TODAY'S VIDEO

TODAY'S VIDEO
Equipment, Setup, and Production
Second Edition

Peter Utz

P T R Prentice Hall, Englewood Cliffs, New Jersey 07632

Library of Congress Cataloging-in-Publication Data

Utz, Peter.
    Today's video : equipment, setup, and production / Peter Utz. --
2nd ed.
       p.   cm.
    Includes bibliographical references and index.
    ISBN 0-13-925033-6
    1. Television--Handbooks, manuals, etc. 2. Video tape recorders
and recording--Handbooks, manuals, etc. 3. Television--Production
and direction--Handbooks, manuals, etc.   I. Title.
TK6642.U88 1992
778.59'9--dc20   91-43759
                                                    CIP

Editorial/production supervision: *Brendan M. Stewart*
Cover design: *Andrew J. Lipinski*
Prepress buyer: *Mary McCartney*
Manufacturing buyer: *Susan Brunke*
Acquisitions editor: *George Z. Kuredjian*

 © 1992, 1987 by P T R Prentice-Hall, Inc.
A Simon & Schuster Company
Englewood Cliffs, New Jersey 07632

The publisher offers discounts on this book when ordered in
bulk quantities. For more information, write: Special
Sales/Professional Marketing, Prentice Hall, Professional &
Technical Reference Division, Englewood Cliffs, NJ 07632.

Printed in the United States of America

10 9 8 7 6 5 4 3

ISBN 0-13-925033-6

Prentice-Hall International (UK) Limited, *London*
Prentice-Hall of Australia Pty. Limited, *Sydney*
Prentice-Hall Canada Inc., *Toronto*
Prentice-Hall Hispanoamericana, S.A., *Mexico*
Prentice-Hall of India Private Limited, *New Delhi*
Prentice-Hall of Japan, Inc., *Tokyo*
Prentice-Hall of Southeast Asia Pte. Ltd., *Singapore*
Editora Prentice-Hall do Brasil, Ltda., *Rio de Janeiro*

To my wif BARBara who has been learningthe   word processor processor
processor.

CONTENTS

PREFACE

The trouble with television textbooks is that they all seem to be written for a field that existed 15 years ago but doesn't exist today. They teach the theory and practices of big-time TV studios with expensive equipment and large production crews.

Most of the television action today is in small studios (or no studio) with relatively inexpensive equipment and small production crews (sometimes one person). It occurs in college media centers, cable companies, corporate training centers, hospitals, churches, and small TV production houses. That is one reason why I wrote this book: to help teach the kind of video predominant today.

Another problem I've found with TV textbooks is that they focus on the theory, aesthetics, and terminology for producing and directing *network* broadcasts. I'm sorry to break this to you, but NBC isn't hiring fledgling TV directors. Glamorous network jobs are as rare as pandas and dodo birds.

If you want a broadcast TV job, you're not going to get it pumping gas and waiting for a break. The formula for success is to start low (perhaps in industry or in a small-town TV station) and get the experience to include in your resume. Building a good reputation with your employer can't hurt either. But before this process can start, you have to land that all-important first job in television.

Which brings me to the second reason for writing this book: Video employers aren't looking for people who can talk about TV and its impact on society, etc. They want people who can *do* things, people who are masters of their machinery, people who can connect and run video equipment and make a TV production, single-handedly, if necessary. This book is designed to teach you how to do everything. Yourself.

According to ITVA (International Television Association) and *Video Manager* (a trade journal), video employer surveys show that they are more inclined to hire applicants with hands-on equipment experience. No book by itself can give you that, but this one aims to come as close as it can.

In short, most of the video activity today is in education, industry, and small-scale studios and production shops where one needs to know how to do almost everything oneself. And if you long for a big-time broadcasting career, you'll still have to start small somewhere and work your way up. In either case, you'll have to master the technology, and that's what this book is all about.

Who Will Understand This Book?

This book will start simply and build slowly. You'll need no previous knowledge of TV to understand it. You don't even need to be mechanically inclined. You will encounter no heavy mathematics, almost no electronics, and only a small amount of necessary terminology to memorize.

This book is geared to a moderate reading level (I won't snow you with long words) and has lots of pictures to help you make sense of new concepts.

I'm an educator first and a TV producer second. That means I'll be using every trick in the book to help you understand and remember the things you've read. Here are a few of the tricks:

- *All subjects are divided into small, bite-sized chunks,* easy to understand and remember.

- When things make sense, they are easier to grasp and remember. For this reason, you'll see brief, *simple explanations of how equipment works* and why it behaves the way it does. Knowing what makes something tick equips you to learn quickly how to operate a device and to troubleshoot problems, using simple common sense.

- There are *numerous headings and subheadings* to help you locate information quickly.

- *Technical words* that are commonly used in the profession and should be learned *will appear in SMALL CAPITALS*, so they'll stand out.

- *Small, manageable groups of technical words* to be learned will appear at the beginnings of chapters and amid chapters, before they are used. Some words are more important than others. Those of crucial importance that should really be learned will be preceded by an asterisk (*) in the miniglossaries.

- In case you need to look up something, the *index is very complete.* It will also direct you to the miniglossaries should you need to look up the meaning of a word.

- I use a *lighthearted approach.* Just when I think you're falling asleep, I'll crack a joke. There's no reason why video can't be fun.

Additional Features of This Book

- *Home video included.* Nearly all educational and industrial TV producers (and many broadcasters) are using VHS and 8mm VCRs for some program distribution. Because it's so popular, it is important to know how to set up and use home video gear. Unfortunately, most professional books omit this subject, leaving you to learn it from the home video magazines and books or, worse yet, the operation manuals. This book integrates home video into the discussion of all video

equipment and techniques. Consider it one-stop shopping.

- *Aesthetics included.* There's more to video than knob twiddling. Woven into the equipment care and feeding rituals will be artistic guidelines and creative suggestions. Video is an art as well as a technical skill.

- *Studio and portable TV are both covered.* Some TV production occurs with crews, lights, several cameras, and control rooms and in studios having everything *including* kitchen sinks. ENG (electronic news gathering), EFP (electronic field production), and other portable video operations are performed on the run. The well-rounded TV person needs to know both styles and this book covers both.

- *Equipment ailments and cures listed.* The Twenty-third Law of Technology states: ''Machines should work; people should think.'' Neither happens with regularity, so Murphy's Law comes into effect: ''If anything can go wrong, it will, and at the most critical time.'' For every major type of equipment, I have included a section on how to troubleshoot the inevitable malfunctions.

- *Trade tips and shortcuts included.* Scattered throughout this book are numerous tricks and shortcuts used by television professionals.

- *Tables and charts provide handy reference.* Not only will this book serve you as an introductory text to video technology, but it will remain a handy reference. Numerous tables review accessories and their uses, tape costs and consumption, recording formats, and useful definitions. Many drawings show possible ways to connect video equipment.

In short, this book prepares you to handle the planning, production, and equipment of *today's video.* If this book seems heavy and expensive, it is because you are holding *three books* in your hand:

1. An introduction to television (often called TV I) textbook

2. An intermediate TV production (often called TV II) textbook

3. A video reference handbook, a place to look up the answers to video questions

All three texts are combined into this one book. As a textbook, it will carry you through several semesters of formal video study. As a reference book, it's organized for easy research or review.

ACKNOWLEDGMENTS

Many thanks, models: Patricia Magee, Donna Insinga, Dorothy Van Duyne, Carolyn Miller, Marcia Earl, Bob Morley, Joseph Sauder, Jim O'Rourke, Wynn Petrakian, and Maureen McGrath.

Photos were the work of Jean de Geus, Chuck Brehm, Colleen Harrington, and Deborah Murray and myself.

No thanks at all go to my model Ugly Al, whose pictures came out so badly that no photo lab would print them. Sorry about what Al did to your camera, Jean.

TEACHER'S MANUAL AVAILABLE

Instructors are invited to contact Prentice Hall to get a copy of the instructor's manual containing:
Course outlines
Teaching objectives and study questions and answers

ABOUT THE AUTHOR

Dr. Peter Utz received his bachelor's degree in physics and his doctorate in instructional technology from the University of Massachusetts. He has produced and directed more than 500 instructional TV productions for the City University of New York. He has also published *Video User's Handbook*, an industrial TV reference guide, *The Complete Home Video Book*, a comprehensive handbook on the setup and operation of home video equipment, *Do-It-Yourself Video, A Beginner's Guide to Home Video, Create Excellent Video* and *Making Great Video*, handbooks for camcorder enthusiasts, and *Making Great Audio*, a manual for recording and reproducing sound. He has published over 170 articles in media and television journals and is video editor for *AV Video* magazine. He presently supervises the media department at the County College of Morris in Randolph, NJ where he also teaches professional and home video courses.

There are lesser-known jobs out there for people who have mastered the equipment.

TODAY'S VIDEO

HOW TV WORKS

Don't give up now—you are only on the first page. Learning how TV works may sound complicated, but it isn't going to be. This chapter aims to satisfy your curiosity while at the same time showing you how this electronic chain of machinery all fits together. *Knowing how* things work will help you make sense of all the knobs and meters and cables you will be introduced to later on.

TV SETS

A TV set is a box of electronics and a big, empty glass bottle called a PICTURE TUBE. Some people call this element a CATHODE RAY TUBE or CRT. You've come to know the flat part of the bottle as the TV screen. The neck of the bottle is hidden beneath the bump you see in the middle of the back of every TV set. Inside the neck of the bottle is an electronic gun, sort of a machine gun that shoots electricity. This gun shoots a beam of electrons at the inside face of the TV screen, which is covered with phosphor dust. Where the electrons hit the phosphor, it glows. To make a picture, this gun sweeps across the screen from side to side, much as your eyes sweep across each line of this page and eventually cover the entire page. Special SYNC circuits tell the gun how fast to sweep and when to sweep as well as when to stop shooting at the end of a line (just as you stop reading at the end of a line and zip your eyes back to the beginning of the next line). SYNC also tells the gun to stop shooting when it reaches the bottom of a TV screen and then reaims it at the top and starts it shooting again. As the beam zigzags across the screen, another signal (VIDEO) tells the gun to shoot harder or weaker depending on whether it is tracing a lighter or darker part of the picture. By turning up the BRIGHTNESS control on your TV, you tell the gun to shoot harder at the screen, thus lighting up the phosphor more brightly. By adjusting the VERTICAL hold or the HORIZONTAL controls on your TV set, you adjust the electronics which tells the gun when to sweep and how fast to sweep.

This gun zips its beam across the screen 15,735 times each second, and it starts at the top of the screen nearly 60 times each second. The phosphors keep glowing until the next time the beam comes around and zaps them again. Thus the screen appears smooth and flicker-free. European TVs, incidentally, retrace themselves only 50 times per second rather than 60, and, as a result, European TV flickers more noticeably than U.S. TV. If you look very closely at your TV screen, especially at the light part of the picture, you will be able to see all those tiny horizontal sweep lines that make up the picture. Figure 1–1 diagrams the process.

Actually, when the electron gun sweeps its lines across the screen, it doesn't do all the lines at once. First it does the odd-numbered lines, leaving empty spaces for the even-numbered lines. Then it goes back and fills in the even-numbered lines. In the first sixtieth of a second, it will sweep

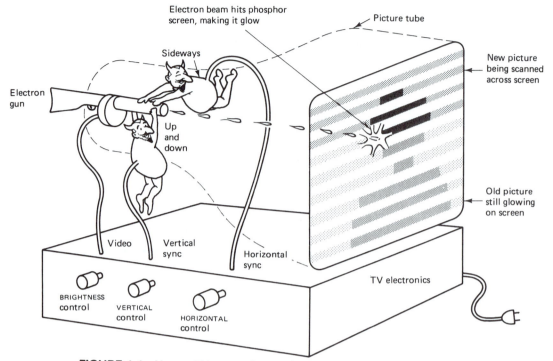

Electron beam hits phosphor
screen, making it glow

Picture tube

Sideways

New picture
being scanned
across screen

Electron
gun

Up
and
down

Old picture
still glowing
on screen

Video   Vertical
           sync

Horizontal
sync

TV electronics

BRIGHTNESS
control

VERTICAL
control

HORIZONTAL
control

FIGURE 1-1   How a TV set makes a picture.

lines 1, 3, 5, 7, etc., making a total of 262½ lines on the screen. This is called the ODD FIELD. In the next one-sixtieth of a second, it draws lines 2, 4, 6, 8, etc., making another 262½ lines on the screen. This is called the EVEN FIELD. Therefore, it really takes one-thirtieth of a second to draw each *complete* picture. The complete picture is called a FRAME. Each of these frames is motionless, but because they go by so quickly (30 per second), they make the picture appear to be moving. The process is much like that of a movie projector.

TV CAMERAS

A TV camera works like a TV set in reverse. Instead of making light, it senses light.

The camera lens focuses the image of a scene onto a light-sensitive PICKUP CHIP or, in older or top-of-the-line professional models, PICKUP TUBE. The chip is a postage stamp-sized circuit covered with several hundred thousand

microscopic light sensing "eyes" made of transistor-like material. These "eyes" are lined up like pockets in a giant egg crate, and when light strikes a pocket, it generates electrons. The brighter the light, the more electrons are amassed. Thus, the light creates a tiny image made of electrons.

Another circuit in the camera measures the electric charge in the pockets, and reads the data out as a varying electric signal. SYNC circuits assure that the electric charges stream out at exactly the right speed so that one row of pockets makes the video signal for one sweep of the electron gun in the TV set. Next, SYNC reads out the next row of pockets, and the next, until the 262½ *odd* rows have been read. SYNC then clocks out the 262½ *even* rows of data, making a complete FRAME, 525 lines of video picture. Before the process starts over again, another circuit empties the chip's pockets, ready to measure more light.

So, following the process from camera to TV: The lens focuses an image on the camera's PICKUP CHIP. The chip converts the thousands of dots of light into electrical

---

MINIGLOSSARY

*CRT or cathode ray tube   A vacuum tube with an electron gun at one end and a phosphor screen at the other which glows when struck by electrons from the gun. TV picture tubes are CRTs having the familiar TV screen at one end.

*Sync   A circuit or a signal which directs the electron gun in a camera or TV picture tube to hold a TV picture steady on a screen. Sync also synchronizes the electronics of other TV equipment.

*Field   The TV picture created in one-sixtieth of a second by scanning an electron gun over *every other* line in the picture. In the United States there are 262½ odd-numbered lines in a field,

followed by 262½ more even-numbered lines making the next field one-sixtieth of a second later. The two fields together make a frame, a complete TV picture.

*Frame   A complete TV picture lasting one-thirtieth of a second, composed of two fields or 525 scanning lines (in the United States).

*Pickup tube or chip   The light-sensitive part of a TV camera which "sees" the picture and turns it into video signals.

*Video   The picture portion of a broadcast TV signal; an electronic signal making a TV picture.

charges. A SYNC signal reads out these charges one row at a time as a varying voltage, a VIDEO signal. Thus, bright and dark parts of the picture become strong and weak voltages in the VIDEO signal.

This VIDEO signal goes to your TV set and tells its electron gun to "shoot, don't shoot," thus creating light and dark areas on the TV screen.

High-speed clock circuits in the camera make the SYNC signal (which becomes part of the VIDEO signal) that starts

the camera chip's readout at the top of the picture, feeding out one line at a time. When the TV receives the signal, it uses SYNC to start the TV tube's electron gun at the top of *its* picture and makes the beam zigzag back and forth corresponding with the chip's readout. Thus, the zigging and zagging are both SYNCHRONIZED together, and the TV set creates a picture similar to what the camera saw. Figure 1–2 diagrams the process.

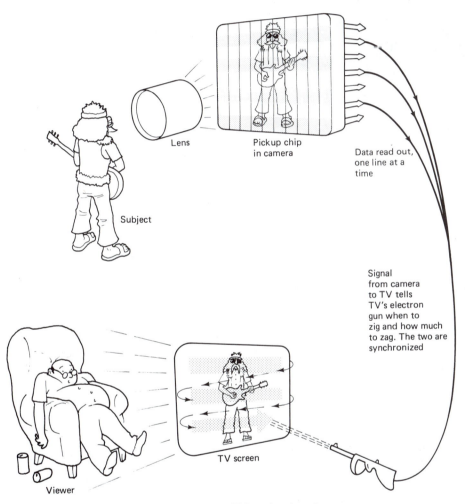

Lens

Pickup chip in camera

Data read out, one line at a time

Subject

Signal from camera to TV tells TV's electron gun when to zig and how much to zag. The two are synchronized

TV screen

Viewer

FIGURE 1-2 How the camera and the TV make the picture.

MINIGLOSSARY

*Single-chip camera A black-and-white camera or a color camera with a pickup chip sensitive to all colors at once.

*Three-chip camera A TV camera with three pickup chips inside, one sensitive to the red parts of the picture, another sensitive to green, and the third sensitive to blue.

*NTSC video National Television Standards Committee method used in the United States for electronically creating a colored TV signal.

*RGB video Video signals traveling on three separate wires. Red parts of a colored picture go on one wire; green, on the second; and blue, on the third.

*Monochrome Black and white (not color).

*Primary colors Three colors which can be combined together to create all the other colors. TVs in the United States use red, green, and blue as primary colors.

Circuit chip replaces pickup tubes in today's color cameras. (Courtesy of RCA Broadcast Systems).

COLOR

In a darkened room, aim a red flashlight at a white wall, and you'll see red; turn it off, and you'll see black. Shine a blue light on the wall, and you'll see blue. Shine green, and you'll see green. Shine red and blue together, and the colors will mix to create a new color, magenta. Shine red and green, and you'll see yellow. Shine all three, and you'll get white, as illustrated in Figure 1–3. All the other colors can be made from shining various proportions of these three PRIMARY COLORS.

A monochrome (black-and-white) TV set makes its picture by electronically projecting black (which is really just the absence of white) and white onto a screen in various proportions. A colored TV makes its color picture by creating three pictures on its screen: one red and black, another green and black, and the third blue and black. Where only red is projected, you see only red. Where red and green pictures overlap on the screen, you see yellow. Where all three pictures are black, you see black. Where all three colors converge with proper strength, you see white.

The picture tubes in color TV sets have three electron machine guns, one for each color. The face of the color TV screen is made of dots or bars of phosphor which shine red, blue, or green when hit by electrons. Look really closely at your TV screen while it's displaying a white picture, and you'll be able to see the tiny colored dots or stripes which make up the whole colored picture. As shown in Figure 1–4, the electron gun for the blue color is arranged so that it can only hit the blue phosphors on the screen. The red gun can only hit red, and the green gun, only green. The three guns independently scan a picture onto the screen, and the three pictures overlap to create dazzling views of sea rescues and loose dentures alike.

In the back of some colored TV sets, there are adjustments for these three electron guns. They are used by TV repairpeople to weaken or strengthen a certain color, balancing all three so that one doesn't overpower the others and tint all your pictures.

Color TV cameras work much like color TV sets in reverse. Somehow the camera has to make a red, a green, and a blue picture and combine them into a single VIDEO signal.

One way to do this would be to have three black-and-white TV cameras, side by side, all take a picture of the same subject. One camera would look through a red lens

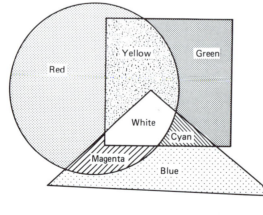

When you shine colors on a wall or light up color phosphors near each other, their colors add, making new colors.

The red circle + the green box = yellow.

The green box + the blue triangle = cyan.

The blue triangle + the red circle = magenta.

Red + green + blue = white. Different proportions of red, green, and blue make up all the other colors.

When mixing paint (rather than projecting light), the colors *subtract*, resulting in colors completely different from the above when they mix (usually various shades of brown).

FIGURE 1-3 Additive color mixing. (See also back cover)

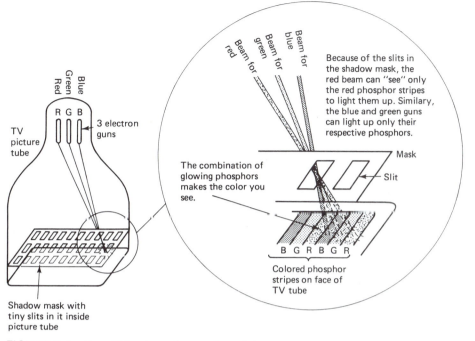

FIGURE 1-4 How TV sets make a color picture.

and see only the red parts of the picture. Another camera would look through a blue lens and would see only the blue parts of the picture. The third camera, looking through a green lens, would see only green. Thus the three black-and-white cameras could give you a red, a blue, and a green VIDEO signal. To simplify matters, manufacturers have put the three cameras into the same box where they can share some of the same electronics. They also share the same lens and are called THREE-CHIP cameras. So that each PICKUP CHIP can "see" a particular color, the lens image is split into three images using either two-way mirrors or glass prisms (Figure 1–5). Each image passes through a colored filter so that one PICKUP CHIP sees red; another, blue; and another, green. These "pure" VIDEO signals are called RGB (red, green, blue) VIDEO and are used by computers, TV projectors, and other devices where supersharp colored pictures are necessary. But RGB VIDEO requires three wires to carry the three colored signals. To make things simpler, most cameras have a circuit which combines the three colors into a single VIDEO signal which requires only one wire. In the United States, this VIDEO signal is called NTSC (National Television Standards Committee) VIDEO. The TV set at the other end of the wire separates this signal back out into its red, green, and blue components, which it sends to the red, green, and blue guns in its picture tube. Because the three

picture signals had to be combined to run down a single wire and then had to be separated again, some of the picture sharpness was lost in the process. For this reason, NTSC colored pictures are not as sharp as RGB or black-and-white TV pictures.

Less expensive TV color cameras use only one PICKUP CHIP to "see" all three colors. The chip has colored stripes on it much like the stripes of a color TV screen. Red parts of a picture would activate the red stripes on the camera chip. Blue parts would activate the blue stripes, and green parts would activate the green stripes. The electronics in the camera then senses which stripes were activated and turns this information into a color VIDEO signal. Because of the space taken up by the stripes on the face of the PICKUP CHIP, SINGLE-CHIP color cameras don't give as sharp a picture as THREE-CHIP color cameras.

AUDIO, VIDEO, SYNC, AND RF

When a television program is produced, whether "live" or by video tape recording, a camera takes the picture, changing it into an electrical signal called VIDEO. A microphone takes the sound and makes another electrical signal called AUDIO. And a special device called a SYNC GENERATOR

MINIGLOSSARY

*Audio The sound part of a TV broadcast. Sound, turned into an electrical signal.

*Sync generator An electrical device which makes sync (timing) signals which synchronize TV equipment and keep TV pictures stable.

Composite video Video (picture) signal with the sync (timing) signal combined.

Noncomposite video Video (picture) signal without sync combined.

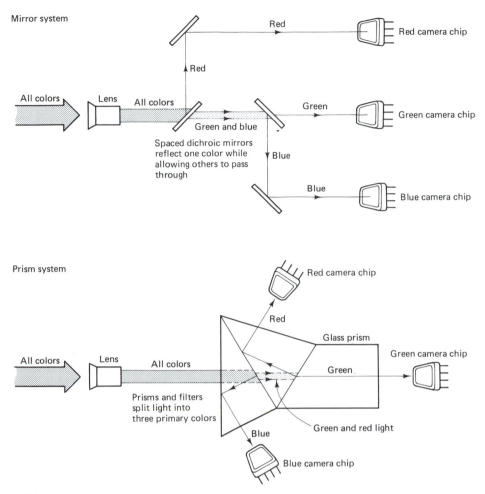

FIGURE 1-5   Three-chip color camera systems to break colored light into its primary colors. (See also back cover)

creates a third electrical signal called SYNC that keeps the picture stable. When SYNC and VIDEO are electronically combined into an electrical pulse, the signal is called *composite* VIDEO, but most TV people refer to this simply as "VIDEO." The TV broadcaster then combines the AUDIO and VIDEO and SYNC using a device called a MODULATOR, which codes the three into another signal called RF (which stands for radio frequency). The RF is transmitted, travels through the air, is picked up by your antenna, and goes into your TV receiver. By tuning your TV receiver to a particular channel (the same one that was broadcasted), the tuner circuit in the TV set decodes the RF signal and separates it back into VIDEO, AUDIO, and SYNC as shown in Figure 1–6. The VIDEO goes to the TV screen, the AUDIO goes to the speaker, and the SYNC goes to special circuits which hold the picture steady. By adjusting your TV's BRIGHTNESS, CONTRAST, HUE, and COLOR INTENSITY knobs, you adjust the TV's VIDEO circuits. By manipulating the VOLUME control, you adjust the AUDIO circuits. By turning the VERTICAL or HORIZONTAL controls, you adjust the SYNC circuits in your TV.

Television signals may sound confusing until you realize that all you have are just a few basic signals and that

MINIGLOSSARY

*Modulator or RF generator   Electronic device which combines audio and video signals, coding them into RF, a TV channel number.

*RF or radio frequency   The kind of signal which is broadcast through the air and comes from a TV antenna. RF is a combination of audio and video signals coded as a channel number.

Chrominance   The color part of video signal.

Luminance   The black-and-white (brightness only) part of a video signal.

Horizontal sync   The part of a sync signal which aims the TV's electron gun left and right. This holds the picture from jittering or straying sideways.

Vertical sync   The part of the sync signal which controls the up and down motion of the TV's electronic gun. This holds the picture steady and keeps it from rolling vertically.

*Audio mixer   Mixes audio (sound) signals perhaps from several microphones and combines them into one audio signal.

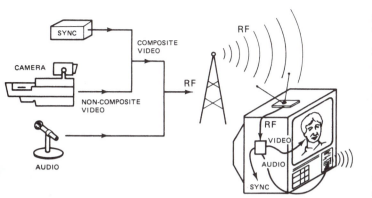

FIGURE 1-6 AUDIO, VIDEO, SYNC, and RF.

the rest are merely combinations of these basic signals. Figure 1–7 diagrams how these signals are combined.

The color camera's signal, you'll remember, was really three signals, one red, one green, and one blue. To make them fit on one wire, they were combined to make one VIDEO signal. Coded into that one signal are two kinds of information. The color information is called CHROMINANCE. The black-and-white part of the information is called LU-MINANCE. The CHROMINANCE part of the signal tells your TV set what colors to paint on its screen. The LUMINANCE part of the signal actually creates the light and dark parts of the picture.

Just as the color VIDEO signal has two components, the television SYNC signal has two main parts: VERTICAL SYNC, which aims the electron gun in your TV set up and down, and HORIZONTAL SYNC, which zigzags the gun from

side to side. If you misadjust the VERTICAL control on your TV set, you'll see the VERTICAL SYNC signal as a black bar rolling through the picture. Misadjust the HORIZONTAL SYNC control on your TV set and you'll see a vertical bar or diagonal lines going across your picture. Those lines are the HORIZONTAL SYNC signals. The combination of these two signals holds the TV picture steady on your screen. If either signal becomes weak or is missing, your picture will roll up or down or will tear sideways.

SYNC is made by an electronic clock circuit called a SYNC GENERATOR. SYNC GENERATORS can be built into cameras or can be in a separate box that stands alone. Professional ones make not only VERTICAL and HORIZONTAL SYNC signals but some other SYNC signals with bizarre names like BURST, BLANKING, and SUBCARRIER. Relax; these last three are beyond the scope of this text.

As mentioned before, the VIDEO and the SYNC signals are usually combined to form what we commonly call VIDEO. All this can fit on just one wire.

Compared to VIDEO, AUDIO sounds simple. A microphone or some other audio device makes an electrical signal which can pass down a wire and perhaps be tape-recorded. When several microphones are used at once, their signals can be mixed together (using what else?—an AUDIO MIXER, naturally) to travel down a single wire.

The next trick is to find a way to combine the AUDIO and VIDEO signals so that the sound and picture can share the same wire. This is where RF comes in. An RF GENERATOR or MODULATOR can combine the AUDIO and VIDEO signals and change them into a TV channel number. All home video

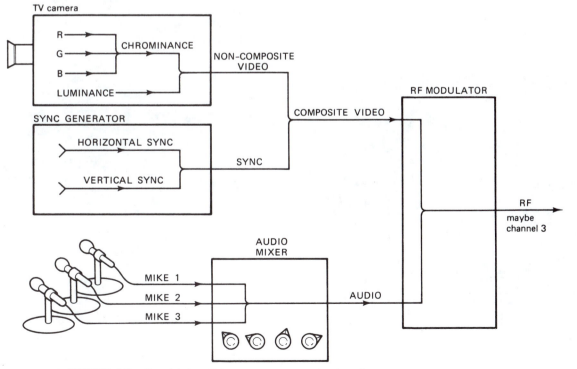

FIGURE 1-7 Combining TV signals onto a single wire.

tape recorders have one of these MODULATORS built in and generally change the VIDEO and AUDIO signals into a channel 3 signal which now can be fed to the antenna terminals of any TV set and be tuned in by switching the set to channel 3.

One of the great features of RF is that more than one channel can be transmitted on a wire at the same time. You're already familiar with how one antenna wire going to your TV set at home can give you channels 2, 4, 5, 7, etc., with no problem.

The bad thing about RF is that in the process of MODULATING the AUDIO and VIDEO signals together, a little of the picture sharpness is lost. When your TV tuner separates an RF signal back into VIDEO and AUDIO again, the picture sharpness decreases further still. About 10% of your picture sharpness is lost by MODULATING and DEMODULATING it.

A LITTLE HISTORY

Now that you know a little about how TV works and what the TV signals do, we will focus on particular items of TV equipment. As we study each item, we may examine in further detail how each machine goes about the magic of making television.

If this chapter sounded a bit complicated, perhaps you can now appreciate how hard it was to build the first TV. Television wasn't invented all at once; it was the result of one man making something *like* television and others improving upon the idea until it became TV as we know it today. The invention of television probably traces back to 1843 when a Scottish watchmaker, Alexander Bain, received a patent for an "automatic telegraph" which was able to print out pictures of letters over a telegraph wire. In 1883 a German science student, Paul Nipkow, proposed a method of creating images using a spinning disc with holes in it. It was not until 1925 that John Logie Baird of Great Britain demonstrated a spinning disc system for mechanically transmitting pictures. The same year, C. Francis Jankins was independently doing the same thing in Washington, D.C. But these were awkward mechanical devices, able to recreate only fuzzy silhouettes and basic shapes. Meanwhile, in 1923, Vladimir K. Zworykin was busy developing the first TV camera tube (called an iconoscope). Philo T. Farnsworth used a form of this device in 1927 when he demonstrated the first *electronic* (as opposed to a mechanical spinning disc) picture transmission system. So who invented TV? Baird and Farnsworth get most of the credit, but both gentlemen were using combinations of machines invented by somebody else.

In 1928, Felix the Cat was one of the first TV images transmitted. This Felix was viewed by TV experimenters from New York to Kansas.

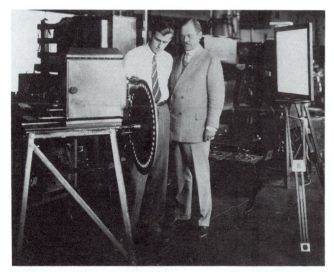

Mechanical spinning disk TV system, circa 1927. (Courtesy David Sarnoff Research Center)

TV MONITORS
AND RECEIVERS

TV RECEIVER

According to a recent study, there are more TV sets in the United States than there are bathtubs. Another study showed that the average three-year-old spends 4 hours a day watching TV.

If television is such a big part of our lives, then why do so few of us know how to connect up and use them? Why is it that half the color TV sets you see are displaying green or magenta faces? And why do so few folks watch UHF TV stations?

Considering the billions of dollars that go into producing and broadcasting crisp pictures to our homes, and considering that most of this book will involve producing or recording crisp pictures of your own, it would be a shame to have all that work turn into mush in a horribly adjusted TV set. So here come the basics on how to connect and adjust TV sets followed by some lesser known secrets of "the tube."

Adjusting the Controls on a TV Set

Since there are so many kinds of TV receivers, the descriptions in this chapter will not work for everybody. Usually the difference will be in the labeling and/or the location of the knobs and switches. When in doubt, read the TV's instruction manual (which I bet you can't find).

Getting Turned on. Okay, let's turn the set on. If you can't find the POWER, ON, ON/OFF, or VOLUME switch, try turning, pulling, or pushing a few. There are no external knobs on a TV receiver that when maladjusted can permanently harm the set, so go ahead and explore.

Now that you've switched the set on, you'd like to know if the set is really working. You can wait a minute for the TV to warm up and come on, or you can immediately tell that the set is on by

1. Seeing the light illuminating the channel numbers come on (most receivers, except the small ones, have channel selectors that light up)
2. Hearing a faint, high-pitched squeal that tells you that the TV circuits are working

If your set has an "instant-on" feature, you will also get a picture immediately. Sets without this capacity require a warm-up period before the picture appears.

If you have waited about 30 seconds and nothing happens, turn to the section on common TV ailments and cures in this chapter.

Channel Selector. Television and radio signals travel through the air as invisible electromagnetic waves. Some waves vibrate very fast and are said to have a high FREQUENCY, while others oscillate more slowly and have a lower FREQUENCY. As you tune your radio up from the lower numbers to the higher numbers on the dial, you are tuning in stations with higher FREQUENCIES. Unlike radio dials, which show you the actual FREQUENCY you are turned to, television dials just give you channel numbers. Channels numbered 2–13 are called the VHF or very high frequency channels. Channels 14–83 are called the UHF or ultra high frequency channels. UHF channels are usually selected on a separate tuning knob from the VHF channels, and the UHF channel selector is usually activated when the VHF channel selector is turned to ''U'' or ''UHF,'' a spot on the dial between channels 13 and 2. The UHF and VHF channels have their own separate FINE-TUNING knobs. To tune in a station properly, assuming your antenna is correctly aimed (more on that later), you adjust FINE TUNING as follows:

1. First switch off any automatic controls if your set has them, as they will confuse the issue. They may be labeled AUTO, AUTO COLOR, or AUTOMATIC FINE TUNING (or AFT), or AFC.

2. Next adjust the FINE-TUNING control, which is probably a knob concentric with the channel selector as in Figure 2–1. On some sets, you have to push it in and turn it to make the dial work. When adjusted in one direction, the picture will look soft and a little fuzzy. When adjusted in the other direction, the sound may become raspy and cause wavy lines in the picture and it becomes rough edged and grainy as in Figure 2–2. If you adjust to this second position and then back up just a little until the picture is sharp and the sound is good, you will have a well-tuned signal.

3. Now turn on the AFT or AUTO control (or whatever), and this will reactivate automatic circuits which will further tune the set to the station. Your picture should now look the best it can be.

Some TV sets have push-button tuning, with perhaps 12 or 14 little windows with numbers behind them. To tune in channel 4, you would simply push button 4. If channel 4 looks a little soft or a little grainy, it may mean that channel 4 needs to be fine-tuned. To do this, first look for a trapdoor (usually near the channel buttons) and open it. Inside you will see tiny thumbwheels, one for each channel. Find the channel 4 thumbwheel and give it a tweak (this is a *highly technical* term which means make a small adjustment) in one direction or the other.

FIGURE 2-1   Channel selector and concentric FINE-TUNING dial.

FIGURE 2-2   FINE TUNING at one extreme, showing a grainy picture, sometimes with slight wavy lines in it.

MINIGLOSSARY

*Frequency   The number of times a signal or sound vibrates each second, usually expressed as cycles per second or hertz (Hz).

*VHF   Very high frequency; TV channels 2–13.

*UHF   Ultra high frequency; TV channels 14–83.

Preset   A control which is set once and is not readjusted each time it is used. TV channel presets are used as fine-tuning ad-justments. Once set, you may switch to a channel without fine-tuning it again.

Express or direct access tuning   TV tuner which selects the channel and fine-tunes it after you punch the channel number into a calculator-type keypad.

*Saturation   The vividness of a color. Colors lacking saturation look pastel.

Little plastic channel numbers
can be removed and changed.

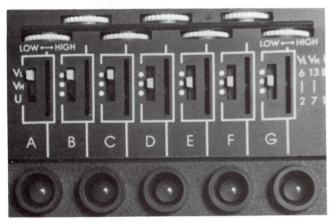

Behind door are thumbwheels for
tuning each button to a channel.

Push button TV set with trapdoor.

Brand-new TV sets of this type and sets which have been moved from a different town may need to have all their channel "whoojies" adjusted to pick up your local stations. You do this by pressing the first button and adjusting the first thumbwheel until you've received channel 2 (or whatever is the lowest-numbered channel received in your area). You may also find little plastic numbers somewhere which can be inserted in the windows so that when you press the first button the light will light up behind the first window and illuminate the little channel 2 number that you inserted. Now press the second button, adjust its knob, and tune in the next higher station in your area. Do this for all the stations you can. Sometimes it's handy to have another (already-tuned) television nearby to help tell you which station is which.

Many push-button models have a tiny switch labeled L, H, and U next to each thumbwheel. To tune in the low-numbered channels 2–6 with this knob, you have to throw the switch in the L position. To tune in the high-numbered channels, 7–13, you have to throw the switch in the H position. To tune in UHF stations, you would turn this switch to the U position. Once the push buttons are tuned to various channels, they are called PRESET. Sometimes the factory or local distributor PRESETS the channels for you. Usually you have to reset them for your own locale.

Sometimes a TV set will have only 12 channel buttons, but you may receive 20 stations in your area. You will have to select the 12 stations you watch most often and sacrifice the others. But if you ever want to watch a program on one of the "other" stations, nothing stops you from picking one of your PRESETS and readjusting it to the other station. It can always be tuned back to normal later.

Some TVs have push buttons 0–9, like a calculator. On these TVs there is no muss, no fuss, no fiddling with FINE TUNING—you just press the channel number you want, and it appears. *Note:* If the channel number is a single digit, like a 6, then you have to press 06 to get that channel. But to get channel 41, simply press 4 and 1. Being able to punch in the channel number you want and go directly there is called EXPRESS or DIRECT ACCESS tuning. Other names for this feature are QUARTZ-CONTROLLED tuning and PHASE LOCK LOOP tuning.

In Chapter 4, we will examine some TV channels which are not listed on your dial and can only be received through a CABLE CONVERTER BOX or a CABLE READY TV.

Brightness and Contrast. Black-and-white TVs don't make black-and-white pictures—they just make pictures of nothing and white. The *nothing* part is the same darkness you find on the screen when the set is turned off. A TV that is turned off and is sitting in your root cellar will have a darker screen than the same TV sitting on your balcony on a sunny day. When the set is turned on, its basement picture will have rich black blacks, while the balcony picture will be washed out and gray. Since you can't make the blacks on the screen blacker (without throwing a heavy blanket over you and your patio TV), you make the whites whiter by turning up the BRIGHTNESS and CONTRAST. This makes the blacks *look* blacker by comparison. But when the environment is dark again, it pays to turn these controls back down. Not only is the excess contrast unnecessary, but it robs your picture of sharpness and subtlety.

All of the preceding is true for colored TVs too: Their pictures are made of nothing and color, and their blacks are only as black as the face of the picture tube when the set is off. If you examine a number of color sets in a TV showroom, you may notice that some have light-faced picture tubes and others have dark-faced ones. The dark-faced ones will yield blacker blacks than will their counterparts in a brightly lit room, so they are more desirable TVs to have in this respect. Some color sets have a PICTURE control which adjusts the CONTRAST and BRIGHTNESS together. To simplify this day-night accommodation, others have an electric eye sensor which measures the brightness in the room and automatically adjusts the picture accordingly.

FIGURE 2-3 Vertical roll—adjust VERTICAL control.

Vertical, or Vertical Hold.   This is not the position lovers take when kissing good night. It is the knob which adjusts the VERTICAL SYNC circuits in your TV so that your picture doesn't roll, flop over, or jitter. Normally the electron gun stops shooting at the bottom of your picture and starts again at the top. Misadjust SYNC, and the gun will errantly think it's at the end of the picture when it's right at the middle. That black bar you see rolling across your screen (see Figure 2–3) is the gun ending one picture, turning off, and starting another, but in the wrong place. VERTICAL HOLD usually needs adjustment when the TV is getting a garbled signal. Video tape players, weak TV stations, or other interference sometimes mess up the SYNC signal, requiring you to adjust VERTICAL. Some newer sets lack a VERTICAL HOLD control.

Horizontal, or Horizontal Hold.   No, this is not the way you pin your wrestling opponent. This adjustment, hidden on many sets, moves the picture left and right, centering it on your screen, and keeps it from tearing into diagonal lines (as in Figure 2–4). It adjusts the HORIZONTAL SYNC circuit that tells the electronic gun that it's at the end of a line and should turn off, zip back, and start shooting a new line across the screen.

FIGURE 2-4 Diagonal lines—adjust HORIZONTAL HOLD.

What throws the HORIZONTAL HOLD off is the same thing that messes up VERTICAL HOLD: a bad signal from somewhere (unless the TV set is defective).

Color and Hue.   To adjust the colors on your TV set correctly, follow these steps:

1. First make sure your TV set has "warmed up" for 2–3 minutes before you start adjusting unless it is an instant-on type. "Cold" TVs have a tendency to change colors while warming up.

2. Tune in a station as best you can. Turn off the automatic controls and turn down COLOR all the way, and you will display a black-and-white picture on your color set.

3. Now adjust your BRIGHTNESS and CONTRAST so that you have black blacks, white whites, and smooth grays in between. Now that you have a good black-and-white picture, we are ready to put the color back. Incidentally, if a station is weak or you are playing a defective tape through the TV or your antenna signal is fouled up, the TV's picture may flash color on and off or do other distracting things. By turning the COLOR all the way down as described here, you will get rid of the distracting color mess and may enjoy the show more even if it can be seen only in black and white.

4. Turn up the color control to increase the SATURATION or vividness of the color until it looks normal.

5. Turn the HUE or TINT control to change all the colors from a greenish cast to a reddish cast. To have a good color picture, you want *enough* color and the *right* color.

Let's focus some more on getting the *right* color. Dresses, scenic canyons, and extraterrestrials can be any color. You can't use them as a guide to proper color adjustment because you never know if you've made them the color they are really supposed to be. Somehow, though, you can always look at a Caucasian face and tell if the color is right. Therefore, we use "flesh tones" to guide our HUE adjustments. To do this, first tune to a station with lots of flesh. Daytime soap operas always have a lot of facial close-ups and so do newscasts. Next turn the COLOR *way up* too far, so you can see the colors better. Then adjust HUE until the face colors look right. Last, turn COLOR down until the overall picture looks good. You're done. The theory here is that if the faces look good, then everything else will look good.

Newer TVs have a special circuit which makes this adjustment for you. This feature may be called by various names (like Colortrak), but what it does is lock the circuits onto a signal broadcast by most stations as part of their TV picture. This signal, called the vertical interval reference signal (VIRS), makes your TV circuits strive to reproduce the color and contrast exactly the way they left the studio.

Connecting up a TV Set

In a fit of extravagance, your company has decided to retire its 1954 black-and-white round-screened Zenith in favor of an up-to-date model. Feverishly you uncrate the glass and chrome beauty and throw away the packaging. As usual, the instructions were with the packaging. Now you'll have to set it up just using common sense. After plugging it in and turning it on, your next endeavor is to get the TV signal into the set so that you have a clear picture and good sound. For this, you need an antenna unless you are receiving your TV signal via "the cable" in which case you connect the cable TV wire to the back of the set in the same place that your antenna would go. If you are in a neighborhood where TV signals are strong (such as 20 miles from the TV station with no mountains or tall buildings between you and the TV transmitter), you can usually get a satisfactory TV signal from the TV set's own built-in antenna if it has one.

Because sometimes a TV set will use a separate external antenna and sometimes it will use its own antenna, there is a place behind the set where the antennas are connected and disconnected. To use the set's own antenna, one must check to see that it is connected. This is sometimes achieved by flipping an ANTENNA switch to the INT (or INTERNAL) position, which means that the built-in antenna is automatically connected to the TV's circuits when the switch is in that position. Most times, however, you connect the set's antenna to its circuits by finding the appropriate wires (see Figure 2–5), slipping the metal ends of the wires under the two screws, and tightening the screws with a screwdriver. It doesn't matter which wire goes under which screw so long as the wires from the RABBIT EARS (the VHF antenna, which is the one that telescopes) go to the antenna connection marked VHF. Your TV set may also have come with a little BOW TIE or loop-shaped antenna, which is for picking up UHF stations. This antenna connects to the screws marked UHF. Now all you do is tune in the stations you want (see the channel selector section earlier in this chapter).

If you are distant from a TV transmitter, the RABBIT EARS will give you a grainy or snowy picture. To improve the picture, you need to disconnect the RABBIT EARS and connect the wires from a larger, more sensitive antenna in place of the RABBIT EARS. It doesn't matter which wire goes under which screw as long as the two bare wire ends are not touching each other or any other metal. Once the antenna

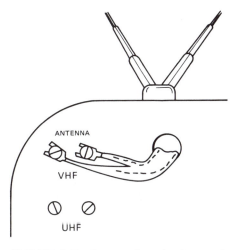

FIGURE 2-5 Connecting the internal or "rabbit ear" antenna.

is connected as in Figure 2–6, you are ready to use the set. For receivers with the INT/EXT switch, be sure the switch is set for EXT (there will be more about antennas in Chapter 3).

Usage Tips

The TV receiver is only a machine. Under appropriate viewing conditions, it can be a powerful tool. In the hands of a skillful user, the impact of television is greatly reinforced. Unfortunately, the programs that cost $1 million to produce and $200 to rent and require $3000 worth of video equipment to display to a class or a group are degraded when the folks in charge of setting up TV monitors botch the job. People often treat TV setup as a trivial matter. It is not.

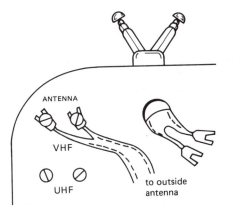

FIGURE 2-6 Disconnecting the internal antenna and connecting an outside antenna wire.

MINIGLOSSARY

Antenna switch Selects whether a TV's internal (monopole or rabbit-ear) antenna or external (rooftop) antenna is to be used.
Rabbit ears A portable TV set's two telescoping rod antennas.

Also called dipole antenna.
Bow tie Portable TV antenna which looks like a bow tie, used for UHF stations.

This last step in the video process is as important as the last step in manufacturing a car: polishing it. Have you ever seen anyone buying a new car that wasn't polished?

Share the following viewing hints with colleagues, teachers, students, conventioneers, salespeople, and other users who wish to get the most out of this medium:

1. Avoid bright lights in the room, or tilt the set slightly forward to reduce reflection and glare on the screen surface.

2. If the room is cursed with an unshaded window (or other bright light source), place the TV in front of the window. That way the screen will be shaded from the light and easier to see. Placing the TV screen perpendicular to the window also helps cut down reflections.

3. If the building is made of brick and steel and the TV has its own built-in antenna, place the TV near a window. TV signals travel fairly well through glass but poorly through brick and steel.

4. A 9-inch (TV screens are measured diagonally) screen can be comfortably seen by 5 or fewer viewers. A 21-inch TV screen is good for *up to* 20 viewers. For more than 30 viewers, definitely use a TV projector or more than one receiver.

5. If echoes are a problem, angle the set out from a corner of the room. If the room has cloth curtains, draw them closed. If the TV set has a TONE control, turn it to "sharp" (not bassy) to improve the sound of speech.

6. Television shows are most effective and most easily remembered when *preceded* and *followed* by some discussion of the subject shown.

The best TV watching environment is a cool, quiet room with dim lights and comfortable chairs clustered from 4 to 10 feet away from the TV screen.

Common TV Receiver Ailments and Cures

Here's what to do when things go wrong, go wrong, go wrong.

1. *Nothing happens when you turn the set on.* Most likely, the set isn't getting power.

a. Make sure it is plugged in.
b. Or, unplug it and plug it in somewhere else.
c. Or, make sure the wall plugs aren't controlled by a wall switch which is turned off.
d. Press the red CIRCUIT BREAKER (an electronic resettable fuse) button on the back of the set. If the set comes on shortly after you push the button, fine. If it blacks out again, however, something is wrong; don't use the set.
e. If the set is a portable and designed to run on batteries, look for a power selector switch labeled BATTERY/AC or CHARGE, or AC/DC, and be sure it is set to the AC position, or *away* from CHARGE or DC.
f. If the set can be remote-controlled, it may have a switch on it somewhere labeled MAN/REM for MANUAL or REMOTE. In the MANUAL mode, the set is responsive to the controls on the set. In the REMOTE mode, it responds to a hand-held keypad or some other remote controlling device.
g. Are you sure the set is not on? Some TVs, if they're not getting a signal, display a silent, black screen which makes the set look like it's turned off. Are any lights lit on the set indicating that it is "live?"
h. If none of this works, read the instruction booklet (the one that is in the dumpster along with the rest of the packing material that came with the set).

Now that your set is on, you may still not get good picture and sound. Here's what can be wrong and what to do:

2. *Good picture, no sound*
a. Check the volume. Is it turned up enough?
b. Make sure there are no headphones plugged into the set (they may automatically cut off the sound).
c. Flip channels. If none of the channels have sound, the problem is in the receiver. If only the one channel has no sound, the problem is in the transmission from the station or your videocassette recorder.

MINIGLOSSARY

*Circuit breaker   An electronic resettable fuse found on TVs and other electronic devices. Pressing the red button resets the fuse.

Mute   A control which cuts out the sound but leaves everything else going.

SAP or supplementary audio program   Technique for broadcasting a third, additional sound track along with stereo TV signals.

*Copy protected   A signal recorded on a video tape renders the tape uncopyable.

Convergence   The precise overlapping of a color TV's three primary colored pictures to make one multicolored picture.

Degauss   Demagnetize (remove residual magnetism).

Purity   An internal color TV adjustment done to a picture tube to make the screen colors even and uncontaminated by other colors or patches of colors.

d. Newer sets with remote control have a MUTE switch (handy for cutting out the sound when you answer the telephone). Try the switch and see what happens.

e. Some newer stereo televisions have a switch which allows you to select monaural, stereo, or SAP (supplementary audio program). SAP is an extra audio channel often used for receiving bilingual broadcasts. If you find that switch in the SAP mode and a second language isn't being transmitted, you'll probably get no sound. Switch it to another position.

3. *There is sound but no picture*

a. If the picture is black, turn up the BRIGHTNESS control.

b. If the picture is gray and washed out, turn up the CONTRAST.

c. Again, switch channels. Something may have temporarily stopped the picture transmission from the station.

d. If your set is listening to a videocassette recorder, also try switching channels. Maybe the set isn't listening to the VCR's channel. Also check to see that the VCR is playing and is connected to the TV.

4. *Bad picture, good sound*

a. If the picture is not centered on the screen, as in Figure 2–7, adjust the HORIZONTAL control. If you cannot find it on the front of the set, look on the side, in back, or behind a trapdoor below the screen. Sometimes it's hidden behind a hole in the cabinet and must be adjusted with a long screwdriver (although a tall whiskey sour will do).

b. The picture has diagonal black lines through it (as in Figure 2–4). Again, adjust the HORIZONTAL control. If this doesn't help, try another station. The problem may be in the transmission.

c. If the picture flips or rolls vertically as in Figure 2–3, adjust the VERTICAL control until the picture is stabilized in the center of your screen.

d. If the picture jitters and shakes or tears into diagonal lines as in part b or rolls easily as in part c, you may be getting a very weak antenna signal or interference from another signal. Check your antenna. If your signal is from a video tape recorder, perhaps the tape has been COPY-PROTECTED (has an anticopying signal on it) and is too unstable to play. Some TV sets will tolerate COPY-PROTECTED TV signals, and others do not. Try another TV. COPY PROTECTION signals will be discussed further in Chapter 13.

e. The screen looks murky (as in Figure 2–8). Dark places fill in, and everything looks very black or very white with no grays. Adjust the CONTRAST control.

f. Just the opposite of part e: The picture looks faded and gray (as in Figure 2–9). Turn CONTRAST up.

Some TVs have an "electric eye" which senses the brightness in the room and turns the brightness and contrast of the TV screen up or down to adjust for it. The electric eye can be tricked, however, if a bright light happens to be shining on it or it gets covered with something (mozzarella from last night's pizza party, perhaps). When the electric eye sees too much light, it will make the picture too bright, and if it sees too little light, it will make the picture too dim. If you can't solve the lighting problem (or the electric eye has gone "kaflooey"), simply readjust the BRIGHTNESS and CONTRAST controls to overcome the effect.

5. *The program and receiver are in color, but the picture is only black and white*

a. Adjust the COLOR and COLOR INTENSITY control that governs the amount of color in the picture. If it is too low, the picture will look washed out. If it is too high, the picture will be too vivid to watch comfortably.

b. Adjust the FINE TUNING. For best results, first switch off the AUTO or AUTO COLOR or AUTOMATIC FINE TUNING or AFT or the AFC button (if your set has one of these) before adjusting the FINE TUNING. Then adjust the FINE-TUNING control, which is probably a knob concentric with the channel selector (as in Figure 2–1). On some sets you have

FIGURE 2-7 Picture not centered—adjust HORIZONTAL.

FIGURE 2-8   Picture needing CONTRAST turned down.

FIGURE 2-9   Picture needing CONTRAST turned up.

to push it in and then turn it to make the dial work. When adjusted in one direction, the picture becomes soft and fuzzy. When adjusted in the other direction, the sound may become raspy and cause wavy lines while the picture becomes rough-edged and grainy (as in Figure 2–2). Sometimes TV sets lose their color when they are mistuned. FINE-TUNE the set to where the picture is slightly grainy and then back up just a little until the picture is sharp and the sound is good. Now you have a well-tuned picture. The color may even come on. Now activate the AFT or AUTO control (or whatever), and your picture should look the best it can be.

6.   *The show is black and white, but the color re-*

ceiver keeps flashing colored splotches over the picture. Ask your TV repairman to adjust your COLOR KILLER control on the TV. Meanwhile,

a.   Turn the COLOR INTENSITY knob down all the way.

b.   Or, turn off the AFT.

c.   Or, turn off the AFT, and then turn the FINE TUNING farther toward the soft fuzzy picture position until the color flashing stops.

7.   *The receiver and show are both in color, but the color looks terrible.*

a.   First turn off the AFT, AUTO COLOR, or whatever.

b.   Then adjust HUE and COLOR to suit your tastes. Use someone's face as a guide of good color. When the flesh tones are just right, everything else generally looks good.

c.   If this doesn't help, make sure the FINE TUNING is properly adjusted.

d.   Finally, turn the AFT back on.

e.   Do not mess with any color knobs in the back of the set. They will undoubtedly confuse the issue.

8.   *Objects in the picture on your color TV seem to have colored ridges around them.* This problem is especially noticeable when white lettering appears on a black background, and the problem is often more pronounced in the corners of the picture.

a.   Try adjusting the TV antenna. You may be getting "ghosts" (reflected TV signals interfering with your main signal).

b.   If this happens on all channels, call a TV repairperson. Your TV set is most likely out of CONVERGENCE, which means the set is not painting its red, green, and blue pictures precisely atop one another.

9.   *When showing black-and-white pictures on a particular TV, part of the screen area seems tinted one color or another.*

a.   Make sure there are no magnets, loudspeakers, or big motors near the TV screen or behind the TV.

b.   Try turning the TV on and then off. Wait a minute and repeat. Do four times. Is the problem gone? If not,

c.   Call the TV repairperson to either DEGAUSS the screen (remove residual magnetism) or adjust the TV picture tube's PURITY.

10.   *The picture has grain, snow, or ghosts as in Figure 2–10.*

a.   Adjust the FINE TUNING as previously described.

b.   If the set is operating on its own antenna—

FIGURE 2-10 Snowy or grainy picture is usually caused by a weak antenna signal.

whether one or two VHF RABBIT EARS or a clip-on UHF BOW TIE or LOOP antenna— aim the antenna around in various directions to see what happens. If the antenna is a telescoping one, be sure it is *pulled all the way out* to full extension. If your TV set is receiving its signal from an outside antenna or from a cable TV company, there may be an antenna or a cable problem making the antenna signal weak. Grain and snow are usually signs of a weak TV signal.

   c. If a videocassette recorder is feeding the signal to the TV set, the recorder may be having problems. They will be discussed later.

11. *Picture squirms badly on the screen, often with a wide white bar wiggling through the picture as in Figure 2–11. The sound is often bad too.*
   a. You may have tuned into a SCRAMBLED pay TV station. You'll need to rent a DESCRAMBLER (described further in Chapter 4) to make the programs viewable.

12. *On some pictures, especially with ones with white lettering on them, the TV sound buzzes.*
   a. Try adjusting the FINE TUNING.
   b. The antenna signal may be too strong. Adjust the RABBIT EARS or look in the back of

the set to see if there is a DISTANT/LOCAL switch near the antenna terminals. Switch it to LOCAL to weaken the signal.
   c. If the preceding doesn't help, your TV set needs an adjustment by a TV repairperson.
   d. If the buzzing happens on one particular station, try calling the station. Sometimes the broadcasters pump out too strong a picture signal (called OVERMODULATING THE VIDEO) which drives everybody's TV crazy.
   e. If the buzzing comes when you're playing a videocassette into your TV, perhaps your tape player's RF GENERATOR is OVERMODULATING its VIDEO. See if a technician can adjust it (more on this in Chapter 5).

13. *There is a lot of static in your TV's sound, especially on the letter S.*
   a. Try everything in part 12.
   b. Your TV speaker is probably torn.

14. *People complain of getting a minor electric shock off the TV screen.*
   a. This is harmless static electricity, which normally builds up on the TV screen.

FIGURE 2-11 Squirmy picture; often bad sound. Are you trying to watch a SCRAMBLED pay-TV station?

MINIGLOSSARY

Loop antenna Circular portable UHF antenna.

*Scramble Distortion of the TV sync signals by a broadcaster or a cable company to render the picture unwatchable (it contorts and wreathes) without a descrambler.

*Descrambler An electronic device (usually rented from a pay-TV company) used to convert scrambled TV signals to viewable ones.

Overmodulating Using too much video signal when making an RF signal, which results in buzzing from the TV speaker when white lettering appears on the screen.

*RF generator or modulator Electronic device which combines audio and video to make RF (radio frequency), a channel number. Usually built into VCRs so that signals may be fed to TV antenna terminals.

Hum bar Hum is usually 60-Hz (60 cycles per second) electrical interference from power lines. When seen on a TV screen, this interference creates a soft dark bar across the screen.

Power supply Circuit in electronic equipment which converts household electricity to the kind of power (correct voltage and frequency) the equipment needs in order to run.

b. If anyone complains of a jolt from a TV's *metal* parts, pull the plug immediately; the set may be dangerous. Have it checked out.

15. *A soft dark horizontal bar occupies part of the screen or rolls through it. It may look like Figure 3–38.*
    a. This is called a HUM BAR and may be caused by your AC (electric) power "leaking" a signal into your picture. Check your cables and move the electric wires away from your antenna wires.
    b. Your TV set's POWER SUPPLY circuit may be dying. Call your TV service.

16. *Circles are egg shaped. Boxes have curved sides.*
    a. If you can find the TV set's HORIZONTAL LINEARITY (or HOR LIN) control, that will adjust parts of the picture that get too fat.
    b. The VERTICAL LINEARITY (or VERT LIN) control adjusts parts of the picture which get too tall.

Both of the preceding adjustments are likely to be hidden behind the set and require tweaking through a labeled hole with a skinny screwdriver or plastic hex wrench. Perhaps ask a TV techie to do it.

17. *TV picture looks too big for the screen. Parts of the picture disappear off the edges.* The picture is what is called OVERSCANNED. To reduce its size,
    a. Look for a VERTICAL HEIGHT (or VERT HT) control to reduce the height.
    b. Look for a HORIZONTAL SIZE (or HOR SIZE or WIDTH) control to shrink the width.

As in part 16 above, these controls are usually tucked away behind labeled holes in the back of the set and may require a technician with a long skinny screwdriver.

18. *The picture appears too small for the screen. You see a black margin around it.* This is called UNDERSCANNED. Studio and control room TV sets are often UNDERSCANNED on purpose so the total picture, edges and all, can be seen. If it's a problem for you,
    a. Adjust VERTICAL HEIGHT and/or

b. Adjust HORIZONTAL SIZE as described in part 17.

19. *People complain of eyestrain while watching TV and ask if they are getting radiation from the set.* Today's black-and-white TVs make no significant radiation. Color sets make some X-rays, but they are trapped inside the set. Color sets also have an automatic disabling circuit which shuts the set down if it has the potential for making too many X-rays. In short, they are safe. Eyestrain usually results from watching a set too long or watching from too close to the screen. Viewers should sit five times the picture's *width* away from the screen; this would be 7 feet for a 19-inch (diagonally measured) TV screen.

TV MONITOR

In contrast to a TV receiver, a television MONITOR does not play AUDIO and does not change channels. All it does is display a picture that it receives directly from a TV camera or recorder via a wire or cable. Like TV receivers, MONITORS have ON/OFF switches and controls for CONTRAST, BRIGHTNESS, VERTICAL, and HORIZONTAL HOLD and perhaps also for PICTURE HEIGHT and WIDTH, but they have no AUDIO controls. Who would ever use such a thing as a TV without sound and without a tuner? Professional TV people, security companies, people who *only* are interested in a picture from a camera, etc. Why should they pay to have a speaker and tuner take up space in the box when they have no intentions of using them? Also, TV MONITORS, being specialized equipment, generally have better circuits and give sharper pictures than regular TV sets. They also cost more. But there's another reason why professionals prefer MONITORS to regular TV receivers. To get VIDEO into a home TV receiver, you have to convert it to a channel number using a MODULATOR, so you can feed it into the antenna of the TV. Once this RF signal goes to the TV's antenna, the TV reconverts it back to VIDEO. All this converting and reconverting makes the picture fuzzier, as you can see in Figure 2–12. It would be nice to skip all this RF foolishness and send the VIDEO from the videocassette machine or camera *straight* to the TV MONITOR, and thus have no picture degradation.

There was a time when the only people who used TV MONITORS were television professionals. The general public

MINIGLOSSARY

Horizontal linearity   TV adjustment which controls how a TV reproduces shapes without stretching or distorting them in the horizontal direction.

Vertical linearity   TV adjustment controlling how a TV reproduces shapes in the vertical direction without stretching or distorting them.

*Overscanned   A TV picture blown up too big on the screen,

causing the edges of the picture to be cut off and hidden from view.

Vertical height control   TV set control which stretches or squashes the picture vertically.

Horizontal size or width control   TV set control which makes the picture skinny or fat.

(a) RF Signal

(b) Video Signal

FIGURE 2-12 Comparing an RF signal with a direct VIDEO signal.

has recently become more aware of the potential of home video and has demanded television equipment of increasing quality. There are now numerous models of high-performance television MONITORS designed for home use. Video connoisseurs would use a supersharp TV MONITOR for their picture, their home stereo for the sound, a fancy tuner to pick up a zillion stations from far away, and a videocassette machine to record the shows that they missed while working a second job to pay for all this equipment.

Back to business. To feed a signal to the MONITOR, simply plug the VIDEO cable into the socket in the back of the set marked VIDEO IN, and a picture should appear on the TV screen (if there's a VIDEO signal in the cable).

Connecting a TV Monitor

First you have to find a wire that has some VIDEO in it. If the socket on the back of the MONITOR matches the plug on the end of the wire, you're in business. If not, you may have to get an ADAPTER to marry the two together (there will be more on cables, plugs, sockets, and adapters later).

Sometimes a MONITOR will have two connections in the back, one for SYNC and one for VIDEO. That means this MONITOR can be used in studios which run SYNC and VIDEO separately. Simply plug the VIDEO cable into the VIDEO input and SYNC input, and you're cooking—almost.

If the MONITOR has a switch that reads INT/EXT SYNC, it will accept SYNC supplied by a cable *only* when the switch is in the EXT position. INT tells the MONITOR to disregard

the extra SYNC coming from the cable and deal only with the VIDEO. If you are not sure which position is correct, try each and check the picture to see when it looks more stable. If the picture collapses, rolls, tears, jitters, or looks like the examples in Figures 2–3, 2–4, or 2–11, this switch may correct the problem.

Using Several TV Monitors to Show the Same Picture

You may notice a lot of other sockets and switches on the back of the MONITOR. Some of these are for connecting several MONITORS together. Before discussing how this is done, let's consider the gas company. It runs a big gas pipe to your house, by your house, and to the next house. You use a tiny amount of gas, and the remainder goes to your neighbor. He uses a bit, and what's left goes to his neighbor. At the end of the line the gas company puts a cap on the end of the pipe to plug it up. They don't want any unused gas leaking out.

Similarly, VIDEO is run to your MONITOR by plugging a VIDEO cable into the VIDEO input of the MONITOR. The MONITOR samples a tiny bit of the signal to make the picture. If the signal is to go to another MONITOR after that, it is LOOPED or BRIDGED, which means that another cable is connected to the back of the first MONITOR and runs to the VIDEO input of the second MONITOR as shown in Figure 2–13. Most of the video signal enters the first MONITOR and then exits through these wires in the "looping through" process. It is possible to show the same picture on five or

MINIGLOSSARY

Monitor A TV set which has no tuner and usually has no speaker (as opposed to a TV receiver, which has both). Such a TV displays video signals but not RF signals. Any device used to observe or hear the quality of a signal (i.e., audio monitor).

Modulator RF generator.

Adapter A connector which allows one type of plug to fit into another type of socket.

Loop or bridge An electrical connection which allows most of the signal to enter one socket and continue out an adjacent socket to be used elsewhere.

Loop through The act of sending a signal to and through a device (using a bridged input) and then on to another device.

Terminate Inserting a resistance (usually 75Ω for video) at the

so MONITORS this way, with each taking a bit of the signal and passing the rest on.

The gas company can't put an unlimited number of customers on the same gas line (they wish they could!). Similarly, you cannot string too many TV MONITORS on the same video cable. The signal will get weak, and the pictures will become dim and grainy.

Similar to the gas company which had to plug the gas main after the last house, something has to be done to the last MONITOR on the line. If it is not LOOPING THROUGH to somewhere, the last MONITOR must be TERMINATED. This is a sort of electronic "plugging up."

There are two ways to TERMINATE this line at the end. The first way is to plug a 75-OHM TERMINATOR into the back of the MONITOR in the place where the cable LOOPING to the next set would have gone, as shown in Figure 2–14. This 75-OHM TERMINATOR looks like a video plug without a cord. TV studios keep them around because every time they use a piece of VIDEO equipment with a VIDEO input that can be BRIDGED or LOOPED, they have to TERMINATE the device (unless they are LOOPING somewhere).

The second way to terminate is to use the switch provided for this purpose (if the equipment has one). It will probably be near the socket and will be marked 75Ω (Ω is the symbol for OHM) or 75-OHM TERMINATOR. The switch positions may be marked "TERM" and "BRIDGED," or 75Ω

FIGURE 2-14   MONITOR properly terminated.

and HI Z. Use the BRIDGED or HI Z position when the signal is to LOOP THROUGH to another piece of equipment. In the TERM or 75Ω position, the switch TERMINATES the signal. Don't use *both* the switch and the terminator plug—use one or the other. Using both is called DOUBLE TERMINATING and causes a picture problem.

How do you tell when a MONITOR or another piece of equipment has the capability of being LOOPED or BRIDGED? Usually the sockets come in pairs set close together and have the words LOOPED or BRIDGED printed near them.

In short, the law reads like this: *If the* MONITOR *(or other device) is made so that it can* LOOP THROUGH, *it must be either* LOOPED *to somewhere else or* TERMINATED.

So if, for instance, you are running VIDEO to only one MONITOR and if it has a terminating switch, the switch must be in the 75Ω position as in Figure 2–15. If, conversely, you LOOP a MONITOR to something, the MONITOR switch should then go to the HI Z position.

SYNC works the same way video does insofar as LOOPING is concerned (thank goodness!). SYNC can be LOOPED to several MONITORS and must be TERMINATED at the last one. Figure 2–16 shows a proper setup.

One last note regarding TERMINATING VIDEO and SYNC lines. If you don't do it properly, it won't damage any equipment. A wrongly terminated cable could mess up a

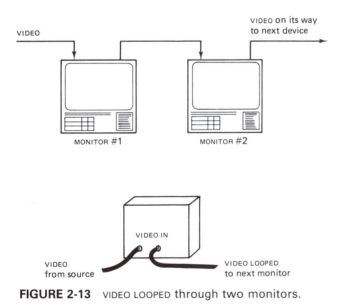

FIGURE 2-13   VIDEO LOOPED through two monitors.

MINIGLOSSARY

end of a cable carrying a signal. A signal may be looped through several devices but must be terminated at the last one by plugging in a terminating plug or throwing a switch to the 75Ω position.

*Ohm   A measure of electrical resistance or impedance. Things must be the same impedance to be electrically compatible with each other. In video, 75 ohms (75Ω) is standard for cables, inputs, and outputs.

Hi Z   High impedance, *not* terminated; *not* 75Ω. An input ready to loop a signal to somewhere else.

Double terminating   Installing two 75Ω terminators on a video cable which should only have one (usually by throwing a 75Ω switch *and* adding a 75Ω terminal plug to the socket).

*TV monitor/receiver (or receiver/monitor)   TV set that can act either as a monitor (using video) or a receiver (using RF).

FIGURE 2-15 Another MONITOR properly TERMINATED.

video tape recording as well as make some studio video equipment misbehave. If, however, all you are doing is sending a video signal to a TV MONITOR, you may hardly notice the difference whether the MONITOR has been properly terminated or not.

VARIOUS KINDS OF MONITORS

In a broad sense, the term MONITOR means any kind of device that allows you to check on the quality of your program. You would MONITOR something by observing it, paying particular attention to the quality. An AUDIO MONITOR could be a speaker allowing you to check the quality

of your sound. The VIDEO MONITOR allows you to check the quality of your picture. There are various kinds of VIDEO monitors. The TV RECEIVER, discussed earlier in the chapter, is often called a MONITOR, but video people are very precise about calling them RECEIVERS. If the TV set *doesn't* have a TUNER in it, we call it a MONITOR. Here are some other kinds of "whichamacallits."

TV Monitor/Receiver

A TV MONITOR/RECEIVER (or RECEIVER/MONITOR) does the job of either a TV MONITOR or a TV RECEIVER. It is two machines in one. It accepts either RF from which it derives picture and sound, or it accepts VIDEO and AUDIO separately and displays them.

The most common MONITOR/RECEIVERS lack SYNC inputs and may look exactly like home TV sets except for an extra switch or two and a couple extra sockets on the side or back. The switch changes the TV from a MONITOR to a RECEIVER.

The simplest MONITOR/RECEIVERS have rectangular sockets with eight little holes. You'll find that classroom TVs have such sockets. The eight-pin plug that goes into this is called, strangely enough, an 8-PIN (Figure 2–17), and a video tape machine generally lives at the other end of the cable. Instead of connecting a VIDEO cable from the tape machine to the MONITOR and then connecting a separate AUDIO cable (to get the sound from the tape player to come out of the TV set), this 8-PIN connector is used. It has both VIDEO and AUDIO wires in it. One convenient cable does the whole job.

To attach the cable, line up the pins on the plug with the holes of the corresponding socket on the MONITOR/RE-

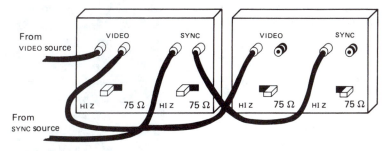

FIGURE 2-16 MONITORS properly wired and TERMINATED for separate SYNC and VIDEO.

MINIGLOSSARY

*8-pin Rectangular plug with eight pins and a cable that goes with it used to carry audio, video, and other signals between a monitor/receiver and a videocassette recorder.

Line An external auxiliary input often used for a video signal. Can also be the final video or audio output signal from a device.

*RGB Red, green, blue. An RGB monitor displays a picture from three video signals—one for the red parts of the picture, one for green, and one for blue. Also the name given to the kind of

video signals which represent component colors, rather than the combined colors.

*RS232-C Standardized multipin computer connection.

T connector Video connector which allows three wires to connect together.

*Characters Letters, numbers, spaces, or punctuation marks which can be printed or displayed on a TV screen.

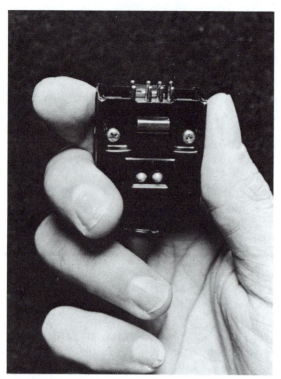

FIGURE 2-17 8 PIN plug.

CEIVER. Then push the plug in until it clicks. Because two of the eight pins are spaced differently than the others, there is only one way the plug can go into the socket (although I have seen some King Kong types who have managed to insert the plug other ways). To unplug it, squeeze together the two buttons on either side of the plug to unlatch the connection and then withdraw the plug as shown in Figure 2-18.

If someone has been using the MONITOR/RECEIVER as a regular TV RECEIVER and you now wish to use it to play a video tape, all you have to do is the following:

1. Turn the MONITOR/RECEIVER power on. Doing this assures that the device will be "warmed up" and ready to go by the time you finish the other steps.

2. Plug in the 8-PIN plug and be sure the other end of the cable is plugged into the video tape player as shown in Figure 2-19.

3. Find the switch on the MONITOR/RECEIVER that says AIR/LINE or TV/VTR and switch it to the LINE or VTR position. This switch converts the TV RECEIVER into a TV MONITOR with sound. That's all there is to it.

If you wish to use the MONITOR/RECEIVER as a straight RECEIVER again, simply flip the switch back to TV or AIR. Done. You don't even have to disconnect the 8-PIN plug.

Remember that in the AIR position the TV set is a RECEIVER and requires an antenna signal.

What if your MONITOR/RECEIVER only has separate VIDEO and AUDIO sockets in the back? As shown in Figure 2-20, just run the AUDIO cable from the video tape player's LINE OUT or AUDIO OUT socket to the MONITOR's LINE IN or AUX (for AUXiliary) IN or EXT (for EXTernal) AUDIO IN, or simply AUDIO IN. Then run the VIDEO cable from the video tape machine's VIDEO OUT socket to the MONITOR's VIDEO IN or EXT IN socket. On some MONITOR/RECEIVERS the label reads simply EXT IN, and they leave it to your good sense to tell which socket is AUDIO and which is VIDEO from the socket shape. Most MONITOR/RECEIVERS, however, are well labeled so you can't go wrong deciding where to plug things in. Figure 2-21 shows a rear view of several MONITOR/RECEIVERS.

Once your machines are connected properly, set the TERMINATING switch to 75Ω or TERM unless you are LOOPING THROUGH to another MONITOR or MONITOR/RECEIVER.

FIGURE 2-18 Pinch the buttons on the 8 PIN plug to unlock it before removing it from a socket.

MINIGLOSSARY

*__Underscanned__   TV picture which is smaller than the screen, showing the black edges of the picture on the screen.

*__Underscanned monitor__   TV monitor which can shrink its picture to display the edges.

*__Pulse-cross monitor__   TV monitor which can shift the picture down and sideways so that the corners of the pictures are in the center of the screen. This professional monitor is used to observe picture and sync problems.

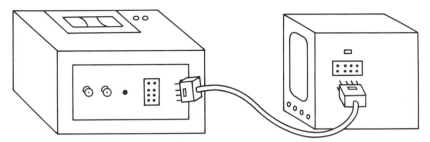

FIGURE 2-19 Connecting a TV MONITOR/RECEIVER to a video tape machine via an 8 PIN cable.

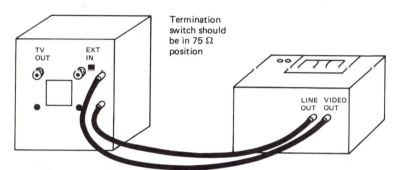

Termination switch should be in 75 Ω position

FIGURE 2-20 Wiring separate AUDIO and VIDEO to your MONITOR or MONITOR/RECEIVER.

(a) Sony Portable

(b) Classroom Style

RCA SelectaVision used by home videophiles.

FIGURE 2-21 Rear view of some popular TV MONITOR/RECEIVERS.

If you are LOOPING to another set, connect the wires as shown in Figure 2–22 and make sure the first TV is *not* TERMINATED and that the second TV *is* TERMINATED (switched to 75Ω).

Most TV MONITOR/RECEIVERS having remote control have no TV/VTR switch. To access the audio/video inputs, you punch in a special channel number, like 91 (for AUX INPUT 1). To return to regular TV channels, just type in the channel number.

Have you noticed that it seems strange to be taking a signal out from a place called EXT IN? The IN sort of implies that the signals go in, not out. Any of them will accept a signal into the TV set. The reason why you get a signal coming out of the adjacent IN socket is that the sockets are BRIDGED or electrically connected together. As described earlier, almost everything that goes in one of those sockets is free to come out of the adjacent socket to be used elsewhere. The TV set isn't *creating* any of the signal that is LOOPING to the next set; it is only sampling a little bit and allowing the remainder to pass on. In fact, you can turn off, unplug, or throw darts at the first set, and it will still pass the signal on to the second set if the two are properly connected.

Some MONITOR/RECEIVERS have additional AUDIO and VIDEO sockets labeled TV OUT, or AUDIO OUT, or VIDEO OUT (see Figure 2–21). Here's what they do: RF (remember RF?) comes down the antenna wire or cable TV wire and into the TV. Depending on where the CHANNEL SELECTOR is set, the TV decodes the RF signal into VIDEO, SYNC, and AUDIO. The combined VIDEO and SYNC come out the VIDEO socket labeled TV OUT, while the AUDIO comes out the adjacent AUDIO socket. These two signals could be sent to a video tape recorder or to another MONITOR, thus allowing you to watch a TV program on two sets at the same time. So to convert a broadcast TV program into VIDEO and AUDIO for recording it or redisplaying it, just connect to the TV OUT sockets.

Now back to the 8-PIN connector for a moment. Remember how the 8-PIN plug carried AUDIO and VIDEO to the

set only without the mess of connecting separate wires to the VIDEO IN and AUDIO IN? The 8-PIN does more than that. It not only *sends* signals to the MONITOR/RECEIVER, but it will *receive* AUDIO and VIDEO signals *from* the RECEIVER. Use the 8-PIN, and you don't have to bother with the TV

Front

Rear

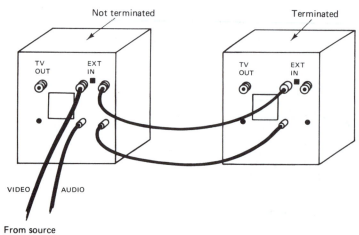

FIGURE 2-22 Connecting two MONITOR/RECEIVERS.

FIGURE 2-23 RGB VIDEO MONITOR.

OUT sockets. You'll see the convenience of this later when we discuss recording TV programs off the air.

RGB Monitor

In Chapter 1 we saw how some color TV cameras could make VIDEO signals for the red, green, and blue (RGB) parts of the picture. We learned that when the three signals were combined to go down a single wire, some of the picture sharpness was lost. Some higher-quality video equipment, in order to keep pictures as sharp as possible, uses these separate COMPONENT video signals rather than the combined video. When computers are used to create video pictures, sharpness is essential because words are often on the screen, and the lettering has a tendency to become fuzzy and illegible if the pictures aren't sharp. Therefore, computers that can make color TV signals, some color cameras, and some high-quality video production equipment use RGB video signals. To view those signals, you need an RGB video MONITOR, like the one shown in Figure 2–23. It looks the same as any other color video MONITOR except that it has some extra sockets and may have an extra switch to change it from a regular MONITOR to an RGB MONITOR. Although not awfully necessary for regular video applications, the RGB MONITOR is almost a necessity for displaying computer text, as you can see in the example in Figure 2–24.

Figure 2–25 shows how to connect an RGB MONITOR to a color video device which uses RGB signals. The video outputs of such devices are labeled R, G, and B. Simply connect the R output to the R input on the MONITOR, the G to the G, and the B to the B (connect these up differently, and your reds could all appear as blue). Some computers have multipin sockets on them to feed RGB video to their color TV MONITORS. Similarly, some color TV MONITORS designed to work with computers will have matching multipin computer connectors. This fairly standard computer connection is called an RS232-C and is shown in Figure 2–26.

Many RGB sources and RGB MONITORS use a separate SYNC connection (SYNC isn't part of their RGB signal). If your picture rolls or tears into diagonal lines, you'll need to connect this SYNC cable, too, and switch the MONITOR to EXT SYNC.

Here's a trick, sometimes useful in a pinch: If you don't have an RGB MONITOR but do have a TV MONITOR (preferably a black-and-white model), try sending the G

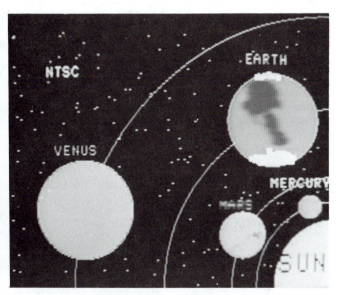

Regular NTSC signal used.   Parts of scene which were colored (the continents, the word "MARS") became fuzzy.

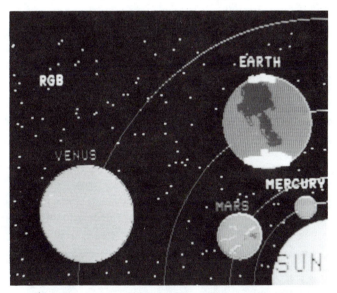

RGB signal used

NTSC text

RGB text

FIGURE 2-24   RGB VIDEO and regular (NTSC) VIDEO compared. (See also back cover.)

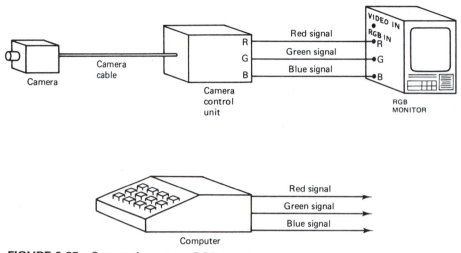

FIGURE 2-25   Connecting up an RGB monitor.

(green) signal to the MONITOR's VIDEO IN. If that doesn't work, try R or B. Often, text-generating computers make white letters which emit the same VIDEO signal at all three outputs. And what if you get a picture but it rolls uncontrollably, and your MONITOR doesn't have a SYNC input? Then connect a cable to the source's SYNC and VIDEO (or G) outputs, attach them together using a T CONNECTOR (Figure 2–32), and send the combined signal to your MONITOR.

Computer Monitor

Some TV MONITORS are designed *only* for computer use. Most models create a green picture on the screen, although some create amber and a few create a white image. The screens on these MONITORS aren't designed to show moving images but to display text. Where normal TV sets can display a maximum of only 40 characters (letters) per line on the screen, these special MONITORS (often called CRTs for cathode ray tubes) can easily display 80 sharp characters on a line. Computer MONITORS have picture tubes with different kinds of phosphors in them which glow for a long time after

FIGURE 2-26   RS-232C connector used to attach computers with RGB monitors.

they've been zapped by the electron gun. This keeps computer text from flickering on the TV screen. Regular TVs, on the other hand, flicker much more when displaying computer text and may cause eye fatigue. Some computer MONITORS have video inputs just like regular TV MONITORS. Others have multipin inputs and are designed to connect with a specific computer. Figure 2–27 shows a monochrome computer MONITOR.

Underscanned and Pulse-Cross Monitors

These monitors are specialized TV sets used to observe technical aspects of the TV signal. The UNDERSCANNED MONITOR, as mentioned before, shows the black edges of the picture on purpose. The PULSE-CROSS MONITOR puts the *edges* of the picture at the screen's center to make them easier to see. Chapter 15 will go into more detail.

VIDEO CONNECTORS AND PLUGS

VIDEO and RF signals, because they have high frequencies, need special wire to travel through efficiently. VIDEO signals require 75Ω COAXIAL or COAX wire, which is round, has a single conductor in the center, and has a braided shield around the outside which helps keep the signal in and keep interference out. Like most wire, COAX is covered with a rubber-like insulation to protect it from moisture.

VIDEO signals are almost always carried through COAX. RF signals are often carried through COAX but are sometimes carried through flat, ribbon-like antenna wire (more on this in Chapter 3). In a pinch, you can try using any kind of wire to carry an RF or video signal. When doing so, try to keep the wire short, as a lot of the signal is lost traveling through the wrong kind of wire. Figure 2–28 shows some TV cables.

FIGURE 2-27 Computer MONITOR.

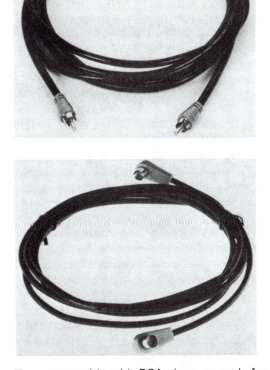

Top, VIDEO cable with RCA plugs on ends for use with most home videocassette recorders. Special COAXIAL wire is used for carrying VIDEO signals efficiently. Common PATCH CORDS used for your home stereo equipment also may have RCA plugs and can work in a pinch, but degrade the VIDEO signal. Bottom, RF cable with push-on F plugs on ends for use with cable TV connections, TV ANTENNA INPUTS, videocassette recorder ANTENNA INPUTS, anywhere RF signals need to go. (Photos courtesy of Comprehensive Video Supply Corp.)

FIGURE 2-28 TV cables.

There are three popular kinds of VIDEO plugs and sockets. Home video equipment almost always uses the PHONO or RCA plug shown in Figure 2–29. What is confusing about this plug is that it is also used for AUDIO. If you are not familiar with the ''feel'' of the cable, you can accidentally pick up an AUDIO PATCH CORD, which is used to connect turntables and audio cassette decks and *also* has RCA plugs on the end. VIDEO cables are stiffer and thicker than AUDIO cables, and that should be the tip-off as to which kind you have in your hand. Incidentally, you can use VIDEO cables with RCA plugs for AUDIO with no problem. In the industrial and professional video fields, two other kinds of plugs are used. One is the PL259 connector shown in Figure 2–30. The PL stands for plug. Another name for this is the UHF connector. It goes into a socket like those shown in the Sony portable and classroom TV pictured in Figure 2–

21. To plug it in, simply press it into the socket and then turn the collar clockwise. The collar screws onto the socket, making a firm connection. To remove the plug, unscrew the collar until it is free and then pull on the plug.

Incidentally, the following is true for *all* electronic equipment: When unplugging it, *always pull by the plug— never pull by the cord*. The cord is weak, and the plug is strong.

MINIGLOSSARY

*Coax or coaxial wire Stiff, round wire about ¼ inch in diameter, used to carry video, sync, or RF (antenna) signals.

*Phono or RCA plug Small connector used to carry audio signals and, in home video equipment, video signals and sometimes RF signals.

*Audio patch cord Wire with audio plugs on each end for feeding signals between two audio devices.

*PL259 or UHF connector A male industrial plug used for video and sometimes RF. Goes into SO259 socket. Rarely used today.

*BNC The most popular industrial connector used for video or sync. Sometimes used for RF.

*Barrel connector An adapter with a socket at each end which allows two cables having male plugs to be connected together.

FIGURE 2-29 PHONO (or RCA) plug and socket. Normally used for AUDIO, most home videocassette recorders use these connectors for VIDEO.

FIGURE 2-32 VIDEO adapters.

FIGURE 2-30 PL239 or UHF connector used in professional video.

FIGURE 2-31 BNC plug used in professional video.

The second popular professional video connector is the BNC (bayonet nut connector), shown in Figure 2–31. Figure 2–15 shows the back of a TV monitor with some BNC sockets. You plug in a BNC by pushing the plug into the socket and rotating the collar one-half turn to the right until it "clicks." To remove it, turn the collar to the left and pull on the plug.

Video Adapters

What if the cable has a PL259 plug, but the socket in the back of the MONITOR is a BNC type? Such incompatibility occurs all the time because different manufacturers use different kinds of sockets. Anyone who does serious TV work stocks a bunch of ADAPTERS like those shown in Figure 2-32. These ADAPTERS permit you to convert from one kind of connector to another. Learn the names of these VIDEO connectors and ADAPTERS and how to recognize them as they will become part of your everyday video language.

VIDEO connectors are either male or female. Examine a few, and you will readily deduce the origin of the sexual connotation. Almost always the plugs will be male, and the sockets in the TV devices will be female. The PL259 (male plug) goes into an SO259 (female socket). The BNC male plug goes into a BNC female socket. The BARREL connector allows two males to connect, and the T CONNECTOR puts together two males and one female.

TV ANTENNAS
AND WIRES

Entire books are written about the many different kinds of TV antennas and methods of connecting them. *Since the antenna is the eyes and ears of the TV receiver, it is important that the signal it delivers to the set be the best possible. Furthermore, the programs you videotape-record off the air will only be as clear as your antenna signal was.*

HOW TV ANTENNAS WORK

The broadcaster pumps his or her RF signal out of the antenna at a certain frequency or channel number. This signal travels in a straight line except when it bounces off something like a mountain, a large building, or a water tower. The signal will bend slightly to reach a little ways over the horizon. As the signal travels farther, it spreads out and gets weaker. If your antenna is within line of sight of the TV transmitter and close enough to it, the antenna will sense this tiny RF signal and funnel it down the antenna wires to your TV set.

Figure 3–1 shows some outdoor TV antennas. Those long probes (called ELEMENTS) that you see on rooftop antennas are a certain length in order to sense certain channels. One pair of probes called the DIPOLE acts much as your rabbit ear antenna flattened out. It sends the signal to the antenna wire. In front of the DIPOLE are some shorter ELEMENTS called DIRECTORS. Like a magnifying lens, they strengthen the signal and make the antenna more directional. Behind the DIPOLE are some longer ELEMENTS called REFLECTORS bouncing more of the signal back for the DIPOLE to pick up. They also increase the directionality of the antenna while blocking out signals which may be coming from behind.

The more ELEMENTS an antenna has, usually the more sensitive it will be (good for distant stations) and the more directional it will be (rejecting interfering signals coming from the side or the back). Unfortunately, the more ELEMENTS it has, the more accurately the antenna will have to be aimed to work properly and also the heavier and more expensive it will be.

Different channels have different frequencies, and the invisible electromagnetic waves have different sizes. It takes

MINIGLOSSARY

Elements   The long probes on a TV antenna.
Dipole   Antenna with two elements. A rabbit ear antenna is a portable dipole.

Power line antenna   TV antenna connection which uses house wiring as the antenna.
*Twin lead   Flat ribbon-like antenna cable containing two wires, nearly always 300Ω.

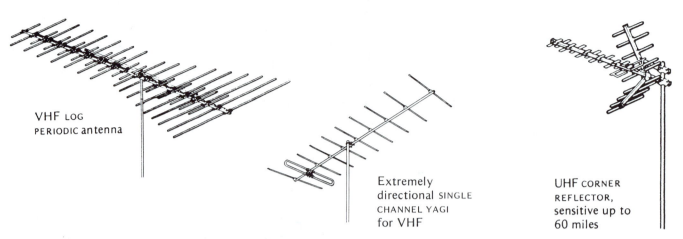

VHF LOG PERIODIC antenna

Extremely directional SINGLE CHANNEL YAGI for VHF

UHF CORNER REFLECTOR, sensitive up to 60 miles

FIGURE 3-1   Outdoor TV antennas (Drawings courtesy of Winegard Co.).

long antenna ELEMENTS to pick up channel 2 because the waves are long. Channel 13 has shorter waves, thus requiring shorter ELEMENTS. UHF antennas have tiny ELEMENTS to accommodate the tiny wavelengths of those ultrahigh frequencies.

MONOPOLE ANTENNA

The simplest of the built-in antennas is the MONOPOLE (see Figure 3–2) for use only when you are close enough to the TV transmitter to receive a strong signal. To work, it must be fully extended (telescoped out), and on small receivers it picks up both VHF and UHF stations. Point the antenna around the room using trial and error to discover its best position.

DIPOLE OR RABBIT EAR ANTENNA

The V-shaped DIPOLE antenna (see Figure 3–3) is a little better than the MONOPOLE because it is more directional and

FIGURE 3-2   MONOPOLE antenna.

more sensitive. It is used under the same conditions as the MONOPOLE, but aiming the DIPOLE is a little more critical. For example, if Bugs Bunny were sniffing in the direction of a TV station, his bunny ears, spread in a V, would be oriented for the best reception of that station.

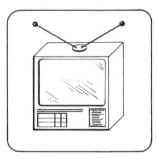

FIGURE 3-3   DIPOLE or RABBIT EAR antenna.

LOOP AND BOW TIE ANTENNAS

If your receiver has a separate antenna for UHF, it will probably look like a little loop (see Figures 3–4 and 3–5). For best results, aim it so the hole of the loop faces the direction of the signal transmitter. Some UHF antennas look like bow ties and clip to the MONOPOLE or DIPOLE antenna as in Figure 3–6. If you were wearing the bow tie antenna, you would get the best reception while facing the direction of the TV transmitter. The UHF antenna must be connected to the UHF screws on the back of the set.

PLACEMENT OF PORTABLE TVS OR TV ANTENNAS

Television signals do not penetrate brick or steel well, so if you are in a building with such construction, position the receiver with its MONOPOLE or rabbit ears accordingly, such as in front of a window. If you still have trouble and if you happen to have some insulated (rubber-coated) wire handy, stick about 3 feet of the wire through the window to the

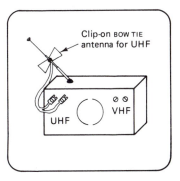

FIGURE 3-6 BOW TIE antenna.

advantages of the portable DIPOLE over your built-in TV antenna are the following:

1. It's a bit more sensitive than a MONOPOLE.
2. It's easier to move the antenna to where the signal is best than to lug the whole TV around, perhaps balancing it on a window ledge. A television falling from the fourteenth floor is rough on passersby.

The POWER LINE antenna shown in Figure 3–8 is something you see ads for. "Turn your whole house wiring into a giant TV antenna," they'll say. Bull pudding! Sometimes they work; usually they don't. Don't buy one unless you can return it.

IMPROVISED TV ANTENNAS

You've rushed a TV set to the chairman of the board's office so he can see himself be torn to shreds on "60 Minutes." When you get there, you discover the set doesn't have a built-in antenna and you forgot to bring one with you. "60

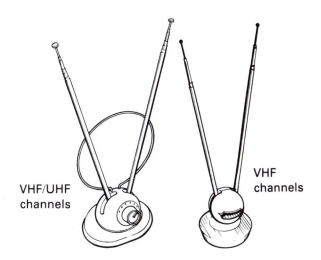

FIGURE 3-7 Indoor portable TV antennas (Courtesy of Winegard Co.).

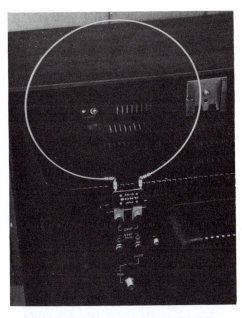

FIGURE 3-4 LOOP antenna for UHF.

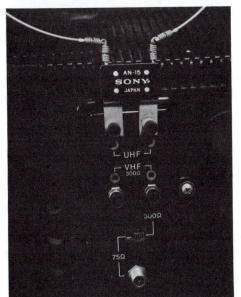

FIGURE 3-5 Connection for LOOP antenna.

outside. Strip 1 or 2 inches of the insulation off the indoor end and wrap the bare wire around the MONOPOLE or DIPOLE antenna on your set. The wire may help carry the TV signal in from the outside and transfer it to your antenna for you.

If you are in a basement or other enclosed area and have trouble getting a good picture, try touching the antenna to something metal: a curtain rod, a metal wall, a bookcase, a steam pipe, or whatever. This may improve the reception. *Caution*: Do *not* touch the antenna to any "live" electrical wiring. Your TV receiver will become an instant hot plate, and you may be the first one cooked.

Portable TV antennas, like those in Figure 3–7, only work well if you are pretty close to a TV station. The

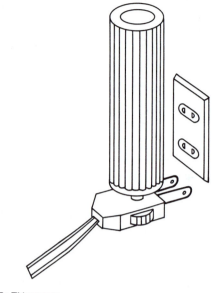

To TV antenna
terminals

FIGURE 3-8 POWER LINE antenna.

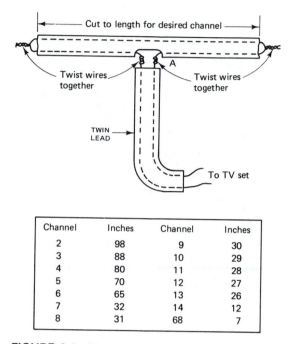

FIGURE 3-9 Homemade FOLDED DIPOLE
antenna.

Channel	Inches	Channel	Inches
2	98	9	30
3	88	10	29
4	80	11	28
5	70	12	27
6	65	13	26
7	32	14	12
8	31	68	7

Minutes'' comes on in 60 seconds, and there's no time to run out for one. If you can't dream up something fast, you may be unemployed in 55 seconds.

Try to locate a clothes hanger, or better yet, a few feet of wire. The clothes hanger you would spread open and hook to one of the antenna screws on the set. Try aiming it around like a MONOPOLE. With the wire you could hang one end out of a window (if possible), and after removing the insulation from the other end, you could attach it to one of the antenna terminals at the back of the set.

Experiment—who knows, you may discover a revolutionary new shape for the superantenna of the future.

Improvised antennas

If you don't have an antenna but you do have some time and wire, you can build one yourself and impress your boss at the same time. It's not hard to manufacture your own FOLDED DIPOLE out of inexpensive antenna wire called TWIN LEAD (more on this wire later). Such an antenna could be strung up in an attic, on the wall behind a TV receiver, under the carpet, or stuck to cardboard so you could aim it. Figure 3–9 shows how to make one. Although this FOLDED DIPOLE is most sensitive to only one specific channel, it will also pick up other channels to some degree. Attach the wires to the proper terminals (UHF or VHF) behind the set as shown in Figure 2–6. Cover the bare twisted wires with tape, keeping the two wires at point A from touching each other.

BIGGER ANTENNAS

If you live well away from the television transmitter, you will need a separate exterior antenna. Generally speaking, the bigger the antenna, the more directional and more sensitive it will be. If your neighborhood gets only one VHF channel, mount a big VHF antenna in a high place, aim it toward the transmitting station, run the antenna wire down to your receiver, and connect it to the screws marked VHF on the back of the set. To receive a UHF station, put up one of the various UHF antennas and connect the lead-in wire to the UHF terminals on the back of the set.

Directionality is the antenna's ability to ''see'' in just one direction, so if you have trouble with a weak signal or with interference or ghosts, you will probably need a large antenna, because they are more directional and able to reject unwanted signals.

KINDS OF TV ANTENNAS

Figure 3–1 showed various outdoor TV antennas. If you live in the boondocks with only one channel to watch, you need but a simple antenna with elements a certain length for that station. Your best bet might be a SINGLE-CHANNEL YAGI because it's relatively inexpensive and directional, giving it high sensitivity with minimal interference.

If you can receive several TV stations, you have a choice of hooking up several individual antennas or mounting a single ALL-CHANNEL antenna. ALL CHANNEL antennas are bigger, heavier, and more expensive because they must contain all the different-sized DIPOLES for all the different channels. With separate antennas, if you receive one station from the north and another from the northeast, you can aim each in the right direction. The ALL-CHANNEL antenna presents a problem, however. If both stations are very strong and the antenna is not too directional, you might try aiming halfway between the two. If the north station is very strong and the northeast station is very weak, you might aim toward the northeast. If the stations are both weak or too far apart or if your antenna is very directional, then you may need an ANTENNA ROTATOR (or ROTOR).

The ROTATOR is an electric motor which fits on the antenna mast and turns the antenna following guidance from

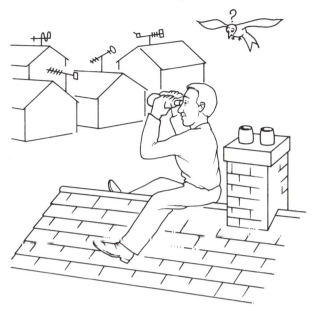

FIGURE 3-11 Check out neighborhood to see who gets good reception. Their kind of antenna may work for you too.

a console near your TV. (See Figure 3–10.) A ROTATOR needs to be pretty heavy duty to support a large antenna in windy places. It is likely to break when the antenna is covered with ice, so buy good-quality mountain-climbing gear to scale the roof in the winter.

Remember, if all your TV channels are arriving from a single direction, you do not need a ROTATOR, so don't let your TV dealer or installer talk you into one of these expensive and sometimes troublesome contraptions.

When selecting a TV antenna, one rule of thumb is to look around the neighborhood, find out who gets good TV reception, and see what kind of antenna they use and how high they mounted it. Often, *but not always,* what worked for them will work for you. Also note: Although any antenna will pick up both color and black-and-white signals, color TVs require a better signal for clear color pictures. Thus, color generally requires bigger and better antennas than black-and-white.

INSTALLING A TV ANTENNA

Buy a paperback guide on this subject for all the details. Here are some important tips the guide may forget to tell you:

FIGURE 3-10 ANTENNA ROTATOR (Courtesy of Channel Master).

MINIGLOSSARY

Yagi A type of outdoor TV antenna.

Single-channel antenna Antenna designed to pick up one channel only.

*All-channel antenna Antenna designed to tune in all TV channels.

*Rotator Motorized rooftop device used to steer a TV antenna in different directions following a command from a console near the TV set.

Standoff A hook which fastens twin lead wire to a building but holds the wire a distance away from the building or from metal.

1. Don't mount the antenna until you've marched around the rooftop (or wherever) with the antenna seeking out the best signal. Sometimes a matter of 6 feet can make a big difference.

2. Wear sneakers or you'll leave long fingernail scratches down the roof and a dent in the ground.

3. Be aware that chimneys, a popular place to mount antennas, expel acids along with the smoke. These acids will accelerate the corrosion of your antenna and connections.

4. Build your antenna mast *too strong*, especially if it must withstand much wind and ice. You don't want it coming down and "shish-kabobbing" the neighbor's dachshund.

AIMING THE ANTENNA

There are several methods for aiming your antenna if you don't have a ROTATOR. First, you could bring a small portable TV up onto the roof with you and connect it to the antenna and watch the results as you aimed hither and thither. The second method requires a walkie-talkie and a helper. The helper with the brains sits downstairs deciding when the picture looks best. The one without brains teeters on the

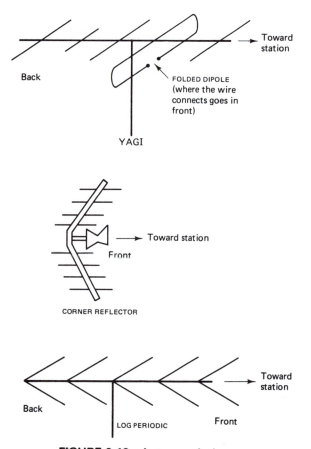

FIGURE 3-12   Antenna aiming.

rooftop, rotating the antenna. Figure 3–12 shows which direction antennas are pointing when you aim them.

Sometimes things get in the way of the signal on its journey to your antenna. Trees and wooden houses block the signal to some degree. If the signal is very strong to start with (like when you are within 30 miles of the transmitter), there's usually plenty of signal left over even when some gets trapped in the trees. Otherwise, you may need a tall mast to get the antenna over the tree line. You may also need a "Piece of the Rock" for when the whole works impales a "Piece of the Roof" in the next ice- and windstorm.

Mountains are a big problem. You might have to mount your antenna on a tower to "see" over the mountain. Sometimes the TV signal will hit a mountain, building, or large metal object and bounce back to your antenna. This is a blessing in cases where a mountain blocked off your original signal. Instead of building a 347-foot tower to peek over the mountain at the original signal, you simply aim toward where the signal is "bouncing" from, as in Figure 3–13.

This "bounced" signal is not a blessing when you receive the original signal *and* the reflected signal together. The result is two signals making two pictures side by side, almost superimposed. One may look like a ghost. Solution: Get a very directional antenna and aim it at the strongest signal; it will reject the weaker signal.

ANTENNA WIRE

TV signals are very fussy about the wires and connectors they'll pass through. And the weaker the station, the fussier they become. A coat hanger will pick up a powerful station, but to pick up a weak station, you need a good antenna and proper wire. There are basically three kinds of proper wire:

1. Flat 300Ω (the Ω symbol stands for OHM, pronounced like *home*) TWIN LEAD
2. Sort of rounded 300Ω SHIELDED TWIN LEAD
3. Stiff, round 75Ω COAX (short for COAXIAL CABLE)

Figure 3–14 diagrams the three types of wire. Each has its own advantages and disadvantages.

Twin Lead

Commonest, cheapest, and worst of the three types of wire is simple TWIN LEAD. It looks like narrow ribbon with thick rounded edges. Some types have heavy-duty insulation (plastic covering) for outdoor use, and others have light flimsy insulation for indoors. Indoor TWIN LEAD will last a long time indoors but will survive only a year outdoors. The insulation

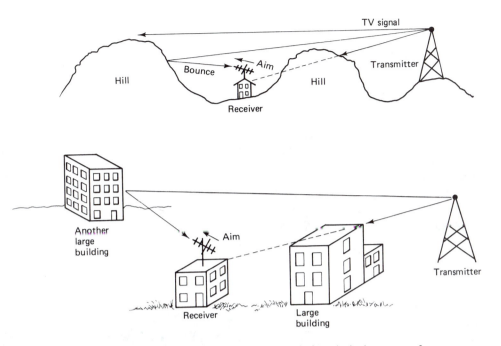

FIGURE 3-13 Sometimes you get a good signal aiming away from the TV transmitter.

will crack, letting water in. TWIN LEAD works abominably when new and wet and even worse when cracked and wet. Outdoor TWIN LEAD is hardier and lasts 4–6 years.

The bare wire connectors can be wrapped around the screw terminals (clockwise for best tightness) on the TV and on the antenna and the screws tightened (see Figure 3 –15). Sometimes, the wire comes with crimp-on SPADE LUGS which fit easily under the screws.

TWIN LEAD cannot be taped to the antenna mast, nor can it *ever* touch metal like rain gutters or aluminum win-

dows. Doing so will weaken the signal, cause ghosts, or make your color flash on and off. Nor should TWIN LEAD lie against a wet roof or the side of a building.

So how *do* you string such wire from place to place? On telephone poles? Yep, almost. You thread it through STANDOFFS such as the one diagrammed in Figure 3–16 which holds the cable away from all surfaces. Inside the home, you can be a little sloppier, fishing the TWIN LEAD under a carpet or behind a couch. Whatever you do, don't try to staple TWIN LEAD to the wall. Staples are metal, a no-no. TWIN LEAD tends to pick up interference easily. The interference could be ghosts (multiple images), the tick-tick-tick of car ignitions, or buzzing from nearby motors. To reduce this interference, try the following:

1. Anchor the antenna wire where it leaves the antenna (using a special STANDOFF made for antenna masts) so that it doesn't flap in the breeze.

2. Run your antenna wire along the "quiet" side of the building away from interferences such as the street, electric wires, or the neighbor's CB radio antenna.

3. Twist the wire once every foot or so as you thread it along its route as shown in Figure 3–17.

4. Do not coil up excess TWIN LEAD. Figure out how much you need, leave a little slack near the TV, and then *cut off the excess.*

One last indictment of TWIN LEAD: It works dreadfully when wet, even when new. When it gets old and cracked, it may fail altogether when wet.

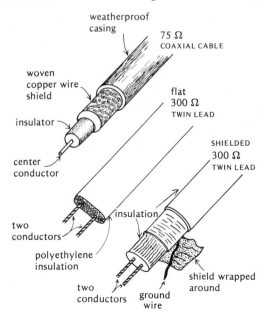

FIGURE 3-14 Kinds of antenna wire.

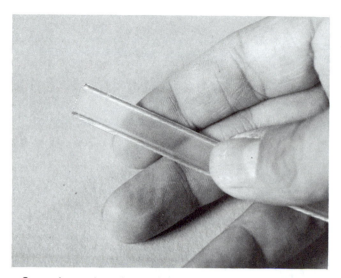

Sometimes the wire ends have no metal showing . . .

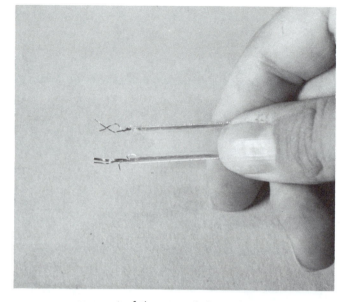

. . . or most of the strands have broken off.

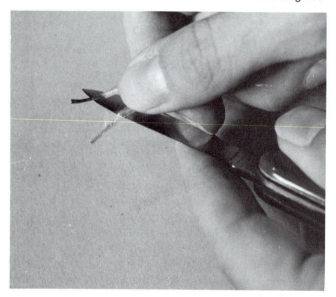

Shave off the insulation revealing bare wires.

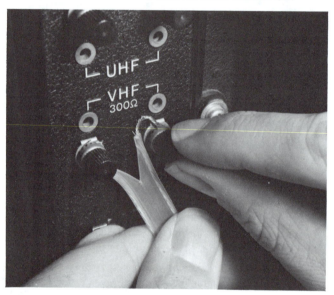

Wind wire around antenna screw terminals and tighten.

300Ω TWIN LEAD with SPADE LUGS for easy connecting.

FIGURE 3-15   Connecting TWIN LEAD antenna wires.

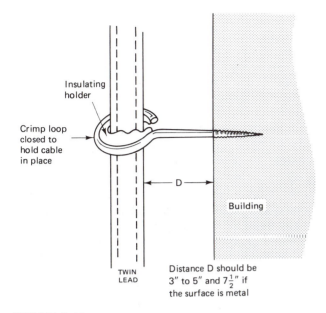

FIGURE 3-16 STANDOFF.

Professionals call it by its technical names:

1. RG-59U for common use.
2. RG-6U for stronger, more interference-proof cable and medium-length runs.
3. RG-11U for heavy-duty use with weak signals and/or long antenna wires. This wire costs about twice as much as the RG-59U.

Like SHIELDED TWIN LEAD, COAX rejects interference and can be attached anywhere and coiled fairly indiscriminately. Coil and tie COAX in 1½-foot diameter circles behind your equipment to hide excess. Tighter coils run the risk of kinking the cable and harming its electrical properties. COAX is not bothered by water and should last outside for 10 years.

Because COAX antenna wire ends with a screw-type plug rather than bare wires or strands, the connection is

Shielded Twin Lead

SHIELDED TWIN LEAD costs twice as much as its humble cousin but is immune to most interference, doesn't require STANDOFFS, can be run next to metal, and can be coiled if desired. It connects up like TWIN LEAD (wrapping the two wires around the two screws) but requires one extra step: A third wire wrapped among the foil shield is a GROUND wire that must be connected to something GROUNDED. Some examples of GROUNDED things include:

1. A copper cold water pipe
2. A metal stake driven deep in the ground
3. The metal conduit or metal armor which covers electrical wires between outlets or light switches or the metal switch-boxes themselves, if exposed (but don't go digging inside the boxes where "live" electrical wiring is present)
4. Another wire connected to any of the preceding

Coax Cable

A little more expensive than SHIELDED TWIN LEAD is 75Ω COAXIAL CABLE (usually abbreviated to COAX). It looks like a stiff ¼-inch rubber hose and has F CONNECTORS on the ends (see Figures 3–18 and 2–28).

FIGURE 3-17 Twist TWIN LEAD along its route to cancel interference.

MINIGLOSSARY

*Shielded twin lead** A "fat" ribbon of TV antenna cable with two wires inside and a metal foil shield surrounding them to keep out interference. Usually 300Ω.

Ohm The symbol for ohm is Ω. TV antenna wires are generally 300Ω, 75Ω, or 72Ω. Ohm is an electrical property of the wire indicating its resistance to certain signals passing through it.

Spade lugs Small metal half-circles bonded to the ends of some twin lead cables and accessories to simplify their connection to TV antenna terminals.

Ground A wire or connection leading into the earth's soil. Cold water pipes are often a good ground.

*F connector** A small socket or plug used for RF or TV signals.

RG–59U Technical name for commonly used coax antenna and video cable.

Interference Unwanted signals which "leak into" your wires or devices and compete with your desired picture and sound, often causing grain, snow, or diagonal or wavy lines on a TV picture.

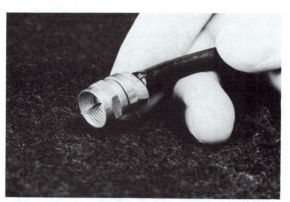

Crimped-on F PLUG

F SOCKET on the back of a TV set

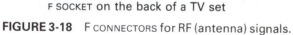

FIGURE 3-18   F CONNECTORS for RF (antenna) signals.

more solid and durable. Crimped-on F CONNECTORS are less likely to fray at the antenna end. At the set end, they allow the cable to be disconnected and reconnected to the set without significant wear. Because of its durability, 75Ω COAX with F CONNECTORS is used in all cable TV and industrial applications.

COAX conducts TV signals slightly less efficiently than TWIN LEAD, which means that if your signal is very weak, TWIN LEAD or SHIELDED TWIN LEAD will do a better job. Some of this loss is in the COAX wire itself, but much of it is due to the transformers.

Impedance Matching Transformers

Impedance? Transformers? Where did they come in? Let's back up.

IMPEDANCE, measured in OHMS (Ω), is an electrical attribute of electronic devices and cables. Some devices let electricity through easily—they have *low* IMPEDANCE. Other equipment holds the current back—they have *high* IMPEDANCE. Things of the same IMPEDANCE go together; things of unlike IMPEDANCE do not—at least not efficiently.

Most TV antennas are 300Ω. Nearly all TV sets have 300Ω antenna inputs. Connect the two with a 300Ω TWIN LEAD, and everybody's happy. There is also such a thing as a 75Ω TV antenna. Some TVs also have 75Ω antenna inputs (that's the F SOCKET you saw in Figure 3–18). Take a look back at Figure 3–5 and notice that next to the F SOCKET is a switch (in the 300Ω position) which can switch the set to the 75Ω mode. Switch it to 75Ω, connect the system together with a 75Ω cable, and you are in business. So what if your TV set has only 300Ω antenna inputs? What do you do with your 75Ω cable?

You plug your 75Ω COAX into a MATCHING TRANSFORMER (sometimes called a BALUN) like the one shown in Figure 3–19. This converts 75Ω to 300Ω and vice versa.

If your antenna and TV were both 300Ω and your antenna wire were 75Ω COAX, you'd install a MATCHING TRANSFORMER at both ends of the COAX to mate the COAX to the antenna and TV. Note that every time you insert something extra in your antenna line such as a TRANSFORMER, even though it's needed, you'll lose maybe 10% of your signal. This INSERTION LOSS can't be helped. Sometimes you can avoid unnecessary insertions. For instance, if you have a 75Ω COAX antenna wire going to the TV set shown in Figure 3–5, plug it straight into the 75Ω input; do not transform it to 300Ω and connect to the 300Ω terminals. Also, try to use 75Ω antennas if you're using 75Ω cable.

Do not try to use CB (citizens band radio) antenna cable in place of TV cable. It looks the same, but CB cable is 52Ω and will often have the numbers RG-58U or RG-8U on it. It will work in a pinch but won't work well.

MINIGLOSSARY

*Impedance matching transformer   A small adapter which allows a cable with one impedance to connect to an input, output, or cable of another impedance. For TV antenna signals these transform 75Ω to 300Ω and vice versa. Also called a balun.

*Impedance   Measured in ohms (Ω), it is an electrical property of a circuit involving its resistance to electrical current. Devices and cables of the same impedance can work together. Those of differing impedances have difficulty.

Insertion loss   A decrease in signal strength when a device is connected into a circuit. Accessories with low insertion losses are desirable to preserve signal strength.

*Antenna preamplifier   An electrical device usually connected near the antenna which makes a weak antenna signal stronger.

Antenna joiner   Electrical device which connects to two or more TV antennas and sends the combined signals to your TV set.

75 Ω COAX
cable connects
here

300 Ω TWIN LEAD
to TV set

FIGURE 3-19   IMPEDANCE MATCHING TRANSFORMER or BALUN (Courtesy
of Winegard Co.).

Table 3–1 summarizes the advantages and disadvantages of each kind of TV wire.

TABLE 3–1
Comparison of TV antenna cables

Cable Type	Efficiency Dry	Efficiency Wet	Rejection of Interferences	Cost
300Ω TWIN LEAD	75%	6%	Fair	Low
300Ω SHIELDED TWIN LEAD	60%	60%	Good	Medium-high
75Ω COAX	40%[a]	40%	Good	High

[a] Including losses by transformers.

OTHER ANTENNA GADGETS

You might want to boost a weak antenna signal, split it and send it to several places, or mix signals from several antennas. Here are the gadgets that help you do it.

Antenna Preamplifiers

What if you are using one antenna to serve only one receiver but you still have a grainy, snowy picture, the sign of a weak or distant signal? Perhaps you make sure the antenna is aimed right. Perhaps you buy a bigger, more sensitive antenna. If those tricks don't work (assuming nothing is wrong with your antenna wire or your set), you could buy an ANTENNA PREAMPLIFIER (or, as it is sometimes called, an ANTENNA AMPLIFIER or ANTENNA BOOSTER). This device usually mounts on the pole near your antenna and connects to your antenna wire. Powered by house current, it boosts the antenna signal, thus improving your set's picture.

Don't try to use an ANTENNA PREAMPLIFIER in place of a good antenna. The antenna itself is the place to start. Get the best signal you can first, and *then* amplify it if necessary. Amplifying an inferior signal with ghosts, snow, or interference from a misaimed or too-small antenna will just give you a stronger, inferior signal with ghosts, snow, or interference. For the ANTENNA PREAMPLIFIER to really help, it must have a relatively ghost-free, clean signal to start with. Once it has this "pretty good" signal, it will boost the quality to "good" or maybe even to "excellent." Also, the ANTENNA PREAMPLIFIER will help get back some of the signal that's lost as it travels down long antenna wires. Figures 3–20 and 3–21 show ANTENNA PREAMPLIFIERS and how they are connected up.

It is best to boost the antenna signal at a point as close to the antenna as possible. That is why the outdoor PREAMPLIFIER is better (though more expensive and less convenient) than the indoor type. Here's the reasoning behind this: If you have a weak antenna signal running down a long wire and other interfering signals "leak into" that wire, then the PREAMPLIFIER will boost the weak antenna signal and the weak interference signal, making them *both* strong. If, however, you PREAMPLIFY the weak antenna signal near the antenna before interference can "leak into" the wire, then you have a *strong* antenna signal and *weak* interference signals.

Antenna Joiners

You have a couple of antennas on the roof and you need to run the wire from them to your TV set. All these signals can travel over the same wire, but *you can't simply hitch the same wire to all the antennas*, at least not without losing

---

MINIGLOSSARY

Band   A set of related frequencies. UHF (ultrahigh frequency) is one band 470–890 MHz (megahertz).

Band separator/joiner   Electrical device which separates combined bands (like VHF, UHF, FM) into separate bands (like FM alone) or combines separate bands so the signal can travel on a single cable.

*Passive   Electrical device which doesn't need electrical power (i.e., batteries or power from the wall outlet) to operate.

High band   Channels 7–13.

Low band   Channels 2–6.

*TV coupler   A small electrical device which allows two or more TVs to share signals from the same antenna wire.

Active   Electrical device which requires electric power to operate. TV antenna preamplifiers and amplified TV couplers are active.

*Signal splitter   Small electrical device which divides a TV signal into several components. A TV coupler could split an antenna signal into two parts for two TVs. A band splitter could divide a multichannel antenna's single signal into separate bands such as UHF, VHF, and FM.

Transformer splitter   A combined device which does the job of a matching transformer and a band splitter.

Type that goes on antenna pole

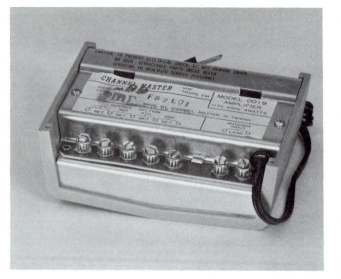

Type that goes inside the home

FIGURE 3-20   ANTENNA PREAMPLIFIERS (Courtesy of Channel Master).

a lot of your signal. The right way to connect antennas is through an ANTENNA JOINER or BAND SEPARATOR/JOINER like the ones shown in Figure 3–22. The wire from one antenna goes to one input of the JOINER, while the wire from the other antenna goes to the other input. The JOINER efficiently combines the signals and sends them out its output. The unit is PASSIVE, which means that it requires no batteries or power to make it work. ANTENNA JOINERS are usually designed to combine UHF and VHF antenna signals or HIGH-BAND (channels 7–13) and LOW-BAND (channels 2–6) signals.

Figure 3–23 shows how ANTENNA JOINERS are connected.

TV Couplers

What happens if you want to connect two receivers to the same antenna? In a pinch, you could attach a second piece of wire to the antenna and run this second piece to the second receiver. But this method wastes a lot of your precious TV signal. Instead, spend a few dollars on a TV COUPLER, which will efficiently divide the antenna signal into two (or more) signals for two (or more) separate receivers.

Match the coupler to your needs. If you have only two TVs, buy a COUPLER with only two ports (outputs), not three or four. If for some reason you end up with an unused port on your COUPLER, that port must be TERMINATED with

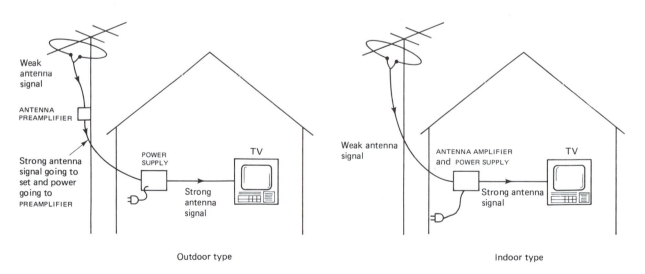

FIGURE 3-21   Connecting up an ANTENNA PREAMPLIFIER.

300 Ω
BAND JOINER
**(Courtesy of
Blonder Tongue)**

75 Ω
BAND JOINER
**(Courtesy of
Winegard Company)**

FIGURE 3-22 ANTENNA JOINERS

a 75Ω TERMINATOR like the ones mentioned in Chapter 2. Also, a four-port COUPLER will send one-fourth the signal to each port *whether a TV is connected there or not.* If only two TVs are connected to a four-part COUPLER, each will still only get one-fourth of the signal strength.

TV COUPLERS come in two varieties: PASSIVE and AMPLIFIED. The PASSIVE COUPLER costs only a few dollars and is easy to install, but all it does is share the signal among the sets it serves. If there are two sets, each gets half the signal strength; if three, each gets one third; and so on. When you have more than enough signal to begin with, the degradation in picture quality may not be noticeable. But

if your signal is weak to start with and provides a picture that is already slightly grainy, using a PASSIVE COUPLER will make that picture even worse. Figures 3–24 and 3–25 show PASSIVE COUPLERS and how they are connected.

When signals are weak, you need an AMPLIFIED COUPLER (Figure 3–26). This accessory costs about $50, is powered by house current, and is connected in much the same way as its passive brother, but it provides much better results. Instead of simply sharing the available signal among receivers, the AMPLIFIED COUPLER boosts the signal to each receiver so that each gets full signal strength. Also, it doesn't matter if some of the ports go unused. Each TV will get its

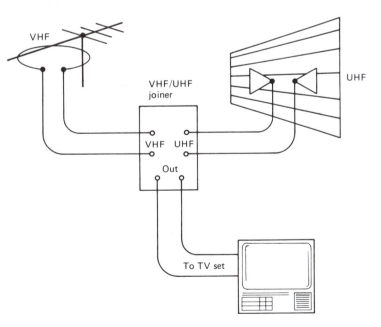

FIGURE 3-23 Connecting up an ANTENNA JOINER.

FIGURE 3-24 PASSIVE TV COUPLER.

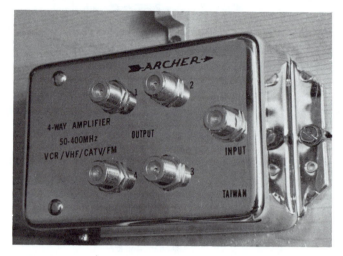

FIGURE 3-26 AMPLIFIED TV COUPLER.

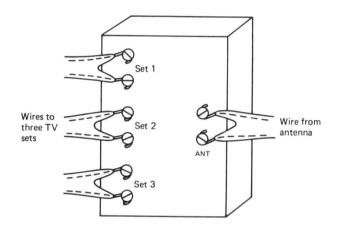

FIGURE 3-25 Connecting TV COUPLERS.

Signal Splitters

A TV SIGNAL SPLITTER is a device that takes the TV signal in a single wire and divides it into several signals. It could either divide the TV signal into equal parts for several TVs (in which case the splitter is often called a TV COUPLER) or divide the signal into UHF, VHF, and FM signals to go to specific places (in which case the SPLITTER is technically called a BAND SPLITTER). Here's how the BAND SPLITTER is used.

Your ALL-CHANNEL antenna picks up UHF, VHF, and usually FM radio signals. All three signals travel through a single wire from the antenna. To separate the signals so that you can send each one to its proper terminals on the TV set, you use a VHF/UHF/FM SIGNAL SPLITTER. If FM is not important, you use a VHF/UHF SIGNAL SPLITTER and connect the wires as shown in Figure 3–27.

If the ALL-CHANNEL antenna is sending you its signal via 75Ω COAX *and* you happen to have the appropriate 75Ω input on your TV, there's no need to split the signal. Simply plug the COAX into the 75Ω input, switch the TV antenna's switch to the 75Ω position, and you're done.

What do you do if the antenna has a 75Ω COAX wire and your set doesn't have a 75Ω socket in the back? You

full share of signal. In general, the AMPLIFIED COUPLER is a must for institutions that connect multiple sets to the same antenna.

If you buy an AMPLIFIED COUPLER, make sure it has UHF capability if you want to boost UHF channels too. Also, make sure it has cable TV capabilities if you are using those signals. Some AMPLIFIED COUPLERS only boost VHF.

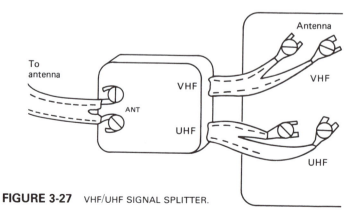

FIGURE 3-27 VHF/UHF SIGNAL SPLITTER.

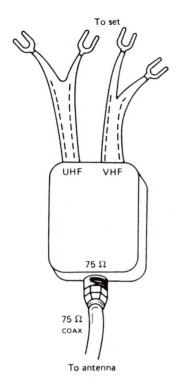

FIGURE 3-28 MATCHING TRANSFORMER which converts 75Ω TWIN LEAD. This model also acts, as most do, asa SIGNAL SPLITTER.

FIGURE 3-29   300Ω antenna wall connection used mostly in homes.

buy a MATCHING TRANSFORMER/SPLITTER like the one in Figure 3–28. It does two jobs at once. It converts the signal from 75Ω to 300Ω for your TV, and it also splits the signal into UHF and VHF for you to connect to those terminals on your TV. These devices are very common and inexpensive.

TRANSFORMER/SPLITTER combinations like this are often used in institutions and in buildings serviced by cable TV or where subscribers share a community antenna. Sometimes you never see the cable at all, only a socket in the wall. If the socket has screw threads on it like the 75Ω socket on the back of your receiver, you will need a 75Ω COAX cable with an F CONNECTOR on each end to join the set and socket. No SPLITTER will be necessary. On the other hand, if the wall socket has three holes about a ¼ inch apart, as in Figure 3–29, it requires a 300Ω TWIN LEAD with a matching plug. This kind of setup is usually found in private homes and will require a SPLITTER if you want both UHF and VHF.

ANTENNA JOINERS and TV COUPLERS do very similar jobs. One combines signals together, and the other divides them. The same goes for BAND SPLITTERS and JOINERS. A

BAND is a bunch of related frequencies. VHF is a BAND, UHF is another BAND, the low TV channel numbers can be considered a low BAND, and the high channel numbers are a high BAND. An ANTENNA JOINER can join a UHF BAND antenna and a VHF BAND antenna to make one signal. At the other end of this wire, a BAND SPLITTER can separate the UHF and VHF signals again. BAND JOINERS and SPLITTERS are so similar that they are often used interchangeably, but because the ANTENNA JOINERS often reside outdoors, they need to be weatherproof.

Adapters

The most popular adapter is the MATCHING TRANSFORMER, discussed earlier in the section on antenna wire. Figure 3–30 shows some more adapters.

The PUSH-ON adapter is very handy when you have to frequently connect and disconnect an antenna cable. F CONNECTORS are normally plugged in by pushing them into the socket and rotating the nut-like collar a zillion turns. Removing an F CONNECTOR requires unscrewing and unscrewing. The PUSH-ON connector simply pushes on and pulls off. In Figure 3–30, can you tell which adapter is for video and not for RF?

Ghost Eliminators

Large metal structures like water towers and buildings sometimes bounce back to you a delayed TV signal which interferes with the original TV signal and causes a double image, or "ghost." Inexpensive GHOST ELIMINATORS,

MINIGLOSSARY

Push-on F connector   Special F connector which pushes on and pulls off rather than screws onto an F socket.

Ghost eliminator   Electrical device to remove double images (ghosts) from a TV picture.

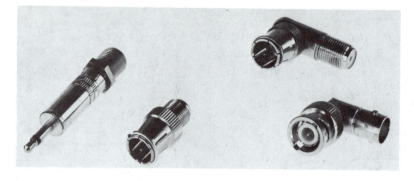

FIGURE 3-30 TV adapters. From left to right: female F to male mini (used in some video recorders), female F to male F PUSH-ON, female F to right angle F PUSH-ON, female BNC to right angle male BNC. (Courtesy of Comprehensive Video Supply Corp.).

sometimes built into portable RABBIT EAR antennas, are often advertised as the solution. Don't expect much from these devices. Short of hiring an exorcist or holding a seance, the best way to get rid of ghosts is to buy a good directional antenna and aim it carefully.

Some recent DIGITAL televisions have circuits in them which can recognize a ghost and remove it from the picture.

Attenuators and Filters

You *can* have too much of a good thing. If your TV antenna signal is too strong, your picture may bend or be "contrasty" or you may hear buzzing in your sound. Some TVs have DISTANT/LOCAL switches on the back which when switched to LOCAL will ATTENUATE the signal coming in from a strong local station. You can also buy an ATTEN-

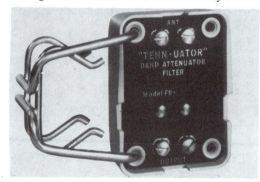

FIGURE 3-31 ATTENUATOR (Courtesy of Blonder Tongue Labs).

UATOR to connect between your antenna and your TV set to weaken the signal. Figure 3–31 shows an ATTENUATOR which clips onto the antenna mast.

An ATTENUATOR can be made selective so that it will block out just one TV channel. Such ATTENUATORS are then called FILTERS. Cable TV companies often use FILTERS to stop pay-TV signals from coming to your house when you're not paying for them. Hospitals, for instance, would use a FILTER to keep their patients from viewing a particular channel showing live surgery and meant just for doctors and nurses.

Antenna Switches

If you had two antennas and wished to switch your TV from one to the other, you could use an ANTENNA SWITCH (or A-B SWITCH or RF SWITCH or COAX SWITCH), such as the one shown in Figure 3–32. It is a special kind of switch designed for RF signals. These switches also allow you to switch a signal to one of two TV sets. Buy an extra-good-quality one if you plan to switch between a very strong signal (like cable TV) and a weak one (such as from an antenna). Cheap RF switches will "bleed" some of the strong signal into your weaker signal, interfering with it.

Lightning Arrester

The LIGHTNING ARRESTER protects your TV system somewhat from static electricity and lightning strikes. It connects to your antenna wire, usually near the ground somewhere, where you attach a "ground" wire which leads to a metal

MINIGLOSSARY

Attenuator Small electrical device which reduces the power of an electrical signal. Useful for weakening too strong TV stations.

*Filter Small electrical device which can remove a certain frequency (i.e., a certain channel) from a signal. Some filters can remove many frequencies, leaving just the desired ones. Also called a trap. Audio filters remove certain tones from a sound signal.

Antenna switch An electrical switch specially designed for antenna signals.

Lightning arrester Device which clips onto an antenna or cable TV cables, and connects to a grounded wire. It is designed to divert the shock of a lightning bolt so that the current doesn't damage your equipment.

Antenna system accessories

Type	Where Used	Function
Antenna amplifier	Near antenna	Increase strength of signal from antenna
Balun, or matching transformer	Antenna and receiver	Match 75Ω coaxial cable to 300Ω twin lead
Joiner	Antenna	Combine the outputs of two or more antennas into one signal
Two-way coupler	Near TV sets	Enable use of same antenna by two receivers
Four-way coupler	Near TV sets	Enable use of same antenna by four receivers
High-pass filter	Input to TV set	Reduce or eliminate interference from radio transmitters
Lightning arrestor	Transmission line at point of entry into building	Protect receiver against damage by lightning and heavy static charges
VHF/FM splitter	Near TV set or FM radio receiver	Separate VHF TV, and FM radio signals and channel them to the appropriate receivers
VHF/UHF splitter	Near TV-antenna terminals	Separate VHF and UHF signals and channel them to the proper receiver inputs

post driven into the dirt outside the building. Hopefully, if lightning strikes the antenna, it flows down the antenna wire, through the post, and harmlessly into the earth, rather than toasting your TV tuner.

COMMON TV ANTENNA AILMENTS AND CURES

The most common TV antenna problems don't involve the antenna at all. The trouble is generally with the antenna wire and more often involves TWIN LEAD than other types

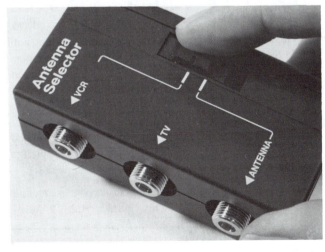

FIGURE 3-32 RF switch

of wire. Check to see if it is disconnected, loose, touching metal, cracked, wet, corroded at the connections, or otherwise in bad shape.

Also, you need to decide whether your problem is with the antenna signal or with your TV set. One way to do this is to disconnect your outdoor TV antenna and try a portable rabbit ear antenna in its place. If the trouble disappears (taking into consideration the limited sensitivity of the little rabbit ear), your outdoor antenna system is probably faulty. If the symptoms are unchanged, perhaps the problem is with the TV set.

Let's start with TV interference problems:

Sporadically you get interference in your picture and sound as follows: The interference may be in the form of a crosshatch or herringbone pattern on the screen like in Figure 3–33. There may be bars of color floating through the picture. In extreme cases, your picture may even turn negative like a photograph negative. The sound may buzz or sound like a garbled Donald Duck. If this is your problem, you probably have CB interference, good buddy.

Citizens band radios transmit on frequencies much like TV. If a strong transmitter is near your TV, its signal can sneak into your "M*A*S*H" reruns.

You can be more sure your interference is from CB if:

1. The problem is worse on channels 2, 5, and 6.
2. The problem occurs only at certain times, like a

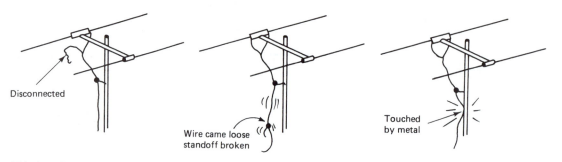

TV signal problems are often caused by a faulty antenna wire.

certain hour of the evening (CBers are creatures of habit too).

3. You can hear the voice and you can recognize CB language like "breaker, breaker," or "10-4" or at the end of every other sentence the words "fer sure." The voices seem preoccupied with someone called "Smokey." You may also hear call letters like WBS7341 or a slangy identification like "Square Dance Lady" or "Roadrunner," and you may even pick up a clue to the speaker's address.

4. You take a stroll around the neighborhood and notice strange antennas looming over rooftops or out windows. You may see superlong antennas attached to cars nearby.

Whose fault is TVI (TV interference)? If it's only on channels 2, 5, or 6, it's usually the CBer's fault. CB transmitters broadcast a strong signal at their proper frequency, and a tiny bit of signal at some multiple of that frequency is called a *harmonic*. Your TV channel 2 happens to be just twice one of the CB frequencies. Channel 5 is three times another CB frequency. Your TV, when tuned to these channels, is sensitive to the CBer's harmonic and is receiving that signal in competition with your desired TV channel.

FIGURE 3-33 Interference from a CB transmitter.

The remedies for CB-caused interference are as follows:

1. Locate the culprit.
2. Ask the CBer nicely to install a LOW-PASS FILTER on his or her antenna. This should filter out those nasty harmonics which bother your TV.
3. If you can't locate the CBer or you get no co-operation, try
    a. Checking the phone book for the local CB club. Some have committees that try to resolve problems between the CBers and the TV viewers.
    b. Contacting the FCC (listed under U.S. Government, Federal Communications Commission in the phone book). They are pretty busy, so writing them a letter may be faster than getting through by phone.

If the interference is on other channels besides 2, 5, and 6 or if the CBer was cooperative but it didn't help, the fault may be with your antenna or your TV. First see if it's the antenna's fault. Disconnect the antenna during the next CB transmission and see if you still pick up the interference (don't worry about your TV station at this point). If the interference remains, it's not your antenna that is picking it up—it is your TV set. It needs modification to reject the interference.

This can often be corrected by adding a special FILTER to circuits in your TV set. Many TV manufacturers will provide the FILTER free or at nominal charge. A technician could install it, but it's simple enough to do yourself. Your TV dealer may even contact the manufacturer in your behalf to order the FILTER. Then again, could you move the set to some other room better shielded from the interference?

If, in the preceding experiment, the interference stopped when you disconnected your antenna, then it's the antenna that's the culprit. There are several possible remedies:

1. Install a HIGH-PASS FILTER between your antenna cable and your TV. This "passes" the TV signal

---

MINIGLOSSARY

Low-pass filter   Filter placed between a radio transmitter and its antenna to reduce the harmonics from being broadcast. The harmonics interfere with people's TV reception.

High-pass filter   An antenna filter which allows normal TV channel frequencies to go to the TV set but stops interference from lower frequencies.

FIGURE 3-34 HIGH PASS FILTER
(Courtesy of Radio Shack).

to your TV while rejecting the interfering signal. Sometimes these devices are called TRAPS (because they "trap" the offending signal). Radio Shack and other TV parts stores sell them. If one TRAP helps but not enough, try two at the same time. Figure 3–34 shows a TRAP.

2. Another remedy for TVI is to switch from the flat TWIN LEAD that you may be using to the SHIELDED TWIN LEAD or to COAX. These are well shielded from outside interference.

Sometimes a very close AM radio station will interfere with your TV, perhaps throwing vertical, red, wavy lines into your TV picture. There's not a lot you can do (other than move away or switch to cable TV). The station's chief engineer might give you some suggestions if you call.

To read more about this subject, consult the *Consumer Electronic Service Technician Interference Handbook—TV Interference* (catchy title) available from the Electronics Industries Association, 2001 Eye Street N.W., Washington, D.C. 20006, or Sony's *Interference Handbook* from Sony's Technical Publications Department, 4747 Van Dam Street, Long Island City, N.Y. 11101.

Interference in the form of black horizontal jagged lines ripple or roll through the picture most often on channels 2, 3, and 4: This may be caused by electrical arcing of a nearby machine or appliance such as a photocopier, a compressor, a large fan, a loose three-way light bulb, a sputtering fluorescent fixture, a vacuum cleaner, a floor polisher, electric pencil sharpeners, razors, engravers, etc. Try turning off everything around (except the TV) and watch.

If the problem stops, switch things back on again one at a time until you find the offender.

Sometimes a snowy buzzing accompanies an electric motor running somewhere. Often power tools or electric mixers spit "spikes" (electrical power surges when switched on or off) and buzzing signals down their electrical cords and into the building wiring. They then pass into your TV set. Try plugging a POWER LINE FILTER (see Figure 3–35) between the wall socket and the TV set or between the wall socket and the appliance. POWER LINE FILTERS also quiet down audio interference and may even prevent computer glitches too.

Sporadic interference that goes tick-tick-tick, speeds up, slows down, is often accompanied by dashes of colored snow (or black-and-white snow), and sounds sort of like a car's engine: This is static from auto ignitions. Try:

1. Making sure your TWIN LEAD antenna wire is twisted about once every 2 feet throughout its journey from antenna to TV set.
2. Relocating your antenna and antenna wire as far as possible from the street. Try to put a building between the antenna (and wire) and the road.
3. Installing SHIELDED TWIN LEAD or COAX antenna wire.

Constant herringbone pattern on a certain channel sometimes affected by your program's sound: That channel's TV signal is too weak. It needs a better antenna, better wire, or a BOOSTER.

Suddenly one night, your local channel is driven off the screen by another, perhaps snowy image. This new image superimposes itself over yours so much that you can

FIGURE 3-35 Installing a LINE FILTER either at the offending appliance or set itself will reduce interference (Courtesy of Radio Shack).

actually read words and decipher the picture. In more moderate cases, you see persistent diagonal lines or ripples (much like Figures 3–33 and 5–47) over your show: This is called CO-CHANNEL INTERFERENCE. There are, for instance, lots of channel 2s. Your TV only picks up the closest one—normally. Sunspots and other atmospheric conditions sometimes ionize the air about 60 miles up, causing this air layer to reflect TV signals from distant stations. Thus, another channel 2 which is over the horizon and perhaps 400 miles away comes bouncing in to compete with your local channel 2.

There's nothing you can do. This unusual phenomenon will go away in a couple of days.

Sometimes you see this problem if you receive cable TV and there is a strong TV station broadcasting on the same channel nearby. The broadcast TV signal "leaks" into your TV set and interferes with the cable signal. In this case, try to shield your TV from the radio waves by moving the set to where metal and brick are in the way. Perhaps try constructing a shield out of tinfoil and partly wrapping the set (leaving the vent holes open so the TV can stay cool).

Picture displays "windshield wiper effect" and writhing herringbone pattern: This effect, shown in Figure 3–36, is called ADJACENT-CHANNEL INTERFERENCE and results when a strong channel's signal "leaks into" the adjacent channel that you are watching. This is most noticeable between channels 3 and 4 and between channels 10 and 11.

Like earthworms, many antennas are nearly as sensitive out their rears as they are out their fronts. In such cases, if you're between two stations broadcasting on ad-

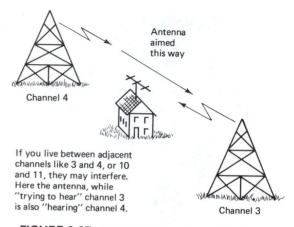

If you live between adjacent channels like 3 and 4, or 10 and 11, they may interfere. Here the antenna, while "trying to hear" channel 3 is also "hearing" channel 4.

FIGURE 3-37   ADJACENT CHANNEL INTERFERENCE.

jacent channels, your antenna will pick up both (see Figure 3–37). The solution is buy a TRAP (a FILTER) tuned to the unwanted station (many are adjustable for various channels) and attach it between your antenna cable and your TV. "Windshield wiper" and constant herringbone problems can also occur when you have several antenna wires running in a jumble close to each other, especially if one of the wires is carrying a strong TV signal and another wire is carrying a weak one. In the mess of wires, the strong signal sometimes leaks into the weak signal's wires. The solution to this problem is to "dress" your wires neatly and try to

FIGURE 3-36   "Windshield wiper" kind of interference from adjacent channel.

FIGURE 3-38   Dark, soft, horizontal bar across picture, caused by bad ANTENNA BOOSTER.

MINIGLOSSARY

*Power line filter   Removes unwanted signals, electrical pulses ("spikes"), and other interference mixed in with your power from your wall outlet.

Co-channel interference   Wavy lines or other interference appearing on the TV screen caused when a TV set receives more than one signal at a time on the same channel (i.e., two channel 3s at once).

Adjacent channel interference   Wavy lines or two TV images simultaneously appear on the TV screen. A problem appears when you're viewing a weak station while another strong station, one channel number higher or lower, is broadcast nearby or from the same direction as the weak station.

keep strong RF signals from getting too close to wires with weak signals. Antenna signals are typically weak. Cable TV signals are typically strong. So if you are running TV wires near cable TV wires, this problem is likely to occur. It also occurs in cheaply made ANTENNA SWITCHERS which allow you to switch between an antenna signal and a cable TV signal.

So much for interference. Here are some specific antenna and antenna wire problems:

Snow, ghosts, no color, color flashes on and off, picture rolls, black ridges around faces, etc.:

1. Replace antenna wire if cracked, frayed, weather-beaten, or old.
2. Check for poor connections, especially between the antenna wire and the antenna where wind has a tendency to loosen the connectors.
3. Replace old or broken STANDOFFS.
4. Install more STANDOFFS, especially if the problem shows up on windy days.
5. Straighten bent antenna elements.

6. If using COAX, retighten the screw-on F CONNECTORS at the TV, SPLITTER, COUPLER, or elsewhere.

If the picture suddenly becomes snowy after you've disconnected and reconnected your COAX, check the F CONNECTOR; perhaps you've bent that thin center wire in the plug. It should be straight like in Figure 3–18.

Ghosts, or loss of color, or loss of detail in picture: If you are using TWIN LEAD, it is probably touching metal somewhere.

Your antenna has a booster and you suddenly get a grainy or snowy picture or you see a wide, soft, dark, horizontal bar across the picture, perhaps sliding up the picture like in Figure 3–38: This is a HUM BAR. Your BOOSTER is probably shot. Fix or replace it.

The lessons you've learned in this chapter about cabling and interference apply to all video equipment and computers, too, not just antennas and TV sets. You will find that the video business is a constant battle for clean interference-free signals, not just from antennas but from everywhere.

4

CABLE TV

HOW CABLE TV WORKS

Gather 'round, boys and girls, and I'll tell you the story of cable TV. Once upon a time (about 1948), a man lived on a mountaintop and got fabulous TV reception. The Valley Villagers below, blocked by mountains, got no TV reception. Poor villagers—no ''Flintstones,'' no ''Wheel of Fortune,'' no ''Family Feud,'' no violence, no sex (well, maybe sex), no Carson reruns, zippo. The Mountain Man, whose wife was the overworked village librarian, hit upon an idea to lighten her work load. He offered to split his antenna signal and run a long antenna wire down the mountain to the village. He would boost the signal so that there would be enough to go around and would run high-quality coax wire on the telephone poles past all the village homes. Since the wires, boosters, the telephone pole rentals, antenna, and upkeep all cost money, the Mountain Man would charge a nominal fee to all subscribers on his COMMUNITY ANTENNA.

The village fathers, after a noisy debate and a quiet payoff, granted the Mountain Man an exclusive FRANCHISE

to be the only antenna in town. After all, it wouldn't make sense to have a cluster of competing antenna wires strung from pole to pole. Besides, the resulting monopoly would guarantee the Mountain Man that his large initial investment would not be endangered by competition. About half of the homes that could be reached by the cable chose to subscribe for $2 per month. The Mountain Man sent a truck around to connect them up. Local youths, good at shinnying up telephone poles, also connected up a few homes, bypassing the $2 fee of course.

Everyone was happy, except the village librarian, who now, for lack of readership, was cut from the payroll.

To increase subscribership, the Mountain Man added a few more antennas for more channels, pulling in some stations from 100 miles away. The viewers loved it. Local broadcasters hated it. They felt that people watching distant stations would dilute local viewership, reduce the station's rating, and decrease the amount the sponsors would pay for ads.

While in the process of seeking out distant stations,

MINIGLOSSARY

Community antenna Large antenna, receiving good reception, feeding its signals to many homes at once. Also called MATV for master antenna TV, often used in apartment buildings where one antenna feeds all apartments.

***Franchise** Contract between a municipality and a single cable

company whereby the company has exclusive rights to market cable TV services to the population for a specified number of years. In return, the cable company promises to provide certain channels and services.

the Mountain Man came up with a channel 4 from the next state. But he already had a channel 4 (a different one) on his system from a local station. How could he send this new station to his subscribers? Meanwhile, he received no stations on channels 6, 8, and 10.

Merlin, the village video wizard, came to the rescue with two solutions to the problem: (1) A CONVERTER which would simply translate the channel 4 signal to say 6, and pump it down the wire like any other channel (2) A DE-MODULATOR (a TV tuner) which would separate the channel 4 RF signal into video and audio. Then he might run the separate video and audio to a MODULATOR that combines them into a channel 6 RF signal like any TV transmitter normally does.

Thus, in goes channel 4, out comes 6, and ring goes the Mountain Man's phone as his subscribers call wondering why the channel 4 news appears on channel 6. While the new subscribers signed on, the profits poured in. The FCC, sensing cable's mighty impact on the population, set up regulations to guarantee ''community service.'' One rule required the Mountain Man to build his own TV studio and make available an ACCESS CHANNEL where anybody could air their views, run video tapes of their school plays, run college courses via TV, or even sneak in a little ''adult entertainment'' on the uncensored, unrestricted channel. Not a lot of people watched (except for the ''adult'' stuff). Eventually, the courts told the FCC that they had exceeded their authority, and the cable industry was essentially DE-REGULATED.

Meanwhile, the village fathers, aware that the Mountain Man was making a bundle on their FRANCHISE, pressured him to keep the unprofitable but public-service-minded ACCESS CHANNEL open. So the previously empty channel 8 was filled with telecourses, local sports, public affairs, and some local news produced in the Mountain Ministudio.

Channel 10 was still empty, so the enterprising Mountain Man aimed a camera at a drum with some cards on it and sold advertisements. Nobody watched. Improving on the idea, he bought a CHARACTER GENERATOR which could electronically type words on a screen and roll them by. He teamed it up with mininews and weather services so that now the viewer could watch the weather report repeatedly crawl across the bottom of the screen, hear a 24-hour radio-type newscast, and see a progression of local ads on the screen.

One day the Mountain Man, while attending his local theater, hit upon an idea: show a popular movie over TV and have everybody pay extra if they wanted to see it. Better yet, show a couple of movies per week with lots of reruns *and* no ads.

But how do you get people to pay extra for something already coming into their home? And what channel do you put it on now that all the channels are filled up?

Merlin came to the rescue again with another CON-VERTER (or DECODER) box. The box would be labeled with channels 2–13 and A, B, C, etc. Depending on where the dial was set, the incoming signal would be converted to, say, channel 3. People would connect the box between the cable and their TV antenna input and would turn their TV to channel 3—forever. The box would do the tuning. Because it had this special tuner, the box could tune in channels a TV couldn't get. Not only did this provide a range of more channels that could be cablecast, but it made it necessary for subscribers to rent the box if they wanted to pick up those ''special'' channels with PAY-TV movies. Merlin made a bundle manufacturing DECODER boxes for the Mountain Man to rent. Merlin's shifty nephew Fletcher also made a bundle copying the circuit and selling the boxes directly to the public, who for the one-time cost of the box could now pick up PAY-TV for free.

The viewers loved it. The movie theaters hated it. The broadcasters, afraid that the cable would syphon off their prime movies, lobbied against it. Merlin's Aunt Haddie could never figure out how to work it, least of all the part where she had to tune her set to channel 3 to pick up the converter box's channel 7 which really started out in the next state as channel 4.

Movie rentals are pretty expensive. The Mountain Man was approached by Mr. City Slicker with the following proposal: Mr. Slicker would lease popular movies at phenomenal expense for cable use. Mr. Slicker would then play the movies over his own cable system, feeding the signal to *other cable systems*—for a price of course. In cases where distribution by wire was too expensive, Mr. Slicker would beam the signals through the air using MICROWAVE transmitters (which instead of *broadcasting* in all directions,

MINIGLOSSARY

*Converter   Electronic device which translates one channel number (one frequency) into another (another frequency). Often rented from cable TV companies, a converter (or decoder) box connects to your TV and does the tuning instead of your TV tuner. The box usually puts out channel 3, and your TV remains tuned to channel 3.

*Demodulator or tuner   Electronic device which changes channel numbers (RF) into video and audio signals.

*Access channel   Cable TV channel set aside for local community use, like town meetings, school sports, local affairs, and news.

Deregulated   Removal of laws and restrictions imposed by Congress, the FCC, or some regulatory body.

*Character generator   Electronic device with a typewriter keyboard which electronically displays letters, numbers, and symbols on a TV screen.

Microwave   An extremely high band of radio/TV frequencies used with satellites to relay TV signals. On earth, used to transmit TV signals in beams about 5–10 miles long between mobile TV vans and the broadcasting station or between cable TV companies.

Network   Group of TV broadcasters or cable TV companies wired together to share signals from each other.

narrowcast in a focused beam at a particular target). The Mountain Man could put a MICROWAVE receiving antenna on his tower to pick up the signals.

Mr. Slicker showed the Mountain Man a briefcase full of maps with crisscrossing lines. These lines represented a web of connections passing his movie signal along from cable company to cable company, eventually covering the map with a NETWORK. Mr. Slicker pulled out more diagrams showing how he would eventually beam his movies up to a satellite 22,300 miles in the sky. All the participating cable companies would be able to point giant ''dish'' antennas at the ''bird'' and get his movie signal without messing with the NETWORK. The Mountain Man couldn't wait to sign on the dotted line. By the time the ink was dry, Fletcher (remember Merlin's nephew?) had somehow acquired the plans for MICROWAVE and satellite receivers and was selling do-it-yourself kits to crafty consumers who wanted to pick up the PAY-TV signals without paying.

By now the Mountain Man was cablecasting 25 channels to his subscribers, some stations in Spanish, some specializing in sports, some religious, and some giving courses over TV. About 30% of his subscribers paid an extra $10 per month for PAY-TV. The cash rolled in.

Then Merlin, in cahoots with Mr. Slicker, came up with an idea the Mountain Man didn't like: put a high-power DIRECT BROADCAST SATELLITE (DBS) in space that would beam pay movies *directly* to the homeowner, skipping the Mountain Middle Man. The villagers would simply rent a small dish-shaped antenna and converter box from Mr. Slicker and aim it at the bird to pick up Slicker's movies directly.

Suddenly one night, lightning struck Mountain-Visions' array of antennas, lighting up the whole sky and $50,000 worth of video gear and the first 120 sets along the cable's path. The mountain cable switchboard also lit up and didn't quiet down until everything was fixed.

Mountain Man sold out to an MSO (MULTIPLE SYSTEM OPERATOR, a conglomerate that buys up FRANCHISES), moved to Aruba where his wife got a job as a librarian, and they lived happily ever after without the companionship of Fred Flintstone, Wonder Woman, Johnny Carson, or Kermit the Frog.

DECODING CABLE'S SIGNAL FOR YOUR TV

To get a better picture of how cable TV packs all those channels into the same wire and how you get them to play through your TV, let's study frequencies for a minute.

Frequency Explained

When we listen to music, we are sensing sound vibrations through our ears. High notes are made of many vibrations per second (sometimes called cycles per second) and are said to have a high pitch or high frequency. Low notes have fewer vibrations per second for a low frequency.

Instead of saying vibrations per second, we use the more popular term HERTZ, named after the German physicist H. R. Hertz and abbreviated Hz.

The lowest note we mortals can normally hear is 20 Hz. The highest note we can hear is about 20,000 Hz or 20 KILOHERTZ (KILO means 1000), abbreviated 20 KHz. Normal speech occurs at about 1000 Hz (1 KHz).

The frequencies of the invisible electromagnetic waves used to transmit TV programs are nearly a million times the frequencies of speech. The million vibrations per second are called a MEGAHERTZ (MEGA means million) and is abbreviated 1 MHz.

Why are these TV frequencies so high? Back in Chapter 1, in the section on how TV works, we saw that the electron machine gun in the picture tube zigzagged across the screen 15,735 times per second, tracing out its picture. During each of those zigs, the gun had to shoot, stop, shoot, and stop many times to create the light, dark, light, and dark parts of the picture. The TV signal had to have coded in it an electrical vibration for every one of those shoots-and-stops. It takes millions of these vibrations per second for the TV signal to carry the message to the gun to shoot-and-stop millions of times per second. There's really more to it than this, but we'll save that for the engineering books.

Frequencies Used by Cable Companies

So that one TV channel doesn't interfere with another and so that broadcasters of other communications don't interfere with each other, the FCC has allocated certain frequencies to each. These are shown in Table 4–1. Notice that channels 2–6 take up one set of frequencies and then there's a gap between 6 and 7. Then come channels 7–13. TV broadcasters can't use the frequencies in this gap because FM radio, police, and airplanes use these frequencies. Your average TV can't even tune these frequencies in.

But cablecasters don't have to be concerned with encroaching on the police and air traffic frequencies because the cable signal stays *in the cable*. Therefore, the cable

MINIGLOSSARY

*Pay-TV** Cable or broadcast television for which you pay extra, usually for the rental of a descrambler which makes the programs visible on your TV. Pay-TV usually consists of movies and sports without commercials.

*Direct broadcast satellite or DBS** High-powered orbiting satellite which receives signals from earth and beams them back down, blanketing a part of the country so that they are easily tuned in with a 3-foot dish antenna and a special (usually rented) receiver which feeds up to four channels to your TV set.

Multiple system operator Large cable company that owns and operates many little cable companies in various municipalities.

*Kilohertz** One thousand cycles (vibrations) per second, represented by 1 KHz, which is near the sound frequencies of speech.

*Megahertz** One million cycles (vibrations) per second, represented by 1 MHz, which is near the frequency of video signals.

TABLE 4–1

Cable and regular TV frequencies

	Frequency in MHz										
	5	54	88	108	120	174	216	318	402	470	890
Broadcast TV channels		2, 3, 4, 5, 6				7, 8, 9, 10, 11, 12, 13				14, 15, 16, . . ., 81, 82, 83	
Kinds of broadcasts	CB radio	VHF TV	FM radio	Maritime, weather satellite, air traffic, police, public service, two-way radio, two-meter amateur radio		VHF TV	Amateur radio, government, mobile, air traffic	Coast Guard, satellite, mobile, air navigation		UHF TV	
Names given to these frequencies		Low-band TV			Mid-band TV	Highband TV	Superband TV	Hyperband TV		UHF band	
Cable TV converter channels	T_1, T_2 T_3, . . ., T_{14}	2, 3, 4 5, 6 (unless converted to a different channel number)	FM radio or A_3, A_4, A_5	A_1, A_2 A, B, C, D, E, F, G, H, I		7, 8, 9, 10, 11, 12, 13 (con-verted to a different channel number)	J, K, L, . . ., V, W	W + 1 through W + 17			
Cable ready TV channels		2, 3, 4, 5, 6		14, 15, 16, 17, 18, 19, 20, 21, 22, when in "cable TV" mode		7, 8, 9, 10, 11, 12, 13	23, 24, 25, 26, . . ., 41 when in "cable TV" mode			14, 15 16, 17, . . ., 83 when in "normal" TV mode	

companies can use these frequencies as they please. So they fill these gaps with more TV channels. Of course for your TV to be able to receive these frequencies, you need to rent the cable company's CONVERTER, which translates these MIDBAND and SUPERBAND frequencies to another frequency your TV can receive.

You may be wondering why the cable companies didn't just convert their TV channels to UHF (which your set *can* receive) and avoid messing with a CONVERTER box. There would seem to be plenty of UHF channels available. There are two reasons for their decision:

1. Ultrahigh frequency (UHF) signals are extremely hard to transmit cheaply through long cables. Lower frequencies are easier.

2. If you could pick up the stations easily without a CONVERTER, then how could cable companies make money renting them to you?

In short, the cable company charges you to hook up to its "antenna." To discourage folks from climbing the phone poles and hooking themselves up for free, and for technical reasons related to transmitting certain frequencies, the cable companies transmit some TV signals on channels your average TV set doesn't get. You then must rent a CONVERTER from them to translate these frequencies to channels you *can* get. Figure 4–1 shows how to hook up the CONVERTER.

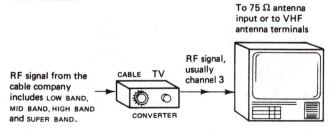

FIGURE 4-1   CONVERTER hookup to TV.

MINIGLOSSARY

Midband   Cable TV channels A–I.

Superband   Cable TV channels J–W.

Some Cable TV Channels

Most cable TV companies today have the capacity to offer up to 37 TV channels. Many are upgrading to 54 channels. Some, they receive locally by antenna, and others by satellite. Here are a few satellite channels:

C-SPAN	Cable Satellite Public Affairs Network	House of Representatives in action (or inaction)
MTV	Music Television	Rock music
DIS	Disney Channel	Original films and Disney favorites
CHN	Cable Headline News	News updated every 30 minutes
CNN	Cable News Network	In-depth news, world, national, financial, sports
WC	Weather Channel	24-hour weather, regional and national
USA		Family entertainment and sports
A&E	Arts and Entertainment	Programs specializing in visual/performing arts
LIF	Lifetime	Talk TV, health, life styles, entertainment
NIK	Nickelodeon	Revived old TV shows, family & children's entertainment
TLC	The Learning Channel	Educational programming
TC	Travel Channel	Information on business and leisure travel
DSC	Discovery Channel	Family programming
TNT	Turner Network TV	General programming
TBS	WTBS (Turner Broadcast)	Channel 17 Atlanta, movies, sports
BET	Black Entertainment TV	Programs of interest to black community
CMTV	Country Music Television	Country music
EWTN	Eternal World TV Network	Religious broadcasts

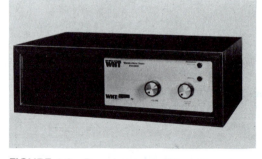

FIGURE 4-2 Broadcast pay-TV DESCRAMBLER. (Courtesy of Wometco Home Theatre)

Pay-TV Channels

Besides your basic TV channels, you can subscribe (at extra charge) to a variety of other special channels offering sports, movies, children's programs, etc., depending on your cable company. Here are some of the PAY-TV services generally available:

HBO	Home Box Office	Movies
SHO	Showtime	Movies
MC	Movie Channel	Movies
Cinemax		Movies
Playboy		Adult films
ESPN	Entertainment and Sports Programming Network	Sports

People living near large cities are sometimes able to receive "wireless cable TV" whereby signals are transmitted through the air rather than through cable. Perhaps 12 channels of PAY-TV signals would be broadcast on superhigh microwave frequencies. Subscribers, for a monthly fee, would rent special antennas and DESCRAMBLERS to turn these special signals into regular TV channels.

If cable TV is connected to your home and you're already renting the CONVERTER box, what stops you from watching PAY-TV channels without paying? The answer is a simple FILTER (or TRAP) like those described in Chapter 3 for reducing adjacent-channel interference. When the cable company rents you their CONVERTER, *without* the PAY-TV option, the CONVERTER has in it a FILTER which stops the PAY-TV channel from getting through. What keeps a cable TV subscriber from renting the CONVERTER, opening it up, and removing the FILTER?

1. You've got to understand some electronics to know which part to remove.
2. It's illegal. You're *renting* the box; it's not yours to tamper with. Such electronic piracy is considered theft-of-service, a crime.
3. Crafty cable companies sometimes don't put the FILTER in the CONVERTER but instead put it in the junction box on the telephone pole where the cable branch splits off to your house. Thus, the PAY-TV signal never reaches your home. Equally crafty subscribers sometimes climb the pole, open the box, and remove the tiny plug-in FILTER. But, again, they are tampering with the cable company's property in order to steal service.

Section 605 of the 1934 Federal Communications Act prohibits the unauthorized reception and use of *over-the-air* PAY-TV broadcasts. Unauthorized reception of *cable* signals is generally a misdemeanor under state law.

The FILTER is *one* way cable companies trap their PAY-

TV signals. A second method of keeping their signals from being intercepted for free is to SCRAMBLE them. They SCRAMBLE their signal and then rent you a DESCRAMBLER (sometimes also called a DECODER) to make the channel watchable. They SCRAMBLE the picture by messing up the sync (remember sync?) so your picture bends and tears across the screen as in Figure 2–11.

Most of these CONVERTERS and DECODERS have one common fault. If you have a remote control on your TV, forget about it. The channel switching is done *manually* on the CONVERTER box (unless *it* happens to be remote controlled). All programs, once switched at the box, are viewed on the same channel on your TV, rendering your TV's remote control useless. As far as your TV is concerned, there is only one station in town, whatever the box is putting out. This problem also bugs you if you are using a home video recorder. You may want to program your recorder to

record one show on one channel and then another show later on another channel. Your recorder will switch channels as programmed, but your CONVERTER won't be switched, *and it's the CONVERTER that tunes the stations*. A further irritation involves one of the reasons some people buy home videocassette recorders in the first place—to watch one show while recording another. The CONVERTER translates only one station at a time, so unless you have two CONVERTERS, both your TV and videocassette recorder will be receiving the same channel.

PROGRAMMABLE CONVERTERS can be set to switch channels while you're away. Meanwhile, your VCR could be programmed to record the results. ADDRESSABLE CONVERTERS are able to automatically activate their PAY TV DE SCRAMBLING circuits. For instance, you'd call the cable folks requesting Showtime; they'd send a signal down the TV cable activating the DESCRAMBLER in *your* box alone (billing

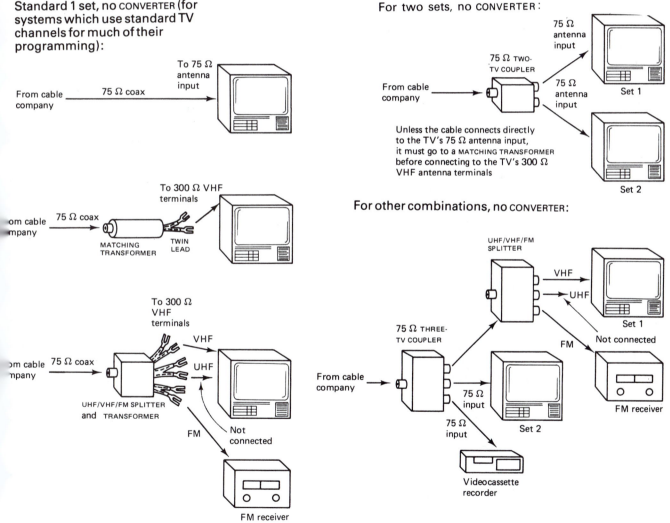

FIGURE 4-3 Cable TV connections without CONVERTER using regular TV sets. TVs will pick up only channels 2–13.

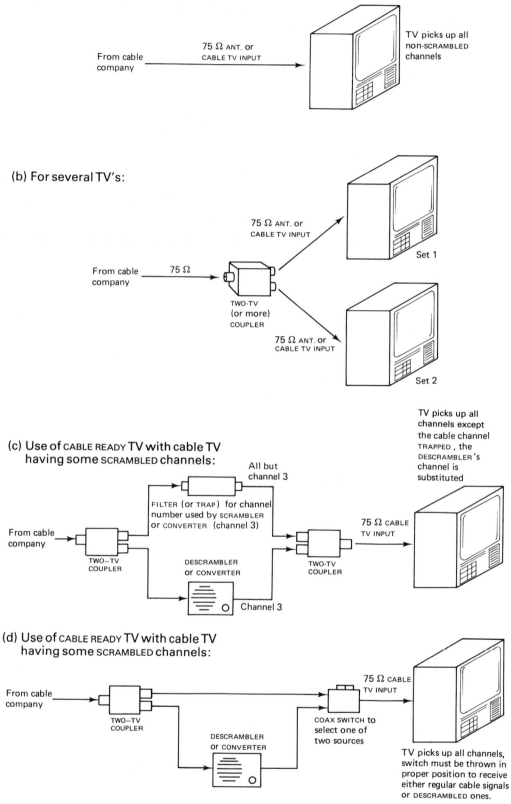

(a) Use of CABLE READY TV with cable TV:

From cable company → 75 Ω ANT. or CABLE TV INPUT → TV picks up all non-SCRAMBLED channels

(b) For several TV's:

From cable company → 75 Ω → TWO-TV (or more) COUPLER → 75 Ω ANT. or CABLE TV INPUT → Set 1 ; 75 Ω ANT. or CABLE TV INPUT → Set 2

(c) Use of CABLE READY TV with cable TV having some SCRAMBLED channels:

From cable company → TWO-TV COUPLER → FILTER (or TRAP) for channel number used by SCRAMBLER or CONVERTER (channel 3) → All but channel 3 → TWO-TV COUPLER → 75 Ω CABLE TV INPUT

DESCRAMBLER or CONVERTER → Channel 3

TV picks up all channels except the cable channel TRAPPED, the DESCRAMBLER's channel is substituted

(d) Use of CABLE READY TV with cable TV having some SCRAMBLED channels:

From cable company → TWO-TV COUPLER → COAX SWITCH to select one of two sources → 75 Ω CABLE TV INPUT

DESCRAMBLER or CONVERTER

TV picks up all channels, switch must be thrown in proper position to receive either regular cable signals or DESCRAMBLED ones.

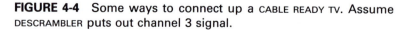

FIGURE 4-4 Some ways to connect up a CABLE READY TV. Assume DESCRAMBLER puts out channel 3 signal.

you for the extra service). Requesting access to just *one* show is called PAY-PER-VIEW.

Ways of Connecting Cable TV to Bypass the Converter

There are several ways around this "only one station in town" syndrome. You could use *no* CONVERTERS and hook up the cable as if it were a standard TV antenna, but you'd only receive channels 2–13 (assuming the cable company used those frequencies). You could use CABLE READY TVs, which do not need CONVERTERS and can connect directly to the cable and pick up all the cable TV stations (except SCRAMBLED and TRAPPED ones). You could use a separate CONVERTER for each TV. You could buy an UP-CONVERTER or BLOCK CONVERTER which will change all the cable TV signals to UHF signals which the average TV set can receive. In the figures that follow, you will see some possible cable configurations. Not everything may work; you may have to experiment.

Everything you learned in Chapter 3 applies to cable as well as to antennas. You use matching transformers, splitters, and couplers the same way as before. Essentially, the cable is like your antenna wire, only you're renting the signal in it from someone else. You can also use boosters and amplified couplers the same way as before, but with one wrinkle: These devices must be designed for cable signals, or else they will make channels J–W a bit snowy. Look for a label on the device indicating "for cable use" or "bandwidth 50–400 MHz" to show it can handle the cable TV frequencies.

Using No Converter. Figure 4–3 shows some possible cable connections which do not employ a CONVERTER. Note that on most systems you can only receive channels 2–13 on standard TV sets using the "raw" cable signals.

Cable Ready TVs. A CABLE READY TV can pick up not only the regular VHF and UHF stations but can pick up the cable TV frequencies as well. No CONVERTER is needed. Figure 4–4 shows several hookups.

Using Several Cable Converters. Somewhat cumbersome (but preferred by many cable companies) is the technique of splitting the signal into several signals and sending each through its own CONVERTER on the way to each TV set. Figure 4–5 diagrams some possible hookups using CONVERTERS.

Up-Converters or Block Converters. UP-CONVERTERS (also called BLOCK CONVERTERS) are little boxes

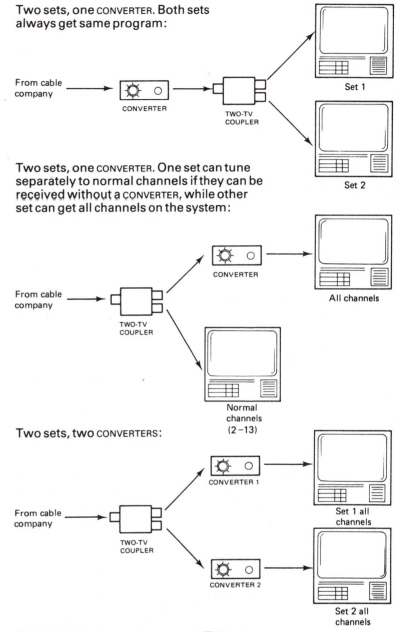

Two sets, one CONVERTER. Both sets always get same program:

Two sets, one CONVERTER. One set can tune separately to normal channels if they can be received without a CONVERTER, while other set can get all channels on the system:

Two sets, two CONVERTERS:

FIGURE 4-5 Connecting several TVs to one or more CABLE CONVERTERS.

which take all the incoming cable frequencies and convert them to UHF frequencies. See Table 4–2. Cable channel 2 would become, say, a UHF channel 36. Channel 5 would be changed to channel 40. Channel 13 may become channel 62. Cable channel A would become 47, I becomes 55, and R becomes 71 and so on. Using an UP-CONVERTER is like having only a UHF antenna hitched to your TV, except this UHF antenna pulls in *lots* of stations.

Using an UP-CONVERTER can be very confusing unless you make a chart showing what UHF channels you tune to

---

MINIGLOSSARY

*Cable ready A TV or VCR with a special tuner able to pick up the cable TV channels directly without a converter box.

TABLE 4–2

Sample table showing how an up-converter changes cable TV channels to UHF channels

Cable channel:	2	3	4	5	6				
UHF channel from up-converter:	36	37	38	40	41				

Cable channel:	A	B	C	D	E	F	G	H	I
UHF channel from up-converter:	47	48	49	50	51	52	53	54	55

Cable channel:	7	8	9	10	11	12	13	
UHF channel from up-converter:	56	57	58	59	60	61	62	

Cable channel:	J	K	L	M	N	O	P	Q	R
UHF channel from up-converter:	63	64	65	66	67	68	69	70	71

FIGURE 4-6   UP-CONVERTER.

for the stations you want. The big advantage of UP-CONVERTERS is that they convert *all* the channels at once, not just one channel at a time.

Figures 4–6 and 4–7 show UP-CONVERTERS and some ways to wire them to your TV system.

Once you've discovered where to tune the UHF dial to pick up your familiar stations, you are ready for a bonus. You can now use your TV's remote control to select the various converted-to-UHF stations, and your programmable videocassette recorder can simply select the UHF channel it's supposed to record.

Having a CABLE READY TV or an UP-CONVERTER can often save cable fees. Many cable TV companies will reduce your subscription fee by about $2 per month if they don't have to provide a CONVERTER box to you.

TWO-WAY CABLE

Most cable systems are designed to send TV signals in one direction—to your home. TV programs aren't the only thing that can travel through a cable. Computer data, alarm system signals, and all kinds of coded information can traverse the wire—*in both directions*—if the system is so designed.

With INTERACTIVE cable systems the TV signals travel in one direction from the HEAD END (the cable company's transmitters) *to* the customer, while computer data or house alarm signals may go *from* the customer to the cable company's HEAD END.

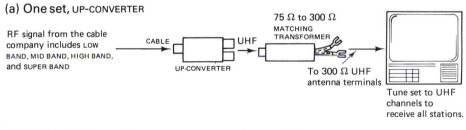

(a) One set, UP-CONVERTER

RF signal from the cable company includes LOW BAND, MID BAND, HIGH BAND, and SUPER BAND

CABLE → UP-CONVERTER → UHF

75 Ω to 300 Ω MATCHING TRANSFORMER

To 300 Ω UHF antenna terminals

Tune set to UHF channels to receive all stations.

FIGURE 4-7   Connecting an UP-CONVERTER to your TV system.

MINIGLOSSARY

Interactive cable   Cable TV that not only sends shows *to* your home but receives signals *from* you such as a fire/burglar alarm, questionnaire responses, and orders to purchase goods.

Head end   The place where the cable TV company sends its signals from. This is not necessarily where its offices are or where its studio is. It is the center where the signals start their journey down the web of wires to homes.

(b) One set, UP-CONVERTER with built-in COUPLER
(for systems which use standard TV channels
for much of their programming):

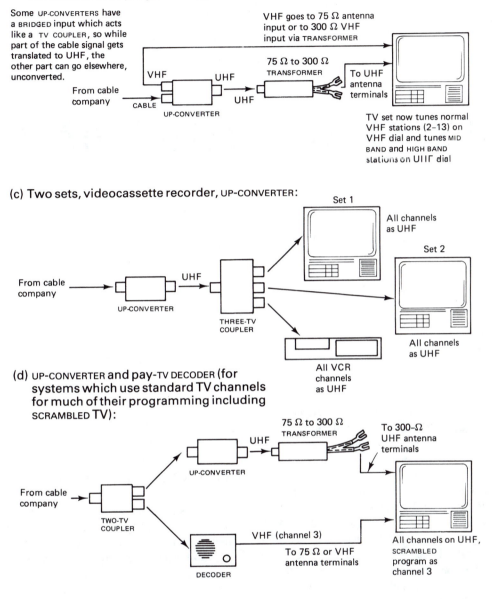

Some UP-CONVERTERS have a BRIDGED input which acts like a TV COUPLER, so while part of the cable signal gets translated to UHF, the other part can go elsewhere, unconverted.

From cable company

VHF goes to 75 Ω antenna input or to 300 Ω VHF input via TRANSFORMER

75 Ω to 300 Ω TRANSFORMER

To UHF antenna terminals

TV set now tunes normal VHF stations (2–13) on VHF dial and tunes MID BAND and HIGH BAND stations on UHF dial

(c) Two sets, videocassette recorder, UP-CONVERTER:

Set 1
All channels as UHF

Set 2
All channels as UHF

From cable company
UP-CONVERTER
THREE-TV COUPLER

All VCR channels as UHF

(d) UP-CONVERTER and pay-TV DECODER (for systems which use standard TV channels for much of their programming including SCRAMBLED TV):

From cable company
TWO-TV COUPLER

UP-CONVERTER

75 Ω to 300 Ω TRANSFORMER

To 300-Ω UHF antenna terminals

DECODER

VHF (channel 3)
To 75 Ω or VHF antenna terminals

All channels on UHF, SCRAMBLED program as channel 3

FIGURE 4-7 Cont'd

VIDEO TAPE RECORDERS

So far we have been talking about television signals that originate in somebody's studio and arrive at your TV set either via over-the-air broadcasts or through a cable. Today, however, very few television shows—professional or otherwise—are "live." Most are recorded and stored on video tape for playing back whenever you wish. This chapter will deal with this play/record process.

Once you've mastered the video tape recorder, you'll have taken the biggest, most important step toward becoming video literate. Of all the video machinery, the video tape recorder is the one device which you should know backward and forward, inside and output. Furthermore, it is not a simple machine. That explains why this is such a large and detailed chapter. There's too much here to swallow in one gulp. Study the subject in sections and take your time.

KINDS OF VIDEO TAPE RECORDERS

Video tape recorders are like critters in the ocean; they come in all sizes. There are professional video tape recorders which cost more than a home (with pool, sauna, and Mercedes in the garage). Some home models cost just a little more than dinner for four at a swanky restaurant.

VCRs, VTRs, and Camcorders

A VTR is a video tape recorder. It can both record and play back video tape. Technically, "VTR" means only reel-to-reel recorders (as opposed to cassette), but people commonly use this term to mean any video recorder, even videocassette recorders.

A VCR is a videocassette recorder. It can record a videocassette or play it back. The CASSETTE or VIDEOCASSETTE is a small case that holds a full reel of video tape and an empty take-up reel. After the cassette has been inserted into the VCR, the machine automatically draws the tape from the cassette, threads it, plays it, and winds it back onto the take-up reel. You can remove a cassette in the middle of a program (the machine automatically unthreads the tape before it ejects the cassette to you) and come back later to pick up where you left off. Rewinding is not required

---

MINIGLOSSARY

*Video tape recorder or VTR   A machine which can record picture and sound on a tape. Nearly all can also play back a tape. Although a videocassette recorder is also a video *tape* recorder, VTR usually implies that reel-to-reel tape is used rather than cassette.

*Videocassette recorder or VCR   A video tape recorder which uses cassettes rather than open reels of tape.

Videocassette player or VCP   A machine which can play a videocassette but cannot record one.

*Camcorder   A VCR and camera in one box.

unless you want the tape to start at the beginning again.

The VTP is a video tape player, and a VCP is a videocassette player. Neither record; they just play. They are handy for libraries and learning centers where tapes are to be played and not accidentally recorded over and erased.

In the early days, all the videotape machines were reel-to-reel. Now, only high-level professional and commercial stations use them. Nearly everything else, nowadays, is recorded on videocassettes.

CAMCORDERS (if you just arrived from Mars) are VCRs too. Not only can a CAMCORDER record sound and picture from its built-in microphone and camera, but many models can also record audio and video signals sent to them over a wire. Nearly all CAMCORDERS can also play back their tapes to a TV monitor or receiver. There is more on CAMCORDERS later in this chapter.

FORMAT AND COMPATIBILITY

A teacher calls you saying that she wants to play a tape to her class tomorrow. Before spending an hour setting up the equipment in the classroom, there is an important question you should ask: What kind of tape has she? In order to play her tape, you must have the right kind of machine. A VHS tape won't play in a ¾U machine; the cassette is a different shape and the tape a different size. The two are INCOMPATIBLE. Fortunately, most tapes are distributed in the VHS FORMAT and VHS is the most common type of VCR. Next time someone asks for a video setup, first find out what kind of tape they want to play. If they say ¾U and you have ¾U VCRs, you are in business.

Maybe 90 percent in business. There are still some things that can go wrong . . . go wrong . . . go wrong . . .

For one videotape machine to play tapes made on another machine, the following criteria must be met:

1. Both machines must be the same format. Both must use COMPATIBLE reels, cartridges, or cassettes.
2. If the recording is in a SUPER format (Super VHS or HI8), it must be played back on a SUPER type of VCR.
3. Both recorder and player must be color in order to display color. Both must be "hi-fi" models to provide high-fidelity sound. (If one is hi fi and the other is not, you'll still get sound, only "lo-fi" sound).

4. Both must use the same television standard (one cannot be a foreign country's standard).

Let's attack these criteria one by one:

Format

Tape comes in different widths to fit various video tape machines. Every tape machine can work with only one width of tape. Generally (but not always), the wider tapes give you higher picture quality, and the video tape and VTRs will be more expensive. There is ¼-, ½-, ¾-, 1-, and 2-inch tape. But there is more to FORMAT than just size. There is tape speed and other electronic differences which make not all ½-inch tapes playable on all ½-inch machines.

To simplify matters, various manufacturers and video associations have agreed to standardize on several FORMATS. *So if a tape is recorded on a certain FORMAT machine, it should play back on the same FORMAT machine* regardless of manufacturer (with a couple of minor exceptions to be discussed shortly).

For the common business, industrial, and school video user, there are three primary low-budget FORMATS:

1. VHS: Most popular FORMAT, found in most colleges, most public schools, many industrial media centers, a few commercial TV production shops, and about 75 percent of U.S. homes.
2. ¾U: Second most popular FORMAT, found in nearly every college and industrial media center, commercial and professional TV production shop, and many public schools.
3. 8MM: The third most popular FORMAT found in the same places as VHS but used mostly in CAMCORDERS. 8MM CAMCORDERS use 8-millimeter wide tape (about ¼ inch) in cassettes able to record 2 to 4 hours each.

Figure 5–1 shows some popular home and industrial videocassette recorder types and the cassettes that go into them. For the high-end professional and commercial producer, there are three popular FORMATS:

1. *1-INCH TYPE C:* Reel-to-reel 1-inch tape. Studio FORMAT heavily used by U.S. commercial broadcasters and others who require the highest technical quality (and can afford the expense).

MINIGLOSSARY

Video tape player or VTP   A machine which can play a video tape but cannot record one.

*Videocassette   A box containing video tape connected to an internal supply reel and a take-up reel. Used in VCRs.

*Format   The way the tapes, cassettes, and video recorders and

players are designed so that one machine can play another machine's tapes. Machines of the same format should be able to play each other's tapes.

Compatible   The ability to play a tape on any same-format machine and get good picture and sound.

VHS (Courtesy of RCA Consumer Electronics)

VHS cassette

MII (Courtesy of JVC)

MII Cassette

¾U cassette

¾U (Courtesy of Sony Corp. of America)

8mm (Courtesy of Sony Corp. of America)

8mm cassette

FIGURE 5-1 Different FORMAT VCRs and tape.

2. *Betacam:* Most popular commercial and professional CAMCORDER FORMAT, often used in news. High quality but expensive. Uses tape similar to (but not the same as) Betamax recorders, at six times the speed and records the individual component colors separately for very high-quality pictures.

3. *MII:* Second most popular commercial and professional CAMCORDER FORMAT, also used in news. High quality, but less expensive than BETACAM, this FORMAT uses cassettes similar (but not the same as) VHS, but records them at a different speed; tapes last 20 or 90 minutes instead of 2–6 hours. MII also records the individual component colors separately for higher-quality pictures.

Here are some less popular and older FORMATS you may come across:

1. *BETA:* This excellent ½-inch videocassette FORMAT comprised 10 percent of the home video market in the 1980s, but lost the battle to VHS at the end of the decade.

2. *½-inch EIAJ:* Popular reel-to-reel FORMAT used in school in the 1960s.

3. *1-inch:* This size tape was used in several different reel-to-reel FORMATS popular in colleges and in industry in the 1960s and 1970s but has largely been replaced by 3/4U. The only 1-inch FORMAT to survive the 1980s is TYPE C mentioned earlier.

4. *2-inch QUAD:* Older high-quality reel-to-reel FORMAT used almost exclusively by professional and commercial broadcasters.

5. *VHS-C:* Another super-light portable FORMAT, the cassette is the size of a deck of cards. With an adapter, the VHS-C cassette can be played in a standard VHS player. This FORMAT competes primarily with 8MM.

6. *Extinct and almost extinct:* Other FORMATS are Phonovision (waxed video discs, circa 1928), Cartivision (using ½-inch tape in a cartridge), Philips VCR, InstaVision, V-Cord and V-Cord II, SVR, VX-100, Akai ¼-inch reel-to-reel, CVC (compact videocassette about the size of an audiocassette marketed by Technicolor), and the high-quality professional IVC9000 2-inch helical recorder. These obscure FORMATS are beyond the scope of this book (thank goodness!).

In short, if a home or school friend brings you a tape, most of the time it will be VHS. Occasionally it will be one of the other formats, ¾U (mostly schools and industry), 8mm, or VHS-C (mostly vacationing camcordists), or maybe an old BETA.

If a professional brings you a tape, it's likely to be BETACAM, MII, 1-inch TYPE C, or one of the SUPER FORMATS (described shortly).

Tape Speeds

As for cassettes, all VHS cassettes are compatible and will play interchangeably if played at the proper speeds; all BETA cassettes are compatible and will play interchangeably if played at the right speeds. Not all VHS and BETA machines will play at all speeds, though. The three speeds on a VHS machine are VHS-2, VHS-4, and VHS-6, which represent 2-hour recordings, 4-hour recordings, and 6-hour recordings on a standard T-120 cassette. Other names for these three speeds are SP (standard play), LP (long play), and EP (extra

---

MINIGLOSSARY

* VHS   Video home system. The most popular of the ½-inch videocassette formats.

VHS-C   VHS compact format using VHS tape in a minicassette.

VHS-HQ   Slightly improved version of VHS format, totally compatible with it.

*S-VHS   Super VHS, much improved version of VHS, downwardly compatible with VHS.

Beta   Introduced by Sony, nearly extinct, ½-inch videocassette format.

Superbeta   Slightly improved totally compatible version of beta.

ED beta   Extended definition. Much improved version of beta, downwardly compatible with it.

*¾U   Popular industrial videocassette format. Uses ¾-inch tape in cassettes.

¾U-SP   Superior performance version of ¾U, totally compatible with it.

*8mm   Eight millimeter. Nearly ¼-inch wide tape used in popular lightweight home camcorders. This is also the width of home movie film which is also called 8mm.

*Hi8   Much improved version of 8mm, downwardly compatible with it.

*Type C   Popular professional reel-to-reel recorder format using 1-inch tape.

*MII   Professional camcorder format using VHS-like cassettes, recording separate colors at high tape speed for high quality.

*Betacam   Most popular professional camcorder format using beta-like cassettes, recording separate colors at high tape speed for high quality. Expensive.

*Betacam-SP   Improved version of betacam, downwardly compatible with it.

*D2   Very high-quality composite video digital recording format. Expensive.

D1   Very high-quality component video digital recording format. Expensive.

Super   SVHS, Hi8, ED Beta or any improvement to a VCR format that increases the picture sharpness above 400 lines of resolution. Sometimes called High Band.

High Band   High-resolution (over 400 lines) VCR format.

long play). The three speeds on a BETA are called BETA-1, BETA-2, and BETA-3 for 1-, 2-, and 3-hour recordings on a BETA L-500 cassette (longer and shorter cassettes, of course, allow you to record longer and shorter times). A VHS-2, 4, 6 machine can play any VHS recording. A BETA-1, 2, 3 machine can also play any BETA tape. If a tape were recorded at the VHS-4 speed and your machine didn't have that speed, you would not be able to play that tape. In short: *VHS cassettes will play on VHS machines but not on BETA. BETA cassettes will play on BETA machines but not on VHS.* Furthermore, *to play the tape, the machine must be able to play at the same speed the tape was recorded at.*

All ¾U videocassettes are interchangeable with one noteworthy exception: portable ¾U VCRs take smaller cassettes (like the Sony KCS-20). These minicassettes will play okay in standard-sized ¾U VCRs, but standard-sized cassettes won't fit into the portable VCR's tiny mouth.

SUPER Enhancements

Manufacturers have been struggling with ways to improve the picture quality on VHS, ¾U, 8MM, and BETA VCRs without sacrificing compatibility with older models. The first round of improvements made the picture look about 20 percent sharper. In order to see this improvement, the tape had to be recorded *and* played back on an improved machine. The tape, however, was completely compatible with older machines; that is, a tape recorded on a new machine would play back just fine on an old machine, but would lack improved quality.

For VHS, there is VHS-HQ (High Quality). Special circuits make the picture *look* sharper, even though it isn't. No special tape needs to be used.

The enhanced version of ¾U is called SP for SUPERIOR PERFORMANCE. A tape recorded on an SP machine will display a sharper, smoother picture. Normal tapes can be played on an SP machine and SP tapes can be played on a normal machine, but with no improvement in picture quality.

The old BETA VCRs were supplanted with SUPERBETA capable of displaying a 25 percent sharper picture. As with the others, SUPERBETA tapes are totally COMPATIBLE with regular BETA machines, but you lose the improved picture quality with such a combination.

Next came the "SUPER" FORMATS, SUPER VHS, HI8, and ED BETA. In all three cases the improvements are similar: the picture detail is improved from 240 lines of resolution (a measure of picture sharpness) to 400+ lines. Further, the video signals can be displayed in two different ways: normally, as COMPOSITE video, or Y/C, where the color parts of the signal are carried on separate wires from the brightness parts of the signal making colors sharper. To achieve the sharper picture, special video tape must be used. When a SUPER machine makes a recording on SUPER tape, the results look dazzling if played back on another SUPER machine. *The recording, however, is unviewable* when played on a "regular" machine. In this regard, the SUPER models are INCOMPATIBLE with their regular brothers. By throwing a switch, the SUPER machines can make regular tapes to play back on regular machines. Further, the SUPER machines can play back regular tapes made on regular machines. This is called DOWNWARD COMPATIBILITY whereby newer machines can make improved tapes playable only on improved machines, but can still make and play unimproved tapes for common machines.

Hi-Fi Sound

Some VHS, BETA, and all HI8 VCRs can record high-fidelity sound. The sound is invisibly coded into the picture and becomes part of it. If a HI-FI recorded tape is played back on a HI-FI VCR, you will hear the extraordinary fidelity. If, however, the tape is recorded on a regular "lo-fi" VCR, or a "lo-fi" tape is played on a HI-FI VCR, then only low fidelity will be heard.

Foreign Standards. A tape recorded in Europe on a VHS recorder will not play on a VHS VCR here in the

MINIGLOSSARY

*SP Standard play—The 2-hour speed of a VHS VCR.

LP Long play—The 4-hour speed of a VHS VCR.

Downward compatibility Ability of an improved VCR to play back and record older format tapes, even though it can also make "improved" tapes that the older machines cannot play.

Hi Fi Ability of some VCRs to record high fidelity sound.

Resolution Picture sharpness, measured in "lines."

*EP or ELP or SLP Extra play or extra long play or super long play—the 6-hour speed of a VHS VCR.

*T-120 Standard size of a VHS videocassette. Plays 2, 4, or 6 hours.

L-500 A common size for beta videocassettes. Plays 1, 2, or 3 hours.

*NTSC National Television Standards Committee. Experts who developed the NTSC video standards which ensure that all TV signals in the United States are compatible.

PAL Phase alternate line—a European video standard incompatible with the U.S. NTSC system.

SECAM SEquential Color And Memory—a video standard used in much of Asia, incompatible with our NTSC system.

*Tri-standard A TV or VCR which can work with an NTSC, PAL, or SECAM TV signal.

Standards converter Expensive electronic device which changes one TV standard into another (i.e., PAL into NTSC) for use with other TV equipment.

United States; their electricity is different, and their TVs make the picture in a slightly different fashion. Even though the FORMAT is the same, tapes made using different TV standards are not interchangeable. Blank tapes, purchased overseas, can be used on USA equipment, however.

The United States uses the NTSC (National Television Systems Committee) standard, which broadcasts 525 interlaced scanning lines every $\frac{1}{30}$ second, puts the sound at a certain frequency, and encodes the color a certain way.

In most of Europe, Australia, Italy, and Singapore, the PAL (Phase Alternate Line) system is used, with 625 scanning lines repeated every $\frac{1}{25}$ second along with other changes. There are variations of PAL like PAL-M in Brazil.

Russia, France, Iran, Poland, Saudi Arabia, and Issas use SECAM (SEquential Color And Memory), which scans 625 lines (except for France, which uses 819 lines) every $\frac{1}{25}$ second plus further variations from PAL.

In short, you can't play most foreign tapes on your VCR even if they're both the same FORMAT.

You can buy TRI-STANDARD videocassette recorders which will record and play back in the three main standards, but require use of special TRI-STANDARD TVs to display the signals. Exceptions: Instant Replay (Miami) makes special VCRs that play NTSC, PAL and SECAM tapes into regular TVs. Panasonic's $2500 AG-W1 VCR will play or record in NTSC, PAL, and SECAM, *and* will convert one standard to another while copying tapes.

There are professional machines called STANDARDS CONVERTERS which will change one country's standard of TV signal into another country's standard. Because STANDARD CONVERTERS are so expensive, there are companies who will convert your tapes for you (for a price).

Table 5–1 reviews video FORMATS.

Time Lapse

Another kind of video recorder is the TIME LAPSE recorder (Figure 5–2). It can make a tape last 8, 24, and sometimes up to 200 hours depending on the model VCR used. Security organizations use them for surveillance, and scientists use them to document how something changes over a long period of time.

You don't get something for nothing by packing all that time onto one little tape. Something gets left out. The something missing is smoothness of action. Watching a TIME LAPSE tape is like watching someone project slides or a filmstrip as fast as they can. For the TV producer, the TIME LAPSE VCR is almost useless (except when used in the *normal* mode like a regular VCR).

FIGURE 5-2 Javelin XL-6001 TIME-LAPSE VCR. Connects to regular TV camera and TV monitor and uses regular VHS cassette, but can record up to 200 hours on a cassette. (Photo courtesy of Javelin Electronics, Inc.).

PLAYING A TAPE

Your boss says, ''I have an important client coming over in five minutes. Play him this tape. The equipment is set up. I'll be back in half an hour.''

Stay cool. Here's what you do:

1. *Turn on the TV* you will use to watch this presentation. While it is warming up, *turn on the power for the tape player*. A pilot light will probably come on to tell you the machine is getting power. If it isn't, look for a TIMER switch and flip it (some timers must be on for the VCR to work; others must be off).

2. *Put the tape on the machine.* If it's a reel-to-reel tape player, look for a threading diagram printed on the face of the player somewhere. Usually the full reel of tape goes on the left spindle, and the empty take-up reel goes on the right spindle. The tape is then threaded around all the doodads and guides shown on the diagram. Wrap the tape around the take-up reel hub several times, and you are finished. If using a VIDEOCASSETTE, press the machine's EJECT button, and its mouth will open. Remove the videocassette from its cardboard box. Holding the cassette so that its label is right side up (readable), insert the cassette into the machine's mouth. The little trapdoor on the cassette goes in first. See Figure 5–3.

3. *Press down the cassette compartment* (closing its mouth) until it clicks, locking it into place (gulp!). Some FRONT LOADING videocassette recorders don't have a trapdoor on the top. Instead they

---

MINIGLOSSARY

Time lapse Method of compressing time by taking a picture every few seconds (or minutes) and playing them back 30 per second, to speed viewing. Time lapse VCRs can record many hours on the tape.

TABLE 5–1

Popular video recorder formats

Format	Tape Size	Notes
VHS	½-inch cassette	Comprising almost four-fifths of the home VCR market, VHS (video home system) is also spreading among industry and schools. Actually three formats, VHS-2, VHS-4, and VHS-6 (or SP, LP, and EP, as some call it), represent three speeds yielding 2, 4, or 6 hours of playing time (depending on tape length—these playing times are for a common T-120 videocassette). A tape recorded at the 6-hour speed must be played at the 6-hour speed. A VHS-2, 4, 6 machine will play all three speeds (it's switchable), but a VHS-2 machine can't play a tape recorded at the VHS-4 or VHS-6 speed. The VCRs cost about $200–$1000.
VHS-C	½-inch minicassette	Stands for VHS compact. Ultraminiature (5.3 lb) VCR recording in the VHS format but using smaller ⅓–1 hour cassette, about the size of a deck of cards. The mouth of a VHS-C machine is too small to hold a normal VHS tape, but a VHS-C cassette can be played in a normal-sized VHS player if you stick the tiny cassette in a special adapter first.
8MM	¼-inch cassette	8mm tape in minicassettes, popular for its small size, records 2 or 4 hours, depending on tape speed.
BETACAM	½-inch cassette	Sony's professional camcorder using 30 minute cassettes.
MII	½-inch cassette	Matsushita's professional camcorder and studio format using 20-90 minute cassettes.
SVHS, HI8, BETACAM-SP, ¾U-SP	various	Improved versions of the above formats.
D-2	various	Digital recording techniques used by high-end professionals. Tapes can be copied without degradation. Expensive.
¾U	¾-inch cassette	Widely used in schools and industry. VCRs cost about $2500. A cassette plays up to 1 hour. Picture is about 20% sharper than that from ½-inch home VCRs.
C	1-inch reel-to-reel	High quality, but expensive, professional studio format. Tapes play 1½ hours.
QUAD	2-inch reel-to-reel	Stands for "quadruplex" (the other recording configurations are called *helical*). It's an older format used by professionals. VTRs cost about $90,000. Tape plays 1–3 hours.

have a slot in the front. Holding the cassette with the label facing up where you can read it and with the trapdoor on the cassette facing the VCR, push the videocassette gently into the VCR's mouth. It will swallow the cassette and lick its lips.

4. *Rewind the tape to the beginning if necessary.* Do this by pressing the REWIND button. You do not have to hold the button down.

5. *Play the tape.* Press the PLAY (or FORWARD or FWD) button on the VCR to set things into motion.

6. *Tune in the TV set.* If the VCR is sending its signal to the *antenna* input of the TV set, then tune the set to channel 3 or 4. If the VCR is sending *video* and *audio* to the TV set, make sure the set is switched to LINE or VTR (*not* AIR). At this point you should be viewing the tape. If you see a regular TV show playing,

a. Maybe you are on the wrong channel.

b. Maybe the antenna signal instead of the VCR signal is going to your TV. Look for an OUT-

---

MINIGLOSSARY

Front loading   Cassette goes into a slot in the front of the VCR instead of into a trapdoor or pop-up mechanism atop the VCR.

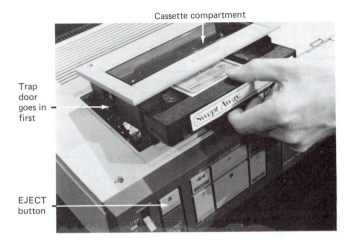

Cassette compartment

Trap door goes in first

EJECT button

FIGURE 5-3 Inserting videocassette.

PUT SELECT (or TV/VIDEO or TV/CASSETTE or TV/VTR) switch on your VCR and flip it to VCR (*away from* TV).

7. *Adjust the TV.* Loudness, brightness, color, and so forth are adjustments made only on the TV receiver. Stopping, playing, rewinding, and fast forwarding (winding ahead) are all functions of the VCR.

Assuming everything is working, rewind the tape to the beginning again so that it's ready for the client (who shows up 4 hours later and 2 minutes before you are ready to go home).

Working the Tape Player's Controls

Not all machines work the same way, but here are some generalizations:

Most VCRs today are automatic, you can push any button in any order without fouling up the tape. If the machine is on PLAY and you wish to rewind, merely push REWIND, and the machine does the rest. If the machine does not obey your order, press STOP, wait for the tape to stop, and then press REWIND. The same goes for changing between other modes: The machine will either obey you or make you press STOP first and wait until *it* is ready for your next command.

On moldy oldie manual machines whose levers go "snap" or "clunk" when you move them, you must be more careful. Switching directly from PLAY to REWIND, from FAST FORWARD to PLAY, or from FAST FORWARD to REWIND can stretch or break the tape. You must *move the control one step at a time* and wait a moment at each step to allow the motors and wheels to stabilize.

Finding Things Quickly on a Tape

As the boss ran out the door, he told you to show the client the part of the tape on "folding bathtubs," a new product the company has just perfected. Here's how you'd find that section of the tape: You could play the tape straight through until you came to that part, but that would be very slow.

You could switch the player to FAST FORWARD, let it wind for a minute, and then switch it to PLAY to see if you were there. If you were lucky enough to catch the middle of the folding bathtub exposition, you might then switch the machine to REWIND for a few seconds and then turn to PLAY again. You could repeat this process until you were at the beginning of the bathtub segment. In most cases, however, trying to find something on a tape this way is very haphazard. You could skip past something in FAST FORWARD and not even know it.

If you have some idea of where something is on a tape, you may try this: On many machines, 1 minute of fast forwarding equals about 20 minutes of regular playing time. So if you think the section you want is 40 minutes into the tape, let it FAST-FORWARD for about 2 minutes. This is also very haphazard because some tape machines wind faster than others or wind faster at the end of a tape than at the beginning.

Another way to gauge how far you are into a tape is to look through the window on the cassette. If something is near the middle of the tape, then the cassette take-up reel should be about half full.

Some video tape players have FAST SCAN or PICTURE SEARCH capabilities. It's like a FAST FORWARD or REWIND *with the picture showing.* If you FAST-FORWARD to near where you think the segment starts, you can then switch to the FORWARD SEARCH mode and watch the tape at 3, 5, or 10 times the normal speed until you get to the point you want. Note that FAST FORWARD and REWIND are faster than the SEARCH speeds, so it's best to get close to your target using FAST FORWARD or REWIND.

All these methods are cumbersome. The best way to find something on a tape is to use the INDEX COUNTER.

Index Counter. Nearly all tape machines have an INDEX COUNTER. Working like the odometer on your car,

---

MINIGLOSSARY

*Scan, fast scan, or picture search Playback of a VCR at several (up to 20) times the normal speed. Useful for skimming through a show in search for a particular event.

*Index counter An indicator on a VCR like the mileage meter on a car which changes numbers as the tape moves through the machine. It is handy for locating events on a tape or estimating the length of a production.

it keeps track of how much tape you've used. Expensive machines count minutes or feet of tape and are quite accurate. Inexpensive machines count turns of the take-up reel. Such counters are not especially accurate but at least give you an idea of where you are on the tape.

Had your boss INDEXED his tape, he would have written something like the following on a sheet of paper and put it with the tape or in the tape box:

Footage	Contents
000	Violin-tuning machine
225	Water hardener
340	Dandruff vacuum
405	Folding bathtub
560	Underwater bicycle
780	Shag carpet shears
900	Spider web remover
950	Infant repellent
1070	End

To use this index, you would

1. REWIND the tape to the beginning.
2. Locate the INDEX COUNTER or TAPE FOOTAGE COUNTER. By pressing a button or turning a dial nearby, RESET the numbers on the counter to 000.
3. To view the segment on folding bathtubs, FAST-FORWARD the machine to about 400. STOP. Then PLAY and watch for the dandruff vacuum section to finish and the folding bathtub section to begin.

The index allows you to jump ahead to 900 and examine the spider web remover and then go back to 225 and see the presentation on water hardeners.

The counters on most inexpensive video tape equipment aren't too accurate and may err by 10 percent. And if you are using a JVC machine but the tape was recorded on a Panasonic machine, the difference may be as much as 30 percent. So don't put *too much* trust in the counter.

Most INDEXES do not measure time, making it hard for you to correlate the machine's index numbers with the length of programs on a tape. If you do a lot of searching for things on tape, you may want to make a little table showing elapsed time verses INDEX COUNTER numbers.

INDEX COUNTER vs. time

Counter Number	Elapsed Time
000	
100	
200	
300	
400	
.	
.	
.	

There is an easy way to make such a chart. Aim a TV camera at a clock and hook it up to the video recorder. Walk away and let it record, at the two-hour speed, the clock face. Then rewind the tape. RESET the COUNTER to 000. FAST-FORWARD the tape to 100. Play it a little and see what time the clock shows. Write down the elapsed time next to the COUNTER number. Then zip ahead to 200 and repeat the process. Do this until you've filled your chart with COUNTER numbers and elapsed times. Now whenever you want to find where one ½-hour show ended and the next one began, you would look for the 30-minute mark on your chart and forward your machine to the corresponding COUNTER number. Note that if your cassette player works at three speeds, the 1-hour mark at the fast speed will be the same as the 2-hour mark at the middle speed and will be the same as the 3-hour mark at the slow speed.

Professional and editing video tape recorders often have more accurate INDEX COUNTERS which can count right down to the thirtieth of a second. We'll see more of those in Chapter 14 (editing).

SETTING UP A VIDEO TAPE PLAYER

Suppose your boss didn't have time to set up the tape player for you before he fled the office, but it is still your job to play his tape for the client.

Before getting into specifics, here are five general steps to setting up any video tape player and TV:

MINIGLOSSARY

*Reset   Setting something back to the beginning. Resetting an index counter sets all the numbers to zero.

1. Check to see that you have a tape player that is compatible with your tape (another look at the cassettes shown in Figure 5–1 may help). Also, make sure you are using a SUPER type VCR if playing an SVHS, HI8, or ED BETA tape.

2. Somehow you have to get the picture and sound from the tape player to the TV set.

3. Plug the tape player and the TV into a wall outlet and switch their POWER on.

4. You have to get the TV to "listen" to the tape player.

5. Load the tape and play it. If using a home VCR, the tape player will automatically determine the right speed to play the tape and will play it. You don't have to switch anything. If the tape plays at the wrong speed, that means the machine *can't* play that speed and you'll have to find another machine.

You may find it handy whenever you load a cassette to rewind it to the beginning and RESET your INDEX COUNTER to 000.

The rest involves the quirks in the machinery you are using and is a matter of connecting the right cables and finding the right buttons.

Using a TV Receiver

The only way to get a signal into a TV receiver is through its antenna input. Find an output on the tape player labeled RF OUT or VHF OUT or ANT OUT or TO TV. Connect the appropriate wire between the tape player's output and the TV's VHF antenna input as in Figure 5–4. If using coax, one F connector will plug into the tape player while the other F connector plugs into the TV. Other systems use flat twin lead between the two.

Nearly all home videocassette recorders (VHS and BETA) have built-in RF GENERATORS that send out their signals on channels 3 or 4. The channel number is probably marked on the back of the video player. Large, professional tape players generally do not have an RF GENERATOR built in and cannot work directly with a TV receiver. Wiring in a separate RF GENERATOR is possible but will be covered in Chapter 15. ¾U and other industrial VCRs may or may not have RF GENERATORS built into them. Often the RF GENERATOR is a separate option which plugs into the VCR as shown in Figure 5–5. This option costs about $50 and comes in a variety of channel numbers. For best results, find out what

TV channels are received in your area and purchase an RF unit that works on one of the *unused* channels. For instance, if your neighborhood receives channel 3 and if your RF GENERATOR is also for channel 3, when you use the unit, the two signals will interfere with each other. But by using a vacant channel, there is no competition between the signal from your RF GENERATOR and the one from the broadcasting transmitter.

Assuming you have an RF GENERATOR, let's go on to the next step, getting your TV to "listen" to the tape player.

Tune the TV set to the proper channel, the one the RF GENERATOR is designed for. Play the tape. Sound and picture should appear on the TV receiver, perhaps after some adjustment of the fine tuning.

Home VCRs (VHS and BETA) often have a switch on them which tells them to send their signal from the tape to the TV. Switch this OUTPUT SELECT or PROGRAM SELECT button to the VTR or VCR mode (*away from TV*).

Using a Monitor/Receiver

You can't use a true MONITOR for TV playbacks as they don't have speakers. Many manufacturers make monitors with speakers, and as long as you aren't using RF, you can wire them up and treat them the same as monitor/receivers. If you *had* to use a straight TV monitor for playing back a tape, you might get by wearing headphones or an earphone connected directly to the VCR if it has a PHONE or EAR output.

A monitor/receiver can be connected just like a TV receiver by sending RF from the tape player to the monitor/receiver's antenna terminals and switching the receiver to channel 3 or 4. But this does not give as sharp a picture as sending direct audio and video from the tape player to the monitor/receiver. Besides, some tape players don't have RF GENERATORS built in. Figure 5–6 diagrams the two processes. Figure 2–12 showed the difference in sharpness between RF and video.

With 8-Pin Connector. Figure 2–19 showed how to connect a TV monitor/receiver to a video tape machine via an 8-pin cable. This cable alone carries both sound and picture. Monitor/receivers have to be told which of their inputs to listen to. The switch for this is probably labeled TV/VTR or TV/VTR/LINE. When using the TV set as a RECEIVER, the switch should be in the TV position. To make it listen to the tape player, turn it to the VTR position. It should now be ready to play the tape.

MINIGLOSSARY

*TV receiver A TV set that tunes in channels but doesn't have audio or video inputs.

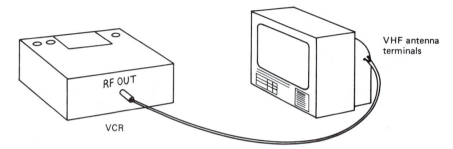

FIGURE 5-4 Playing a tape into a regular TV set.

With Separate Audio and Video Cables. Figure 2–20 showed how to connect separate audio and video cables from a tape player to a TV monitor/receiver (or monitor with sound). Essentially, you connect a coax video cable from the tape player's VIDEO OUT or LINE OUT to the monitor/receiver's VIDEO IN or EXT IN. Similarly, you connect an audio cable from the tape player's LINE OUT or AUDIO OUT to the monitor/receiver's AUDIO IN or EXT IN. As before, the monitor/receiver has to be told to listen to those inputs. If the selector is marked TV/VTR/LINE, switch it to LINE. If that doesn't work, try VTR. You're ready to play.

Using an S Connector

SVHS, HI8, and ED BETA VCRs make two kinds of video signals, COMPOSITE and Y/C. The COMPOSITE video output works fine with common TV monitors as described earlier. If your TV monitor accepts a multipin S CONNECTOR, then use this wire in place of the video wire. By sending the Y/C signals into the monitor, you will get sharper, cleaner colors. You still connect the audio cable in the usual way (described previously).

MAKING A VIDEO TAPE RECORDING

All the procedures that you learned about playing a tape will apply to recording a tape. You load the tape cassette the same way. Also the tape must be the same FORMAT as the machine.

Before we get to specifics, here are some generalities which apply to nearly all video recorders:

FIGURE 5-5 Installing RF GENERATOR. RF GENERATOR feeds TV through channel 3 or 4 depending on unused channel in your area.

MINIGLOSSARY

*S connector Small multipin connector carrying Y/C video signals from or to "super" VCRs.

*Y/C Video signal separated into two parts: brightness (Y) and color (C). Such signals yield sharper, cleaner color than composite video signals.

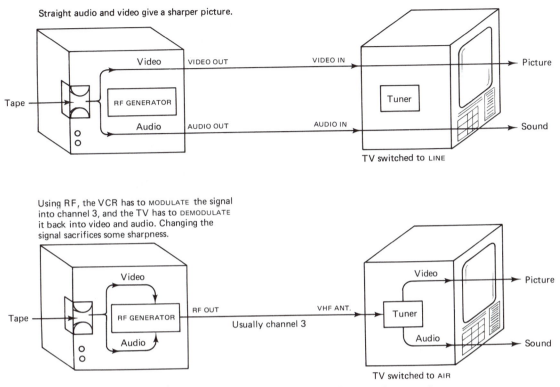

Straight audio and video give a sharper picture.

Using RF, the VCR has to MODULATE the signal into channel 3, and the TV has to DEMODULATE it back into video and audio. Changing the signal sacrifices some sharpness.

FIGURE 5-6 RF vs. video between a tape player and TV monitor/receiver.

1. You have to tell them which of their inputs to "listen" to. The INPUT SELECTOR does this. Often you may choose between a TV or tuner input, a camera input, or a general video input (good for about anything else).

2. You always want to monitor (watch) your recording. You have to tell the recorder what signal to send out to the TV; this is done with the OUTPUT SELECTOR. Home recorders can send either the antenna signal or the tape signal to the TV. Some industrial models can display the signal *going to* the tape or a signal *played off* the tape.

Industrial VCR with Automatic Controls

If the VCR has automatic video and audio circuits (as most do), here's what to do:

1. *Do a video and audio check before starting to record.*
   a. Press the big red RECORD button.

   b. Glance at the monitor/receiver connected to the VCR and notice if it is displaying a reasonably good picture and has acceptable volume for the sound. If there is no picture or no sound, something is wrong. When it is connected to the VCR, the monitor/receiver displays what the VCR "sees." If there is something wrong with the signal going to the VCR or if the VCR is doing something wrong, the difficulty will become evident on the monitor/receiver. Stop everything right there until the problem is located and corrected.

2. *To begin the recording:*
   a. About 10 seconds before the show actually begins, push the RECORD button down. While holding it down, switch the machine to PLAY or FORWARD.
   b. After the program starts, keep an eye on the monitor/receiver to make sure the picture and the sound remain okay.

MINIGLOSSARY

*Output selector Switch determining which of several signals will be fed to a VCR's output for viewing.

*Input selector Switch which determines which input (which source) a VCR will "listen" to.

A ROSE BY ANY OTHER NAME		
Different machines often use different names for the same switch or the same function. Rather than listing all the possible switch names each time one is mentioned, I will select one popular name and call it that. Here is a list of some of the other names for "whichamacallits" on a video tape machine:		
Various Names	*Names I'll Use*	*Definitions*
INDEX COUNTER TAPE FOOTAGE COUNTER	INDEX COUNTER	Shows how much tape has been used
RF OUT VHF OUT ANT OUT TV OUT TO TV	RF OUT	Output which sends RF (usually channel 3) to a TV's antenna input
INPUT SELECT PROGRAM SELECT REC SELECT TV/CASS TV/TUNER SOURCE ANT/VTR TV/CAM/LINE TV/AUX	INPUT SELECT	A switch determining which of a VTR's inputs it will "listen to" and record
OUTPUT SELECT PROGRAM SELECT TV/VTR TV/VCR TV/VIDEO TV/CASSETTE ANTENNA/VTR	OUTPUT SELECT	Switch on a home VCR which sends to the TV set either 1. Antenna signal 2. Signal from video-cassette while it's recording or playing back
VIDEO OUT LINE OUT	VIDEO OUT	Output sending a VTR's video signal to another device; signal is from the tape when in PLAY and is what is being recorded when VCR is in RECORD
AUDIO OUT LINE OUT HI LEVEL OUT	AUDIO OUT	HIGH-LEVEL (strong audio signal) sent from VTR to another device; signal is from tape when VTR is in PLAY; when VTR is RECORDING, presents the signal being recorded
AUDIO IN LINE IN AUX IN	AUDIO IN	Input for a HIGH-LEVEL (strong) sound signal (not a microphone)
PAUSE STILL FRAME	PAUSE (sometimes) STILL FRAME (sometimes)	VCR function which stops the tape from moving and displays a "frozen" picture
SEARCH SCAN	SEARCH	Playback of the tape at several times normal speed, forward or reverse, useful for skimming through programs quickly
SKEW TAPE TENSION	SKEW	Control on a VCR which keeps the top of the TV picture from bending or wobbling

3. When the program ends, let the machine run an extra 10 seconds and then press the STOP button. The reason for the 10-second space at the beginning and end of the tape is to provide a LEADER. Why bother with a LEADER?

   a. When VCRs thread and unthread the tape, they may scratch it a little, degrading the picture. The LEADER assures that only unimportant things get scratched.

   b. Some VCRs automatically advance the tape a bit before they begin playing it. If other people with machines different from yours borrow your tape, you don't want them to miss the first 10 seconds of your presentation. The LEADER acts as a safety margin before the beginning of the show.

   c. If the tape is to be edited later, the editing VCRs need a picture to "lock onto" as they get a running start before performing the edit.

   d. A LEADER at the beginning or end of a reel-to-reel tape provides room for people to thread and handle the tape without damaging the actual recording.

Industrial VCR with Manual Controls

Some VCRs can control audio and video automatically or manually. Look for switches called VIDEO LEVEL: MAN/AUTO and AUDIO LEVEL: MAN/AUTO. Switching these to MAN will give you manual control over the video and audio recording circuits.

If you wish to take manual control over the recording levels or the VCR has only manual audio and video controls instead of automatic circuits, the video and audio checks become more tedious. For these machines, you do the following:

1. *Check video level.* First you need to "get" a video level; that is, you need to adjust the video circuits to record a picture that turns out not too faded, not too bright, just right. To do this, press the red RECORD button. This starts the recorder's "record" circuits going so that you may observe how they are doing and adjust them as necessary.

Have your video source—the camera or whatever you are using—send you a sample picture so your VCR can "see" a typical, normal signal. Next, look at the meter labeled VIDEO, VIDEO LEVEL, or just LEVEL. If the VCR has only one meter, that means the meter can show *either* audio *or* video levels. A nearby VIDEO/AUDIO switch will make

the meter display the video or audio level. Switch it to VIDEO.

This meter should be pointing in the green area. If it points below the green, turn up the VIDEO or VIDEO LEVEL control until the meter needle goes into the green. Otherwise the picture will come out too faded. If the needle points into the red area, turn the VIDEO control down, or else the picture will have too much contrast or may even have streaks in it. The VIDEO control actually adjusts the brightness and contrast of your recording.

Some of the more expensive recorders allow you to monitor (that is, examine or look at) the recorder's picture and sound in the STOP mode as well as in the RECORD mode. You can tell when you are using that kind of VTR because the TV monitor connected to the VTR will display a picture while the VTR is in the STOP mode but will lose the picture if you turn the VTR's power off.

Some VCRs only display a picture in the RECORD/PLAY mode. In this case hit RECORD/PLAY and PAUSE to hold the tape still while monitoring.

Take a glance at the TV monitor or monitor/receiver to see how the picture looks. If your VIDEO meter says your picture is okay but it looks rotten on the TV screen, it could mean either

   a. Your video signal is bad.

   b. The TV monitor is misadjusted.

If the monitor picture is faded, too contrasty, too light, or too dark but the VIDEO meter reads in the green where it belongs, suspect monitor misadjustment. Although it can be fooled, the meter usually tells the truth. Note that *adjusting your TV set will not affect your recording at all*. A monitor doesn't *make* the pictures; it only *displays* them for your convenience. If you really want to take the guesswork out of this process, use a WAVEFORM monitor (discussed in Chapter 15).

2. *Check audio level.* With the record button still down, have someone speak normally into their microphone (or play a sample of the audio source which is going to be recorded). At the same time, observe the AUDIO or AUDIO LEVEL or LEVEL meter (remember to switch it to AUDIO if it is a dual-purpose meter). The needle should wiggle around just below the red area.

It is okay if the needle occasionally dips into the red area—the audience will accept momentary loud noises. But if the needle is too low or if it barely moves, the sound is too weak. Turn up the AUDIO control. On the other hand,

---

MINIGLOSSARY

*Leader Unrecorded space (from 10 seconds to 3 minutes) at the beginning of a tape, often used to protect the actual program from threading damage.

*Video level How strong a video signal is. On VTRs, the video level control will adjust the contrast of your video recording.

*Audio level How "loud" a sound signal is. Adjusting the audio level on a recorder determines the recording's loudness.

if the needle is in the red area too much, or "pins" (goes as far as it can go and holds there), the audio is too loud; turn it down.

Now hear what it sounds like on the monitor/receiver. If the meter says the sound is right but the monitor/receiver is blasting you in the ear, probably the sound on the monitor/receiver is too loud and needs to be turned down.

Incidentally, some VTRs don't need a monitor/receiver for sound because they have their own speakers. On these machines, you may use a TV monitor for viewing the picture and the VTR's own speaker for listening to the sound. Also, some VCRs have a headphone output for listening to the sound.

Some cassette machines have an AUDIO LIMITER control. When this control is *off*, you have total manual control of the sound volume. But when it's *on*, you have only partial control. When on, the AUDIO LIMITER allows you to control the sound volume manually *except* when a loud sound peak comes along. Then the LIMITER momentarily knocks the volume down to minimize the audio distortion from the loud passage.

To use this feature, you first set the AUDIO LIMITER to OFF and do a sound check, adjusting the audio manually. When the adjustments are complete, just switch the control to ON; you're now in manual control of the sound except during loud volume peaks.

3. *Begin the recording* as described before by pressing RECORD and PLAY and allowing a 10-second lead-in.

4. *During the recording*, watch the video and audio levels and adjust them whenever necessary to keep them at their proper settings.

5. *Finish the recording* as described before switching to STOP after a 10-second lead-out.

If you have a choice, when should you use *manual* and when should you use *automatic* controls? If you are going to be too busy during the production to pay attention to your audio and video levels, set them to AUTO. If, however, you have the time to watch them, set them to MANUAL. MANUAL is sometimes better than AUTO because automatic circuits can easily be fooled by loud noises or bright lights. A human, being more aware of the content of the show, can make recording level decisions much better than a circuit can.

RECORD/PLAY/PAUSE Dangers

As mentioned before, some VCRs require you to press RECORD and PLAY in order to monitor your picture. Since you don't *really* want to be recording at this time, you can press PAUSE to suspend tape motion until you are ready to begin.

When you do this the video recording heads start spinning and wearing against the tape. You can usually hear them. This is okay for a couple of minutes, but most VCRs protect the tape and their heads by switching themselves to STOP after three minutes. Don't let this surprise you, the automatic shutdown is normal. In fact, if you have finished making your audio and video checks, you will save wear and tear on the machine and tape by switching it to STOP manually. Incidentally, the same is true when playing a tape: if you switch the VCR to PAUSE, the heads will be wearing against the tape. Try to avoid unnecessary PAUSES. Again, the VCR will automatically shut down if you leave it displaying a STILL FRAME for more than three minutes.

Home VCRs

Professional VHS 8MM, SVHS, and HI8 VCRs operate very much like industrial ones. Because "industrial-strength" VCRs are so expensive, schools and industry often buy home-type VCRs. Home VCRs have a little less picture and sound quality and are less durable than the industrial models. They're also a little more complicated because they are designed to do *several* jobs:

1. Record video and audio signals from its video and audio inputs
2. Record TV programs off the air from its antenna input
3. Allow antenna signals to *pass through it* (as if it weren't there) on the way to a TV set
4. Tune in a particular channel off its antenna and send that channel's signal out its antenna output to a TV set

To do these extra jobs, home VCRs have extra buttons. Typically, you'll find them connected to an antenna and a TV set for recording programs off the air. To make sense of all the wiring and switches, let's study the home VCR in detail for a moment.

Normal TV Viewing. VCRs, with the exception of some portables, are designed to connect *between* your TV antenna and your TV set as shown in Figure 5–7. While watching normal TV, the antenna signals travel from the antenna, pass *through the VCR untouched* (the VCR could even be off), and then travel to your TV, which you can watch normally. To watch a tape played on the VCR, you flip a switch which electrically disconnects the antenna and substitutes the VCR's signal on channel 3 (or 4). Tune your

---

MINIGLOSSARY

*Audio limiter   Automatic control on a recorder that reduces volume during a recording if the sound becomes too loud. The

audio limiter doesn't affect the quiet and medium parts of the recording.

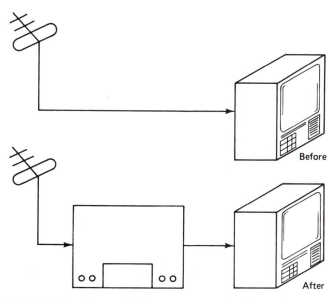

FIGURE 5-7 Home VCRs connect up between the antenna and the TV set.

Playing a tape:

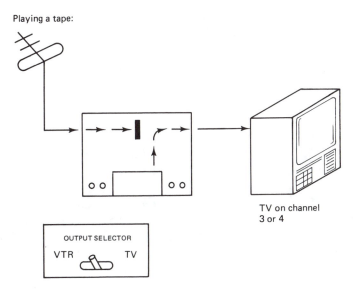

TV on channel 3 or 4

Watching normal TV:

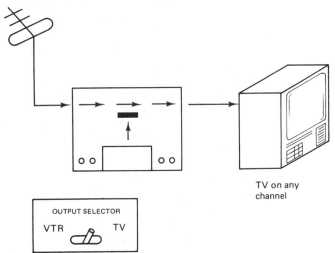

TV on any channel

FIGURE 5-8 Playing a tape vs. watching TV.

TV to channel 3 to watch the tape. Some machines make the substitution automatically (flipping their own switch when you play a tape).

Here are the steps you take to watch normal TV with a VCR hooked up:

1. *Turn on the TV set.*
2. *Switch the* OUTPUT SELECTOR *to TV.* The VCR does *not* have to be turned on.
3. *Select the desired TV channel* on your TV set.

That's all.

Playing a Tape. To play a tape, switch the VCR's OUTPUT SELECTOR to VTR (*away* from TV). This feeds the cassette program to your TV and inhibits the antenna signal. Tune the TV to channel 3 (or 4). Again, some machines flip their own OUTPUT SELECTOR switches when you press PLAY. Figure 5–8 illustrates the process. Press PLAY.

Recording Programs off the Air. Let's assume the VCR has already been properly hooked up to the antenna and TV (you'll see how to do this later). Essentially, you have to tell the VCR what *source* to listen to (camera, another VCR, or the TV antenna) and what *output* to send to your TV (the show the VCR is recording or *some other shows* coming in on the antenna).

You also have to tell the VCR which channel to record and at what speed it should record it.

One thing all recorders *will* do by themselves is ERASE the tape automatically as they record.

Warning: If you have something on a tape that you wish to keep and you accidentally record a new program

over it, your old program will be erased and gone forever. If you have the 2-hour movie *Indiana Jones and the Temple of Doom* recorded from last week and this week you start the tape from the beginning with a new recording of the half-hour program *Doogie Howser,* you'll lose the first quarter of *Indiana Jones.* You may get something like ''*Doogie Howser . . . and the Temple of Doom.*''

Back to business. Here are the steps you follow to record shows off air:

1. *Turn on the VCR and TV set.* Tune the TV to the VCR's channel, usually 3. Deactivate the VCR's timer by pressing the button marked TIMER. The timer light should now go off. (With the timer engaged, the VCR won't listen to your commands; it'll just sit there like a lump.)

2. *Load the cassette into the VCR* as described earlier. Make sure you have the proper length tape for the desired recording time.

3. *Tell the VCR which source to listen to.* The INPUT SELECTOR should be turned to TV.

4. *Tell your VCR what signal to send to your TV.* It's best to monitor the beginning of your recording to make sure everything is okay, so you'll want your TV to receive the VCR's signal that's being recorded. Switch the OUTPUT SELECT switch to VTR.

5. *Set the recorder's SPEED.* For highest-quality picture and sound, use the fastest speed (shortest play time). On VHS machines, this may be labeled SP (standard play—2 hours on a standard T-120 tape). On older beta machines this may be the X-2 (2-hour recording on an L-500 tape). If the program is too long to fit on the tape at that speed or if you're a cheapskate who likes to cram as much on a tape as possible regardless of picture quality, switch to a slower speed (longer playing time). On VHS machines, this may be labeled LP (for long play—4 hours on a T-120) or EP (extra long play—6 hours on a T-120). Longer tapes, of course, will allow longer recording times. Avoid recording at the 4-hour (LP) speed. Picture quality will be no better than the 6-hour speed, and some VCRs can't play back the 4-hour speed tapes. Further, many VCRs can only record at the 2- and 6-hour speeds.

6. *Switch the VCR's tuner to the channel you want to record.* Tuning a VCR is like tuning a TV. The procedures described in Chapter 2 for tuning a TV apply just as well here.

Note: If your antenna signal was poor or your VCR was mistuned, your recording will look bad. Once you have recorded a show poorly, it will remain that way—forever, or until you erase the tape and record something new.

If your TV set isn't perfectly tuned to start with, you will have difficulty knowing when you have properly tuned your VCR. You may think your signal is bad when it really isn't. Here's how to avoid this problem: First play a tape which you know is good. Tune your TV to channel 3 (or 4) to view the tape. FINE TUNE the TV until the picture comes in best. Now your TV is *calibrated.* Remove that tape and insert the tape you want to record. Press RECORD, if necessary, to get a picture from the VCR's tuner. As you FINE TUNE the VCR, you will see the TV's picture get worse or better. Make it better. From here on, do all of your FINE TUNING adjustments on the VCR. Incidentally, the tuning of the TV set *will not* affect your recording. The TV could

even be turned off. Your TV is just a monitor, showing results. Your VCR's tuner makes the results.

7. *Reset the tape INDEX COUNTER to 000* by pressing the button next to it. Later this can help find things on your tape.

8. *Start recording* if the TV says your picture and sound are okay. Press the RECORD and PLAY buttons simultaneously. Some machines have a single wide RECORD button which sets everything in motion.

9. *Check your recording* if you have time. Would you rather know that something was wrong right away or after 2 hours of recording *nothing*? Before the show starts, do steps 1–8, recording just 15 seconds of any show. Then play it back. If it's okay, rewind and get ready to record for real. This precaution may sound silly now. I guarantee it won't after you've lost a few shows.

10. *Rewind the tape* to the beginning when you've finished the recording. *Label your tape!*

Recording One Program While Viewing Another.

Your school wants you to record an important political debate for review in class, but you'd rather watch wrestling on another channel. With home VCRs it is possible to do *both*.

1. *Carry out steps 1–9 from the previous section on recording TV programs off the air.*

2. Once satisfied that your recording is successfully underway, *send your antenna signals straight to your TV* by switching the VCR's OUTPUT SELECTOR to the TV position.

3. *Select your desired broadcast TV channel on your TV.* This will not affect your recording. The VCR is recording the channel *it* is tuned to, while the TV is showing the show *it* is tuned to.

If you want to check on your recording from time to time, simply flip that OUTPUT SELECT switch to VTR and tune your TV to channel 3. This will not affect your recording. Figure 5–9 diagrams the process.

Recording with an Automatic Timer.

Automatic timers all work a little differently. Some are so complicated that you need a doctorate in electrical engineering just to program them. The simplest timers, called ONE EVENT timers, will record one event in one day. This means the recorder will tune in *one* channel and record it sometime within the next 24 hours. These timers are simplest to set up.

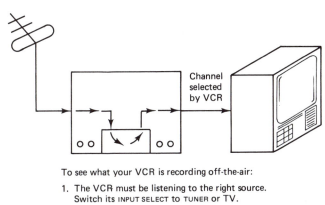

To see what your VCR is recording off-the-air:

1. The VCR must be listening to the right source. Switch its INPUT SELECT to TUNER or TV.

2. Your TV must be listening to the VCR. Switch the VCR's OUTPUT SELECT (or VTR/TV SELECT or ANTENNA switch or whatever) to VTR (or VCR or CASSETTE or VIDEO, or whatever). Tune the TV to channel 3 or 4.

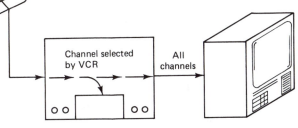

To watch other TV broadcasts while your VCR records one of them:

1. The VCR must *still* listen to the antenna. Switch its INPUT SELECT to TV.

2. Your TV must listen directly to the antenna. Switch the VCR's OUTPUT SELECT to TV. Tune the TV to the regular channels.

FIGURE 5-9 Watching your recording while you record vs. watching normal TV.

1. *Prepare the VCR to make the recording.*
   a. Is it listening to the right input? The INPUT SELECT should be switched to TV.
   b. Have you tuned in the channel you want to record?
   c. Have you selected the recording speed?
   d. Did you put a long enough (or any) tape in the machine to record the *whole* show?

2. *Tell the VCR when to start and when to stop.* Setting the start and stop times is much like setting a digital electric clock. Somehow you need to press a START or ON button and set the clock for the time the recording is to begin and then press an OFF or STOP button and set the clock again for the time the recording is to finish.

3. *Give control of the VCR over to the timer.* Often you do this by switching the timer ON. At this point, the other VCR controls stop working, and the timer is in command (switching the timer off returns control back to you).

If you're using a cable TV converter, it must be tuned to the station you wish to record, and left turned on. Your VCR, meanwhile, would be tuned to the converter's output frequency (channel 3 or 4). PROGRAMMABLE converters can be programmed to tune in one channel at one time, and another at another.

Most home VCRs have PROGRAMMABLE timers which can record several events over several days. They can record one channel at one time, stop, switch channels, and record another channel at another time. Some will record up to 14 different programs (which could be on 14 different channels) over a 2-week period. Telling these timers exactly what you want them to do is fairly complicated. That's why the instruction manuals look so hairy.

To record a show, your PROGRAMMABLE timer needs to know six things (all those instructions in the manual are showing you how to *tell* the timer these six things):

1. Which of its eight (or so) memories you wish to store your instructions in. Each memory is like a pocket in which you place a message. When the right time comes, the message pops out of the pocket, and the machine reads it to find out what to do next.

2. What week (this week or next week) the recording will start.

3. What day (Monday through Sunday) the recording will start. Some machines will permit you to select *all* days, so that you can record the same show at the same time each day.

4. What time (hour, minute, a.m., or p.m.) the recording will start.

5. What channel is to be recorded.

6. What week, day, and time the recording will stop.

Most timers help guide you by flashing on and off some part of their display to tell you what it is they want to know next. Others may display a menu of choices on your TV screen.

In short, somehow you have to tell the timer these six things *for each show* you want to record.

As a note of encouragement, when I got my first

MINIGLOSSARY

*Programmable   Ability to tell a machine to do something on its own. A VCR's programmable timer stores instructions for when to start and stop recording and what channel to record. It may remember several such instructions covering a period of days or even months.

programmable VCR, it took me an hour to learn how to set the timer. I somehow recorded a lot of Saturday morning "Bugs Bunny" when what I wanted was Saturday evening "Nature of Things." Fret not if it takes a while to get the hang of programming these things. I'm now 95 percent accurate, and it takes me one-half minute.

Remember, when you finish programming the timer, switch it ON *(press the* TIMER *button) to give the timer control over the VCR.*

Recording from the Audio and Video Inputs. Most home VCRs are designed to accept straight video and audio signals as well as RF. Connect the audio and video signals to the VCR just as you would an industrial tape machine. Throw the INPUT SELECT switch to CAMERA, AUX, or VCR (away from TV) so that it listens to those inputs.

Wiring Several TVs and a VCR Together. Connecting one antenna to one VCR to one TV is fairly straightforward. If you have cable TV, the wiring gets more complicated but looks a lot like the examples in Chapter 4. If you have several VCRs, the wiring starts looking like spaghetti. In fact, in honor of this nest of wires, an organization called Videofreex published an excellent little book called *Spaghetti City Video Manual* (Praeger, New York, 1973). If you have several TVs, then you add a whole new tangle to the ball of wires. Let's save multiple VCR wiring for Chapter 13 and concentrate on a few common setups for one VCR and several TVs. Figure 5–10 diagrams some ways to connect your VCR to several TVs and lists the advantages and drawbacks of each setup. If using cable TV, you are better off using cable ready equipment or installing up-converters to allow all devices to receive all channels all the time rather than the one channel a decoder box is tuned to.

Always try to keep cable runs as short as possible to avoid TV signal interference and to maintain a lusty signal.

Ensuring a Good Recording

You are not likely to heed this warning until the inevitable happens to you, but here's the warning anyway. *Before making your recording, make a short test recording and play it back to confirm proper functioning of the equipment.* It is much easier to invest this moment of prevention than it is to redo an hour of production.

The easiest way to check recording quality is with monitors. Figure 5–11 shows a handy setup. Monitor 1 receives the signal directly from the camera, switcher, or other video source. If the signal is bad to start with, this monitor will show it. There is no reason to believe that there's something wrong with your VCR or monitor 2. If monitor 1 shows a good picture and monitor 2 shows a bad one, then your problem is in your VCR or in monitor 2 or the cables between them. A bad picture on monitor 2 is generally a warning that the VCR is not receiving its picture properly.

A good picture on both monitor 1 and monitor 2 tells you that the VCR "sees" a good picture. *A good picture on monitor 2, however, does not guarantee that you are getting a good recording.* Here's why:

Before it records the picture, the VCR samples a tiny bit of the signal and sends it to monitor 2. Monitor 2 isn't showing what the VCR *recorded*, just what the VCR was "*seeing*." If the tape were bad (or even if there were no tape at all), the VCR would "see" a good picture and send it to monitor 2.

The only way to be absolutely sure that a recording came out okay is to play it back. While playing, the VCR sends the signal *recorded on the tape* to monitor 2. If the tape is good, monitor 2 will show it.

Things which ruin recordings and show up only on playback are dirt or scratches on the tape, wrinkles or creases in the tape, or defective video recording heads.

Some higher-quality video tape machines have a special mechanism in them to play back a tape *at the same time it's being recorded.* This is called INSTANT VIDEO CONFIDENCE. A switch on the video recorder allows you to select whether monitor 2 sees the signal *going to* the VTR or the signal *playing back* from the VTR (a fraction of a second after it is recorded).

Because most VTRs *do not* show you the condition of the actual recording, it is wise to record and play back a sample before starting actual taping. Whenever you are between taping sessions, it is also wise to check a little of the recording to make sure it is coming out.

Yet even if you play a sample and find that the tape is recording perfectly, this is no guarantee that a speck of dust won't clog up the machine or that the tape won't crease after you start recording again. Checking just minimizes the losses. Some studios, especially where productions are expensive and hard to repeat, record each session with two recorders; the second recorder makes a "safe" copy, an insurance policy against the failure of the first machine.

MINIGLOSSARY

A/B switch   Electrical switch which selects either the signal from cable A or the signal from cable B and feeds the results to a TV, VCR, or other destination.

Instant video confidence   Feature on some VTRs which allows them to play back the picture hundredths of a second after it is recorded while the VTR is still recording it. Handy for assuring the video heads are not clogged.

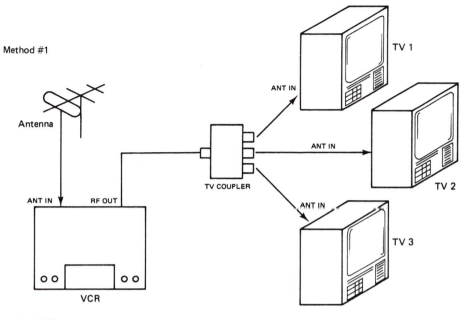

Method #1

Capabilities:

1. With VCR's output switched to TV, all TV's can view any channel. The VCR can simultaneously record any channel. The recording cannot be viewed simultaneously.
2. With the VCR output switched to VCR, all TV's see only what the VCR is recording or playing. They must also be tuned to the VCR's channel (usually 3 or 4).

Note: If the antenna signal is weak, there may not be enough signal to feed three TV's especially if their wires are long. In such cases, instead of using a PASSIVE TV COUPLER (uses no power), buy an ACTIVE COUPLER (which amplifies the signal while splitting it). Or, insert an RF AMPLIFIER or ANTENNA AMPLIFIER or ANTENNA BOOSTER *between* the VCR and the COUPLER.

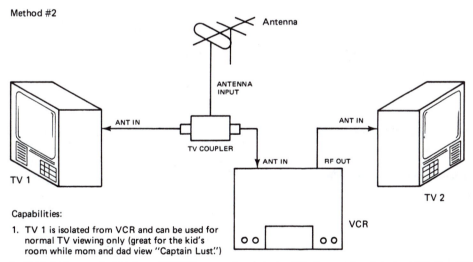

Method #2

Capabilities:

1. TV 1 is isolated from VCR and can be used for normal TV viewing only (great for the kid's room while mom and dad view "Captain Lust!")

2. TV 2 displays regular TV channel if VCR's output is switched to TV. If output is switched to VCR, then TV views only what the VCR is recording or playing. TV must be tuned to VCR's channel to display the VCR's signal.

FIGURE 5-10   Connecting your VCR to several TVs.

Method #3

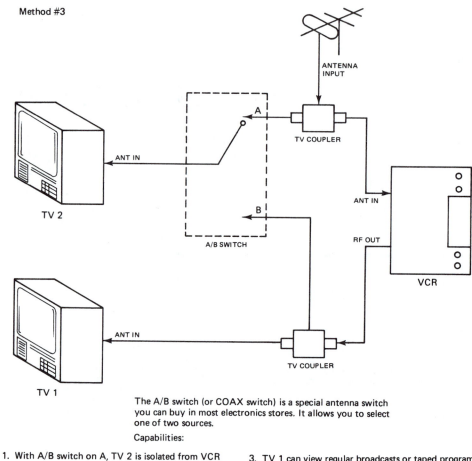

The A/B switch (or COAX switch) is a special antenna switch you can buy in most electronics stores. It allows you to select one of two sources.

Capabilities:

1. With A/B switch on A, TV 2 is isolated from VCR and views regular broadcasts.
2. With A/B switch on B, TV 2 watches what the VCR is putting out, which could be (a) regular TV channels if the VCR output is switched to TV, or (b) whatever the VCR is recording or playing if the output is switched to VCR.

3. TV 1 can view regular broadcasts or taped program depending on position of VTR's output switch.

In short, anything can be viewed on either TV with this setup, depending on where the switches are set.

Don't forget to monitor audio, too. In Figure 5–11, you do this by:

1. Plugging headphones into the VCR headphone output. This way you can hear what the VCR is hearing.

2. In place of monitor 2, use a monitor/receiver and send it both video and audio or send it RF. Either way, the sound will come out the monitor/receiver's speaker. If the sound bothers others, you may be able to plug an earphone into the monitor/receiver if it has such a socket.

In both preceding cases, you are monitoring *what the VCR is hearing*: You still don't know if good sound is really being recorded on the tape. As before, the only way to check this is to make a sample recording and play it back and hear how it sounds.

CONNECTING A VTR OR VCR FOR A RECORDING

The following will apply to most industrial video recorders. Home video recorders have a few twists of their own which

MINIGLOSSARY

*Low level or microphone level   Weak audio signal, as from a microphone.

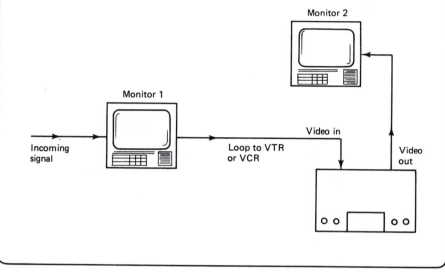

FIGURE 5-11 Setup showing the first monitor displaying the incoming signal and the second monitor displaying the VCR output.

will be mentioned later. Fancy editing video recorders will be discussed further in Chapter 14. Here are some generalizations on how to set up the equipment.

1. *Identify your signal sources* (camera, microphone, or whatever) and run video and audio cables from them to your VCR. If possible, run the video sources through a monitor first so that you can confirm for yourself that there is indeed satisfactory video coming over those cables (as in Figure 5–11). The signals, whether connected directly to the VCR or connected indirectly by looping through the monitor, must go to the VCR inputs.

2. *Plug the sources into the VCR inputs.* A VCR can handle only one video and only one audio source at a time (except for two-channel or stereo VCRs, which can record two audio tracks at once). If you are using two cameras or two microphones, they must be connected into some intermediate device that can select or mix the signals and send out only one video signal and only one audio signal. Figure 5–12 shows such a setup.

Audio. Most VCRs have a socket labeled MICROPHONE where an appropriate mike can be plugged in. An adapter may be necessary if the plug doesn't fit the socket. This socket is appropriate for weak audio signals like

Microphones

Most record turntables

Telephone pickup coils

Electric guitar pickups

The MICROPHONE OUTPUT of an audio mixer

In general, if your audio source has no power supply of its own, doesn't use batteries, and doesn't need to be plugged into the wall to work, it should be plugged into the MICROPHONE INPUT or MIKE IN of the VCR.

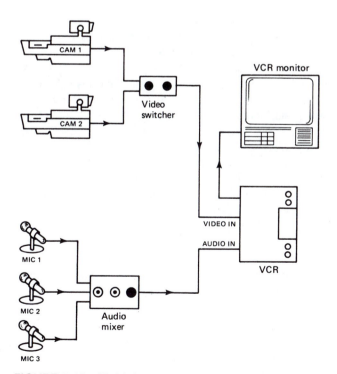

FIGURE 5-12 Multiple sources going to VCR inputs.

VCRs are likely to have another audio input labeled AUX (for AUXILIARY) or LINE or LINE IN or HIGH LEVEL IN or AUDIO IN. These inputs are not so sensitive and can take the stronger audio signals you would get from

AM or FM tuner

Audio tape deck

Audiocassette player

Monitor/receiver

Movie projector

Turntable with built-in preamplifiers

VTR

VCR

Mike mixer

You can expect a strong audio signal from a device if its output is labeled

LINE OUT

HIGH LEVEL OUT

PROGRAM OUT

AUX OUT

PREAMP OUT

AUDIO OUT

PHONE

EAR

MONITOR OUT

In most cases, if the audio source needs electricity to operate, the audio signal should go in the AUX IN or LINE IN of the VTR as shown in Figure 5–13.

Some home VCRs lack AUDIO INPUTS and have just the supersensitive MICROPHONE or MIC inputs. So how do you get HI LEVEL signals into the machine without over-driving the input? Use an AUDIO ATTENUATOR (Figure 5–14).

Video. Nearly all VCRs except home types have a BNC or an SO259 socket labeled VIDEO IN or LINE IN. (SO259 is the socket that the PL259 plug goes into. Review Figures 2–21 and 2–30 if you don't remember what these connectors look like.) Some may also have another connector labeled CAMERA. In addition you may also find the familiar 8-PIN connector.

If your video source is a camera that sends its signal over a cable with a round multipin plug at the end, you use the VCR input marked CAMERA. If the video source (which

---

LO LEVEL AND HI LEVEL EXPLAINED

A microphone turns sound into a tiny electrical signal. A preamplifier changes the tiny signal into a medium-sized electrical signal. An amplifier turns a medium-sized signal into a big electrical signal. A speaker changes a big signal into sound. Medium-sized signals are easiest for electronic equipment to handle, so most audio devices have PREAMP or HI-LEVEL outputs for sending medium-strength signals to other devices. They also have LINE, HI-LEVEL, or AUX inputs to receive medium-strength signals from other devices.

Some audio devices have earphone, headphone, or speaker outputs. In some cases, signals from these outputs are too loud even for the AUX IN of a VTR: The recorded sound may come out raspy and distorted. The cure for this is to connect an AUDIO ATTENUATOR between the source and the input. The ATTENUATOR cuts down the strength of the signal. If you don't have an ATTENUATOR, and the source's signal is too strong; try turning its volume control down very low.

---

may still be a camera) uses the more common 75Ω coax and a BNC plug, use the VIDEO IN connection to the VCR. If the video source is a monitor/receiver which uses an 8-PIN connector, use the 8-PIN socket. These connections are diagrammed in Figure 5–15.

Somewhere on the VCR will be a switch called INPUT SELECTOR or CAMERA/TV/LINE or something similar that tells the machine which particular input to use for its recording. If you are using a camera plugged directly into the CAMERA input, switch the selector to CAMERA. If your source is a video tape player, or a camera, or the TV OUT from a monitor/receiver, use the LINE input and switch the selector to LINE. If you are recording from a monitor/receiver through an 8-PIN cable, switch the selector to TV.

3. Plug a monitoring device into the VCR (as in Figure 5–16). Somehow you must be able to find out how the VCR is handling the video and audio. If you are using a monitor/receiver with an 8-PIN cable, plug one end into the VCR and the other into the monitor/receiver and switch the monitor/receiver to the VTR mode.

If you are using separate audio and video cables, plug them into the VIDEO OUT and AUDIO OUT from the VCR. The other ends of these cables go into the VIDEO IN, AUDIO IN, or EXT IN sockets of the monitor/receiver. Switch the monitor/receiver to the VTR mode so it "listens" to those inputs. Don't forget to flip its terminator switch to the 75Ω position.

If using only a video monitor to check the picture, then run a cable from the VCR's VIDEO OUT to the monitor's VIDEO IN. Remember to terminate. To check the sound, use

---

MINIGLOSSARY

*High level or hi level Strong audio signal typically sent from an aux out or a line out of a device.

*Attenuator Small electronic device that reduces the strength of an audio signal.

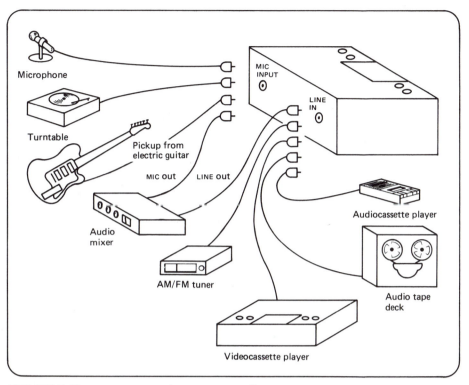

FIGURE 5-13   HIGH LEVEL and LOW LEVEL audio sources.

either the VCR's built-in audio system, earphone, or headphone output (if it has one) or a separate loudspeaker system. For the latter, run an audio cable from the AUDIO OUT of the VCR to the AUX or HIGH LEVEL or LINE input of the loudspeaker system.

You can also monitor the audio and video of your recording with any TV receiver by using RF (if the VCR has an RF generator). Connect the recorder's RF output to the antenna terminals of the TV receiver. Turn the receiver to the proper channel to observe both picture and sound.

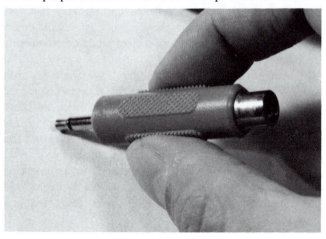

FIGURE 5-14   AUDIO ATTENUATOR connects between a "strong" audio source and a "sensitive" audio input, like a MIC IN to cut down the signal strength.

As mentioned earlier, it is always better to monitor the signals *going into* the VCR as well as those *coming out* of the VCR so that if something goes wrong, you'll have some idea about where the problem is. Figure 5–17 reviews how a VCR can monitor both its sources and its outputs.

4. *Plug everything in and turn it on.*

5. *Thread the tape or insert the cassette.*

6. *Do your video and audio checks.*

7. *Record and play back something to make sure everything is working.*

8. *Now you are ready to record* (Ta dah!).

Figure 5–18 reviews the kinds of things which can be plugged into a VCR's inputs, outputs, and "whatputs."

Recording a TV Broadcast "Off Air"

Home videocassette recorders record off air with ease, as we saw earlier in this chapter. Industrial video recorders lack tuners, making the process more cumbersome. Somehow we have to feed an antenna signal to a tuner which can make the audio and video to send to the recorder as diagrammed in Figure 5–19.

There are several ways to do this:

1.   Send the antenna signal to a separate tuner which makes audio and video to send to the VCR. A monitor/receiver connected to the VCR will help

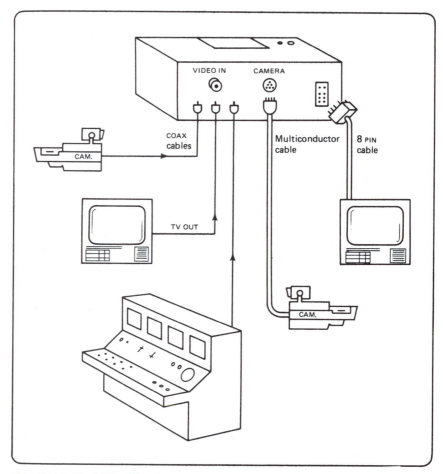

FIGURE 5-15   VCR video inputs.

you tune the tuner and check the signal quality.

2. Find a monitor/receiver with TV OUT connections and use *it* as the tuner. The antenna signal goes into the monitor/receiver, which then makes video and audio, which you send to the VCR. A second monitor/receiver connected to the VCR's outputs will assure that the signal is getting through okay.

3. The most elegant setup of all involves only one monitor/receiver, but it must be connected up just right or else it can become a trap:

You connect the antenna to the monitor/receiver, which in turn is connected to the VCR via an 8-pin cable. Switch the monitor/receiver to the TV mode, find the station, and get the picture to come in clearly. Now you know you have a good signal to send to the VCR. You switch the VCR to the TV mode (or whatever mode listens to that input), press RECORD, and check your audio and video levels.

Although you could make the recording now, *don't*. There's one little step to do that may save your recording someday. Remember how your monitor/receiver was in the TV mode? That means it is showing the signal the TV set is receiving off the air. But what about the signal the VCR is actually recording? Your monitor/receiver, which customarily showed you what your VCR was producing, is now showing you only the TV broadcast signal. Seeing a picture on the screen may lull you into thinking that you are monitoring a recording. You're not. Your VCR could be turned off and the monitor/receiver's picture would still be there.

Here's what to do: Switch your TV monitor/receiver to the VTR mode. In most sets, the following will happen: The TV will change the RF into audio and video and send the two signals over the 8-pin cable to the VCR for recording. The VCR, when in the RECORD mode, sends the signal to the heads for recording on the tape and also sends a sample of the signal out the 8-pin cable for monitoring. The monitor/receiver picks up the signals and displays them. Here the TV is showing *what the VCR is seeing*, not just what the tuner is seeing.

In short, when in the VTR mode, the monitor/receiver displays how good a job the VCR is doing at recording the show. In the TV mode, it shows only what the set is receiving directly off the air. The picture may look exactly

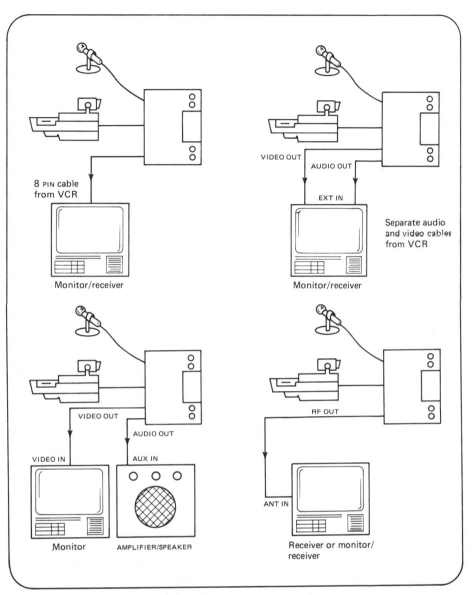

FIGURE 5-16 Ways of monitoring a recording.

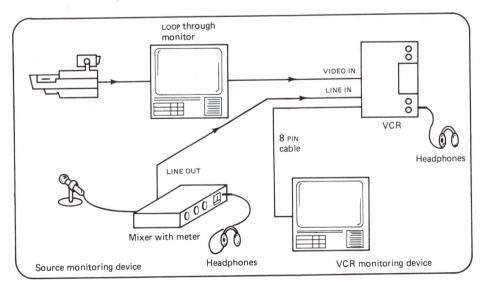

FIGURE 5-17 Monitoring both the sources and the VCR.

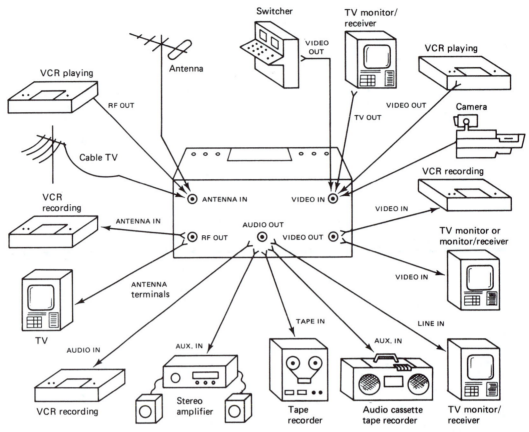

FIGURE 5-18   Reviewing the ins and outs of RF, video, and audio between a VCR and other equipment.

the same in both cases, but in the VTR mode if your recorder poops out, you'll see the problem on the screen.

There's nothing magic about using an 8-pin cable between the two as described. You could send the TV OUT from the TV to the inputs of the VCR and outputs of the VCR to the inputs of the TV.

If recording shows off cable TV, the connections are very similar to those shown in Chapter 4. In each case, simply substitute a home VCR or a VCR-connected-to-a-TV for the TV set shown. The problems and joys you faced using descramblers, converters, up-converters, splitters, etc., are all the same when feeding your signal to a VCR.

OTHER VCR CONTROLS AND FEATURES

As you can see from Figure 5-20, a VCR teems with inputs, outputs, and switches. Not all features and sockets will be

found on all VCRs. Simple videocassette players may have very few features. Inexpensive home and industrial VCRs may also be featureless. Top-of-the-line machines, however, have more bells and whistles than you'll ever find time to ring or blow.

Remote Control

A REMOTE CONTROL is a keypad which allows you to operate a VCR's controls at some distance (Figure 5-21). Industrial models often connect to the VCR via a wire with a multipin connector at the end. The plug goes into a socket on the VCR labeled REMOTE. Many home VCRs replace the wired REMOTE with a wireless, battery-powered keypad which sends infrared light signals through the air to a sensor on the VCR. To work, the WIRELESS REMOTE must be aimed like a flashlight in the general direction of the VCR.

MINIGLOSSARY

Remote control   Control of a device such as a VCR with a keypad held in the hand. Some require a wire connection to the recorder; others use infrared light signals and are wireless.

Unswitched   Some VTRs and other devices have convenience

power outlets in the back, handy for feeding power to other devices. If this power ceases when the VTR's power is turned off, it's called "switched." Unswitched outlets are on all the time as long as the VTR is plugged in.

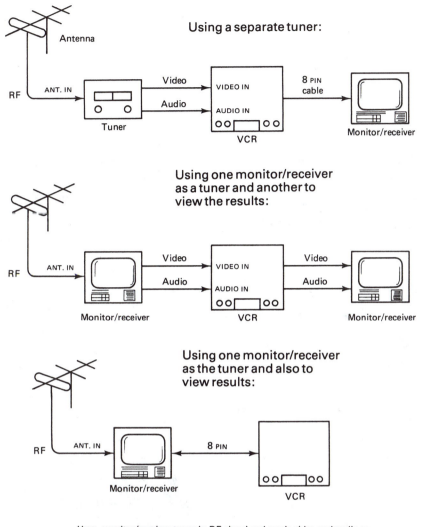

Using a separate tuner:

Using one monitor/receiver
as a tuner and another to
view the results:

Using one monitor/receiver
as the tuner and also to
view results:

Here, monitor/receiver tunes in RF signal and sends video and audio to
VCR. VCR sends its video and audio *output* back to TV set, which in
the VTR mode, displays it.

FIGURE 5-19 Recording a program "off air."

Most REMOTES have functions like REWIND, STOP, PLAY, FAST FORWARD, PAUSE, and PICTURE SEARCH in the forward or reverse directions. Home models often include CHANNEL CHANGE and VARIABLE SLOW MOTION controls.

In Chapter 14 we'll see how EDITING CONTROLLERS can plug into the REMOTE sockets of several VCRs at once, controlling all of them.

Convenience Outlet

In the back of some VCRs you'll find an electrical outlet. This AC power socket is handy for plugging in small accessories such as separate timers, converters, or decoders or a small lamp. If the outlet says UNSWITCHED, it provides power whether the VCR is turned on or not (handy for timers which must keep going after the VCR is turned off). If the outlet is labeled SWITCHED, then it turns on and off with the VCR (handy for activating decoders during the time a show is being recorded).

These outlets are usually labeled something like "300W MAX," which means they can feed up to 300 watts of power to an appliance (like three 100-watt light bulbs). Plug your 1000-watt steam iron into it, and you risk zapping your circuits or blowing a fuse inside your machine. Depending on your TV's wattage, it's probably safe to plug your TV into this outlet.

Dew Indicator

For the VCR to operate properly, the tape has to slide smoothly through its internal mechanism. When the machine is cold and the air is humid, the surfaces become

Features galore.

"sticky," and the tape risks getting hung up and damaged. The DEW INDICATOR senses this dampness (illuminating a light to notify you of this) and makes you wait until it's safe. If the DEW INDICATOR brings your TV production to a standstill, just leave the machine *on* and wait for the light to go out. Sitting and staring at the machine will not shorten this process. When the light goes out, your VCR will function normally again.

S CONNECTORS

SUPER VHS, HI 8MM, and ED BETA VCRs, as well as high-quality TV monitor/receivers and other video equipment can handle their video signals two ways. The normal way, COMPOSITE VIDEO, uses one wire to transmit the color and brightness signals. This is the method you've seen diagrammed throughout this book. The improved way is to use the S (SUPER) connector, if your equipment has it, to transmit the brightness (Y) and color (C) signals separately. For the sharpest, smoothest colors, use the S (or Y/C, as they are sometimes called) connectors whenever possible.

Counter/Reset/Memory

We've already seen how the INDEX COUNTER can help us find things on a tape. Some counters have a MEMORY button which makes locating things even easier. The MEMORY button will stop the machine from rewinding whenever the INDEX COUNTER reaches 000. Here's how to use it.

Say you are playing or recording "Star Trek" and you think you'd like to come back to the exciting scene where aliens burn up all the circuits on the Enterprise (for the sixty-seventh time). First you set the COUNTER RESET to 000. Next, switch MEMORY on. Let the tape continue playing. Later, when you want to find that spot, simply press REWIND. When the counter reaches 000, the machine will automatically STOP. You are there.

One disadvantage of this system is that when you press RESET midtape, you are no longer keeping chronological track of elapsed time. This may or may not be a problem, depending on whether you're trying to keep a log of what's where on the tape.

Speed Select

¾U and industrial VCRs are designed for one standard speed only. TIME LAPSE VCRs can be switched between various slow speeds.

VHS recorders are equipped with one, two, or three speeds: SP (2-hour standard play), LP (4-hour long play), and SLP or EP (6-hour super long play or extra play). Beta

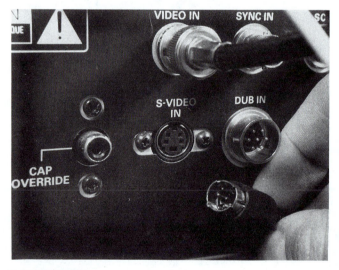

S CONNECTOR

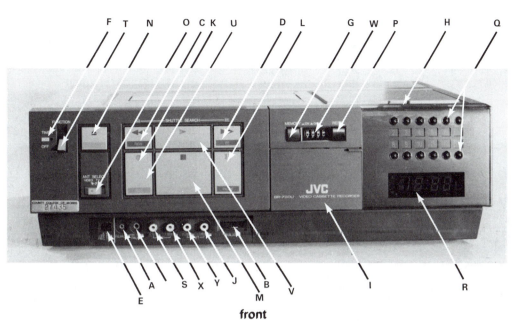

front

(A)	wired REMOTE CONTROLS	**(M)**	STOP button
(B)	TRACKING control	**(N)**	EJECT button
(C)	REWIND (REW) button	**(O)**	OUTPUT SELECT (VIDEO/TV) switch
(D)	FAST FORWARD (FF) button	**(P)**	tape/COUNTER reset button
(E)	recording speed selector	**(Q)**	channel selectors
(F)	POWER/TIMER lamp	**(R)**	timer display
(G)	memory SEARCH switch	**(S)**	AUDIO INput
(H)	automatic fine tuning (AFT) switch	**(T)**	POWER/TIMER switch
	and FINE TUNING controls behind trap door	**(U)**	RECORD (REC) button
(I)	timer controls behind trap door	**(V)**	PLAY button
(J)	VIDEO OUTput	**(W)**	tape COUNTEF
(K)	RECORD lamp	**(X)**	VIDEO INput
(L)	PAUSE button	**(Y)**	AUDIO OUTput

rear

FIGURE 5-20 Location of inputs, outputs, and controls on one VCR.

FIGURE 5-21 Remote control.

recorders similarly can have up to three speeds: X-1 (1 hour), X-2 (2 hour), and X-3 (3 hour). Of course tapes of different lengths will change how long the VCR can record at a given speed.

You select the tape speed when recording. VCRs automatically sense and adjust themselves to the correct speed when playing back.

Many industrial and editing VHS and 8MM VCRs, in order to provide the best picture possible, work at only one speed, the fastest speed. If recording with a three-speed machine, you should select the fastest speed (the shortest recording time) for the highest-quality recording and to assure that your tape will be playable on industrial machines.

Tracking

Generally, you shouldn't have to adjust the TRACKING control. Leave it in the FIX or center position. When playing a tape, if your picture shows a band of hash across it as in Figure 5–22 or jiggles a lot, it is time to adjust TRACKING. Turn it until the picture clears up.

This control compensates for minor differences between similar FORMAT recorders and tapes. You are most likely to have to use it when playing a tape recorded on someone else's machine or even when playing your own tapes when your machine gets worn and out-of-adjustment. This control does nothing when you are recording—only when you are playing tapes.

FIGURE 5-22 TRACKING misadjustment.

Another control, SLOW TRACKING, smoothes out your picture while your VCR is in STILL or SLOW MOTION play.

Some VCRs (all 8MM types), have AUTOMATIC TRACKING, making their adjustments electronically with no help from you.

Audio Dub

When you record a program, you automatically record both picture and sound. With AUDIO DUB you can go back and rerecord *new* sound, erasing the old sound as you go and leaving the picture untouched.

To activate the AUDIO DUB feature:

1. Plug a microphone into MIC IN, or some other audio signal source into AUDIO IN as described earlier. If using a portable VCR and camera, you can even use the microphone built into the camera.
2. You may wish to check your sound before you start recording by listening to it through your system. You could
    a. Plug an earphone into the EARPHONE jack if your VCR has one.
    b. Run RF to your TV and monitor your sound there.
    c. Connect your hi-fi to the AUDIO OUT jack on the VCR.

On some decks, you will automatically be able to hear your audio source with the VCR on STOP. On others you must press AUDIO DUB to send the sound through the monitoring outputs. On a few, press AUDIO DUB/PLAY/PAUSE simultaneously to hear your source.

MINIGLOSSARY

Slow tracking Adjustment on a VCR used to attain a clear picture while the tape is playing slowly or is still framed.

*Audio dub Feature on video recorders which allows you to record new sound (erasing the old sound) on a tape while leaving the picture untouched.

*Noise bars Bands of snowy hash across the TV screen usually evident in still and scan modes and when mistracking occurs.

3. Find where you wish to start dubbing new sound.

4. Press AUDIO DUB and PLAY together.

5. You will hear the new sound being recorded. The old sound is being erased as you go.

6. When done, press STOP.

Some VCRs have STEREO sound and features such as SOUND-ON-SOUND or SOUND-WITH-SOUND. You can make AUDIO DUBS on hi fi VCRs, but the new sound will be low fidelity. We'll get into all these features in Chapter 10.

Still Frame

When playing a tape, STILL FRAME "freezes" the picture on the screen. Some VCRs do a better job than others at holding a clean still picture on the screen. Sometimes the picture will stop with a horizontal band of hash across it (which looks exactly like the tracking problem shown back in Figure 5-22). This hash is called NOISE or a NOISE BAR. No, a NOISE BAR is not a loud tavern. Jus as in audio, where "noise" is unwanted sound, in video, "noise" can be unwanted stuff in the picture like hash, grain, snow, or specks.

Home videocassette recorders with *two video heads* generally leave NOISE BARS in their STILL FRAMES. Home VCRs with three, four, or five video heads can make a NOISELESS STILL FRAME. Some VCRs automatically make the picture NOISELESS. Others require fidgeting with the SLOW TRACKING control.

Adjusting SLOW TRACKING while the machine is in STILL FRAME will not improve the picture you see at that moment. Jump to another STILL FRAME, and the adjustment you made will then go into effect. Generally, if you adjust SLOW TRACKING so that the picture is NOISE-free while playing in the slow motion mode, the picture will be clean in the STILL FRAME mode too.

¾U VCRs may have this adjustment on the back or bottom, perhaps labeled V-LOCK ADJUST.

Expensive editing and professional tape machines have a similar feature called AUTO SCAN TRACKING or DYNAMIC TRACKING. On these machines the video recording heads are specially built to provide a superclear STILL FRAME.

DIGITAL STILL FRAME is a high-end home VCR feature that electronically "grabs" and displays an excellent STILL FRAME from the tape while it's playing. It will hold the STILL indefinitely; the tape doesn't have to be PAUSED.

FRAME ADVANCE is a STILL FRAME subfeature. Once in the STILL FRAME mode, a second control allows you to advance to the next video picture and stay there, or on some models, by holding the button down, you can continuously advance the tape slowly, picture by picture, like viewing a movie one frame at a time. On professional tape machines, this feature is often called JOG.

Unless the VCR is especially designed for NOISELESS STILL FRAMES, those familiar NOISE BARS will roll across the screen as the picture progresses in the FRAME ADVANCE mode.

Variable Speed Slow Motion

Sometimes combined with STILL FRAME or FRAME ADVANCE features, VARIABLE SPEED SLOW MOTION allows the tape to proceed slowly forward at various speeds.

Double or Triple Speed Play

With DOUBLE or TRIPLE SPEED PLAY the tape passes through the machine at twice or three times the normal speed while it is playing. Although the picture moves at the fast pace, funny things happen to the audio. Some machines mute the audio so you can't hear it. Others allow you to hear the sound. Like a speeded up record, the Donald Duck sound is nearly incomprehensible. More expensive VCRs include a special SPEECH COMPRESSOR circuit which renders the sound much more intelligible. Another name for this circuit is LIPLOCK. Is there someone in your family who you wished had LIPLOCK?

Double and triple speed play are very useful features on editing VCRs where many hours of tape have to be reviewed and indexed.

Search

SEARCH or SCAN or HIGH-SPEED PICTURE SEARCH shuttles the tape *forward* or *reverse* at 5, 10, or maybe 20 times normal speed with its picture still showing.

Automatic Backspace

The AUTOMATIC BACKSPACE feature is found on the better camcorders and VCRs. With this feature, when you stop recording, the VCR automatically backs the tape up a little ways. When you start recording again, the machine *first* starts the tape moving and gets it up to speed before actually

---

MINIGLOSSARY

Dynamic tracking Professional VTR feature which allows the tape to be played at various speeds including still frame while making a clear picture.

Frame advance VCR feature allowing the tape to be moved forward one video picture at a time.

Speech compressor Electronic circuit which changes the pitch of fast playing tape so that it sounds normal (not like the "Chipmunks").

Varactor Electronic tuning of TV channels using 10 buttons allowing you to punch in any channel number directly.

switching into RECORD. This attribute ensures clean, glitch-free edits from scene to scene.

This feature is nice, but it can also foul you up. In Chapter 14 you will learn how to PAUSE your recording at crucial places to edit out commercials, change scenes, or change shots or angles. However, every time you interrupt the smooth motion of the tape, such as stopping it or pausing it, you cause a glitch. Different manufacturers have attempted to minimize these unsightly blemishes from folks' backyard epics with AUTOMATIC BACKSPACE. Some VCRs have a switch called EDIT to activate this feature.

These systems endeavor to match the new incoming sync signal with the old recorded sync signal and thus maintain a smooth flow of sync pulses recorded on your tape. When played back, the smooth sync flow yields a steady picture on your TV. Also, every time you hit PAUSE while recording, the tape backs up a little. When you UNPAUSE, the circuits start the tape moving, line up the sync signals, and record the new picture over the tail of the old one.

This method does have some disadvantages:

1. Precise editing is difficult. You have to allow a 2-second ''tail'' at the end of each scene so that the VCR doesn't start the next scene on top of something you wanted to keep.

2. Since there's an overlap of video signals, the color sometimes gets messed up for a couple of seconds. This shimmering spectrum of color is called VIDEO MOIRE.

Stereo

Many home and a few professional VCRs will record and play back in stereo. There are two audio tracks, one for the left channel and one for the right channel.

There are several ways to get signals onto those two channels. One way is to send the signal to the left and right AUDIO INPUTS on the VCR. Another way is to use a stereo microphone connected to the camera or the VCR while making your recording.

A stereo VCR cannot necessarily receive stereo TV broadcasts *off-air*. To do so, it needs a stereo TV *tuner* to

decode the stereo broadcasts and make two-channel sound. This feature is called MTS (multichannel television sound) and is found on newer stereo TVs and VCRs. An MTS switch on your VCR will make it sensitive to these broadcast stereo signals so that they can be recorded. Many TV broadcasts are now in stereo.

If a program is monaural (single channel rather than stereo), then that is all your VCR will record. Also, unless your TV set is able to play stereo, you will get monaural sound even if the VCR and the tape are stereo. VCRs are designed so that stereo tapes will play on monaural machines and monaural tapes will play on stereo machines. In both cases, the sound will be monaural.

The word *stereo* does not imply *high fidelity*; it only means that there are two audio channels, one for the left ear and one for the right ear. Those two channels can be stereo but still deliver mediocre sound quality, as is usually the case with home VCRs (except the hi-fi versions, which will be described shortly).

Dolby

DOLBY is a method of improving sound quality, mostly by reducing background hiss in a tape recording. When making a recording, switch DOLBY *on* to put the improved sound on the tape. When playing the tape back, also switch DOLBY *on* to extract the best sound from the tape. A DOLBY tape can be played on a non-DOLBY machine and vice versa, but the sound will be slightly worse than normal quality.

If you are making a recording for someone who does not have DOLBY, then switch your DOLBY *off*. If you are playing a tape which was not recorded in DOLBY, also switch DOLBY *off* for best results.

Audio-1, Audio-2, Mix

¾U VCRs have two independent channels of audio called AUDIO-1 and AUDIO-2. You can record on the AUDIO-1 channel by plugging your source into the AUDIO-1 INPUT of the VCR. You can record on AUDIO-2 by plugging into the AUDIO-2 input of the VCR. When playing back the tape, you flip the AUDIO/MONITOR switch to CH-1 to listen to the AUDIO-1 channel or to CH-2 to listen to the AUDIO-2 channel

---

MINIGLOSSARY

*Stereo   Two separate audio channels are used at the same time. One represents what the left ear would hear and the other, the right.

MTS   Multichannel television sound, technique for stereo TV broadcasting.

*Backspace   Act of moving a video tape backward slightly. Helpful in producing glitchless (clean, smooth) edits.

Moire   Faint diagonal lines across a TV screen sometimes seen when you record over unerased tape.

*Monaural   Single-channel audio. Opposite of stereo.

*Dolby   Method of improving recorded sound quality on consumer audio and video equipment.

Audio-1, audio-2   Names given to the two audio channels on a ¾U VCR. Audio-2 is often the main channel. Some home VCRs may have two-channel audio or stereo audio.

*Mix   Switch on a two-channel or stereo VCR which allows both channels to be mixed and heard together.

or to MIX to listen to both channels at the same time. Although some VCRs will record and play back in stereo, two-channel audio does not mean the same as stereo. With two-channel audio you can listen to the first sound track or the second track or a mixture of both sound tracks, but the sound you get is still monaural.

Dub

DUB is not the same as AUDIO DUB. This is a feature found in editing and high-quality ¾U and professional VCRs. With the help of a special DUB CABLE connected between two VCRs, it allows extra sharp copies of videocassettes to be made. You'll see more about this in Chapter 13, "Copying A Video Tape."

Automatic Rewind

AUTOMATIC REWIND or REPEAT is a feature which allows a tape machine to play the same show over and over all day long. When the machine reaches the end of the tape, or a certain point programmed into the tape, the machine will stop and rewind to 000 on its INDEX COUNTER and will switch to PLAY.

Skew or Tape Tension

SKEW or TAPE TENSION is an adjustment on a VCR which tightens or loosens the tape as it plays. SKEW does not affect recording. Home VCRs have this control factory preset, making it inaccessible to the user.

When tape shrinks or stretches or is pulled too tight by the VCR playing it, the picture bends at the top or flutters back and forth. This is called FLAGWAVING and is shown in Figure 5–23. In extreme cases, the picture collapses altogether into a mass of diagonal lines like in Figure 2–4. Adjusting SKEW may straighten out the problem. If it doesn't, or if your VCR lacks this control, try adjusting HORIZONTAL HOLD on your TV.

FIGURE 5-23   FLAGWAVING—adjust SKEW.

Insert Edit

The INSERT EDIT feature is found on editing videocassette recorders and will be described more fully in Chapter 14, "Editing A Video Tape." Briefly, an INSERT EDIT is done in the *middle* of a program. It cuts a piece out of a show and puts a new piece in. Home VCRs with this feature will make a fairly neat jump from one scene to the other at the edit point, leaving a very tiny glitch on the picture. Professional video editors will do the same without leaving any glitch at all.

Sync in

To make it possible to mix the pictures from a VCR with the pictures from cameras and other sources in the TV studio, they all must be synchronized electronically. SYNC does this, and the VCR's SYNC INPUT is where the studio's SYNC signal enters the tape player.

Hi Fi

One-inch VTRs have good sound fidelity. ¾U VCRs have average sound fidelity. Many VHS VCRs, because of their

---

MINIGLOSSARY

*Dub   To duplicate, as in "please dub this tape." Also, the name for the copy of a tape, as in "the dub is on the shelf." Dub cables assist in the process of sending signals from a VCP (video cassette player) to a VCR. *Audio* dub means to replace the present recorded sound with new sound.

*Skew   The control which adjusts the tape tension on a video tape machine to hold the picture steady during playback. A skew error appears as a fluttering, pulling to the side of the top a TV picture.

*Flagwaving   The sideways pulling and fluttering seen at the top

of a TV picture caused by a skew misadjustment or some other tape tension error.

*Insert edit   Feature allowing a VTR to record a new segment in the *middle* of a program, erasing what it's replacing.

*Hi-Fi   High Fidelity or true-to-life sound. VHS and 8MM hi-fi VCRs record the sound with almost "perfect" sound quality.

*Simulcast   TV program having its sound *also* broadcast over FM radio in hi fi.

slow tape speed, have fairly poor sound fidelity. VHS manufacturers have overcome this fidelity problem by changing the way they record the audio on the tape. Instead of placing it on a linear track of its own, much like an audiocassette player does, the hi-fi models encode the audio in with the video on the tapes. You can't see it in the picture, but the sound is here. The process results in excellent stereo audio fidelity (as good as the best audio tape recorders) yet doesn't degrade the picture.

Hi-fi home VCRs are compatible with their "low-fi" brothers. A tape recorded on a hi-fi machine will play back on a regular machine and vice versa; it just won't have high fidelity.

Feeding low-fidelity audio into a high-fidelity recorder will not make high fidelity. Broadcast TV programs are medium fidelity. Recordings made in the studio could be medium or high fidelity, depending on the microphones, mixers, and room acoustics. Prerecorded hi-fi tapes are one way to get hi-fi sound out of your hi-fi recorder. Another way would be to record a TV SIMULCAST off your FM receiver, a technique which will be further described in Chapter 10, "Audio."

GETTING MORE OUT OF A VCR

In the coming chapters, you'll see dozens of tricks for making your video recordings better looking and easier to produce. Here are a few now.

Erasing Part of a Tape

Say you recorded the program *60 Minutes* at home, viewed it, and didn't want it any more. Now say you used that same tape to record a documentary for your school, "Teachers' Salaries: Why Are They So Abysmally Low?" The documentary runs only one-half hour (maybe no one could answer the question), leaving the last half of *60 Minutes* still on the tape. Since it is rather unpolished video etiquette to play a program to a class that ends with a sudden jump to unrelated trivial video leftovers, you would like to erase the last 30 minutes of *60 Minutes*.

The process is simple; just record a new program over *60 Minutes*—a program of *nothing*. There are several ways to do this:

1. Switch the VCR to a vacant channel (snow) and set the timer to record for half an hour.

2. Better yet, switch the INPUT SELECT to CAMERA or VIDEO and make a ½-hour recording. Since

nothing is connected to those video and audio inputs, nothing will be recorded.

3. If you want the recording to be smooth black and silence, connect a TV camera to the VCR but leave the lens cap on and don't plug in a microphone. Switch your INPUT SELECT to CAMERA, and you will record a picture of black.

Drying out a VCR When the Dew Lamp Goes on

When you move your VCR from a cool place to a warmer, humid place, water condenses inside it, making the mechanism sticky. This could damage your tape if the DEW sensor inside your VCR didn't shut your machine down, forcing you to wait until it dried out.

So what do you do, twiddle your thumbs while your clients mix their ninth round of Chivas Regal? No way! Just dig out your handy blow drier, switch it to its *lowest heat*, open the cassette compartment, and remove the cassette. Then, holding the drier about 6 inches from the hole, blow the mechanism dry. One minute may do the trick, and just in time—some of your clients have moved to your boss's 90-year-old Scotch, and one is wearing your wicker wastebasket.

Note: I did not say *high* heat with the nozzle *stuck into* the cassette hole. You don't want to melt anything, just dry it.

Squeezing out the Highest-Quality Picture and Sound

Record your master tapes with the best machine you've got. BETACAM, MII, D2, and 1-INCH are preferable to ¾U-SP, SVHS, HI8, and ED BETA, which are all preferable to ¾U, which is preferable to 8MM and VHS.

Start with a good picture and sound while recording. Use a strong enough antenna, properly aimed, good antenna wire, and tight connections. FINE-TUNE your stations carefully. To really be sure your VCR is receiving a good signal, monitor the beginning of the recording through your TV.

Use the fastest speed (shortest recording time) for highest picture and sound quality on home VCRs.

If playing back through the antenna input of a TV, FINE-TUNE the TV carefully.

For better sound, run the VCR's AUDIO OUT to the AUX inputs of your good $300 sound system rather than your TV's $6 sound system.

For a 10% sharper picture, use a monitor or monitor/

MINIGLOSSARY

Gyroscopic error   Sideways bending or breakup of the TV picture as it plays back, caused by movement of a VCR while it was recording the picture.

*Y adapter   Used in audio, a Y-shaped connector or wire which combines two signals or splits one signal into two.

receiver and use the VCR's VIDEO OUT directly. If using SVHS or HI8 with a TV having an s input, use that connection rather than VIDEO. Definitely use the s or VIDEO cables rather than RF when playing your VCR through a TV PROJECTOR. The RF fuzziness shows up even more when the picture is blown up.

Leave a 10-second LEADER at the beginning and end of every recording.

If you plan to make an *important* recording, first make a sample recording and play it back to make sure that everything is working okay. If you are recording long stretches of material with short breaks (perhaps commercials), use these respites to rewind and play back a little of your recording just to see how it is coming. *Simply viewing your TV screen as the recording is being made doesn't completely guarantee that the recording is going onto the tape.*

VCRs (and this includes camcorders) don't like to be tipped, rocked, rotated, or spun around while running. The spinning VIDEO HEADS inside the machine act like a gyroscope and fight the motion. This shifts the tape speed momentarily, causing the picture to waver in what's called a GYROSCOPIC ERROR.

If a show is going to run past the end of one tape and onto another, don't bother rewinding the first tape just then. EJECT it, pop in the new cassette, and start recording immediately. This will save you 3 minutes and 40 seconds of missed program while you rewind the tape. You can rewind it later; it won't hurt the tape to wait.

Every time you record a tape, *label it.* I can't count the tapes I've seen accidentally erased because they weren't labeled and subsequently were misplaced or recorded over.

If you plan to swap home videocassettes with others, *again*, use the fastest speed for recording. The faster speeds make more stable tapes. Because tapes recorded on one home VCR will play less solidly on another VCR, it helps to start with the most stable signal you can get.

Y Adapters. A Y ADAPTER (Figure 5–24) is a wire which mixes two audio signals together. It's handy for mixing the stereo outputs from a VCR and sending them to a monaural amplifier. It can also mix the outputs from a stereo source and send the result to a monaural input or one channel of a ¾U VCR. The Y ADAPTER is also handy for taking a monaural signal and sending it simultaneously to channel 1 and channel 2 of the ¾U VCR so that it may be recorded on both channels simultaneously.

An alert reader may notice that it is possible to plug a Y ADAPTER into a VCR's VIDEO OUT and feed the signal to two places—or, conversely, use the adapter to feed two sources into one VIDEO IN. Feeding the signal to two places with a Y ADAPTER will work, but it will weaken your signal badly, creating a faint or snowy picture. Feeding two sources (like two VCRs) simultaneously into your VCR's VIDEO IN by the Y ADAPTER won't work at all; you'll just get mush. The only proper way to share a video signal between several devices is to

1. LOOP it from one to another
2. Send the signal to a DISTRIBUTION AMPLIFIER (discussed in Chapter 15)
3. Send the signal to a VCR which has several outputs

The same goes for the T ADAPTER in Figure 2–32. Why do they make it if you shouldn't use it? That's a good question. I've never heard a good answer. Perhaps a T ADAPTER is for hooking things up wrong when you can't hook 'em up right.

If using a stereo home VCR for a monaural recording, record the sound on *both* channels and play it back from *both* channels. Recording on two tracks spreads the magnetism over twice the area of tape, making the total recorded

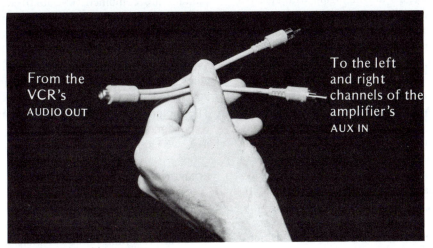

From the VCR's AUDIO OUT

To the left and right channels of the amplifier's AUX IN

FIGURE 5-24 Y ADAPTER.

signal stronger. When played back, this stronger signal will have less background hiss and will have better fidelity.

If using a ¾U VCR, record your sound on both channels if you can (for the same reason as above). If you can only use one channel, make it channel 2. First reason: Channel 1 is recorded at the outside edge of the tape. If the edge is manufactured inaccurately or gets damaged later, the channel 1 sound will suffer. Channel 2 is farther from the edge and is safer. Second reason: It's an industry-wide practice to use channel 2 as the primary channel. If you use channel 1, some day an unknowing person may play your tape and think it's defective because there's no sound and they won't think to switch their machine from "standard" channel 2 to channel 1.

Clean Environment

VCRs don't like liquids, chalk dust, cat hairs, cigarette smoke, high humidity, beach sand or salt spray, bumps, vibrations, high heat, or sustained direct sunlight. A VCR is a delicate princess who should live in a cool, clean, dry, dust-free room, be fed clean video tape, and receive an annual checkup from a video doctor.

Recording a Series of Segments—Stop Versus Pause

You'll see more on this subject in Chapter 14, (Editing), but here's just a brief view of how to record several segments onto the same tape neatly.

If you have recorded a scene and expect to record another scene within three minutes, use PAUSE to stop your VCR. If you expect to take more than 3 minutes to prepare for your next scene, then use STOP on your VCR. Here's why: When you hit PAUSE on a VCR or CAMCORDER, the tape stops in its tracks waiting for you to UNPAUSE. When you UNPAUSE, the tape begins moving again with barely a glitch in the picture. The problem comes when you wait too long in PAUSE and the machine automatically shuts down (protecting the spinning heads from wearing out the tape).

If you press STOP, the tape unthreads from the VCR, losing its place. When you press RECORD/PLAY again, the tape rethreads itself *but not to where you left off.* Depending on your machine, the VCR may begin recording over the *tail* of your last scene, or may begin recording a few moments after your last scene ended (perhaps leaving a moment of blank tape between the old and the new scenes).

A few VCRs try to solve this problem with an AUTOMATIC BACKSPACE feature. On such machines when you press RECORD/PLAY, the VCR backs itself up a little way, switches to PLAY (showing you the tail of the previous scene in your viewfinder), and then switches to RECORD automatically when it reaches the end of the previous scene.

Although this AUTOMATIC BACKSPACE feature insures "clean" edits (without glitches), it can throw your timing off a bit: You must remember to start the action when the *VCR* switches itself to RECORD, not when *you* switch it to RECORD.

What happens if you are recording a scene and it runs afoul, and you wish to rerecord just the end of it? Here you would back the tape up a little way, and then play it, poising your finger over the PAUSE button. When you come to the flub, hit PAUSE, next hit RECORD/PLAY, next UNPAUSE, and begin recording. The VCR will erase the old stuff while recording your new scene. Depending upon the VCR's circuitry and whether it has FLYING ERASE HEADS (described further in Chapter 14), you may see a few squiggly lines, some rainbow color for a moment, or maybe no glitch at all.

Some portable VCRs have a POWER SAVER or RECORD LOCK feature which is very helpful when recording a series of segments with a long wait between them. This switch powers down the VCR *without* unthreading it. This not only keeps your place, but saves your batteries too. When ready to resume shooting, power the machine back up in the PAUSE mode where you left off and UNPAUSE to continue your recording.

HOW A COLOR VTR WORKS

Before continuing, it is time for a little science lesson. The way a VTR records color determines to a large degree the sharpness of the picture and the cost of the machine.

It is fairly easy to make a sharp black-and-white picture. It is difficult, however, to make a sharp color picture, send it down a wire, and record it. Let's look at several methods for doing this, starting with the high-quality methods and working our way down.

In Chapter 1 we learned that cameras make three video signals: red, green, blue (RGB). The three primary color pictures are combined (ENCODED) so that they can travel down a single wire. The ENCODING process degrades the sharpness of the colors (CHROMINANCE), but manages to maintain most of the sharpness in the black-and-white (LUMINANCE) parts of the picture. Put another way, COMPONENT color video has sharp colors, but requires three wires. COMPOSITE video has fuzzier color, but requires only one wire. The picture will still look sharp to us because our eyes respond mostly to the black-and-white variations of the picture; we hardly notice that the colors are fuzzy.

COMPONENT Video Recorders

There are video tape recorders capable of recording the three color COMPONENTS in their entirety. You might say these VTRs are recording three complete video signals at once. The task is difficult and the machines are expensive. They are used primarily for recording computer-generated animation. Most COMPONENT recording today is done on DIGITAL VTRs employing the D1 format. DIGITAL VTRs slice the video signal into tiny chunks, measure them, give each measurement a number, and record the number digitally. Thus, instead of recording vibrating video signals, DIGITAL VTRs record long streams of numbers, ones and zeroes. Upon playback, the ones and zeroes are converted back into video signals for display.

One great advantage of DIGITAL VTRs is that you can make copies of a DIGITAL video tape without losing any quality. With ANALOG VTRs, every time you copy a tape, signal errors and noise add up to cause graininess, fuzziness, and color aberrations in the picture. When DIGITAL VTRs copy tape, they are copying ones and zeroes. Even if the numbers are a little fuzzy, they are still ones and zeroes that can be accurately read by the machine. Thus, a copy of a DIGITAL video tape is *exactly* the same as the original. This has great advantages when you are editing many layers or generations.

Other COMPONENT Recording Methods (such as Y/I/Q or Y/R-Y/B-Y)

It is difficult to record all three color components in their entirety. It would be easier to record the LUMINANCE portion of the picture well and record the colors only half well. The color degradation would be small and your eye would barely see the difference.

Figure 5–25 shows how this is done. The RGB signals are combined to create the Y (LUMINANCE) signal. (You will recall that if you add red, green, and blue together, you get white. Thus, the color COMPONENTS combine to make a

black-and-white image.) Other parts of the circuit subtract the LUMINANCE (Y) from red (R) to make a signal called R-Y. Another circuit subtracts LUMINANCE from blue to make B-Y.

You started with three colors on three wires and you ended up with three colors on three wires, only they are different colors now. The COMPONENT VTR now records the Y signal in its entirety, but it compresses and combines the two color signals (also reducing their sharpness) as it records them. Different machines do this differently. Sony's BETACAM, for instance, records the Y signal with one swipe of the spinning video head and then records the combined color signals with the next swipe, then repeats the process. Some manufacturers, instead of using R-Y and B-Y, use similar signals called I and Q. Still, the process is the same.

You might say that these COMPONENT VTRs are recording two video signals at once. To do this, the machines must run rather fast and are quite expensive.

BETACAM and MII are COMPONENT VCRs.

HI BAND COMPOSITE Recorders

Recording three signals at once (R, G, B) was very tough. Recording two signals at once (Y, colors) was tough. It would seem simple to record plain old COMPOSITE NTSC video, using one wire and one signal only. It isn't. High-quality, expensive HI BAND VTRs are needed to record the extremely high frequencies in the NTSC signal to convey all the detail in the picture. ONE-INCH TYPE C ANALOG VTRs and D2 COMPOSITE DIGITAL VCRs make HI BAND recordings. The TYPE C VTRs do an excellent job of recording all of this detail by running one-inch tape quickly through them. The D2 DIGITAL recorders run a little slower but still are

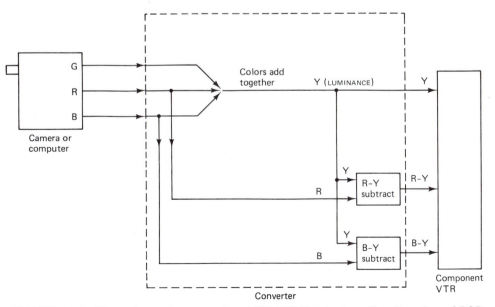

FIGURE 5-25 The color COMPONENT signal Y/R-Y/B-Y is just another iteration of RGB.

very fast. The D2 machines, like the D1 VTRs, digitize the video image into ones and zeroes and record the numbers. On playback, the numbers are converted back into video. In the meantime, if a tape is copied it is only the numbers that are copied and the digital copies are "perfect." You can copy down many generations without any degradation to the picture.

COLOR UNDER VCRs

The fine detail in a picture involves high frequencies. So does the color. It is difficult to record high frequencies, especially when using slow tape speeds; the magnetic vibrations are hard to pack close together. It would be easier if the VCR didn't have to deal with the high-color frequencies (3.58 million vibrations per second—3.58 MHz). Common VCRs (¾U, ¾U-SP, VHS, SVHS, 8MM, HI8, BETA, ED BETA, and others) solve this problem using a technique called COLOR UNDER. The VCRs filter the color out of the COMPOSITE VIDEO signal making two signals, Y (LUMINANCE), and C (CHROMINANCE). The LUMINANCE is recorded directly while the CHROMINANCE goes through a special process: The high-frequency 3.58 MHz is downconverted to a much lower and easier-to-record frequency. VHS and SVHS machines change to 3.58 MHz to 629 kHz. 8MM and HI8 change the high-color frequency to 743 kHz. BETA, ED BETA, ¾U and ¾U-SP change the high-color frequency to 688 kHz. All of the machines then record this lower (hence the word *under*) color signal along with the LUMINANCE signal.

When the tape is played back, the color signal is boosted back up to 3.58 MHz and is combined with the LUMINANCE, making COMPOSITE video again. Unfortu-

nately, once you've lost the color detail by converting the color frequency down, it's gone forever: You don't get it back when you boost it back up to 3.58 MHz. For this reason all COLOR UNDER VCRs have pretty smeary color. While the LUMINANCE part of the picture may have a RESOLUTION of 240 to 400 lines, the color parts of the picture have a RESOLUTION of about 50 lines. (Chapter 6 describes RESOLUTION more fully). Fortunately, your eye is very forgiving and still hardly notices the color fuzziness. This fuzziness does become more pronounced when the tapes are copied.

The COLOR UNDER technique has additional disadvantages: The process of separating the color from the LUMINANCE part of the picture degrades the picture quality. DOWNCONVERTING the color frequency degrades the color a little more. Boosting it back up again degrades it. Combining it with the LUMINANCE contaminates the picture further.

When a common VCR is used to copy a tape, these problems are compounded: The videocassette player must upconvert the color and combine it with the LUMINANCE to make COMPOSITE video, which it would send to a videocassette recorder whose job it is to *reseparate* the color, and *redownconvert* it. All these conversions take their toll. Higher-quality VCRs designed for copying and editing have special DUB CONNECTORS which allow the CHROMINANCE and LUMINANCE signals to travel *unconverted* over separate wires from the player to the recorder. Thus, the player may send its LUMINANCE signal over one wire and its 688 kHz downconverted color signal over another wire, to another VCR ready to record these signals without any conversion. This feature is called Y/688 DUB (named for the COLOR UNDER frequency). On VHS equipment it is called Y/629 DUB.

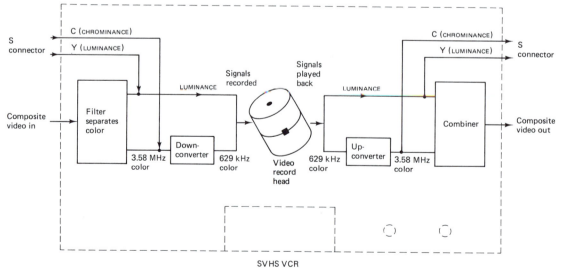

COLOR UNDER

SUPER VCRs

SUPER VCRs (SVHS, HI8, ED BETA) use a technique of pre-serving their color quality similar to the DUB feature men-tioned previously. Instead of using their COMPOSITE video connectors, they use their S (or Y/C) connectors. The LU-MINANCE (Y) travels over one wire while the CHROMINANCE (C) at the 3.58 MHz frequency travels over a separate wire. Because the two signals don't travel on the same wire, they don't interfere with each other. When a SUPER VCR receives Y/C signals directly from a camera or other source, it doesn't have to separate the color from the LUMINANCE. As an added bonus, the LUMINANCE without the color interference can be recorded at a higher frequency than normal, preserving more picture detail (400+ lines of resolution). The VCR must still downconvert the 3.58 MHz color signal, a dam-aging but necessary evil.

When the tape is played back, the low-color frequency is multiplied back up to 3.58 MHz again. If a SUPER VCR is playing the signal into a TV set using an S connector, the color signal goes straight to the TV's color circuits without going through any unnecessary conversions. This permits the LUMINANCE parts of the picture to remain unprocessed and very sharp. The same is true for copying a tape from a SUPER VCR to another SUPER VCR using the S connector. The 3.58 MHz color signals travel on a separate wire from the LUMINANCE. The LUMINANCE is recorded directly while the CHROMINANCE is downconverted to the lower recordable frequency. Even though upconverting and downconverting occur in this process, this isn't as damaging to the color and sharpness as the full-blown conversion to COMPOSITE video would have been.

Now you see the reason why higher-quality, more expensive VCRs do a better recording job. VHS and 8MM VCRs are inexpensive, use tape slowly, but make fairly fuzzy pictures with even fuzzier color. These tapes do not copy well. The same is true for ¾U tapes, but to a lesser extent because the signal is spread over a larger amount of tape. SUPER VCRs use special tape and special circuits to sharpen the picture and when S connectors are used, improve

the picture even more. Still, all of these formats are COLOR UNDER resulting in fuzzy color even though the black-and-white aspects of the picture may be sharp.

D2 and 1-INCH TYPE C machines overcome the color problems (at significant expense) by using fancy electronics and using high tape speed. COMPONENT VCRs such as BETA-CAM and MII overcome the color problems by recording the color *completely* separate from the LUMINANCE. Recording each COMPONENT color *individually* (as with the D1 VTR) affords the ultimate quality but at tremendous expense.

PORTABLE VCRS AND CAMERAS

Although most people use camcorders nowadays, there are still quite a few separate VCR/camera ensembles in exis-tence. There are also numerous portable VCRs in use where smallness is important. Here's how to operate them.

Portable VCRs work like console models do. They are usually smaller, have more automatic circuits, may have a multipin input for a camera, and will run on batteries. The features, functions, and feeding of portable recorders are nearly the same as their tabletop brothers.

Operating a Portable VCR

Once someone else has set it up, operating a portable VCR is easy:

1. Find something to shoot, preferably not some-thing in the dark.
2. Remove the camera's lens cap. Thereafter, don't aim the camera at anything too bright, like the sun. If your brother-in-law, Harold, isn't too bright, aim the camera at him.
3. About a minute before you are ready to begin taping, switch the VCR's power to *on*. Since either the camera can PAUSE the VCR (with its trigger) or the VCR's PAUSE button can PAUSE it, let's fix things so that the camera does the

MINIGLOSSARY

*Component video   Video signals carrying separate colors on separate wires. RGB, Y/I/Q, Y/R-Y/B-Y are component video signals.

Encode   To combine component video signals into a composite video signal.

*Composite video   The combination of three color video signals traveling on one wire. NTSC video is composite video.

NTSC   National Television Standards Committee. United States organization that determines standards for video signals.

Hi band VTR   Video recorder capable of recording full-fidelity color signals (as opposed to color under signals).

*Color under   Video recording method where color is separated from luminance and converted to a lower frequency for inexpensive recording.

Analog VTR   Video recorder that records the continuously vary-ing video signal onto the tape (as opposed to digital).

Digital VTR   Video recorder that converts the video signal to ones and zeroes (digits) and records the numbers. Upon playback, the numbers are converted back to video.

Y/688 dub   Method of sending separate luminance and down-converted color signals between ¾U editing-type VCRs to preserve color quality.

PAUSING; it's more convenient that way. To do this, you will first want the VCR's PAUSE *off* and the camera's PAUSE *off*. A methodical way to coordinate this is to press PLAY and look to see if the tape is moving. If it is, you are all set; the camera's trigger will PAUSE for you. If it *isn't*, pull the camera trigger. If this UNPAUSES you, you're all set. If it doesn't, UNPAUSE the VCR's button. If that doesn't get things started, pull the camera trigger again. Eventually you get the machine UNPAUSED and playing.

4. Now pull the trigger to PAUSE everything (after all that). Next, press RECORD. The camera's viewfinder will light up either when you switch the VCR's power to *on* or when you press PLAY or RECORD. Either way, the camera will take 30 seconds to "warm up" and give you a picture. If the camera is an MOS (metal oxide semiconductor) or CCD (charge-coupled device) type, it will give you an instant picture.

5. While in the PAUSE mode, focus the lens and adjust the iris (f-stop). More on this in Chapter 6.

6. When ready to record, pull the camera trigger, and the VCR will switch from PAUSE to FORWARD and will begin recording.

7. To stop recording temporarily, pull the trigger again, and the VCR will switch to PAUSE.

8. To start recording once more, pull the trigger again.

9. To *finally* stop recording, switch the VCR to STOP (if you wish, you may PAUSE the recording first by pulling the camera trigger). In the PAUSE mode, VCRs will consume electricity, wearing down your battery even though you are not making a recording. In the STOP mode, some VCRs and cameras keep consuming electricity; others don't. To save your batteries, you may wish to switch the POWER to OFF if the machine will be idle for a while. As mentioned before, some VCRs have a POWER SAVER switch which "powers down" the VCR while in the PAUSE mode.

10. To play back a sample of what you recorded, just press REWIND and back up a ways. When the tape is rewound, press it to STOP.

11. To play, switch it to PLAY and look into the tiny viewfinder on the camera. The image will appear there *if* your camera happens to have an electronic viewfinder (a tiny black-and-white TV set) in it. To hear sound, find the earphone. It usually lives in a little pocket in the carrying case. Stick the plug into the earphone socket on the VCR or the camera; the other end sticks into your ear (or is it the other way around?).

Setting up a Portable VCR for Use

Recording from a Camera.   The camera connects to the VCR via a multipin plug. Just line it up, push it in, and screw the tightening collar to hold the plug in.

Next, check the VCR's INPUT SELECT switch. When operating with a camera, this switch must be in the CAMERA position.

On most portable TV cameras, sound is picked up by a sensitive mike built into the front of the camera and recorded automatically.

Not all portable cameras will work with all portable VCRs. First, the plugs may be different. Even if the plugs match, there is no guarantee that the right signal is traveling on the right wire. Recently made portable cameras and recorders generally put their audio and video and power signals on the same wires, making it likely that your camera will work but will not be able to use all its fancy features. When buying different make or model cameras or VCRs, check first to see if they are compatible.

Recording Audio.   Instead of using the mike built into the camera, you can substitute your own (say a LAVALIER or a SHOTGUN mike, described further in Chapter 10). These mikes plug into the MIC input of the VCR. Home VCRs usually take a MINI PLUG for their mikes. Professional VCRs take a three-pronged XLR mike plug. You may need an adapter to match your mike's plug to the machine's socket.

Using the mike built into the camera is very handy but tends to cause echoey sound, because if the camera is 8 feet from the performer, so is the microphone. Microphones pick up sound best when they are close to the performer. The only way to move the mike close to a performer without moving the camera too is to use a separate microphone.

If the sound source is prerecorded and requires a HI-LEVEL INPUT, look for an AUX or AUDIO IN socket on the VCR for this.

In most cases, the audio level is automatically adjusted. More professional models have manual adjustments too.

AUDIO DUBS are possible by pushing the DUB button on the deck while switching the VCR to FORWARD. The mike in the camera may be used in this process, or you may plug in a second mike.

Recording Off-Air Broadcasts.   The big difference between console and portable VCRs is that consoles *include* the TV tuner while the portables, for weight reasons, don't. The tuners are separate and connect to the VCR through a multipin umbilical cable or through several cables to the VIDEO IN, AUDIO IN, and REMOTE inputs. To pull in distant stations, tuners have ANTENNA inputs which connect to your rooftop antenna or cable TV.

Sometimes these tuners have PROGRAMMABLE TIMERS with all the features the console models have. Also included in most units is a power supply that runs your VCR without discharging your battery. In fact, most units *will charge* your VCR's battery. Of course they must be plugged into a wall outlet, connected to a VCR, and turned *on* to do this.

To really be sure that you've accurately tuned the tuner for a recording, you may wish to review the results on a TV set beforehand. Here's how:

Attaching a TV Set. Most portable VCRs have a socket labeled RF OUT. On older and industrial models, connect this to the TV's ANTENNA terminals. Switch the VCR's OUTPUT SELECTOR (if it has one) to VCR to send the VCR signal (rather than the antenna's) to the TV. Switch the TV to channel 3 (if that's what the VCR's RF GENERATOR puts out). That's all. It's a lot like your console VCR/TV.

Newer models do it a slightly different way. The antenna runs to the tuner. The tuner runs a multipin cable to the VCR. The VCR sends its RF back to the tuner over the multipin cable. You plug your TV to the ANT OUTputs on the tuner. The OUTPUT SELECTOR is also on the tuner. Not much is really changed except the location of the TV connection and the OUTPUT SELECTOR. The tuner may also house the TIMER and an infrared REMOTE control sensor.

What's handy about both of the setups in the adjacent figure is that only one or two cables have to be disconnected before you can run off with your portable VCR. The TV and antenna wires stay put.

If using a ¾U VCR, double-check that the RF GENERATOR is installed. If it's missing, you'll get zippo out your RF OUTPUT.

With portables, as with console VCRs, it's possible to run both audio and video to an audio amplifier and speaker and a TV monitor, or you could run both signals to a TV monitor/receiver for viewing.

Sometimes the portable VCR isn't used with a tuner. Instead, it's used to make tapes out in the field (or the woods), and then it comes home to connect to the TV for viewing. Industrial VCRs and older and cheaper VCRs were designed to work this way and use an external ANTENNA/ VTR switch when feeding a TV. Figure 5–26 shows the hookup. In fact, this kind of hookup is used for video games and many other devices which connect to your TV. With the ANTENNA switch, you can select whether your TV will "see" the antenna signal (ANT position) or "see" the VCR's signal (VCR position).

Why bother with the switch? Why not simply connect the VCR's RF cable to the TV's antenna terminals and then connect the TV antenna to the same terminals? Such a connection *would work*, although not awfully well. You could get a grainy, ghosty signal on your TV. Here's why: RF doesn't care where it goes when it comes through the RF cable from the VCR. If the TV is still connected to its antenna, some of the RF signal will go into the TV, as it

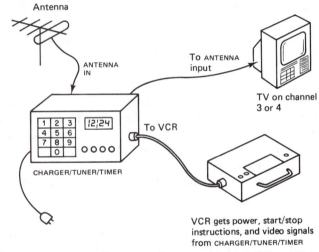

Older method:

Old method:

Tuner/power supply connections to portable VCR and TV.

should, and the rest will detour out the antenna wires to the rooftop antenna. In fact, the RF going up the antenna will actually broadcast out the antenna a little and may interfere with other people's TVs. The Federal Communications Commission frowns on renegade TV broadcasters scattering signals willy nilly over the airways.

Cable TV and master antenna systems face the same problem. If your TV set is still connected to your building antenna system when you pump RF into your set, everyone on your system will get interference from your signal.

Time for a true story. A young man and his wife bought a portable VCR and camera. After taking it back to their apartment, they decided to set up the camera to record themselves making love. When finished, they eagerly played the tape back on their home receiver. How exciting! The next morning, the couple noticed funny stares from other tenants in the lobby. Someone asked, "Haven't I seen you

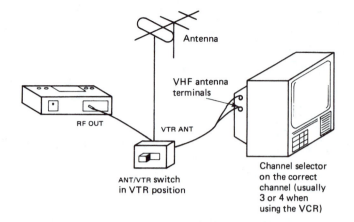

FIGURE 5-26   Receiving RF on a TV receiver from a portable VCR.

somewhere before?'' In their exuberance, the couple had neglected to disconnect the master antenna cable from their TV before attaching their VCR to it. It is not known exactly how many tenants had watched this X-rated gem on their TV sets that evening or had recognized their neighbors as the main characters.

The same danger exists (but not to the same degree of embarrassment) with video games that you connect to your TV antenna terminals. The games send out an RF signal that must go into the TV set *only* and not detour up the antenna wires for all to see.

In short, when sending RF to a TV set, make sure

1.   The set is disconnected from its rooftop antenna,
2.   The set is disconnected from any master antenna or cable system, or
3.   If you have an ANT/VCR junction box with a switch, the switch is in the VCR position.

Powering the VCR

For short on-location shoots, a rechargeable battery which fits inside the VCR will power the VCR for up to 1 hour. See Figure 5–27. These batteries usually slide into a hole in the VCR somewhere and, when clicked tightly into position, will make their own electrical contact. Other VCR batteries have a plug at the end which has to be inserted in the BATT or DC input after the battery has been inserted.

If you press RECORD and FORWARD but do not pull the camera trigger, you still are using power even though the tape may be PAUSED and not moving. Add this ''standby'' time to the time you spend actually shooting, and you can estimate that your battery will probably power you through only half an hour of actual recording.

As the battery gets older, it will serve shorter and shorter duty cycles. To check your battery power (and to estimate how much longer you can shoot before the battery dies), glance at the battery meter somewhere on the VCR. *The meter may register only when the VCR is in RECORD, not when it is in STOP.* In these cases, if the VCR is just sitting somewhere and you wish to check the battery, merely push the RECORD button and look at the meter. The meter will give the answer in about a second. You may then release the lever, and the machine will return to STOP. If the meter reads way in the white, the battery has a lot of life left. If the meter reads near the red, the battery may have 5 minutes or so of life left. When the meter reads in the red, the battery is not sufficient to power the VCR. *Even though the motor runs, if the battery meter reads in the red, don't record;* you'll be wasting your time. The VCR motor speed starts to drop (imperceptibly at first) when the battery is nearly expended, rendering the recorded passage unplayable. If you are recording and suddenly notice that your viewfinder picture begins to jiggle or roll, that may also be a sign that your battery is weak. Check the meter. Many portable TV cameras are equipped with LOW BATTERY POWER lights inside the viewfinders to warn you of impending power loss.

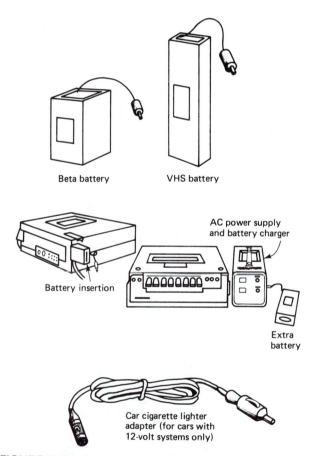

FIGURE 5-27   Powering the VCR.

Some VCRs have automatic circuits inside which shut off the VCR when the battery power becomes too weak to make a good recording.

There are two popular kinds of VCR batteries. One is called GEL-CELL and works much like a car battery. The other is NICAD, which stands for NIckel CADmium battery. NICADS are smaller, lighter, hold more power, recharge faster, and last longer than GEL-CELLS. They also cost about twice as much. When GEL-CELLS run low on power, the process is gradual, and by watching your meter, you can see the battery get weaker. This helps you predict how long your battery is likely to last. NICADS, however, fail precipitously. One minute they're great, and the next minute they're dead. To really be sure you have enough power for a shooting session, bring along an extra battery (preferably a charged one).

The VCR's battery must power the VCR *plus* the camera. How long the battery lasts will depend greatly on how much juice the camera uses.

Professional ENG (electronic news gathering) and EFP (electronic field production) cameras can work off the VCR's battery or can have a battery of their own attached. This battery takes some of the load off the VCR.

To operate a VCR on location for up to 3 hours, there are optional external chargeable battery packs that connect to the VCR. They plug into the VCR socket labeled DC or 12V IN or EXT BATT.

To operate near vehicles using their battery power, there is a "cigarette lighter" plug and cable which can suck power from your car while you're shooting. Your car has to use a 12-volt battery (most do). If your car battery is old, it may be wise to start your car and let it run every so often unless you enjoy being stranded on-location.

For extended shooting, bring along your ac adapter, which can power the VCR from a wall outlet when your battery is used up. The adapter can also recharge the VCR's internal battery. Some ac adapters are shaped like VCR batteries and slip into the VCR without a trace except the ac cord.

None of the preceding power supplies is yet designed to power anything *other than* the VCR and its camera. TV monitor/receivers and other equipment must be powered separately either by batteries or by ac.

If you wish to play back some of your recording "in the field" (a good idea just to make sure everything is being recorded okay), nearly all cameras which have electronic viewfinders will allow you to view your tape through the viewfinder. Plug in an earphone for sound, and you can see and hear your recordings without dragging along a separate

TV. Take note that this playback also takes electricity and drains batteries.

You'll find more about VCR batteries and how to charge them in Chapter 16. One interesting note on how to squeeze more life out of a battery bears stressing. When you "power up" your VCR and camera, they start draining your battery—even if you are not actually taping. "Power down" the system whenever possible to conserve juice.

Nonprofessional Camcorders

A CAMCORDER is a camera and VCR in one box. The convenience of not having two separate items with a dangling cord between them is obvious. Also, great efforts have gone into making CAMCORDERS weigh 6 pounds or less. The popular types of consumer CAMCORDERS use VHS, SVHS, VHS-C, 8MM, or HI8 tape.

Videotaping in the Winter

Portable video equipment, if it could talk, would beg to stay inside where it's dry and warm, but ice skating, skiing, snowball fights, snowmobiling, and traffic-snarling blizzards occur outdoors (you needed this book to tell you that). What problems will you and your equipment face in the cold?

Shortened Battery Life. Above all others, this may be your biggest problem. Up to 50% of your battery's life is lost when the temperature reaches freezing. Expect your 1-hour battery to last about 30 minutes. This goes for NICADS as well as GEL-CELLS.

To avoid this problem, keep the battery warm, like in your pocket or wrapped in a blanket until you're ready to use it, and *then* chuck it into the machine.

You may even keep the whole VCR (and thus the battery) warm with a blanket while it operates. By all means, don't set the recorder down on the ice or frozen ground; this will cool it and its battery very quickly.

Try reducing power demands on the battery. Don't stand around in PAUSE or STANDBY any length of time. Either shoot or kill the power. Remove the plug-in RF GENERATOR if it is not being used; save-a-watt.

Bring two batteries for more shooting time. Keep the second battery warm until used.

Condensation. Ever notice how moist the outside of a glass of iced-tea becomes on a humid day? VCRs and cameras suffer the same problem. Whenever they are cooler than their environment, dew collects on them. This isn't

---

MINIGLOSSARY

*Nicad NIckel CADmium. Lightweight, high-power battery for VCRs.

*Gel-cell Battery designed like a miniature car battery, only with gelatin inside rather than liquid. Less expensive than nicad; used in VCRs.

8mm (Courtesy of Sony Corp. of America)

VHS (Courtesy of Hitachi Sales Corp. of America)

Consumer CAMCORDERS.

Betacam (Courtesy of Sony Broadcast Products Company)

Professional CAMCORDER.

such a big problem when you move outdoors to shoot in the cold, but when you come back in, with frigid gear, it starts collecting humidity from its warm surroundings. The dampness is devastating to the electronics, makes the tape stick (instead of slide) inside the VCR, and will most likely trip the DEW SENSOR, a safety device in the VCR which turns it off (wisely) when there's excessive condensation.

When bringing cold equipment inside, immediately cover it with a plastic bag and let it warm up for an hour or more. The bag will seal out the humidity-laden air so that water can't collect on the goodies. When the equipment reaches ambient temperature, it's ready for use. Take it out of the bag first.

One way to hasten the drying out process is to open the cassette tray (removing the cassette) and blow warm (not hot) air from a hair dryer into the VCR. In a few minutes, the machine should be defrosted. Let the inside temperature restabilize for a couple more minutes before chucking in your cassette.

Don't play a cold tape in a warm machine and vice versa. The tape will act "sticky" if it is not the same temperature as the machine.

Videotaping in the winter.

None of this applies to batteries. It is perfectly fine to use warm batteries in a cold VCR.

Extreme Cold. If it gets *really* cold (below −40°F), don't use your VCR. Its circuits could become damaged. Go out in a blizzard, and your lens might ice up. Slightly damp skin will freeze to very cold metal. If it is really cold, wear gloves. Professionals use specially made electric blankets to keep their equipment warm.

Glare. Snow and ice reflect a lot of sunlight. Shoot at a high f-stop or attach a neutral density filter to your lens. Use a polarizing filter to reduce snow and ice surface reflections. Keep the sun at your back lest your subjects are backlit so strongly that they look like silhouettes. (More on all of this in Chapter 6.)

Video equipment (except batteries) works quite well when cold. It works poorly when it is changing temperature. Your camera may give poor color. Your tape may stick. Your lens may fail to zoom smoothly.

Videotaping at the Beach

Video is an action medium, and what better place for action than at the beach? The beach, with its games, kids, and bathing beauties, can be a recording gold mine—if you know how to insulate your equipment from the harsh environment. Here are your enemies and how to avoid them.

Salt Air. Ocean beaches are bathed in an invisible spray of salt air. The salt corrodes the switches, gums up VCR mechanisms, and clouds lenses.

Wrap your equipment in plastic bags and tie them closed. White or clear plastic is better than black. Black will get hot in the sun. Since your wrapped equipment can't "breathe" and cool itself on hot days, run it for only 15 minutes at a time. Then let it cool awhile. Keep it shaded for best cooling. Wrap the camera so the lens sticks out (it has to "see") and make a little hole and tape it tight to your viewfinder opening so that you can "see" too. Later, you can wipe the salt spray off the lens and viewfinder glass with a wet swab. Use *fresh* water for this.

Dampness. Strangely, your greatest danger isn't a tidal wave or your boat capsizing. It's the "dry land" danger of dripping swim suits, kids with water buckets, and half-empty Budweisers spilling into your gear.

Next in line comes the sea dampness which permeates your equipment and collects. Then comes the total dunking—a total disaster at best.

Like before, wrap everything while shooting. Store your gear in waterproof cases. A waterproof ice chest makes a fine temporary home for your little treasure (and it may even float in a capsize). Suitcases or other containers will work too. Avoid cardboard; it gets soggy when wet. Pack towels or whatever around the equipment to cushion it during travel. Don't, however, bring the equipment home wrapped in *sandy* beach towels.

For long-term storage in damp environments, buy some packets of SILICA GEL from your photo store and pack them in with your equipment. *It* will absorb the humidity rather than your machinery. When the packets get damp, simply bake them on low heat in the oven to dry them out (if they are cloth covered, wrap them in perforated tin foil to keep the cloth sacks from scorching).

Sand.   You know how sand feels in your bathing suit, and you know how it tracks into your car, into your home, and somehow into your bed and magically into your breakfast cereal the next day. Similarly, that abrasive, destructive grit will find its way into your machinery unless you wrap it and store it tightly. Also keep your cables off the ground—they'll transfer sand to your storage box.

Heat.   Store everything out of the sun. Heat damages tape and makes VCRs "sticky" inside.

Above all, don't move your equipment quickly from a cool air-conditioned environment to a sun-baked beach. (Where have you heard this before?) It will collect water like a cold drink can, and the salty condensate will eat battery contacts, switches, everything. If you are changing from cold to hot places, put the equipment in plastic (keeping out the moisture) and let it "warm up" to outside temperatures.

After the Damage is Done.   Even after wrapping the equipment, you may come home and find sand or salt-water film on your equipment. Here's what to do:

1. With a freshwater-moist sponge, wipe down your VCR and camera casings.
2. Decake dried salt with undiluted alcohol on a soft lint-free cloth. Avoid scented cleaners; they leave a film.
3. Take off any removable covers and open any trapdoors or compartments on your VCR and camera, and using an alcohol or freshwater-dipped swab, clean the seals, gaskets, and edges of the covers.
4. Clean around switches and movable parts.
5. If dampness gets into your camera lens, fungus will grow, causing a fine, spidery, light-colored buildup on the glass. This can be minimized by

"drying out" the lens in a warm dry place (like under a lamp) after shooting. Otherwise, if you notice the fungus is growing as you hold the lens up to the light, bring it to a camera store for cleaning. If the fungus builds up too much, the camera store may be unable to remove it.

6. If your boat capsizes and "deep sixes" your gear,
    a. Turn it off.
    b. Remove the battery.
    c. Swim—don't paddle—to your nearest bucket of fresh water and dunk it. Slosh it around to get out all the salt—immediately. Repeat the rinse five more times, each time using a new batch of fresh water.
    d. Take it in for service, or if you're handy, open it up and dry it out in a warm place. Blow a fan on it for a day. Still it will be wet *somewhere*. Check carefully for moisture. Don't try to use it until you're sure it's dry. With luck, it may work. Then again, your drowned machine may still need resuscitation by an expert.
    e. Last, go back to the scene of the accident and look for survivors (your family maybe).

HOW VCRS WORK MECHANICALLY

Open up a VCR someday and look inside. You'll see a marvel of circuits, wheels, pulleys, brakes, sensors, and other gadgets. VCRs are so complex that it is amazing they work at all (sometimes they don't). If the opportunity comes your way, watch a VCR operate with the lid off. It's a high-tech thrill to see tiny robot fingers extract the tape from the videocassette and thread it around the guides and the spinning VIDEO RECORD HEADS. Figure 5–28 diagrams this process for ¾U and VHS recorders.

The VIDEO RECORD HEAD is the heart of the system. To understand it, let's go back to a simpler system and see how audio tape is recorded.

To make an audio recording, sound vibrations go in a microphone and are changed to small electrical vibrations. These are sent to an AUDIO RECORD HEAD, which is a block of metal with an electromagnet in it which changes the electrical vibrations into magnetic vibrations. In the recorder, a rotating spindle called a CAPSTAN pulls the tape through

---

MINIGLOSSARY

*Video heads   Electromagnets attached to a spinning drum inside the VCR, responsible for recording the picture.

Audio head   Stationary electromagnet inside a VCR which rec-

ords the sound on the tape or plays it back.

Capstan   Shiny spinning rod inside the VCR which pinches against the tape and draws it through the mechanism.

FIGURE 5-28 Video-cassette threading patterns.

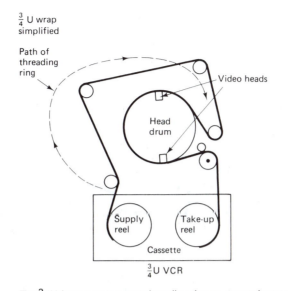

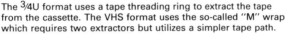

The 3⁄4U format uses a tape threading ring to extract the tape from the cassette. The VHS format uses the so-called "M" wrap which requires two extractors but utilizes a simpler tape path.

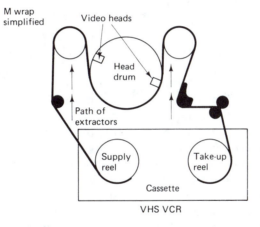

the machine. The tape is a plastic ribbon impregnated with metal particles which can be magnetized as the tape slides over the HEAD. Figure 5–29 diagrams the process.

When audio tape is played back, just the opposite happens. The magnetic vibrations on the tape slightly magnetize the head as they pass by it. This *generates* a tiny electrical signal in the head. The tiny electrical vibrations from the head are amplified in the tape recorder and can vibrate a speaker, making sound again.

An audio tape recorder has to deal with thousands of vibrations per second, but a video signal is millions of vibrations per second. If the tape didn't move fast enough, all those vibrations would overlap each other and erase each other. It's sort of like writing so small on a piece of paper that the letters all overlap and you can't read them. The solution to this problem is to move the tape very quickly to make space between the magnetic vibrations. The tech-

nical term for this is HIGH WRITING SPEED. Primitive longitudinal video recorders (LVR) ran at 200 inches per second. A reel of tape didn't last very long, making LVRs unpopular. There's a more elegant way to make the VIDEO HEAD move quickly across the tape, which allows the tape to move slowly through the cassette. This technique requires a *spinning* VIDEO HEAD. As the tape moves slowly through the machine, it passes over a drum which rotates 30 times per second. Attached to the drum opposite each other are the VIDEO HEADS, as shown in Figure 5–30.

The VIDEO HEAD DRUM is tilted so that each head takes a diagonal swipe across the tape as the drum rotates. Since the tape is moving, one head makes one swipe, and the next head makes another swipe right next to it. Each swipe records one picture. Notice that even though the tape is moving slowly, the VIDEO HEAD is moving quickly across it at a high WRITING SPEED.

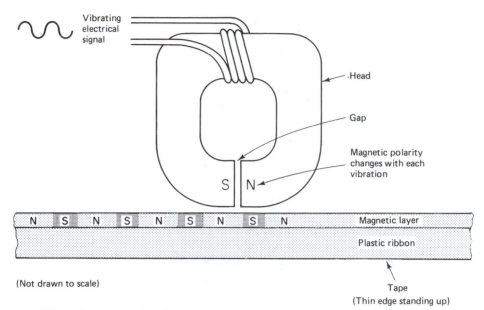

FIGURE 5-29   Record head magnetizing the tape.

Because of the way the tape is wrapped around the drum, this process is called HELICAL SCAN. 8mm, VHS, ¾U, 1-inch type C, and many other recorders use HELICAL SCAN.

While the recording is being made, an ERASE HEAD upstream of the VIDEO HEAD is erasing old material from the tape. At the same time, a stationary AUDIO HEAD is recording audio at the edge of the tape. At the opposite edge of the tape, a CONTROL HEAD is placing 60 pulses per second on the tape. These pulses correspond with the VERTICAL SYNC pulses on your TV picture. When the tape is played back, these pulses act like the sprocket holes in a movie film. They keep the picture centered so it doesn't roll up or down. They also help guide the VIDEO HEAD so that as it plays back a tape, the head follows exactly the same path that the record head took. The recorded control track is sort of a "drum beat" which is used by the VCR's motors so they know how fast to play the tape and how fast to spin the VIDEO HEAD drum. The TRACKING control on a VCR adjusts where the VIDEO HEADS take each swipe. Adjusted perfectly, the heads perfectly follow the recorded paths on the tape. When TRACKING is misadjusted, the VIDEO

HEADS may play a little of the unrecorded space *between* the paths or even stray into the adjacent path.

Figure 5–31 shows the path left by the VIDEO HEADS across the tape. The slower VHS and 8mm speeds crowd more picture into the same space. Each swipe of the VIDEO HEAD is now closer to its neighbor. Reducing the distance between each swipe (this blank space is called a GUARD BAND) makes it hard for the VIDEO HEAD to "listen to" one swipe and not get interference from the adjacent swipe. One way to make more space between each swipe is to make the VIDEO HEAD trace a narrower path. But narrower VIDEO HEADS make a poorer picture than fat heads do.

In short, industrial VCRs use a fast tape speed, leave a large GUARD BAND between the video tracks (reducing interference), and use fat VIDEO HEADS to make high-quality recordings (one case where "fat heads" are desirable). As we slow down the speed of the tape as is done with VHS and 8mm formats, the traces left by the wide VIDEO HEADS start to overlap (Figure 5–31b). This degrades the quality of the picture. If we make the heads narrower (Figure 5–31c), we regain some space between the tracks, but now we are using skinny VIDEO HEADS, which can't make as

MINIGLOSSARY

Writing speed   The speed of the video heads relative to the tape.

Video head drum   Spinning cylinder inside a VCR with video heads attached to it.

*Helical scan   The method of recording a video signal diagonally across the tape by winding the tape in a spiral around a rotating drum with video heads on it.

Quad   Older recording format used on professional VTRs.

*Erase head   Electromagnet inside a VTR upstream from the video head. The erase head demagnetizes the tape prior to the video head recording on it.

Control head   Electromagnet in a VTR which records timing pulses on the tape and plays them back. These pulses precisely guide the speed of the tape.

Guard band   Unrecorded strip of area between audio or video tracks on a tape.

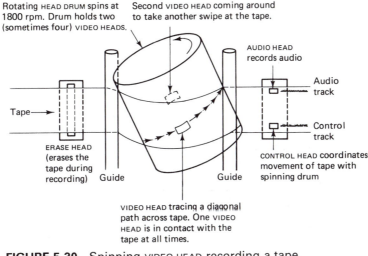

Rotating HEAD DRUM spins at 1800 rpm. Drum holds two (sometimes four) VIDEO HEADS.

Second VIDEO HEAD coming around to take another swipe at the tape.

AUDIO HEAD records audio

Audio track

Tape→

ERASE HEAD (erases the tape during recording)   Guide   Guide

Control track

CONTROL HEAD coordinates movement of tape with spinning drum

VIDEO HEAD tracing a diagonal path across tape. One VIDEO HEAD is in contact with the tape at all times.

FIGURE 5-30   Spinning VIDEO HEAD recording a tape.

good a picture. If we now slow down the tape some more, we use less tape, but look at the overlap between the traces in Figure 5–31d and 5–31e. Lousy narrow heads, lousy narrow overlapping, lousy picture. And the slower tape speed crowds the audio vibrations too, making the sound weaker, hissier, and muffled. This is why you should use the fastest tape speed possible with 8mm and VHS machines.

Now go back to Figures 5–30 and 5–31 and think about a STILL FRAME. Can you see how it works? The tape stands still; the heads keep spinning, tracing the same picture path over and over again. Move the tape forward a tad, and the *next* picture will show. Move the tape halfway between the pictures, and the heads will play the mush between the tracks. Voilà, a TRACKING problem like in Figure 5–22! Let the VCR STILL-FRAME too long in the same spot, and the heads start to eat through the tape. The same phenomenon happens with the soles of shoes.

ALL ABOUT TAPE

Tape is the principal diet of your VCR. Since you are what you eat, it is important to know a little about this durable yet delicate ribbon.

What Is Video Tape?

Video tape is a thin plastic film impregnated with a fine powder (cobalt-doped gamma-ferric oxide or chromium dioxide) which, when sliding over the HEADS, gets magnetized. You can't see the magnetism: Recorded tape looks no different from blank tape. The magnetic vibrations on the tape are transformed electronically into a picture and sounds by the magnetism-sensing playback HEADS of your VCR, in conjunction with various electronic circuits.

Video tape must be manufactured to such exacting standards that it's incredible that there's so much of it around. Imagine manufacturing a ribbon of tape about 800 feet long *exactly* ½ inch wide and .75 mil (.75 thousandths of an inch—about one-fourth the width of a human hair) thick. The plastic ribbon must neither stretch nor shrink significantly yet be strong enough to withstand the wear and tear of threading and winding. The tape must be smooth enough to slip easily through the mechanism and not abrade the delicate spinning VIDEO HEADS inside the machine. The magnetic powder on the tape is specially formulated to hold the kinds of magnetism used in video recorders and must be very pure, uniform, and resistant to flaking or shedding from its base. Even the video cassettes themselves have to be built to exacting tolerances with 36 parts to meter the flow of tape from one hub to the other without sticking or jamming.

Like audio tape, video tape can be erased (demagnetized) and used over. With healthy machines and a clean environment, this rerecording process may be repeated hundreds of times.

There is no such thing as "color" video tape. The *signals* that are recorded on the tape make the color, not the tape itself. All video tape can record hi-fi sound *if* the VCR is equipped with that feature. Some video tape formulations make better hi fi recordings than others.

Kinds of Tape

As shown in Figure 5–32, video tape comes in a variety of sizes. Professional quad video tape is 2 inches wide and comes on large reels. Professional/industrial tape is 1 inch wide and usually comes on 6½–11¾-inch reels. Tape used in industrial and educational applications is ¾-inch-wide and comes in U-matic cassettes. Home VCRs (VHS, SVHS,

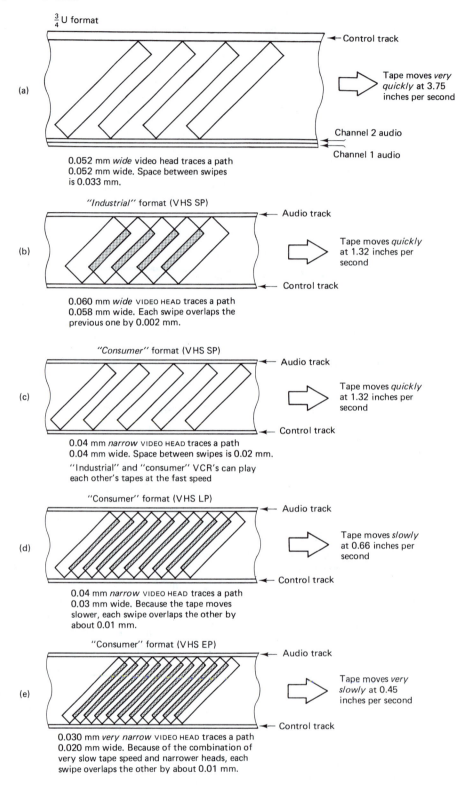

(a)

$\frac{3}{4}$ U format

← Control track

Tape moves *very quickly* at 3.75 inches per second

Channel 2 audio
Channel 1 audio

0.052 mm *wide* video head traces a path 0.052 mm wide. Space between swipes is 0.033 mm.

(b)

"Industrial" format (VHS SP)

← Audio track

Tape moves *quickly* at 1.32 inches per second

← Control track

0.060 mm *wide* VIDEO HEAD traces a path 0.058 mm wide. Each swipe overlaps the previous one by 0.002 mm.

(c)

"Consumer" format (VHS SP)

← Audio track

Tape moves *quickly* at 1.32 inches per second

← Control track

0.04 mm *narrow* VIDEO HEAD traces a path 0.04 mm wide. Space between swipes is 0.02 mm.

"Industrial" and "consumer" VCR's can play each other's tapes at the fast speed

(d)

"Consumer" format (VHS LP)

← Audio track

Tape moves *slowly* at 0.66 inches per second

← Control track

0.04 mm *narrow* VIDEO HEAD traces a path 0.03 mm wide. Because the tape moves slower, each swipe overlaps the other by about 0.01 mm.

(e)

"Consumer" format (VHS EP)

← Audio track

Tape moves *very slowly* at 0.45 inches per second

← Control track

0.030 mm *very narrow* VIDEO HEAD traces a path 0.020 mm wide. Because of the combination of very slow tape speed and narrower heads, each swipe overlaps the other by about 0.01 mm.

FIGURE 5-31 Path of VIDEO HEAD across the tape.

VHS-C) use ½-inch-wide-tape, and professionals use similar ½-inch tape t 6 times the speed (MII, Betacam, ED beta). Some home and semiprofessionals also use 8mm-wide-tape (8mm, Hi8).

Packing all that information into a tape isn't easy, and the less tape used per hour, the tighter the signal gets squeezed and the lower the quality of the picture and sound. Notice in Table 5–2 how many square feet of tape are used per hour by the professional and industrial recorders. By comparison, home machines use very little tape.

Dropouts

This economy of tape usage has its price, DROPOUTS. No, DROPOUTS aren't students who failed their TV I midterms. DROPOUTS are specks of snow on a TV screen caused by the VIDEO HEAD passing over a "bare spot" on the tape where the magnetic powder has flaked off, leaving no signal (and no picture) behind. DROPOUTS also occur when flecks of dirt, or tape debris, cover the magnetic powder in places, impairing contact with the spinning VIDEO HEADS. Figure 5–33 shows what a DROPOUT looks like on the TV screen. Some DROPOUTS appear as a blink on the screen for a thousandth of a second and are gone before you see them. Others may last longer and are quite distracting. Better tape has fewer DROPOUTS, but all tape has some, especially at the

beginning where the tape gets more abused from threading. Slower, denser recordings are more sensitive to these scratches, defects, and debris and therefore display more DROPOUTS.

Length

Tape manufacturers are still attempting to pack more minutes into a cassette. They can't put a bigger reel in the cassette; it won't fit. But they can make the *tape* thinner, putting more feet on a reel. Table 5–3 lists some of the common cassette sizes and their typical costs.

Superthin tape may give you more playing time, but it has its disadvantages too. Some tape is so thin that it stretches and snags easier than regular tape. What does this all boil down to?

1. If you want the best picture and sound, use standard thickness tape (T-120, or lower numbered) and the fastest VCR speed. The same goes if you plan to copy, edit, or play your tape through a TV projector.

2. If you plan to do much winding, rewinding, pausing, scanning, starting and stopping, editing, or frequent ejecting or inserting of your tape, use the standard thickness tape.

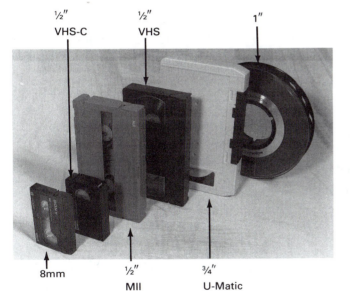

FIGURE 5-32   Kinds of video tape and cassettes.

TABLE 5–2

Tape speeds and consumption

Format*	Width (in.)	Speed (in./sec.)	Sq ft/hr
8MM (LP)	¼	.28	2.25
8MM (SP)	¼	.56	4.5
VHS-EP	½	.44	5.5
VHS-LP	½	.66	8.2
VHS-SP	½	1.31	16.4
MII	½	2.66	33.25
BETACAM (Broadcast professional)	½	4.67	58.25
¾U (Industrial, educational)	¾	3.75	70
TYPE C (Broadcast professional)	1	9.61	240
QUAD (Broadcast professional)	2	15	750

MINIGLOSSARY

*Dropout   A speck or streak of snow on the TV screen seen when a video tape player hits a fleck of dirt or a "bare" spot when the tape is playing. Dust or scratches can also cause a dropout to be *recorded* on a tape.

*Safety tab   A button or tab on a videocassette that can be removed to render the tape unrecordable (thus unerasable).

FIGURE 5-33   DROPOUT as it appears on a TV screen.

3. If you plan to record a long passage from beginning to end and play it back nonstop from beginning to end, feel comfortable in using the thinner tapes (numbers higher than T-120 or KCA-60).

If you play a cassette all the way through, you are likely to get a pleasant surprise. The manufacturers generally put more tape in the cassette than advertised. A VHS T-120 cassette generally has as much as 15 extra minutes in it at the SP mode. 8mm's are known to have 5 extra minutes in their P6-120s. KCA-60s have about 2 minutes extra.

When stocking cassettes, longer is not always better. Sometimes, shorter cassette lengths help you organize certain scenes or subjects together. Also, if you have numerous segments recorded on the same tape, it won't take as long

to wind from one to another if they aren't a mile apart on a long, long tape. Short tapes are most handy when you're shooting scenes in the field. All "takes" of a particular scene can go on one cassette, while shots of another scene appear on another cassette. Editing the scenes together is simplified as you organize the cassettes in the order of the story.

Avoiding Accidental Erasure

Picture yourself recording a "Dukes of Hazzard" rerun and looking down at the cassette box to discover that you're recording over priceless sequences of your college president receiving an award from Michael Jackson. Or you loan your school's only copy of "Emergency Medical Care" to your neighbor and get it back with his kids' recording of Saturday

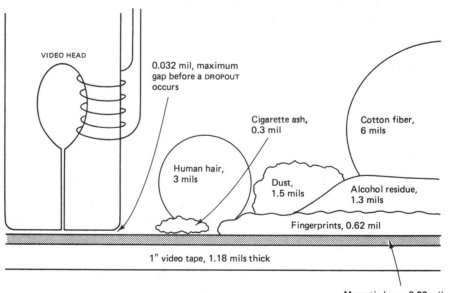

Tape debris ruins intimate contact between VIDEO HEAD and tape.

TABLE 5–3

Cassette running times

Type	Length (ft)	Play Time (hours:minutes)			Common Price ($)	Dollar Per Minute at Fast Speed
8mm metal particle		SP		LP		
P6-60MP		1:00		2:00	7.00	.12
P6-90MP		1:30		3:00	7.50	.08
P6-120MP		2:00		4:00	8.00	.07
Hi8 metal particle		2:00		4:00	13.00	.10
SP6-120						
Betacum metal		One speed only				
BCT 10M	233		0.10		20.00	2.00
BCT 20M	466		0:20		22.00	1,10
BCT 30M	700		0:30		24.00	.80
BCT 60ML (large)	1400		1:00		33.00	.55
BCT 90ML (large)	2100		1:30		52.00	.58
VHS-C		SP		EP		
TC-20		0:20		1:00	5.00	.25
TC-30		0:30		1:30	6:00	.20
VHS		SP	LP	SLP		
T-15[a]	115	0:15	0:30	0:45		
T-30	225	0:30	1:00	1:30	3.25	.11
T-60	420	1:00	2:00	3:00	3.60	.06
T-120	815	2:00	4:00	6:00	3.80	.04
T-150[b]	1075	2:30	5:00	7:30	10.00	.05
T-160[b]	1147	2:40	5:20	8:00	10.50	.06
T-180[b]	1260	3:00	6:00	9:00	11.00	.06
¾U		One speed only				
KCS-10[c]	235		0:10		8.20	.82
KCS-20[c]	423		0:20		9.00	.45
UCS-20	423		0:20		8.60	.43
KCA-30	610		0:30		9.60	.32
UCA-30	610		0:30		9.60	.32
KCA-60	1174		0:60		13.00	.22
UCA-60	1174		0:60		15.00	.27
¾U-SP		One speed only				
KSP-10	188		0:10		10.00	1.00
KSP-20	376		0:20		12.00	.60
KSP-30	564		0:30		13.00	.43
KSP-60	1128		1:00		17.00	.28

[a] Used by cassette-dubbing houses or available on special order.

[b] Extended length tape which uses a thinner base than standard length tapes, allowing a greater amount of tape to be spooled onto the cassette reels.

[c] Minicassette for portable ¾U VCRs.

Note: Most manufacturers include a little extra tape in the cassettes, yielding a little higher running time than listed here.

morning cartoons on it. Murphy's 108th Law of Recording states that junk recordings *never* get accidentally erased; prized ones *always* do.

Nothing short of copying all of your tapes and keeping them in a vault will completely protect them. But two easy precautions will save you a lot of disappointments:

1. *Label everything as soon as it is recorded.* Don't guess what is on your tapes. You're likely to record over your wrong guesses.

2. *Remove the SAFETY TAB from the videocassette so the tape cannot be recorded upon.* Figure 5–34 shows how.

¾U: Pop out the red button on the bottom of the cassette. Save the buttons. You will need them if you change your mind someday and wish to record over your tapes (erasing them).

VHS: Break off the plastic safety tab from the edge of the cassette (don't bother keeping the tab; covering the hole with a piece of scotch tape will

¾U

Turn tape over and pop out red button to protect cassette from being recorded over. Reinsert button to restore recording ability to prevent erasure.

to prevent erasure   to record again

VHS

To protect a recording from accidental erasure, break off the SAFETY TAB on the rear edge of the videocassette using a screwdriver. With the tab removed, the RECORD button won't activate and the tape is "safe" from erasure. If you change your mind and wish to record, cover the hole with a piece of adhesive tape.

tab   to prevent erasure   to record again

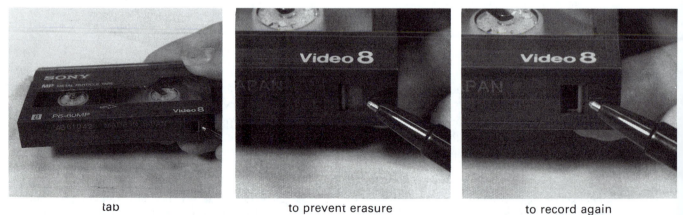

tab   to prevent erasure   to record again

FIGURE 5-34 Preventing accidental erasure.

restore the cassette's recordability).

8mm: Throw the switch on the rear of the cassette. This procedure does not affect playback or any other feature of the VCR or tape. It just defeats the VCR's RECORD function.

Care of Video Tape

Longevity. Magnetic tape is not an archival medium. Although the magnetism recorded on it will last easily 50 years, the plastic ribbon itself stretches and contracts with heat and cold and "relaxes" with age, causing the picture to bend at the top. This bending (shown in Figure 5–23) is called FLAGWAVING (the top of the picture flutters back and forth), or SKEW ERROR (the top of the picture bends to the side), or TAPE TENSION ERROR (the machine is misadjusted and is stretching the tape as it plays it).

Even under ideal environmental conditions, your tape may begin to suffer the "bends" after a year or so. Playing the tape once or twice a year will help, perhaps stretching this figure to 10 years. When the problem gets bad enough, the whole picture may twist into diagonal lines like in Figure 2–4. Much early broadcast video tape was lost this way.

Conceivably, a technician or a knowledgeable VCR owner could open the machine and readjust the TAPE TENSION on the VCR, making a shrunken tape play again, and could then copy it. Once a playable *copy* existed, the VCR could be adjusted back again to its proper TAPE TENSION. Superthin tapes (numbers higher than T-120, P6-120, or KCA-60) are more likely to stretch than normal tapes.

A tape can be played between 100 and 200 times before the picture becomes noticeably degraded. PAUSING wears out the tape quickly.

Storage. The lifetime of your recording is directly related to how carefully it is *stored* and how gently it's *used.* Here are the rules of storage:

1. *Store cassettes upright, in a vertical position.* This prevents the edges of the tape "ribbon" from getting "roughed up" against the inside of the cassette as it lays on its side.

2. *Keep the cassettes in their boxes.* This keeps out dust and dirt which cause DROPOUTS.

3. *Store tapes at average temperature and humidity*—about 70°F (20°C) and 50% or less humidity. Temperatures above 140°F (60°C) will permanently damage the tape (the cassette warps). A car trunk, passenger area, or glove compartment on a hot day may easily exceed the 140°F maximum. Leaving a tape in the sun may also overheat it. Mildew will grow on damp tapes.

4. *Keep tapes away from magnetic fields,* such as hi-fi speakers, amplifiers, transformers, magnets, or big electric motors. The magnetism from these devices can partially erase your tape.

5. *Wind or rewind tapes before storage.* Don't store them partially rewound because you want the DROPOUT-causing threading process to occur at the beginning or end of your tape, not in the middle of your show. Tape archivists advise playing valuable tapes all the way to the end for long-term storage. This provides very uniform tension on the tape. Also, when you rewind the tape before playing it back, you "air it out," dehumidifying it so it will slip easily through the machine when it plays.

6. *Don't leave recorded tapes threaded in the VCR* for days or weeks; this creases them where they are bent around the tape guides.

7. *Remove the SAFETY TAB on any tape you wish to "erase-proof."*

8. *Keep cassette labels smooth and tight.* Dog-eared, loose, wrinkled labels, or labels-over-labels, tend to snag in the VCR.

Handling

1. *Treat cassettes gently.* Dropping them rubs the delicate tape edges against the cassette housings, abrading the tape. If you damage a cassette, don't try to play it. It may jam in your VCR. You can buy new cassette shells as kits and transfer the tape to the new housing.

2. *With VHS machines, use PAUSE rather than short STOPS.* It stretches the tape a little every time the VHS machine unthreads and rethreads on STOP. On PAUSE, the tape stays threaded.

3. *Don't PAUSE a tape more than a couple minutes.* After a while, the spinning VIDEO HEAD will wear out the tape or crease it, causing DROPOUTS in that part of the program.

The list could go on, including things like "don't smoke near VCRs or cassettes because smoke particles will make a film on the tape and machinery and will clog the VIDEO HEADS," but the precautions, though accurate, start to get ridiculous. How many "rules" you follow will depend on how much of a perfectionist you are. Many small shops knock their tapes around, storing them half-wound on their sides atop a loudspeaker in damp basements. Their tapes will not last for years or look great, but many users don't seem to notice. Two things *no* tape will tolerate even if you're an indiscriminating viewer are *heat* and *dirt.* Radiators, hot cars, sunny windowsills, dusty workshops, and sandy, salty beaches can wreck a videocassette.

Larger production houses who use 1-inch reel-to-reel recorders and create costly TV programs take tape care very seriously. Here are some further rules which apply more specifically to professional reel-to-reel tapes.

Special Care for Professional Reel-to-Reel Tapes

1. Smoking, drinking, or eating should not be allowed within a 10-foot radius of the tape machine or tape storage areas. *Never* should ashtrays, coffee, etc., be placed on a table that is used as a tape work area. Wash and dry hands after eating and before handling tape.

2. Carry the reel by the hub, not by the flanges (see Figure 5–35). The reel flanges are to protect the delicate edge of the tape. If you squeeze the flanges, they will bend inward and will press against the edge of the tape and damage it. When a tape gets older and "looser," some tape machines will rewind the tape unevenly, leaving

uneven layers of tape on the reel as shown in Figure 5–36. These uneven layers cause concern. Squeeze the reel flanges, and you'll damage these edges. Lay a reel on its side rather than its edge, and these tape edges will rub against the reel. Another reason to store tape boxes on their edges is that a reel or box lying on its side invites people to set something on it. The weight can damage the tape edge.

3. The windows in the tape reel are for threading the tape and for viewing how much tape is on the reel. Keep your fingers out of those windows so you don't soil the edge of the tape.

4. Do not mark on or write on the tape while it is still wrapped on the reel. The pressure of your writing can deform several layers of the tape.

5. Before storing a tape, the free end of the tape must be secured so that it doesn't flop around or the reel doesn't unravel and become loose. This is best done with approved HOLD-DOWN tape (Figure 5–37). *Do not* try to secure video

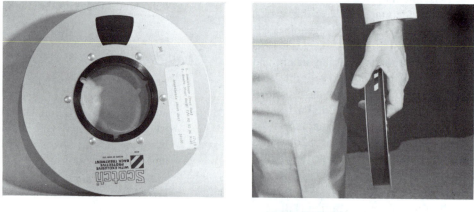

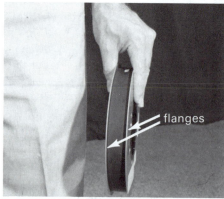

FIGURE 5-35   Carry the tape by the hub, not by the flanges.

MINIGLOSSARY

Flanges   On a reel of tape, the circular walls which protect the tape's edge.

Hold-down tape   Special adhesive tape used to keep the loose end of a reel of video tape from flopping around.

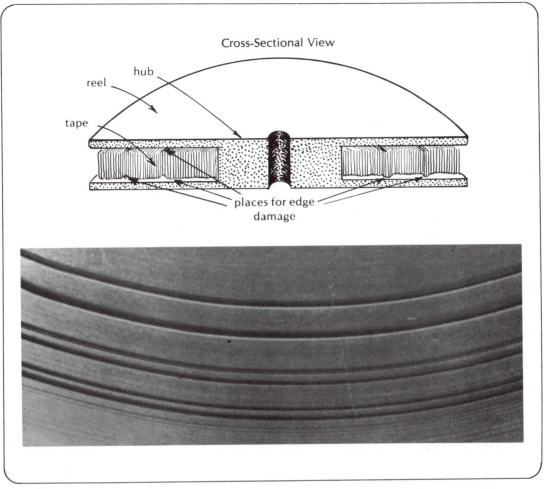

Cross-Sectional View

reel

hub

tape

places for edge
damage

FIGURE 5-36   Uneven layers of tape (Courtesy of 3M/Magnetic Media Division).

tape with masking tape or scotch tape. They leave a gooey residue that spreads everywhere.

6. If a tape has a wrinkled beginning (probably from threading, etc.), the wrinkled area should be snipped off. Now you can see why it's good not to start recordings at the very beginning of a tape.

7. When threading a tape around the take-up reel, wind it around at least six turns. This will ensure that the reel won't slip and spin free and then "catch" on the tape, snapping it.

8. If water (or other crud) gets onto the tape, remove it immediately. Both flanges should be unscrewed, and all water should be dabbed off with a lint-free cloth.

9. If a reel is dropped and the flange is severely bent, remove the bent flange and replace it with a good one. Don't simply wind the tape off the

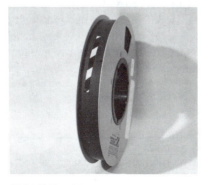

FIGURE 5-37   HOLD-DOWN tape.

reel onto a new reel if the tape is rubbing against the flange. Once the flange problem is solved, the tape *should be* wound onto another reel and back again to eliminate tape stresses that the dropping may have caused.

10. If a contaminated or damaged reel of tape is run on a machine, the machine should be cleaned before playing another reel of tape.

11. If for some reason (such as splicing) you *must* handle tape while it is threaded in the machine, use special lint-free gloves so you don't leave smudgy fingerprints on the tape.

12. In case of fire around magnetic tape, use CO_2, Freon, Halon, or some other inert chemical. Water will damage the tape.

13. Watch for CINCHING (Figure 5–38). CINCHING occurs when layers of tape in the middle of the pack have folded or washboarded as a result of layer-to-layer slippage. You can usually see CINCHED tape from the side by looking through the reel windows. CINCHING could be caused by tape machines with misadjusted brakes, clutches, or tape tension controls. Another cause for CINCHING is very loose tape, the result of storing it or transporting it without securing the loose end with HOLD-DOWN tape. If you see CINCHED tape, try to get rid of the wrinkle before it becomes permanent. Wind the tape all the way to the end and then rewind it to the beginning again. It should now lie flat. If your VTR won't wind the tape tightly and smoothly, use another machine. Take the first machine out of service until it gets fixed. Permanently CINCHED parts of tapes are unviewable.

Organizing a Tape Collection

Schools and home recording enthusiasts all have a tendency to record too much and store too much. Almost every program seems too valuable to be erased. So another box of videocassettes gets purchased, and another batch of "priceless" recordings goes on the shelf. The programs most likely to be recorded and never watched include graduations, guest lectures, school events and concerts, school plays, and

speeches made by faculty. For political reasons you often cannot refuse to make these recordings, but there is no good reason why they must remain on the shelves for 10 years unwatched. So regardless of your filing system, it pays to examine your shelves once a year and "weed out" the unnecessary recordings. People who fail to "weed" their tape inventory tend to get buried in their recordings and have a hard time locating the important ones.

It is better to have a short pencil than a long memory. You will need some method of logging your recordings on paper. The most common method is logging and shelving by ACCESSION NUMBER.

Number your cassettes on the box and on the cassette shell itself (in case they get separated) from 1 to whatever. Thereafter, each new tape gets the next higher number. Get a loose-leaf or other type of notebook and number each page. List each tape's contents with COUNTER numbers on the corresponding page. You may wish to include length of show, date, time of broadcast, channel, tape speed, whether or not the program had the commercials edited out, original or rerun, the year the movie was made (usually seen at the movie's end), the cast, original recording or duplicate, recording quality, black-and-white or color, or other details on the page. For off-air recordings, you could easily slice out the *TV Guide* listing for the show and paste it on the page.

Later, you can sort the entries by title or subject and print up tape catalogues with alphabetized titles and tape numbers.

For large collections, computerized data bases are very handy for sorting and listing the program titles alphabetically. The computer may also sort by subject or even by kind of tape (i.e., VHS, ¾U).

Storage Boxes and Labels. The boxes videocassettes come in are fine except that when you shelve all the various brands side by side, the mix of colors and trademarks starts to make your shelves look messy. Also the space available for titling is a bit cramped. Furthermore, some cardboard tape boxes fall apart rather easily.

¾U videocassettes can be purchased with "bookshelf" cases (Figure 5–39) which are plastic and strong and will lock closed. Some have hooks on them so that they may be hung on a rail and don't require a shelf.

Half-inch videocassette boxes are usually plastic; some snap shut and are sturdy enough to be mailed. They cost about $2.

MINIGLOSSARY

*Cinch A washboard-shaped wrinkle in a video tape that occurs when the end of a loose roll of tape is pulled and some tape slides off the roll without the roll actually turning.

Accession number Numerical order (1, 2, 3, etc.) assigned to tapes as they are acquired or recorded.

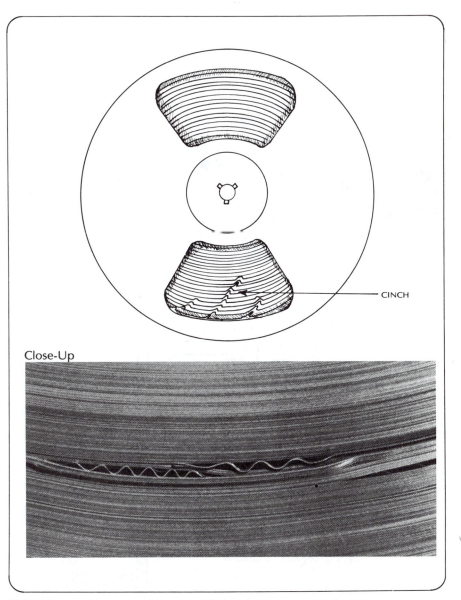

Close-Up

FIGURE 5-38 CINCHING (Courtesy of 3M/Magnetic Media Division).

Many stationery stores sell blank ¾-inch by 3-inch labels which can be typed and then stuck onto your present cassette boxes and cassettes. Placing same-sized labels on the boxes will keep your tape library looking uniform. Also, when you make a new recording over an old one, you can neatly affix a new label rather than erasing, scraping, or blowtorching off the old label.

Remember, when numbering cassettes, to put your number on both the box *and* the cassette. Place the cassette's number so it will be visible through the recorder's window while the tape is in use. This way, you'll not lose track of what tape you're playing *now*.

And here's a special hint for fitting more tapes on those cramped shelves of yours. Most people store cassettes upright (long side faces out, like books on a library shelf).

This way the cassette labels on the spines of the boxes are easy to read. The cassettes, however, take up a lot of shelf height, and you get fewer shelves per cabinet or per range. If, instead, you set the cassettes on their long edge (spine facing the ceiling), they are not so tall, and more shelves can be used for the same height. There may no longer be room enough to show the cassette's title, but you can put the cassette ACCESSION NUMBERS on ¾-inch square labels and affix these to the short edge of the tape, which now faces out. Figure 5–40 shows an example of such an arrangement.

Never mail video tapes in those paper-filled cushioned envelopes. The envelopes invariably tear inside, releasing head-clogging dust and flakes. Use special mailing boxes or plastic "bubble" wrap packaging.

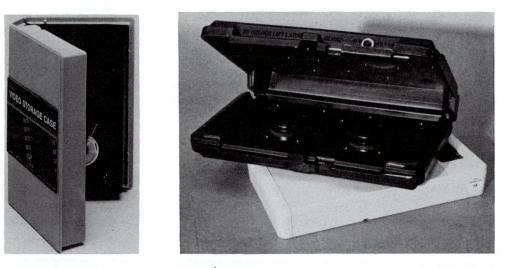

FIGURE 5-39   Videocassette storage boxes.

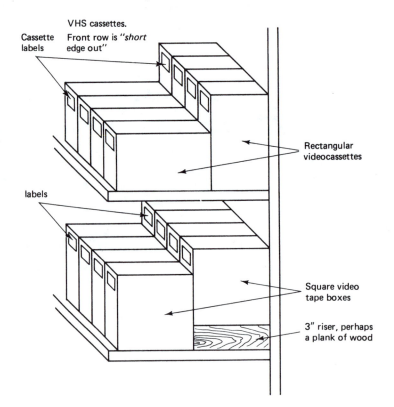

FIGURE 5-40   Consolidated typical "industrial" shelving of videocassettes.

COMMON VCR AILMENTS AND CURES

You followed the directions (you thought). Everything is connected and rarin' to go. Only it doesn't. Before you run whimpering to the repair shop, there's a lot you can try in order to diagnose and solve your VCR's problem yourself.

After a while you will learn the eccentricities of your machines and will intuitively head right for the trouble when your VCR goes kaput. Until that time, you're going to have to use logic and patience.

You will find that 80 percent of your problems are because you failed to throw a switch, turn a dial, or connect something up the way it should be. Another 10 percent of your problems will be minor defects, maybe in one of your connecting cables, perhaps a broken wire. Another 5 percent of the time it's a dirty or broken switch somewhere—still an easy thing to replace or fix. Then 5 percent of your time the machine fails because *it* has a problem and needs technical attention.

Let's look at some common VCR problems and how to solve them.

Machine Control Problems

In these examples, the machine's buttons don't seem to do what they're supposed to.

1. *VCR fails to operate;* POWER *light does not come on:*
    a. Check to see if the VCR is plugged in.
    b. Is the socket it's plugged into operated by a wall switch somewhere? If in doubt, check the outlet using a lamp.
    c. TIMER may be switched to AUTO, or OFF, or ON and waiting for a scheduled time. Check which position manually activates the VCR, or try flipping it to another position and watching for signs of life.
    d. Turn TIMER/SLEEP switch off.
    e. Look for a REMOTE switch. Perhaps the machine only wants to listen to its remote control.

2. POWER *light comes on but tape doesn't move when you switch to* PLAY:
    a. Check PAUSE; it should be off.
    b. Check the REMOTE socket; if anything is connected there, *it* may be PAUSED. (An earphone accidentally plugged into the RE-MOTE PAUSE socket will PAUSE the machine too.)
    c. Is the DEW light on? If so, you'll have to wait until the machine dries out before it will work. Try the hair dryer trick mentioned in the section on getting more out of your VCR in this chapter. Is the machine cold (thus condensing humidity)? Move to Arizona where it's warm and dry.
    d. Is the cassette properly loaded? Are you trying to use a beta cassette in a VHS machine? That will never do.
    e. On beta VCRs, is the STANDBY light on? If so, wait till it goes off.
    f. Maybe the tape is loose in the cassette; it should be taut. Often VCRs sense loose tape and switch themselves to STOP. EJECT the tape and give one of the tape hubs a turn with your fingers to tighten the tape. Or press REWIND, STOP, WIND, STOP and then PLAY. The rewinding and winding may tighten the tape.

g. Could it be that the tape is at its end and needs to be rewound? Some cassette machines have an AUTO OFF feature which stops the tape at the end of play. Some also have an AUTO OFF indicator that lights up to apprise you of this condition.
h. Are you using a portable VCR? Perhaps your battery died, or got too weak. When a 12-volt battery drops down to about 10.4 volts, the VCR senses the weakness and shuts itself down. Try another power supply. If using your portable in very cold weather, the oil in it could become gooey, slowing down the VCR's wheels. The VCR may sense this and shut itself down to avoid damage. The VCR may also sense whether its transistors are too cold to function and will shut itself down.
i. Do you have a camera connected to the VCR? Perhaps its PAUSE trigger is pulled.
j. If using a reel-to-reel recorder, perhaps you misthreaded the tape. Many machines sense this error and automatically shut down if the tape is misthreaded or too loose.

3. *Buttons won't depress:*
    a. Switch the POWER on.
    b. Make sure you have a cassette in the VCR.
    c. Check everything in example 2.

4. RECORD *button won't depress or activate:*
    a. Check the cassette SAFETY TAB. If it's been removed, the tape cannot be recorded. See the section on avoiding accidental erasure earlier in this chapter.
    b. Check everything in examples 2 and 3.

5. AUDIO DUB *button won't depress or activate:*
    a. Same problem as example 4; most likely SAFETY TAB.

6. FAST FORWARD *does not function:*
    a. Maybe you're already at the tape's end. Rewind.
    b. Check everything in examples 2 and 3.

7. *VCR stops in* REWIND *before end of tape:*
    a. Turn the MEMORY or SEARCH switch *off.*

8. *Cassette won't insert into VCR:*
    a. Is it the right FORMAT (VHS, 8mm, or ¾U) cassette?
    b. Are you holding it right? The trapdoor part points away from you; the reel hubs face down, the window, up.

9. *Cassette* EJECT *doesn't work:*
    a. Switch POWER *on,* press STOP, wait a moment, and then try again.

10. *The tape moves slowly or moves backwards when you press* RECORD *or* PLAY:
    a. The belts in your VCR are tired, slippery, or broken. Take two aspirin and call your repairperson in the morning.

11. *The tape rewinds or winds very slowly (it should take less than 3½ minutes to completely wind or unwind a tape):*
    a. Tired belts; same as example 10a. Air pollution and smoke tend to age belts quickly, causing them to glaze (get slippery), crack, or stretch.

12. *While playing or recording, the VCR mysteriously stops itself. It may even rewind itself or rewind and begin playing/recording again:*
    a. Does the VCR have a REPEAT switch? If it's turned *on,*
       1. When the tape reaches its end, it will rewind and repeat.
       2. When the INDEX COUNTER reaches 000 on some machines, they will rewind and start over.
    b. The tape may be scratched or perforated. VCRs have automatic shutoffs which sense this and stop. If the problem always occurs at the same spot on the tape, you might try removing the cassette and examining the tape for damage (explained further in Chapter 16).

Signal Control Problems

In these examples, you can't get a picture or sound where you want it to go. Often the culprit is a misflipped switch. If you do much connecting and disconnecting of cables, it may be a cable plugged in the wrong socket.

If the cables are moved often, the connectors sometimes get loose or the wires fray inside the plug. After trying the obvious (switches, correct connections), try wiggling the cable at the plug end or rotating the plug in its socket (except for multipin plugs) and watch your TV for flashing and listen for crackling. A frayed wire or dirty or bent plug may make contact intermittently when wiggled, cutting your signal off and on. When you doubt a cable, try another if you have one. A successful substitution will prove the first cable defective. They are usually easy to fix (if you're the handy type) or fairly cheap to replace (except for the multipin cables—they're worth fixing because they're expensive).

Sometimes switches get dirty inside. When you switch, the TV shows a lot of breakup and hash, and the audio may crackle. Flipping the switch back and forth a dozen times may clean it somewhat, but a deteriorating switch will become more and more intermittent and unreliable. If you're handy, there are sprays you can squirt in there to clean a dirty switch. Otherwise, a technician may have to clean it for you.

Now let's see what else can go wrong, go wrong, go wrong. . . .

1. *Can't get regular TV programs on the TV set:*
   a. Switch the VCR's OUTPUT SELECT, or ANTENNA, or VTR/TV switch to TV.
   b. Is the VCR connected to the antenna? Is the TV's ANTENNA input connected to the VCR? They should be.
   c. If you're using the VCR's tuner to feed the TV, make sure *it* is FINE-TUNED, and your TV is FINE-TUNED to the VCR's channel (3 or 4), and your VCR's INPUT SELECT is on TV and its OUTPUT SELECT is on VCR.

2. *Can't get VCR to play through the TV:*
   a. The VCR's OUTPUT SELECT must be in the VCR position.
   b. Check the connections between the VCR and the TV.
   c. Is your TV tuned to the same channel as your RF CONVERTER?
   d. If you're sending video and audio straight to the TV monitor/receiver, make sure the monitor/receiver is switched to LINE or VTR or AUX or VIDEO, not to AIR or TV.
   e. Check to see if the TV even works by switching the VCR's OUTPUT SELECT to TV or ANTENNA and tuning the TV to various broadcast channels. If nothing comes in, perhaps the VCR-to-TV connection is bad or the TV is bad. If stations are received, then the TV and its connections are okay. Switch the OUTPUT SELECT back to the VCR mode and check channel 3 or 4. Check to see that the RF MODULATOR isn't missing.
   f. Wiggle the cables while playing a tape and watch the TV for flashing to check for loose connections.

3. *Unable to record TV programs:*
   a. Flip the VCR's INPUT SELECTOR to TV.
   b. Make sure the cassette's SAFETY TAB is still there.

MINIGLOSSARY

Repeat switch   Switch on a VCR that tells it to play a tape over when it reaches the end.

c. Unplug the camera and microphone from the deck. They sometimes automatically override the TV inputs.

4. *No sound while recording a TV broadcast:*

a. You left the microphone plugged into the VCR, overriding the TV.

b. Check any separate AUDIO IN connection for the same thing.

c. If your portable VCR uses a separate tuner with separate AUDIO and VIDEO connections, the AUDIO wire may be loose.

d. Maybe you have sound but you can't hear it. Can you get sound while playing other tapes or from off-the-air broadcasts? If not, maybe your TV has an earphone plugged into it, cutting off its speaker.

e. Are you using a TV monitor instead of a receiver? TV monitors don't have sound.

f. Turn up the volume control on the TV. If the sound crackles on and off as you do, work the knob back and forth a couple of times or tap it. You may have a volume control that is dirty inside or has a "dead" spot. If you get the crackling volume working for the time being, leave it. Later, have a technician look at it. It's generally easy to fix.

g. If you're using a monitor/receiver and sending separate audio and video to it, check your audio wire.

5. *No picture while recording a TV broadcast; sound is okay:*

a. The camera or a separate video input is still connected to the VCR.

b. On portable VCRs, check that the tuner's video cable is connected to the VCR's VIDEO IN.

6. *Camera or microphone signals won't record:*

a. Switch the VCR's input switch to CAMERA or VIDEO (away from TV or TUNER).

b. Are the microphone and camera plugged in? Is the camera turned on? Is the mike turned on? Some have ON/OFF switches.

Using Your Monitors to Help Track Down Signal Problems

If you're using a small studio setup like in Figure 5–17, you are well equipped to track down picture and sound problems using a methodical step-by-step process. First locate the problem. Is it the monitor/receiver, its cables, the VCR, the cables to the VCR, or the source? Let's start with the source.

1. *Source.* If you loop the signal from the video source (camera, etc.) through the monitor on the way to the VCR, you will know if you have video to start with. Similarly, you can verify the presence of audio if it goes from the sound source (microphones, etc.) into a monitoring device like an audio mixer with a meter or headphones on the way to the VCR.

If the camera viewfinder has no picture, the camera is probably at fault. If the viewfinder has a picture but the camera's monitor does not, it's probably the fault of the cable between the two. If the monitor has a good picture but the VCR's VIDEO LEVEL meter shows no signal, then the cable between the monitor and the VCR is probably at fault. If the VCR's VIDEO LEVEL meter reads in the green area but the VCR's monitor shows no picture, the fault probably lies with that monitor or its cable to the VCR.

On the audio side, if you are sending a signal to a mixer but the mixer's meter doesn't wiggle, you should check the microphone, its cable, its connection to the mixer, the switches on the mixer, and the volume controls. If the meter does wiggle but you hear nothing over the headphones, there may be something wrong with the headphones. Try another pair. If the headphones and the mixer's meter read okay but the VCR's AUDIO LEVEL meter doesn't wiggle, your problem is probably in the cable between the mixer and the VCR. If the VCR's meter wiggles but you get no sound from the VCR's monitor/receiver, then it's probably the fault of the monitor/receiver or its cables. Try plugging headphones directly into the VCR to check the sound.

In general, always check to see if the sources are turned on. Are the cables connected to the source's outputs? Do the cables go to the right inputs on the monitoring device? Are the plugs in tight? Are the monitoring devices turned on? Do the monitoring devices have switches on them which need to be flipped to put them into the right mode for operation? If you suspect a faulty cable, replace it with another cable. Eventually, something is bound to work.

2. *VCR.* Make an audio and video check. Do the audio and video meters show that you have a signal getting to the VCR? If not, turn up the audio and video controls. If the meters still don't move, check the INPUT SELECTOR. Is the VCR "listening" to the right inputs? If so, and if there is still no signal on the meters, check for exotic switches on the VCR which may be fouling up the signal.

Try playing a tape. If the tape doesn't move, something is wrong with the video recorder. If it's a reel-to-reel VTR, could the tape be misthreaded? Is the tape taut as it should be? Does the machine have power? Something should light up if it does. Is the VCR listening to a built-in timer instead of you? Are you at the end of a cassette? Is there a PAUSE button pressed somewhere?

If you can play a tape on the machine and see good picture and sound, at least you know that the VCR monitor and cable are good.

3. *Monitor.* How are you getting the signal to the

VCR monitor or monitor/receiver? If you are using an 8-pin cable as in Figure 5–17, try unplugging it and any other cables from the TV monitor. You should now get a smooth gray or black screen (assuming the set is turned on and getting power). If you see and hear snow or ''Wheel of Fortune'' reruns, your monitor/receiver is errantly switched to TV or AIR. Switch it to VTR and watch for that blank screen. Now reconnect just the 8-pin cable. Start playing a prerecorded tape. If it appears on the monitor, then your monitor and cable are okay. If not, the cable or monitor is bad. Try wiggling the cable and the plug while watching the TV screen for flashes of picture.

If sending RF to a monitor/receiver from a VCR, the checkout is quite similar. Play a prerecorded tape, and if you see ''Cosby'' reruns on the screen instead of your taped program, your TV is probably listening to the wrong channel. Switch it to channel 3 or 4. Make sure the VCR's OUTPUT SELECT switch is turned to VCR. If all you get is snow in all cases, then check your RF GENERATOR and your cable connecting the VCR's RF OUT with the ANTENNA IN.

When all else fails, unplug and reconnect all of the cables. Sometimes mistakes are made in the tangle of wires, and the only way to straighten things out is to do it over from scratch.

Special Audio Problems

Chapter 10 will sound off on this subject, but here are a couple common gremlins to chase away:

1. *A screech or howl comes out of the TV and you're not watching ''Ghostbusters'':*
   a. Turn down the volume on your TV receiver or sound monitor. You've got what they call FEEDBACK. No, FEEDBACK isn't the chef's position on a football squad. What happens is this: Sound goes in the microphone, goes into the VCR, gets sent to the TV receiver, and comes out the speaker loudly enough to go back into the microphone again. Around and around the sound goes until you break the cycle by turning the volume down or moving the mike farther from the TV. Figure 5–41 diagrams this process.
2. *Echoey, hard-to-hear sound:*
   a. Move the mike closer to the subject.

   b. Move to a room with more carpet, curtains, and furniture. You have too many echoes from hard bare walls.
   c. Turn the volume down on your TV; some of the sound is getting back to your mike.

3. *Microphone sound is weak, tinny, or has hum or hiss:*
   a. The mike is not right for your VCR. Read Chapter 10 and try another mike.
   b. The mike may be defective or of poor quality.
   c. The connection may be bad, or you may be using the wrong kind of mike wire.

4. *Using AUDIO IN, the sound distorts or sounds raspy: your audio signal is too strong; the sound may also have lots of hum and hiss:*
   a. Reduce the volume from the source.
   b. Attenuate the signal using methods described earlier in this chapter or in Chapter 10.

5. *While playing a tape, you get no sound, the wrong sound, two soundtracks at once, or what seems like half the sound:*
   a. Video tapes can have two sound tracks. You can listen to channel 1, 2, or a mixture of both. The trick is to play back the correct sound track. If you hear no sound, perhaps you're listening to the silent track. Try the other. If you're hearing Spanish and you want English, perhaps the tape is bilingual and you're listening to the wrong channel. Try the other. If you hear both languages together and you want only one, perhaps your AUDIO SELECT switch is on MIX. Switch it to CH1 or CH2. If you hear just sound effects when you should hear narration *and* effects, switch the selector to MIX. If using RF or the VCR's headphone output, the AUDIO SELECT switch will probably take care of everything. If, however, you're running an audio cable from the CHANNEL 1 AUDIO OUT to a separate amplifier and loudspeaker, then CHANNEL 1 may be all the amp will hear. You may have to unplug the cable from the CHANNEL 1 output and stick it into the CHANNEL 2 socket to hear CHANNEL 2. The AUDIO SELECT switch won't make any difference here.
   b. If, with a VHS or beta VCR, you record a

MINIGLOSSARY

Feedback   A loud screech coming from a loudspeaker when sound enters a microphone, gets amplified, and then comes out the speaker only to be picked up again by the microphone and amplified more.

Distortion   Poor quality sound, usually raspy and loud, often caused by too strong an audio signal.

Bulk tape eraser   Large electromagnet used for erasing (demagnetizing) an entire reel or cassette of audio or video tape at once. The procedure takes about 4 seconds.

AFC time constant   An internal circuit design on a TV set which determines how much it jitters and flagwaves when playing tapes.

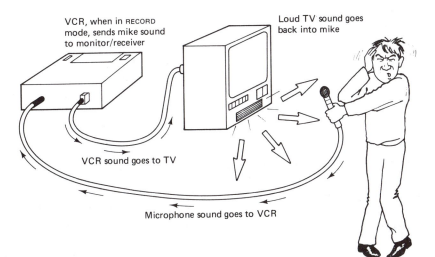

VCR, when in RECORD mode, sends mike sound to monitor/receiver

Loud TV sound goes back into mike

VCR sound goes to TV

Microphone sound goes to VCR

FIGURE 5-41 Audio FEEDBACK.

tape in monaural, then record over it in stereo, and *then* play it back on a monaural VCR, the old sound from the first recording may faintly linger, because the monaural VCR recorded the whole width of the sound track, while the stereo machine only erased and replaced the top 45% and the bottom 45% of the track, leaving an unerased gap in the middle (Figure 5–42). When played on the monaural VCR, the *whole* track is played, which is the top track plus the bottom, *plus the*

unerased leftovers in the middle. The solution is to BULK-ERASE the tape (Chapter 15) or to erase the tape with the monaural machine before reusing the tape in stereo.

6. *While playing a home-edited tape, the sound cuts in and out, or doesn't quite go with the show:*
   a. Switch the VCR from HI FI to LINEAR or NORM. If someone edited new pictures on a tape with HI FI sound, they obliterated the old HI FI sound along with the picture, but may have left the LINEAR sound intact. Also AUDIO

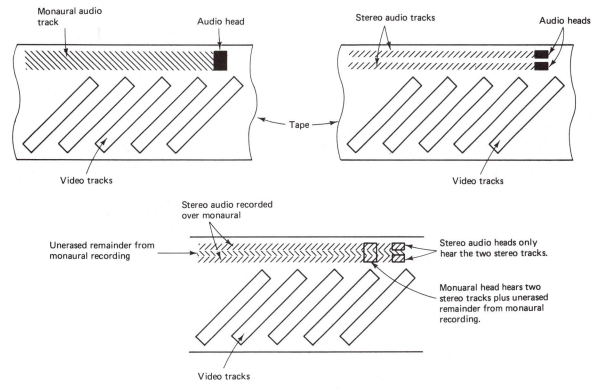

Monaural audio track

Audio head

Video tracks

Stereo audio tracks

Audio heads

Video tracks

Tape

Stereo audio recorded over monaural

Unerased remainder from monaural recording

Stereo audio heads only hear the two stereo tracks.

Monaural head hears two stereo tracks plus unerased remainder from monaural recording.

Video tracks

FIGURE 5-42 Stereo audio versus monaural audio on a home VCR.

DUBS would only affect the LINEAR tracks, not the HI FI. More on this in Chapters 10 and 14.

Picture Quality Problems

1. *Picture is distorted, usually with a band of snow or NOISE through it. Figure 5–22 shows a mild case. Worse cases may look much like Figure 2–10. The picture may jiggle a lot, even though you're not watching "Babes" reruns:*

   a. Adjust the TRACKING control. This is perhaps the most frequent adjustment you will need to make on a VCR. Usually the picture is best when the TRACKING control is set in the middle at FIX, but tapes made on another recorder are most likely to require that the TRACKING knob be turned to some position other than FIX. Turn the TRACKING control until the breakup moves off the screen at either the top or bottom and the picture remains stable. Old, stretched, hot, cold, or damp tapes won't track well at all.

2. *On PAUSE or STILL, VCR's picture has a band of noise through it as in Figure 2–10 or 5–22:*
   a. Similar to example 1, adjust the VCR's SLOW TRACKING control. Some manufacturers call this a V POSITION control and hide it behind or under the VCR. The effect of your adjustments may not appear until you PLAY and PAUSE the tape again.

3. *No color or poor color when playing a tape:*
   a. Adjust the TV's FINE TUNING.
   b. Perhaps the VCR's FINE TUNING wasn't properly adjusted when the recording was made. If so, the program will never play in color.
   c. Are you sure this is a color program?
   d. Turn up the COLOR control on the TV set.
   e. Check to see that the TV's ANTENNA input switch is set to 75Ω if you're using coax cable or 300Ω if you're using twin lead.
   f. If you're using straight video into a TV monitor, make sure the set is properly terminated (as explained in Chapter 2).

4. *Using AFT worsens reception.* You've tuned your VCR manually with the AFT *off*, and everything looks great. You switch the AFT *on*, and things get worse, or perhaps you lose color:
   a. Your AFT circuits need adjustment by a technician.

   b. In the meantime, leave it *off*, and you'll get by.

5. While playing a tape, the picture blanks out or temporarily distorts or shrinks:
   a. Is there a power-hungry motor, heater, microwave oven, or other appliance on your circuit sapping power? If this power interference occurred during the recording, the glitch is there to stay. If it happens only during playback, the tape is okay; go back and view again.
   b. If there's an air conditioner or big motor nearby, it can interfere with your video signal.
   c. If it looks like TV interference is being picked up by your VCR's tuner, check Chapter 3 for ways to cure TV interference.

6. *Tape plays at the wrong speed.* The picture may roll, bend, tear, or collapse into diagonal lines, looking like Figure 2–4:
   a. Is your SLOW MOTION feature engaged?
   b. Perhaps the machine is damp inside and the DEW SENSOR failed to stop the machine. Let it warm up a bit and try again.
   c. Your VCR's insides may be dirty. Run a CLEANING CASSETTE through (process described in Chapter 16) and see if it helps.
   d. Is your portable VCR's battery dying? Double-check the ac power adapter connection. Does the power meter say it's getting enough juice?
   e. Are you using a cheapie or COUNTERFEIT cassette? It could be rubbing inside.
   f. Try another cassette tape. If the problem persists, perhaps the VCR's drive belts are old, tired, slippery, or in need of replacement.
   g. You would get speed problems if your ac power fluctuated in frequency. Do you have weird power? (I have a grandmother with weird powers.)
   h. Was the tape recorded at a speed your VCR can't play? If the sound plays fast or slow, that's your tipoff that the machine and tape are different speeds.

7. *The picture bends or flutters at the top during playback.* FLAGWAVING (or TAPE TENSION ERROR or SKEW ERROR) is shown in Figure 5–23. It is usually caused by the tape being too tight, too loose, or too shrunken as it plays through the machine:
   a. Adjust the HORIZONTAL HOLD on your TV set.

b. Adjust the SKEW or TAPE TENSION control on your VCR if it has it (most home models don't).

c. If the problem happens on all tapes, have your VCR adjusted inside. If your TV is very old or has tubes, the picture will always look worse than it has to. A technician can modify the TV's AFC TIME CONSTANT to improve the picture. If you have another TV, try that one.

d. If the tape and/or VCR are cold, let them stabilize to room temperature.

e. If the tape was recorded long ago, it probably has shrunk. Without adjusting the VCR, it will probably never play right. One thing you can try is winding the tape all the way to the end. Then, rewind it all the way to the beginning. Now try to play it. The winding may stretch or relax the tape a little, making a marginal tape playable.

f. Superthin tapes stretch easily, especially in hot weather, causing FLAGWAVING.

g. The problem may be similar to example 2 or 6 earlier. Try those maneuvers.

8. *The picture looks very grainy or even snowy on playback only. It looked fine when it was recorded (eliminating the possibility that you recorded a snowy picture to start with) and the sound is okay. Perhaps it looks like Figure 2–10:*

a. If the picture is very snowy, it could be due to a worn or dirty VIDEO HEAD. You can't easily see the dirt on a VIDEO HEAD, but a tiny speck of it can clog this very sensitive device. Usually a clogged HEAD will not completely obliterate the picture—some faint image may show through the snow. If this is what you see, you can be fairly sure your problem *is* a clogged HEAD. What do you do for a clogged HEAD (besides taking nasal decongestants and getting lots of bed rest)? You clean the HEAD with a special "cleaning cassette." Following the manufacturer's instructions, insert the cleaning cassette into the machine just as you would a normal cassette. Play the special cassette for about 30 seconds and then remove it without re-

winding it. The HEADS should now be clean. Chapter 16 gives further details on this procedure.

9. *The TV used to work all right before you hooked up the VCR, and the VCR works okay elsewhere, but together they display interference or herringbone patterns on the TV screen or in the recording like in Figure 3–33 or 5–47:*

a. The TV or the recorder is sending out a weak signal which is interfering with the other's tuner. Move the two farther apart, maybe 3 feet. If moving them apart isn't possible, try placing a grounded metal plate between the two. Sometimes tinfoil wrapped around part of the TV works (but don't cover its vent holes). Also, avoid long antenna cables coiled up behind the equipment. Use good-quality, well-shielded cables and avoid excessive lengths.

10. *You used to get sharp cable TV pictures before you added the VCR and a few other video devices like switchers. Now you get a ghost on some channels:*

a. Nearby TV stations are broadcasting strong signals *directly into your equipment* while the cable company is sending you the same programs on the same channels slightly delayed (because their signals traverse so much wire). The unwanted direct signal "leaks into" your wires and into poorly shielded equipment to cause ghosts. Seal these invading signals out by using COAX or SHIELDED TWIN LEAD everywhere and well-shielded, high-quality switchers and other accessories.

11. *The TV and program are both color. While playing your ¾U or 1-inch tape, the picture displays vertical bands of color, like a rainbow (see Figure 5–43):*

a. Some color video players have an extra control called COLOR LOCK or COLOR PHASE or COLOR HOLD. This control is generally clicked into its middle position, where most color tapes play well. When vertical bands of color appear or when the picture suddenly loses its color or doesn't maintain the correct hue,

---

MINIGLOSSARY

Color lock Control on a VCR which allows tapes to be played with good color and without shimmering rainbow effects on the TV screen.

Head switching noise A small horizontal discontinuity at the bottom of the TV picture (usually off the screen) caused when each spinning video head leaves the tape and the twin head takes over.

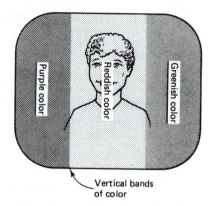

Vertical bands
of color

FIGURE 5-43   COLOR LOCK adjustment.

it's time to adjust the COLOR LOCK control. Turn the control slowly to the left or right (on some videocassette machines, you will have to make the adjustment with a screwdriver) until a proper picture is restored. The control may have to be returned to its original position to play normal tapes after you are finished.

12. *You press* RECORD *and look into the VCR monitor only to find the picture contrasty, faded, bending, or streaking in the bright places, as in Figure 5–44:*
   a. First check the signal you're feeding to the VCR. Is it good? If not, Chapter 6 tells how to fix it.
   b. If the picture going to the VCR is good but the picture coming out is bad, maybe it's the VCR's fault:
      1. If the picture looks too contrasty, streaks in the bright places, or bends in the white places, turn the VIDEO LEVEL on the VCR down.

2. If the picture looks pale, faded or rolls easily, or jitters on the TV screen, turn the VCR's VIDEO LEVEL up.
   c. The VIDEO LEVEL METER on the VCR is usually accurate. If it reads in the green, where it belongs, yet the monitor shows a bad picture, try readjusting the monitor's controls. Maybe the VCR's picture was good but the monitor was off a little. Reviewing the VCR's VIDEO LEVEL problems once more:

Problem	Looks Like	Cure
Contrasty	Figure 2–8	Turn VIDEO LEVEL down
Faded	Figure 2–9	Turn VIDEO LEVEL up
Bending, squirming sideways	Figure 2–11 or 5–23	Turn VIDEO LEVEL down
Bright parts of picture have streaks of black	Figure 5–44	Turn VIDEO LEVEL down
While monitoring the VCR's signal using RF, the sound buzzes; picture may also be contrasty, streaky, or bendy		Turn VIDEO LEVEL down

13. *While playing a tape, there are excessive* DROP-OUTS *or thin bands of snow across the screen as in Figure 5–45:*
   a. If the band slides down the screen and eventually disappears off the screen as the tape plays, this is a tape defect. It could be a wrinkle, a tape splice, a dent in the tape, or a spot where the magnetic surface has

FIGURE 5-44   VIDEO LEVEL too high. White parts of picture streak and sometimes turn black or gray.

FIGURE 5-45   Thin bands of snow across screen, usually a sign of damaged tape.

flaked off. There is no cure; the tape will always play like this. If the problem is severe and you can actually *hear* the bad part as it runs through the machine, you should consider not playing the tape again. The damaged tape could "snag" on the delicate video heads and damage them.

b. If the band stays at the same height on the screen as the tape plays, the tape probably has a scratch on it.

    1. The tape cassette could be scratching the tape. Perhaps it can be opened and fixed so it doesn't scratch *more* of this tape (the damage already done is irreversible).

    2. The VCR could be scratching the tape. If so, it could scratch *all* your tapes. Try another tape of little importance. If the problem shows there too, it's the VCR's fault. Get it fixed before it damages more tape.

c. If tapes play okay on other machines but show the snowy streaks only on this one, there may be a dirty contact in the video head drum assembly. Have a technician clean it.

14. *Off-air TV signal looks okay when fed to a TV only, but when the VCR is wired into the system, the picture gets snowy or grainy or color is lost:*

a. The connections or the antenna wire between the VCR and its TV monitor might be bad. Check out the ends or try another wire.

b. If the TV signal is weak, giving it *two* machines to feed (the VCR and TV) may be enough to deplete it altogether. Try BOOSTING the signal first (see Chapter 3) and then send it to the VCR and TV.

15. *While playing a tape, a thin horizontal line appears across the TV screen, looking like someone sliced the picture with a knife. See Figure 5–46:*

a. If while playing a ¾U VCR into an UNDERSCANNED TV monitor (one with the black edges of the picture showing) you see one or more of these creases sitting at the *bottom* edge of the picture *and it stays there*, that's normal. This is called HEADSWITCHING NOISE, a tiny glitch in the picture occurring when one video head leaves the tape and its twin brother begins to take over. When

FIGURE 5-46 Horizontal slice across screen could be a sync abnormality.

a ¾U tape gets copied, another horizontal slice gets added to the picture. The experts can often tell what generation (master, copy, copy of a copy) a ¾U tape is by counting these lines (if they don't overlap too much). Usually these lines sit about ¼ inch from the bottom of a 12-inch UNDERSCANNED TV picture. On normal TVs, they usually don't appear at all; they're hidden. They often show up when

    1. The TV is misadjusted.

    2. A third- or fourth-generation copy is played (the lines stack up and begin to appear).

    3. The recorder is misadjusted and is placing the line too high in the picture. This requires a technician to fix.

b. If the line rises and passes out of the picture in a second or two, there was probably a bump in the tape. Leaving the tape threaded and idle for a long time can "bend" the tape where it goes around guides and rollers. Dirt particles can also dent the tape, as can writing on it. Bad cases cause DROPOUTS and a band of snow as in Figure 5–45.

c. Someone may have switched from one picture to another without the proper synchronized equipment (more on this in Chapter 6). The switch causes a blip which rises off the screen in a few seconds.

d. A brief power disruption during the recording caused the recorder to "blink" and momentarily lose sync. The sync blip will rise off the screen and disappear. There's no way to get rid of the line at this point other

---

MINIGLOSSARY

*Copyguard, Macrovision   Antipiracy techniques employed by prerecorded tape producers to thwart tape copying.

than recording over it. A frequent cause of such power disruptions is a large motor connected to the same circuit as the VCR. When the motor turns on, the VCR hiccups. Some examples of machines with large, blip-causing motors are a large Xerox copier, air conditioner, electric heater, furnace, freezer, or large lights.

16. *You play a prerecorded videocassette and the picture jitters and rolls:*
    a. Adjust TRACKING.
    b. Adjust your TV's VERTICAL HOLD.
    c. If you bought the tape commercially, it may be COPYGUARDED—its sync signals are messed up to keep people from duplicating it. These signals are messing up your VCR or TV. Send the tape back and ask for a non-COPYGUARDED edition.
    d. If a friend gave you this copy which he made from a COPYGUARDED tape, naturally it shouldn't play. That was the tape producer's whole idea.
    e. Cheap, COUNTERFEIT, PIRATED, or nth-GENERATION (copies of somebody else's copies which were previously copied from somebody else's copy) tapes are likely to be of low tape quality, low-cassette-shell quality, or low recording quality, and they just won't play well.
    f. Clean the HEADS with a CLEANING CASSETTE (process described in Chapter 16). Your CONTROL HEAD, which synchronizes the motors and tape movement, may be dirty. Also the CAPSTAN (the part that pulls the tape steadily through the machine) may be dirty. If you're handy, open the machine and clean these parts manually to do a better job (process also shown in Chapter 16).

17. *If when playing a tape, the picture flashes bright and dark, maybe even blanks out:*
    a. Again, you may have PIRATED tape, this time with MACROVISION anticopying signals.

18. *When playing a tape, the picture looks smeared, blotchy, or overcontrasty, much like Figure 5–44.*
    a. You're trying to play a SUPER (SVHS or HI8) recording on a "regular" VCR. Only SUPER VCRs can play SUPER tapes.
    b. The VIDEO LEVEL was too high when the tape was recorded. There is nothing you can do.
    c. Check to see if your monitor is misterminated. This affects contrast.

19. *When playing a tape into a TV receiver, the sound buzzes when white words of very white parts of a picture are shown:*
    a. The tape could have been recorded that way. Some TV stations OVERMODULATE their video; that is, they mix too strong a video signal into their RF, and some of this signal "leaks into" the sound. The problem is aggravated by misadjusted FINE TUNING. If the TV station caused the problem, your VCR simply recorded the signal warts and all. If you *don't* have the problem with other stations and *do* have the problem with other VCRs or TVs, call the station TV engineer and ask him to check on it.
    b. Your VCR's RF GENERATOR may be OVERMODULATING the video signal like the TV station in example a. Does the problem always happen with this VCR, regardless of what is played or what TV is used? If so, have a technician adjust the RF GENERATOR for lower video level.
    c. Your TV could have misaligned IF (intermediate frequency) circuits. Have a technician fix it if the same TV gives you all the problems.

20. *When playing a VCR using RF into a TV receiver, you see a ghost (a faint second image) or light or dark bars floating through the picture. The picture may look like Figure 3–33 or 3–36 or Figure 5–47:*

FIGURE 5-47   VCR's channel 3 is competing with a local station's channel 3 causing faint diagonal lines. Try switching RF converter to channel 4.

a. The RF GENERATOR in your VCR is using the same frequency (channel) as a nearby TV station, giving you CO-CHANNEL INTERFERENCE. Switch the RF GENERATOR and TV to another channel.

b. You have a tangle of antenna wires, picking up interference. Organize them neatly and place them farther apart.

c. Your antenna cables aren't well grounded or well shielded and are picking up interference.

d. Your TV and VCR are too close together. One is interfering with the other. Review problem 9.

e. If you're using boosters, splitters, switchers, or other gadgets on your antenna wires, they may be poorly shielded, and interference may be "leaking" into or out of them.

21. *Portable VCR mysteriously turns itself on when in its carrying case:*

a. You probably bumped the case hard enough to push one of the "feather-touch" function controls. Some buttons are mighty sensitive.

Programmable Timer Problems

Here you've set your timer to catch Tim Allen improving his home with a 300-horsepower ceiling fan, and when you come back to play your gem,

1. *You play your tape only to find a blank recording.* You forgot to switch the INPUT SELECTOR to the TV or TUNER position.

2. *You play your tape only to find a picture with no sound.* You left a microphone or some other audio source plugged in.

3. *You play your tape only to find sound without a picture.* You left the camera plugged into the VCR.

4. *You program the VCR but frequently miss recordings:*

a. Is your timer's clock set to the right time—exactly? How about its AM/PM setting?

b. Are you being careful about programming in all details—which week, which day, a.m/p.m.? Are you using the *TV Guide* channel number or the channel number your *cable company* happens to be using for this station? Often you have to translate the listed channel into the channel *you* get.

c. Are presidential announcements, public station fundraisers, or extended sporting events throwing off the broadcast schedule?

d. Do you have power outages? Some VCRs have backup power to keep their clocks going during momentary blackouts; others do not. Digital timers usually flash to let you know their timers are "kaflooey." Often a power outage will deprogram (lose from memory) the programming instructions you gave the VCR. Most programmable timers allow you to check what instructions are stored in memory.

5. *Some programmed recordings come out nice but others snowy, grainy, or in black and white when they should be in color:*

a. Do you get good reception of all channels? When viewed normally, a weak station will still be weak when you record it. It will then play back snowy or lack color.

b. If you have an aimable TV antenna (with a ROTOR, perhaps), is it aimed toward the station you'll be recording? Remember that your programmable recorder won't reaim your antenna for each station you choose to record.

6. *Watching the last 20 minutes of "The Right Stuff," the picture flips and suddenly you're watching the wrong stuff:*

a. The TV broadcast was probably delayed 20 minutes and your VCR shut off before it was over.

b. You miscalculated the time when the VCR should switch itself *off*.

6

TV CAMERAS

KINDS OF CAMERAS

The first television cameras were as big as a St. Bernard's doghouse and took two strong men to aim and operate them. The St. Bernard hasn't changed over the years, but today you can fit the camera *and* VCR into a birdhouse.

Manufacturers have designed specific kinds of cameras for various applications. There are basic inexpensive black-and-white cameras used for surveillance. These cameras perched atop telephone poles in parking lots and stationed at warehouse back doors do not include VIEWFINDERS. Imagine, who would climb up there to look through them? Cameras which are aimed and operated manually usually include the VIEWFINDERS so that the camera operators can view what they are shooting.

Some cameras are configured only for use in the studio. Many of their controls are operated remotely in the control room. The cameras are tethered to the control room by a thick umbilical containing many wires. These cameras are often heavy and big, but it doesn't matter since they only have to be wheeled around the studio floor. Studio cameras often sport many knobs and special features and allow the attachment of large lenses, teleprompters, and special pedestals.

Portable cameras, on the other hand, need to be as lightweight as possible. They often have fewer features and fewer external controls. Some portable home video cameras are so automated that they have almost no visible controls at all. You simply point and shoot.

Color cameras range in price from $500 to $80,000. Naturally, the $80,000 camera gives a much sharper, purer, and more stable picture than its bargain basement home video cousin. The quality of the picture depends on many factors in the camera's makeup. Where a home video camera would use a single ½-inch or ⅓-inch wide CCD chip to create a full-color picture, an industrial camera would use three pickup chips, one for each primary color, and the chips would be ⅔-inches wide to produce sharper pictures. The professional camera chips would have more PIXELS, photosensitive transistors, able to ''see'' the picture in more detail.

There are several ways to describe a camera. It could be a 1-CHIP, 2-CHIP, or 3-CHIP color camera. Some cameras (top-of-the-line models and older models) use PICKUP TUBES

MINIGLOSSARY

*Viewfinder   The part of a TV camera you look through to see where the camera is aimed.

132

rather than CHIPS. Some separate their colors with mirrors while others use prisms. Some cameras are DOCKABLE to VCRs; they can stand alone or they can have a VCR connected directly to them turning them into CAMCORDERS. Other CAMCORDERS can only work as a team; the camera sends its signal *only* to the built-in VCR. Most CAMCORDERS have video inputs and outputs allowing the camera to send its signal elsewhere or to have other video signals sent to its VCR. Let's explore some of these differences in more detail.

IMAGE SENSORS· TUBES AND CHIPS

A TV camera "sees" with its PICKUP TUBE or its CHIP. The larger the face of the TUBE or CHIP, the sharper the picture is likely to be (and the larger, more power hungry, and expensive the camera is likely to be).

MONOCHROME cameras make only black-and-white images and need only one CHIP or TUBE to sense the brightness in the picture. Professional color cameras have three CHIPS or TUBES, one dedicated to each of the primary colors red, green, and blue (review Figure 1–5 and also the back cover). Home video and inexpensive industrial cameras use one CHIP or TUBE with fine colored stripes on its face which allows it to sense all three primary colors at once, but at reduced sharpness. Some mid-priced industrial cameras and camcorders use two CHIPS, one for luminance (black-and-white parts of the picture), and the other to sense chrominance (colors). With such cameras, the colors may be a little fuzzy but the more important brightness details are very sharp.

The "perfect" TV camera would: (1) have excellent picture sharpness, (2) be sensitive in low light, (3) resist SMEAR and BURN-INS, (4) reproduce color faithfully, and (5) be affordable.

The RESOLUTION (picture sharpness) of today's CHIP cameras depends largely on the number of PIXELS in the CHIP. Inexpensive cameras may have only 250,000 PIXELS on a ½-inch CHIP where expensive cameras may have 450,000 PIXELS on a ⅔-inch CHIP. These PIXELS sense dots of light, and the more dots you can use to make up your picture, the sharper it will look. Use a magnifying glass some day to compare a newspaper photograph with a magazine photo. You will see why the newspaper photo appears fuzzier; it is made of fewer, larger dots than the magazine

picture. The larger size camera CHIPS have room for more PIXELS able to recreate a sharper picture.

The "perfect" camera would be sensitive in low light, allowing the camera to produce a good picture without the need for many extra lights. This is handy in ENG (Electronic News Gathering) situations where normal room lighting, or perhaps one light, is all that is available. Some cameras do a better job in low light than others.

TUBE cameras have the nasty habit of getting BURN-INS. BURN-INS occur when TUBED cameras are aimed at a bright light or the sun or a contrasty scene for too long. The bright parts of the scene "burn themselves into" the picture and remain there no matter where the camera is aimed. Mild BURN-INS go away in time, but serious ones caused by very bright lights may damage the TUBE forever. Figure 6–1 shows examples of BURN-INS. Cameras which are to be used by novices or outdoors need to be BURN-IN resistant.

CHIP cameras have no trouble with BURN-INS; you can aim them into lights with no fear. It is not something that you should do on purpose, however, because a bright light or the sun focusing through the camera's lens for a long period of time could still singe the CHIP.

Some CHIP cameras have a tendency to SMEAR when viewing bright lights. You'll see a vertical streak through bright objects running from the top to the bottom of your picture. The SMEAR remains until the bright object moves off the screen. Figure 6–2 shows an example of SMEAR.

If you are buying a new camera, which should you get, TUBE or CHIP? Unless you are a broadcast professional, you will want a CHIP camera. Tables 6–1 and 6–2 describe TUBES and CHIPS further.

Professional TV cameras use the more expensive PLUMBICON tubes which give excellent pictures. Among the CHIP cameras are the MOS (Metal Oxide Semiconductor) and its more expensive cousin the CCD (Charge-Coupled Device) used in higher-quality industrial cameras. Both are extremely rugged and tiny compared to camera TUBES.

MIRRORS AND PRISMS

Single-chip color cameras have no mirrors or prisms; they don't need them. The light from the lens is focused directly on the color-striped pickup CHIP which creates the red, green, and blue signals. Because one CHIP is doing three jobs, the picture is not as sharp as in a THREE-CHIP camera.

MINIGLOSSARY

Pixels   Picture elements, tiny dots that make up the picture. In a camera, pixels represent the tiny light-sensitive transistors that store the image.

*Chip   Miniature electronic circuit consisting of thousands of transistors. A TV camera chip senses the image.

One-chip camera   One image pickup chip senses all the colors plus black-and-white aspects of a TV picture.

Three-chip camera   More expensive TV camera which has separate chips to sense each primary color in the picture.

Pickup tube   Vacuum tube light sensor on a TV camera. Plumbicon, saticon, and vidicon are three types.

Dockable   Camera/VCR feature whereby the two can work independently or can be joined into a single unit becoming a camcorder.

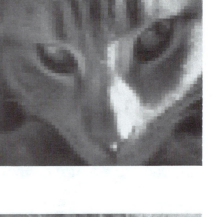

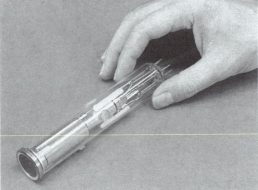

Camera CHIP and TUBE.

Compare the detail in the two pictures. One is made of 3658 PIXELS, the other 14,632.

FIGURE 6-1 BURN-IN (top) from aiming at ''THE END'' sign too long; (bottom) from shiny object.

FIGURE 6-2 SMEAR, the result of aiming a CCD camera at a bright light.

TABLE 6–1

Tubes versus chips

Tube advantages
- Higher resolution (700 lines vs. 240–550)
- Slightly smoother, richer colors

Tube disadvantages
- Complex electronics
- Larger size and weight
- Higher cost ($50,000 vs. $25,000 for pro models)
- Uses more power (drains batteries faster)
- Sensitive to electromagnetic fields (can't use near high-tension lines)
- Sensitive to vibration (bumps misalign it; very loud noises affect the picture)
- Nonlinear (straight lines bend a little)
- Sharp at center but image becomes fuzzier at the edges, especially corners
- Tubes need replacement after approximately 3,000 hours of use
- More technical attention needed to maintain excellent picture
- Subject to BURN-INS from bright light

Chip advantages
- Simpler electronics
- Small and lightweight
- Uses very little power
- Very rugged
- Needs very little maintenance
- Not permanently damaged by bright light
- Perfect linearity
- Image sharp all over, including edges
- Capable of electronic shuttering (1/125, 1/250, 1/500, 1/1,000 sec for "catching" motion while maintaining sharp images)
- Chips seldom "wear out"
- Instant on (no warmup time)

Chip disadvantages
- Smear (vertical STREAKING of bright areas in some types of chips)
- Aliasing (straight diagonal edges look slightly jagged)
- Occasional pixel defects (causing tiny dots or unevenness to smooth colors)
- Grainy signal at high temperatures

TABLE 6–2

Kinds of TUBES and CHIPS

Photosensor	Cost	Features
Plumbicon tube	high	High resolution, excellent in low light, excellent color.
Saticon tube	medium	Very high resolution, good in low light, excellent color.
Vidicon tube	very low	Medium resolution, color, low-light sensitivity. Tends to burn-in easily. Old technology.
IT (interline transfer) CCD (charge-coupled device)	low	Medium resolution, sensitive to smear from bright lights, good color.
FIT (frame interline transfer) CCD (charge-coupled device)	medium–high	Medium resolution, minimal smear, good color, slight lag (blurring of moving objects), preferred by professionals.
FT (frame transfer) CCD	medium	Medium resolution, reduced smear, good in low light, good color. Infrequently used.
MOS (metal oxide semiconductor)	low	Medium resolution, poor in low light, slightly noisy (grainy) pictures. Not commonly used except in consumer cameras.
HAD (hole accumulated diode) IT (interline transfer) CCD	medium–high	Medium resolution, no smear, good in low light.

What the SINGLE-CHIP camera loses in picture quality, it gains in low cost and small size.

THREE-CHIP cameras (seen in Figure 1–5) divide the image into separate colors using mirrors or prisms. Prism cameras are physically larger, cost more, but perform better.

Mirror optics are less expensive to manufacture than prism optics so these cameras are cheaper. Because of distortions in the mirrors, these cameras do not produce as sharp a picture as their prism pals. Because the mirrors "waste" some of the light, these cameras also need more light to operate.

MINIGLOSSARY

*Monochrome   Black and white (as opposed to color).

*Smear   A temporary white vertical streak passing through bright objects in a CCD camera's picture.

Burn-in   In tubed cameras, a spot, streak, or blemish on the TV screen which remains even when the camera is focused on a new scene. Burn-ins are usually caused by aiming at a bright object for too long or momentarily aiming the camera at a very bright object like the sun.

Resolution   Picture sharpness.

*CCD   Charge-coupled device, transistorized light sensor on some tubeless TV cameras.

MOS   Metal oxide semiconductor, transistorized light sensor on some less expensive tubeless TV cameras.

VIEWFINDERS

The VIEWFINDER is what you look through to see where your camera is aimed. Not all TV cameras have VIEWFINDERS. Many security and industrial cameras do not. Most portables have a VIEWFINDER that fits up against your eye. Studio cameras have larger VIEWFINDERS that are easy to see at some distance. Some VIEWFINDERS are detachable, handy for tight spaces.

Electronic Viewfinder

Nearly all TV cameras have ELECTRONIC VIEWFINDERS, small TV monitors which display the picture exactly the way it is being sent to the VCR. Focus, framing, iris, and zoom all manifest themselves in the picture you see.

Color is the one thing you don't see on your ELECTRONIC VIEWFINDER. With rare exceptions, all ELECTRONIC VIEWFINDERS are black and white, even on color TV cameras. That's not a mistake. Black-and-white VIEWFINDERS give a sharper picture than color (of prime importance when focusing). They are smaller, lighter, cheaper, and adequately perform their basic mission—to display what you are viewing. ELECTRONIC VIEWFINDERS may also display messages such as battery power, pause, light level, time, and other functions of the camera and VCR.

The tiniest camcorders have the VIEWFINDER built into the camera, streamlining the package. Industrial camcorders and cameras have separate VIEWFINDERS which are detachable. Advantages of a detachable VIEWFINDER are:

1. Once the VIEWFINDER is removed, the camera and VIEWFINDER are more easily packed and transported.
2. Detachable ELECTRONIC VIEWFINDERS are handy when in a pinch you need to conserve battery power—simply disconnect to save-a-watt when shooting.
3. Many ELECTRONIC VIEWFINDERS hitch up to either side of a portable camera (for left-eyed or right-eyed people) and some can be adjusted in many directions, freeing the camera operators to hold the camera above or below themselves. Some finders can be removed from the camera and attached to an extension cable for remote viewing—handy if your camera needs to be up in a tree or in the middle of a cattle drive.

The heavier, higher-quality camcorders are designd to sit on the shoulder. They have their VIEWFINDERS attached to the front of the camera, improving its balance and steadiness. Tiny camcorders are light enough to hold up to your eye and use the built-in VIEWFINDERS. Figure 6–3 shows camcorders with built-in and detachable VIEWFINDERS.

Streamlined built-in VIEWFINDER. *(Courtesy Quasar)*

Adjustable, detachable VIEWFINDER. *(Courtesy Chinon America, Inc.)*

FIGURE 6-3 Camcorder VIEWFINDERS.

MINIGLOSSARY

*Electronic viewfinder   Tiny TV monitor mounted on a camera showing the image the way the camera sees it. It can also be used to view tapes played back in the field.

Return   Switch on a studio TV camera that displays the final program image (as opposed to the camera's own image) to help the camera operator position objects in the viewfinder. Useful for coordinating with other camera shots.

One great advantage of the ELECTRONIC VIEWFINDER on a portable camera or camcorder is that it can display the image *played back* from your VCR. Thus, after recording a sequence, you can rewind the tape and play it through your VIEWFINDER to see how you are doing.

Portable TV cameras and camcorders use miniature TV monitors with 1½-inch screens. You would view the screen by placing your eye against a rubber cushion and looking through a magnifying lens built into the VIEWFINDER. This will make the picture big and easy to see. Often these VIEWFINDERS have a hinged lens/eyecup mechanism that can be flipped out of the way, allowing you to view the TV monitor directly with both eyes or with a friend. You'd better be close friends, as the picture isn't much bigger than a postage stamp.

ELECTRONIC VIEWFINDER eyepieces sometimes have focusing adjustments of their own. The eyepieces, like binoculars, often have to be adjusted to your eye (otherwise you may think your camera's picture is blurry when really your VIEWFINDER is out of focus). To use this adjustment, aim the camera at a distant object, focus the lens at infinity (∞), and zoom out all the way. That will make a picture that *should* be sharp in the VIEWFINDER. Now adjust the eyepiece to make the VIEWFINDER picture sharp (and comfortable) for your eye. Done.

Studio TV cameras, unrestrained by size and weight, use larger, more easy-to-see VIEWFINDERS. Most sit atop the camera and can be tilted and turned in various directions. They usually have a visor around the screen to block out reflections from studio lights. They have the usual monitor controls for brightness, contrast, horizontal, and vertical, and some have switches to allow the camera operator to view another camera's picture or the picture the director has selected. This feature called RETURN is handy when a camera operator must keep his part of the picture in the left half of the screen while another camera operator is keeping his picture in the right part of the screen, while the director splits the screen in half. Professional TV cameras have VIEWFINDERS which can also display test signals, electronic waveforms, and messages about camera circuit operation.

Remember that the camera VIEWFINDER only *displays* the picture, it doesn't *make* it. Adjusting the VIEWFINDER's brightness, contrast, or any other controls will have no effect on the camera's recorded picture.

Some TV cameras are designed for both studio and portable use. In the studio the camera with its full-sized VIEWFINDER is placed on a tripod, and is connected to a multiwire umbilical cable which remotely controls the cam-

era's electronics. To use the camera in portable situations, the large VIEWFINDER is detached and replaced with a miniature one. The multiwire umbilical is replaced with a battery pack and a cable leading to the VCR. The tripod is replaced with a cushioned shoulder mount. Figure 6–4 shows such a camera.

OPTICAL-SIGHT VIEWFINDER

The OPTICAL-SIGHT VIEWFINDER, an inexpensive little range finder, shows you approximately where your camera is aimed. OPTICAL-SIGHT VIEWFINDERS are found on industrial surveillance cameras which are often aimed once and then left. The OPTICAL-SIGHT VIEWFINDER's greatest advantage is its price, as it costs next to nothing. Its disadvantages are:

portable configuration

studio configuration

FIGURE 6-4 Industrial color camera designed for portable or studio use.

MINIGLOSSARY

Optical-sight viewfinder Inexpensive, simple lens scope, gunsight or crosshairs used to help aim a TV camera.

Parallax How the positions of objects change relative to one another as you move by them. If your camera looks at a subject from one position and you look at the subject from another, the two of you will see slightly different pictures.

1. It is not very accurate.

2. You don't automatically know if you are focused or if the picture is bright enough.

3. You can't tell where you are zoomed.

4. If you forget to uncap the lens before shooting, your OPTICAL-SIGHT VIEWFINDER won't remind you.

5. You can't check your recordings by playing them through the OPTICAL VIEWFINDER.

6. Since the OPTICAL-SIGHT VIEWFINDER is a way off from the camera lens, it is not looking "straight at" subjects that are close to the lens. This sighting difficulty is called a PARALLAX error.

OPTICAL-SIGHT VIEWFINDERS are sometimes found on superminiature camcorders and on underwater camcorder housings. Since you can't put your eye up to the VIEWFINDER under water, you simply look through the gun-sight range finder to roughly aim your camera.

MICROPHONES BUILT IN

All CAMCORDERS and some of the lower-cost industrial cameras have microphones built into them. These are usually ELECTRET CONDENSER microphones, which are very sensitive and give good sound quality, especially for conversation.

They do have one fault. Since your camera is likely to be 6 feet or more away from your subject, that means that your built-in mike is also 6 feet away. This results in echoey speech and a distracting amount of background noise such as doors slamming, dogs barking, traffic, wind, and even the camera operator snorting and sniffing.

You could solve all these problems by recording in a soundproof tomb using a corpse for a camera operator (no breathing sounds). This wouldn't work if your show had to be "live," would it? When really good sound is necessary, it is possible to override the built-in mike and use a separate microphone such as a lavalier which you hang around your talent's neck or attach to his or her lapel. The plug goes into the MIC input of the VCR. Because the subject speaks directly into the mike (a foot away), the sound is clearer. If the talent moves around too much for an attached mike, you could use a SHOTGUN MIKE.

A SHOTGUN MIKE, so named because of its long shot-

FIGURE 6-5 Some microphones can attach to the ACCESSORY MOUNT atop the camera and can be switched from a wide angle of pickup, to SHOTGUN. *(Cam Ear Courtesy Silver Creek Industries, Azden Courtesy Electronic Mailbox, Azden, ECZ-770)*

MINIGLOSSARY

MIC Short for microphone input, a highly sensitive, low-level audio input.

Electret condenser Type of microphone, usually built into portable TV cameras. Sensitive and inexpensive, they have good sound fidelity.

*Shotgun mike Microphone shaped like a gun barrel, which "listens" only in the direction it is aimed.

Accessory mount Threaded hole on top of camera or camcorder for attaching a light, microphone, or other accessory.

gun-like barrel, "listens" in one direction only. Someone standing nearby could aim the SHOTGUN MIKE at the talent as he or she moved around.

Some cameras have SHOTGUN MIKES built into them. This is handy because the microphone automatically aims wherever the camera aims.

What if you don't want your SHOTGUN mike to "listen" in only one direction? Use a *switchable* SHOTGUN mike that works either as a SHOTGUN, or as a normal mike. Some camcorders come with switchable SHOTGUNS, but separate switchable SHOTGUN mikes can be attached to the ACCESSORY MOUNT atop many cameras (Figure 6-5).

The more professional cameras allow different microphones with different characteristics to be plugged into their VCRs as needed.

OPERATING CAMERAS

There are so many different kinds of TV cameras that generalizing about how to operate them is difficult. Let's start with the simplest camera and work our way up to the more complex types.

Monochrome Surveillance Cameras

If you are using your camera to keep an eye on the warehouse at night or to observe traffic flow in the parking lot, you do not need an elaborate, expensive color camera with a VIEW-FINDER, built-in mike, and other bells and whistles. A plain vanilla color camera is likely to cost $800. A simple black-and-white industrial camera costs as little as $300 and gives a picture *twice* as sharp as its color counterpart. *Twice* as sharp! *And* it will work in much dimmer light, *and* it's generally smaller, lighter, and more rugged than a color camera.

The disadvantages of black-and-white industrial cameras are

1. No color (ugh!).

2. They're small but because they use house current they are not as portable (imagine running a 7-mile extension cord down to the street corner).

3. Today's research dollars are going into color cameras. Black-and-white technology is stagnating.

The simplest camera consists of a lens, a box of electronics with built-in automatic controls to give you a good picture, an electric cord for power, and a socket called VIDEO OUT. You need a cable to connect the VIDEO OUT to the VIDEO IN of the TV monitor or a VCR (Figure 6–6).

There is not much to using this camera other than connecting it, plugging it in, turning it on, letting it warm up, uncovering the lens, aiming the camera, and then focusing the lens. How can you tell if you are focused? One way is to look at a TV monitor somewhere. You could have one connected either directly to the camera or to a VCR that is connected to the camera. When the VCR is in the record mode, its monitor will display your camera's picture. If the monitor is connected directly to the camera, it should display a picture whenever the camera is on and the lens is uncovered.

Sometimes, to simplify running wires to a surveillance camera, manufacturers will use a single cable which sends power to the camera and receives video back from it. Figure 6–7 shows a possible connection.

A surveillance camera may come with a FIXED FOCUS lens, which does not have to be adjusted, or it may come with a simple STANDARD LENS, which is focused by rotating a collar on the lens. The lens may have a mark on it and numbers etched along the collar to tell you how close the lens is focused. Since most surveillance cameras are used to take pictures of things 10 feet away or more, it is best to focus the lens for 10 feet or infinity (∞).

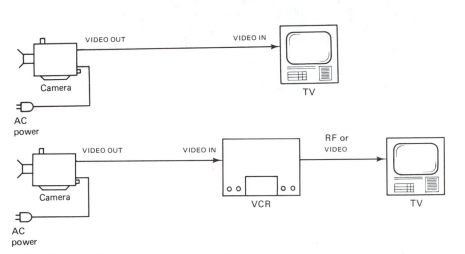

FIGURE 6-6   Connecting up a surveillance camera.

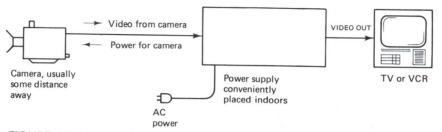

FIGURE 6-7 Connecting a surveillance TV camera which uses one cable.

Simple Industrial Cameras

One step above surveillance cameras are the basic black-and-white cameras used in classrooms and industry. They generally make a better-quality picture and have more features. Figure 6–8 shows one without the viewfinder, and Figure 6–9 shows the same camera with the optional viewfinder attached.

The Lens.   The lens typically found on even a simple TV camera would be a ZOOM LENS (Figure 6–10). Although Chapter 7 will tell more about lenses, here are a few basics to get you started. There are three moving parts that can be adjusted on a ZOOM lens. When using a simple camera, you reach around to the front of the camera and make your adjustments by rotating parts of the lens. Always check what you are doing by watching the VIEWFINDER or a TV monitor. The three adjustments are

1. FOCUS. Turning this part of the lens makes the picture sharp or blurry.
2. ZOOM. Turning this part of the lens makes the picture look closer or farther away.
3. IRIS. Turning this part of the lens in one direction allows lots of light to pass through the lens and increases the contrast in your picture. Turning it in the other direction restricts the amount of light allowed through the lens and decreases the contrast, making the picture look grayer. In general, you adjust the IRIS so that the picture looks good.

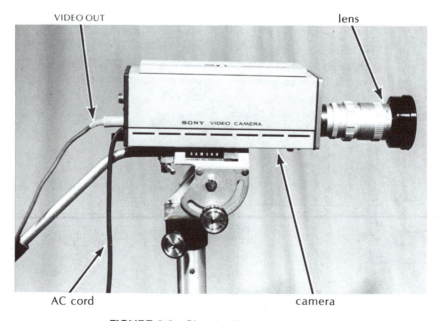

FIGURE 6-8 Simple classroom camera.

MINIGLOSSARY

Fixed focus   Lens which cannot change focus from near to far.
Standard lens   Inexpensive, nonzooming lens which gives a "normal" (not close-up, not wide angle) field of view.

*Zoom lens**   A lens which can "zoom in" or "zoom out" to give a closer-looking picture or a wider angle of view.

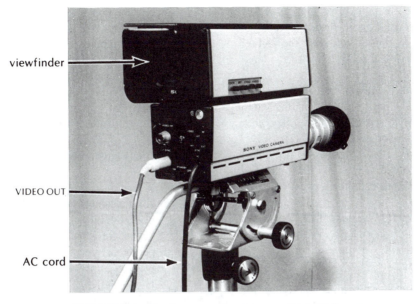

FIGURE 6-9 Basic monochrome camera with VIEWFINDER.

FIGURE 6-10 ZOOM lens.

You will be using focus and zoom all the time. Usually you set the IRIS at the beginning of the show and don't touch it thereafter.

Correct Focusing. There is only one way to correctly focus a zoom lens. The method takes about 5 seconds and should be done before the show actually starts. During the show, you may not have time to use this proper method and will have to focus as best you can. Proper method of focusing:

1. Zoom in all the way on your central subject, making it look as close as you can.
2. Focus the lens.
3. Zoom back out to the kind of shot you want.

If you use this method, you can zoom in and out, and your subject will stay in focus. If you don't use this method, your picture will go out of focus as you zoom in or out, and you will have to keep refocusing repeatedly. In any case, if the distance between your subject and your camera changes—the subject moves or you move the camera or you pick another subject at a different distance from the camera—you will have to refocus if you want the picture to stay sharp throughout the entire zoom range of the lens.

Viewfinder Controls. VIEWFINDERS are just tiny TV monitors. They sample a little of the camera's video signal before sending it out to its destination. Like other monitors, they have controls for BRIGHTNESS, CONTRAST, HORIZONTAL, and VERTICAL. These controls do not affect the picture that is being recorded. They only make the picture comfortable for the camera operator's eyes.

Camera Controls and Connectors. You may discover extra knobs, buttons, and sockets on your TV camera. These add to the flexibility of the camera's use by allowing it to do special tasks.

Sync: The SYNC switch has positions marked IN (for INternal) and EXT (for EXTernal). In the IN position, the camera generates its own sync (INside itself), mixes it with the video, and sends both to the VIDEO OUT. You would connect the VIDEO OUT via coax cable to a TV monitor or VCR. Use INternal SYNC when using just one camera alone, perhaps feeding a VCR or TV monitor.

When the SYNC switch is set at EXT, the camera does *not* generate its own sync signal—something else, usually an external SYNC GENERATOR, must make the signal for it. Use EXTERNAL SYNC when you have *several* cameras, a switcher, and an external SYNC GENERATOR.

Since it is inconvenient to run a separate set of wires from a SYNC GENERATOR, most cameras are equipped with a multipin socket labeled EXT SYNC for the cable that carries the sync *to* the camera and the video *from* the camera. This round multipin connector is called a DIN connector and is standard through Europe and is fairly popular in the United States. Figure 6–11 shows a male and female DIN connector. This cable is used when several cameras are connected to a switcher.

Why does anyone bother with EXTERNAL SYNC?:

1. When making their own SYNC internally, most simple cameras do a mediocre job of it. The sync gives the picture a tiny jitter. If your shop has an expensive SYNC GENERATOR, sending this superstable sync signal to the camera will make its pictures superstable too.

2. If you're using several cameras at the same time and are switching from one to another, all the cameras must have their electronics synchronized (the electron guns must be zipping and zapping in unison). EXTERNAL SYNC does this. How it all happens will be explained in Chapter 11.

Genlock: GENLOCK is very similar to EXTERNAL SYNC. It allows your camera to synchronize its electronics with another camera so the pictures are mixable and cleanly switchable.

With EXTERNAL SYNC, the camera requires an EX-

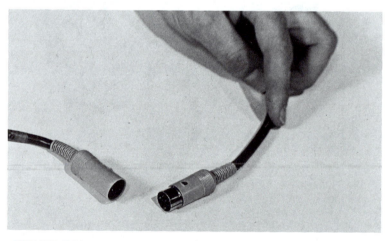

FIGURE 6-11   DIN connectors.

MINIGLOSSARY

*Din connector   Round multipin plug or socket, about ¾ inch in diameter.

*External sync   Electronic pulses, coming from outside the TV camera, which tell it when to zig and zag its electron gun. Syn-chronizes the camera's picture with other cameras in the studio so the pictures can be mixed or switched.

Sync generator   Electronic device that makes sync signal used to synchronize the electronics of several cameras so their pictures can be mixed together.

TERNAL SYNC signal to drive its electronics. With GENLOCK, the camera requires any *video* signal to drive its electronics. The GENLOCK circuit "locks onto" the video signal and synchronizes its electronics with it.

Studio cameras usually use EXTERNAL SYNC inputs, because SYNC is readily available around the studio. Portable cameras often have no SYNC or GENLOCK inputs because they usually work alone. Portable cameras designed to *sometimes* work with other cameras have GENLOCK inputs, making them easily synchronizable with other cameras in the field (where video may be handy but SYNC is not). Also, some home TV cameras, designed to work with other cameras or with computers, have GENLOCK inputs.

To GENLOCK two cameras, you would make camera 1 the "master." It would run on its own (INTERNAL sync). As shown in Figure 6–12, you would send camera 1's signal to a switcher and also send some of camera 1's signal to camera 2, probably via a 75Ω coax wire. Switch camera 2 to the "GENLOCK" mode so it "listens" to the other camera's video. Send camera 2's signal to the switcher and *voilà*, you can now switch or dissolve neatly between the two pictures. Note that you may have to make some minor adjustments (probably called HORIZONTAL PHASE) on camera 2 or at the switcher to *exactly* match the two, but this little knobtwiddle will be described later.

Gain: GAIN adjusts the strength of the video signal from the camera. If your picture seems a little too dark or faded, you should first throw more light on the subject. Second, you should open the lens IRIS. Third, if all else failed, you would turn up the video GAIN from the camera to boost brightness and contrast.

Intercom: The INTERCOM is a headset that the camera operator wears in a studio, making it possible to listen to and speak with the director in the control room. The INTERCOM generally plugs into a socket in the TV camera, and the signals travel through the camera's multiwire umbilical cable. If the camera is working alone (without the umbilical), the INTERCOM circuit will not work.

Tally light: The TALLY LIGHT is a red light in the front of the camera that goes on when the TV director selects your camera's picture to be recorded. Your TALLY LIGHT will go out and another camera's TALLY LIGHT will go on

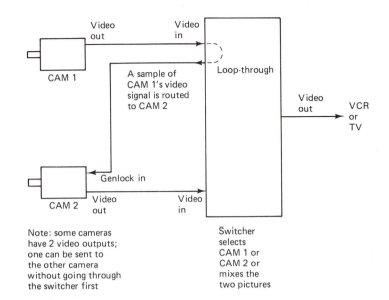

FIGURE 6-12   GENLOCKING two cameras.

when the director switches from your camera to the other camera. If the director is creating a special effect, like a split screen which uses two cameras at once, both their TALLY LIGHTS will be on at the same time.

When a studio camera is used alone (away from the studio and switcher), the TALLY LIGHT does not operate.

A second TALLY LIGHT is usually built into the studio camera's VIEWFINDER to let the camera operator know his or her camera is "on."

A message to fledgling camera operators: Don't goof off when your TALLY LIGHT is "on."

Camera adapter: Portable TV cameras do not *have* to be used portably. They can be placed on a tripod and used in the studio just like any other camera with the help of a CAMERA ADAPTER. The CAMERA ADAPTER is a box which gets its power from a wall outlet and makes all the signals the camera needs to operate. The CAMERA ADAPTER box has a standard VIDEO and AUDIO output and generally has a cable which sends a REMOTE PAUSE signal to a console VCR. Figure 6–13 shows how a portable camera can be connected to a VCR.

MINIGLOSSARY

*Gain   Camera adjustment which controls the strength of the camera's video signal, altering contrast and brightness of the picture.

Genlock   Ability of a camera or other TV device to receive an external video signal and synchronize its own video signal to it, so the two videos can be neatly switched or mixed.

*Intercom   An earphone/microphone headset that allows the director in the control room to speak to the camera operators in the TV studio.

*Camera adapter   Box of electronics that a portable camera can plug into (instead of directly into a VCR) that powers the camera and distributes the camera's video and other signals via standardized outputs.

*Tally light   Lamp on the TV camera which goes on when the camera's image has been selected by the TV director. It tells both the performer and the camera operator that the camera is "on."

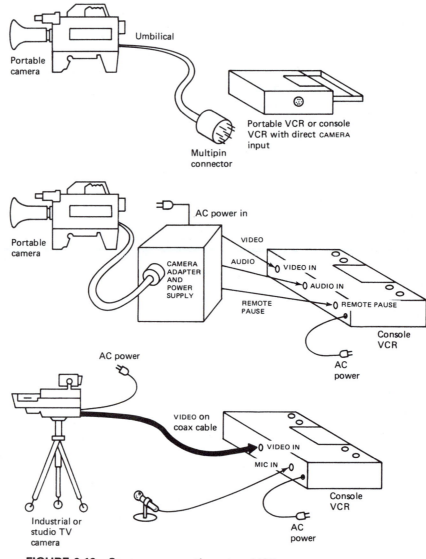

FIGURE 6-13   Camera connections to a VCR.

Controls on a Portable Color Camera

Although many portable cameras and camcorders come festooned with automatic controls, the older or professional models do not. Some have automatic controls with manual overrides allowing you to second-guess the built-in computers. The two most important and universal controls on a color camera are COLOR TEMPERATURE and WHITE BALANCE. These adjust how the camera "sees" colors.

Color Temperature.   COLOR TEMPERATURE describes the warmth (redness) or chill (blueness) of a scene. For example, have you ever noticed how cold and sterile

offices lit with fluorescent lights look? Or have you looked into a darkened room illuminated only by the light from a black-and-white TV set and noticed how stark and bluish everything seemed? On the other hand, have you noticed the warmth in a home lit by incandescent lamps or the warmth of a supper lit by candlelight, or the richness of the whole outdoors during an August sunset? These differences are caused by the COLOR TEMPERATURE of the light.

Under different lighting conditions, the color of things changes drastically even though you may not be aware of it with your naked eye. The eye of the camera, however, sees these differences and makes them even more pronounced. A face that looked red and rosy when lit by a

MINIGLOSSARY

*Color temperature   The redness or blueness of a scene, the result of the kind of light used to illuminate the scene. Also the

name given to the color TV camera control which adapts it to these varied lighting conditions.

COLOR TEMPERATURE wheel on a JVC camera.

sunset will look deathly pale when photographed on a foggy day. Somehow the color camera must be adjusted to compensate for these differences in lighting so that colors will look familiar and proper. This is called COLOR BALANCE.

Some color cameras have a built-in set of colored glass lenses that counteract the ''coldness'' of the light and bring it into proper balance. The COLOR TEMPERATURE control on most cameras is a four-position thumbwheel. Next to the wheel is usually a chart listing which positions to set the wheel for various lighting conditions. Sometimes these conditions are described by icons (little pictures), and sometimes they are described in more technical terms. The precise measurement of COLOR TEMPERATURE will be discussed in detail in Chapter 9, but for now just follow this chart of where to turn the wheel for various lighting situations.

Position	Color Temperature (°K)	Lighting Situation
1	3200	For shooting scenes under studio lamps or outside during a sunrise or sunset—all ''warm'' light conditions
2	4500	For fluorescent lamp lighting
3	6000	For bright or hazy sunshine
4	8000	For shooting outdoors in cloudy or rainy weather or with a clear blue sky without direct sunshine

Not all cameras will use the same filters or numbers for their COLOR TEMPERATURE controls. Some may simply have a two-position switch marked INDOORS/OUTDOORS. Other cameras may not have a wheel with built-in lenses but will have separate colored filters which screw onto your lens. Whatever the situation, it is important to adjust your camera for the correct COLOR TEMPERATURE *first*, before making any other camera adjustments.

White Balance. Every time you use your camera or change lighting conditions (like moving from indoors to outdoors or even from scene to scene sometimes), you have to ''teach'' your camera what color *white* is.

You remember how a color picture is the composite of three pictures, one green-and-black, one red-and-black, and one blue-and-black. A certain mix of these three primary colors is needed to make pure white. If the mix is off, you get tinted white. Sometimes things that are supposed to be white (a white piece of paper on a desk) turn out not to be white at all (the desk is in an orange room, casting orange light on the paper). Still you'd like the paper to look white on camera, so by adjusting WHITE BALANCE, you adjust that mix of primary colors to *make* it white. WHITE BALANCE, or WHITE LEVEL SET, or WHITE SET is often adjusted as follows:

1. Always adjust your COLOR TEMPERATURE filter *first*. Next:
2. In the area you plan to shoot (and in its light), place a white card in front of the color camera close enough to fill the camera's VIEWFINDER screen. If you don't have a white card, then aim the camera close-up at a white T-shirt or some other white surface. Some cameras come equipped with a milky white camera lens cap which you

WHITE BALANCE controls.

MINIGLOSSARY

*White balance The mix of primary colors which results in pure white light. On color cameras, the controls which strengthen the blue or red colors so that none overpowers the other, allowing white objects to appear pure white, not tinted.

can place over the lens and use that as your white surface.

3. Adjust your IRIS to its proper setting.

4. Find the WHITE BALANCE meter either on the camera or in the VIEWFINDER.

5. With the card still filling the screen, adjust the WHITE BALANCE control (or controls—there may be two, one perhaps labeled RED and the other BLUE) to move the WHITE BALANCE meter needle to its proper position. Some manufacturers want the meter needle centered; others, as low as possible; and others, high—check your instructions.

Sometimes the COLOR TEMPERATURE and WHITE BALANCE controls are coupled into one control called by either name or something else like COLOR BALANCE, COLOR TONE, or TINT.

If you used your camera instructions to wrap your last trout catch and don't know which way to set your needle, then try this:

1. Hook up a color TV to your camera/VCR system.

2. Adjust your TV's color so it looks nice while viewing a strong TV station with lots of flesh tones.

3. Switch the VCR to feed the camera signal to the TV and now observe the color of the white card on the TV. If it's not white, adjust your WHITE BALANCE controls until it *is* white.

Automatic White Balance.   AUTO WHITE BALANCE adjusts the controls for you at the push of a button. Still you have to hold the white card in front of the camera while pushing the button.

Some cameras don't even require a WHITE BALANCE button to be pressed. They adjust themselves totally automatically but don't do as good a job of it as you can do with a white card.

Automatic Gain Control (AGC).   Just as you can adjust the brightness and contrast of your TV's picture, a camera can adjust the brightness and contrast of its picture. This feature, sometimes abbreviated AGC is common to all

color portable cameras and many other kinds too. In bright scenes, it will darken the image (lower the GAIN); in dim scenes, it will add contrast to the image (increase GAIN). There is a limited range over which the camera's AGC can make adjustments for scene brightness. Beyond these limits, one must "open" or "close" the lens IRIS to increase or reduce the light allowed into the camera.

Automatic Iris.   The lens IRIS controls how much light the camera "sees." On some cameras you manually adjust a collar on the lens to adapt to bright or dim lighting conditions. Cameras with AUTOMATIC IRIS will perform this adjustment for you, sensing the amount of light admitted through the lens and opening or closing the lens accordingly.

There are times when AUTO IRIS can get fooled. For such cases, most cameras will allow you to manually override this control. A typical situation is when someone is standing against a light background or in front of a window washed with daylight. The AUTO IRIS will adjust the lens to give an excellent rendition of the background, while the subject comes out dark and murky. By manually "opening" the lens further, you can get the subject (the important part of the picture) to look good while overexposing and sacrificing the unimportant background. Another name for this manual override is BACKLIGHT.

Akin to the BACKLIGHT control is the DYNAMIC CONTRAST CONTROL found on more expensive professional cam-

6dB, 12dB BOOST.

MINIGLOSSARY

Automatic white balance   By pressing one button and holding a white card in front of the camera, this will automatically adjust the camera's circuits to make pure white.

AGC or automatic gain control   Automatically adjusts the brightness and contrast of a camera's picture.

Automatic iris   Camera circuit which senses the amount of light in a scene and opens or closes the lens iris to adapt to it.

Backlight   Light coming from behind a subject. Also a control

on a TV camera which improves a backlit picture (keeps it from looking like a silhouette).

*Boost   Camera control which makes it extrasensitive in dim light.

Dynamic contrast control   Camera circuit extending its contrast ratio beyond the normal 30:1, allowing very bright and very dark areas to exist in the same picture.

eras. This feature allows you to aim your camera at something dark with a bright background and have *both* look good. The camera adjusts part of its picture for the bright areas, rendering the highlights evenly, but less bright, and adjusts for the dark parts of the picture, brightening it up some so the detail doesn't become murky.

+6dB, +12dB Boost. Cameras are designed to work with a certain amount of light. When the light gets too dim, the image from the camera gets dim and murky. The colors will look especially dingy. The BOOST control will make the camera more sensitive in dim light. This is handy when you get into situations where you have no choice but to use the existing light or one small portable light to shoot with.

You never get something for nothing. When you boost the sensitivity of the camera, you also boost the NOISE or graininess in the picture. The BOOST control should always be left *off* unless it is absolutely needed.

Usually, the switch is labeled with a number (like +6dB) to tell you how much BOOST you are getting. The bigger the number, the bigger the BOOST.

Auto Fade. AUTO FADE is a control found on some camcorders allowing you to fade the picture out by pressing a button. This is often a neat way to end a scene, fading to black, or white. AUTO FADE can also be used the other way to open a scene fading up from black to the full picture.

Character Generator/Timer. One way to put titles on the TV screen is to painstakingly letter them onto a piece of paper and aim the camera at it. A quicker way to put words on the TV screen is to use a CHARACTER GENERATOR, built into some of the more expensive home video cameras and camcorders. By pressing buttons on the camera, you can electronically select letters to appear on the TV screen. On most models, these letters will even appear over your picture. Some cameras even have a built-in clock which will record the date and time along with your picture. This would be handy, for instance, if you were documenting when a factory was polluting the air. Your video tape would show the smoke coming out the stacks while the time and date would be shown on the screen.

Using these CHARACTER GENERATORS is fairly easy. You would switch on the CHARACTER GENERATOR circuit, then select the position for the first letter, and then select the letter. Repeat the process for all the letters of your title. Then you store the title in "memory" and call the entire

title out at the push of a button when you want it to appear on your tape. Some generators are easier to use than others. They have a separate button for each letter. Others have only two buttons, FORWARD and BACKWARD. If the first letter you wanted was a "D," you would press the FORWARD button 4 times. You would then advance the CURSOR (positioner) to the next position and then press the FORWARD button 11 more times to advance to "O." You'd then move the cursor ahead one more step and press the REVERSE button 9 times to get back to "G." Voilà! It may sound hard, but remember that magic markers and paper weren't easy either.

Such CHARACTER-GENERATED text looks a bit chunky. An example appears in Figure 12–23. The unadorned block letters may be satisfactory for amateur and some school uses but wouldn't look right in a professional production.

Electric Zoom. It is difficult to reach around in front of a portable TV camera and grasp the lens to zoom it. To simplify matters, manufacturers have built ELECTRIC ZOOM controls into their cameras, usually in the form of a ROCKER SWITCH. Press one end of the ROCKER SWITCH, and a miniature motor will zoom the lens in for a close-up. Press the other end of the switch, and the lens zooms out. On the better cameras, a gentle press of the ROCKER SWITCH zooms the lens slowly, while a firm press of the button zooms you quickly.

Automatic Focus. Some cameras take all the guesswork out of focusing by doing it for you. These cameras are handy when novices or children and some grand-

ROCKER SWITCH to control ELECTRIC ZOOM.

MINIGLOSSARY

Auto fade Control on some cameras which fades the picture to black at the end of a scene or fades up from black at the beginning.

*Electric zoom Electric motor on a lens or camera which zooms the lens at the touch of a button.

Rocker switch Lever which rocks back and forth. Pressed one

way, it could make an electric zoom lens zoom in. The other way would zoom the lens out.

*Automatic focus Electronic system in some cameras that senses whether the picture is sharp and electrically focuses the lens to correct blurry pictures.

mothers will be using the camera. The feature is also handy when you will be too busy to focus your shots (like getting ready to jump out of an airplane for a skydive).

AUTO FOCUS cameras are not 100% accurate. They cannot tell whether it was the flowers in the foreground that you wanted sharp or the bride and groom in the background that you wanted sharp. Some will focus on whatever is in the center of the picture. Others will focus on whatever makes up most of the picture. AUTO FOCUS cameras use one of three different methods to measure the distance to your subject—INFRARED, ULTRASONIC, and ELECTRONIC. Although each system can sometimes be fooled, all systems have a manual override allowing you to switch off the AUTO FOCUS control and focus manually. Sometimes it's handy just to press the AUTO FOCUS button for a moment to focus the picture and then let the camera stay at that setting until you are ready to change the focus again.

Infrared: The INFRARED AUTO FOCUS camera emits a brief burst of invisible INFRARED light from one of two windows near the lens. The light bounces off the closest object in the center of the picture and is detected by a sensor behind the second widow. The angle between the outgoing and incoming beam is measured electronically sending a signal to an electric motor to focus the lens.

This method of auto focusing works only with surfaces that reflect light. A black curtain, because it absorbs light, screws up the system. A second impediment to INFRARED auto focusing is that it can't focus on small objects. Something 4 inches wide and 10 feet away is too small for the mechanism to "see." The lens housing, because it holds the INFRARED apparatus, is large compared to other auto-focus methods.

The INFRARED system focuses quickly and accurately (except when you zoom in to extreme telephoto) and because the INFRARED beam is its own light source, the system works well in dim light.

Ultrasonic: These AUTO FOCUS camers use inaudible sound waves for focusing. They emit a short burst of high-frequency sound from a horn transmitter/receiver near the lens. The sound bounces off the subject and is "heard" by the receiver on the camera. A circuit in the camera

measures the time it takes for the sound to make the round trip and calculates the distance.

ULTRASONIC AUTO FOCUS cameras have their problems too. Anything that stops the sound waves from echoing back confuses the camera. If you aimed the camera through a window, for instance, the sound would bounce off the window and come back to the camera and make the camera focus on the window glass rather than the object behind the window. Similarly, an open smooth beach or a large flat floor can also fool the camera because the sound reflects away from the camera and never bounces back to get measured. Some atmospheric conditions such as strong winds, fog, snow, hail, and heavy rain also confound the camera. Even shrill metallic sounds will interfere with the ULTRASONIC sound and disrupt the focusing. ULTRASONIC AUTO FOCUS, since it doesn't rely on light, works well in dim light.

Electronic: ELECTRONIC AUTO FOCUS systems electronically "look" at the picture to determine whether it is fuzzy or sharp. The contrast between objects is greater in a sharp picture than in a fuzzy one, so the camera adjusts the lens to maximize that contrast.

This system can be fooled too. Smooth, foggy pictures with very little contrast give the machine very little to "lock onto." It also works poorly in low light.

PIEZO TTL (through-the-lens) AUTOFOCUS is one type of ELECTRONIC AUTOFOCUS. It looks for contrast in the camera's picture and then focuses the lens to maximize that contrast. One drawback to PIEZO is RACKING, the constant refocusing, or "hunting" by the lens that causes the picture to go in and out of focus until it eventually gets sharper. The action is much like a pendulum that swings back and forth several times before coming to rest. The RACKING process is slow and sometimes disconcerting to the viewers.

Another drawback to PIEZO is that it needs to "see" well to work, making it a poor performer in low-light situations.

FUZZY LOGIC is another ELECTRONIC AUTOFOCUS technology. Where normally a lens motor would spin the lens a turn this way and a turn that way in search for "perfect" focus, FUZZY LOGIC rotates the lens in tiny, imperceptible

---

MINIGLOSSARY

Infrared   Invisible light. When used in autofocus cameras, it reflects light off a subject to sense the distance to the subject in order to focus the lens.

Ultrasonic   Inaudible sound. When used in autofocus cameras, it bounces sound waves off the subject to calculate the focusing distance.

Electronic auto focus   Circuit that "looks" at a camera's picture to determine if it is sharp and focuses the lens appropriately.

*Resolution   Picture sharpness, usually measured in "lines." The greater the number of lines, the sharper the picture.

dB or decibel   A measure of the strength of one electronic signal compared to another. The higher the dB number, the greater the signal strength.

Piezo autofocus   Electronic autofocus method that focuses by maximizing contrast in the picture.

Fuzzy logic   Autofocus technology that increases focusing accuracy by rotating the lens by tiny amounts, not noticeable to the eye.

steps. These minute adjustments hold the image in focus without "overshooting" the mark and without making changes large enough to be noticed by the eye.

AFT (AUTO TRACKING FOCUS) is another technique for autofocusing pictures. This system uses PIEZO to speedily bring the picture into rough focus, then uses FUZZY LOGIC to perfect the picture. Unlike FUZZY LOGIC alone that sometimes gets "fooled" into focusing on a contrasty foreground or background, AFT homes in on the subject (even if it isn't in the center of your picture) and focuses on it.

No AUTOFOCUS system is as accurate as the human brain. Only *you* would know that Johnny Jr. is the main subject that belongs in focus, not Barbecue Bill in the next yard or Hog-the-show Hester waving in the foreground. Switch the lens to MANUAL to override the automatic controls when shooting groups of people or scenery. Both shots require that *you* decide whether the people or the scenery is most important. MANUAL is also a good way to guarantee sharp shots of recitals, speeches, and other static images; once you get the shot focused right, you don't need to change it and don't want some computer circuit diddling with it.

Finally, with lens attachments (described in the next chapter) and auxilliary lenses, MANUAL focusing may be the only option as AUTOFOCUS lens motors are not built to handle the extra size and weight of these add-ons.

Studio Color Cameras

Studio color TV cameras start at about $3000 and go out of sight at $80,000. Many good ones can be purchased for $10,000.

What is it that the "good" cameras do that the cheaper ones don't?:

1. They make a sharper picture. The sharpness may achieve 400–700 LINES OF RESOLUTION as opposed to 240–440 for home-type SINGLE-CHIP cameras.

2. They give a smoother picture with less graininess. Graininess is technically measured as so many dB SIGNAL-TO-NOISE RATIO. The bigger the dB number, the better. 62 dB is appropriate for

Studio camera (Courtesy Sony Broadcast Products Co.).

professional cameras, while 46 dB is common for home cameras.

3. The professional cameras use BROADCAST GRADE CHIPS, circuit chips with few, if any, flaws such as dark pixels. Cheaper cameras display more blemishes.

4. Finer color adjustment is possible. SINGLE-CHIP cameras make color only one way. If two similar SINGLE-CHIP cameras are looking at the same subject and one is a little off, then you will see a difference in the color as you switch from one to

MINIGLOSSARY

*Signal-to-noise ratio (S/N ratio)   A number describing how much desired signal there is compared to undesirable background noise. The higher the S/N ratio, the "cleaner" the signal.

*Genlock   A circuit which "listens" to another video signal and manufactures its own video signal in step with it.

*Color bars   Vertical bars of color used to test cameras and other video equipment.

*CCU or camera control unit   Box of electronic circuits which can remotely adjust the operation of a camera as well as provide power and signals to it.

*Pedestal   Electronic control on a camera which adjusts the

brightness of the picture. Proper adjustment yields blacks which are the right darkness.

Black balance   Color camera adjustment which makes blacks pure black (not tinted one color or another).

Video reverse   Camera switch which makes blacks white and whites black and changes colors into their complementary colors (opposites).

Sweep reverse   Button on a camera which switches the camera's picture left to right or flips it upside down.

*Switcher   Push-button device which selects one or another camera's picture to be viewed or recorded.

the other. There's not much you can do to correct the color in these cameras. THREE-CHIP color cameras, however, allow a wider range of correction so that cameras can be more perfectly matched together.

5. They are more easily synchronized. Studio cameras are designed to work together. Their SYNC circuits must all be driven by a single source so that all paint their pictures in unison. Some cameras designed for both studio and portable use can be GENLOCKED. GENLOCK is a circuit in the camera which will "listen" to the video from another camera, VTR, or some other source and will lock its SYNC circuits to it. This makes it possible for the camera to work in harmony with other cameras.

6. They include test signal generators. The better cameras will create COLOR BARS and other test signals which assist technicians in adjusting their circuits and adjusting their colors so they match other cameras in the studio.

7. They come with remote CAMERA CONTROL UNITS. A CAMERA CONTROL UNIT (CCU) is a box of electronics which can adjust the camera's operation from the control room. This allows engineers to adjust the many controls on the cameras from one place without running into the studio. This also allows the camera controls to be adjusted during a show.

8. They have numerous specialized features such as
   a. PEDESTAL: a control which adjusts how black the blacks are.
   b. BLACK BALANCE: a control which makes sure that blacks are not tinted (similar to what WHITE BALANCE does).
   c. VIDEO REVERSE: makes the camera's pictures negative (blacks will be white, etc.), a handy feature for turning photographic negatives straight into positive video pictures.
   d. SWEEP REVERSE: makes the picture swap its left and right sides, giving a mirror image. This is handy if the camera is viewing a movie shown on a REAR SCREEN PROJECTOR which makes the pictures reversed. This unreverses them.
   e. VERTICAL SWEEP REVERSE: flips the picture upside down. This is handy when you aim the camera in a mirror to look straight down on something. Everything ends up upside down until you throw this switch, which makes it rightside up again.

Connecting Studio Cameras. A portable TV camera would connect directly to a VCR. Studio cameras,

on the other hand, are designed to work together and connect to a SWITCHER which can select which camera's picture is used and may even create some special effect between those pictures such as a FADE or WIPE (more about these in Chapter 11).

Portable TV cameras have all of their electronics built into them. Studio cameras have much of their electronics built into their CAMERA CONTROL UNITS housed in the control room. Studio/portable cameras may have all of their electronics in them but will disregard the built-in circuits and listen only to the CCU circuits when used in the studio.

The camera and the CCU are connected by the multiwire umbilical cable which sends signals back and forth between the two. The CCU also gets a SYNC signal from the studio's master SYNC GENERATOR. Since all the cameras' CCUs receive the same SYNC signal, they all paint their pictures in unison. Each CCU sends its video to the switcher. Figure 6–14 diagrams the process.

For simplicity, I have diagrammed the sync signal as just one wire. In some systems, this is adequate. In advanced color TV systems, sync is carried on more than one wire. The specialized types of sync may be called VERTICAL DRIVE, HORIZONTAL DRIVE, BLANKING, BURST FLAG, and SUBCARRIER. Although it sounds complicated, it just means that several wires come out of the SYNC GENERATOR and connect to the first CCU and loop through it to the next CCU and to the next. Just what these signals do and how they interact are beyond the scope of this text (thank goodness!). Suffice it to say that all these wires should be the same length so that these synchronizing signals all arrive at their destinations together. If one wire were longer than the others, its signal would arrive late because the signal had more wire to go through. These delays cause color shifting when you switch from one camera's picture to another's.

Controls on a Studio Camera. The more expensive professional studio cameras have more buttons than an admiral's dress uniform. The newer cameras have more bells and whistles than you could ever hope to ring or blow. Here are some of the more common features found on studio cameras.

Sweep reverse: HORIZONTAL SWEEP REVERSE will swap the left and right sides of the picture. Writing will appear backwards. I recall using HORIZONTAL SWEEP REVERSE once with an exercise instructor who could never remember that the students' left was not the same as her left. By switching the TV camera to HORIZONTAL SWEEP REVERSE, I was able to make it possible for her to lift her left arm, say "Lift your left arm," and have the students lift their left arms in synchrony with her.

VERTICAL SWEEP REVERSE creates a picture which is upside down. Outside of special effects, this feature is handy for correcting shots made through a mirror. For instance, if you wanted to shoot a bird's-eye view of someone cooking, you could place a large mirror over the stove about a foot above the chef's head. Aiming the camera into the mirror would allow the viewers to see straight into the pots and pans. The image would seem upside down, however. By switching the camera to VERTICAL SWEEP REVERSE, the picture can be made upright, as if the camera were looking over the chef's shoulder.

Negative/positive: NEGATIVE/POSITIVE or POLARITY REVERSE will make blacks white and whites black. It can also change the camera's colors into complementary colors. This is handy when you have shot black and white movies or colored film negatives. Instead of printing the positives for use on television, you may project the raw negative images into the camera, switch the camera to NEGATIVE, and have the pictures come out right. This saves one step of photo processing.

NEGATIVE has another application. When a company logo or a subtitle is to be superimposed on the TV screen, it is easiest to read if the words are white. They tend to disappear into the picture if they are black. Black artwork and black words on white pages, however, are easiest to produce. One could photograph the logo or subtitle turning it into white words on black background, but this is a cumbersome extra step. A camera with NEGATIVE capability can make the switch for you so that your easy-to-make black words will appear white.

Return: Some TV cameras have one or more buttons which allow the camera operator to select another camera's picture or some other image to be displayed on the viewfinder. If, for instance, the director were creating a CORNER INSERT, where your picture was inserted into the corner of some other camera's picture, you would need to know just what part of your picture was being used and what part wasn't. By pressing RETURN or LINE, your camera's viewfinder would show the final effect.

Controls on a CCU. The CAMERA CONTROL UNIT (CCU) is a box of electronics which can remotely control the circuits in a TV camera. Many if not all of the adjustments mentioned so far may be found on the CCU. Generally, a TV engineer would adjust these controls before the program begins and may tweak the controls a bit during the production.

To recap a bit:

1. IRIS: opens and closes the camera's electric IRIS to create more contrast and adjust DEPTH-OF-FIELD (described further in Chapter 7)

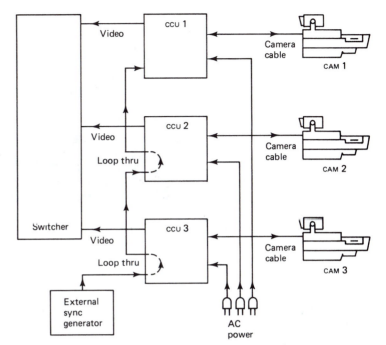

FIGURE 6-14 Cameras and camera control units.

2. GAIN: increases the strength of the video signal, also improving contrast

3. PEDESTAL: makes the entire picture lighter

Many color CCUs have additional controls (Figure 6–15.

R gain: Adjusts the redness of the camera's picture.

B gain: Adjusts the blueness of the camera's picture. By aiming the camera at something white and adjusting the R GAIN and B GAIN, you can tint the white image redder or bluer. By turning both controls down, you tint the white image greener. Under normal circumstances you balance these two controls to make the white image perfectly white with no tint whatsoever.

R pedestal: Adjusts the amount of redness in the dark parts of a camera's picture.

B pedestal: Adjusts the amount of blueness in the dark parts of a camera's picture. Turning both R and B PEDESTALS down leaves only green in the dark parts of a picture.

Bars: This allows the camera to put out COLOR BARS, a test signal used to technically judge the colors coming from a camera.

MINIGLOSSARY

Corner insert A wipe effect where one corner of the TV screen shows one camera's picture while the rest of the screen shows another's.

Return Feature on some cameras which allows another camera's picture (or some other video image) to appear on the first camera's viewfinder.

FIGURE 6-15 Color camera CCUs.

Subcarrier phase: SUBCARRIER PHASE or SC PHASE or simply SC is like the hue control on your TV set. It adjusts all the colors at once. Turned in one direction, your reds will start to look a little orange. Turned in the other direction, this knob makes your reds look a little magenta. This control affects the green and the blue at the same time.

Cable length: A camera with a longer cable sends signals to the switcher that are slightly weaker and fuzzier than those of its brothers. This weakness causes the picture to appear softer and more faded than the others, especially the colors. Adjusting the CABLE LENGTH control on your cameras matches their picture strengths and sharpness.

H phase (or horizontal phase): A camera with longer cables also gets its sync signals slightly later than the others and sends its picture back to the switcher later. This is because it takes longer for a signal to travel through a longer cable. This time shift causes the picture to be horizontally shifted on the TV screen. H PHASE is the control which shifts the picture left or right to match up with the others.

Setting up Cameras and Matching Colors.
Before you begin a studio TV production, you must always adjust your cameras so that they all produce the proper brightness, contrast, and color. If you don't, the viewer will see a distracting change in the brightness or color of the picture as you switch from one camera shot to another. Industrial and professional TV studios usually employ a TV engineer or technician who makes these adjustments before the program begins. Two essential tools for making these adjustments are a WAVEFORM monitor and a VECTORSCOPE;

both will be described further in Chapter 15. Smaller studios don't have the luxury of a TV technician to tweak the cameras, and you may have to do this yourself. You may even have to do it without the help of the WAVEFORM monitor and VECTORSCOPE. Although a detailed procedure for adjusting cameras will appear in Chapter 15, here is a brief general approach which requires only a color TV monitor and a TV SWITCHER to switch from one camera's picture to another.

Let's presume that all the cameras are connected and the studio has been set up for the production. Here's what to do to prepare the cameras for the show. Carry these steps out in the order listed.

1. *Getting turned on*
    a. Turn on house sync generator.
    b. Turn on pulse distribution amplifiers (if your system has them).
    c. Turn on switcher.
    d. Turn on camera control units.
    e. Turn on cameras (unless the CCUs turned them on).
    f. Turn on TV monitor.
    g. Let the system warm up for 5 minutes. Some equipment is not too stable when cold and will "drift" during the first 5 minutes. There's no sense making adjustments that are going to change by themselves anyway.

2. *Perfecting the image*
    a. Adjust the TV monitor. Send the TV monitor a signal that you know is good. This could

---

MINIGLOSSARY

*Depth-of-field   The span of distance from a lens which appears in focus at one time. Wide depth-of-field means far and near objects in the picture appear sharp.

Subcarrier phase   Adjustment on cameras and other video equipment which alters all the colors coming from them.

Cable length   CCU control which adjusts the strength of signals

coming from a camera, matching them to the strengths of other cameras with longer or shorter cables.

Horizontal phase or H phase   Control on a camera's timing circuits which adjusts the picture sideways to line up with other cameras' pictures.

be an off-the-air TV signal, a tape, or COLOR BARS from a camera or a test signal generator. Don't try to use the signal from one of your yet-to-be-adjusted cameras. As described in Chapter 2, turn the monitor's color down and adjust the brightness and contrast. Then turn the color up while viewing a TV program with good "flesh tones" and adjust the hue so the skin looks the right color. There are more precise ways to adjust the colors on a TV monitor which will be described in Chapter 15, but this should at least get you started with a picture you can trust.

b. Prepare the set for shooting. Light the stage area the way it will be lit for the actual production (more on this in Chapter 9). Move the cameras to the positions they will take during the production and focus them. Have someone (perhaps a performer) remain on the set long enough to make your camera color adjustments.

c. Adjust the COLOR TEMPERATURE wheel on all the cameras to match the kind of lighting used. Generally, studio lighting is 3200°K.

d. Adjust each camera's IRIS. If your cameras have an electronically controlled AUTO IRIS, then switch them to that mode. If not, you could adjust each IRIS to make the camera's picture look good, but sometimes other camera controls are so fouled up that even a proper IRIS setting makes a poor-looking picture. If you can't seem to get a good picture, just try f4 as a safe estimate. You can go back and change it later if f4 was a poor choice. The IRIS affects both contrast and brightness. Color is unimportant at this point. To simplify matters, you may wish to turn your TV monitor's color control down to make a black-and-white picture. Generally speaking, we wish to create a good black-and-white picture from the color cameras before we deal with making the color right.

e. Adjust GAIN and PEDESTAL. Like IRIS, GAIN and PEDESTAL affect the contrast and brightness of your picture. On CCUs which have these controls, adjust the GAIN so that your picture has enough contrast. Adjust PEDESTAL so that dark parts of the picture do not merge together into murky shadows. When PEDESTAL is adjusted too high, you see no blacks in the picture, only shades of gray and white. Figure 6–16 shows some GAIN and PEDESTAL adjustments and misadjustments.

f. Adjust WHITE BALANCE. For this you will want to see colors, so if you turned your TV monitor's color control down earlier, turn it back up now. Some cameras have an AUTO WHITE BALANCE control. In many, it works like this:

1. Place a white card in the area where you will be shooting, *or* place a milky white lens cap over each camera's lens.

2. Press the AUTO WHITE BALANCE button and hold it down for at least 5 seconds.

3. You're done. A peek at your TV monitor should show a pure white screen. If the screen appears tinted, either your cameras are off or your TV monitor is off. It is hard to tell which is at fault, so if you wish to be 100% sure, you'll need to have a technician examine the monitor and certify that it is telling you the truth. Once you can trust your monitor, you can tell whether your AUTO WHITE BALANCE is working as it should. Now that you have pure white, you may remove the card or white lens caps. Most studio cameras will allow you to adjust their WHITE BALANCE manually. As before, you focus the cameras on a white card or have them look through milky white lens caps. If your monitor shows you a pure white picture, you need go no further. If the picture looks a little pinkish, then turn the R GAIN down a little. If it looks a little bluish, turn the B GAIN down a little. If the picture looks greenish, turn both GAIN controls up. Sometimes your eyes seem to glaze over with a rainbow of colors, and you can't tell what you're looking at anymore. One way to straighten yourself out is to turn the color control on your TV monitor down and observe what color the screen is. That should be *genuine* white. Now turn the color back up on the monitor and see if you can match the cameras to that kind of white. Some cameras may only have one knob which you turn in one direction to make the picture redder and the other direction to make it bluer. Simply twiddle it back and forth until the picture looks neither reddish nor bluish.

g. Adjust BLACK BALANCE. Just as you don't want whites to be tinted, you also don't want the blacks in your pictures to be tinted. Not all cameras have an adjustment for BLACK

GAIN high

GAIN low

GAIN and PEDESTAL right

PEDESTAL high

PEDESTAL low

FIGURE 6-16   GAIN and PEDESTAL adjustments and misadjustments.

BALANCE, but if yours does, here is how to adjust it:

1. Make your cameras' pictures dark either by placing lens caps over the lenses, or by turning out the lights, or by turning the lenses' IRISES past f16 all the way to C (closed).

2. Look at the picture on your TV monitor. If it is totally black, temporarily turn up its brightness a little so that it is dark gray. If the gray isn't tinted, then you're finished; set everything back to normal. If the dark gray appears greenish, reddish, or bluish, then continue with the following steps:

3. Turn R PEDESTAL down to remove a reddish tint. Turn B PEDESTAL down to remove a bluish tint. Turn both PEDESTALS up to remove a greenish tint. Some cameras may have a single knob for this adjustment which you turn one way to boost red and the other way to boost blue. *Note:* Adjusting PEDESTAL may affect GAIN, so after doing this step, you may have to go back and redo step 6.

h. Adjust SUBCARRIER PHASE. Like the hue control on a TV set, this control changes all the colors from your camera at once. To properly adjust the colors:

1. Aim the camera at one of your performers (who by now has probably melted under the lights).

2. Adjust SUBCARRIER PHASE (or SC PHASE or SUBCARRIER or BURST or BURST PHASE) so that the skin looks the right color. Do this in turn with each camera.

i. Adjust CABLE LENGTH controls. Using one camera as a standard, you will be trying to match the other cameras one by one to it. Line your cameras up side by side and have them take close-ups of a performer. Try to make all three shots have equal brightness and color vividness. You can compare the colors by switching back and forth from one camera's picture to another while observing the TV monitor. If your switcher has a SPLIT SCREEN capability, you may even be able to view two cameras at once on the same TV screen. Adjust your *second* camera's CABLE LENGTH control until its flesh tone matches the color of your first camera. If you have a third camera, make *its* flesh tone match the first camera.

j. Adjust H PHASE. This control needs adjust-

ment only when the camera cables have been changed. If your cameras never get swapped around, this adjustment will stay the same. Also, you need an UNDERSCANNED monitor to make this adjustment correctly. You need to see the HORIZONTAL SYNC PULSE, the black bar on the left and right sides of the picture. While observing the bar, switch from one camera to another. Does the bar shift sideways? It shouldn't. Turning H PHASE will move the bar and picture horizontally. Adjust the cameras so that their bars line up as you switch.

Now that your brightness, contrasts, and colors are matched, you're ready to shoot. What do you mean your performer has dissolved into a puddle of sweat in the middle of the floor?

Studio/Portable Color Cameras

Portable color TV cameras are getting so good nowadays they perform better than studio color cameras did only a few years ago. So if a portable camera works better than a studio camera, then why not simply carry it into the studio and use it there?

This is not so simply done. If you plan to use only one camera and send its signal directly to a VCR, then there's not much of a problem. The moment that you take two portable cameras and send their signals to a switcher, you've got big problems unless those cameras can accept EXTERNAL SYNC. Cameras which aren't designed to convert to studio use don't have EXTERNAL SYNC inputs. Without EXTERNAL SYNC or GENLOCK inputs the portable cameras can't synchronize their electronics so that the switcher can switch from one camera's picture to another's smoothly.

Cameras which are strictly portable have other limitations in the studio. What camera operator wants to do an entire show with his or her eye pressed up against a hot rubber eyecup? And who wants a camera sitting on his or her shoulder for an hour with one hand strapped to the camera's lens barrel?

In short, even though they're good, portable cameras can't automatically be transplanted into the studio. This limitation applies to nearly all of the home video cameras.

There are industrial color cameras, on the other hand, which are designed for studio/portable convertibility. In the field they use a tiny viewfinder held inches from your eyeball. The viewfinder may have little lights showing you whether your VCR is recording or on PAUSE, or showing how your battery is doing, or indicating whether your lighting is adequate. In the studio, you hitch up a larger, easier-to-see viewfinder which has a TALLY light and perhaps a switch to select another camera's image. In the field, the camera may dock with your VCR or send its video signals over a coax wire to the VCR. In the studio, an umbilical connects the camera to the CAMERA CONTROL UNIT which provides the power to the camera, receives video from it, and sends regulated SYNC to the camera as well as numerous remote control signals for the camera's electronics. In the field, the camera sits atop the operator's shoulder while the lens is manually controlled or operated by push buttons or automatic controls. In the studio, the lens is usually controlled by a switch on the tripod handle. The internal CCU may take charge of the camera's IRIS, WHITE BALANCE, and GAIN controls.

Because of their broad flexibility, studio/portable color cameras appear to be the trend for the near future.

CAMERA CARE

1. Lock your camera pedestal controls when your camera is idle. A camera with a heavy lens might suddenly tilt down, slamming its lens against the tripod.

2. Cap your lens when the camera is idle. This protects not only the lens glass but protects a tube camera from bright lights.

3. Don't knock the camera around. It is fragile and easily misaligned.

4. Avoid extremes in temperature. CCD cameras make grainier pictures above 115°F. The heat in the trunk of a parked car on a sunny day can also damage a camera's circuits. In superfrigid weather, the zoom lens may get "sticky" and may fail to rotate smoothly.

5. Avoid dampness.

6. If traveling by air, don't check your camera with your baggage. Remember the American Tourister luggage ad with the ape tossing around the suitcases?

CAMERA RESOLUTION AND SENSITIVITY

Perhaps the two most important measures of a camera's capabilities are its RESOLUTION (sharpness) and its sensitivity (ability to make a picture in dim light).

Resolution

To explain picture RESOLUTION, think of telephone poles. If you had a fuzzy picture, 100 telephone poles side by side would merge into one mushy mess. If you had a real sharp picture, each telephone pole could be easily discerned. One way to measure the sharpness of the TV picture is how many telephone poles can you see before they blend together. Since shots of telephone poles are hard to come by, technicians use TEST PATTERNS of black lines on white paper (see Figure 6–17) or electrically generated signals called MULTIBURSTS. You would keep squeezing more and more vertical lines onto the screen until they no longer could be counted separately, and call that the RESOLUTION of the device (camera, monitor, or whatever was being tested).

This is where the term LINES OF HORIZONTAL RESOLUTION came from: You are looking horizontally across the screen and counting the vertical lines visible on the screen. Notice that this has nothing to do with the *horizontal scan lines* that create the picture. Regular NTSC TV always has 525 horizontal scan lines. Those lines affect *vertical resolution*. Camera sharpness, however, is measured in LINES OF HORIZONTAL RESOLUTION.

The preceding description is oversimplified a bit for clarity. To be precise, if you had a vertical white bar, a black bar, a white bar, and a black bar filling the TV screen, this would calculate as 3 LINES OF HORIZONTAL RESOLUTION. Why? Technically, you call the white bar one line and the black bar one line. Then why isn't the answer four lines? Because LINES OF HORIZONTAL RESOLUTION are professionally calculated *per picture height*. Since TV screens are always ¾ as tall as they are wide, you count only ¾ of the bars showing. Since ¾ of 4 is 3, the answer is 3 LINES OF RESOLUTION.

Inexpensive TV cameras generally reproduce 240 LINES OF HORIZONTAL RESOLUTION, and professional models up to 750. Broadcast NTSC video goes up to 330 lines, while VHS VCRs can record only 240. Note that in each of these cases we are talking about *luminance* RESOLUTION, black telephone poles over a white background. Colors, on the other hand, are much fuzzier. Fortunately, our eyes tend to perceive sharpness from the black-and-white parts of the picture and "forgive" the fact that the colors seem to be smeared on over the top. The only time when color resolution becomes a serious consideration is when we are viewing electronic graphs or trying to read colored text.

MINIGLOSSARY

Test pattern   Chart used for measuring a camera or other video device's performance such as resolution.

Lux   A measure of illumination, the amount of light needed to make a 1-volt video signal.

Footcandle   A measure of illumination, the level of brightness found 1 foot from a candle; about 10 lux.

High-speed shutter   Feature on chip camera or camcorder allowing the camera to electronically capture each image more quickly, "freezing" the action with less blur.

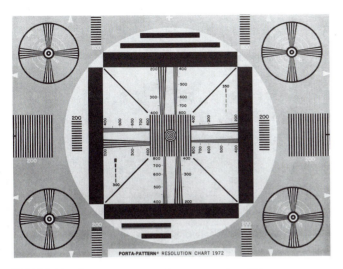

FIGURE 6-17 TEST PATTERN for measuring TV camera RES-OLUTION. (Courtesy Porta Pattern)

So why buy an expensive 600-LINE TV camera if you know the extra sharpness is wasted on 330-LINE broadcasts or 240-LINE VHS tape? Might as well use a 330-LINE camera, right? Wrong. The 600-LINE camera will still look better than its 330-LINE brother. Here's why:

1. The 600-LINE camera is making a stronger, more robust 330 LINES than its 330 counterpart, even though everything above 330 LINES is lost.

2. Expensive cameras tend to make better signals in other ways. These enhancements make the picture *look* better for reasons other than RESOLUTION.

Chip-type cameras are often described by the number of PIXELS in the chips, rather than by the more common standard LINES OF RESOLUTION. As mentioned earlier, the more PIXELS a chip has, the sharper the picture. Industrial CCD cameras often have chips measuring 610 × 492 PIX-

ELS. The bigger number is always the horizontal measurement and the smaller is the vertical. Multiply the two together to get the total, 300,120 PIXELS. To convert horizontal PIXEL numbers to LINES OF HORIZONTAL RESOLUTION, multiply by .53. Thus a chip 610 PIXELS across yields about 323 LINES OF HORIZONTAL RESOLUTION.

Camera manufacturers have recently been able to squeeze more RESOLUTION out of the preceding formula by offsetting the three camera chips (rather than lining them up pixel for pixel). It's sort of like lining up three big windows (representing three big PIXELS) and then looking through them. If you now slide the middle window a bit to the left and raise the last window up some, the "hole" through the three windows is smaller, representing a "smaller" or "sharper" PIXEL triad. This trick squeezes 700 *apparent* LINES OF RESOLUTION out of chips designed for only 550 LINES.

Camera sensitivity

Camera sensitivity is measured in LUX or FOOTCANDLES (one FOOTCANDLE equals about 10 LUX). A camera that can make a picture in very dim light may be rated at 1 LUX, where a less sensitive camera needs maybe 10 to 20 LUX and perhaps an extra lamp or two. Note that a camera rated at 1 LUX *minimum illumination* barely makes a picture at 1 LUX. It still may require 40 LUX of illumination before the picture *looks good* and renders vibrant colors. Also note that manufacturers determine camera specifications in different ways, some stretching the truth. For instance, a camera rated for 3 LUX in the United States would be rated 7 LUX in Japan where standards are more strictly controlled.

HIGH-SPEED SHUTTERS

One feature in cameras and camcorders that affects apparent picture sharpness and camera sensitivity is the high-speed shutter. Common TV cameras "shoot" 30 TV frames (pic-

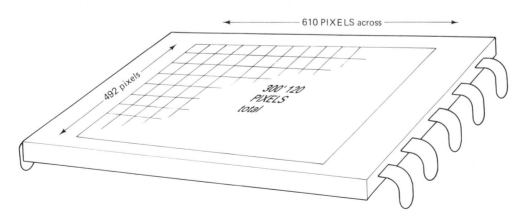

PIXELS on a CCD camera chip.

tures) per second. As any camera buff knows, if you snap a photo of Grandpa hitting a golf ball at $\frac{1}{30}$-second shutter speed, you'll get a blur (even for Grandpa). To catch fast action, you need to use faster shutter speeds, such as $\frac{1}{250}$, $\frac{1}{500}$, or maybe $\frac{1}{1000}$ second. Note that these faster shutter speeds allow less light to enter the camera, so to keep a bright picture, plenty of sunlight (or lamps) may be needed. The same is true for TV cameras with fast shutters. Although they can ''freeze'' the action making a sharper picture (especially noticeable when still-framed), they require more light for the faster shutter speeds. *Note:* Even though the shutter may be set to $\frac{1}{250}$ second, the camera is still making 30 pictures per second. The pictures are just electronically captured more quickly. Figure 6–18 shows the HIGH-SPEED SHUTTER control on a camcorder.

FIGURE 6-18 HIGH SPEED SHUTTER and other enhancements on a camcorder. (Courtesy Chinon)

COMMON TV CAMERA AILMENTS AND CURES

The first step, as always, is to home in on the problem. If your camera has a viewfinder and sends its signal to a TV monitor on the way to a switcher, you are equipped to track down the elusive camera gremlins. Although there are 100 things that can go wrong with a camera, there are probably only 10 things that go wrong 90% of the time. Figuring out *where* the problem occurs tells you a lot about *what* the problem might be. Try to narrow the problem down to whether it's caused by the camera, the viewfinder, the cables, the sync system, the switcher, the VTR, or your monitor. Now let's attack some specific problems.

Picture Problems

1. *No picture on ELECTRONIC VIEWFINDER or TV set*:
   a. Is the camera turned on?
   b. Is the VCR turned on?
   c. If using an AC POWER ADAPTER with your camera, is *it* turned on?
   d. Is the lens uncapped? This isn't as silly an oversight as you may think.
   e. Is the lens IRIS open enough? Turn it to low f-numbers.
   f. Is the camera designed to work with this VCR?
   g. Check the camera for REMOTE/MANUAL switches and throw them into the proper position. If a CCU is being used, check its switches and knobs. Are they in logical positions?
   h. Is the camera's FADE button depressed?
   i. Is the camera switched to the CHARACTER GENERATOR mode, showing you a screen of text that hasn't been made yet (black screen)?

   j. Is there a dark lens FILTER attached to the lens from a prior shoot?

2. *Good picture in VIEWFINDER; no picture on the monitor*:
   a. Wiggle each end of the camera cable near the plug. If the picture flashes on and off the screen, you have a broken wire—very common on simple camera setups. Substitute another camera cable for now and have a technician examine the plug for the broken wire.
   b. Are the VCR's switches (INPUT and OUTPUT selectors) set to the right positions?
   c. Is this camera designed to work with this VCR?
   d. If using an AC POWER ADAPTER, are all its wires connected properly to the VCR?
   e. If using an industrial camera, is its GAIN turned up? If using a CCU, are its GAIN and PEDESTAL controls turned up?
   f. Your monitor may be malfunctioning. Check its switches or substitute another monitor.
   g. The monitor cable may be malfunctioning. Substitute another cable.

3. *Good picture in the VIEWFINDER; the picture on your monitor rolls*:
   a. Check the VERTICAL HOLD and the HORIZONTAL HOLD on the TV. The VERTICAL HOLD controls roll, but if the HORIZONTAL HOLD is also messed up, you may need to adjust them both for a stable picture.
   b. Try flipping the SYNC switch at the rear of the monitor to the INT position.

4. *Pictures roll on both the TV and the VIEW-FINDER:*
   a. Check VERTICAL HOLD on both the ELECTRONIC VIEWFINDER and TV set. If that doesn't help, check battery power—is it low?
   b. Does this VCR go with this camera?
   c. You may have a SYNC problem. If the camera is the industrial type, check for a switch marked SYNC. If your camera is working with others, these switches should all be in the EXT position. If your camera is alone, feeding its signal directly to a VCR, its switch should probably be in the INT position.
   d. Perhaps your SYNC system has failed or has been turned off. Check the master SYNC GENERATOR.
   e. If using a gas-powered generator or if using an INVERTER (a device that changes dc, like from your car's battery, to ac, like from the wall outlet) to power your equipment, make sure these sources
      1. Can provide at least 110% of the wattage your total video equipment needs.
      2. Control the frequency of their alternating current to 60 cycles per second plus or minus ½ cycle.

5. *Picture on VIEWFINDER or monitor screen "draws in" on all sides, yielding a shrunken picture:* Your voltage is low.
   a. Check your battery power.
   b. If using ac power and long extension cables,
      1. Shorten the extension cords or get heavier ones.
      2. Turn off any unnecessary power using equipment on the extension cord.
      3. Turn off any high-wattage lights plugged into the same circuit.

6. *Picture is faded on both the monitor and the VIEWFINDER:*
   a. Perhaps there is insufficient light on the subject.
   b. Is the lens IRIS closed too much? Turn to the lower f-numbers.
   c. Is the lens covered with water or salt spray, or did you leave a dark FILTER attached to your lens?
   d. Is the camera's GAIN turned too low?
   e. Does the camera work at different shutter speeds? Try the slowest.

7. *Picture is good in the VIEWFINDER but faded on the TV monitor:*
   a. On industrial cameras the GAIN or PEDESTAL may be low.
   b. The TV's BRIGHTNESS or CONTRAST may be too low.
   c. The video signal may have been improperly split, LOOPED, or TERMINATED, losing some of its "punch."

8. *Camera's picture is fuzzy on both the VIEWFINDER and the TV monitor:*
   a. Focus the lens.
   b. Are you using a MACRO lens or a close-up lens attachment while viewing a distant subject (more on this in Chapter 7)?
   c. Is your lens screwed into the camera tightly?

9. *Ghost of same image stays on TV and VIEWFINDER screens no matter where you point the camera (Figure 6–1):* You have a BURN-IN from having a tubed camera "see" the same contrasty picture for too long, or it has been aimed into a bright light. To remove mild BURN-INS try:
   a. Turning the camera off and waiting a day or so before you use it again.
   b. Turning on the camera and aiming it at a smooth white (not shiny) object, like an out-of-focus close-up at a well-illuminated sheet of dull white paper. The white image *must fill* the screen. The BURN-IN may go away in about an hour or so.
   c. Have the technician remove the camera tube and replace it with a new one. The old tube may "heal" itself if left on the shelf for about a year. Label the tube's box with a message "burn-in" and the date.
   d. Check your lens to see if there is any crud on it. If you cover the lens and a mark or streak remains, it's a BURN-IN. If the streak disappears, your lens, tube, or chip is probably dirty. Chapter 16 tells how to clean them.

10. *Picture won't stay in focus throughout the zoom range:*
    a. Have you focused correctly (zoom in, focus, then zoom out)?
    b. Do you have a close-up lens attachment, or is your lens in the MACRO mode? Return the lens to normal.

MINIGLOSSARY

Inverter   A device which changes DC electricity (from a battery) into AC electricity (from your wall outlet).

c. Are you trying to focus on something too close for your particular lens? Read the focus etchings on the lens; the lowest distance marked is the closest distance to which the lens will normally focus.

d. Has the camera been bumped or knocked around? If so, the tubes, chips, or mirrors may have jarred out of alignment. A technician can readjust this alignment.

e. Have you changed lenses on your camera? If for some reason one lens doesn't screw in as far as another, the image will not be properly focused.

11. *THREE-TUBE color camera makes pictures with colored ridges around the edges. The ridges are especially noticeable around white objects in front of black backgrounds*: Your camera is out of CONVERGENCE. To make a proper picture, it must precisely lay the green, red, and blue pictures atop one another. If CONVERGENCE is misadjusted, then the three pictures don't match up. Repair requires a technician.

Other Problems

1. *When you pull the camera trigger, the separate, portable VCR fails to start*:
   a. Have you pressed RECORD/PLAY as you should?
   b. Check to see that the VCR *itself* isn't PAUSED. Its PAUSE button must be off for the camera's PAUSE to work.
   c. Is something (an earphone, perhaps) plugged into the REMOTE input on the VCR? Unplug anything from the REMOTE jack.
   d. Does this VCR go with this camera?
   e. Is there a CAMERA or REMOTE switch on the VCR flipped to the wrong position? Sometimes the VCR has to be told to listen to its CAMERA input.

2. *The camera trigger and its little PAUSE light seem to work opposite from the way they're supposed to; everything else works okay*: Your camera apparently goes with the VCR *except* the PAUSE part is wired up backward. Some cameras have a switch to correct this incompatibility. The camera still can be used; you just have to relearn what the PAUSE light is now telling you.

3. *Weak or no audio is being recorded from the camera; picture is okay*: To monitor your audio, connect up a TV to the VCR while the camera and VCR are in RECORD. If you don't hear sound from the TV speaker, double-check the sound by trying an earphone in the EARPHONE or HEAD-

PHONE OUT socket of the VCR. Once you're sure the sound is deficient:
   a. Check the MIC IN or AUDIO IN. Nothing should be connected there if you are recording from the camera's built-in microphone.
   b. Does this camera go with this VCR?
   c. Is your VCR's INPUT SELECT in the VCR or TV mode? Switch it to VCR.
   d. Are you using an external microphone? If so,
      1. Make sure its ON/OFF switch is turned ON.
      2. Some mikes have built-in batteries. Is your battery dead?

4. *You can play a picture into your camera's ELECTRONIC VIEWFINDER, but the sound is missing*:
   a. Make sure the tape has sound recorded on it to start with. Try playing the VCR into a TV.
   b. With the earphone-equipped cameras and different-brand VCRs connected via adapter cables, sometimes the adapter cables aren't wired to send sound from the VCR to the camera. Plug the earphone into the VCR directly.
   c. Have you selected the correct audio channel? On a ¾U VCR, you would normally record your sound on channel 2. Are you listening to channel 2?

CAMERA MOUNTING EQUIPMENT

Really smooth camera movements require a tripod. A tripod also alleviates arm fatigue during long shooting sessions. During "trick" shots, it holds the camera still enough for edits to be unobtrusive (more on this in Chapter 14, Editing).

Figure 6–19 shows a tripod with a HEAD and DOLLY. The HEAD part at the top (not to be confused with the heads on a VCR, which are something else altogether) connects to your camera and allows you to aim it using a long handle. The tripod part usually has an adjustable PEDESTAL raised and lowered by a crank to allow for high and low shots. (And PEDESTAL is another word with two meanings. The camera height PEDESTAL has nothing to do with the camera's PEDESTAL control, a knob which electrically adjusts the camera's brightness.) The tripod legs often telescope so that the camera can be raised an additional 3 feet overhead. On the bottom is the DOLLY, a set of wheels which allows the tripod to glide smoothly over the floor.

Tripods

Figure 6–20 shows a less expensive and less flexible portable tripod used primarily with home video equipment. The typical portable tripod has three-part telescoping legs which

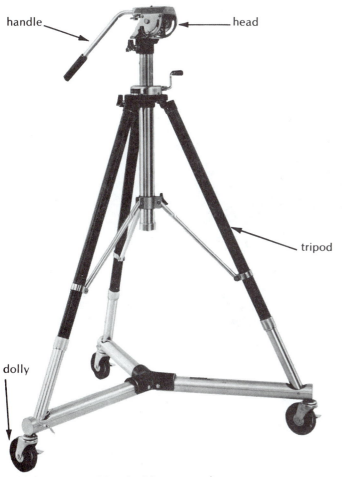

handle — head

tripod

dolly

FIGURE 6-19 Tripod with HEAD and DOLLY.

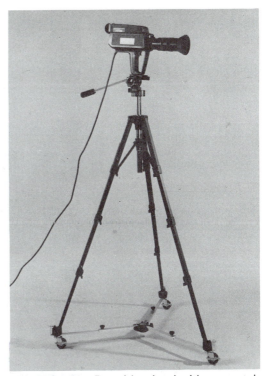

FIGURE 6-20 Portable tripod with HEAD and DOLLY (Courtesy of Comprehensive Video Supply Corp.).

extend to 4 feet or shrink down to about 18 inches when closed. Smaller, flimsier models may lack the PEDESTAL and crank, and the HEAD may be wobbly. They're really meant for still photography and don't smooth out camera moves a heck of a lot, but they are better than nothing and are very light and compact. Unless you're the type who would mount a diamond in a plastic setting, you may not want to trust your $2000 color camera to a $29 ''precarious pod.''

Most small industrial users would find the tripod shown in Figure 6–19 to be most adequate. Studios using heavier cameras may wish to employ a sturdier, heavy-duty tripod like the one shown in Figure 6–21. Not only can its legs be lengthened, but its legs may be spread apart varying distances, allowing the tripod to lean, squat low, or conform to unusual terrain.

The heavy-duty crank allows the PEDESTAL to be elevated with ease. Make sure you tighten the elevator lock when finished cranking it so that gravity doesn't wind it down with a ''zwoop!'' When you *do* decide to elevate the camera, be sure to *loosen the controls first*; otherwise the flimsy parts of the crank mechanism will strip and wear out quickly.

Some tripods have telescoping legs which permit the elevation of the camera to be nearly doubled. Lengthening the legs can be a tedious task for one person, so find a buddy to help you. Loosen the collars on all three tripod legs and also the collar which braces the legs to the PEDESTAL. Then both of you should be able to lift the camera straight into the air with the tripod legs extending automatically to the floor. While one of you holds the camera, the other runs around to tighten the three leg collars and the brace that goes to the PEDESTAL. Before leaving the camera, check it to make sure it is sturdy. Sometimes a half-tightened leg collar will squirm loose, and the camera will come crashing down like a mighty tree in the forest.

MINIGLOSSARY

*Head Top part of a camera tripod that holds the camera.

*Dolly Bottom part of a camera tripod that has wheels. Also the act of moving the camera toward or away from a subject.

*Pedestal The elevation control on a camera tripod. ''Pedestal up'' means raise the camera higher.

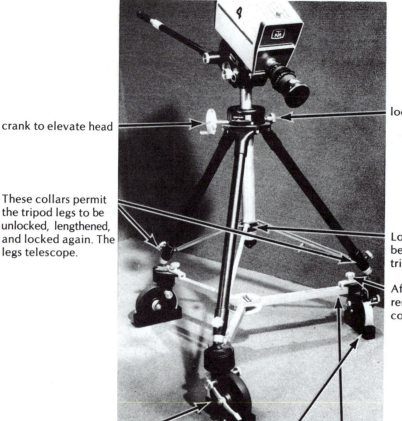

crank to elevate head

lock for elevator

These collars permit the tripod legs to be unlocked, lengthened, and locked again. The legs telescope.

Loosen this collar, too, before lengthening tripod legs.

After these clamps are removed, the tripod comes off the dolly.

Each caster can lock so the wheel won't roll.

Another lock forces the wheels to aim in a specific direction. This is helpful when you dolly in and don't want to be bothered aiming the wheels around before you can dolly back out.

The adjustable cable guards sweep cables out of the way. If they're too high, you'll roll over the cables. If they're too low, you'll scrape the floor.

FIGURE 6-21 Tripod controls.

Lighter, more portable, and cheaper than the tripod is the MONOPOD, a one-legged tripod (Figure 6–22). Although it can't stand up by itself, it takes a lot of weight off your arms and steadies your shots quite well as you hold it upright. It moves with you quickly with a minimum of setup (for when you are shooting wild scenes of that tornado ripping the town apart a block behind you).

And then there's the CENTIPOD, a furry little worm with 100 legs. Although capable of holding tiny cameras, the CENTIPOD is the only model that turns into a butterfly.

The COUNTERWEIGHT STUDIO PEDESTAL (Figure 6–23) is a heavy-duty monster used in professional television studios where the cameras are massive. Often there is a TELEPROMPTER attached to the camera adding even more weight. Naturally it wouldn't be feasible for a camera operator to crank this whole shebang up and down manually. Instead, the PEDESTAL has springs, pulleys, counterweights, or pneumatic pressure to help raise and lower the camera. The movement is so smooth that it can be done while the camera is ''on,'' unlike the industrial tripods which wobble as they are cranked.

FIGURE 6-22 MONOPOD (Courtesy of Comprehensive Video Supply Corp.)

Heads

Figure 6–24 shows a basic FRICTION HEAD appropriate for small studios and schools.

To TILT and/or PAN your camera, first loosen the controls on the head while holding onto the handle you use to aim the camera. When PANNING or TILTING, you may want to keep these controls slightly taut so they provide a little drag and thus mask some of your jars and shakes. Professionals consider anything but smooth-flowing camera movements to be very ungainly.

If you leave the camera, *tighten the HEAD controls.* The weight of the camera often makes it want to TILT. If it squirms loose while unattended, it could TILT down abruptly and smash its lens against the tripod. Also, if someone kicks the camera cable, that could swing the unattended camera

around and PAN it into a bright light, causing a BURN-IN on a tubed camera.

The FRICTION HEAD in Figure 6–24 has a spring built into it to help hold the camera upright, making it easier to aim. Amateur and portable HEADS like the one shown in Figure 6–20 have no way of counterbalancing the camera, making camera moves awkward and unstable.

Figures 6–25 and 6–26 show two heavy-duty tripod HEADS. The CRADLE and CAM LINK HEADS are very stable and are good at smoothing out jiggles as you tilt. If the camera is balanced correctly, you should be able to let go of the camera handle and have the camera settle to a safe horizontal position rather than nosediving or falling backward. These HEADS are found on semiprofessional and the better industrial equipment.

These heavy-duty HEADS often have two TILT adjustments and two PAN adjustments. The first TILT adjustment locks the head so that won't TILT. The second adjustment is called DRAG, and it creates a resistance as you TILT the head. This reduces little camera jiggles, especially while

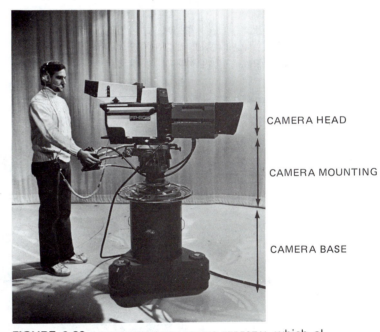

CAMERA HEAD

CAMERA MOUNTING

CAMERA BASE

FIGURE 6-23 COUNTERWEIGHT STUDIO PEDESTAL which allows the camera to be raised and lowered smoothly while on the air. (Courtesy of Cablevision Systems Development Co. and *Understanding Television Production—* Prentice Hall)

MINIGLOSSARY

Monopod One-legged tripod.

Counterweight studio pedestal Heavy duty studio camera pedestal and dolly that allow the camera to be raised and lowered smoothly with ease.

*Teleprompter Device that sits near the camera lens and allows the performer to read the text while his eyes appear to be looking at the camera lens (the viewer).

*Friction head Inexpensive tripod head with locks to impede unwanted camera movement.

*Cradle head Heavy-duty camera support to keep the camera stable when it's free to tilt; i.e., the camera won't suddenly tilt down.

*Cam link head Heavy-duty camera support to keep the camera from tilting down abruptly when free to move; the camera simply comes to rest in a safe horizontal position.

TELEPROMPTER (Courtesy KPHO and *Single Camera Video Production*—Prentice Hall)

Pneumatic pedestal. (Courtesy of Listec and *Understanding Television Production*—Prentice Hall)

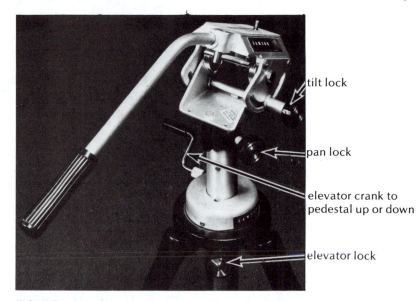

FIGURE 6-24 Simple FRICTION HEAD for tripod. Basic tripod controls and movements.

you are trying to hold the camera still. If you try to TILT the camera quickly, however, the DRAG control will slow you down. If you anticipate such a move, it is best to loosen the DRAG control. Similarly, the PAN lock control stops the head from PANNING, and the PAN DRAG control adds resistance to this camera movement. Inexperienced camera operators who don't realize that there are two controls for

both PAN and TILT often find themselves struggling with their cameras while trying to aim them.

The better HEADS also allow you to adjust the position of the camera handles. If the camera is to be very low, then you will want the handles to come straight back so that you don't have to stoop over to aim the camera. If the camera is PEDESTALED up high, you may want to bring the handles

BASIC TRIPOD CONTROLS AND MOVEMENTS

TILT: To aim the camera up and down on a vertical axis, like nodding your head "yes." *Tilt up* means to shoot higher, toward the ceiling. *Tilt down* means to aim lower, toward the floor.

PAN: To aim the camera back and forth on a horizontal axis, like shaking your head "no." *Pan left* means to turn the camera to your left. *Pan right*, of course, means to turn it to the right.

To *DOLLY*: To travel forward or backward across the floor with the camera. *Dolly in* means to move the camera forward, tripod and all, closer to the subject. *Dolly out* means to pull back.

To *TRUCK* or *CRAB*: To travel from side to side across the floor with the camera, tripod and all. *Truck right* means to travel to your right, and *truck left* tells you to go in the other direction.

To *PEDESTAL*: To adjust the elevation of the camera above floor level. *Pedestal up* means to make the elevation greater. *Pedestal down* means to decrease the height of the camera.

DOLLY, TRUCK, PEDESTAL: When used as nouns, these words refer to parts of the camera tripod mechanism. The dolly or truck is the base with wheels which supports the actual tripod. The pedestal is the vertical shaft that raises or lowers the camera.

nearly straight down so that your arms aren't waving overhead to aim the camera.

The CRADLE and CAM LINK HEADS are too heavy for portable use. Professional film- and videomakers will spend several hundred dollars on a sturdy tripod and a FLUID HEAD (Figure 6–27). The HEAD has oil-damped components inside it, making its movements smooth and precise.

Before using a FLUID HEAD, loosen its controls and move the camera up and down and back and forth about 10 times. The fluid gets ''stiff,'' and it needs to be ''worked'' a little to allow smooth movement.

Attaching the Camera to the Head. Attaching the camera to the HEAD is sometimes tricky. There is a threaded hole in the base of the camera or the base of the trigger handle. In the HEAD is a captive bolt which shouldn't fall out (if you can't find the bolt, it probably fell out). Somehow that bolt has to screw into the hole in the camera's base. It's not easy to get the bolt started straight. Starting it crooked will strip the threads, so be patient.

Cheaper HEADS have a single bolt. Industrial HEADS usually have a bolt with a TIGHTENER RING. As shown in Figures 6–28 and 6–29, the bolt goes through a slot in the

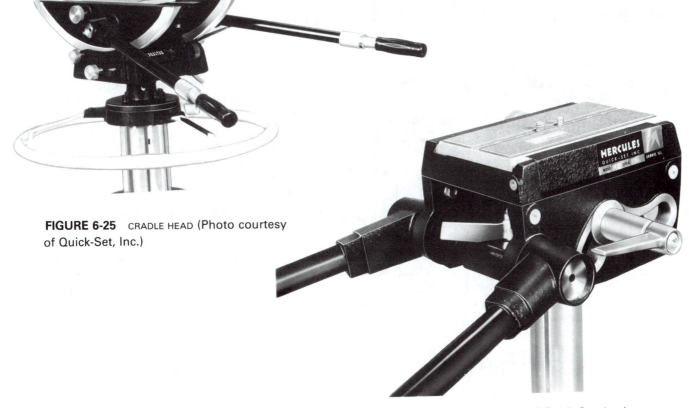

FIGURE 6-25 CRADLE HEAD (Photo courtesy of Quick-Set, Inc.)

FIGURE 6-26 CAM-LINK HEAD (Photo courtesy of Quick-Set, Inc.)

MINIGLOSSARY

*Fluid head Camera support that dampens the tilting and panning movement of the camera, smoothing out jerky movements.

DRAG control

FIGURE 6-27   FLUID HEAD for portability and smooth moves.

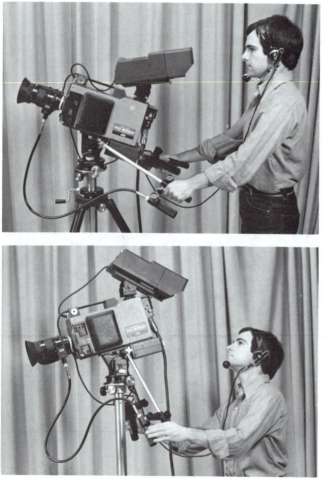

Handle adjustment for comfort.

HEAD and into the hole in the camera's base. The bolt is tightened as far as it will go into the camera. If the big, free-turning TIGHTENER RING gets in the way while you're doing this, screw it down toward the head of the bolt, away from the camera. Now loosen the bolt a bit so that both bolt and camera can slide backward and forward on the head (the bolt slides in the slot and takes the camera with it).

Find the place where the camera feels balanced and does not try to TILT up or down on its own. To do this, carefully loosen the TILT LOCK on the tripod HEAD *while holding the camera* so that it doesn't get dumped off the free-tilting HEAD. Shift the camera forward and backward until it feels balanced and then tighten the TILT LOCK. Next, tighten the TIGHTENER RING against the HEAD by screwing it up toward the camera as far as you can. Do not use pliers—that could strip the threads. Your fingers should get it tight enough.

Some heads don't have a slot for balancing. Without this little luxury of good balance, the camera is harder to aim.

Because this method of attaching cameras to HEADS is so awkward, some manufacturers have simplified the process. Some HEADS come with a removable plate which slides firmly into a groove in the HEAD and locks in place. This is called a QUICK RELEASE or a WEDGE MOUNT. The WEDGE plate can be attached to the base of the camera using the normal bolt and tightener ring method. Once the plate is attached to the camera, the pair can be slid easily onto the camera head and locked in place. Figure 6–30 shows such a QUICK RELEASE mount.

MINIGLOSSARY

Tightener ring   A large round nut on the mounting plate of a camera head for tightening the camera to the mount.

*Tilt lock   Camera head control to lock the camera in place so it can't tilt.

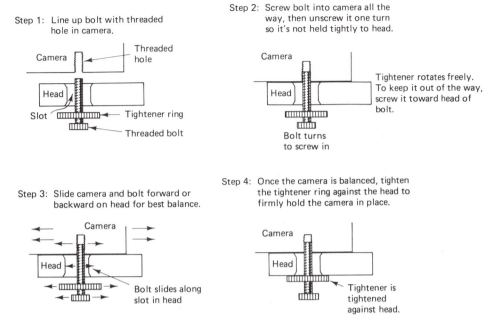

Step 1: Line up bolt with threaded hole in camera.

Step 2: Screw bolt into camera all the way, then unscrew it one turn so it's not held tightly to head.

Step 3: Slide camera and bolt forward or backward on head for best balance.

Step 4: Once the camera is balanced, tighten the tightener ring against the head to firmly hold the camera in place.

FIGURE 6-28 Steps for fastening camera to head.

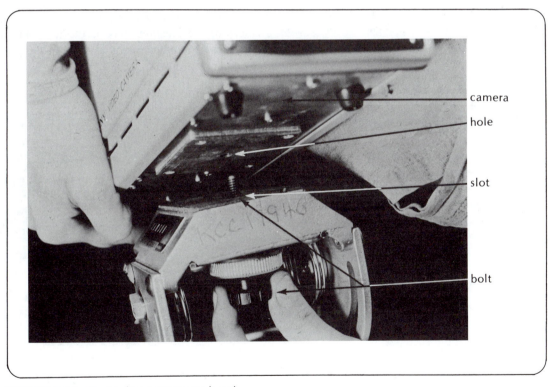

FIGURE 6-29 Fastening camera to head

MINIGLOSSARY

*Quick release A plate attached to the base of the camera by a bolt. The plate can then clip quickly onto the tripod head.

Wedge mount A type of quick release camera mount that slides into a wedge-shaped groove in the tripod head.

FIGURE 6-30   QUICK RELEASE camera mount allowing the camera to be easily removed from the tripod head.

Dollies

DOLLIES are the casters that hold up the tripod. Almost all DOLLIES will glide smoothly over a smooth floor. For an uneven floor or a carpeted floor, bigger wheels are necessary to smooth the bumps. The DOLLY in Figure 6–19 would work pretty well, but the one shown in Figure 6–21 would work very well. The simplest DOLLIES have casters which rotate in all directions, and usually the wheels will lock so that the camera doesn't roll. The DOLLY in Figure 6–21 also has CABLE GUARDS for each caster to sweep cables out of the way as the camera traverses the floor. If these guards are set too low, they'll snag on minor obstructions like the edge of a carpet. If set too high, they won't sweep the cables, and you'll roll bumpity jiggle over them. Another whoojit locks the casters, aiming them in a specific direction. This is handy when you want to DOLLY IN and don't want to be bothered aiming the wheels around before you can DOLLY back out. When this dingus is unlocked, the casters will rotate freely in whatever direction you pull the camera.

Before you DOLLY or TRUCK, make sure the wheels of the DOLLY are unlocked so they can turn. You'll be able to move the camera more easily if you anticipate and prepare by

1. Orienting the wheels in the direction you want to go
2. Getting cable and other obstacles out of your path
3. Preparing your camera cables so that they can follow you easily

Moving cameras smoothly from place to place is difficult enough without having to wrestle with tangled and twisted cables or sticking casters as you go.

Body Mounts

ENG (electronic news gathering) and EFP (electronic field production) cameras are pretty heavy, especially when you have to carry them around all day. It's especially difficult to walk and run with them. Special BODY MOUNTS are sometimes used to distribute the camera's weight and stabilize the picture. Almost all ENG cameras come with a SHOULDER POD which holds the camera balanced on the shoulder of the operator. Some also serve as a stand when the camera is set down.

The more elaborate BODY BRACE distributes the camera's weight to the operator's waist. They help balance cameras without a SHOULDER POD or cameras which need additional support. One disadvantage of the brace is that it picks up the breathing motion of the operator. Unless the cameraperson holds his or her breath, the picture tends to sway up and down.

SHOULDER POD

MINIGLOSSARY

*Cable guards   Metal shields that sweep cables out of the way so the camera dolly doesn't roll over them.

*Shoulder pod   Cushioned device connected to the base of a camera, allowing it to rest on the operator's shoulder.

Body brace   A cushioned framework worn by the camera op-

erator; it acts as a mount for a TV camera.

Steadicam   An elaborate framework of levers and springs used to hold a camera steady while the camera operator walks or climbs. A harness straps to the camera operator while the camera attaches to the other end of a movable arm.

STEADICAM JR camcorder stabilizing system. (Courtesy Cinema Products)

MERLIN CRANE ARM (Courtesy of Listec TV Equipment Corp.)

Steadicam

A more elaborate body mount is the STEADICAM. It permits the camera operator to move at will while the camera remains quite steady. For instance, a cameraperson could climb stairs with a STEADICAM and his or her body would bounce up and down with each step. The camera, however, would glide smoothly up the incline with the gentle push of a hand. The STEADICAM performs its magic with an elaborate framework of springs, levers, and hinges all attached to a padded body brace tightly strapped to the camera operator. STEADICAMS are much handier than tripods but cost a lot (they are usually rented), take some skill to operate and are quite heavy, tiring out the poor camera operator quickly.

If a $40,000 STEADICAM seems too pricey for you, and you're using a miniature camcorder weighing less than 3.5 pounds, you might try a STEADICAM JR. This little 2-pound camera mount includes an LCD monitor and batteries, and holds your picture steady for about $550. A delicate balancing system keeps your camera upright while you tilt and rotate the gimballed pistol grip as you move. Holding

this contraption away from your body for more than 10 minutes becomes a pain in the elbow, however.

Professional Mounts

Larger, professional studios employ fancier ways to hold the camera.

The STUDIO CRANE is usually a motor-driven device with three or four wheels and a movable boom which can raise the camera from close to the studio floor (about 1 foot) to about 10 feet high. Hydraulic pistons allow the camera to move smoothly from one position to another, carrying the camera operator along for a fun carnival ride.

The CRANE ARM is a smaller, lighter version of the heavy-duty STUDIO CRANE but without the carnival ride. It doesn't carry the camera operator with the camera. Most versions allow the camera to be operated at one end of a counterbalanced boom arm, perhaps 10 feet up in the air, while the camera operator stands at the other end viewing a TV monitor and aiming the camera remotely from the floor.

MINIGLOSSARY

Studio crane   Large studio device able to smoothly lift camera and operator high into the air.

Dolphin crane arm   Device for lifting cameras high into the air and aiming them while the camera operator remains on the ground.

Improvised Mounts

When shooting on location, you usually don't have room to carry much mounting equipment. Most camera mounting equipment is quite bulky. Here are a few shortcuts.

Wobbly camera shots will make your scenes look amateurish. Steady your camera by leaning it against the hood of a car, a fence, or a notch in a tree. If that's too hard, then hold the camera and brace yourself against something solid. Perhaps you can set your elbows on a desk or lean your shoulder against a doorway.

A pillow or bean bag is indispensable when doing a lot of on-location shooting. It does wonders for cradling a camera for steady shots. You can lie prone on the ground and use it to steady your low-angle shots. It's great for sinking your elbows into when you've propped yourself over the hood of a car. If you're shooting children (with a camera of course), you'll find yourself taking many medium-low-angle shots, probably while standing on your knees. A pillow or bean bag saves untold bruised kneecaps.

There are a lot of ways to move without a DOLLY. Wheelchairs glide smoothly, even over irregular surfaces. One person pushes while the other sits and holds the camera.

If you're shooting near shopping centers, the ubiquitous shopping cart gives a smooth ride over indoor floors (find one with casters that aren't square).

How would you shoot a bowling ball as it rolled down the alley? Try lying on an upside-down carpet remnant while two helpers push you with brooms.

You can shoot from the window of a moving car, from the tailgate of a pickup truck, from an electric company cherry picker (a bucket on a long mast capable of lifting a person high into the air), or from a fire truck ladder. Speedboats, roller coasters, and hot-air balloons are all possible sources of creative camera angles. If the bumpiness of the ride makes your picture jiggle too much, try to zoom the lens out as far as you can, hiding most of the wobbles.

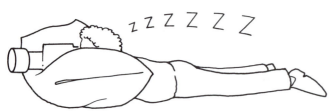

A pillow or bean bag will help you steady your shots.

Tai Chi Stance

If tripods are too cumbersome for you or you are moving a lot, there are several methods for steadying a hand-held camera. Figure 6–31 shows the Tai Chi stance (from Oriental martial arts), which minimizes natural body sway while putting you in excellent position for smooth moves in most any direction. Stand with your legs 18 inches apart, slightly pigeon-toed, with knees slightly bent. Keep your neck and the camera in close to your body. From the Tai Chi position, you can PAN by turning at the waist, you can TILT using your whole torso, you can PEDESTAL DOWN by further bending your knees, and you can DOLLY in or out by proceeding to walk, bent over in the Groucho Marx style, sliding your feet along, letting your knees absorb all the ups and downs while your torso glides smoothly through the air. Perhaps it sounds more like Oriental torture than good camera posture, and indeed it gets uncomfortable over time. But with a little practice, this silly-looking stance will deliver nice-looking pictures.

And while we are on the subject of walking with the camera, here's a good habit to get into, one practiced by professional microscope and telescope users: Stick one eye

FIGURE 6-31   Tai Chi position for holding camera.

MINIGLOSSARY

*Tai chi stance   Body position for hand-holding a camera steady.

EIS, ELECTRONIC IMAGE STABILIZATION steadies your shots inside your camera. *(Courtesy Panasonic)*

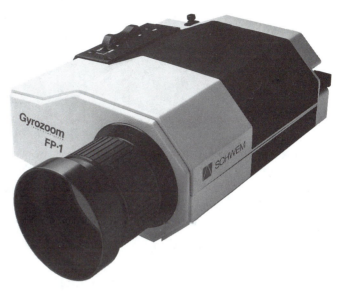

GYROZOOM professional image stabilizer, allows you to ride a horse or a helicopter while keeping steady shots. *(Courtesy Schwem Technology)*

to the eyepiece but *keep both eyes open*. It may be awkward at first, but a half hour of practice will teach your "unused" eye not to see. Thereafter, you don't have to tire your face squinting one eye closed, *and* you reap an unexpected benefit when you try to walk. Your unused eye starts keeping track of curb sides, low limbs, trip cables, and horse droppings as you march hither, thither, and yon.

A camera that rests on your shoulder will give a more stable picture than one with the common pistol grip. With the shoulder cameras, the camera gains stability by being pressed against you at several points: the eyecup (or forehead), the side of the face, the shoulder, the trigger grip, and the lens. The pistol grip camera, on the other hand, is held at the end of your bobbing arm. If using a pistol grip camera, stabilize your arms by pressing your elbows against your chest. Press the eyecup against your brow and hold onto the lens. Again, uncomfortable but stable.

As mentioned before, it will be more restful and more stable if you can brace yourself or your camera against something while shooting. Sit down, lean against a wall, brace the camera against a pole, or hold it against a car hood or a rock for rock solid shots.

For really low shots, the movable ELECTRONIC VIEWFINDER will make it easy to look straight down into the finder while cradling the camera beneath you, perhaps at knee height. If you can connect up to a TV set to view your shots, you may not need the VIEWFINDER at all, freeing you to hold the camera anywhere. Low shots are also easily done by hugging the camera to your waist and aiming it while viewing the TV screen.

ELECTRONIC IMAGE STABILIZATION

ELECTRONIC IMAGE STABILIZATION, or EIS, is an electronic way to get the bumps and jitters out of your shots. Sensors in the camera lens of some camcorders react to your body motions, reaiming the camera lens slightly to compensate for your movement. The system adds a small amount to the cost of a home camcorder and is more compact than the slightly unwieldy STEADICAM JR.

A professional model called the Schwem GYROZOOM does the same thing as described earlier, only better. The $15,000 zoom lens device attaches to a professional camera and using gyros, stabilizes the picture as the operator walks, runs, rides a horse, or shoots from a helicopter.

Both of the preceding mechanisms have one fault: They can't tell an accidental shake from a legitimate pan or tilt of the camera. The internal inertia will resist the pan at first and then overshoot the mark as the device starts to "catch up." Both take some getting used to, and like the STEADICAM and STEADICAM JR., require practice and concentration to achieve the desired results.

MINIGLOSSARY

Electronic image stabilization (EIS)   Electronic mechanism used in cameras to reduce shakiness in the picture.

Gyrozoom   Gyro stabilized zoom lens used with professional cameras to steady pictures.

MORE ABOUT TV CAMERA LENSES

Anybody who wears glasses knows how important lenses are. The lens is the camera's window to the world. This window could be a $50 STANDARD or FIXED FOCAL LENGTH (it doesn't zoom) lens or a $2000 ZOOM lens. The price depends on the complexity and quality of the lens. The principles for using and caring for both kinds of lenses will be the same.

HOW LENSES WORK

For both FIXED FOCUS and ZOOM lenses, light enters the lens, becomes concentrated or magnified by the optical elements in the lens, passes through an IRIS which reduces the light, and is then focused onto the chip or tube (see Figure 7–1).

The glass parts of the lens are called ELEMENTS, and their shape and position determine how much a picture will be magnified. The amount of magnification a lens gives you is called FOCAL LENGTH.

The positioning of the outermost ELEMENT of a lens generally determines whether far or near parts of a scene will appear sharpest. This is called FOCUS.

The third thing a lens can do is reduce the amount of light permitted through it. The IRIS or F-STOP handles this job.

All three factors are quantifiable and are used in describing the lens and its application. Let's look at each in more detail.

The Iris or F-Stop

The F-STOPS measure the ability of a lens to allow light through it. The IRIS ring, a rotatable collar somewhere on the lens, has numbers etched on it—typically these are 1.4, 2.0, 2.8, 4, 5.6, 8, 11, 16, and 22—F-STOPS. The F-STOPS represent the speed of the lens. SPEED is a photographic term describing how much light is allowed through the lens. The lower the f-number, the "faster" the lens—or the more light it allows through for a bright, contrasty picture. In general, you'd keep the lens "wide open" (set at the lowest f-number) if you're shooting inside with minimal light. Out-

---

MINIGLOSSARY

Element   Each glass part of the entire lens.
*Focal length   The distance between the optical center of a lens and the surface where the image is focused when the lens is focused

on infinity. The apparent magnification or angle of view of a lens.
Speed   A measure of how much light a lens can transmit. A "faster" lens has a lower f-stop number.

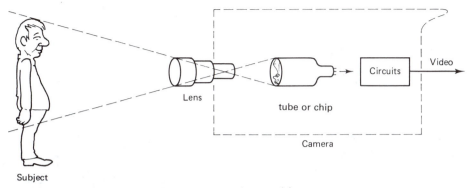

FIGURE 7-1 Lens focuses image onto tube or chip.

doors, in bright sunshine, you'd "stop down" to a small lens opening (a high f-number) to allow limited light.

Why aren't the numbers simply 1, 2, 3 instead of weird decimals? Because the F-STOP numbers are a result of a mathematical formula:

$$\frac{\text{focal length}}{\text{diameter of lens}} = \text{F-STOP}$$

Figure 7–2 diagrams this relationship. Common lens sizes make the F-STOP numbers come out weird, plus there's a lot of tradition mixed in here.

An F-STOP of 1.4 allows in twice as much light as f2. In fact, each time you "click off" an F-STOP on the IRIS— 1.4, 2, 2.8, 4—you double the amount of light to the camera. Thus, f1.4 admits eight times as much light as f4.

So where does IRIS come into the story? The IRIS is the mechanism inside the lens which cuts out the light. It performs much like the automatic IRIS in your eye, the part that makes your eyes blue or brown. Tiny flaps in the IRIS intersect to create a larger or smaller hole, permitting more or less light to pass through the lens. See Figure 7–3.

In summary, the IRIS is adjusted to the f-number that allows in *enough light* for your camera to "see" to make a good picture but *not too much light*, which will result in too contrasty a picture or poor DEPTH-OF-FIELD. So what's DEPTH-OF-FIELD?

Depth-of-Field

DEPTH-OF-FIELD is the range of distance over which a picture will remain in focus. Good DEPTH-OF-FIELD occurs when something near you and something far from you can both be sharply in focus at the same time. Poor DEPTH-OF-FIELD is the opposite. Things go badly out of focus when their distance from you is changed. Figure 7–4 diagrams this relationship.

Generally, one wants to maintain good DEPTH-OF-FIELD so that all aspects of the picture are sharp. There are times, however, when for artistic reasons, one would prefer to have the foreground picture (the part of the picture which is up close) sharp while the background is blurry. Such a condition focuses the viewer's attention on the central attraction in the foreground or middleground while making

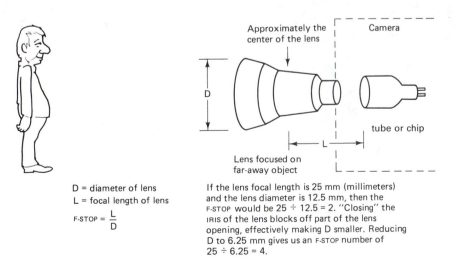

D = diameter of lens
L = focal length of lens

$$\text{F-STOP} = \frac{L}{D}$$

If the lens focal length is 25 mm (millimeters) and the lens diameter is 12.5 mm, then the F-STOP would be 25 ÷ 12.5 = 2. "Closing" the IRIS of the lens blocks off part of the lens opening, effectively making D smaller. Reducing D to 6.25 mm gives us an F-STOP number of 25 ÷ 6.25 = 4.

FIGURE 7-2 F-STOP measurement.

IRIS ring adjusts lens opening.

inside the lens

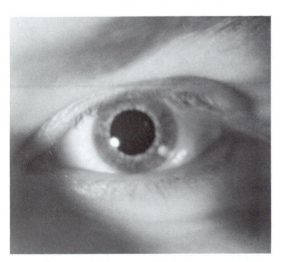
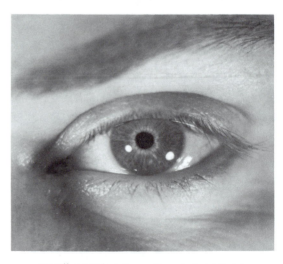

eye

large opening needed in dim light

small opening to reduce bright light

FIGURE 7-3   IRIS.

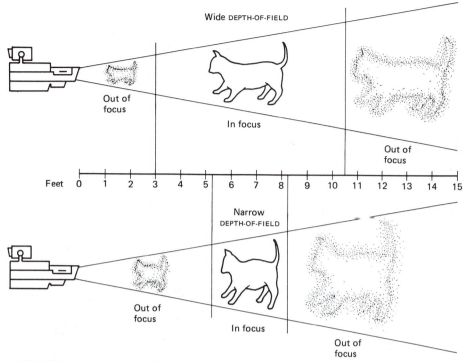

FIGURE 7-4   DEPTH-OF-FIELD.

the fuzzy background unobtrusive. And it's especially handy for hiding the inevitable scratches and smudges on your TV scenery. In such cases, poor DEPTH-OF-FIELD is an advantage.

The mechanism for adjusting the DEPTH-OF-FIELD is our old friend the IRIS. As you see from Figure 7–5, low f-numbers give poor DEPTH-OF-FIELD, while high f-numbers give excellent DEPTH-OF-FIELD.

Notice how you never get something for nothing. As you improve your DEPTH-OF-FIELD by increasing your f-number, you simultaneously reduce the amount of light permitted through the lens. A film photographer would make up for this loss of light by slowing down the shutter speed of his camera. This would expose the film to the light longer. But in TV cameras, you can't do this. TV cameras electronically take their pictures at ⅟₆₀ second, *always*. So while f22 may give you excellent DEPTH-OF-FIELD, it will let in very little light for a gray, dull picture. f2.8 would let in plenty of light for a brilliant contrasty picture, but your DEPTH-OF-FIELD would be very limited. Figure 7–6 shows the image from a TV camera at different IRIS settings. What can one do to get the best of both worlds?

1.   Try to get as much light on the subject as possible. This will make up for some of the light that the high f-numbers cut out.

2.   Decide where to compromise. Usually, people go for the bright enough picture and sacrifice DEPTH-OF-FIELD.

3.   One can make the camera work harder to compensate for the insufficient light associated with the high f-number. On color studio cameras you can increase GAIN. Some portable/studio color cameras have a +6dB, +12dB BOOST switch. Avoid the faster camera shutter speeds, if your camera has them. All these adjustments will enhance the contrast of your picture. Some cameras with automatic controls will make these adjustments for you. While these GAIN and BOOST adjustments will give you a brighter, more contrasty picture, the picture will become more grainy with the increase in these levels.

4.   One can make the VCR work harder and increase the contrast of its recorded picture by turning up its VIDEO LEVEL control. Most home VCRs will do this automatically. As with the camera, increasing these levels also increases the graininess of the picture.

You still never get something for nothing. Excellent DEPTH-OF-FIELD gives you poor contrast. Boosting the contrast electronically makes your picture grainier. The only way to get excellent DEPTH-OF-FIELD *and* a smooth and contrasty picture is to pour lots of light onto the subject. The subject had better not be a dish of ice cream.

A good compromise in the outdoor setting is to provide plenty of light and set the lens at f4 for moderate DEPTH-OF-FIELD and moderate contrast. Indoor shooting may gen-

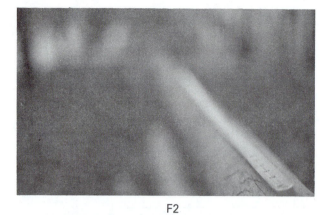

F2

F4

F8

F2.8

F5.6

F11

F16

F22

FIGURE 7-5 DEPTH-OF-FIELD at different IRIS settings. Camera is focused at the 2-inch mark.

F2.8: Contrast to spare but poor DEPTH-OF-FIELD

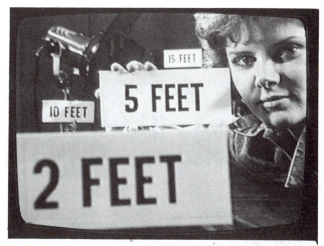

F4: A good compromise

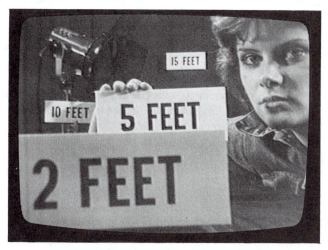

F8: Sometimes okay

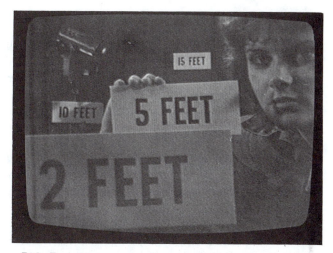

F16: Excellent DEPTH-OF-FIELD but picture is too faded

FIGURE 7-6   How a TV camera responds to different IRIS settings. The camera is focused at 5 feet.

erally require f1.4 to f2.0. Bright daytime shooting outdoors may permit f8 to be used.

Focal Length and Zoom Lenses

The FOCAL LENGTH of any lens is measured in inches or millimeters (mm) and describes two attributes of the lens:

1. How far from the optical center of the lens the image is focused (length L in Figure 7–2). Actually, video users don't give a hoot about this measurement.

2. How wide an angle the lens will cover. This factor gets plenty of hoots.

A lens that doesn't zoom is called a FIXED FOCAL LENGTH lens. (A lens which has been run over by a truck is called a *broken* focal length lens.)

A WIDE-ANGLE FIXED FOCUS LENGTH LENS like the one shown in Figure 7–7 is likely to have a FOCAL LENGTH of 12mm or less. It will display a wide field of view like the LONG SHOT shown in Figure 7–8. It is useful for surveilance of large areas or for shooting in cramped quarters where you can't easily back up far enough to "get it all in."

FIXED FOCAL LENGTH GENERAL-PURPOSE lenses for TV cameras, like the one shown in Figure 7–9, typically have a FOCAL LENGTH of about 25mm and will display a

---

MINIGLOSSARY

*Wide-angle   The opposite of telephoto, a wide-angle lens takes in a broad panoramic view. The lens "gets everything in," but everything may appear small in the picture.

*Telephoto   The opposite of wide-angle, a telephoto lens magnifies the view, like binoculars. It also has a narrow field of view, concentrating on one part of a picture and cutting out the rest.

FIGURE 7-7   FIXED FOCAL LENGTH WIDE ANGLE lens.
(Courtesy of Century Precision Optics)

medium field of view like the medium shot shown in Figure 7–10.

A TELEPHOTO lens like the one in Figure 7–11 will have a FOCAL LENGTH of 50mm or more and will display a narrow field of view like the CLOSE-UP shown in Figure 7–12. The TELEPHOTO lens gives close-ups of objects far from the camera. These are handy when you want to shoot action shots of a raging bull without actually climbing into the bull ring.

You'll notice from Figures 7–7, 7–9, and 7–11 that you can almost guess the FOCAL LENGTH of a lens just by looking at it. Long FOCAL LENGTH lenses are physically long, and short FOCAL LENGTH lenses tend to be short.

A zoom lens has a variable FOCAL LENGTH (as opposed to a FIXED FOCAL LENGTH) and may range from WIDE-ANGLE to TELEPHOTO. One example of how a zoom lens's range

of FOCAL LENGTH can be expressed is as follows: 12–72mm. The lens can give a WIDE-ANGLE shot like a 12mm lens and can be zoomed in to give a shot like a 72mm lens. This lens has a ZOOM RATIO of 6:1 (6 to 1). You can zoom it in to six times its lowest FOCAL LENGTH and make something look six times closer. This ratio can also be expressed as 6×.

Such lenses cost around $300 and are typically found on the better home cameras and the lower-priced industrial cameras. A lens with a ZOOM RATIO of 3:1 can be had for $200 and is common on inexpensive home portable TV cameras. A 10:1 zoom lens would be found on the the better cameras and has a broad enough zoom range to permit the camera to be used in a variety of situations. It can take WIDE-ANGLE shots in a cramped studio as well as TELE-PHOTO shots of hungry lions in the jungle. Generally, the greater the range of the zoom, the higher the cost will zoom (like $700–$2000).

Zoom lenses with broad 10×, 12×, or 15× ZOOM RATIOS tend to be long and heavy, making them a little hard to work with when portability is paramount. Lenses with a 6× ZOOM RATIO are relatively small and light.

Changing the FOCAL LENGTH of a lens does more than simply change the magnification of the picture. Other subtle changes occur. *The greater the FOCAL LENGTH, the greater the magnification of the picture, the narrower the field of view, the flatter the scene looks, and the narrower the DEPTH-OF-FIELD becomes.*

You are probably familiar with the telephoto shots of baseball games where the players in the outfield look like they are standing on top of the pitcher, who himself looks inches away from the batter. If you want to compress distance so that far-away things look closer to nearby things,

FIGURE 7-8   WIDE ANGLE lens (low FOCAL LENGTH number) gives a LONG SHOT.

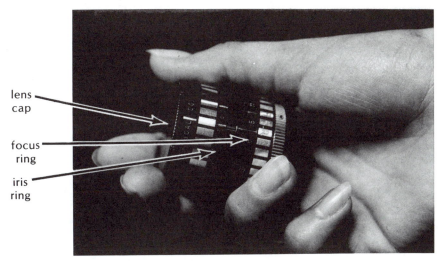

lens
cap

focus
ring

iris
ring

FIGURE 7-9 FIXED FOCAL LENGTH STANDARD lens.

FIGURE 7-10 STANDARD lens (about 25mm) gives a MEDIUM SHOT.

FIGURE 7-11 FIXED FOCAL LENGTH TELEPHOTO lens. (Courtesy of Tamron Industries)

FIGURE 7-12 TELEPHOTO lens gives a CLOSE-UP picture.

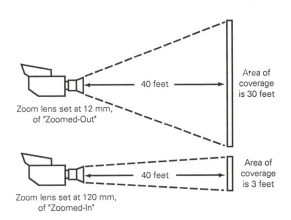

40 feet

Area of coverage is 30 feet

Zoom lens set at 12 mm, of "Zoomed-Out"

40 feet

Area of coverage is 3 feet

Zoom lens set at 120 mm, of "Zoomed-In"

Angle of view from a zoom lens. 12-120 Zoom Range (from *Single Camera Video Production*—Prentice Hall)

TELEPHOTO

WIDE ANGLE

TELEPHOTO lens compresses distance while WIDE ANGLE lens emphasizes it. In both cases, the man and the tunnel are the same distance from the car.

WIDE-ANGLE shots have excellent DEPTH-OF-FIELD, and TELEPHOTO shots have poor DEPTH-OF-FIELD. This sometimes makes it hard to shoot moving objects with a TELEPHOTO lens; they are constantly wiggling out of focus on you. A WIDE-ANGLE lens, on the other hand, hides the focusing errors. WIDE-ANGLE shots are useful when you have to run with the camera or shoot without taking time to focus.

FOCUSING

You have learned that focusing is done by rotating part of the lens. This process always involved looking through the viewfinder (or a TV monitor) to see your results. If you have no viewfiner or are forced to guess when things are properly focused, do the following:

Observe the focus ring on the lens. You'll see numbers etched in it representing the distance in feet or in meters at which the object will be in focus. Estimate the distance to your subject and turn this ring to the appropriate number, and you will be roughly focused.

then use a TELEPHOTO LENS. If you'd like to stress the difference between nearby and far-away things, use a WIDE-ANGLE shot. If a vehicle or person is coming toward you and you wish to make their motion appear slower or wish to hold them in the scene longer, then have them start from far away and use a TELEPHOTO lens. If however, you wish to exaggerate their speed or progress, use a WIDE-ANGLE lens. Once near you, the subject will appear to whip by.

TELEPHOTO shot gives narrow DEPTH-OF-FIELD.

How FOCAL LENGTH affects DEPTH-OF-FIELD.

WIDE ANGLE shot gives broad DEPTH-OF-FIELD.

As you think back to the relationship between DEPTH-OF-FIELD and IRIS settings, you appreciate the fact that it is much harder to get something into focus if the DEPTH-OF-FIELD is very narrow. But there are times when this narrow DEPTH-OF-FIELD can *help* you focus.

Take the example in which you have good DEPTH-OF-FIELD and are trying to focus. You turn the focus ring, and the blurry picture becomes less blurry, less blurry, pretty good, fairly sharp, sharp, maybe sharper, maybe not sharper, still fairly sharp, a little blurry—so you start turning back in the other direction. There is a range where the picture doesn't change much while you turn the focus ring. Which position is right?

Now take the example of poor DEPTH-OF-FIELD. You turn the focus ring, and the blurry picture becomes blurry, blurry, less blurry, good, perfect, good, blurry again—so you start turning back. Here the picture zaps into focus and out again. There is no guesswork as to where the right focus position is; there is just one narrow range where the picture is good.

The moral of the story is: *For accurate focusing, open the lens to the lowest f-numbers, focus, and then return to the higher f-number.*

You also remember that a zoom lens has its poorest DEPTH-OF-FIELD when it is zoomed in. Therefore, if you wish to focus on something precisely, you zoom in first in order to exaggerate your focusing errors.

Teaming up these two methods together gives you a superaccurate, microprecise, but takes-a-while-to-do regimen for focusing that goes like this:

1. Zoom in all the way on the subject.
2. Open the IRIS all the way.
3. Focus the lens.
4. Zoom out to the shot you want.
5. Reset the IRIS to the appropriate f-number.

Obviously, this list of steps is only useful when you have plenty of time and a subject that is either dead or tied down. Nevertheless, this is the only way to get a sharp picture of things that *have* to be perfectly focused such as graphics or close-ups of small items.

What do you do when you have a subject who moves around a lot? How do you stay in focus? Here are some possibilities:

1. If you are a good focuser, just stay alert and adjust for every movement. Most of us are not good focusers, however.
2. Flood the subject with light so that you can use high f-numbers for broad DEPTH-OF-FIELD.
3. Stay zoomed out. Focusing inaccuracies are most

Lens close to face

Face distortion

Lens close to car

Car distortion

WIDE-ANGLE lens distortions.

noticeable in close-ups. When you are zoomed out, nearly everything appears sharp.

4. Try to get the subject to move *laterally* to you, not toward or away from you. Since the subject stays roughly the same distance from the camera, you will not have to refocus, just aim.

5. Try to use big subjects so that you can zoom out or can stay farther away from them. Why are big, zoomed-out subjects easier to focus? To fill a TV screen, little objects must be magnified. You do this by zooming in or by moving the camera closer to them. A zoomed-in lens blows up all the little focusing inaccuracies, especially if an object is itself deep or is moving toward or away from the camera. A camera close to an object also exaggerates the focusing problems. When an object 3 feet away from a camera moves 1 foot, you get a very noticeable 33 percent focusing error. When an object 30 feet away moves 1 foot, the error is a minor 3 percent. Combining both the zooming and the nearness concepts, we find that zooming in on a postage stamp held in somebody's hand 3 feet away will display formidable focusing problems as the hand moves. But zooming out on a giant poster of a postage stamp held 30 feet away poses no focusing problems even if the poster moves a foot.

6. Last and least, try to confine your subjects by seating them, tranquilizing them, encumbering them with microphone cables, or marking a spot on the floor where they must stand.

Selective Focusing

A tricky way to move the viewer's attention from one object to another is by SELECTIVE FOCUS. First you focus on an object in, say, the foreground, leaving the background blurry. Next you refocus for the background, blurring the fore-

Your attention is drawn to the alarm clock.

Your attention is drawn to the sleeper.

FIGURE 7-13   Selective focus.

ground. Your viewer's eyes will stay riveted to whichever part of the picture is in focus at the time. Figure 7–13 shows how your attention can be moved from the time on the clock to the person asleep in bed. Another SELECTIVE FOCUS technique is used with long rows of objects (soldiers, flowers, toys on an assembly line). Focus first on the closest items and slowly refocus to the farthest ones. One by one, each will pull into focus and then retreat into a blur as your eye follows down the line to the end. This is a technique called PULL FOCUS, focusing from one object to another while the camera is "on."

To intensify the SELECTIVE FOCUS effect, it is good to have a narrow DEPTH-OF-FIELD. Use a long FOCAL LENGTH lens and a low F-STOP (if possible) and try to keep the nearby objects as near as possible and the far ones as far as possible.

LENS CONTROL SYSTEMS

Figure 6–10 showed a manual zoom lens. To focus one, you rotate the outside ring. To zoom, you rotate the middle ring. To change F-STOPS, you rotate the ring nearest the camera.

The IRIS ring is seldom adjusted while you are shooting. The focus and zoom rings need almost constant adjustment. It's not easy to do this with one hand. Only practice will teach your fingers which ring does what so that focusing and zooming become automatic for you.

No one ever has trouble telling amateur camera work from professional camera work. Amateurs invariably wiggle the camera while zooming, make jerky zooms, change zoom or focus too late, or zoom a little in the wrong direction before discovering their error and then have to zoom in the

right direction. All of these bad moves are very obvious on the screen and make a TV production look amateurish. There is only one way to wipe out all of these lens control problems at once, and it's summarized in one word, practice.

Learn which ring is focus and which is zoom. *Learn* which way to turn the zoom ring to zoom in or zoom out (it is inexcusable to "try" one direction of the ring and then the other to find out which way zooms you in; you should *know* this; it should be automatic). *Learn* which way to turn the focus ring when your subject is coming toward you.

None of this is easy, especially if you have to do it with one hand. If the camera is on a tripod and one hand is aiming the camera, then you only have one hand left for focusing and zooming. If you're using a home-type camera

---

FOCUSING TIPS REVIEWED

1. Practice, practice, practice, practice, practice—or you'll be no darned good!

2. Remember, if time permits, that you focus on a subject by (1) zooming in all the way first, (2) focusing for a sharp picture, and (3) zooming out to the shot you want.

3. If your camera is on (its tally light is lit) or if time doesn't permit proper focusing, you omit steps (1) and (3) and just focus as best you can without extra zooming.

4. Objects closer than 5 feet (or the closest distance etched on the lens barrel) probably cannot be focused clearly, so keep your distance.

5. When you're trying to focus in step (2), use this method: (1) Turn the focus knob until the picture changes from blurry to sharp to a *little* blurry again. (2) Since you've gone too far, turn the knob back again slightly. Once you're used to it, this technique becomes fast and accurate.

6. Practice, practice, practice. Step 2 should take 3 seconds, and you will have a picture that is crystal perfect.

MINIGLOSSARY

*Selective focus   Adjusting the focus of a lens so that one part of the picture is sharp and other parts fuzzy, useful for directing attention.

Pull focus   Adjusting the focus of a lens, often to keep a moving subject sharp, while your camera is "on."

on a trigger handle, you still have only one hand with which to focus and zoom because the other is holding up the camera. To make camera handling easier, the manufacturers have devised some ingenious ways to zoom and focus lenses.

Pump Zoom

Figure 7–14 shows a PUMP ZOOM. It has a large ring on the lens barrel which can be rotated to focus the lens or slid in or out to zoom the lens. Thus, one hand can focus and zoom at the same time. One disadvantage of this type of lens is that pumping the barrel in and out may jiggle your picture.

Cable Drive

Common in older studios is the manual CABLE DRIVE (Figure 7–15). A focusing knob is mounted on the camera handle, and a zoom crank is mounted on the side of the camera. Each is connected to the lens by a flexible cable. On some models, there are two camera handles, and the zoom crank is mounted on one of them.

The crank allows you to perform very smooth, slow zooms with one hand while the other hand constantly guides the aiming of the camera.

With a little practice, you can learn to focus at the same time you're aiming the camera by placing the camera

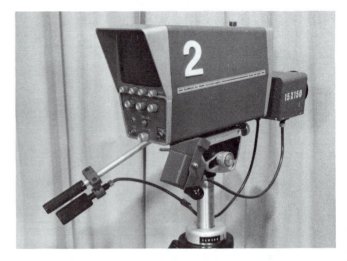

FIGURE 7-15   CABLE DRIVE.

handle in the palm of your hand and running the tips of your fingers over the rotating focus knob nearby. See Figure 7–16.

It takes some time to get the hang of cranking and focusing and aiming the camera in one smooth move, but with practice, you can achieve very professional-looking results.

One disadvantage of cable drive cameras is that the cables, which hang down somewhat, tend to snag on things like parts of the camera's tripod when the camera is PANNED, TILTED, or PEDESTALED UP. Also, some cable drives, when they get old, get "sticky," yielding jerky movements instead of smooth ones.

FIGURE 7-14   PUMP ZOOM.

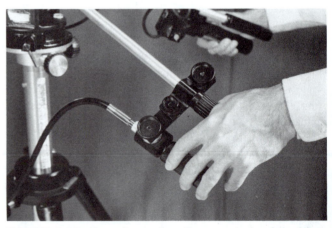

FIGURE 7-16   Focusing a handle-mounted cable drive with your fingertips.

MINIGLOSSARY

Pump zoom   Zoom lens with a sliding barrel that slides in or out to zoom, and focuses the lens when rotated.

*Cable drive   Cranks or knobs, mounted on or near the tripod handles, are connected to the lens via cables and remotely control the lens's zoom and focus.

Rocker switch   A two-way switch that pressed one way zooms a lens in, and if pressed the other way zooms it out.

FIGURE 7-18   Electric zoom, manual focus on a portable camera.

Push left to zoom in.

Push right to zoom out.

FIGURE 7-17   Electric zoom, manual focus on a studio camera.

Electric Zoom

Instead of a crank to zoom your studio camera's lens, some modern cameras come with an ELECTRIC ZOOM control. Figure 7–17 shows one mounted on the camera handle. It is a ROCKER SWITCH. You operate it by placing your thumb in the groove on the switch as shown in the figure and moving your thumb to the left to zoom in and to the right to zoom out. Pushing your thumb farther makes the camera

zoom faster. Meanwhile, the rest of your right hand firmly grips the camera handle, allowing you to maintain steady aim. As before, the left hand grasps the left handle (further steadying the aim) while the fingertips focus the lens.

As mentioned in Chapter 6, portable cameras may also have an electric zoom with a ROCKER SWITCH. As shown in Figure 7–18, one hand slides under a strap (to help you hold the camera), and two fingers press the ROCKER SWITCH, one way to zoom in and the other way to zoom out. The other hand grasps the lens barrel to manually focus the lens. This two-handed operation allows you to focus and zoom at the same time. This is the most popular setup for industrial and ENG portable cameras.

ATTACHING AND CLEANING LENSES

Home video cameras and camcorders typically have lenses built into the camera body itself; they can't be removed. Usually the autofocus, autoiris, zoom motors, and sensors are wrapped around the lens and tied inextricably to the camera's circuits in one compact package.

Industrial and professional camcorders are generally the opposite; their lenses are removable and replaceable with various types suited to the task at hand. They usually lack autofocus, and the zoom and iris motors (if there are any) come attached to the lens. Figure 7–19 contrasts these two kinds of camera lenses.

Many industrial and professional cameras, and most 16mm-film cameras come with a C-MOUNT, which indicates a standard hole size, lens base size, and thread size. Any C-MOUNT lens can be screwed into the face of any C-MOUNT camera as shown in Figure 7–20. Another called the BAY-ONET MOUNT is attachable with a half twist. Figure 7–21 shows a mount popular with JVC. There are also adapters to convert from one mounting type to the other. C-MOUNTS and BAYONET MOUNTS are generally found on cameras having 1-inch or ⅔-inch image sensors.

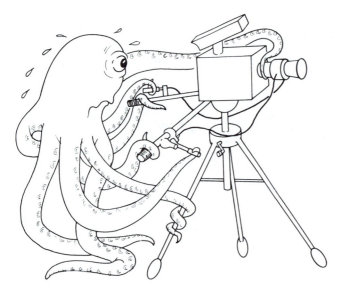

Manipulating all the camera controls at once takes some practice.

FIGURE 7-19   Built-in vs. exchangeable lenses.

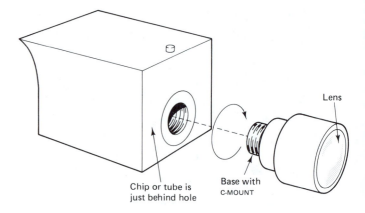

FIGURE 7-20   Attaching lens to camera.

Semiprofessional camcorders with ½-inch chips may use a VL-MOUNT, a bayonet-type of lens mount with a special twist—the lenses, even though they contain autofocus, autoiris, and zoom controls, *can* be swapped with other VL-MOUNT camcorders. The mounting connection as well as the camcorder and lens circuits are standardized to make them all interchangeable.

When changing lenses, be very careful not to touch or soil the chip or tube just inside the hole in the camera's face. Dirt on the chip will show up in your picture. If the chip is dusty (and the tiniest flakes of dust make a big difference, especially in bright scenes), gently brush it off with a soft CAMEL'S-HAIR BRUSH like the one shown in Figure 7–22. A stream of dry air may also blow off lint and dust particles.

The same care is appropriate for TV camera lenses. The outside lens, however, always seems to have a way of attracting fingerprints. Since oily prints don't brush off easily, one must resort to gentle wiping with lens tissue. If that doesn't work, try a clean rag dampened with soapy water as a last resort. Take care to rub gently lest you scratch the thin blue or amber coating on the lens. The coating on the lens helps the light go through the glass in the lens.

Wipe the lens with a rotary motion to reduce the chance of scratching it. Some manufacturers advise against using eyeglass lens tissue to clean camera lenses because it may scratch the lens.

Lenses are pretty delicate and like to be pampered. Like most TV equipment, they abhor bumps, water, sand, oil, smoke, and food. The lens cap protects them from some of these hazards when the camera is not in use.

MINIGLOSSARY

C-mount   Standardized connection between TV camera lenses and TV cameras, used in industrial cameras.

Camel's-hair brush   Brush of soft camel's hair, often with bellows in the handle, for blowing dust off lenses.

Bayonet mount   Lens-to-camera connection popular with JVC and others; lens removes with ½ twist.

VL-mount   Standardized lens mount for ½-inch chip camcorders assuring automatic lens controls are compatible with camcorder, thus allowing lenses to be swapped.

FIGURE 7-21   JVC camera mount.

How can you tell whether that fleck of dust you see on the monitor screen is from the camera's lens or from its chip? Rotate the lens, unscrewing it slightly from the camera mount. If the dirt rotates also, it is on the lens. If the spot of dust remains stationary, it is on the chip (or else you have a tube with a burn-in).

If the dirt is on the lens, is it on the outside glass or on the glass at the base of the lens? After snugging the lens back into its mount, try rotating the focus ring of the lens. If the dirt remains stationary, it is probably on the base glass. If it rotates, it is on the outside glass.

If you can't tell whether you have a burn-in or a spot of dust on the tube, try capping the lens. A burn-in will remain even if the picture is totally dark. A piece of dust will only show in a light scene. In fact, the brighter the scene, the more the dust will show. Bad pixels in a chip could show up as white dots in your scene (with your lens capped or not), and sometimes dark dots in a bright scene.

Lens Format

Now let's go back for a closer look at the C-MOUNT and lens interchangeability. Any C-MOUNT lens will fit any C-MOUNT camera including movie cameras. There are even adapters which allow C-MOUNT cameras to be used with the P-type, T-type, or "bayonet" lenses from 35mm slide cameras. Just because the lenses fit, however, doesn't mean that they all work equally well with all cameras. There's a little complication called FORMAT which is of interest to you if you swap lenses around between dissimilar cameras. FORMAT refers to the size of the image the lens projects onto the sensitive face of the camera's chip or tube. This little image must be the right size to fit the face of the tube.

You'll remember from Chapter 6 (of course) that there are different-sized camera chips. Home video cameras gen-

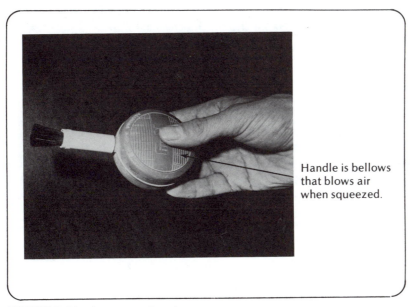

Handle is bellows
that blows air
when squeezed.

FIGURE 7-22   Lens-care brush.

MINIGLOSSARY

Lens format   Describes the size of the lens's focused image. It should match the size of the camera's pickup chip.

Vignetting   A condition where the picture's edges (usually the corners) show the dark edges of the lens, often because the lens format is too small for the pickup chip or tube.

erally use ½-inch chips. Industrial cameras use ⅔-inch chips. Professional cameras use ⅔- and 1-inch tubes. Cameras with ⅔-inch chips need lenses that make ⅔-inch pictures. These are called ⅔-inch FORMAT lenses.

To work properly, the lens FORMAT must match the chip size, but there is room to cheat a little. If you wanted to, you could successfully use a larger FORMAT lens on a smaller camera if the lens mount fit. There would be image to spare. The only strange thing you'd notice is that the lens would act a little more "zoomed in" than it should. This is because you are using only the middle part of the picture, and throwing away the outside and this tends to magnify the picture, giving it that "zoomed-in" look.

For example, here's the FOCAL LENGTH for a "normal" lens on three different kinds of cameras:

Type of Camera	Normal lens FOCAL LENGTH
35mm slide	50
⅔-inch chip	15
½-inch chip	10

All three mentioned in the table make a "normal" picture with *their own* camera. Put the bigger lens on a smaller camera and you get a TELEPHOTO (greater FOCAL LENGTH) effect.

A larger FORMAT lens will work on a smaller FORMAT camera, but the opposite is not true. If you tried a ½-inch lens on a ⅔-inch camera, the image would be too small,

and the camera would "see" the dark edges of the lens in the corners of its pictures, a phenomenon called VIGNETTING (see Figure 7–23).

CLOSE-UP SHOOTING

Normally, if you try to shoot something closer than 4 or 5 feet from your camera, the picture will be blurry. If you must get closer to your work, the following options are available.

Close-up or Macro Lenses

CLOSE-UP or MACRO lenses are made especially for close work. Some are able to focus on things only an inch away. Figure 7–24 shows a MACRO ZOOM lens.

In their "normal" (nonMACRO) mode, most of these lenses work like any other lens, focusing down to 4 feet or so. They also zoom normally. To change them to MACRO, you generally throw a "safety catch" (which prevents the lens from going into this mode accidentally) and then refocus on a very near object. Sometimes the zoom control becomes the MACRO *focus* control. Many MACRO lenses can be refocused easily from a super-close-up back out to normal distances, opening the possibility of "arty" or creative transitional effects (see Chapter 8).

Unfortunately, while in the MACRO mode, zoom lenses cannot zoom. Also, many MACRO zoom lenses have a "dead area" where they cannot focus. The MACRO mode may work from 1 inch to 3 inches and the non-MACRO mode may work from 2 feet to infinity, but you can't focus anywhere between 3 inches and 2 feet.

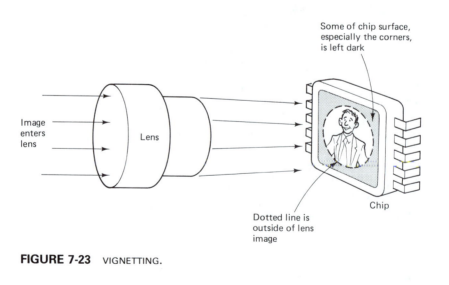

FIGURE 7-23   VIGNETTING.

MINIGLOSSARY

*Macro lens   A lens that can be focused on a very close (sometimes touching) object.

*Close-up lens attachment   A lens element that screws onto your existing lens, allowing it to focus closer than normal.

Diopter   The measure of a close-up lens attachment's strength. The larger the number (+1, +2, +3) diopter, the closer the lens can focus.

FIGURE 7-24 MACRO ZOOM lens
(Courtesy of Vivitar Corp.)

Close-up Lens Attachments

Your regular zoom or FIXED FOCAL LENGTH lens can be made to focus on closer subjects by the mere addition of a CLOSE-UP LENS ATTACHMENT as shown in Figure 7–25. Attach these by unscrewing the lens shade and screwing on the CLOSE-UP LENS ATTACHMENT in its place. The curved surface of the attachment should face *away* from the camera. *Do not* screw close-up lens attachments down tight; they easily seize up and become hard to remove. The lens shade may now be screwed onto the close-up attachment. Leave this slightly loose too.

Incidentally, here's a trick for removing a seized up CLOSE-UP-LENS ATTACHMENT. Screw a lens shade onto the attachment, and if you have a lens cap that goes over the

lens shade, put that on too. Now try twisting the three off together. The lens shade helps act as a handle. And here's another trick while we're at it. If a lens shade is hard to remove, place the lens cap on it first and then try unscrewing it. The cap holds the shade perfectly round so that it doesn't bind while unscrewing. Now back to business.

With a close-up lens attachment on your zoom lens, you can zoom the full range without going out of focus, assuming that you are the right distance from your subject and that you followed proper focusing procedure to start with. This is the greatest advantage of close-up lens attachments over MACRO lenses.

To select a proper CLOSE-UP LENS ATTACHMENT, you need to know two things:

1. The magnification of the lens
2. Its compatibility with your present lens

Magnification. CLOSE-UP LENS ATTACHMENTS don't appreciably magnify an image per se. They merely permit you to bring your camera closer to your subject without going out of focus. They essentially make your camera nearsighted.

The power of a CLOSE-UP LENS ATTACHMENT is measured in DIOPTERS. The bigger the DIOPTER number, the stronger the lens and the closer your camera can "see" with it on. The weaker ones are +1 and +2 and the stronger ones, +5 and higher. Where normally we could shoot from infinity up to 4 feet, a +1 DIOPTER CLOSE-UP LENS ATTACHMENT lets us shoot ranges from about 3 feet to 1½ feet. A +2 DIOPTER attachment gets us from 1½ feet to 1 foot, and +3 gets us from 1 foot to ¾ foot. Up to +3 diopters, your image will stay sharp throughout the zoom range. Above +3, it becomes increasingly necessary to refocus as you zoom in order to keep a sharp picture.

CLOSE-UP LENS ATTACHMENTS are generally threaded so that after one is screwed onto your regular lens, another CLOSE-UP LENS ATTACHMENT can be screwed onto the first attachment—sort of a piggyback arrangement. In such cases, the DIOPTERS are additive: A +2 lens added to a +3 lens is equal to a +5 DIOPTERS in power. When stacking lenses, the highest-DIOPTER lens should be nearest the camera. The curved part of the glass should always face out.

Adding lenses tends to decrease sharpness and brightness, especially in the corners of your picture. To combat the blurriness caused by using several lenses at once, flood the subject with light and stop your IRIS down to f16 or more if possible.

Compatibility. Most zoom lenses are threaded so that the lens shade can be screwed off and a CLOSE-UP LENS ATTACHMENT screwed on.

Photo equipment stores sell such attachments for about

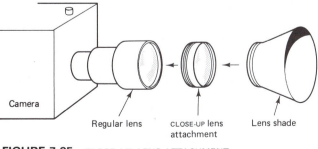

Camera — Regular lens — CLOSE-UP lens attachment — Lens shade

FIGURE 7-25 CLOSE-UP LENS ATTACHMENT.

$25. Wider-diameter lenses cost more. TV cameras generally use wider lenses than those used on film cameras sold in photo stores. You may need to try a photo store with a broad lens selection in order to meet up with an attachment large enough for your camera's lens.

For a CLOSE-UP LENS ATTACHMENT to fit on your lens, two things must match:

1. The diameter of the attachment must be the same as the diameter of your camera lens. This number is usually printed on the lens. Home camcorders typically use 46, 52, and 55mm diameters.
2. The thread sizes must match.

Usually, if you can bring your camera lens to a photo equipment store, you can try out the CLOSE-UP LENS ATTACHMENTS for size.

What do you do if you already have a CLOSE-UP LENS ATTACHMENT but it's just a little too small for your lens? You buy a STEPPING RING which will adapt one size lens to another size lens. STEPPING RINGS do not span great differences in sizes but will handle small differences. Figure 7–26 shows a STEPPING RING.

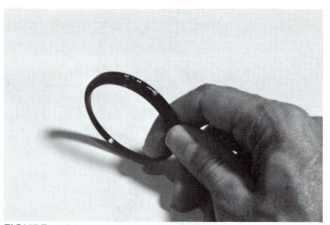

FIGURE 7-26 STEPPING RING adapts one size lens to a slightly different sized lens attachment.

TELEPHOTO AND WIDE-ANGLE CONVERTERS

Close relatives to the CLOSE-UP LENS ATTACHMENT is the TELEPHOTO and WIDE-ANGLE CONVERTER. Whatever focal length your STANDARD or ZOOM lens is, these attachments

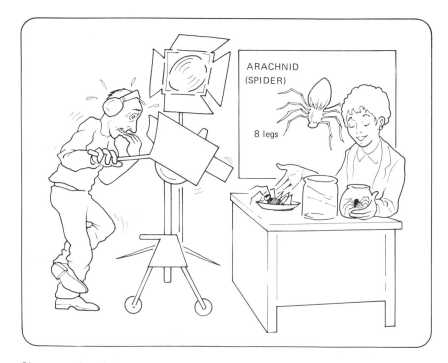

Close-up shooting.

MINIGLOSSARY

Stepping ring   Adapter to allow a lens attachment to fit a lens of a different size.

*Telephoto converter   Lens attachment to increase the focal length (magnification) of a lens for a narrower angle of view.

*Wide-angle converter   Lens attachment to decrease a lens's focal length, giving the image a wider angle of view.

will make it greater or smaller, thus yielding a wider view or a more magnified view of your subject. Figure 7–27 shows a WIDE-ANGLE and TELEPHOTO CONVERTER and the image you get when using them.

Some CONVERTERS are reversible; they're threaded at both ends. Attach the CONVERTER one way and it multiplies your FOCAL LENGTH 1.5X, making your lens more TELE-PHOTO. Attach it the opposite way and it multiplies your FOCAL LENGTH .6X, making it more WIDE ANGLE. If, for instance your 6:1 zoom lens had a range of 20mm to 120mm, the 1.5X CONVERTER could make your lens act like it was 30mm to 180 mm (better for bird watching and lion safaris). The other way around, the .6X CONVERTER makes your lens 12mm × 72mm (better for spelunking and indoor shots). Figure 7–28 shows a reversible CONVERTER.

Beware that some CONVERTERS may cause VIGNET-

Normal view

Normal view

"2X TelXtender®" TELEPHOTO CONVERTER

"Curvatar™" WIDE-ANGLE CONVERTER

View with "2X TelXtender®"

With "Curvatar™"

FIGURE 7-27 TELEPHOTO and WIDE-angle converters (Courtesy of Spiratone, Inc.).

FIGURE 7-28   REVERSIBLE CONVERTER. (Courtesy Electronic Mailbox)

TING, especially in the WIDE ANGLE modes. Try before you buy.

LENS FILTERS

A FILTER blocks something unwanted and lets through something wanted. Lens FILTERS attach to the outside of your lens and block out unwanted colors, brightness, or glare.

Sometimes, lens attachments which are not really FILTERS are called FILTERS simply because they look like FILTERS and attach to the outside of the lens like filters. They don't filter out anything; they may distort or add some special effect to the image from the lens such as star patterns, a multiple image, a halo effect, a fog effect, etc.

There are two ways to buy FILTERS. One is to buy round glass LENS ATTACHMENTS, threaded and ready to screw onto your lens. These are convenient but expensive. The other is to buy a FILTER HOLDER which attaches to the outside of your lens and a kit of FILTERS which drop into the holder. This method is cheaper but more cumbersome.

Polarizing Filters

Like your Polaroid sunglasses, this lens attachment reduces glare and reflections from water, snow, highways, windows, and sky. It very slightly darkens the whole scene, doesn't affect color rendition, yet specifically knocks out shine and reflected light.

Once attached to your lens, *it must be rotated to the right position to work properly.* You can tell what's right by looking through your viewfinder and watching the highlights disappear and come back as you rotate the lens.

POLARIZING FILTERS do remarkable jobs at making lakes and streams appear more transparent (less reflective) so that you can look *into* the water. POLARIZING FILTERS remove the shine from tarred highways, making them look a rich black. They also make the sky look bluer, richer, and less hazy.

Aim a tube type camera at the sun, and you'll "burn" the tube, blinding it. Reflections from water, chrome, glass, and wet beaches can be dangerous too. Using a POLARIZING FILTER or NEUTRAL DENSITY FILTER (described shortly) on your lens and/or shooting at high f-numbers will give you some protection. If none of this is possible, try to shoot on cloudy or hazy days or wait for the sun to fall lower in the sky for reduced brightness and glare. Chip cameras are rather immune to the above dangers.

When shooting far-away vistas or doing aerial photography, the air generally lends a blue haze to your picture. Using a POLARIZING FILTER will remove much of that haze.

Colored Filters

Amber FILTERS help sharpen aerial and long-distance outdoor photography (as do POLARIZING FILTERS).

A red FILTER will make a blue sky very dark. On a black-and-white camera, a red FILTER can make day seem like night.

Neutral Density Filters

NEUTRAL DENSITY FILTERS are lens attachments that are like dark glasses for your camera. They reduce the amount of light through the lens much like the IRIS does. They don't change color rendition and are very handy if you do a lot of beach or ski slope shooting where the light might be too

---

MINIGLOSSARY

*****Filter**   A lens attachment to eliminate glare or certain colors or modify the image in some way.

Filter holder   Small carrier to clip onto a lens and accept slide-in filters.

*****Polarizing filter**   Like Polaroid sunglasses, these lens attachments cut out glare and reflections.

*****Neutral density filter**   A gray glass lens attachment that cuts light coming through the lens, thus reducing picture brightness.

bright for your camera even with the F-STOP at its highest number.

What would you do if you had a very bright scene but you wanted to have a narrow DEPTH-OF-FIELD (perhaps to emphasize something of importance)? Using low f-numbers would give you a narrow DEPTH-OF-FIELD but would let in too much light. Here is where a NEUTRAL DENSITY FILTER comes in handy. It reduces the light without affecting DEPTH-OF-FIELD.

Special Effect Filters

Figure 7–29 shows a few SPECIAL EFFECT FILTERS and their resulting images. These FILTERS can make the center of the image sharp and the outside blurry for a dreamy effect or can make glistening parts of the picture have star twinkles or can create multiple images. Other SPECIAL EFFECT lenses can focus half your picture on a nearby object and the other half of your picture on a distant object—sort of like bifocal eyeglasses. Another attachment can give you a foggy effect, handy for doing portraits.

Table 7–1 lists some of the more popular FILTERS and summarizes their uses.

LENS AILMENTS AND CURES

1. *Camera is out of focus all the time*:
   a. Is the lens still wearing a CLOSE-UP LENS ATTACHMENT? Look very carefully. Some-times these attachments look so much like the lens that you can't tell they're there.
   b. Is your lens switched to the MACRO mode?
   c. Is the lens not screwed tightly into its C-MOUNT?
   d. You wouldn't be trying to focus on something closer than the lens can "see," would you?

2. *Lens can be focused but goes out of focus as it is zoomed*:
   a. Make sure the lens is screwed tightly into its mount.
   b. Sometimes a lens designed for one camera doesn't work exactly well with another. Usually the problem is caused by the lens not being the perfect distance from the chip or tube. Somehow the lens has to be screwed closer or further from the chip, or the chip needs to be moved inside the camera.
   c. Related to the preceding, are you using an ADAPTER to make your lens fit the camera? If so, it may be fouling up the lens-to-chip distance.
   d. Is your lens in the MACRO mode?
   e. Are you trying to focus on something too close?

3. *Corners of picture seem dark*:
   a. Are you using the right FORMAT lens? If the lens is designed for a smaller FORMAT camera, it will VIGNETTE.

"Center Sharp" lens attachment   "Multimage 5C" prism attachment   "Crostar ISQ" lens attachment

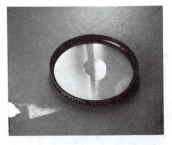

Resulting CENTER FOCUS image   Resulting multiple image   Resulting STAR PATTERN image

FIGURE 7-29   Camera lens effects (Courtesy of Spiratone, Inc.).

TABLE 7-1

FILTERS

Color camera filters

Skylight (1A) or haze (UV)—Nearly clear lens absorbs ultraviolet light. Removes excessive blues from open shade. Excellent for protecting camera lens from blowing sand, water, fingerprints, etc., and may be left on all the time.

Polarizing—Reduces outdoor reflections in glass, chrome, water. Deepens blue sky and whitens clouds. Reduces bright light. Take off when not needed.

#85—Orange FILTER makes camera think it's seeing indoor incandescent light even though it's outdoors. Use on cameras without indoor/outdoor COLOR TEMPERATURE adjustments.

Graduated—Half the lens is tinted, then gradually changes to clear glass for the other half. A blue tint at the top half of the lens, for instance, would create a blue sky while the lower foreground looked normal.

Neutral density (ND-X1, X2, X3, . . .)—Simply reduces the light in a bright scene. The higher the "X" number, the darker the FILTER.

FILTERS must be bought in sizes which fit the diameter and threads of your present lens. If this is a difficulty or you already have FILTERS from your photo equipment, buy a STEPPING RING which adapts your lens to your FILTER.

---

   b. Are you using CONVERTERS or ATTACH-MENTS? Sometimes these too cause a VIGNETTING problem.

   c. Are you using several CLOSE-UP LENS ATTACHMENTS at once? They tend to make the corners dark.

4. *White spots or octagonal figures float through the picture as you move the camera (Figure 7–30):* Light is shining into your lens.

   a. Reposition your camera so that light doesn't shine into your lens.

   b. Use a bigger LENS SHADE to protect your lens from the direct light.

   c. Hold a square of cardboard near the lens so that it casts a shadow on the lens.

   d. Move the light farther away from the subject.

   e. Use a very expensive lens with professional lens coating. These lenses are less susceptible to FLARE from lights.

   f. Use a NEUTRAL DENSITY FILTER and open the IRIS to a low F-STOP.

5. *Specks of dust appear in your picture:*

   a. If they move only when you rotate the outer lens surface, then that is the dirty element. Brush it clean with a CAMEL'S-HAIR BRUSH or by blowing a dry stream of air on it. Cans of compressed air are available for this purpose (Figure 7–31).

   b. If the dust only moves when you unscrew the entire lens, then it is most likely on the camera end of the lens. Remove the lens and brush or blow it clean.

   c. If the blemish disappears when you cover the lens, it is probably dust on your chip. Unscrew the lens and take a "looksee." Brush or blow the chip clean. Incidentally, the tiniest flakes of dust seem very large to a tiny chip or tube.

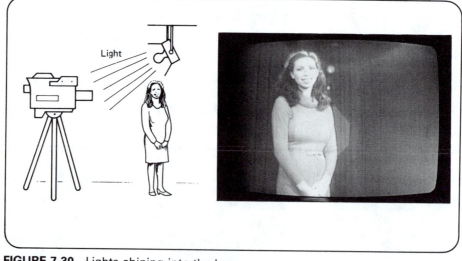

FIGURE 7-30 Lights shining into the lens.

MINIGLOSSARY

Lens shade A funnel-shaped visor that attaches to the outside of a lens shading it from lights and the sun.

Flare A bright spot, streak, or geometric pattern seen in the picture, caused by light streaming directly into a lens and reflecting off its internal glass elements.

FIGURE 7-31 Cans of compressed dry air for blowing dust off lenses etc.

You have to look very closely or use a magnifying glass to see them.

d. If the blemish doesn't go away when you cover the lens, then it may be a tube burn-in or a dead pixel in your chip. Your tube or chip may have to be replaced.

e. Industrial TV camera tubes sometimes have inherent blemishes—imperfections in the tube created during its manufacture. Even CCD and MOS image sensors sometimes have blemishes, usually in the form of tiny white dots. These blemishes cannot be fixed outside of replacing the tubes or sensors with higher-quality models (at higher expense).

f. If the dust or dirt cannot be blown off or brushed off, try wiping the lens or chip with a lint-free cloth dampened with window cleaner. Be sure not to leave any soapy film behind.

SPECIALIZED LENSES

There are PEEPHOLE lenses for "totally hidden videos" (check with your lawyer before slipping it under your neighbor's fence). There are probe-type lenses that can move easily around train models, miniature cities, even through donut holes. Some used by proctologists have flexible tubes (although they don't feel so flexible if you've ever been on the receiving end of one).

Miniature camera/lens ensembles can be attached to a surgeon's eyeglasses or to a robot's arm.

Special image intensifying lenses attached to supersensitive cameras can almost see in the dark. One model equipped with a built-in infra-red laser illuminator can "see" in total darkness.

Probe lens makes little worlds life-size (Courtesy Innovision Optics)

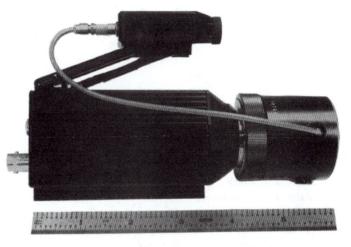

Zero-light CCD "Laser Cam" can see in the dark, using tiny infrared laser illuminator (Courtesy Prime Lasertech, Inc.)

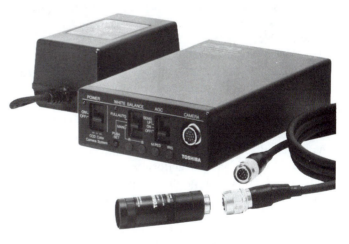

Miniature camera used in medicine (Courtesy Toshiba)

Specialized cameras and lenses.

CAMERA TECHNIQUES AND PICTURE COMPOSITION

So far, we've studied only how to operate the equipment. Now that we can work our machinery, let's focus on how to use it to make professional-looking pictures.

Anyone can pick up a camera and make a boring, fuzzy video tape that plays like the typical home movie. Audiovisual departments by the score stockpile dust-laden videocassettes of programs so dull and dry that nobody wants to watch them. It's easy to get so involved in the mechanics of producing a TV program that you forget to make it interesting, forget to make it visual, forget to make it memorable.

Many people may share in the production of a TV program, but the buck stops with the camera operator. What *you* get is what everyone will see. (Unfortunately, the spendable bucks don't stop anywhere near the camera operator.)

Sometimes the camera person is but one cog in the giant wheel of teleproduction. There are other times when the camera operator writes the script and directs the show. It is hard to tell where straightforward camera aiming ends and creative production planning begins. A good cameraperson is part scriptwriter, part director, part acting coach, and part technician all at once. For this reason, you will see more than just camera angles and cute effects covered in this chapter. The best shots don't just happen, they are planned, so you will learn how to plan them. The camera

operator is not a robot, a remote control responding to the verbal commands of the director. The cameraperson needs to appreciate how his or her image is to be woven into the fabric of the entire production.

In short, as you become more expert at camera work, you will be developing skills in lighting, audio, directing, and stage management. These crafts are all related, and proficiency at one adds to your competency at the others.

Our stairway to expertise will have three steps:

1. Mechanics of skillful camera handling
2. Aesthetics of picture composition
3. Camera ''tricks'' which add visual interest or create a mood

It is the nature of art that sometimes you will balance on several steps at once, using a camera ''trick'' to compose an appealing shot or using raw mechanical skill to create a desired illusion.

TV is all illusion anyway; making your audience see something that really isn't there is part of the magic of television. *A bigger part of the magic, however, is making something look like it really is. It doesn't just happen.*

You'll be shown a lot of ''rules'' on the aesthetics of picture composition and camera movement. These ''rules'' are meant to be broken at times—but *after they've been learned*. You'll be master of your tools if you can "do it right" whenever you need to and then "do it your way" whenever you *want* to.

Once you become aware of these "rules," "tricks," and techniques, a funny thing will happen to you; you will never be able to watch TV again without becoming conscious of the camera angles and shot composition. You'll see the "rules" being followed and sometimes broken. You'll have become a gourmet of visual craftsmanship. You'll be "video literate."

CAMERA MOVES

First let's review the lingo. Figure 8–1 shows the fundamental moves the camera can make. They apply whether you are using a tripod or holding the camera by hand.

If you are operating the camera in a studio, there will probably be other cameras in use besides yours. To help the director to keep track of them all, the cameras are numbered. Learn your camera number. Your director won't say, "Teddy, tilt up" or "Zelda, zoom in." The director will command you by number saying, "Camera 3, tilt up" or "Camera 2, zoom in."

A good director will generally give you a "ready" command before asking you to do something. He or she will probably say something like, "Get ready to zoom in —zoom in." This will give you time to place your hands on the proper controls and psych yourself for action.

Many actions require no commands from the director but just common sense. For instance, unless there is a special reason for not doing so, *your pictures should always be in focus and centered*. You generally should follow the motion on the screen and zoom appropriately to keep the important action in the picture.

In more complicated shots where your camera would be taking only part of the final picture, you may have to consciously break these "rules." For instance, you may have to keep the performer in the right-hand half of your screen so that another camera can place something else in the left-hand part of the viewer's screen. Put another way, your viewfinder picture looks strange, but the final picture from both cameras looks fine.

Keep the tops of people's heads near the top of your viewfinder. Don't leave an airy space above them when they sit down or decapitate them when they stand up.

When two people are on your screen (a TWO SHOT), zoom out enough to keep them both in the picture most of the time. However, if two people are in the picture and one is leaving, let the person walk out of the picture. Don't try to zoom way out and keep them both visible. It is very natural for a person to leave someone and walk out of the

picture. To the viewer, it's as if the person simply left the room.

The Steady Camera

Every move you make during taping will be seen and perhaps unconsciously become part of your message. A picture that bobs around, even a little, betrays amateurism or implies that you are looking through somebody else's eyes. The camera, for instance, would follow someone down a flight of stairs, becoming the pursuer. Jumpy, hand-held camera work can also imply peril, reality (as in newsreels), and frenzy. Unless you intend to portray these moods, you'll want to keep the camera rock solid while taping.

For this reason, studio cameras spend most of their time sitting on tripods. Here are some tips on using them.

Tilt and Pan

If using a tripod, think ahead to loosen the controls so that your camera will move effortlessly. If you expect frequent, fast moves such as with sports, loosen the pan and tilt locks all the way. If you're doing slow, gentle moves, leave a little drag on the controls to dampen some of the jiggles.

If, however, you find yourself shooting a motionless scene for a long time, your arms will get awfully tired holding the camera steady. It will be much easier to lock the controls and let the camera hold itself. You must forever remain alert for an upcoming pan or tilt so that you can get your controls unlocked in time to carry out the move.

If working hand-held, use the Tai Chi position (Figure 6–31) and move your whole torso when you tilt or pan.

If you expect to move the camera somewhere while recording, figure out where you want to go first. This avoids zigzagging and "searching" kinds of shots. If you can, PAUSE the VCR first and try out the move to practice it and also decide on the picture composition. Then UNPAUSE and carry out your practiced move.

I recall, as an amateur, filming panoramic scenes of the Grand Canyon. I started at the ridge, moved down the wall of the canyon to the base, panned right for a ways, tilted back up to the rim, panned right some more across the rim, and then panned left for a long sweep. I was so awed by the scenery that I constantly felt as if I were missing something. I therefore crisscrossed and zigzagged my poor viewers' eyeballs out. The proper technique would have been to *select* the part of the canyon I was going to show first. Next, plan a strategy for moving from one part of the panorama to the other. Next, carry out the plan in one smooth, leisurely sweep. Be patient. Don't ruin the shot you are taking now in order to "get it all in." Strangely,

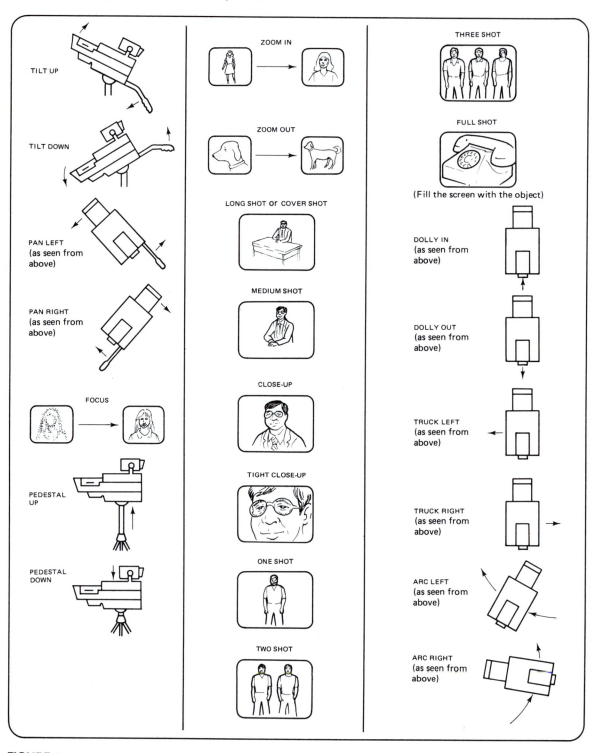

FIGURE 8-1   Basic camera moves.

your audience will never miss the shots you don't take, but they will abhor the hasty pans and the plummeting tilts you take as you try to photograph every weed and boulder.

It sometimes helps to pick an interesting beginning and an end to a sweeping panoramic shot. Perhaps starting with a view of one canyon ridge, framed by some nearby tree limbs would do. In fact you could start with the limbs in focus and the background blurry and PULL FOCUS, bring-ing the canyon into focus. The out-of-focus tree nearby will actually add depth and dimensionality to your picture (an-other case where something out of focus *adds* attractiveness to the picture). Next, carry out a slow pan across the ridge, ending with a shot of tourists by your side, leaning over the guardrail, taking pictures of the canyon. Such a scene pro-vides an ending to the shot, a comfortable place to jump to another shot.

Dolly, Truck, and Arc

If using a tripod on a dolly over a smooth floor, the process of traveling is easy. Do think ahead to unlock your wheels and to sweep cables and obstructions out of the way so you don't drive bumpity-bump over them.

If time permits, aim the wheel casters in the right direction so a gentle push is all that is needed to get the camera moving. A single wayward caster will do wonders at swaying you off course during a dolly or truck maneuver.

To further guarantee a smooth move, lay out your camera cable behind you so that it trails you easily. Professional studios even have FLOOR MANAGERS to assist in this cable handling process. When working alone, it sometimes helps to throw the cable over your shoulder, keeping it out of the way of the tripod. This also makes it easier to kick the cable out of the way when you're backing up.

When your camera is "on" (your tally light is lit), camera travel should be smooth and slow to give a gliding effect. When your camera is "off" and the director asks you to "truck right," then you should do it with haste. Your object is to move from here to there and get reframed and focused as fast as possible. In many cases, the director may have only two cameras taking pictures at a time. If one camera fouls up the shot, the director can quickly switch to the second camera's picture. But if the second camera is in motion or focusing or otherwise not ready, the director is stranded with the first camera with no escape. So you can see how popular you can become with your director if you are able to reset your camera quickly.

If traveling hand-held, it is best to first zoom out to a wide angle. A telephoto shot greatly magnifies camera wiggles, while wide-angle shots hide them. Use the Tai Chi posture to start from and then glide like Groucho.

Everyone likes to take a journey. Travel gives pictures motion and reveals new things to see as we move. The dolly and truck are excellent vehicles for such excursions. Camera motion can seem very natural and does not wear thin too quickly. The camera could, for instance, dolly in and then ARC around a desk as the person behind the desk speaks. This move not only presents us with a new view but also moves us from the formal across-the-desk position to a more casual face-to-face (desk to the side) position.

An ARCING shot is handy for showing two people talking together. By slowly circling the two, you can start with one's face (and the back of the other's head) and ARC around to the other's face, gently swinging the audience's attention from one performer to the other.

Moving the camera behind props, through bushes, or through windows or doors gives a strong three-dimensional feel to the picture.

A shot taken over a steep cliff may not convince the viewer of the canyon's depth. Trucking a camera up to the cliff's edge while looking down (and seeing the edge of the cliff disappear underfoot) can take the audience's breath away. (Take triple precautions so that you don't go over the edge with your camera and end up taking *really* spectacular shots.)

Wheelchairs, shopping carts, or any number of wheeled or sliding vehicles can serve nicely as a temporary dolly. Be creative. Skis, toboggans, ice skates, mechanic's creepers, rubber rafts, parachutes, and construction cranes make fascinating (though dangerous) camera mounts.

Despite all the advertisements that you may have seen showing smiling 115-pound models shooting a scene while carrying the VCR on a shoulder strap, it just ain't so. Not only does that little VCR seem to double in weight every 15 minutes that you carry it, but it unbalances you and hampers your moves. Either set it down when you shoot, strap it to your back, or have a friend carry it. That friend can also serve as a "guide dog" for you while you're walking (eyes glued to the viewfinder) as well as keeping your camera cord from dragging or tangling.

Avoid swinging the VCR around too much as you shoot. It won't hurt the machine (unless you bumped it hard), but it's likely to affect the quality of your picture. Inside the VCR, the spinning heads, etc., act a little like a gyroscope and resist changes in direction. Forced reorientation of the machine changes recording speed slightly, causing a horizontal shift of the TV picture. When possible, set the VCR down, use a long cable, and move only the camera.

As long as it's not bumped, the *camera* doesn't care what position it's shooting in or whether it's moving. So when that curvaceous young gymnast invites you to record her on the trampoline, feel free to hop right up there with her (leaving your VCR behind) and shoot. The visual effect of her seeming to stand relatively still with everything else moving up and down will be too stunning to ever erase.

Focus and Zoom

If you finish this book without learning how to properly focus a zoom lens, we both deserve an *F*. Let's review the basics one more time:

1. If your tally light is off or your VCR is PAUSED, focus by
   a. Zooming in all the way
   b. Focusing for a sharp picture
   c. Zooming out to the shot you want

During a studio production, do this quickly to

MINIGLOSSARY

Floor manager Studio crew member who assists by handling cables or relaying director's cues and commands.

minimize the time that your camera is ''out of service.''

2. If your tally light is on or you're recording the camera's picture, don't use the preceding method; just keep a sharp eye on the viewfinder and focus as best you can.

3. Objects closer than 4 or 5 feet probably cannot be focused clearly, so keep your distance.

4. If your camera is on, zoom gently and smoothly. If zooming in, be constantly ready to correct your focus as you go (just in case something moves).

5. Practice, practice, practice, practice, *practice*— or you'll be no darn good!

Know your lens controls. Know by ''feel'' which part of the lens does what. Know instinctively which way to turn the lens to zoom in. A typical amateurish shot is the ''false zoom,'' a slight in-then-out move made because the camera operator didn't immediately know which way to twist the zoom lens to zoom out. *Know* which way to turn the lens; don't experiment while recording. The same goes for focus; if something comes toward you, *know* whether to turn the lens clockwise or counterclockwise to FOLLOW FOCUS.

What's FOLLOW FOCUS? That's a technique used by professional camera operators as they shoot moving objects. With a little practice you can do it too. Zoom in on someone and have them walk toward you from 30 feet away. Try to keep their image sharp as they move. It isn't easy. With practice, you can develop the skill of following a moving target, keeping it centered, keeping it focused, and even keeping it the right size on the screen (by zooming) as it moves around.

But if you can't get the hang of it, stay zoomed out on fast-moving objects. That way your focusing errors won't be as noticeable, and you'll have less trouble keeping your moving target on the screen.

Assuming you've mastered keeping your zoomed-in shots sharp and centered, *use them; television is a close-up medium.*

LONG SHOTS, although easy to shoot, turn into monotonous mush on the TV screen. CLOSE-UPS capture the expressions, the detail, the excitement of a scene. Check out Figure 8–2 and notice how the CLOSE-UP is more interesting than the LONG SHOT.

And now a word about that zoom lens of yours. Everyone who gets his/her hands on one loves playing with it. In and out, in and out, your audience's eyeballs feel like Duncan yo-yos. Zoom to your heart's content while practicing.

Television is a close-up medium. Get in close.

Then go out and force yourself to shoot without zooming. If you want a close-up of something, then PAUSE, zoom in to a close-up, and then UNPAUSE your recording. Do you want to create a sense of travel, motion, or exploration? Then move the whole camera. That will create a *real* sense of travel, not the overworn zoom. Although a zoom and a dolly both can make things look closer or farther away, only the dolly changes perspective as it happens. Can you sense this difference from the two examples in Figure 8–3? If not, then perhaps you have to see the picture in motion to be fully convinced.

One handy use for the zoom lens is to fill the TV screen with action when you can't move the camera. Picture a youngster up at bat. Pitcher and batter are both on screen. The wind up, the throw—it's a hit. You follow the ball into the outfield, *zooming in* as you go. When the fielder fumbles, you'll be zoomed in close to *see* the fumble rather than seeing a dot surrounded by a whole outfield. Here the zoom lens helps you fill the screen with action. To get that nice shot, however, you had to zoom *and* aim *and* refocus simultaneously, no easy task for the unpracticed.

MINIGLOSSARY

Follow focus   Continually adjusting a lens's focus to maintain sharp picture of subject moving closer to or away from the camera.

Zoomed out, the viewer observes the action.

Zoomed in, the viewer participates in it.

FIGURE 8-2 CLOSEUPS capture the excitement of a scene.

Think Ahead

The following advice may seem too obvious to deserve attention, but it deserves attention. It separates the masters from the mediocre.

As a camera operator, try to anticipate every move. Be ready to tilt up if somebody is about to stand up. Be ready to zoom out if someone is about to move from one place to another. Being zoomed out makes it easier to follow unpredictable or quick movements. If someone is about to move to the left, start moving a little before your subject does. This will make a ''space'' for the subject to move into. Such a camera move also creates an unconscious anticipation in your viewers. They will expect the performer to move when the camera moves.

Reviewing some previous commandments for emphasis: Have your controls unlocked if you expect to pan, tilt, or pedestal. If you'll have to dolly or truck, get the cables out of your path and orient your casters. If shooting panoramas, plan a shot to start with, a route to take, and a shot to finish with.

STUDIO PROCEDURES

In Hollywood or broadcast TV studios, procedures are very regimented. For every job there is a person who does only that job. In fact, union regulations often preclude staff members from doing each other's jobs. There are network stories of how an entire studio crew had to wait around for an hour waiting for the set director to arrive to hang a picture. No one else was allowed to. Only electricians can plug power cords in, and only lighting technicians can change burnt-out bulbs. Only the cue card person can hold cue cards for the performers to read.

Smaller studios are usually nonunion and have fewer crew handling more jobs. In a small studio, the camera operator may need to assist with lights, audio, stage management, and makeup as well as operating the camera and oiling the casters.

Studio procedures vary greatly with the size of the staff. According to International Television Association (ITVA) surveys, the average educational or industrial television studio runs with a staff of four people. So here is what you can expect to do as a camera operator in a studio that size.

Before the Shoot

Let's assume that the set and props are ready, the lights are aimed and turned on, and the performers are ready.

Hopefully, the camera operator will have some idea of what the production is about and how it is to be performed. In complex productions, the camera operator may receive a SHOT SHEET listing the kinds of shots the director will be requesting and their order. This list should be studied and the more difficult shots perhaps practiced before the shoot actually begins. The cameras should be moved to their initial positions and the camera cables laid out in such a way that nobody trips over them yet moves can be made easily.

Unlock your PAN and TILT controls and adjust your DRAG controls for easy motion. How is the camera balanced on the tripod head? If it tries to nosedive or tilt up, perhaps it needs to be slid backward or forward on the head. If you leave your camera to do something else before the show begins, relock your PAN and TILT controls.

Is your lens clean? If not, brush or blow it clean, as described in Chapter 7. Focus and zoom the lens. Does it

---

MINIGLOSSARY

Shot sheet   Brief list of the kinds of shots a camera operator will need to take during a show.

Don't try to imply movement by zooming.

Dolly to give a real sense of movement. The dolly is more three-dimensional and more stimulating.

FIGURE 8-3 Dolly versus zoom. Notice the background.

stay in focus throughout the zoom range? If it doesn't, find out why.

Put your intercom headset on and plug it into the camera. The director or video technician may ask you to uncap your lens and will begin the camera tweaking procedure. Here you'll probably aim your camera at a "typical"

scene so that the technician (or somebody) can adjust iris, gain, pedestal, color, etc. Take this opportunity to adjust your viewfinder to give you good pictures. Try to pick a shot which has something pure black and something pure white in it. The technician will use these extreme colors to calibrate the blackest-black and whitest-white signal from the camera.

Horizontal centering of the camera's picture is sometimes a problem. If the camera's viewfinder is misadjusted, its picture won't be centered, and the camera will be mis-aimed to make it "look" centered for the cameraperson. The director will still see a picture which is not centered. To correct this, you may be asked to center your camera on an object and then be told by the control room to pan left or right a little so that the image is centered on their monitors. Once the picture is centered for them, you should adjust your viewfinder's horizontal control to center the picture for you. Now your picture will be the same as theirs, and they won't be hollering at you later for giving them cockeyed shots.

The camera setup procedure may take as much as 5 minutes per camera, so be patient. Each camera, one by one, may have to be white-balanced, black-balanced, and adjusted in a dozen ways to match the images from the other cameras.

If no floor manager is available, the director may select one of the camera operators to cue the talent and wave other instructions to them.

In many cases, the director may select one camera to follow the action and take mostly CLOSE-UPS. The other camera takes COVER SHOTS, mostly MEDIUM SHOTS and LONG SHOTS. It is hard to keep CLOSE-UPS in focus, especially when you have to follow hands manipulating objects. For this reason, the camera taking the COVER SHOTS must *always* keep an acceptable picture on the screen. This is the director's "safety net" for when the CLOSE-UP camera loses its good shot (which it will frequently).

If shooting a talk show or interview, the director may select one camera to follow one person and another camera to follow the other. Or the director may select one camera to follow whoever is speaking *except* for the interviewer or the master of ceremonies. The other camera stays with the interviewer all the time.

Sometimes a camera has to shoot a title card or mounted photographs or the final credits during the show. The director has to decide which camera does what so that the camera that is "off" has time to move from the graphic to the studio scene and back again. If you happen to be that camera, you may have to set up the graphic and make sure it is level, well lit, not reflecting any shine, and not in a place that will be obscured or bumped during the show. Make sure the graphic is the right distance to be focused on and is oriented perpendicularly to the camera (more on

this in Chapter 12).

Test your headset just prior to the show. Make sure you and the director and the other camera operators can hear each other.

You may have learned the lingo of television, but your performers may understand none of it. If you will be giving them CUES, then explain what the CUES mean and what they will look like. Also consider that the performers are probably nervous, and that you, the only other human contact in the studio, are their only source of confidence and consolation. So be warm and friendly.

If certain camera positions will be critical, try them out before the show. Once perfected, mark the camera position on the floor with masking tape so you can relocate it quickly during the show.

As "show time" approaches, restrict your chatter over the intercom. Don't add to the confusion. Things should become increasingly quiet as the "important" messages become the only communication heard.

During the Shoot

About 15 seconds before the program starts, the director will probably announce over the public address system (or through your intercom system, in which case you have to relay the announcement to the performers) "Stand-by." "Stand by" means be ready to start shooting and remain silent. This goes for *everybody*. You may be asked to relay the countdown and CUE to the talent. Listening to the director over the headphones, you will probably do this: You'll say "Stand-by" and point your finger toward the ceiling. You'll then count aloud "Ten, nine, eight, seven, six, five, four, three, (pause), (pause)" and then point your finger toward the talent, who begins speaking. The purpose of not calling out the last two numbers is so that the sound of your voice doesn't get picked up by the talents' microphones if the mikes get turned up too soon. Also, this avoids having the echo of your last words bouncing around the walls when the production begins. Make sure that the performers have been forewarned that they should count the last two numbers to themselves and not expect to hear them from you. Some studios use another technique of holding up 10 fingers and taking one down at a time to show the countdown. This is fine if you have a floor manager whose fingers aren't busy holding the camera at the time.

During the production, the director can speak to the camera operators, but the camera folks mustn't speak back

---

MINIGLOSSARY

*Cue A signal to performers (or crew) telling them to do something. Usually, the director calls out the cue, which is relayed via hand signals by a studio crew member.

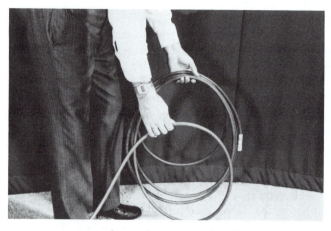

Large loops over hand

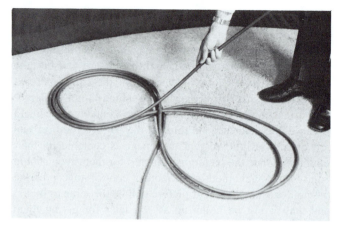

Figure eight

Cable coiling.

to the director—their voices might be picked up by the talents' microphones. One way to communicate with the director is to blow gently into the headset's mouthpiece. This helps to get the director's attention if you think he or she is missing something. The director may also ask you questions to which you respond yes with one blow and no with two blows into your mouthpiece. You could set up a system where three blows means something else, like "I've got to go to the bathroom" and four blows could mean . . .

If you know what shots will be taken and your shot is now finished, immediately move your camera to your next assigned shot. Do not wait until the last minute. Directors like knowing that they can cut to your shot early if necessary.

Note what kinds of shots the other camera operators are taking so that you coordinate well and don't duplicate each other's shots.

At the end of the show, no one should begin speaking until the director gives the "all clear." Sometimes you may think the tape is finished because they have faded the picture out, but the tape may still be running and the sound may still be on. Make sure that the performers know this too.

After the Shoot

Wait a moment until the director checks the end of the tape to make sure the video heads didn't clog and that the show was actually recorded. Sometimes a part of the show needs to be done over again, so you don't want to be too hasty at packing things away.

Once the show is definitely over, cap your lens, lock your controls, and pull your camera to the side of the studio and out of the way. Coil your camera cable, making everything neat. Camera cable, being thick, is hard to coil. It

seems to want to tangle and curl around on itself. Coil it in large loops which you lay onto one hand, while twisting the cable a little with the other hand to get it to circle in the right direction. For really thick cable, lay it in figure-eight shapes on the floor.

Next, run like heck before the director gives you something else to do like dismantling the 29-part set and sweeping the studio floor.

Safety Tips

1. Never aim the camera into a bright light and beware of shiny objects which may reflect a bright light.
2. Lock your camera controls when the camera is idle.
3. Cap your lens when the camera is idle.
4. When you move a camera from place to place, be wary of accidentally aiming the camera into lights.
5. Sometimes TV studio lamp bulbs explode with a white flash and a "pop." Don't look up until *after* the shower of hot glass has finished falling. It's better to catch the glass in your pockets and cuffs or get it down the back of your neck than to get it in your face (such explosions very rarely happen, but play it safe anyway).

Floor Manager Cues

Often the camera operator has to double as a floor manager so you might as well learn the CUES here and now. Figure 8–4 shows some.

Cue		Meaning	Hand signal
Stand by		Show about to start	Finger pointed at the ceiling
Cue		Begin performing.	Point toward the talent. If the talent is to address a different camera, point to that camera.
Speed up.		Accelerate what you are doing. You are going too slowly.	Crank finger forward, much like the "cuckoo" gesture.
Stretch.		Slow down. Too much time left. Keep talking.	Pretend to stretch a rubber band.
One minute left.		Finish what you are doing. One minute 'til the end.	Hold one finger in the air for a while, giving the talent time to notice it (but don't expect them to acknowledge it while performing). Holding up 2 fingers means 2 minutes left, etc.
Cut.		The show is ending now. Stop immediately.	Slice your throat with your finger.

FIGURE 8-4 Floor manager cues.

cue		meaning	hand signal
Walk.		Move in direction shown.	Let your fingers do the walking.
Stop.		Go no further. Stay put.	Hold your hand out like a traffic cop.
Speak up.		You're talking too softly.	Cup hand behind ear.
Closer to mike.		The talent is too far from their mike.	Move palm of hand towards lips.
Look here.		Look into this camera.	Point toward your camera lens. To encourage the talent to speak to another camera, point to *its* lens.
Smile.		You're too serious; smile. (Many amateur performers are nervous and forget to use their charm.)	Smile broadly and point to your mouth.
Relax (not standardized).		Relax. (Many amateurs need help loosening up. Only do this before a show.)	Cock your ears, wave your fingers, cross your eyes, and stick out your tongue.

FIGURE 8-4 Continued

CAMERA ANGLES AND PICTURE COMPOSITION

Basic Camera Angles and the Moods They Portray

You have already seen the results of dollying, zooming, and selective focusing. Figure 8–5 shows examples of some more basic camera angles and describes their impact.

When the camera is low, the performer looks domineering, strong, forceful and authoritative. Political advertisers use this camera angle to strengthen the image of politicians.

When the camera is higher than eye level, the performer looks submissive, docile, unassertive, obedient, weak, or frightened. These camera angles do not have to be obvious. Slight camera height adjustments will leave a subconscious message with the viewer.

To avoid tainting the image with any special meaning, keep the camera at eye level. This is appropriate for newscasts, interviews, panel discussions, and instructional presentations. Sometimes you will find it necessary to elevate the performers on risers so that when they sit down they won't be lower than the camera can go. If this isn't possible, then keep the camera far from the performers so that the angle of view is not steep.

Dominance is also implied by screen position and size. If you have two shots, one of the host and one of the guest, use a slightly larger close-up of the host to portray the host as the more influential of the two. If during an argument between two actors, one is getting the upper hand, frame your TWO SHOTS with that actor closer to the camera. It is also said that the upper right quadrant of the TV screen is more ''powerful'' than the lower left. To symbolically shift power from one TV character to another, frame the dominant one in the upper right-hand part of the screen where his or her presence is stronger.

People walking or running need space to move into; don't center them on your screen. People speaking need space to speak into. If you know someone is going to begin walking, zoom out far enough and start panning to give them a space to walk into before they even begin. This mentally prepares the viewer for the action and makes the shot look very natural. If a person is about to turn to the side and start speaking, you would similarly pan to the side just a moment before he or she started.

Tilted shots imply danger or threat. You would use a tilted shot of someone being chased.

Long wide shots make the performer look insignificant and weak.

There are two ways to show someone how to do something. One is from the viewpoint of an onlooker; the other is from the viewpoint of the ''doer.'' Educational studies have shown that over-the-shoulder shots from the performer's viewpoint are more instructive.

Assorted camera angles can add variety to presentations and make them more enjoyable to watch as long as they *do not distract the viewer from the show*. Strive to balance creativity with singleness of purpose.

Camera Placement and Backgrounds

Two questions should come first to mind as you set up a camera shot:

1. Where is the light?
2. What's in the background?

Lighting. When you are driving into the sunset, it is hard to see the road. The sun glares into your eyes and makes you squint, and it creates reflections on your windshield. For cameras, the same is true, only worse. Bright lights near or behind your subject force your camera's automatic circuits to ''squint,'' creating a very dark picture as in Figure 8–6. Light also reflects off the lens ELEMENTS, creating white dots and geometric shapes (Figure 7–30).

In general, try to keep all the light behind the camera—none behind the subject (with the exception of carefully controlled backlighting, which will be discussed in Chapter 9). In situations where you *must* shoot toward (not at) the light, the following steps might minimize the problem:

1. Use a bigger lens shade. The lens shade (shown in Figure 6–10) is the funnel-like scoop on the outside of the lens that shields the lens from ambient light. Or you could make a shade with some paper and adhesive tape.
2. Zoom in some in order to avoid as much of the extraneous light as possible. A tight CLOSE-UP of the face in Figure 8–6, for instance, would eliminate much of the glare from the window (but it may be easier to close the windowshade than it is to maintain a good CLOSE-UP of a moving face).
3. Using extra lights, throw more light on the foreground (the face in Figure 8–6) to offset the background light.
4. The bright lights are fooling the automatic iris controls into ''squinting.'' So turn the controls to MANUAL and adjust the camera to make the subject look good even though the window behind the subject may appear overexposed or look washed out. Some cameras have a BACKLIGHT button which does just this; it overexposes a light background so that the darker main subject looks best.

Camera low. Subject
looks domineering.

Camera high. Subject
looks weak or
subservient.

Camera at eye level.
Viewer feels neutral,
person-to-person
relationship with
performer.

People walking need space to walk into.

People speaking need space to speak
into.

Tilted shots give an aura of danger,
frenzy, threat. Combined with moving,
hand-held camera, shows a panicked
subject's viewpoint; the viewer is
running, the viewer is searching, etc.

If trying to teach something, favor the
doer's point of view.

FIGURE 8-5 Camera angles.

Zoomed way out. Subject looks insignificant and dominated by surroundings.

Note that if the light from behind is too bright, a tube burn-in can occur. The preceding steps are applicable if the light is bright enough to affect the picture but too moderate to damage the camera tube. Things that are too bright for a tube camera *ever* to look at are

1. The sun
2. A movie or TV light
3. Any bright, bare bulb
4. Any chrome object reflecting light from any of the first three
5. Mirrors or glass objects that are reflecting bright lights

Chip cameras are fairly immune to the above dangers, but shouldn't be aimed at the sun, just to play it safe.

Things that a tube camera can stand to look at (but not for long periods of time) are

1. An open window that looks outdoors (but not into the sun)
2. Fluorescent lamps
3. Table lamps with translucent shades
4. A flashlight or other weak light
5. A dimmed house light if dimmed or diffused enough
6. Shiny automobiles on a hazy day
7. White clothing
8. A TV screen image

Background.   The good camera operator should also take into account what is behind the subject. With just the wrong camera placement, the bush in the background could appear to grow out of the subject's ear, or a desk lamp could become the subject's antenna as in Figure 8–7. Avoid busy or distracting backgrounds unless they serve a purpose in your program. Clocks are especially distracting because the

Window shade closed

Window shade open

FIGURE 8-6 Excessive lighting from behind the subject.

Lamp in background looks as if it were
growing out of the subject's head.

More distracting backgrounds
(Courtesy Imero Fiorentino Associates).

FIGURE 8-7 Distracting backgrounds.

viewer automatically notes what time it is and chuckles at
the fact that the time is wrong.

Watch for window reflections which betray the fact
that a camera and lights are present. Also, make sure that
cameras and microphones don't cast shadows, especially
moving ones, on the set or performers.

When shooting outdoors, be especially aware of the
line of the horizon. You may get so involved in shooting
your subject that you forget to notice that your horizon isn't
level. A level horizon is stable and unobtrusive. Unless
you're looking for special effects, the horizon should always
be level. A slightly nonlevel horizon gives a heightened

energy to the picture. The same applies for vertical objects
like tall buildings. If they are tilted, they lend urgency or
excitement to the picture.

There are times when you want to hide the horizon.
Imagine taking pictures in a rocking boat. Your tripod is
firmly planted on the deck. You thought your pictures would
be stable, but when you view them later, you see the ocean
tilting back and forth. A few minutes of this are enough to
make you seasick. Instead of handing Dramamine tablets
to your audience, simply zoom your camera in on the subject, avoiding the rocking background.

Sometimes the background is the subject of your picture. Backgrounds could be scenes of raging oceans, peace-
ful valleys, or majestic mountains, or they could be giant
machines, thundering waterfalls, towering pyramids, or
sprawling shopping centers. In each case, the actual size of
the background subject is too large to appreciate, especially
on a tiny TV screen. To give your screen depth, include
something in the foreground of your shot like a tree, a wagon
wheel, a person, something of recognizable size. Better yet,
try to include something in the foreground and middleground
of the shot to show even more dimensionality. It doesn't
really matter a lot if the foreground object is a little blurry
(as may be the case if you have focused on the distant
background). The blurriness will further emphasize the farness of the background and *because* of the blurriness won't
distract attention from the background. Figure 8–8 shows
some examples.

Don't's and Do's of Camera Angles

Now we get deeper into the aesthetics of picture composition. Before we merely concerned ourselves with keeping
subjects sharp, close, centered, and level without wiggling.
If you can do that, you score 75%, but Francis Ford Coppola
you're not. To score 100%, your TV screen must portray
precisely the message you want it to. The image must em-
brace the subconscious nuances you desire yet eliminate
unwanted distractions. Figure 8–9 shows some more "rules"
of picture composition. Can you see why the "don'ts"
should be avoided?

You shouldn't cut a person off at a natural body division such as the ankles, the bust, or the chin. It makes
the body look like there is something missing. If, however,
you leave in part of the body leading to the next part, the
mind will complete the rest of the body using a psychological quirk called CLOSURE. Similarly, if talent is sitting on
or leaning against something, a part of that something should
be showing. Again, the mind, by seeing part of the object
will complete the rest of it.

The RULE OF THIRDS suggests that your center of attention should not be in the dead center of the TV screen.

Level, stable, unobtrusive horizon.

Angled horizon attracts attention, heightens energy.

Vertical tall structure appears stable, solid.

Angled tall structure implies danger, action.

Be aware of your horizon line.

The picture looks best when your eye is attracted to areas just off the center of the screen such as one third down from the top, one third up from the bottom, or one third in from either side.

Performers should be taught how to hold items they are displaying to the camera. The talent shouldn't be looking into a monitor and trying to center a picture for the camera operator. Centering and focusing is the cameraperson's job. Instead, the talent should place an object near his or her face and hold it as still as possible. Here the performer can easily see the detail he or she wishes to point to. Another technique is to place the object on a table (which holds it very still for close-up shooting) and to point out details without moving the object.

More Do's of Camera Angles

Seat People Close Together. One of the oddities of television is how it distorts space. A 5-foot-wide SET becomes the entire universe for your TV show. Your audience never sees the mess of lights, cables, and clutter just a foot ''off screen.'' To them, what they see is all that exists. Visit a TV newsroom which you've grown accus-

MINIGLOSSARY

Closure Describes how the TV viewer mentally fills in the parts of an incomplete picture.

*Rule of thirds The center of attention should not be dead center on the screen but one third of the way down from the top, or up from the bottom, or in from the edge of the screen.

background alone

foreground object

middleground and foreground object

FIGURE 8-8 Using a foreground object to emphasize the size or distance of the background.

tomed to viewing only on TV, and you'll be shocked at how little there is to it.

One aspect of this distortion of space involves people in conversation. Normal Americans converse at about 3 or 4 feet from each other and sit even farther from each other. Not on TV. Three or four feet seems like across the room on TV (See Figure 8–10). Squeeze your people in tight to look "normal." Note that it may take some practice before your performers feel comfortable conversing from less than 2 feet apart.

Seat Host to One Side of Guests.   This way you avoid giving the host "tennis neck" as he Ping-Pongs his attention first to the guest on his left and then to the guest on his right. Cameras and viewers alike go bouncing back and forth to follow the discussion.

Placing the host between the guests has another disadvantage. If you later edit together the TWO SHOTS, which include the host and one guest (as in Figure 8–11) and then

the host and the other guest, the host snaps from the left side of the frame to the right side. First he's looking to the right and then suddenly, he's aimed to the left. Figure 8–12 is more comfortable.

Angle the Guests.   This maxim applies to any shot with two or more persons in it. People facing each other nose to nose suggest an adversary relationship. It is a great way to portray an argument. But in a panel discussion, the shots imply disagreement or debate. Conversely, people lined up facing the camera look like a team of contestants. They appear ready to perform individually for the camera but not ready to react with each other. The most comfortable seating arrangement has people angled toward each other as shown in Figure 8–13.

Reverse-Angle Shots.   No one wants to look at the back of somebody's head. But people usually face each other when speaking, and unless you like profiles, you're always going to be shooting somebody's behind.

DON'T   **DO**

eyes, nose, mouth, chin

bust

hands

knees
hem line

ankles

Cut off a person at any of these natural
divisions

Cut off feet

Include feet

Cut off neck

Leave in part of the body leading to the next part.
The mind will complete the rest of the body.

FIGURE 8-9 Don'ts and do's of camera angles.

DON'T   **DO**

Cut a person off where she contacts her surroundings. Here the talent looks like she's leaning on the side of the TV screen.

Provide enough surroundings for shot to explain itself: the talent is in a chair

Provide insufficient headroom.

Leave just a little space so talent doesn't "bump her head"

Change angle without changing shot size. Twists performer without apparent purpose.

Change shot size as you change angle to add variety and interest.

DON'T

Use many LONG SHOTS.

DO

Television is a CLOSE-UP medium.

JUMP CUT between long shots and closeups without changing angle. With a cut, the viewer expects a substantial change in visual information but doesn't get it.

Change angle about 30° when cutting. Adds variety and smooths transition. Builds a fuller perception of subject.

Place most important picture element in the dead center of the screen.

Use the so-called RULE OF THIRDS placing important picture elements one-third down or one-third up the screen.

DON'T   **DO**

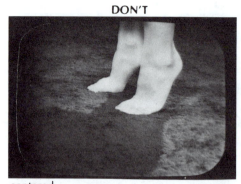

centered

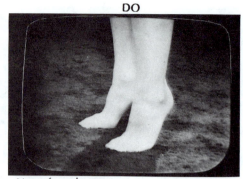

⅓ up from bottom

Place eyes in the middle of the screen.

Place eyes one-third down the screen.

Place the face in the middle of the screen. Too much headroom. Note also how shot cuts body at the bust, also a no-no.

Place face one-third down the screen.

The mouth and eyes are important. Here they follow the RULE OF THIRDS, where the eyes are one-third down the screen and the mouth one-third up. Here the missing chin and forehead will be mentally filled in by a psychological process called CLOSURE.

DON'T

Get zoom happy using your fancy zoom lens to make yo-yos out of your audience's eyeballs.

DO

If possible, cut from shot to shot when necessary

DON'T

Allow talent to wiggle object around while looking at TV monitor, trying to center it for the camera. Also, here the talent can't see object well enough to point to detail.

Hold object to side of face, steadily. Let camera operator do the centering. Detail is easy to see and point to.

Or, set it on something stationary and hold it there.

Talent may look fine in the studio but . . .

FIGURE 8-10   Seat the talent close together.

. . . are too far apart for a good MEDIUM SHOT

First the host is on the left . . .

. . . then the host is suddenly on the right.

FIGURE 8-11   Panel discussion with host between guests.

FIGURE 8-12   Panel discussion with host to one side of guests.

There are lots of ways to get a dialogue with two people facing the camera, like people on a park bench, sitting in a car, or watching TV or a fireplace or with one in front of a mirror. In fact, you'd be surprised at how many natural-looking scenes there are in the soaps showing someone facing *away* from the person they are talking with. Nevertheless, at some point, you're going to have to face the problem of how to shoot faces facing faces.

One way is the over-the-shoulder shot as shown in

Figure 8–14. Shooting over one person's shoulder, we get a close-up of the second person's face as he or she speaks. When the first person speaks, we swap everything around, making a medium shot of the first person over the shoulder of the second.

If you are using a single camera and editing as you go, you'll find yourself getting a lot of exercise traveling back and forth to get a face shot for each actor's lines. If shooting RAW FOOTAGE for later editing, you could save

Nose to nose implies adversary relationship.

Line-up gives no relationship between guests.

Angled is comfortable.

FIGURE 8-13   Angle the guests.

steps by shooting the entire sequence twice, once facing one actor (or actress) and once facing the other. In fact, each sequence could be recorded on a separate cassette. When editing later, all you do is change tapes to change camera angle and get a face shot of the person speaking.

Over-the-shoulder shots are not always easy to do. Performers seem to have a way of stepping to the side and blocking your shot. To counteract this, you must always be ready to ARC your camera to the left or right to unblock the shot.

Reverse-angle shots may be great for two people talking but not for cuts between hands or props and the person holding them. If a performer is angled to the right in a

MEDIUM SHOT, his or her *hands* should be angled to the right when you cut to a CLOSE-UP. To have the hands or prop suddenly change direction would be jarring to the audience. The same is true for the general flow of motion in a scene. If the action is moving to the right in one shot, it should be moving to the right in the next, at least for the start of the shot. If panning left in one shot, the following shot should continue the motion. These are examples of why the camera operator should know *how* the shots are going to be used, so that they may be recorded with later edits in mind.

Avoid Slightly Off-Camera Looks.   Have your performers play directly *to* the camera *or* react to each other or the props, disregarding the camera completely. Direct eye contact is very engaging, making the viewer feel as if the performer is talking to *him* or *her* alone. Profile and off-angle shots make the viewer feel like an observer to the action. Both kinds of shots have their advantages. But avoid splitting the difference. The slightly off-camera shot in Figure 8–15 makes the performer look insincere and shifty-eyed or makes the performer look like she's reading something.

As combination camera operator and FLOOR MANAGER, it may be your responsibility to remind the performers that they are supposed to look *into* your camera; they shouldn't be talking to the floor, the wall, or the studio monitor. You can remind them by reaching out from behind your camera and pointing your finger at your lens. I have found that with amateur talent, 10 such reminders in a single show are not too many. Eventually, the performer learns good eye contact.

If you have a studio monitor which seems to catch your talent's eyes during the performance, then turn it off. They don't need to see what they are doing. In many studios it is part of the preproduction setup procedure to extinguish the studio monitors, just to guarantee the performer's undivided attention.

Use a Familiar Object to Create a Sense of Scale.   Extreme CLOSE-UPS and majestic landscape shots suffer one problem in common. The viewer cannot appreciate the smallness or largeness of the subject without a "visual yardstick" to gauge it by. Figure 2–30 used a hand to define the size of a plug. Figure 8–16 uses a familiar object in the foreground to add dimension and meaning to the landscape in the background. Often a bush, a tree, a person, or an old wagon wheel serves nicely to create scale, balance, and visual variety in a scenic view.

Shooting Children.   . . . And I don't mean with a firearm. Photographing little people requires getting low. If you shoot them from the same height as you do adults, you inadvertently introduce the domination/weakness impression

FIGURE 8-14   Over-the-shoulder shot.

Direct eye-contact engages viewer

No eye-contact. Viewer observes the
actions of others

Slightly off-camera look loses impact,
implies insincerity

FIGURE 8-15   Avoid slightly off-camera looks.

(refer back to the example in Figure 8–5). To keep your shots neutral and unbiased, get down and shoot children at their own eye level.

The same goes when adults interview children. The adult should bend down to the same height as the child so that the camera sees both their heads at about the same level. Otherwise, you get a shot of

1. The child's body and the adult's knees
2. The adult's torso and the child's head
3. An uninteresting LONG SHOT of both of them

When the adult gets down to the level of the child, there is better eye contact, and a warmer relationship is implied. See Figure 8–17.

FIGURE 8-16 Foreground figure to give scale to landscape shot.

As I mentioned before, some of these tasks, like positioning the performers, may not seem to be the cameraperson's job. Strictly speaking, it isn't, but in most industrial settings, the camera operator is responsible for taking a shot which is *technically* good and *also* for assuring that the shot portrays the desired mood. In small TV operations this means being several things at once: camera operator, technician, acting coach, scenic designer, and lighting expert. Then again, this is why many people feel that educational and industrial television is more fun than commercial broadcast television.

CREATIVE CAMERA ANGLES

Popular Alternatives to the Simple Shot

There are lots of ways to show the same thing. The real skill comes in knowing when fancy shots will add spice to the scene and when they will distract the viewer and ruin the scene. Your decision will be a delicate balance between two forces. One is the pressure to ''show off'' as many of the tricks and skills you can perform. I have seen students contrive ways to ring and blow every bell and whistle they can find on their equipment. The opposite force is the ''let's-get-it-done-quick-and-dirty'' approach. You get so involved in cranking out the required shots on your list that you forget to make your program visually interesting. Knowing when to be fancy and when to be straightforward makes you a master of this trade.

May the force be with you.

Focus Shift. Something (or someone) is in the foreground. Something else (or someone else) is in the background. Usually, one of these two has to be out of focus for the other to be sharp. Okay, use this to your advantage. As two people talk, focus back and forth on whichever person is speaking. It's better than panning back and forth and better than a LONG SHOT trying to get both performers into the picture at once. Perhaps the foreground person's expression is paramount (i.e., tears welling in the eyes) while the background person is secondary, even distracting. So focus on the foreground, making the background blurry. Then when the background person's action becomes important (staggering to the cupboard and pouring another drink), *that* is pulled into focus, diverting attention from the face. Yet both characters can be seen at once as you shoot past the tearful face 4 feet from the camera and in the other half of the screen get a LONG SHOT of the drinker 20 feet away. Attention can be focused on one or the other performer, yet the viewer can *keep track of* both.

FOCUS SHIFT is also a popular way to display a long line of something—soldiers, flowers, gravestones, fence posts, etc. Position yourself next to one member of the lineup and shoot toward the farthest member. Focus on the closest (or farthest) at first, and as you change focus, different members will, in turn, become sharp and then blurry as other members become sharp.

Low f-numbers and longer focal length lenses make this focus/defocus effect more pronounced.

If you have a macro lens, you're equipped to carry out a super FOCUS SHIFT from 3 inches out to 20 feet. Picture how this could work with a beginning title. You letter your

MINIGLOSSARY

*Focus shift   Also called ''pull focus''; the act of changing focus to sharpen objects at different distances from the camera to center

attention on them.

camera high

eye level

adult interviewing from above

adult at child's level

FIGURE 8-17 Shooting children.

title onto a foot square piece of clear glass, plastic, or acetate transparency (like those used in schools for overhead projectors). You macro-focus on that title inches away from your lens (make sure you make the title the right size for this). In the background, so out of focus that you can't see it, is your first scene. You refocus from the title to the distant scene as the actions begins. The title will go so far out of focus that it almost disappears, making it a nice cheapie–dissolve effect using only one camera.

The same technique may be used for 35mm slides taped to the front of the macro lens or for FOCUS SHIFTS from miniature soldiers or plane models out to your live actors.

Mirror Shots.   This is another way to slip twice as much into your picture without resorting to dull LONG SHOTS. Imagine this—a CLOSE-UP of a woman putting on lipstick. The camera slowly zooms (or dollies) out, revealing that her face is a reflection in a mirror. You see the side of her head in the foreground. As the camera pulls back farther,

you see someone else reflected in the mirror. It's a man getting dressed across the room. As he speaks, she turns, now facing toward the camera. He steps closer, tying his tie as he walks. Now your camera has two people talking *to* each other, *facing* each other, yet you see *both faces at once.* Nice trick. How else can you get two full face shots at once?

Mirrors are useful any time you want to show both sides of something like an engine, a statue, or some complex device. They're useful for dramatic effects (an actor looks up and in the mirror sees someone climbing through the window). Occasionally, they serve as just another way of formulating a shot.

A mirror hung at an angle over the kitchen stove allows you to get straight-down-into-the-pan shots without the risk of having your camera fall into your tomato sauce. Conversely, a mirror slipped under the car allows you to document defective ball joints on autos awaiting recall.

Mirrors are a handy way to get a bird's-eye view of medical operations without risk of your camera operator fainting onto someone's intestines.

Reflection in a shiny object.

Mirrors are not the only things that reflect images. Imagine a CLOSE-UP of a man's face slowly turning toward the camera. He is wearing those mirror-like sunglasses. You know what he's looking at when the twin likenesses of a sexy sunbather appear in the lenses.

Windows and water reflect too. Windows have an added advantage of being able to mix two pictures together for you, the figure behind the glass and the one reflected by it. In fact, one tricky way to dissolve from one picture to another without an expensive video FADER and two cameras is as follows: First, illuminate the figure behind the glass while keeping the reflected subject in the dark. Then, dim the behind-the-glass scene while simultaneously brightening the reflected scene. The old scene will melt away, being replaced by the new. Makes a nice trick shot for a sci-fi drama too.

Mirrors don't even have to be flat to work. Nice effects can be had by shooting a CLOSE-UP of someone's face reflected in a shiny Christmas tree decoration, teapot, or chrome bumper. Remember all those serving dishes you got as wedding presents but never had occasion to use in the 31 years that you've been married? Hallelujah, the time has come! Use them now for twisting the world into funhouse contortions. It is a good way to start a scene before zooming out and panning to the real world.

Again, be careful not to reflect any bright lights into your camera when you use mirrors. They may cause tube burn-ins or chip smears.

Parallel Movement. Joggers, bicyclists, water-skiers, snowmobilers, and horseback riders all pose the same problem for the camera. They are fun to watch, but they're gone an instant later. To really catch the action, move with them. Video tape that jogger from your car (preferably with someone else driving). Shoot that water-skier from the back of the tow boat or by moving along next to the action; it stays with you while the background slips by.

Adding Movement to Still Objects. Photos and model ships don't move. That doesn't mean that the camera can't make them come alive. Imagine a close-up of a *photo* of a parade. By panning across the picture to the sound of a marching band, the people will *seem* to be marching.

Imagine a slow *zoom in* on a painting, starting first with the whole town but ending with a "stroll" down the main street.

Picture the museum's dinosaur display filling the screen. Cut to a close-up of one of the models. Take the shot from slightly below the creature so it looks domineering and dangerous. ARC the camera around the model, making it seem to turn its head. Add some tension by tilting the camera a little as you go.

Zoom in on a still photo of a horse race with the sound of crowds cheering in the background. Shake the camera as you zero in on the lead horse. You can almost see the mud flying from the hooves.

To the beat of rock music, quickly zoom in-and-out, in-and-out, on the singer's face from the record jacket. Angle your shots of the performer at the microphone, first left and then right, and cut from the first shot to the second. This will heighten energy in the scene, and if the shots are quick enough, the viewer will never notice that the performer isn't moving.

Pan across photos of antique cars. Move the camera so that the cars appear to be traveling forward. You won't fool anybody into thinking that the cars are really moving, but viewers are often inattentive. Leaving them with the *impression* that the cars are moving is all that is necessary.

The same trick works with photographs of birds. Pan across a photo of a bird in flight; it will appear to be gliding through the sky.

Wall Shadow. In the movie *Goodbye Girl*, the young lady is forced to accept a male roommate in her apartment. She's had it with his guitar playing and barges into his bedroom to complain. Seems he strums in the nude. How would you show such a scene without going X-rated?

You could use the tried and true guitar-in-front-of-the-genitals shot. That's okay, but what do you do for your next shot? In the movie they aimed the camera at the flustered lady standing by the door with this shadow of a man projected on the wall. It all seemed very natural, and the scene was creative and novel.

You can use this technique to show sinister shadows, lantern-cast shadows, or just a stylistic view of something before you actually see it (like a shot of the bicycle shadow on the pavement followed by a tilt up to the bicycle).

Shadows make nice surprise revelations. Consider a lady's shadow cast upon a doorway. Then into view comes the shadow's owner, a fellow dressed ridiculously in woman's clothing in order to sneak past a detective.

Low-Angle Cleanup.   If you have a busy background of unwanted cars, bushes, or signs, crouch down and shoot *up* at your subject. The cluttered horizon disappears leaving sky, clouds, mountaintops, or the orderly tops of buildings in your shot.

Such angles can intensify a landscape shot by exaggerating converging lines of perspective. Roads, fences, pipelines, or sandy beaches seem to go *through* you rather than by you.

Arty Tricks

Camera Tricks.   Some consumer camcorders have built-in special effects like MOSAIC, POSTERIZATION, FADE, or STILL FRAME. You could go from one shot to one of these effects, pause, then set up your next shot. Next, unpause, come out of the effect, and now start the new action.

Camcorders with electronic STILL STORE allow you to freeze a "snapshot" of the last frame and dissolve from it to your next scene. This is a cheap but convincing fake dissolve from one shot to another without expensive editing equipment. Camcorders with STILL STORE also allow you to "grab" a title and dissolve from it to your next scene, or vice versa. You can also KEY the letters over a scene or use the letters as cutouts, as if the scene were happening *inside* the letters. More on these effects in Chapter 11.

Lens Flare.   Figure 7–30 showed what happens when you aim too close to a light. The geometric patterns and spots that you get are called LENS FLARE and are usually reduced by the lens shade. But what if you *want* LENS FLARE for artistic effects? Well, remove the lens shade. Play around with your zoom and iris further to enhance the effect, but watch out that you don't shoot directly into a light and burn-in your tube.

If you'd like to use your tubed camera to get one of those dreamy sails-across-the-sunset shots or sun-eclipsed-by-the-glider views, it's still possible—if you are careful.

Place a dark neutral density filter over your lens and stop down your lens to f16. Take a quick zip past the sun with your camera and see if it is leaving a streak. If there's no evidence of temporary burn-in and your pictures is altogether too dark to use, open your lens one stop at a time and repeat the experiment until you notice that the sun is starting to leave a mild *temporary* streak. That's as far as you can go safely. Such a shot should yield nice halos, rings, rainbows, and geometric patterns yet probably a dim silhouetted shot of that sailboat or glider. This technique works well with sun reflections on water, sunsets, tall buildings with the sun-just-behind-them, the sun peeking through the trees as you look up from a stroll on a wooded path, and almost-in-the-sun shots.

If you *must* shoot directly into the sun,

1. Use a film camera for the shots. The film camera won't be harmed, and the film can later be transferred to video tape (see Chapter 12). This way you take no chances.
2. Use a CCD- or MOS-type camera which will be unharmed by bright light.

Gun Barrels, Drain Pipes, Kaleidoscopes.   Remember the title scenes of James Bond 007 where you are looking *through* a gun barrel at James? You could even see the spiral rifling of the barrel's interior. You can do this or something similar by placing tinfoil or cardboard over your lens and making a pinhole in its center. Then aim your camera through the gun barrel or whatever and *flood* your subject with light. Because the pinhole "stopped down" your lens to perhaps f64, you'll have nearly infinite depth-of-field but will need lots of light.

Chrome-plated sink drain pipes are interesting to shoot through. Kaleidoscopes too. You can pull the end off your kid's scope (or wait until three days after Christmas when *he* or *she* has done this) and get kaleidoscopic views of the studio and outdoors.

Silhouette.   Hang a large seamless white sheet from a wall, extending it in a smooth curve to the floor. Perhaps place another on the floor overlapping with the first. Darken the room and flood the sheets with light. Have your performer stand in front of the sheets *but not in the light.* The strong backlighting will create a silhouette effect.

Silhouette effect by brightly illuminating background.

Painting with a Camera. Most of the warm, red sunrises and sunsets you've seen on TV weren't so warm as they looked. The camera shot the scenes with its COLOR BALANCE controls misadjusted to "paint" the scene redder than it really was. You can "cool" a scene by "painting" it blue with the opposite adjustment. Science fiction scenes can be done with this effect, making all the colors unusual and foreign. Faces become green, water becomes magenta. You can imitate the sepia tone of old-time movies by adjusting your camera's color controls for perfect WHITE BALANCE while *aiming it at something light blue.* This will trick the camera into making whites look sepia.

Lost Horizon. We touched on this subject before while discussing how a slightly angled horizon can add energy or frenzy to a shot. Now let's carry angled shots to the extreme.

The viewer who can't see the horizon can be easily fooled much like the way you were fooled at the fun house in that tilted room where *everything* was tilted—except you—and you soon became tilted too.

Shoot straight down at someone climbing on his belly (panting and groaning) over a gently sloped cliff of rocks. The viewer will think the cliff goes straight down. Include the appropriate glances "downward," with an occasional "slip" of the foot and the face-pressed-against-the-rock look to add credibility.

Walls that look like floors and vice versa are splendid sets for such camera tricks. And imagine someone hanging onto a window ledge ready to jump 30 floors—or is it really 30 inches? The camera will never tell.

Picture a rope in the gym backed by a nondescript cinder block wall. Have an athelete climb up a ways, get turned head down, and then slide slowly down toward the floor. Only shoot all this with the camera upside down. Imagine your viewers seeing someone slither to the "top" of a rope and teeter at its free end like an Indian rope trick.

Nail up a sheet of wall panelling—at a 30° angle. Prop a table or desk and chair at the same angle. Stick in a performer holding himself at a 30° tilt and place your camera 30° from level. Now have your actor pour a glass of milk. Guess where the milk will go?

To make a car or runner seem to speed powerfully up a hill, tilt the camera so that the road is at an uphill angle. Shoot with a neutral background, avoiding telltale trees, signs, and houses. This shot is very common in car and truck commercials. The runaway wagon, on the other hand, will look like it is careening down a much steeper slope than it is with the help of a camera tilt in the downhill direction.

A tilted camera can make an airplane "dive" to earth. A back-and-forth tilt can make a boat endure great sea swells.

Reflections and Refractions. Try shooting through a wavy glass for a dreamy or underwater effect. Edmund Scientific, an optics company, sells such glass and all kinds of crazy lenses. So do many photo shops. Kroma Studios in Santa Monica sells Kroma glass, which creates colorful rainbow effects as you shoot through it.

Try shooting through a glass of water or a fish tank or just above a hot radiator or parallel to hot pavement for some nice effects.

How about shooting at a steep angle into a puddle of water, a pan of water, or a lake to catch ripply reflections before you tilt up to the actual subject?

For a dreamy effect, project a scene onto a screen or even a white sheet in a darkened room. Place a pan of water in front of the screen image and shoot the reflected image (with ripples in the water). To make the reflected scene right side up, invert the slide in the projector.

Lens Effects. Back in Figure 7–29 we saw some of the effects available with lens attachments.

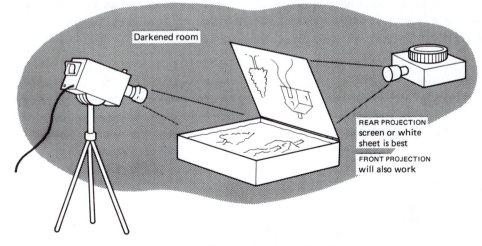

Dreamy effect from reflection off water.

CENTER FOCUS creates a dream-like image where the outside edges of the picture are blurry but the center is sharp.

STAR PATTERNS, also done with lens attachments, create straight lines going through each sharp point of light in the scene. Stylistic views of rippling water in the sun or stage lights above a rock group are the most common applications. The pattern can emphasize the twinkle in an eye or a glint of a tiger's tooth.

Either effect can be used to portray the point of view of a drugged, crazy, crying, frenzied, or dreaming character, showing their distorted view of the world.

A cheapie version of these effects can be improvised. Slip a layer of nylon stocking over the lens; all the focus will go soft, like in a portrait or a semidream. Tape a piece of glass to the lens or screw on a lens attachment. Smear a fine film of vaseline over it, and again you'll get a soft focus effect. Leave the center ungreased, and you have a CENTER FOCUS effect. Wipe the grease up and down, and you'll get sideways shafts of light coming from bright spots in your scene. Half the lens wiped one direction and half the other will create crosses of light emanating from lamps, reflections, and twinkly places. Try a circular pattern. Experiment.

Video Feedback. What do you suppose would happen if you aimed your TV camera at your TV monitor which was displaying your TV camera's picture? What your camera saw your set would show, and your camera would see, and your set would then show and 'round and 'round the signals would go in a FEEDBACK LOOP. This is called VIDEO FEEDBACK.

The visual effects are so limitless that one could sit all night tilting the camera a little, zooming a little, twiddling the TV's color and brightness controls as well as the camera's white balance controls, and then sticking one's hand in front of the TV screen to see what happens. You're unlikely to see exactly the same effect twice and just as unlikely to reproduce a given effect at will.

Tilted cameras produce pinwheel and kaleidoscope patterns which spin, freeze, reverse direction, and break up into separate pinwheels. Iris changes create shrinking and growing blobs of light. Color these blobs with your TV and camera's color controls.

And when you've taped a couple of hours of this fantasy, go back and dub in an appropriate sound track. Hard rock or heavy classical scores set to the same visual piece can create completely different moods.

The effect of watching such a recording is much like gazing into the fireplace while listening to the stereo. Your eyes are riveted, but your mind is free to wander. The name given to these kinds of abstract TV productions is *visual wallpaper.*

It's great with Fritos and wallpaper paste dip. Figure 8–18 shows some examples.

CREATING MOODS AND IMPRESSIONS WITH THE CAMERA

How you show something tells as much of your story as *what* you show. You've learned that by changing camera height you can make a performer look strong or weak. You've seen how tilted shots create suspense and how soft focus and foggy lenses create a dreamy effect. Close-ups, as you've seen, involve the viewer directly with the action. Here are some more mood-creating shots.

Progress vs. Frustration

A jackrabbit is racing to the right. The camera pans along with it but slowly falls behind so that the bunny moves forward in the frame. That's progress. A mountain lion pursues to the right. Here the camera pans slightly faster than the lion can run, leaving more space in front of it. That's frustration. Score: jackrabbit, 1; lion, 0.

A runner in a telephoto shot approaches and approaches and approaches in a vain attempt to reach the camera. Frustration. Shot at a wide angle, the mild-mannered accountant strolls by the camera and appears to loom forward as she nears the camera lens. The look is one of decisive action. Progress.

Suspense

Our hero creeps backwards, and the audience cringes, waiting for him to back into something awful. He looks up slowly; something terrible is bound to drop on him. He pokes his head into an air duct; what's going to grab his sweet face? He draws open the curtains; what monster will leap from behind them? As he sleeps, a large shadow slips across the bedroom wall; what is it? He washes his hands, he looks up, and through the sink window, inches from his face, there is a. . . . He sits at the dressing table and looks up into the mirror; what unholy creature will he see behind him? He casually shaves while we dolly in from behind him, stopping just over his shoulder; what unwelcome surprise awaits him in seconds?

MINIGLOSSARY

Center focus Mood-creating lens effect where the outside edges of a picture are blurry and the center is sharp.

Star pattern Lens effect creating shafts of light gleaming from any bright points of light in a picture.

Video feedback Fantasy effect created when a camera is aimed at a TV monitor displaying the camera's picture.

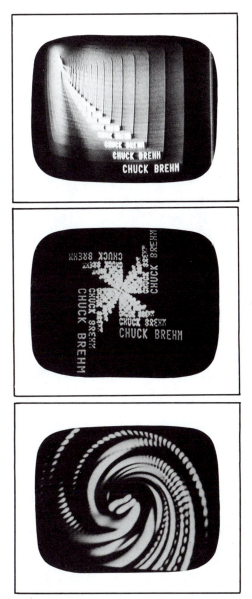

FIGURE 8-18   VIDEO FEEDBACK

In each case, your attention is drawn to what you *don't* see, out of the camera's view, behind the curtain, in the mirror, or behind the camera as it moves closer. For a crash course in suspense shots, see the movie *Alien*. It uses every trick in the handbook.

Anger, Secrets

Pose your performers almost nose to nose and have them shout. Voilà, anger. As they move about, keep their heads very close together.

This closeness *without* the shouting implies intimacy or secretiveness.

Speed

Keep the camera low to the ground (so you can see the ground rushing toward you) and use a wide-angle lens to portray speed. A camera on a skateboard moving at 1 mile per hour looks like it's moving at 60. A camera shooting from the roof of a tall truck at 60 miles an hour looks like it's only traveling at 20.

I once shot a traffic safety tape by mounting a TV camera on the fender of a car and driving it around a congested campus. The scenes made it look as if I had performed speeding maneuvers and daredevil near-misses at every turn. Potholes loomed up like canyons. Casual pedestrians looked like kamikaze jaywalkers. Yet all these shots were taken at a calm 15 miles per hour or less.

Night

Naturally, if you shoot in the dark, you don't get a picture. The object is to make the scene *look like night* but with plenty of light for the camera.

If shooting outdoors in the daylight, pick a cloudless day with blue sky and harsh shadows. Place a red filter in front of the lens to make the blue sky very dark. Either shoot with a black-and-white camera or misadjust the camera's controls to undo the redness you've just added to the scene. Incidentally, night scenes normally have very little color to them anyway. They do not look abnormal in black-and-white. Use a neutral density filter or stop down your lens to darken the scene more. One way to darken the sky without messing up your colors is to use a polarizing filter. If the effect isn't noticeable enough, try using the polarizing filter *and* the red filter together.

Keep performers in shadows or "puddles" of light as they move from place to place. Shoot with the sun halfway or three quarters behind the performer, creating frequent silhouettes. The sun will appear to "moonlight" the scene. Be very careful not to shoot directly into the sun unless you have a sufficient neutral density filter in place and your lens is at a high f-number. The process is very difficult to do well.

If shooting a "night scene" at night outdoors, use one or two harsh lights, creating distinct shadows. Try to BACKLIGHT the performers (described further in Chapter 9) to form a white ridge outlining them. Again, watch out for burn-ins.

If shooting a night scene indoors, create harsh shadows and puddles of light by using one or two bare light bulbs hidden from the camera by props. BACKLIGHT the performers to create silhouettes. If you give your actors lanterns or candles, you have an excuse to beam some light on their faces to catch expressions, etc. The face light doesn't *really* have to come from their lanterns; you could carefully beam some light on their faces from off camera, making it look like the light came from the lantern.

SURREPTITIOUS RECORDING

If something is important enough to risk your life or (more important) your equipment for, that's your business. The camera operator is usually the one attacked when people want to hide what they are doing. Also, you may be the one sued when their privacy has been invaded.

It is generally considered legally permissible to record anyone in public areas (not counting toilets). Streets, sidewalks, parking areas, and public buildings are pretty much unrestricted. One rule of thumb when deciding whether you are illegally invading someone's privacy is to consider the following: Do the unwitting participants have a good reason for believing their activities are totally private and visible to no one else. A football fan making a big gambling payoff in the bleachers is fair game. The same transaction shot with a telephoto lens through the person's living room window is off limits.

The courts give investigative reporters and film crews more leeway than people shooting documentaries or entertainment programs. Using a hidden camera to catch a shop mechanic puncturing someone's tire is treated differently from comedy shots of hospital patients frantically trying to open jammed bathroom doors.

If you're doing a "Candid Camera" type of production, you will be required to secure permission from the participants before you can distribute or publicly perform your production. Warnings over; now let's get on with how to do it.

One way to catch people off guard is to focus quickly for medium distance, zoom out all the way, and hold the camera by your side, perhaps under your shoulder, perhaps at arm's length, held by the camera handle. The VCR is still running, but no one knows it. The camera is still taking pictures (but not of the sun or other bright objects we hope). Since the camera is not up to your eye, people think that it is turned off. Try to divert attention away from the equipment by turning yourself away from the camera. You might set the unit down (running) on a table. You could pretend to switch it off and then button up the carrying case. Just make sure your "dirty tricks" are worth the possible consequences if you are found out.

There are special "peephole" lenses which are very small and permit cameras to hide behind a hole in a wall or behind clutter with just the tiny lens snout sticking out. I remember once investigating storeroom thefts and wondering how to hide the camera and its regular lens. I hit upon the idea of stuffing the camera into a cardboard box of plastic drinking cups and sticking one cup over the lens. The image through the base of the clear plastic cup was

Surreptitious recording.

excellent, and no one noticed the half-open box of cups peering out from the shelf. Lesson: Be creative.

It's best to use black-and-white cameras with wide-angle standard lenses for detective work like this. The black-and-white cameras work well in low light and give sharp pictures. The wide lenses pick up all the action, and if your "spy" gets stolen, you're not out a $1,200 color camcorder and lens.

And after all this, you play back your tape the next day and see the thief in action—wearing a mask.

SHOOTING SPORTS EVENTS

The owner of a small Santa Barbara restaurant won considerable recognition by spending his days videotaping local sandlot football games, city league games, and other sporting events. At night, he played his tapes over a video projector to a packed house. The name of his flourishing restaurant was called The Instant Replay.

Besides entertainment, video can be a useful training tool as the players study their moves and formations. Close-ups can show one player's body position, angle, and form. Team shots reveal the dynamics of the group. Here are some shooting techniques unique to sports.

Camera Angles

For field sports, get as high as you can (without using drugs). Mount your camera atop a school bus or on some kind of tower. Face away from the sun. If the team is practicing moves for later review, the camera's view is of paramount importance. Have the practice moved to where the camera can see best, like on a lawn near a building. Here you can shoot from the window or a roof.

Feel free to zoom in or out on the action while the players are moving. Don't bother zooming in to a close-up of a player standing still.

For football, place the ball carrier to the rear of the frame with the blockers in the front. As the ball carrier passes the line of scrimmage, gradually center him so you get the action both ahead of and behind him. If the runner makes it through the secondary defensemen, gradually let him outrun your pan, thus positioning him in the front edge of the screen so you can reveal as much of the rear action as possible. On passing plays, cover the passer till the ball is thrown. Then pan to the intended receiver while zooming out some. As the ball is caught, continue zooming out enough to get the other players in his zone. On punts, cover the kicker. When the ball is kicked, stay with the kicker a moment to watch for roughing. Then pan to the receiver (you'll probably have time for this as the ball generally stays in the air awhile).

The general technique in sports videography is to catch the *main* event while recording the secondary action as well. No one cares what's happening *behind* the kicker during an extra point attempt, so you frame him off center with space in front. Meanwhile, you're including the charging defense at the other edge of the viewfinder.

Shooting Techniques

If you are shooting games for entertainment and for coaching purposes, don't forget to include a quick shot of the scoreboard after a touchdown or the down indicator or referee signals just to clarify what's going on and to provide a visual respite.

If your VCR makes good PAUSE edits, try to cut out the chaff, like the huddle breaks. Start the tape rolling when the call of the play begins.

When you get near the end of a cassette, eject it during a break in the action and start a new cassette. You don't want to run out of tape in the middle of the game's best play (and Murphy's 109th Law says that the play you missed is the best play of the game).

Shooting sports safely.

Perhaps this is a large enough dose of camera angles and techniques for one sitting. We'll be coming back for another taste of camera and shooting strategies later in Chapter 14 when we cover editing techniques and how to get one shot to flow smoothly into another.

Safety First, or Maybe Second

Remember all those nifty shots on *America's Funniest Home Videos* where the baseball comes flying straight into the camera lens, or the golf ball catches the cameraperson right in the tender places? Good stuff, right? The audience loves it! So do camera repair shops. So do doctors. And funeral directors. So before you tie that bungie cord around your ankles and dive off the Royal Gorge Bridge, you might pause for a millisecond and ask yourself, "Is this shot worth it?"

LIGHTING

The human eye is an amazing thing. It can make wide-angle, crystal-sharp images in color under the worst conditions. The eye is sensitive enough to see by candlelight and tough enough to perform in sunlight 20,000 times brighter than a candle's light.

The television camera is frail in comparison. It needs plenty of light, but too much light can damage the camera tube. It can display only a two-dimensional image which looks flat and dull compared with the 3-D panorama our eyes give us with each glance. Where the eye can discern 1000 different levels of brightness, the best cameras under the best conditions can distinguish only 30 or fewer shades of gray.

Lighting serves two purposes:

1. It illuminates the scene so that the camera can at least "see" it.
2. It enhances the scene to make up for television's visual shortcomings.

THE KIND OF LIGHT THE CAMERA NEEDS

Giant airport snowblowers can chew through 5-foot drifts with ease. Meanwhile, your handy-dandy electric snow shovel gets flooded flinging 40 flakes at a time. Having the right equipment makes a difference. The same is true for cameras. When the networks pay $20,000 or more for a TV camera, they get a machine that is hardly bothered by bright lights and dark shadows. *They'll work better* with the proper lighting, but they can also handle the rigors of a 5-foot snowdrift. Industrial and home cameras, on the other hand, need all the help they can get.

This chapter will list a lot of "rules" of lighting. Some of them matter a lot and some not very much. It all depends on how tolerant your camera and your audience are. No matter what camera you are using, the picture will always look better if you follow the "rules."

Enough Light

Older TV cameras required a fair amount of light to register any picture at all. Rather bright light was required for smooth pictures with vivid colors.

Recent industrial color cameras are much more sensitive, requiring about half the light of their older brothers. Home TV cameras are designed to be even more sensitive in weak light, probably because amateur videophiles generally don't have extra lights or wouldn't bother using them. So, broadly speaking, home video cameras will work in the least light, while older color cameras require the most light.

For industrial cameras, normal home lighting is barely

```
LIGHTING BRIGHTNESS COMPARED
The brightness of a scene can be measured in LUX or FOOTCANDLES (abbreviated fc).
Technically, 1 LUX = 0.093 fc, but it's easier to remember 10 LUX = 1 fc. Although
LIGHT METERS can be used to measure the brightness of a scene, most videographers
use their camera and monitor to determine if lighting "looks" adequate.
    A TV camera, in its specifications, will list the minimum amount of light the camera
needs to produce a picture. Some home cameras work (but not terrifically) in as little
as 10 LUX (1 fc) of light, while some industrial models may need as much as 200 LUX
(20 fc) for good color. It is good to know your camera's light needs so you can predict
whether extra lights will be needed for certain conditions.
```

Brightness (fc) (approx.)	Lux	Situation	Notes
1	10	Candlelight supper	More light needed
3	30	Movie theater during intermission	Poor color
10	100	Public stairwell	OK for some cameras
20	200	Kitchen or living room	OK for some cameras
40	400	Office	OK
60	600	Library or department store	OK
200	2,000	Cloudy day	OK, f5.6 maybe
1,000	10,000	Hazy day	OK, f8 maybe
3,000	30,000	Sunny day, early or late	OK, fill maybe
6,000	60,000	Sunny day, noon	Try ND (neutral density) filter
15,000	150,000	Beach or snow on sunny day	ND filter

sufficient to yield a picture. Although faces and objects will be recognizable, the image will be rough, grainy, or very gray and flat looking.

Office and classroom lighting is generally sufficient for shooting. Depending on the circumstances, you may even be able to "stop down" your lens from its lowest f-number to its next lowest f-number, realizing a little better depth-of-field in the process. Office lighting, though it provides sufficient light to create a picture, doesn't create the shadows and contrast to yield a vivid picture; it will still look somewhat flat and lifeless. The colors may also look drab.

On a cloudy day outdoors, the light is adequate for shooting. You may be able to use f4 to f5.6 for good depth-of-field.

A slightly hazy day is perfect for shooting outdoors. Shiny objects won't be dangerously bright (burning tubes or creating smears), and there will be plenty of contrast at f8, yet shadows won't be too pronounced.

Full sunny days are pretty good for shooting. Avoid highly reflective objects. Use f11 or so. The picture will be bright and vivid but may appear *too* contrasty. Shadows

especially may look too dark, and anything lurking in them may be obliterated.

The sun or its reflection off highly polished surfaces, a welder's torch in action, or direct views of an atomic bomb blast constitute excessive light and must be avoided at all costs if using a tube camera. They might even damage a chip camera.

Home video cameras will work well with office and classroom lighting and will work pretty well in home lighting. In home situations, the color and contrast will always be improved if you can find a way to throw more light on the subject. Notice the difference in the pictures in Figure 9–1 as light is added to the scene.

For special surveillance, police, and military purposes, LOW LIGHT LEVEL CAMERAS have been designed. It is said that the best ones can show you the color of a black cat's eyes in a coal mine. They do this using slower than normal shutter speeds (causing motion to smear), or use infrared light to illuminate the subject. Some use an electronic device called an IMAGE INTENSIFIER to perform in subdued light. Such devices are nearly always black-and-white and are more expensive than standard color TV equipment.

MINIGLOSSARY

Low light level camera   TV camera designed to "see" with very little light—used in military and surveillance.

Image intensifier   Electronic device that brightens the image fed to a TV camera—used in military and surveillance.

LOW LIGHT LEVEL CAMERAS almost see in the dark.

Lighting Ratio

Place something very bright next to something very, very bright, next to something very dark, next to something very, very dark, and you will be able to distinguish one from another readily. A TV camera, on the other hand, will see only two white objects and two black objects. Although your eye can handle something 1000 times brighter than something else in the same scene, and although photographic film can distinguish between an object 100 times brighter than another in the same scene, a TV camera can accept a LIGHT RATIO of only 30. With home video equipment this number may be as low as 15.

The brightest thing in the picture should not be more than 30 times brighter than the darkest object in the same scene. Here's what this means in practice. You wish to tape a person standing in front of an open window during the day. What your camera will see was shown in Figure 8–6. Since the light from the window is very bright, everything else looks black and silhouetted by comparison. The gradations of gray in the clothing and face are all lost. If you close the shade (see the figure again), now the whitest thing in the picture is the wall and some of the clothing. They

are only about 10 times brighter than the hair and other dark parts of the picture. As a result, everything between the blacks of the hair and the whites of the wall gets a chance to be seen as some gradation of gray rather than end up black as they did before. In short, things which are super-bright must be avoided. The brightest part of the scene should be less than 30 times brighter than the darkest part of the scene. Shafts of light coming in the windows, shiny buttons, and chrome hardware (like mike stands) should be avoided or subdued.

Lighting Placement

You want most of the light to come from behind you (the camera). Avoid light coming from behind the subject (bright windows, etc.) as that will silhouette your performer as shown back in Figure 8–6. On the other hand, try not to have all your light coming from *too near* the camera, or you'll lose your shadows, making everything look flat and dull.

Lighting Color

Unless you are after special effects, you'll be using white light. But white light isn't always white. Fluorescent lights are bluish. The incandescent lights in your home are reddish. You don't see this difference with your naked eye, but your camera does. The COLOR TEMPERATURE and WHITE BALANCE controls on your color camera can make up for much of this variation. But your controls can't salvage a scene that's lit on one side with fluorescent or sunlight and on the other side with incandescent or studio lights. The majority of the light in your scene has to be of the same COLOR TEMPERATURE, or else you'll confuse your poor camera into giving you half pink and half magenta faces.

BASIC LIGHTING TECHNIQUES

Existing Indoor Light Only

You're shooting on location and didn't bring lights (maybe you're traveling light, if I may abuse a pun). How do you illuminate your subject?

1. Place your subject where the existing illumination is best, like outdoors (in the daytime, of course) or under office lighting.

MINIGLOSSARY

*****Lighting ratio**   A comparison between the brightest part of a subject and the darkest. If the brightest white in a performer's shirt measured 60 fc (footcandles) and his black hair measured 2 fc, then the lighting ratio would be 60 ÷ 2 = 30.

Lux   A measure of the brightness of an object in a scene. Cameras need a certain degree of scene brightness in order to register a

picture. Ten lux equals about 1 footcandle, another measure of brightness.

Light meter   Electronic device that measures the brightness of a light (incident light) or the brightness of a scene (reflected light) and gives a readout in footcandles or lux.

Scene lit with regular incandescent home lighting

Scene lit by overhead office fluorescent lamps

Scene with just one extra 250-watt lamp added near camera

Scene with three, well placed, extra lamps

FIGURE 9-1 Ambient lighting vs extra lighting.

2. If the camera with its lens wide open (lowest f-setting) still shows a poor picture because of insufficient light, seek out other light sources such as desk lamps. Turn on every lamp in the room. Take the shades off any lamps not appearing in your scene just to boost the lighting more. Move lamps closer to the scene if possible. Replace the light bulbs in lamps with the highest wattage rated for the lamp (if using extension cords and lots of lamps, do not exceed the wattage maximums for the building's wiring and extension cords). Every light that you add will add punch and contrast to your picture. Can you scrounge up any lights nearby? How about an outdoor floodlight? Does anyone have a movie light packed away since their super-8 movie-making days? Be aware while placing such lamps that the closer they are to your subject, the brighter your subject will be illuminated; however, the area covered by the light will be smaller. This can be a problem if your subject is a moving one as he might slip out of the small bright area you have created. Also, moving subjects cause a brightness problem with close lights. If someone moves his head 6 inches closer to a lamp 4 feet away from him, the change in the illumination of his face will be unnoticeable. But if someone 2 feet away from a lamp moves 6 inches closer to the lamp, the illumination on his face increases sharply, causing a pronounced flare or shine on his forehead and cheeks.

3. Avoid bright windows or lights in the background of the shot. If you wish to use light from a window, get between it and the subject so that the subject, not you, is looking into the window light.

Most of the preceding ideas pose COLOR TEMPERATURE problems, but at least you'll have enough light to take your pictures.

Outdoor Lighting

The big challenge with outdoor shooting is shadow control. Picture a bright sunny day. The baby chases the family puss across the green lawn and under a tree. Suddenly the baby's rosy pink cheeks turn muddy gray. Your orange cat turns

KINDS OF LIGHTING

Types of Illumination	Color Temperature
TUNGSTEN bulb for home use	2800°K
TUNGSTEN bulb for photographic use	3200°K
QUARTZ-HALOGEN or TUNGSTEN IODIDE bulb	3200°K
Daylight, early or late	3200°K
FLUORESCENT lamp, warm white bulb	3500°K
FLUORESCENT lamp, white bulb	4500°K
Photoflood bulb for photographic use	5000°K
Daylight, midday	5500°K
HMI (HALOGEN METAL IODIDE) professional lamp	5500°K
FLUORESCENT lamp, daylight bulb	6500°K
Daylight, hazy or foggy day	7000°K

INCANDESCENT   A light with a TUNGSTEN (a metal with a high melting point) filament which glows when electricity passes through it. Common home lamps are INCANDESCENT and have a low COLOR TEMPERATURE. When INCANDESCENT lamps are dimmed, the light becomes redder, but the bulbs will last much longer. Bulbs darken with time (some of the filament boils away and coats the glass), decreasing the light's brightness and lowering the COLOR TEMPERATURE of the light.

FLUORESCENT   A light with a gas inside which glows when electricity passes through it. The COLOR TEMPERATURE is usually high. It is possible, with special hardware, to dim FLUORESCENT lights, but the practice is uncommon. FLUORESCENT lights give off more light per watt than INCANDESCENT lamps, making them more efficient and cooler running. They're not easily aimed or focused. A powdered coating inside the bulb is formulated to glow a specific color; thus FLUORESCENT lights can come in various COLOR TEMPERATURES.

QUARTZ-HALOGEN or TUNGSTEN IODIDE   All means the same thing. Both are INCANDESCENT lamps having a TUNGSTEN filament. The glass bulb is QUARTZ. A gas inside the bulb is IODINE (also called a HALOGEN gas). The gas helps keep the bulb from darkening with age, so the COLOR TEMPERATURE changes very little. Unlike regular INCANDESCENT lamps, QUARTZ lights run very hot, and the bulbs are quite small.

PHOTOFLOOD   A light bulb, sold by photography shops that looks like a normal household lamp but burns with a bright blue-white light, much like daylight. The bulbs darken with age and usually last 10 hours.

DAYLIGHT   That stuff you see outdoors when you're not cooped up inside reading this book.

PHOTOFLOOD bulbs (Photo courtesy of *Single Camera Video Production*—Prentice Hall)

TUNGSTEN HALOGEN studio lamp bulbs

---

muddy gray. The green lawn turns muddy gray. Every once in a while you can see a white flash as the kid's outfit is caught in a stray beam of sunlight piercing the leaves.

The trick here is to fill in the shadows:

1. Shoot on hazy days when shadows are soft.
2. Shoot with the sun mostly to your back so shadows are partially hidden.
3. Stop down your lens as far as possible to reduce excessive contrast.
4. Glue some tinfoil to a sheet of posterboard and "fill in" the shadows with reflected light.
5. Shoot with a bright light near the camera, even in broad daylight. Place the light in such a way as to "fill in" the shadows caused by the sun. To avoid COLOR TEMPERATURE problems, use a "daylight" photoflood, or use an HMI light (described later), or use a colored filter over the bulb to convert the light to 5600°K.

Bright sunlight has a way of driving a camera's AUTO IRIS crazy. The automatic control "locks onto" the brightest part of the scene, making everything else look dark relative to it. Whenever possible, size up the lighting situation and switch to MANUAL IRIS. Let your eye and viewfinder be the judge of what is most important in the scene and make *that* look good, even if something else gets a little over- or underexposed.

This technique is commonly applied when shooting *toward* the sun: In this case, the sun acts as a strong BACK light, rimming the performer, while you use a separate lamp or a reflector to illuminate the talent's face. The background or rim may be too white, darkening the talent's face. Here you would manually open the IRIS a little (or press a BACK LIGHT button on the camera, which does the same thing) to overexpose the background while properly exposing the face. You can only "cheat" like this by one or two f-stops before you overburden the camera's electronics with excessive light. Sometimes it helps to zoom in on the face a bit to decrease the amount of bright background visible.

Outdoor light reflector of aluminum foil over posterboard helps fill in shadows.

If you're shooting early in the morning or late in the afternoon, WHITE-BALANCE your camera often. As the sun nears the horizon, its COLOR TEMPERATURE drops rapidly toward the red. This change can sneak up and bite you if you're too preoccupied with your shoot to notice the changing sun.

One Light Only

You're shooting on location and you brought only one light (perhaps that's all that would fit under your airline seat). Where do you place it?

Don't place it next to the camera because that will give a flat picture without shadows as in Figure 9–2. In most cases, shadows are desirable as they create a sense of depth and texture to the image. Place the lamp at an angle 20°–45° to the right and 30°–45° above the subject as shown in Figure 9–3.

What kind of lighting instrument should be used? Figure 9–4 shows several popular portable lamps used for video and filmmaking. Some can be held by hand, and others clamp to telescoping stands.

If you're practicing penny-pinching teleproductions, all is not lost. Improvise. Buy one (or more) clip-on lights and extension cords at your hardware store. For a daylight color balance, equip them with PHOTOFLOOD light bulbs from a photo store. The light can be clipped almost anywhere, including the top of your camera, to make a roving camera/light ensemble like the one in Figure 9–5. Although that clip-on setup is handy, the shadows it creates aren't too gorgeous.

Dig around in the back of your AV closet or go to a church rummage sale, and you may come up with a movie light (Figure 9–6). They are portable, are very bright, and yield a reasonable COLOR TEMPERATURE.

Shooting with only one light has its liabilities. It is possible that the light could be so bright, compared with other light in the scene, that it "washes out" light-colored parts of the scene and creates harsh shadows. This is called EXCESSIVE LIGHTING RATIO and typically occurs when a very bright light is placed too near the subject. Figure 9–7 shows an example.

The cure for EXCESSIVE LIGHTING RATIO is to find some way to dim or diffuse the light and to create or reflect some light into the shadow areas. Try moving the light farther from your subject. Or place a metal screen, called a SCRIM, in front of the light to diffuse it. Figure 9–8 shows a SCRIM which clips onto the front of a light fixture. Maybe refocus the light from SPOT to FLOOD (described shortly).

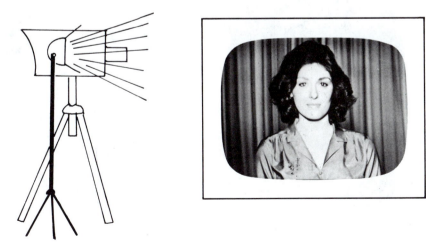

FIGURE 9-2 One light placed near the camera yields a flat picture with almost no shadows on the subject.

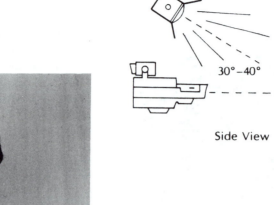

Side View

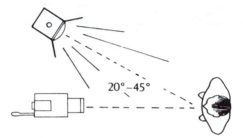

Top View (Lamp could be to left or to right.)

FIGURE 9-3   Optimal placement of single lamp to create depth through shadows.

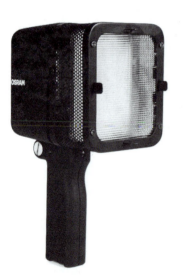

Hand-held light

Camera-mounted light

Brace holds camera-mounted light farther from camera for better shadows.

FIGURE 9-4   Portable TV lights (Courtesy Quality/Universal).

If roving around inside buildings, clamp-on a fixture with a bare 250-watt bulb. You'll also need a long, heavy-duty, extension cord for it. All this can be found for under $6 at your hardware store. A special PHOTOFLOOD light bulb from a photo store would cost extra but would improve color shooting.

FIGURE 9-5 Cheapie roving camera light.

Another alternative is to aim the light at a white ceiling or white wall and illuminate your subject with the fairly shadowless BOUNCE light.

If the room's natural lighting seems to illuminate the left side of the face the most, you should place your professional light on the right side of the face. This way the existing room lighting has a job to do; it fills in the shadows created by your professional light.

Two Lights Only

The professional solution to dark shadows is to use two lights. One light makes the shadows, and the other one decreases their intensity, bringing the LIGHTING RATIO back to where the camera can handle it.

Where should you place your two lights? The first light should go 20°–45° to the side and 30°–45° up as described in Figure 9–3. You place the second light similarly up and to the *other* side of the camera as in Figure 9–9.

The brighter of the two lights acts as the KEY light, providing most of the illumination of the subject, while the

FIGURE 9-6 A movie light makes a handy portable video light.

weaker lamp becomes the FILL light, filling in the shadows somewhat and softening the picture.

If both lights are of equal brightness, one can be made into a FILL light by

1. Moving it farther away from the subject.
2. Placing a SCRIM in front of it to diffuse the light.
3. Aiming the light at something reflective nearby. The diffused reflected light will then fill in the shadows.
4. Refocusing it from SPOT to FLOOD.

Individual taste and circumstances play a large role in setting up lights. There is no law that says a light must

FIGURE 9-7 Excessive LIGHTING RATIO.
Too bright a lamp, too close, causes excessive contrast and deep shadows

MINIGLOSSARY

*Photoflood** Light bulb, available at photo stores, that can screw into normal lamp sockets but gives off proper color temperature for TV or film work.

*Scrim** Glass fiber or metal screen mesh which clips to front of lighting instrument to diffuse and soften light.

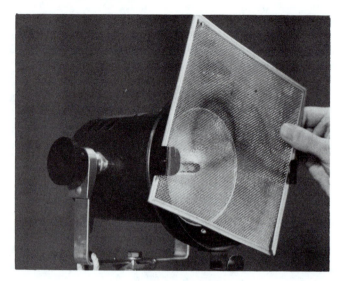

FIGURE 9-8   SCRIM clips onto light fixture to reduce and diffuse light intensity.

be 20° up and 30° over. No laws says one lamp must be brighter than the other so that one is the KEY and the other is the FILL; they could be equal. The ideas set forth here are generalities, not rules.

If working without professional lights, you could make a FILL light out of a clamp-on utility lamp with a large shroud as in Figure 9–10. You can further soften the light by covering the opening with layers of metal window screen. A PHOTOFLOOD bulb will give a higher COLOR TEMPERATURE

light than the standard home light bulb and will also provide plenty of brightness.

If you want all shadows to be harsh, deep, and noticeable, and if you want textures to appear rough and super-three-dimensional, you omit the FILL light. In most cases, however, you don't want *black* shadows, just the dark gentle hint of shadows.

STUDIO LIGHTING

Well-placed, carefully aimed lights of just the right color and intensity are superb for fulfilling lighting's main purposes:

1. To provide enough light so that the camera can "see"
2. To enhance the scene to overcome TV's inherent shortcomings, making the image appear sharp, vivid, and three-dimensional

Such precise lighting adjustments are only possible in a studio where you have total control over all the lighting. Ideally, there should be no windows or fluorescent office lights to contend with.

Figure 9–11 displays a general "typical" TV lighting layout. TV (and film) lighting consists of four basic building blocks, the KEY light, the FILL light, the BACK light, and the SET light. Let's look at them one by one.

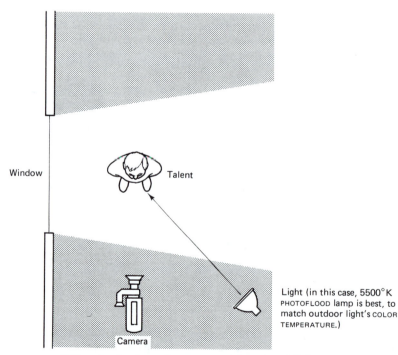

Filling in shadows from a room's natural light.

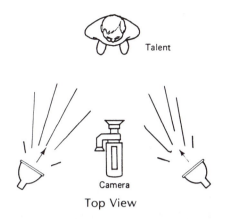

Top View

FILL light made from a utility light with a large reflector covered with metal window screen to further soften the light. The whole works should cost $10 from your hardware store.

FIGURE 9-10   Homemade FILL light.

KEY LIGHTS are usually instruments with a VARIABLE FOCUS control which can be adjusted either to FLOOD the area evenly with light or to concentrate the light in a small SPOT. Generally, if intense light is needed, adjust the instrument to SPOT. If the subject is large or moves around, SPOT may be unsatisfactory because the area illuminated is so small. The solution is either to adjust the instrument— partially or all the way—to FLOOD (sacrificing some of the brightness as the light covers a larger area) or to obtain more instruments to cover the area.

Some studio lighting instruments vary their focus by moving the lamp bulb closer to or farther from the reflector, thus spreading or concentrating the light. Others, such as the FRESNEL lamp shown in Figure 9–14, focus the light by moving the bulb closer to or farther from a lens.

Generally, when lamps are used on a stand, a knob on the instrument may be moved to adjust the lamp's focus. If the lamp is the type that hangs from the ceiling, on the bottom of the lamp there may be a loop that is easily turned by using a pole with a hook on it. See Figure 9–15.

Fill Light

Any light can be a FILL light if it fills in the shadows created by the KEY light. Professional instruments typically have a

FIGURE 9-9   Using two lights.

Key Light

The KEY light (Figure 9–12) is like the sun. It illuminates the subject, creating the main shadows, as shown in Figure 9–13. These shadows help give depth and dimension to the scene.

The term KEY light does not refer to a particular type of fixture but describes the kind of job the fixture does. Almost any lighting instrument can act as a KEY light; however, some are better suited than others.

The KEY light generally has BARN DOORS, metal flaps used to direct the light or shade certain areas of the scene. Like the SCRIM, the BARN DOORS may be clipped onto the front of the fixture.

MINIGLOSSARY

Barn doors   Metal flaps on a lighting instrument that can be closed or opened to direct the light, and shade areas where light is undesirable.

Flood   Broadly focused light that covers a large area evenly.

Spot   Narrowly focused light that concentrates its intensity over a limited area.

Key light   Brightest and main source of lighting for a subject, creating the primary shadows.

Fill light   Soft broad light whose main purpose is to fill in (reduce the blackness of) shadows created by the key light.

Variable focus light   A lighting fixture that can be adjusted from spot to flood or vice versa to direct the light's intensity.

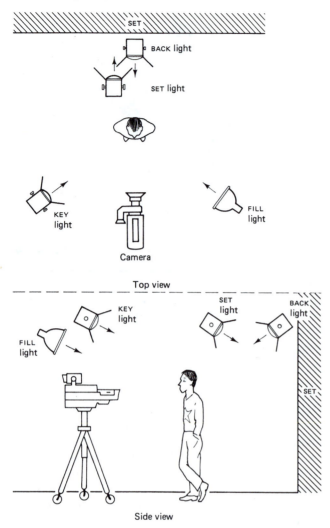

Top view

Side view

FIGURE 9-11 Typical lighting layout.

1000–2000-watt (1–2 kw) bulb surrounded by a large reflective surface. The larger the surface, the more the diffuse the light will be and the softer the shadows will look. Figure 9–16 shows two FILL lights commonly used in studios.

Figures 9–3 and 9–13 show the results of using a KEY light without FILL. The image is harsh and contrasty. Figure 9–13 also shows the result of using a FILL light alone and in combination with a KEY light. The relative brightness of these two lights determines the depth of the shadows created.

While setting up lighting, some people use light meters and measure the LIGHTING RATIO between KEY and FILL and the other lights. If you are not so inclined (and most industrial users are not), a pretty good lighting job can be

done by "eye" if you let the camera do some of the work for you:

1. Set up your lighting the way you think it should be.
2. Aim the camera at the subject to be recorded, and look at a TV monitor to examine the image.
3. Readjust the lighting so that the image looks best *on the TV screen.*

Placement of the FILL light is generally 20°–45° to the side of and 30°–45° above the camera-to-subject axis, just like the KEY light, only on the opposite side of the camera from the KEY light. This placement is flexible, however, and occasionally FILL lights may be found near the floor or near the camera.

Since the KEY light is often thought of as the sun, it is appropriate for this light to come from fairly high up (like the sun), casting shadows downward. Since the FILL light's job is to temper those shadows, the FILL often has to be low (if it were high, it too would be casting shadows downward).

The light from a FILL light can be made harder or softer. Some instruments have a FOCUS control and may also have a holder for a SCRIM, which can further soften the light.

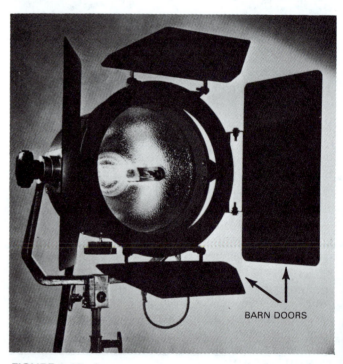

FIGURE 9-12 Instrument generally used as a KEY light.

MINIGLOSSARY

Fresnel Lighting instrument with a circularly ribbed glass lens to focus the light.

Scoop light Funnel-shaped fill light.

Soft light Large lighting instrument with built-in reflector for soft, shadow-free lighting.

KEY light alone

FILL light alone

KEY and FILL together

BACK light alone

KEY, FILL, and BACK together

SET light alone

KEY, FILL, BACK, and SET together

FIGURE 9-13 Various lighting effects in a darkened studio.

241

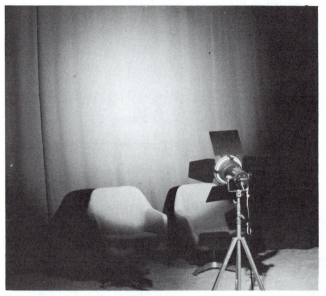

Light focused for SPOT

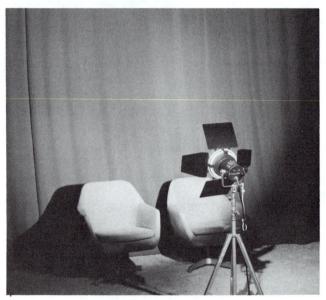

Light focused for FLOOD

SPOT VS FLOOD. SPOT is brighter and a little harder.

Figure 9–17 shows a spun-glass SCRIM used on a SCOOP light for this purpose.

If you don't have a proper FILL light, try softening the light from one of your unused KEY lights to do the job. Adjust its focus to FLOOD and add a SCRIM or two. You could even try aiming the light at the ceiling, a wall, or a white posterboard to diffuse the light.

The softer and more diffuse the light, the sexier and more informal your picture will look. To get the effect

shown in Figure 9–18, you will need to use a large SOFT LIGHT or will have to bounce your light off a large white surface.

When shooting outdoors, the sun usually acts as your KEY light, and you have to dream up something for your FILL. You could bring along an electric light and position it to fill in shadows. Or you could hold up a square of white or tinfoil-covered posterboard and position it to reflect some of the sun's light onto the shaded side of your subject.

Back Light

The BACK light is the third most important of your studio lights (KEY and FILL being the first and second). Its position and effect are shown in Figures 9–11 and 9–13. The BACK light is responsible for most of the dimensionality of the TV picture. Without it, the image is flat and dull; with it, the image stands out from its background and has punch.

The BACK light's job is to rim foreground subjects, separating them from the background. But don't make the BACK light *too* bright, or it will light up the tops of actors' heads and shoulders, distracting the viewer from the actors' faces. Would you rim da Vinci's "The Last Supper" with a frame of blinking neon lights?

The BACK light shouldn't look straight down on the subject (as in Figure 9–19) for it will illuminate the performer's nose (like Rudolph, the Red-Nosed Reindeer) every time the head tilts back. The light should strike from above and behind at an angle 45°–75° up from the horizontal. The

FIGURE 9-14 FRESNEL lamp, generally used as a KEY light (Photo courtesy of Colortran, Inc.).

MINIGLOSSARY

*****Back light** Lighting instrument that illuminates the subject from behind, creating a rim of light around the edges of the subject; usually has barn doors for precise control of light's direction.

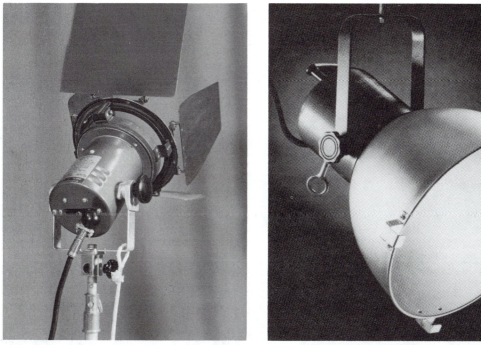

knob for varying focus   loop for varying focus

FIGURE 9-15 Focus controls.

higher and farther back the lamp, the better, because the light being aimed *toward* the cameras has a tendency to shine into the lenses. This causes undesired optical effects like the lens flare in Figure 7–30 and risks burn-ins when careless camera operators tilt too far up. Often the BARN DOORS are a help in shading the cameras from the lights while directing the light only on the performers. Figure 9–20 shows a couple of BACK lights with BARN DOORS.

A cousin to the BACK light is the MODELING light or KICKER light. This light hits the subject from behind and *to the side*, creating a rim of light around the edge of the subject. Its purpose is the same as that of the BACK light—to outline the subject, adding dimensionality. It can be used with the BACK light to strengthen the outlining effect or could be used alone for when the BACK light isn't practical (i.e., when the performer is wearing a straw hat and would

FIGURE 9-16 A SCOOP and a SOFT LIGHT, generally used as FILL lights (Courtesy of Colortran, Inc.).

Sometimes a high KEY light can cause problems, like prominent eye shadows or shadows from eyeglass rims.

Try lowering the KEY light.

Shadows caused by glasses.

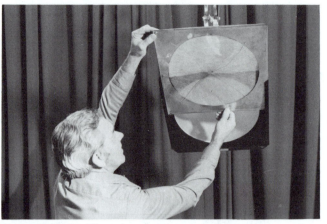

FIGURE 9-17 Fiberglass SCRIM used on a SCOOP to soften the light.

FIGURE 9-18 Soft indirect lighting (Courtesy of Colortran, Inc.).

be in total shadow from a BACK light). The MODELING light substitutes nicely for the BACK light in portable situations where it is inconvenient to erect a BACK light above and behind someone.

Because the BACK light does a more complete job of rimming the subject, it is usually preferred, leaving the MODELING light for special occasions.

Set Light

The SET light illuminates the set or background. Again, Figure 9–11 shows its position, and Figure 9–13 shows its effect. Depending on the brightness of this light, a gray background can be made to look white, neutral, or black

relative to the performer. It is best (usually) to have the background darker than the performers so that they stand out from the background, directing your attention to them. Figure 9–21 shows one result of having too light a background. And if the background is *really* too bright, as in Figure 8–6, you may see only a silhouette of your performers, or their faces may appear very dark.

Not any light can do this job well. The set should be lit fairly evenly. A regular light aimed downward near the set will create a bright spot at the top of the set and will fade off to nothing at the bottom. Placing a lamp farther from the set will light it more evenly, but unless the subject is standing *far* in front of the set, much of the light will spill onto the performer. For good control, you want each

---

MINIGLOSSARY

Modeling light   Light aimed at the subject from behind and to the side, creating a white rim of light to add dimension.

*Set light   Lighting instrument used to illuminate the background or set.

FIGURE 9-10 BACK light aimed straight down on subject.

studio setup

Telescoping light stand elevates lamp to illuminate performer from above and behind.

FIGURE 9-20 BACK lights with BARN DOORS.

KEY and FILL only

KEY, FILL, and MODELING

MODELING light.

light to illuminate one area only so that adjusting one lamp doesn't affect anything else (it will anyway to some extent, but why make things worse?).

Some studios use rows of lights to illuminate the set. One row is above the performer (but not illuminating him or her), and another group may be on the floor aimed up at the set. An easier method of illuminating the set is to use a specially designed SET light. This fixture concentrates the light and throws it *down* to illuminate the bottom of the set while throwing a diminished supply of light at the top of the set near the lamp (where, because of the nearness, the set will get more light anyway). Figure 9–22 shows a SET light.

If you built your set to look like a room or an office, keep in mind that most room light comes from the ceiling and is aimed down. SET lights, in order to duplicate this effect, should not illuminate the tops of your set pieces too

The woman in white fails to stand out from the white background. It is usually best to have the background darker than the subject (Courtesy of Imero Fiorentino Associates, Inc.)

FIGURE 9-21   Light background.

brightly. Walls should look darker at the top. When a camera looks at a wall and the top is slightly darker than the rest of the wall, the viewer mentally assumes that there is a ceiling. We all know, however, that sets are constructed without a ceiling (so that the studio lights can shine down on the set), but your viewers don't need to know that.

The SET light spilling onto the performers is a problem. The KEY and FILL lights spilling onto the set create another problem. If room is available, try to move the set back far enough behind the performers so that this spillage is minimized. Doing so, you'll also reap the benefits of not having to contend with so many shadows from the performers being projected onto the set from the KEY and FILL lights.

Pattern Spotlight

Another instrument used to illuminate sets and backgrounds is the PATTERN SPOTLIGHT or FRAMING SPOTLIGHT like the one in Figure 9–23. The instrument accepts aluminum cutouts and, like a projector, focuses the *pattern* from the cutouts on the studio background. These cutouts could be of shutters, venetian blinds, windows, clouds, church windows, leaves, or stars or even could be custom-made designs and serve to create an illusion or mood. Venetian blinds, for instance, projected onto a paneled studio set, may imply

that the room is an office. A cross projected on a curtain behind a singer would give a religious mood to the set. Often you can avoid the expense and trouble of building complex windows and other set pieces if you simply project the pattern of them on the wall.

The technical term for this metal PATTERN is CUCALORUS, called a COOKIE for short (I can't imagine why).

If you don't have a PATTERN SPOTLIGHT or if you lack a specialized PATTERN to put in it, here's a way to improvise: Use a slide projector for your SPOTLIGHT. Take a slide of

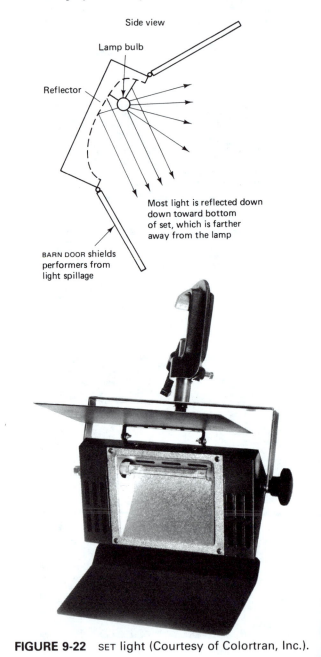

Side view

Lamp bulb

Reflector

Most light is reflected down down toward bottom of set, which is farther away from the lamp

BARN DOOR shields performers from light spillage

FIGURE 9-22   SET light (Courtesy of Colortran, Inc.).

MINIGLOSSARY

*Pattern spotlight   Lighting instrument that accepts aluminum cutouts to project patterns such as venetian blinds, leaves, or other

figures on the background.

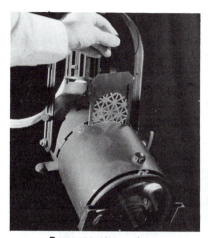

Patterns slip in here.

FIGURE 9-23 PATTERN SPOTLIGHT.

Pattern projected on curtain. (From *Single Camera Video Production*—Prentice Hall)

patterns

the desired black-and-white PATTERN and project that onto your set. Another alternative is to use an X-Acto knife to cut the desired PATTERN into an index card and tape the card to a clear or empty 35mm slide mount for projection. Tinfoil or black slides also make good surfaces to cut PATTERNS into. Flimsy, complex PATTERNS may have to be sandwiched between glass in special custom slide mounts.

Other Kinds of Lights

There are enough specialized kinds of lights and lighting techniques to fill a catalogue. Some are aimable spotlights designed for following an actor on stage. Some instruments are designed for portable use; they are small and fold up compactly. Some are designed for battery use and others for use near explosive gases.

PORTABLE LIGHTING

Name three lights you would take with you on location. Heineken Lite, Michelob Lite, and portable light, right?

Portable lights can be purchased as kits, which may include three or four small-sized 500-watt fixtures with BARN DOORS and adjustable mounts by which they can be attached to portable telescoping lighting stands like those shown in Figure 9–24. Some lighting instruments attach to a heavy-duty clamp which can be clipped about anywhere. Some special clamps will hitch to a drop ceiling like those found in offices and schools. Many kits contain SCRIMS, extension cords, and other handy gadgets, like FLAGS (easily movable flaps for casting shadows and controlling light spillage).

These lights are commonly used indoors and in tem-

porary or quasi-"studios." They are quite popular in schools. Even in larger studios, stand-mounted lights are useful, especially when you want light to come from a low angle. They are a nuisance, however, when you trip over the power cord or when the floor stands get in your way.

Portable lighting kits may also be used outdoors, but they pose two problems:

1. *COLOR TEMPERATURE.* Sunlight usually has a COLOR TEMPERATURE of 5000°–7000°K. Studio lamps usually have a COLOR TEMPERATURE of 3200°K. This means that if the sun were the KEY and a portable light were your FILL, your KEY would look bluish, and your FILL would look reddish.

FIGURE 9-24 Portable lighting kit (Courtesy of Quality/Universal).

MINIGLOSSARY

Pattern An aluminum cutout that fits into a pattern spotlight to create the shapes projected by the light.

Flag Easily movable flap used with lights for casting shadows and controlling light.

Gaffer grip clamp holding a Lowel
Tota-Light on a wall stud.

Another Lowel Tota-Light on a wall bracket.

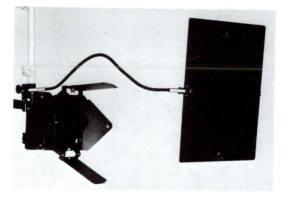

Flexible arm holds a FLAG
on a Lowel Omni-Light.

Some portable lights with clamps (From *Single
Camera Video Production*—Prentice Hall)

2. *Electric consumption.* Normal studio lights and their portable counterparts typically use 500–1000 watts each. This is a lot of power! Often, no more than one or two such lights can be connected to a household wall outlet at once without blowing a fuse. Even the extension cords for such lights must be heavy duty.

For every complex problem, there is always an equally expensive solution. To raise the COLOR TEMPERATURE of the lamp, special filters, the shape and size of SCRIMS, can be inserted into the instrument to cut out some of the redness of the light. These filters also cut light output significantly, requiring almost twice the number of lights to get the same effect. Special bulbs with dichroic coatings will give off light at 5500°K and can be swapped for the bulbs normally used in your lighting instruments. These coatings, however, decrease the brightness of the bulb, and the coatings wear off quickly.

ENG teams frequently use battery-powered lights such as the ones shown in Figure 9–25. Some are held in the hand and get their power from battery belts worn around the waist. Others clip onto the TV camera and aim wherever the camera aims. In both cases, the lights aren't very bright compared with their big brothers, and they drain their batteries quickly. They are good for shooting someone 5 feet away or closer, but they won't illuminate a very large area.

COLOR TEMPERATURE and power problems are *both* solved elegantly and expensively with a recent invention, the HMI light.

HMI Lights

HMI stands for halogen metal iodide, which describes the materials which make up the light bulb. It is a special instrument which creates 5500°K daylight and puts out about

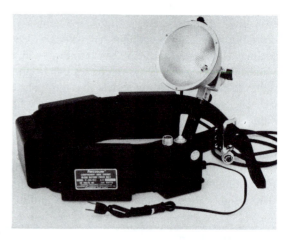

FIGURE 9-25 Frezzi battery belt and light (From *Single Camera Video Production*—Prentice Hall)

three times as much light as a tungsten bulb of the same wattage rating. It also generates about half as much heat.

Because the instrument makes the right COLOR TEMPERATURE to start with, it doesn't need a brightness-robbing filter to correct the COLOR TEMPERATURE. Because of its correct COLOR TEMPERATURE and high efficiency, a 1200-watt HMI light produces just about as much light as a 10-kilowatt tungsten lamp. It can plug into any standard wall socket without blowing a fuse where the 10-kilowatt unit would require special wiring to be installed by a licensed electrician. Also, because the lights don't heat up as much as their tungsten counterparts, they don't turn offices and homes into ovens or overburden a smaller building's air conditioners.

HMI lights have two disadvantages. First, they are expensive. Second, the instruments don't plug directly into a wall outlet. Instead, they plug into a large and heavy BALLAST, which powers the light. The BALLAST plugs into the wall outlet.

In short, HMIs are stingy on power and run relatively cool but require a large and heavy BALLAST. Figure 9–26 shows an HMI light.

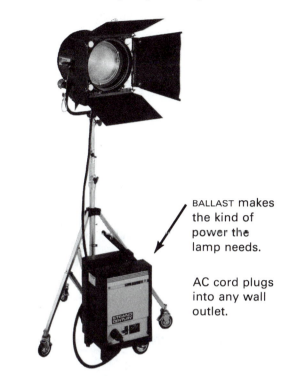

BALLAST makes the kind of power the lamp needs.

AC cord plugs into any wall outlet.

FIGURE 9-26   HMI light.

LIGHTING TECHNIQUES

Lighting Several Areas at Once

The lighting plan for one camera angle may not necessarily work for another camera far from that angle. So guess what you have to do? Check the camera angles before show time to see if lighting problems exist for these various angles. If they do, use the same techniques as before to KEY, FILL, and BACK light for each camera angle.

Sometimes the performers will be moving around. So that they don't walk out of the light, lamps have to be provided to illuminate the path that the performers will take. It could take dozens of lamps to illuminate an entire set. In professional studios it is common to see rows of KEY lights flanked by dozens of FILL lights lined up, crowding the rafters. BACK lights seem to be aiming in every direction. These setups were made one step at a time using the techniques described in this chapter: KEY, FILL, BACK, and SET. The process was repeated for every angle and for every stage position until the entire area was lit.

Sometimes the lighting strategy gets so complex that you need to sketch out your plans on a piece of paper before you try to execute them. This plan is called a LIGHTING PLOT and shows what types of instruments are used and where they are aimed and gives each light a number so that it's easier to refer to later. The LIGHTING PLOT may also include details such as props and furniture and may even show where the talent will be standing or moving. Figure 9–27 is a LIGHTING PLOT for a medium-sized studio production.

The LIGHTING PLOT serves not only to organize your plans (before you start climbing ladders and lugging hardware from position to position), but it allows you to give the plan to another person to execute.

Many studios don't have a great number of lighting instruments to work with. In such cases, one tries to get a single instrument to do two jobs at once. For instance, in the interview setting shown in Figure 9–28, one person's KEY light is the other person's FILL.

If your performer must move from one area to another and you don't have enough instruments to light the entire path, then light the beginning and end points of his or her journey and use low FILL lighting for in between.

Lighting for Color

The basic principles of black-and-white lighting generally hold true for color. Color, however, permits new lighting possibilities while exacting new constraints.

MINIGLOSSARY

HMI light   Halogen metal iodide lighting instrument. Very efficient and uses minimal power, but requires a heavy ballast. Gives off light with 5500°K color temperature.

Ballast   An electrical transformer that properly conditions the electrical power to run HMI lights.

Lighting plot   A drawing to show where the lights are to be aimed. May also include fixture numbers and circuit numbers and show prop and performer positions.

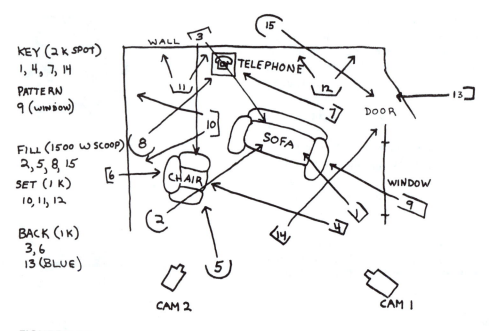

FIGURE 9-27   LIGHTING PLOT.

For color, you need more light than for black and white. With insufficient light, color cameras will produce grainy pictures with muddy color. So for color, you're stuck with introducing more instruments and consuming more power than for black and white. So if you find yourself racing to a classroom with 5 minutes to get set up to record the instructor's lesson, you will probably want to shoot in black and white. You won't have time to set up the lights to make the color look good. If your recording is to show teaching techniques, color isn't important anyway.

Most serious teleproduction is done in color nowadays. With color we can get creative, illuminating our sets and performers with colored lights. Actually, the light bulbs themselves are always white. The lighting instruments have slots in them to hold colored panes of glass or plastic or to hold GELS, a popular theatrical supply which looks like colored cellophane. The light that shines through ends up colored. The GELS slip into the instruments much like a SCRIM does.

Blue lights can give the impression of nighttime, darkness, or cold (this is why in Figure 9–27 instrument 13 was blue). Red lighting may convey warmth and happiness or fire. Lighting an object with different colored lights from different angles offers dimensionality and visual appeal. We pay for this new creative freedom with new headaches. The camera doesn't "see" things exactly the way we do, so something that looks pleasant to the naked eye may look abominable on the TV screen. Lighting adjustments must be made with both eyes on a color monitor. As the talent moves from one area of the stage to another, the relative brightness of various lamps will change, thus changing substantially the vividness, the highlights, the darkness of the

shadows, the overall contrast, and even the color of the subject. To accommodate these problems, the lighting director (the person who does the lighting) must be familiar with all the moves likely to be made on the set and he or she must illuminate each performer for each area.

Background color can be used to add dimensionality to a scene. Where in black and white it was the BACK light's job to make a subject stand out from its background, now the complementary coloration of the background can help to emphasize the subject. Something to watch out for when choosing color backgrounds and props is the effect they may have on other colors in the scene. One colored surface may reflect light of its own color onto an adjacent object, such as an aqua dress casting a sickish blue-green tinge onto the

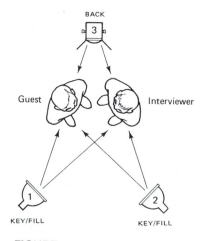

FIGURE 9-28   Let one person's KEY be another's FILL when short of instruments.

Colored GEL.

neck of the TV performer or a yellow detergent box throwing a dingy hue on a nearby stack of clean white undershirts.

Faces are perhaps the hardest things to illuminate correctly. No one will notice if a shirt, table, or backdrop appears more bluish or greenish than it's supposed to. But flesh tones that appear pale, greenish, or reddish brown will not be tolerated (unless you are shooting Casper the Friendly Ghost, the Creature from the Black Lagoon, or the Jolly Green Giant). Not only are skin tones sensitive to outright color changes under various lighting and backgrounds, but these tones are also sensitive to changes in COLOR TEMPERATURE.

COLOR TEMPERATURE made its debut in Chapter 6 where we adjusted the various doohickies on color cameras. Look out, 'cause here comes another dose of COLOR TEMPERATURE. COLOR TEMPERATURE is measured in degrees Kelvin. 3200°K (3200 degrees Kelvin) describes a lamp with a COLOR TEMPERATURE appropriate for color TV cameras. As this number goes up to about 6000°K, the light gets bluer and "colder" (this seems backwards, but as the COLOR TEMPERATURE increases, the scene looks "colder"). Fluorescent lights and foggy days exhibit such COLOR TEMPERATURES. As the number drops down to about 2000°K, the light gets redder and "warmer." Incandescent lamps in the home create such COLOR TEMPERATURES. A face, not to look too red or too pallid, should be illuminated by 3200°K lamps. This COLOR TEMPERATURE is available from studio QUARTZ-IODINE or TUNGSTEN-HALOGEN lamps or from the sun early or late on a clear day.

Rules of Color

Much of what has been written about TV color relates more to artistic taste than to objective principles. Here are some of the more widely held "rules" of TV color:

1. Avoid pure whites. They will be too bright for most color cameras. Avoid pale yellow and light off-whites as these may be too bright for the cameras. Light colors and light gray will probably all reproduce on TV as just "white." Medium-tone colors reproduce best. Dark colors such as maroon, black, and purple may all appear as "black" on TV.

2. Do not mix fluorescent lamps with TUNGSTEN or QUARTZ lamps on your set. This creates COLOR TEMPERATURE problems. If your TV lights are very bright, however, they will overpower the fluorescent lamps, making their effect on the color negligible.

3. The background for a colored object should be either gray or a complementary color. For instance, red looks best before a blue-green background; yellow, in front of blue; green, in front of magenta; orange, in front of green; and flesh tones look best with a cyan background.

4. Bright multicolored subjects look best before a smooth neutral background. Especially avoid "busy" backgrounds as they distract the eye from the main subject.

5. Attention is attracted to items with saturated (pure or solid) color. Pastels attract less attention and are good for backgrounds.

6. Colors appear brighter and more saturated when illuminated by hard light as opposed to those illuminated by soft, diffused light.

7. Black backgrounds make both light and dark colors appear brighter.

8. Use as few colors in a scene as possible—perhaps two or three complementary colors are sufficient.

MINIGLOSSARY

*Gel Colored material that looks like cellophane and can be placed in front of a lamp to color the light. Usually the flimsy gel material is held in a frame which fits the fixture's scrim holder.

*Tungsten-halogen Common TV lamp bulb with a tungsten filament (glowing wire inside) and a quartz bulb filled with iodine (a halogen) gas. It's small, operates at 3200°K color temperature and runs very hot.

TABLE 9–1
Guidelines for complexion, hair color, and attire

Hair Color	Clothing Colors that Accent Complexion	Clothing Colors for Special Effects	Clothing Colors that Are Unflattering
Blonde	Beige Saturated or dark blue Salmon		Yellow upstages the complexion, and it gives violet hue.
Brunette	Saturated or dark blue Medium gray Medium orange	Light gray gives a tanned look.	Yellow makes complexion look pinkish.
Red	Faded pink Medium and light gray		Chartreuse (yellow-green)
White or gray hair	Faded pink Darker reds	Violet-blue accentuates pink flesh tones	Saturated or dark blue gives complexion a sallow look

9. Some colors become indistinguishable when shown on a color TV screen. The colors between red-orange and magenta end up looking about the same. Similarly, blue and violet look about the same on the screen. Your graphic artist should therefore *avoid* highlighting a red apple with red-orange or trimming a blue robe with violet. These nuances in hue will not reproduce.

10. Yellow, gold, orange, red, and warm colors will appear lighter on camera than in real life. Greens look darker on TV than they really are.

Table 9–1 summarizes some guidelines for selecting flattering clothing and background colors for your performers.

Mood Lighting and Special Effects

Now for some lighting trickery.

For the evil look, aim the lamp up from under the chin as shown in Figure 9–29. A weaker low-angle light will give a more subtle effect. Lighting from below builds an unconscious suspense to the shot, perhaps making the performer look untrustworthy or devious.

For the soft, sexy bedroom look, use reflected light only, either by aiming the instruments at white boards or by using something like the Soft-Lite shown in Figure 9–16. Figure 9–18 showed what soft indirect lighting could do.

Hard, direct light does just the opposite; it accents texture and flaws in smooth surfaces. To get hard lighting, like that in Figure 9–30, avoid lamps with big reflectors. Use a typical KEY light as your FILL light. Have the lights hit the subject more from the side than from straight on in order to accent the shadows. The texture of a surface be-

comes more pronounced as the light skims along it from the edge.

With all the other lights in their normal positions, adding a MODELING light aimed from the side of the subject can further accent the dimensionality and texture of the subject. It is especially handy when something, such as a large hat or a prop, blocks the BACK light from doing its job.

A small light placed near the camera's lens and shone into the talent's face will add a sparkle to his or her eyes. Be conservative; too much light will add tears to his or her eyes and complaints to your ears.

Comedies demand upbeat, happy lighting. This is generally done by providing plenty of brightness and FILL lighting. Shadows are minimized, and backgrounds are fairly bright.

Mysteries and dramas are the opposite. Backgrounds are darker, shadows deeper, and overall illumination is lower. The scene may have dark areas and "puddles" of light through which the performers pass. Don't be afraid to allow your performers to pass into and out of the light, to move

FIGURE 9-29 Evil look with lamp from below.

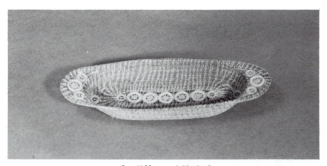

soft diffused lighting

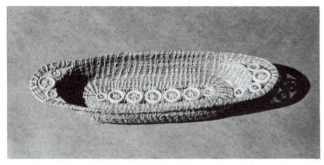

hard, direct lighting from the side

FIGURE 9-30 Hard, direct lighting emphasizes texture.

through shadows, or to be totally backlit, obscuring their features. A drama isn't a newscast; the viewer doesn't expect a perfectly lit view of the talent. The shadows and vagaries improve the scene.

Lighting can sometimes provide a quick-and-dirty way to create the illusion of something being there when it really isn't. Are people supposed to be driving in a car? Park the car somewhere with a blank background (like in an open area) and have somebody wave a light across the car (from front to back) once in a while to imply movement.

Are folks chatting next to a fire? Wave some red or yellow lights around so flames dance across their faces.

With imagination you can project plants, rain effects, explosions, rocket takeoffs, police car flashers, lighthouse beacons, colored spacecraft control panel reflections, green radar scope reflections, all kinds of things, on your performers and their backgrounds.

Shine

Glass, metal, or wet objects pose special lighting problems for television because of the reflection of the studio lights off shiny surfaces on the objects. To minimize these harsh, shiny spots, use soft indirect lighting. Perhaps the Soft-Lite

can act as the KEY light while another instrument aimed away from the subject and toward a white reflector can act as a FILL. If you have a white ceiling, try aiming all the lights at it for glare-free indirect lighting. Where appropriate, use SCRIMS or other semitransparent items to diffuse the light from the lamps. Make sure, however, that your diffusers don't melt or catch fire from the heat of the lamp.

Although this technique solves the shine problem and is appropriate for shooting mechanical objects up close, it is not likely to flatter performers. If both must occupy the same screen, compromises must be made. Can any of the shiny objects like watches or bracelets or shiny buttons be removed? Can the bows of a performer's glasses be raised half an inch to reflect studio lights downward? Can chrome mike stands be traded for ones with a dull finish? Can an actor's face or bald pate be powdered to reduce shine? If a shiny object plays an essential part in the scene, it may be dulled with DULLING SPRAY, a professionally made spray designed specifically to deshine reflective surfaces. Cornstarch, soap, stale beer, milk, or even cloth tape can also be used to make shiny things dull.

Close-up shots of super shiny items like silverware may call for extraordinary dulling efforts. Here, one may erect a tent made of a white sheet over the objects. Lights aimed at the tent from the outside will softly illuminate the area inside the tent. The TV camera can poke its lens in through a hole somewhere to shoot the results.

LIGHTING PROCEDURE

Where do you start?

1. First figure out where the action will take place.
2. Figure out the desired camera angles.
3. Plan which lights you wish to use for KEY, FILL, BACK, and SET lighting. Place and aim these lights approximately.
4. Turn off the HOUSE LIGHTS (general room lights), darkening the room. These lights complicate matters, interfering with the next step.
5. One at a time, test out each light for proper placement, aiming, and focusing.
6. Now, switch on all four lights (KEY, FILL, BACK, and SET if used) for *one of your camera positions*. With your camera at the angle it will be shooting, display the results of your lighting arrangement. Watching a video monitor, adjust the four levels for the proper balance.

MINIGLOSSARY

Dulling spray   Canned liquid to reduce shine when sprayed on objects.

House lights   General overhead work lights used in the studio during rehearsals and between productions.

7. Find the next camera angle or stage position and start the process over.

8. Once all positions are lit, try them all at once and make final adjustments.

DIMMERS

A light DIMMER does just that, it dims lights. The DIMMER makes it very easy to adjust the relative brightness of lights and to fine-tune your LIGHTING RATIO.

You would use the DIMMER in steps 5, 6, and 8 in the preceding section as you raised or lowered the intensity of lights to get the effect you want in the TV monitor.

When working with color equipment, professionals avoid using DIMMERS because a dimmed light gives a reddish glow, affecting the COLOR TEMPERATURE of the scene. To *properly* change brightness in color studios, one either has to change to a lower-power lamp bulb, move the instrument farther away from the subject, place gray filters in front of the lamps, or place SCRIMS on the lamp fixtures.

Because DIMMERS are so convenient, however, most color studios use them anyway, up to a point, to vary the brightness of their lights. The lights can be dimmed about 10% without affecting the COLOR TEMPERATURE appreciably. From 10 to 30%, the light gets redder but is often good enough for all but your "best" productions. The redness really starts to show when you dim the light more than 30%. Since the color shift is most noticeable on faces, it is best to keep the face lights at nearly full brightness. The other lights can be dimmed freely.

When working with black-and-white equipment, feel free to dim to your heart's content. The reddest color of the dimmed light doesn't bother the black-and-white camera image at all.

There are little DIMMERS and big DIMMERS. The little ones are portable and plug into a heavy-duty wall outlet or are wired to a power main by an electrician. Each instrument is plugged into a socket on the DIMMER board as diagrammed in Figure 9–31. Notice that some of your lamps may have their own ON/OFF switches. When these are used with a DIMMER, they may be left in the *on* position (thus you avoid climbing a ladder to turn them on each time you use them). Next, the DIMMER's corresponding power switch is flipped *on* to activate the circuit. Then the dimming control is raised or lowered to adjust the brightness of the light.

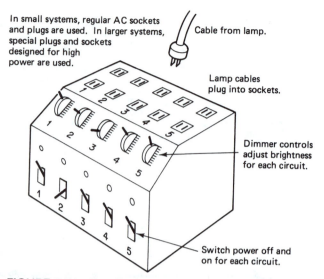

FIGURE 9-31   Small DIMMER.

Caution: Before plugging in a lamp, make sure the power switch for that circuit (or for all circuits if each circuit doesn't have its own power switch) is turned off. If the switch, the lamp, and the dimming controls are all on, giant sparks may jump around each socket as you plug in the lamp. Even bigger sparks may appear if you try to pull out a lamp plug while the power is still on. In short, power down the DIMMER before plugging anything in or unplugging it.

Some DIMMERS allow more than one light to be operated by a single circuit. Be sure not to exceed the power capacity of the unit by putting too many lamps on one circuit. For example, a small five-channel dimmer like the one in Figure 9–31 could be rated for 2000 watts (abbreviated 2000w or 2kw) per channel. That means that each DIMMER control can power no more than 2000 watts of light, such as two 1000-watt instruments, three 600-watt instruments, or one 2000-watt instrument.

Larger DIMMERS designed for permanent studios place a DIMMER REMOTE CONTROL panel in the studio or control room where it is convenient to operate. The actual guts of the DIMMER are large, heavy, and bulky, and they sometimes buzz while in operation. This part of the mechanism is usually placed in another room, out of the way and out of earshot. This big DIMMER then feeds power to the LIGHTING GRID, which is a framework of pipes attached to the studio ceiling from which the lights are hung. The GRID contains

MINIGLOSSARY

*Dimmer   Electronic device to vary the brightness of lamps connected to its circuits.

Dimmer remote control   Control panel with sliders to vary each dimmer circuit's power. The small panel connects via a multiwire cable to the actual large and heavy dimmer circuits. Those circuits

feed power to the lighting grid.

*Lighting grid   Framework of pipes connected to the studio ceiling from which lights are hung.

Master dimming control   A single slider on a dimmer that dims all the lights at once.

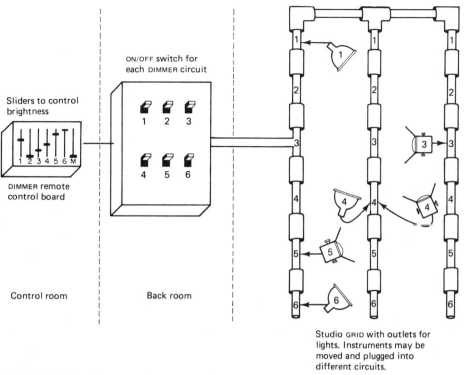

Sliders to control brightness

ON/OFF switch for each DIMMER circuit

1 2 3
4 5 6

DIMMER remote control board

Control room

Back room

Studio GRID with outlets for lights. Instruments may be moved and plugged into different circuits.

FIGURE 9-32 Permanent DIMMER setup.

the electrical outlets for each of the DIMMER circuits. Figure 9–32 diagrams the setup, and Figure 9–33 shows an example of this kind of DIMMER.

Some DIMMERS work like a mike mixer; individual dimming controls adjust brightness on individual circuits, while a MASTER dimming control adjusts all of them simultaneously.

The DIMMER in Figure 9–33 has TWO CHANNELS. The top bank of sliders controls one CHANNEL, while the bottom bank controls the other. Normally, you may do all your lighting with just one CHANNEL, but sometimes you have a complicated lighting change during your show which doesn't afford you the time to diddle the DIMMERS. So what you do is set up your lighting change on one channel (this is called a PRESET) but do the rest of your show on the other channel. Here's an example.

You have a scene in which the electricity is supposed to go out in someone's living room. Naturally, you can't simply turn off the light because the TV cameras won't be able to see. Generally, you create a darkness *effect* by aiming

one hard (perhaps blue) light in from the side and using one BACK light on your performers. This will create dark shadows and the *look* of nighttime. Here's how the lighting setup would go:

1. Turn off the HOUSE LIGHTS.
2. Turn down all the DIMMERS and both MASTER controls.
3. On CHANNEL ONE, turn the MASTER all the way up.
4. Next, turn up the DIMMER controls for the desired side light and BACK light for the "dark room" scene. Once adjusted, fade the CHANNEL ONE MASTER all the way down.
5. Turn up the CHANNEL TWO MASTER all the way.
6. Adjust the remaining DIMMERS for the normal room lighting setup.
7. Start your show by fading up the CHANNEL TWO MASTER, which will light the whole set normally.

MINIGLOSSARY

Channel   On a dimmer, a channel is a set of controls working independently of another set of controls. One channel can be set up for one lighting situation and the second set up for another. Switching channels changes all the lights from one setup to the other.

Preset   On multichannel dimmers, the dimmers can be set up (preset) for one lighting arrangement on one channel, and then the channel is turned off, essentially "storing" the lighting setup for use later when the channel is reactivated.

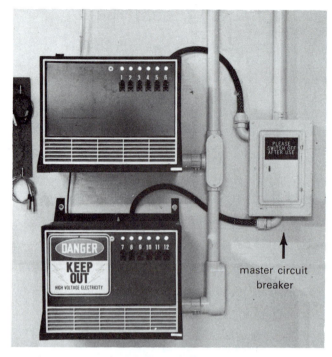

Actual DIMMER circuits

Remote control

master circuit breaker

FIGURE 9-33   DIMMER and remote control.

8.   When the show reaches the point where the lights go out, you quickly fade up the CHANNEL ONE MASTER while simultaneously fading down the CHANNEL TWO MASTER.

9.   If later in the scene the lights come back on again, you reverse the process, dimming ONE while bringing up TWO.

Thus, by moving two faders all the lights will change to the brightness you PRESET them to be. It saves you from adjusting four or five knobs all at once as the scene lighting changes. Larger DIMMERS may have three or more PRESETS, and you might be able to switch from one to another with just one press of a button.

The DIMMER in Figure 9–33 also has switches across the top to permit selected controls to work independently from the MASTER. Thus, you could keep selected lights *on* even while the MASTER dimmer is turned off. Here's how they might be handy:

Say you had only a ONE-CHANNEL-DIMMER board (or a TWO-CHANNEL unit where the second channel was already occupied) and you wanted to create the effect described earlier. You could switch the BACK light and side light *on* independently and leave them on for the entire scene. You'd set up the rest of the lights on the DIMMERS. Before the blackout scene occurs, *all* the lights would be on, but the "nighttime" lights would be drowned out by the "normal" lights. However, when the blackout occurs, you would dim the MASTER, which would take out all the lights but the two independent ones, which now comprise your "nighttime" scene. Thus, you did the whole job with just one CHANNEL.

As before, several instruments can be plugged into the same dimming circuit, but you must always be careful not to overload the circuit. It is wise to know the capacity of each circuit in your dimmer system.

Larger, more professional DIMMER systems have PATCH BAYS, which are like telephone operator switchboards. Instead of climbing a ladder to plug a fixture into this GRID outlet or that, you would plug each lighting instrument into a numbered socket in the GRID. Meanwhile, each DIMMER circuit controls a series of sockets in the PATCH BAY. By connecting a lighting patch cord between one DIMMER circuit and one GRID outlet, you connect one light to that circuit. You can do this conveniently for a large number of lights until you have filled your PATCH BAY with a spaghetti of PATCHES. Figure 9–34 diagrams the process.

It is handy to label the number of each GRID circuit on its corresponding PATCH cable; thus, you know what you're connecting where. It wouldn't help much to label the instrument number on each PATCH cable because the instruments move around from circuit to circuit as you connect them in different places on the GRID. Put another way, when standing in the studio and looking up at the lighting instruments, you don't care what the number of the lighting instrument is as much as you care what circuit it is going into. If you want to illuminate a certain light, you find out

---

MINIGLOSSARY

Patch bay   Like a telephone switchboard, a console of sockets leading to the studio lights and another set of sockets leading to the dimmer circuits. Connecting the two via patch cords allows various dimmers to activate various lights.

Patch cable   This special heavy-duty lighting cable plugs into the patch bay to carry current from the dimmer circuit to the grid circuit and studio lamp.

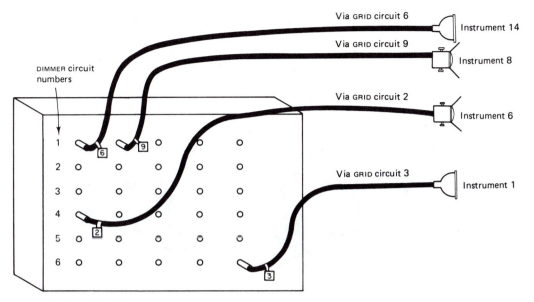

FIGURE 9-34 Lighting PATCH BAY.

its circuit number and then trace it back to find out which DIMMER control that circuit is connected to. Then turning up that DIMMER should turn on that light.

LIGHTING HARDWARE

Lighting Grid

In a studio, the lights must hang from something. Often this something is a combination of crisscrossed pipes suspended from the ceiling and forming the LIGHTING GRID (Figure 9–35). The lighting instruments are connected to C-CLAMPS (Figure 9–36), which attach to the GRID pipes throughout the studio. Sometimes the pipes have electric wires in them and sockets where the pipes intersect. Other systems may have separate wiring conduit, and small power sockets will hang down from it spaced every 3 feet or so. There are usually several sockets for every DIMMER circuit. The lamps may be plugged into these sockets, taking care not to exceed the power rating of any one circuit by running too many lamps from it.

Antigravity Hangers

The height of the lamps may be adjusted using extensions to the C-CLAMP. Extensions with counterbalances that help

raise and lower the lamps are called ANTIGRAVITY HANGERS. Figure 9–35 showed one type of ANTIGRAVITY HANGER, and Figure 9–37 points out another type.

CARE OF LAMPS

Fixtures Get Hot

And boy do they! They make as much heat as a toaster and can toast you if you don't stay clear. Keep the instrument away from anything combustible or meltable. Make sure that the power cord for the fixture isn't draped over the instrument (it could melt). Watch where the lamp is aimed. You can feel the heat of a 1000-watt lamp from 10 feet away, so imagine how hot it gets right in front of it. For instance, aiming the lamp at a wall or curtain less than 1 foot away or so could start a fire in a matter of minutes.

When handling instruments, let them cool before attempting to change bulbs or SCRIMS (unless you go around wearing asbestos gloves). Do not attempt to store instruments until they have cooled adequately. Don't be too surprised if the paint burns off the BARN DOORS sometimes; it doesn't look pretty, but it's common wear and tear.

Try to "warm up" large lighting instruments for a moment at reduced power before dimming up to full power. This prolongs bulb life and reduces the chance of cracking the glass on a cold FRESNEL lens.

MINIGLOSSARY

*C-clamp   C-shaped clamp used to hang lighting instruments from the ceiling grid.

*Antigravity hangers   Spring-loaded mechanisms between the

lights and the grid to allow the lights to be individually lowered (and stay put at various heights) simply by pulling them down or pushing them up.

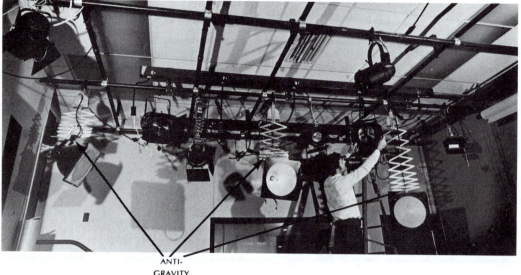

ANTI-
GRAVITY
HANGERS

FIGURE 9-35   LIGHTING GRID.

Moving Lamps

Do not jar, shake, bump, or attempt to move a lamp while it is lit. The filament in the light bulb is white hot and *extremely* fragile. When the lamp is not glowing, the filament is solid again and is fairly rugged. When you turn off a lamp, always let it cool for a few seconds before moving it. Yes, a few seconds is all that is necessary.

Whenever you attach a light to the GRID, *always connect the* SAFETY CORD (Figure 9–38). Sometimes the frequent aiming, turning, and moving of a lighting instrument will loosen it up so that it is hanging only by a whisker. Attaching the SAFETY CORD adds one extra measure of security, so that if the lamp *does* come loose, it doesn't bonk somebody on the head or, worse yet, smash a camera. Some SAFETY CORDS are made of rope, others of wire with a spring-loaded clip, while others might be chain. For lamps with long ANTIGRAVITY HANGERS, you should use one SAFETY CORD to connect the C-CLAMP to the GRID and another to connect the lighting instrument to the bottom of the ANTIGRAVITY HANGER.

Falling lights are not unheard of; this is not an unnecessary safety precaution. I recall seeing someone begin to loosen a C-CLAMP when a 20-pound lamp slipped out of his hands. The hurtling lamp took a long arcing swing by its electric cord, missing the floor by a foot and zipping right between the heads of two performers seated in chairs. No one was hurt probably because the talent had not yet been told that they should sit very close together.

Changing Bulbs

You can assume that a light bulb has burned out when the lamp stops working. To confirm that the bulb has expired,

first *turn off the power to the instrument.* Take a close look at the bulb. If it has a big bulge, if it is blackened, if it is cloudy inside, or if the filament is clearly broken, the bulb is shot. If none of the preceding is true, perhaps the bulb is good and the instrument, switch, DIMMER, cable, or something else is defective.

Bulbs last between 10 and 500 hours depending on the manufacturer and type. When lamps are dimmed, they last much longer than they would at full brightness.

Never touch a good bulb with your fingers, or it won't be much good anymore. Traces of oil from your fingers can

FIGURE 9-36   C-CLAMP for hanging lights from a pipe GRID. (Courtesy of Quality/Universal)

FIGURE 9-37   ANTIGRAVITY HANGER.

While teetering back and forth atop that ladder changing the bulb, here's something else you can do: Check out the lamp's reflector or FRESNEL lens for dust or dirt. The lamp's heat seems to attract dust and film which coat the optics, making the lamp inefficient. *This is no trivial matter!* I've seen lamps which hadn't been cleaned for one or two years, so brown with baked-on dust and film that easily halve the brightness was being wasted. Plain soap and water will generally do the trick.

Power Requirements

AMPS times VOLTS equals WATTS. Homes and schools run on 120 VOLTS, so if a circuit is good for 15 AMPS (as is typical in older homes), then you may use up to 15 × 120 = 1800 WATTS of power on that circuit. If a circuit is rated at 30 AMPS (schools and businesses usually are), then you can use 30 × 120 = 3600 WATTS. In short, the house current you get from the wall socket in your home is good for about 1800 WATTS. Institutional electrical outlets can sustain about 3600 WATTS. So how many 1000-WATT lamps can you use at home without blowing a fuse or burning the house down?

chemically change the glass when the bulb heats up. The glass devitrifies and fails right where the fingerprints were. Handle bulbs with a clean cloth or with the packing that came with the bulb.

Replace bulbs with exactly the same type of bulb or its equivalent. Some lamps can take bulbs of different power and brightness. *Do not exceed the power rating of the instrument.* Removing a 600-watt bulb from a fixture designed for a 600-watt bulb and putting in a 2000-watt bulb will give you more light—until the fixture and its wiring burn up.

To help you find a replacement bulb, hang onto the one you just took out of the lamp. If you examine its base, it will probably have a three-letter code which designates an exact replacement for the bulb. Bulb boxes usually have this three-letter code emblazened on them as well as other details like wattage. Some lighting fixtures have a tag on them telling the bulb type appropriate for the fixture. If yours don't, it might be handy to make your own label and attach it to the lamp.

If you're in charge of purchasing, make sure you buy a few bulbs ahead. I've seen more than one studio running half in the dark because someone neglected to stock up.

Fixtures get hot

MINIGLOSSARY

Safety cord   Loop of chain, cable, or rope that fastens loosely around the lamp and the grid pipe and stops the lamp from falling if its C-clamp becomes undone.

Amp or ampere   A measure of the volume of electrical current. Institutional circuits are usually rated for 20–30 A (amps). Electric wires may get hot as this number is approached.

Volt   A measure of electrical pressure. In the United States, 120 V (volts) is the standard available from common electrical outlets in homes or institutions.

*Watt   A measure of electrical power. Amps times volts equals watts. A studio light may use 1000 W (watts). Institutional wiring may handle 2400–3600 watts per circuit.

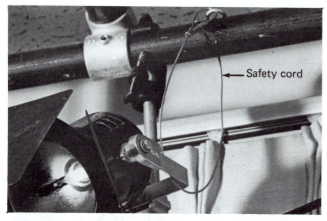

FIGURE 9-38 SAFETY CORD protects folks from falling fixtures.

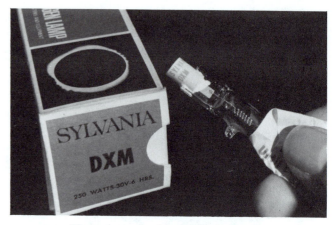

Three-letter code on base of bulb.

Calculating WATTS and AMPS roughly in your head.

The exact calculation goes like this:

watts = amps × volts

Nearly everyone in the United States uses 120-volt power, so

watts = 120 × amps   or
amps = watts ÷ 120

The quickie approximation:
  To change watts to amps, divide by 100.
  To change amps to watts, multiply by 100.
  Or 100 watts equals an amp (sort of).

Before turning on any light, check to see what else is on the same circuit and is also using power. Check also to make sure that you aren't running several lights off one extension cord. An extension cord rated for 15 AMPS (a label

on it may say 15A at 120V, meaning that it can take 15 AMPS of electrical current) can carry only 1800 WATTS of power. Even if you are working in a school whose outlets are rated for 30 AMPS (3600 WATTS), your extension cord can safely handle only 1800 WATTS.

Once you are set up for a remote production and are satisfied that you aren't overburdening the wiring, you're ready to go. Switch the lights on *one at a time* rather than all at once because they use abnormally high amounts of power at the moment when they are just lighting up. Switching all the lights on at once could cause a "surge" of power and blow a fuse. If you switch the lights on one at a time, the smaller surges are spaced out and are less likely to overburden the wiring.

One last note, from experience: Don't operate right at the limit of your power capacity. Murphy's 44th Law of Lighting states that lighting systems will work during rehearsal but will fail during the show. The likelihood of failure increases with the importance of the show.

AUDIO

Half of video is audio, but it never seems to get the attention it deserves. Turn off the picture on your TV set, and the sound will give you a very good idea of what is going on. Turn off the sound, however, and the picture will leave you puzzled.

Television production uses much the same sound equipment and techniques that are used in radio and other audio enterprises. The big difference is that television audio is harder to do. Where radio people can *read* scripts into *nearby* microphones in *soundproof rooms*, television people often have to *memorize* their lines and speak them into *hidden* microphones in a neat-looking environment, on a studio set, or out in the field.

Whole books are written on the subject of audio, and its recording, editing, and reproduction. Just because it's hard, however, doesn't mean that it's impossible. We will start with some basic, easy sound setups and proceed to fancier, more complicated arrangements.

THE BASIC BASICS

The microphone picks up sound vibrations and turns them into tiny electrical signals which travel down a wire into a VCR. The AUDIO LEVEL CONTROL on the VCR (if it has one) allows you to adjust the volume of the sound being recorded. Portable and home VCRs have AUTOMATIC AUDIO LEVEL or AUTOMATIC GAIN CONTROLS (AGC) which adjust themselves.

What kind of microphone should you use? Any kind that has a plug which fits the VCR will probably work. Try it. You can't hurt anything by trying it. If the plug doesn't fit the VCR, use an ADAPTER (shown later) to make it fit. If the mike is designed to hang around the neck or clip to a lapel and the performer wants to hold it or put it on a stand, it will still work. The mike will still pick up sound. If the mike is designated for stand use and the performer wants it to hang around his or her neck, get some string and tie the mike around his or her neck. It will still work.

What if the mike doesn't work? If you get no sound from the mike after having done all the things described in Chapter 5—pressing RECORD, turning up the recording volume, and checking to see if your mike cable is plugged in—try the old standbys of wiggling the wire near the plug, wiggling the plug, or trying another microphone. Also check to see whether the mike itself has an ON/OFF switch on it that is turned to OFF.

MINIGLOSSARY

Audio level control   A volume control. Adjusts sound recording loudness on VCRs.

Automatic gain control or AGC   Electronic circuit that automatically adjusts the loudness of a recording.

Camcorders have built-in microphones which automatically pick up the sound as you make a recording. Although these built-in microphones are handy, they all have one flaw. If the camera is 8 feet from the performer, then your microphone is also 8 feet from the performer and will pick up room echoes and noises galore. For this reason, a separate microphone is often a better way to catch the words of your performer; the mike stays near the performer's mouth and gets good sound while the camera stays at some distance for a flexibility of shots.

These are the basic basics of audio. With them you will be successful at recording the sounds you want most of the time. Audio is somewhat forgiving. Even if not done perfectly, it is still frequently usable.

The rest of this chapter is dedicated to helping you make the sound perfectly right. If your sound is poor, it will distract the viewers from the message. If the sound is mediocre, the entire presentation will appear amateurish and will leave the impression of sloppy workmanship. Professional audio is like paint on a car. The car drives okay without it, but that extra shine is what turns people's heads. How impressive do you want your show to be?

Our quest is for clean, crisp, high-fidelity sound with-

out echoes, buzz, hum, hiss, distortion, or other unwanted noises.

THE MICROPHONE

How a Microphone Works

Sound vibrates a mechanism inside the microphone, making a tiny electrical signal that passes down the mike cable. Many things can be done to that signal before it gets to its destination. Nearly all these things are electrical in nature.

Kinds of Microphones

The words CONDENSER, DYNAMIC, and CRYSTAL refer to what's inside the microphone that makes it work. These components contribute to the microphone's characteristics: its fidelity, its sensitivity, its ruggedness, and its cost.

The words OMNIDIRECTIONAL, CARDIOID, UNIDIRECTIONAL, and BIDIRECTIONAL describe in which direction(s) the microphone is designed to "hear." You'll see later how

	Name	Found on the end of a BALANCED or UNBALANCED LINE	Used with a HI Z or LO Z mike	Used with
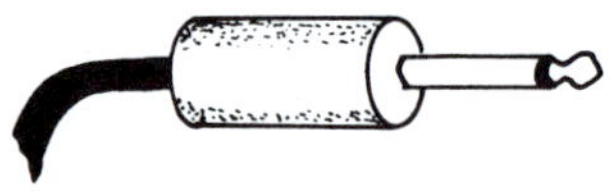	MINI PLUG	UNBALANCED	usually HI Z	Audio cassette tape recorders ½" VCRs Small portable equipment
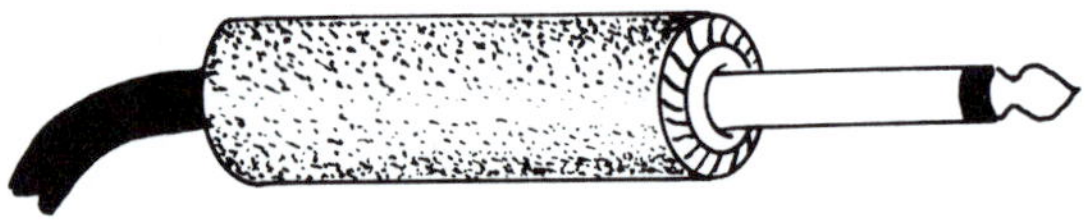	PHONE PLUG	UNBALANCED	usually HI Z	½" VCRs Reel-to-reel audio tape recorders Most school AV equipment
	RCA or PHONO PLUG	UNBALANCED	usually HI Z	Some ½" VCRs Some reel-to-reel audio tape recorders CD and audiocassette players, phono turntables
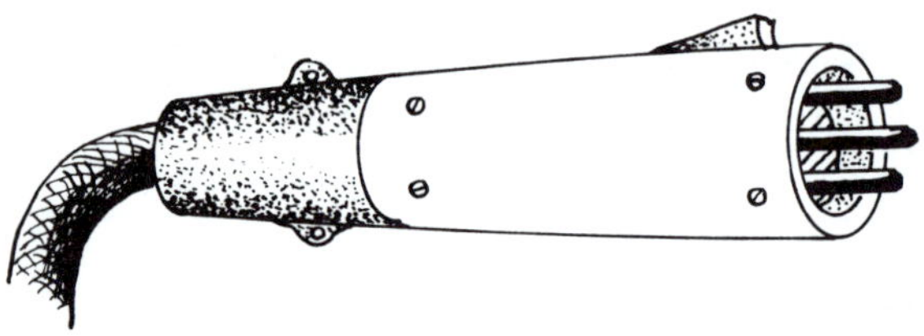	XLR or CANNON PLUG	BALANCED	usually LO Z	1" VCRs Most mike mixers and other audio equipment of high quality Nearly all good microphones

FIGURE 10-1 Various audio plugs.

the various kinds of microphones are selected for different situations, but first browse through Tables 10–1 through 10–4 to become more familiar with audio lingo. These words are important and will be used throughout the rest of this chapter.

Dynamic Microphone. Most common among the semiprofessional and professional microphones is the DY-NAMIC microphone (Figure 10–2). DYNAMIC microphones are quite rugged and reliable, require no special power supply or batteries, and respond to a wide range of audio frequencies. They are usable under severe temperature and humidity conditions. They generally cost $50–$200 depending on their quality.

TABLE 10–1

The ways microphones produce their signals

Condenser microphone A microphone that uses a condenser (a small electrical component) to create the signal. One of the standard microphones in the broadcast industry, it has good sensitivity to a wide range of sound volume and pitch. Disadvantages include fragility, expense, and the fact that this particular kind of microphone operates only with batteries or with some external power supply.

Electret condenser microphone An improved version of the condenser microphone that needs only a tiny power supply to operate. Used alone or built into portable TV cameras, these microphones give good sensitivity to a wide range of sound volume and pitch and give especially clear voice reproduction. Disadvantages include fragility (especially to heat and humidity) and the occasional need to replace the small battery that powers the microphone circuit.

Dynamic microphone Lacks the fidelity of the condenser microphone but is good enough for most video use. This frequently used microphone is rugged and quite trouble-free. Its name comes from the fact that its signal is generated by a moving (hence the word *dynamic*) coil of wire and a magnet.

Crystal microphone Makes its signal when sound vibrates a tiny crystal inside it. Fragile and not very sensitive but very cheap.

TABLE 10–2

Pickup patterns, the directions in which the microphone hears best

Omnidirectional microphone Can hear in all directions regardless of where it is aimed.

Cardioid microphone Can hear very well in front of it, medium well to the side of it, and hardly at all in back of it.

Directional or unidirectional microphone Can hear very well in front of it and hardly at all anywhere else.

Shotgun microphone A very unidirectional microphone that looks like a shotgun barrel.

Bidirectional microphone Can hear well in only two directions: front and back.

TABLE 10–3

Microphone IMPEDANCE and cables

Impedance Is measured in ohms, and 1000 ohms (abbreviated 1000 Ω, or 1 kilohm, or 1 kohm, or 1 kΩ) or more makes something HI Z.

High impedance (abbreviated HI Z, where Z stands for impedance) An electronic term describing certain kinds of microphones, audio inputs, and audio outputs that should be used together. Microphones with 8-foot cables (or shorter) are probably HI Z. Inexpensive audio equipment for small TV setups is usually HI Z. Crystal microphones are usually HI Z.

Low impedance (abbreviated LO Z) An electronic term describing other kinds of microphones, audio inputs, and audio outputs which should be used together. Microphones with 15-foot (or longer) cables are probably LO Z. Most dynamic and condenser mikes are LO Z. Most TV studio equipment and all high-quality audio equipment use LO Z. Low impedance is usually consideered to be less than 500 ohms.

Impedance matcher or impedance matching transformer. A small device which changes something LO Z to HI Z or vice versa.

Unbalanced line An audio cable that has two conductors only. There is a center wire and a braided shield (the second wire) surrounding it. At the end of such a cable is a two-conductor plug, such as an RCA, mini, phono, or phone plug (examples are shown in Figure 10–1). This method of carrying audio is usually found on inexpensive equipment and usually with hi z microphones with short cables.

Balanced line An audio cable with three conductors (two center wires plus a braided shield). It terminates with a 3-pin plug such as an XLR or Cannon plug (shown in Figure 10–1). This kind of wire is usually found on expensive or professional equipment, usually with LO Z microphones. The balanced line is extremely impervious to extraneous electrical interference; or, put another way, unwanted buzzes and hums don't sneak onto your recording.

TABLE 10–4

Audio plugs and sockets

Cannon plug The Cannon company makes plugs, including what it calls the XLR plug shown in Figure 10–1. When Switchcraft makes this particular plug, it calls it an A3M plug. Similarly, other companies will call the plug other names. For some reason, the "Cannon" has become synonymous with this particular kind of plug, much as the words *Kleenex, Xerox, Scotch tape,* and *RCA* plug, though they are company or brand names, have taken on a generic meaning for particular items. This is a common phenomenon. This book uses the most common terms encountered in the field, so don't be surprised to see some items identified by company names.

Mini, phono, RCA, phone Popular types of audio plugs used with unbalanced lines. They all handle both hi and lo level signals, but the RCA (phono) plugs are more often associated with hi or line level signals. Phone plugs are frequently used to handle lo or mic level signals. Phone plugs are very durable and are preferred in educational and institutional settings. Mini plugs are flimsier than their bigger brothers but are small and are used on miniature audio and video gear.

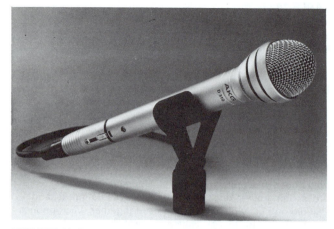

FIGURE 10-2 DYNAMIC microphone (Courtesy of Quality/ Universal).

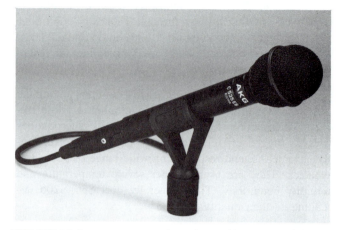

FIGURE 10-4 CONDENSER microphone (Courtesy of Quality/Universal).

One specialized DYNAMIC microphone is the RIBBON or VELOCITY microphone (Figure 10–3). RIBBON microphones are becoming rare, being replaced by condenser mikes. RIBBON microphones are renowned for their "warmth" and smooth bass response. They are preferred in talk shows, giving the host a warm, rich voice. They are BIDIRECTIONAL (listen in two directions) and cost about $400.

DYNAMIC microphones can be made with OMNIDIRECTIONAL, CARDIOID, or UNIDIRECTIONAL pickup patterns. They can be small LAVALIER types, or larger hand-held types, or giant SHOTGUN types designed to be hung from microphone BOOMS.

FIGURE 10-3 RIBBON microphone. (Courtesy of Radio Shack)

Condenser Microphone. CONDENSER microphones (Figure 10–4) have an extremely wide frequency response (pick up a wide range of audio frequencies) and are excellent for voice and music. They can be used indoors or outdoors, but extremes in temperature and humidity may hurt the microphone.

They require an external power supply to make them work and may cost $400–$600. Some have a modular design, allowing you to attach different "heads" to the "body" of the microphone, making it CARDIOID or SHOTGUN or especially sensitive to certain frequencies.

Electret Condenser Microphone. ELECTRET CONDENSER microphones (Figure 10–5) have good frequency response, are much cheaper than CONDENSER microphones, and can be made very tiny. Their smallness makes them excellent choices for TIE CLIP or LAPEL microphones, which should be unobtrusive or nearly invisible.

Like the CONDENSER microphone, they require a power supply, but a tiny battery may be all that is needed.

They generally cost $25–$200, which makes them just the right price for the industrial market. They can have an OMNIDIRECTIONAL, CARDIOID, or UNIDIRECTIONAL pickup pattern.

Although their sound fidelity is better than all but the most expensive DYNAMIC mikes, ELECTRET CONDENSER mikes are not quite as rugged. They should not be left in the sun, heat, or high humidity, but with a little care, they can be used freely outdoors.

Crystal and Ceramic Microphones. CRYSTAL and CERAMIC microphones (Figure 10–6) are supercheap (about $5) and are often included in inexpensive audio tape recorders. They have limited high-frequency response but are okay for speech, especially with public address systems where fidelity is not paramount.

Carbon Microphone. The granddaddy of all microphones is the CARBON microphone. Up until recently,

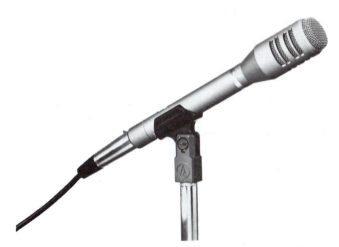

FIGURE 10-5 ELECTRET CONDENSER microphones (Courtesy of Quality/Universal).

these were the typical microphones used in telephones. CARBON microphones have a limited frequency response, good only for speech. They are fairly noisy but put out a strong signal. They work almost anywhere and are nearly indestructible.

Pressure Zone Microphone. The PRESSURE ZONE MICROPHONE (or PZM, Figure 10–7) is quite different from the others. It is usually attached to a desk or flat surface which acts to reflect the sound waves into the microphone. It can also be hung with a flat plate attached to it for funneling the sound.

PZMs have a hemispheric pickup pattern (they listen to everything on the top side of them), which makes them

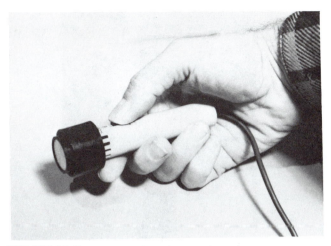

FIGURE 10-6 CRYSTAL or CERAMIC microphone.

ideal for picking up large group discussions and audience reactions. For instance, a PZM sitting in the middle of a table will pick up the voices of all the people sitting around the table. Unfortunately, the PZM also will pick up the sounds of finger tapping and paper rustling and the thumps of people knocking their knees against the table legs.

Stereo Microphone. To pick up stereophonic sound, you can use two microphones. Sometimes it's handier to use a single STEREO microphone, which is actually two microphones built onto the same body so that it looks like a single microphone. Half the microphone listens in one direction and sends the signal to the left channel through a stereo plug at the other end. The other half sends its signal out through another conductor on the same plug. Figure 10–8 shows a STEREO microphone.

STEREO microphones are very handy for stereo recording but are not popular among professionals because

FIGURE 10-7 PRESSURE ZONE microphone (PZM).

MINIGLOSSARY

*Pressure zone microphone Microphone mounted on a flat plate that senses sound reflected from the plate.

Stereo microphone Microphone that ''hears'' in two directions and sends out two separate audio signals to a stereo recorder.

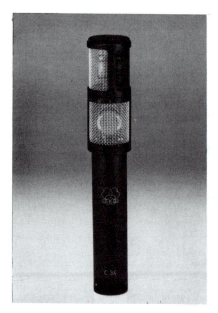

FIGURE 10-8 STEREO microphone (Courtesy of Quality/Universal).

careful microphone placement generally requires one microphone to be in one place while another microphone is in the other. Having them tied together on the same body is considered a compromise.

Pickup Patterns

Microphones are designed to listen in various directions. Sometimes you want them to work like your ears and listen in all directions equally. Sometimes you want them to be like binoculars and listen in only one direction, rejecting noises from behind and to the side. There are four basic microphone pickup patterns: OMNIDIRECTIONAL (listens in all directions), DIRECTIONAL (listens in one direction), CARDIOID (listens mostly in one direction but a little to the sides), and BIDIRECTIONAL (listens in two directions). Figure 10-9 diagrams the different microphone pickup patterns.

Omnidirectional. OMNIDIRECTIONAL microphones are equally sensitive to sound from all directions. No microphone is perfectly OMNIDIRECTIONAL because the body of the microphone interferes with sound reception from the rear. OMNIS are less expensive than their DIRECTIONAL brothers and therefore are the type usually built into home cameras.

OMNIS are excellent for close work. If a speaker turns his or her face to the side or the microphone moves a little, the sound is relatively unchanged because the microphone

can "listen" in all directions. This becomes a disadvantage when the microphone is used to pick up a distant sound because the microphone will hear the room echoes and off-stage sounds as well as the sound of your performer.

Directional. DIRECTIONAL is a fairly general term indicating that a microphone has greater sensitivity in one direction than it does in the other. Its pickup pattern is like half an OMNIDIRECTIONAL microphone or half a BIDIRECTIONAL microphone. PZM microphones are DIRECTIONAL.

DIRECTIONAL microphones are good for picking up small groups of people where you want all of them to be heard but you don't want to hear room echoes and other background sounds. DIRECTIONAL microphones have also been used on podiums where they pick up the sound of the person speaking but reject the sound of the audience or any loudspeakers in the audience area.

Unidirectional. UNIDIRECTIONAL microphones are very sensitive in the direction they are pointed and are insensitive to the sides and rear. They make it possible to move the microphone a little farther from the talent (perhaps getting the mike out of the picture) yet still pick up the sound of the talent loud and clear. They also reject room echoes. They are so sensitive that when used close to a person speaking, if the microphone is turned slightly away from the person, or the person moves from the center of the microphone's domain, or the person turns to the side and speaks, the sound of their voice will drop noticeably. UNIDIRECTIONAL mikes can be used up close only if your talent is tied down and has a stiff neck.

Shotgun. Figure 10-10 shows a SHOTGUN microphone. You can see how it gets its name. It is a very UNIDIRECTIONAL microphone, rejecting all sounds except those coming from where it's pointed. These microphones are especially useful in noisy environments and in situations where you have to pick up sound from quite some distance away.

Short ones (about a foot long) are not as directional as long ones (several feet in length). You can sometimes see a very long shotgun mike at the rear of the room during presidential press conferences.

SHOTGUN microphones are so effective at rejecting off-center sounds that accurate aiming becomes a constant hassle. If a person moves or two people are speaking, the SHOTGUN microphone has to be aimed precisely at the speaker's face or else he or she may not be heard at all.

Just because SHOTGUN microphones are very directional, it does not mean that they are extremely sensitive.

MINIGLOSSARY

*Pickup pattern The areas in which a microphone picks up the best sound. Also a diagram depicting a microphone's sensitivity in different directions.

Directional microphone Microphone that needs to be "aimed" as it is more sensitive in one direction than another.

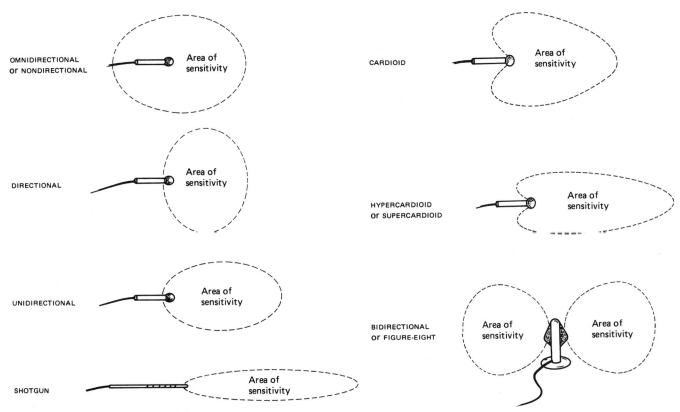

FIGURE 10-9 Microphone pickup patterns.

The SHOTGUN microphone can reject loud sounds to its side, but a weak sound in front of it still remains a weak sound. To pick up weak and distant sounds, you may need a PARABOLIC microphone.

Parabolic. A PARABOLIC microphone (Figure 10–11) has a conventional microphone mounted at the focal point of a large parabolic reflector. The larger the reflector, the more sensitive the device; in fact, these microphones can often pick up the sound of people conversing a block away (what a great way to spy on your friends). You often see PARABOLICS used at football games to pick up the grunts, groans, and bone-crushing blows of the players.

PARABOLICS usually have poor low-frequency response, making them sound a bit tinny. Many also tend to pick up the sounds of the person holding and aiming the microphone as well as distant echoes of cars going by and wind blowing in the trees.

When using a PARABOLIC, remember that light travels much faster than sound. What you see on your screen will happen before the sound gets picked up by your microphone. This delay can be disconcerting, especially if you're taping a marching band from 50 yards away—the marchers will never be in step with their music.

Cardioid. The CARDIOID microphone gets this name from its heart-shaped pickup pattern (review Figure 10–9). These microphones are very popular among professionals. The CARDIOID pattern allows the talent to work at a greater distance from the microphone without having the mike pick up too many room echoes. Yet the microphone allows the talent to move his or her face or the microphone a little without losing that much sound volume. Thus, microphone aiming is not as critical as with UNIDIRECTIONAL mikes.

CARDIOIDS are very insensitive out their rears, making them good for cutting out audience or crowd sounds, ma-

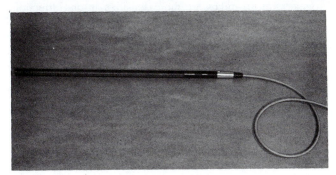

FIGURE 10-10 SHOTGUN microphone. (Courtesy of Shure Brothers, Inc.)

MINIGLOSSARY

Parabolic microphone   Attached to a small bowl-shaped reflector, the microphone picks up weak or distant sounds.

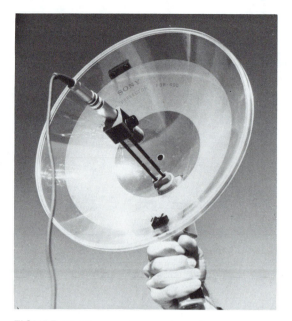

FIGURE 10-11   PARABOLIC microphone.

chine sounds, room echoes, or sounds from loudspeakers in the room.

One disadvantage of some CARDIOID mikes is that the frequency response at the center of the microphone's pattern is a little different from the frequency response at the side. Thus, if a person turns her head or the mike while speaking into it, some of the high frequencies may be lost.

Microphones generally don't have the words CARDIOID, OMNIDIRECTIONAL, or UNIDIRECTIONAL printed on them. CARDIOIDS physically look pretty much like OMNIS. People who work with audio generally memorize the company and model numbers of the various microphones in order to tell them apart and will seek an "Electro-Voice 635A" if they want an OMNIDIRECTIONAL mike or perhaps ask for a "Sure SM53" if they want a CARDIOID.

Hyper- or Supercardioid.   HYPERCARDIOID and SUPERCARDIOID mikes are more sensitive in one direction than their brothers the CARDIOID mikes. You might think of them as a cross between UNIDIRECTIONAL and CARDIOID.

Bidirectional.   BIDIRECTIONAL or FIGURE-8 or BIPOLAR microphones listen in two directions. The term FIGURE-8 comes from the shape of the pickup pattern (review Figure 10–9).

BIDIRECTIONAL mikes are nearly always RIBBON microphones and are quite rare outside of the professional studios. The BIDIRECTIONAL pattern is most useful when

you have a host seated at a desk and a guest opposite the host. The same microphone can pick up both.

Noise-Canceling.   NOISE-CANCELING microphones are used where background noise is extremely loud, such as in a helicopter or on a crowded convention floor. They are often built into headsets with the microphone positioned directly in front of the speaker's mouth. Some of these are simply UNIDIRECTIONAL microphones designed for close talking (made to be insensitive to mouth sounds). Others use an interesting technique of canceling out noise by using a BIDIRECTIONAL pickup pattern. Room noise (or cockpit noise) enters both sides of the microphone equally, canceling itself out. The announcer's voice, because his mouth is so close to the microphone, goes in one side much louder than it goes into the other and is not canceled out. Thus, the microphone listens to him and is quite insensitive to background noise.

Impedance

IMPEDANCE is an electronic term describing how the microphone and its wires transfer its signal to the circuits in the amplifier or tape recorder. IMPEDANCE is measured in OHMS (Ω). There is HIGH IMPEDANCE (HI Z), which is about 20,000Ω, and there is LOW IMPEDANCE (LO Z), which is 100-600Ω. Somewhere in between, there is MEDIUM IMPEDANCE, about 2400Ω.

Microphones have an IMPEDANCE (usually stamped on them somewhere), and audio tape recorders, mixers, and VCRs have a certain IMPEDANCE associated with their inputs and outputs. Even microphone extension cables and other audio cables are designed to work with a particular IMPEDANCE. If not listed on the machine, these IMPEDANCES appear in the equipment specifications.

Technically, LOW IMPEDANCE stuff is designed to work together, and HIGH IMPEDANCE stuff works together. Never the twain should meet except through an IMPEDANCE MATCHING TRANSFORMER, which changes one kind of electrical signal into the other (as we saw with antenna IMPEDANCES in Chapter 3).

When IMPEDANCES are properly matched between a microphone and a VCR (or whatever) input, the microphone is able to transfer the maximum amount of energy and its best fidelity to the VCR.

Generally, small inexpensive audio devices are HI Z(Z stands for IMPEDANCE); large expensive ones are LO Z. The better microphones have a HI/LO Z switch, enabling them to work with either HI Z or LO Z equipment.

MINIGLOSSARY

Hypercardioid microphone   Very unidirectional and slightly cardioid microphone.
Noise-canceling microphone   Microphone that is used close to the mouth, and rejects surrounding noise.

Hi Z   In audio, an input or output having 10,000 or more ohms of impedance (resistance to signal flow).
Lo Z   In audio, any input of 600 ohms impedance or less.

Audio IMPEDANCE MATCHING TRANSFORMER adapter.
(Courtesy of Radio Shack)

To work optimally, a HI Z microphone must be plugged into a HI Z microphone input and a LO Z mike must go into a LO Z input. Some microphone mixers and amplifiers have switches near their input sockets that will change the IMPEDANCE of the input, thus allowing either HI Z or LO Z mikes to be used. Behind those switches (in case you were wondering) are IMPEDANCE MATCHING TRANSFORMERS.

Audio professionals adhere strictly to the laws of IMPEDANCE matching. However, this is not a world where everything is done the "right" way. In the real world, incompatible equipment such as VCRs and a parade of other audio devices have to be interconnected, matching or not, on a daily basis. If you are willing to compromise a *little* on audio quality, it is possible to break some of the "rules" of IMPEDANCE matching and save yourself from carrying around a pocketful of IMPEDANCE MATCHING TRANSFORMERS for the rest of your video lives.

The law was that microphone IMPEDANCE and VCR input IMPEDANCE had to match (or be transformed to match). The compromised law goes like this: A HI Z mike *must* go into a HI Z input, but a LO Z mike can *also* go into a HI Z input. *The IMPEDANCE of the input must always equal or exceed the IMPEDANCE of the microphone.* If it doesn't,

you'll lose your low frequencies, and the mike will sound weak and tinny.

To accommodate the various kinds of microphones, many VCRs and other audio devices have input IMPEDANCES of 10,000Ω or more. Some have at least 1000Ω input IMPEDANCE to accommodate the 150Ω and 600Ω LO Z mikes.

While we are on the subject of IMPEDANCE, let's digress from microphones for a moment. If any audio device (like a mixer) has to send its signal to a VCR or a VCR has to send its signal to something else, we are again obliged to obey the laws of IMPEDANCE matching. For a strong signal and good fidelity, the IMPEDANCE of the source's output must match the IMPEDANCE of the next device's audio input. As before, we can cheat a little. Again the manufacturers accommodate us by making many audio inputs HI Z (10,000Ω or more) and many of their outputs LO Z (600Ω). Because LOW IMPEDANCE things can connect to LO or HIGH IMPEDANCE things, nearly all equipment is compatible.

Nearly all doesn't mean *all*. If you connect two things together and get weak and tinny sound, you may just have a HI Z device trying to feed the input of a MEDIUM or LO Z device.

A word about cables: HIGH IMPEDANCE cables usually have a center conductor and a shield around the outside. If you can't see this, take a look at the plug; it will have only two metal parts (like the MINI, PHONE, and RCA plugs in Figure 10–1). LOW IMPEDANCE audio cables generally have two conductors and a shield and have a three-pronged plug at the end like the CANNON connector shown in Figure 10–1.

Balanced and Unbalanced Lines

BALANCED and UNBALANCED LINES are two ways the mike cable can carry the signal to its destination. The higher-quality BALANCED LINE has two conductors inside a metal shield, making a total of three wires for the cable. The plug

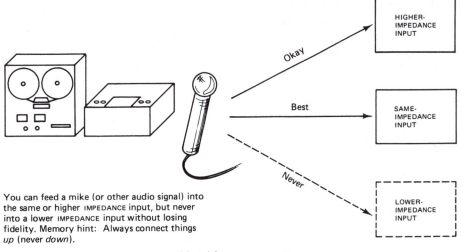

You can feed a mike (or other audio signal) into the same or higher IMPEDANCE input, but never into a lower IMPEDANCE input without losing fidelity. Memory hint: Always connect things *up* (never *down*).

Matching IMPEDANCE

at the end of such a cable would have to have three pins like the CANNON connector shown in Figure 10–1. BALANCED LINES can carry signals a long distance (over 50 feet) yet will pick up very little stray electrical interference (called "noise"). Most professional and industrial audio equipment and the higher-quality VCRs use BALANCED LINES.

Home and semiprofessional audio equipment, home VCRs, and many ¾U VCRs are designed for UNBALANCED LINES. These cables have only one conductor inside a woven shield, making them thinner than their BALANCED counterparts. Two conductor plugs like the MINI, PHONE, and RCA plugs shown in Figure 10–1 go with UNBALANCED LINES.

The mikes, the plugs, and the wires for UNBALANCED systems are inexpensive and simple to maintain. Since many ½ and ¾-inch tape users generally work close to their VCRs, use the mediocre mikes sold with these VCRs, play back their sounds through the tiny speakers on portable TVs, and usually aren't very discriminating about their sound anyway, UNBALANCED LINES are often good enough for these users.

One-inch VTR users generally have made the investment in good mikes, cables, mixers, and other audio equipment, so by popular demand, the 1-inch equipment accepts signals from BALANCED LINES.

There is one sometimes-baffling problem some users encounter when using UNBALANCED LINES—radio interference. While recording, they may hear police calls, CB (citizens band radio), nearby TV stations, or nearby radio broadcasts. These problems are often corrected by using BALANCED LINES—if your mikes and VTRs have that capacity. If not, you may have to call for technical help to get rid of the unwanted interference.

In short, BALANCED LINES are more expensive and more professional and give better-quality sound. UNBALANCED LINES are popular among nonprofessionals, are cheaper, but pick up interference easily.

So what do you do if your mike is BALANCED and your VCR is not? As I'm sure you've guessed by now, there are ADAPTERS which change BALANCED into UNBALANCED LINES and vice versa. To change a BALANCED microphone line into an UNBALANCED LINE for a small VCR, you would use an ADAPTER which had a three-pronged XLR socket at one end and perhaps a PHONE plug or MINI plug at the other. Inside the ADAPTER is a tiny transformer which converts one system into the other.

Transformers are the proper way to make this adaptation. There is also an improper way that *works*, but it loses you some of the benefits of the BALANCED LINE. This shortcut involves taking one of the two center wires in the cable and soldering it to the shield of the cable. Sometimes little ADAPTERS are made which have one of these center wires soldered to the shield wire. In a pinch, they are handy

to have around, allowing you to connect a BALANCED source into any UNBALANCED input via the ADAPTER. The problem with this transformerless ADAPTER is that it renders the entire BALANCED line UNBALANCED. Thus, the whole wire picks up interference, hum, and noise just like any common UNBALANCED line would.

Note: If you start with an UNBALANCED line and go through such an ADAPTER to a BALANCED line, it won't do you any good. Even though the aforementioned ADAPTER is attached to the *end* of the BALANCED mike cord, it makes the whole cable UNBALANCED and thus subject to unwanted noises. The only way to connect a BALANCED line to an UNBALANCED input and preserve its qualities is to use an adapter with a *transformer* in it.

Plugs and Adapters

Figure 10–1 showed the most common audio plugs. Learn the names of these plugs. You will use them frequently. The MINI, PHONE, and PHONO plugs are used mostly on home and semiprofessional equipment. The XLR plug is used on professional equipment. If the plug on your microphone doesn't fit the socket on your VCR, you need an ADAPTER. Figure 10–12 shows a sampling of common audio ADAPTERS. You describe an ADAPTER by telling what kind of socket and plug are on the ADAPTER. Thus, the first ADAPTER in Figure 10–12 is a "MINI plug to PHONO jack" (jack is another name for socket). People who like to talk about sex describe these ADAPTERS by gender, like "RCA female to MINI male" or "female XLR to female XLR."

Although ADAPTERS make things easy for us, they have the potential for being a weak link in our audio system. They sometimes wiggle loose and make poor contact, which results in no audio or crackly sound.

Sometimes ADAPTERS stick out a long way from the device they are plugged into, which places stress on the socket and makes the ADAPTERS easy to dislodge or break. The worst offender is the MINI plug. It is so small and frail that the weight of an ADAPTER plus the weight of an audio cable on the end of the ADAPTER can bend or break off the shaft (the male part of the plug). It is preferable to use the *right* plug when connecting together equipment instead of using ADAPTERS. If your mike doesn't mate with your VCR and you have to use an ADAPTER frequently, you may be better off removing the old plug and installing one that fits. You'll end up with a much more reliable connection.

Microphone Output

Some microphones are more sensitive than others. They create a stronger signal for a given sound volume. The

MINIGLOSSARY

*Balanced line   An audio cable with three wires, two inside a shield. Corresponding connectors have three prongs.

*Unbalanced line   An audio cable with two wires, one of which

is a shield surrounding the other.

*Adapter   An audio device that allows a plug of one type to fit a socket of another type.

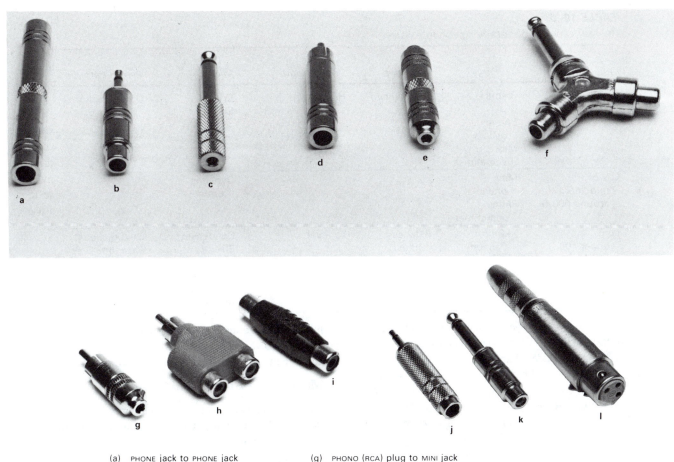

(a) PHONE jack to PHONE jack
(b) MINI plug to PHONO (RCA) jack
(c) PHONE plug to MINI jack
(d) PHONO (RCA) plug to PHONE jack
(e) MINI jack to MINI jack
(f) Y ADAPTER, two RCA jacks to PHONE plug

(g) PHONO (RCA) plug to MINI jack
(h) Y ADAPTER, two RCA jacks to an RCA plug
(i) RCA jack to RCA jack (sometimes called a BARREL CONNECTOR)
(j) MINI plug to PHONE jack
(k) PHONE plug to PHONO (RCA) jack
(l) PHONE jack to XLR jack

FIGURE 10-12 Audio ADAPTERS (Courtesy of Quality/Universal).

sensitivity of the microphone is usually listed as a number, like −56 dBm. The more minus this number (like −60 dBm), the lower the microphone's output or the less sensitive it is. The closer the number is to zero (like −43 dBm), the greater the microphone's output. Generally, audio devices can make up for a microphone's weak or strong output by adjusting their record volume (perhaps automatically). The only time you get in trouble is when a microphone is so weak that you have to turn the volume up all the way and still don't have enough.

Except for specialized microphones which have amplifiers built into them, all microphones put out a LO LEVEL or MICROPHONE LEVEL signal (you may wish to review Figure 5–13 and that part of the VCR chapter to review LO LEVEL and HI LEVEL signals). MICROPHONE LEVEL signals, because they are so weak, must go into MICROPHONE inputs. Any other kind of audio input, like a LINE IN or AUX IN, is too insensitive to pick up the mike's feeble signal.

Table 10-5 reviews the kinds of mikes, cables, and signals that go together.

Frequency Response

FREQUENCY RESPONSE describes the fidelity of a microphone, its ability to pick up high tones and low tones. A mike's FREQUENCY RESPONSE is described by the number of vibrations per second that the microphone can reproduce. Lower-priced industrial microphones used on ½- and ¾-inch VCRs generally have a FREQUENCY RESPONSE of 80–12kHz (80 to 12,000 vibrations per second). Better professional mikes will give a *wider* FREQUENCY RESPONSE of 50–15kHz. True high fidelity requires a frequency range of 20–20kHz. Only the special hi-fi-type VCRs can reproduce this FREQUENCY RESPONSE. Also, only young healthy people can hear this wide range of frequencies. Most of us just hear in the 50–15kHz range, which is what most industrial VCRs can reproduce.

MINIGLOSSARY

*Frequency response   The ability of a device to pick up high tones (high audio frequencies) as well as low tones.

TABLE 10-5
Review of what generally goes with what

Impedance	Plugs	Cables	Signal Strength	Types of Microphones	Pickup Patterns
Hi Z, several kΩ	Phono RCA Phone usually Mini sometimes	Unbalanced lines	Hi level unless its from a mike; then it's lo level	Crystal Electret	Omnidirectional
Lower middle Z, around 600 Ω	Mini often Phone sometimes			Dynamic often	
Lo Z, under 100 Ω	XLR or Cannon	Balanced lines usually	Lo level	Condenser Electret sometimes Dynamic usually Ribbon	Unidirectional Shotgun Cardioid Bidirectional

A microphone should be as sensitive to one frequency as it is to another. Otherwise, when recording music, the bassoon will sound louder than it is supposed to, and the flute might hardly be audible. This evenhanded sound reproduction is called a FLAT FREQUENCY RESPONSE and is preferred for music recording.

Microphones used primarily for voice, such as LAVALIERS, have an uneven FREQUENCY RESPONSE. They're made that way on purpose in order to emphasize the clarity of voice sounds while reducing undesirable low tones such as the rumble of wind or the resonance of a person's chest as he or she speaks (with the LAVALIER microphone hung right next to their chest). Also, since LAVALIER microphones are usually close to people's mouths and bass sounds are stronger there, the LAVALIER microphone needs to have reduced bass sensitivity. Some microphones sometimes are used close and sometimes far from the speaker's mouth and are sometimes used in the wind, or in echoey rooms, or near the rumble of air conditioners and have a switch which reduces the bass FREQUENCY RESPONSE. This switch may be called LOW CUT or BASS ROLLOFF. No, LOW CUT has nothing to do with alluring female attire; it simply means that the low frequencies are cut down, or weakened.

The FREQUENCY RESPONSE of a microphone gives it character. One microphone may sound like the inside of a tin can, while another sounds shrill, while another sounds muffled. When doing a lot of production with microphones, you should be aware that if you change microphones in the middle of a production, your sound character, called SOUND COLORATION, will change noticeably. Edits made later using another microphone may stick out like a sore ear.

Microphone Stands and Mounts

Now that we have our microphone happily plugged in, what do we hold it up with? Unless you are holding the mike in your hand, you will need a stand or something.

Shock Mounts. Let's talk about hand holding for a moment. People who move around with hand-held mikes end up making noises as the microphone tugs against its cord. Also, the fidgeting and the tap, tap, tapping of nervous hands are heard loud and clear along with the voice. The solution is to get a SHOCK-MOUNTED microphone. There's insulation between the microphone's sound-sensitive insides and the shell of the mike. This insulation absorbs much of the hand and mike cable noises.

Some microphones pick up hand and motion sound more than others. The SHOTGUN mike is one example. If you aim such a mike with your bare hands, every finger movement and turn of the wrist will create a creak or rumble in the background of your program. SHOTGUNS need a SHOCK MOUNT or ISOLATION MOUNT (Figure 10-13) to hold them. You grasp the handle, which is attached to a rubber cushion, which in turn holds the microphone.

Desk Stands. DESK STANDS like the ones shown in Figure 10-14 are most convenient for indoor work. The

MINIGLOSSARY

*Low cut filter   An audio circuit, often built into mixers and mikes, that reduces low tones from a sound signal.

Sound coloration   The characteristic tone a microphone gives to the sounds it picks up.

*Shock mount   Protects microphone from picking up noises as it is moved or handled.

Desk stand   A small microphone holder that sits on a desk.

Floor stand   Microphone holder that stands on the floor and reaches up to shoulder height.

Boom   An arm that sticks out, often with a mike hung on the end.

FIGURE 10-13   SHOCK MOUNT for a SHOTGUN mike.

general-purpose one is thinner and less obtrusive, while the adjustable DESK STAND allows more freedom raising or lowering the microphone's height. The VIBRATION ISOLATING stand reduces the sound of drumming fingers, rattling pencils, and scratching pens.

Floor Stands. Standup speakers or singers will probably use a FLOOR STAND like one of those in Figure 10–15. FLOOR STANDS come with an elevation adjustment for performers of different heights. The base is usually quite heavy to keep the microphone from tipping over. Some models have a small BOOM which allows the microphone to get into places where the stand would be in the way (like over pianos and drums and in front of seated people playing instruments). The FLOOR STAND puts some distance between

the microphone and the sounds of heavy footsteps and scuffling feet. If floor vibrations are still picked up, you might consider placing the mike stand on something spongy, like a carpet remnant or a stale pizza pie.

Both FLOOR and DESK STANDS come in shiny chrome or charcoal gray or brown. The dark mike stands are much better for television because they don't reflect the studio lights, blinding your camera. If you are cursed with shiny mike stands, try spraying them with DULLING SPRAY to reduce their shine.

Boom Mike Stands. A BOOM is the sound that you hear when a microphone stand falls over. It is also the name given to the long arm used to hold the microphone out in front of the talent.

One of the simplest, the BABY BOOM, is shown in Figure 10–15. Maybe it was invented in the post World War II era (hence the name "Postwar Baby Boom?"). Joking aside, these are handy when you need to keep the stand out of the way so that your cameras can see the performers better. They're also good for getting a microphone up higher or down lower than a normal mike stand could do.

Another handy BOOM is the FISH POLE. Essentially, it is just a stick with a microphone on the end. Professional "sticks" cost about $130, but at least those models telescope. They are lightweight and portable and allow you to catch the sound up close to a moving target. The problem is that it always takes an extra person to aim the FISH POLE, and that person's arms usually get tired very early in your production.

adjustable height   general purpose

FIGURE 10-14   DESK STANDS (Courtesy of Quality/Universal).

FIGURE 10-15   FLOOR STAND.

MINIGLOSSARY

Dulling spray   Spray-on aerosol that reduces surface shine.
Baby boom   Small boom stand for holding a microphone.

Fish pole   A portable boom in the form of a pole with a mike at the end.

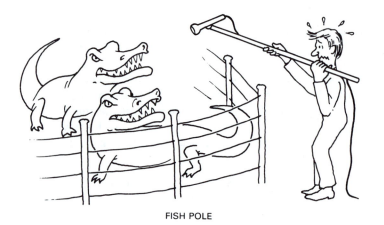

FISH POLE

studio BOOMS.

Baby BOOM

The portable studio BOOM is often a long telescoping rod mounted on casters so that it can be moved about the studio floor. BOOMS can be raised, lowered, and swung left and right. More complex models allow the BOOM arm to be lengthened or shortened as the performer draws nearer to it or farther from it. Better models also allow the microphone to be aimed in various directions, a necessity if using a directional microphone on the BOOM.

Sometimes as the BOOM is swung from one place to another quickly, the breeze gets picked up by the mike. This problem is cured by placing a foam boot, called a WIND SCREEN, over the mike.

Lavaliers and Lapel Clips.   History buffs will enjoy knowing that the LAVALIER microphone (or LAV, as some call.it) is named after Madame Louise Francoise de la Baune le Blanc de la Vallier, mistress to King Louis XIV. She always wore a large single jewel suspended just above her bodice. Although Madame La Vallier ended up in a nunnery, LAVALIER microphones 280 years later still bear her name. The madame's jewel was meant to draw the eye, but today we try to make our mikes invisible.

Older LAVALIER microphones are OMNIDIRECTIONAL and are designed to be worn on the chest about 6 inches below the mouth. Because of the vibration of the chest, bass sounds are very loud. Also, the treble sounds of the lips are some distance from the microphone. A properly made LAVALIER mike will have its FREQUENCY RESPONSE adjusted for this so that it cuts out the excessive bass and boosts the weak treble. Ordinary microphones don't tailor their FREQUENCY RESPONSE this way and will sound bassy when hung around a performer's neck. Similarly, a LAV-

FLOOR STAND, DESK, LAVALIER, and TIE CLIP microphones.

ALIER microphone sounds a bit shrill if hand-held or mounted on a mike stand.

Thanks to the ELECTRET CONDENSER, miniscule LAPEL CLIP microphones the size of a pencil eraser have become available. Clipped to a performer's tie or lapel, the mikes are unobtrusive and almost invisible. Their cables are extremely thin and pliable and run to a battery-fed power supply about the size of a cigar. Regular mike cord runs out the other end and to your VCR.

Coping with Cables

As your control room accumulates video machinery, a jungle of wires will grow behind the equipment, complete with exotic birds and distant drumbeats. There are several ways to make a path through the cable jungle. One way is with a machete. Quick, but expensive. A slower but more civilized way is to *label* the cables with a masking tape tag telling where each goes (or comes from). Then coil up any excess wire (except twin lead) and use tape to hold the loops together. Unravel any kinks or knots.

The most terrifying sound in the video jungle is ''oops—CRASH.'' It's the too-often-heard sound of a shoe snagging a cable and bringing a mike or light terminally to the floor. If you're running long extension cables from cameras, mikes, or lights, slip the cables under a carpet or tape them to the floor using wide duct tape or GAFFERS tape as shown in Figure 10–16. Run them *over* doorways, if possible.

Handy GAFFERS tape is strong, flexible, easily torn by hand, and sticks anywhere.

If the cable you tripped on happens to be a mike cable, the predominant sound in the jungle will become ''bzzz'' or ''crackle, crackle'' or ''(silence).'' You have partially or wholly separated the wire from your plug. If the cable you tripped on happens to be firmly affixed to someone's clothing, the sound you'll hear will be a ''YIPES'' or ''You clumsy $%#@&!.''

When using STAND MIKES, try tying, taping, or clipping the mike cord to the stand's base. Not only will the wire be less of an obstacle resting on the floor, but if tugged, it is less likely to topple the stand.

To keep mike extension cords from becoming unfastened, tie them in an overhand knot, as shown in Figure 10–16. This trick works for electric extension cords too.

AMPLIFIERS AND SPEAKERS

The AMPLIFIER is an electronic device which boosts the strength of a signal. HI LEVEL audio signals from preamplifiers, mixers, VCRs, or whatever go into the AMPLIFIER and get boosted thousands of times more, making them strong enough to drive one or more speakers.

There are several ways to describe an AMPLIFIER. One is by its *power*, how many WATTS of signal it can produce. The more WATTS of power, the louder the sound the AMPLIFIER and speaker can produce. For comparison, portable

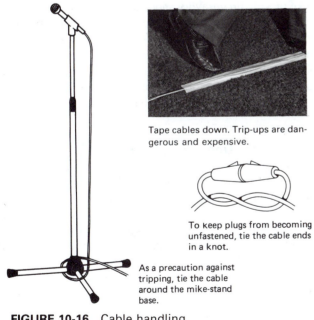

Tape cables down. Trip-ups are dangerous and expensive.

To keep plugs from becoming unfastened, tie the cable ends in a knot.

As a precaution against tripping, tie the cable around the mike-stand base.

FIGURE 10-16 Cable handling.

---

radios and TVs generally contain ⅟₁₀–to–1-WATT (abbreviated W) AMPLIFIERS. Industrial (school) record players and console TV sets may run about 5W, enough to make themselves heard in a classroom. A 30W AMPLIFIER is usually recommended for control room and studio monitoring. It has enough guts to drive a hi-fi speaker and fill a room with sound. Auditoriums and multispeaker public address (PA) systems need 100–500-WATT AMPLIFIERS to feed quality sound to hundreds of ears. For stereo sound, AMPLIFIERS require twice the power, i.e., a 30W stereo AMPLIFIER has 30W *per channel*.

Speakers, like AMPLIFIERS, are rated by power. A 30W AMPLIFIER needs a 30W speaker. Hitch a smaller speaker to the AMPLIFIER, and you could burn out the speaker. It would be safe, however, to hitch up a speaker rated for *more* WATTS than the AMPLIFIER. The extra capacity would never be used.

Speakers also vary in EFFICIENCY. EFFICIENCY describes how much of the electrical energy actually gets turned into sound by the speaker (the remainder is wasted as heat). Most loudspeakers are not very EFFICIENT. A typical bookshelf hi-fi speaker may be only 1% EFFICIENT. A speaker of high quality may be 2% EFFICIENT, giving twice the sound for the same number of WATTS. Some horn-type loudspeakers may be as much as 20% EFFICIENT.

FREQUENCY RESPONSE is another measure of the quality of an AMPLIFIER and speaker. A 20-Hz–20,000-Hz FREQUENCY RESPONSE would be perfect (it matches the range we can hear), but most systems give us a range of 40Hz–12,000Hz or so. This is a MIDRANGE FREQUENCY RESPONSE. WOOFERS are big speakers which extend the lower FREQUENCIES (low bass notes). TWEETERS are tiny speakers which handle the highest FREQUENCIES (shrill high-pitch notes). The wider this range, the better but the more costly the audio system will be.

DISTORTION (sometimes called THD for total harmonic distortion) and noise are other measures of a sound system's quality. They are beyond the scope of this book (thank goodness?), but suffice it to say the less there are of them, the better.

AMPLIFIERS and speakers have IMPEDANCES. As before, the IMPEDANCE of the speaker must match the IMPEDANCE of the AMPLIFIER for maximum signal strength. Nearly all hi-fi speakers are 8Ω. Similarly, nearly all hi-fi AMPLIFIERS have 8Ω outputs. Smaller speakers may be 4Ω, and a few speakers may be found with 16- or 2Ω IMPEDANCES. The more versatile AMPLIFIERS may have several outputs, one for each of the more common IMPEDANCES, 4, 8, *and* 16Ω.

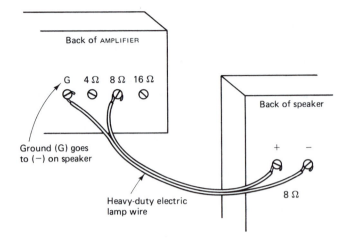

Connecting a speaker to an AMPLIFIER'S output.

CHOOSING AND USING THE PROPER MICROPHONE FOR A RECORDING

General Rules

You want the best sound. You know that LO Z mikes with BALANCED LINES will introduce the least interference and noise into your signal, so use them if you have them.

The closer the microphone is to the subject, the clearer and less echoey the sound will be. However, placing a mike too close to the performer (say 3 inches) often results in the mike picking up unsavory snorts and lip noises. If the letter "p" has too much punch (audio experts call this "popping your p's"), the mike may be too close. One seeks a happy medium between a mike that is too close, is in the way, picks up mouth noises, and distorts, and a mike which is too far from the performer, sounds distant, and picks up room echoes and background sounds.

One Person, One Microphone

Hand and Stand Mikes. If the talent is standing at a podium or sitting at a desk, a CARDIOID or UNIDIRECTIONAL mike on a floor stand or desk stand may be best. If the talent turns to the side while speaking (while addressing a widely spaced audience), the CARDIOID mike is better. It is less sensitive to the fact that the performer may not be standing directly in front of it. An OMNIDIRECTIONAL mike would work but tends to pick up too many room noises.

MINIGLOSSARY

Efficiency   How much of a signal is actually used (i.e., turns into audible sound) compared with how much is wasted by the electronics and simply turns into heat.

Midrange   Middle frequencies of sound, about 40–12,000 Hz.

Woofer   Big speakers, efficient for reproducing bass notes.

Tweeter   Small speaker, efficient at reproducing high notes.

Distortion   The unfaithful reproduction of sound. For example, turning a portable radio up to full volume often causes distorted sound.

The larger hand-held and stand microphones have excellent FREQUENCY RESPONSE and can handle a wide range of volume, making them excellent for singers or instrumentalists.

The *least* desirable way to mike a performer is with an OMNIDIRECTIONAL mike from any distance from the person. The OMNI will pick up echoes galore along with backstage sounds of people shuffling, camerapeople snoring, or studio crew tripping over cables.

Lavalier and Lapel Clips

If the talent is sitting, standing, or walking around, a LAVALIER or LAPEL CLIP mike may be best. They don't give terrific fidelity but are excellent for voice. Being close to the talent's mouth, they reject room noise and provide excellent ''presence'' (intimacy).

Center the LAV or LAPEL mike about 6 inches from the performer's mouth. If the neck cord is too tight, the LAV mike will pick up too many throat sounds, making the speech hollow or muffled. If tied *really* too tightly, the LAV could choke the performer, resulting in a gagging, gasping sound. Connected too loosely, the LAV won't pick up as much sound from the speaker. It may have a tendency to drag on clothing and rattle against buttons.

If the performer expects to move around a lot, it is good to attach the wire to his body or have him hold some wire in his hand so that it will trail along easily.

If the mike mustn't show, it may be hidden under a thin layer of clothing; however, the rustle of the clothing rubbing against the mike becomes a problem. Polyester, chiffon, silk, synthetics, and other stiff fabrics are pretty noisy. Soft fabrics like cotton shirts and wool ties are more silent.

When the humidity is low, you may sometimes hear a *snap*, *crackle*, and popping coming from your LAV mikes. No, it's not the sound of your talent digesting their Rice Crispies breakfast. It's static electricity discharging between the clothing (usually polyester) and the mike. Antistatic laundry spray from your supermarket will quiet this problem.

You'll get clearer sound if the microphone isn't covered. Today's tiny LAPEL mikes can easily poke their heads out through buttonholes or can be camouflaged as tie clips.

Be on the alert for buttons and zippers that go clankity-clank against the microphone when your performer starts moving around. Also watch out for performers who beat their chests during dramatic scenes.

The mike cable may also need to be hidden. The wire can usually be threaded down beneath the shirt or blouse and down a pant leg to exit at the ankle. A mike cable could exit the rear of a blouse or the bottom of a skirt or could run down the back of a leg, under a shoe strap, and across the floor. It is often good to anchor the mike cable wherever it leaves the body so that small tugs on the wire do not jiggle the microphone. The cable could be anchored at a belt loop or taped to an ankle.

Still, as the talent twists and turns, there will be tugging and twisting on the mike cable which may cause creaking and thumping sounds. These can be reduced slightly by tying a loose square knot in the mike cable about 3 inches from the base of the microphone. Somehow the knot absorbs some of the cable movement.

Boom Mikes. If the talent is too active, or if for aesthetic reasons the mike or cable must not show, you may wish to use a BOOM microphone. The term BOOM microphone actually refers to the mechanism which holds the microphone rather than the microphone itself. Various microphones can be used on a BOOM, generally CARDIOIDS and UNIDIRECTIONALS. The BOOM itself can either be a giant wheeled vehicle or a simple FISH POLE with a microphone on the end. BOOMS are expensive (except for the FISH POLE type), and someone must operate them in order to keep the mike close to the performer and yet out of the picture. You also have to worry about the mike BOOM casting a shadow in the picture.

The larger BOOM mikes are used mostly for dramatic productions and the FISH POLES, for on-the-spot news interviews.

Wireless Mikes. For the very active performer, the FM or WIRELESS microphone may be the answer. Instead of sending its signal down a wire to the VCR, the FM mike changes the signal into a radio wave and broadcasts this wave to an FM receiver up to 400 feet away. The FM receiver then sends a regular audio signal to the VCR as diagrammed in Figure 10–17. Good systems are a bit expensive ($500–$1,500) and somewhat more complicated to operate than a simple microphone. They are unbeatable in cases where the talent dances while singing, circulates among other people, interviews members of an audience, rides a bicycle, skydives, or climbs around machinery.

When choosing an FM wireless microphone system, check with your dealer or technician to make sure that the mike is transmitting on an unoccupied frequency in your area. Otherwise, you may pick up commercial FM radio broadcasts, walkie-talkies, or other interference.

FM WIRELESS microphones do have their faults. When used indoors or near metal, their signals sometimes bounce off something before reaching the receiver. These ''bounced''

MINIGLOSSARY

*Wireless microphone A mike transmitting a radio (usually FM) signal to a receiver rather than sending the signal over a wire. Used by performers who need freedom to move without mike cords.

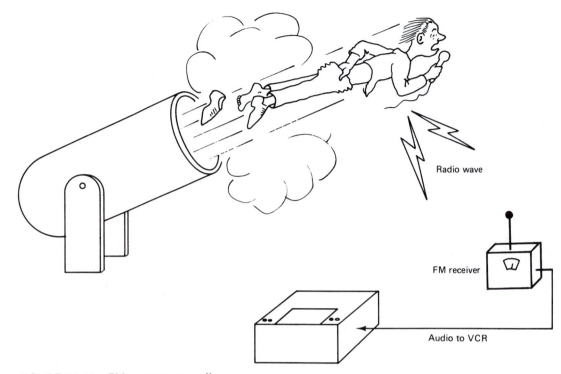

Radio wave

FM receiver

Audio to VCR

FIGURE 10-17   FM or WIRELESS mike.

signals sometimes interfere with the direct signal, weakening it or canceling it out altogether. As the performer moves and turns, you will hear short bursts of hiss or silence in place of your good sound. It sounds like what you sometimes hear on a stereo FM radio in a car traveling through a city, under an overpass, or in the mountains. The sound will be great most of the time, but every so often it fades out. This is especially true of $150 WIRELESS mikes.

The more expensive WIRELESS microphones go to great lengths to reduce these annoying dropouts. Many use two antennas on their receiver—if one antenna can't pick up the signal, the other will. These are called DIVERSITY FM mikes.

Some WIRELESS mikes have the transmitter built into the body of the microphone itself. Others use a separate transmitter about the size of a pack of cigarettes which can take a signal from any standard microphone such as a TIE CLIP mike.

Often, if the mike has to be hidden, you can sneak the "head" of the mike out between the buttons on a shirt. Mikes can be hidden at the cleavage in a brassiere, under a carnation, in lots of places. There are stories to tell about the places that microphones have been hidden. Juicy stories.

You then run the short mike cable to the transmitter, which could be in a coat pocket or strapped to somebody's back under his or her shirt. If taping the transmitter directly to someone's skin, use "ouchless" surgical tape or band aids.

Shotgun.   One problem with the STAND or DESK microphones, is that they are clearly seen in the shot. If your DESK mike must remain out of view of the camera, try

hiding it in a prop, perhaps among the flowers in a centerpiece or camouflaged as part of some tabletop artwork. Otherwise you may need to use a SHOTGUN mike placed just out of view of the camera. If the performer stays in one spot, the SHOTGUN can be mounted on a stand just like a regular mike. If the performer is active, then you'll need to have someone aim the mike, following the performer as he or she moves. Incidentally, anyone charged with this task

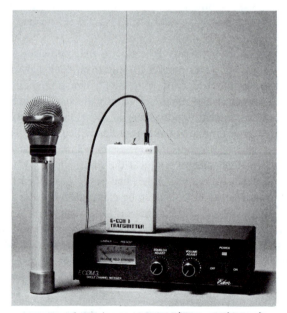

WIRELESS microphone, transmitter, and receiver (Courtesy of Edcor and Single Camera Video Production—Prentice-Hall).

If the talent stays in one place, the
SHOTGUN can be mounted on a stand
outside of the camera's view. . .

Otherwise, someone has to aim
the mike to follow the action.

SHOTGUN mike.

should be wearing headphones and listening to the signal from the microphone. This will improve attentiveness to accurate aiming.

Two People, One Microphone

It is better to have a separate mike for each person; that way you can adjust the volume of each source independently. If this option isn't available, then we try to get one mike to hear two people and the least room noise.

If the two people are sitting across from each other at a desk, the best mike might be a BIDIRECTIONAL one placed on a desk stand between them. The mike is sensitive in two directions only and rejects much of the room echo.

A BOOM mike or a SHOTGUN mike might also work if the mike has to stay out of the picture and if you have an operator to aim it.

As is done on many game shows and news interviews, the emcee or reporter can hold a mike (preferably a CARDIOID type) in his or her hand and hold it up to each of the respondents to catch the replies.

Several People, Several Microphones

A LAV for each speaker is the best situation. This way you can adjust each person's mike volume independently.

If six or more people are speaking, you'll probably run out of microphones and inputs for these microphones, so compromise. Try grouping the people into threes or so and aiming a CARDIOID microphone toward each group. If you are severely limited in the number of microphones you can use, try planting an OMNIDIRECTIONAL mike in the middle of the group with a LAV or CARDIOID delegated to the group leader or to the most important speaker.

PZM mikes are handy for group situations. They will pick up everyone speaking around a table and do a good job at rejecting room echo.

For news conferences where several individuals will be using a podium and others from the audience will be asking questions, place one mike at the podium and place another on a stand in the aisle for the audience. The intent is that the audience members will step up to the mike in the aisle when asking their questions. If a loudspeaker system is also in use for the conference, these microphones should

---

MINIGLOSSARY

Diversity FM microphone   An FM microphone *receiver* that can "listen" to a signal from the mike using two antennas. It picks
the antenna giving the best signal, thus yielding more reliable reception (fewer audio dropouts).

all be CARDIOID or UNIDIRECTIONAL in order to reject as much of the loudspeaker's sound as possible, thus avoiding FEEDBACK. Another way to mike a news conference audience is with a SHOTGUN microphone and an alert assistant to aim the microphone toward each person speaking.

Musical Recording

If the performers are singing, fidelity is paramount. LAV-ALIER mikes are generally designed for speech and are therefore inappropriate. The best fidelity usually comes from the larger HAND-HELD mike (which can also be used on a MIKE STAND or a BOOM). If possible, mike each singer separately for individualized volume control and keep each performer

1 or 2 feet from the mike. If you run out of mikes, group the singers.

Musical recording is a science in itself. If it is necessary to group the musicians, do it so that the lead has a separate mike from the rest, the rhythm gets a mike, the bass gets a mike, the chorus shares a mike, and related instruments share microphones. This way you have independent control of the volumes for each *section* of the band.

Stereo Microphones

Simply plug one mike into the left MIC IN and another into the right MIC IN and shoot. Whatever is picked up by the left mike is recorded on the left channel, and what goes in

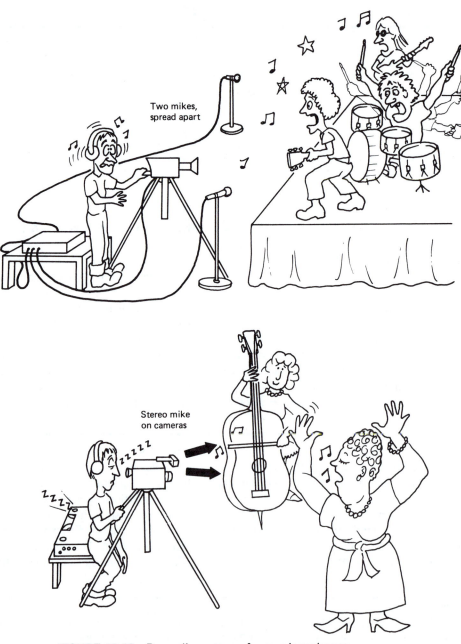

FIGURE 10-18 Recording stereo from microphones.

the right mike is recorded on the right channel (assuming you have a stereo VCR).

Separating the mikes so they each hear something different will improve the stereo effect. Figure 10–18 diagrams the hookup.

A strange problem sometimes occurs when using two microphones spread apart. One mike will hear a sound and another mike will hear the same sound a moment later (perhaps the second mike is farther away from the source and it takes the sound a while to travel to it). If you listen to the effect in stereo, you don't hear any problem. You won't consciously perceive the delay, but unconsciously the delay adds a certain "color" to the sound. The problem arises when you play this stereo signal through a *monaural* system such as a single speaker in a TV. If the sound from the second mike is delayed and OUT OF PHASE with the sound from the first mike, the signals cancel each other. You will hardly hear any sound at all, or it may sound hollow or echoey. To avoid this problem, it is good to always check out your sound on a monaural system just to make sure that it is compatible.

There are also stereophonic microphones that look like a single, fatter microphone. These stereo mikes "listen" in two directions at once to give the stereo effect. Although the stereo SEPARATION (how much *difference* there is between the two signals) is not great, the single stereo mike solves the PHASE problem. Since the mikes are together, sound reaches them both at the same time; one never hears a delay.

Redundant Mikes

Your college president is to give an important speech at graduation. Since you are doing the video recording, you've been asked to do the sound setup too. What happens if the microphone poops out in front of 8000 proud parents, grandparents, fidgeting babies, and the board of trustees? Your career poops out, that's what happens. It's not good to have your career hanging on the thread of a single microphone. So use two. You could just plunk two microphones onto the podium, one to the left and one to the right of the president as in Figure 10–19. They would probably work, but you're taking a chance with PHASE problems. As the president moves from left to right, his voice reaches one microphone before it reaches the other. If that second microphone gets its sound one-half vibration later than the first, its signal will *subtract* from the first mike's signal rather than *add* to it. Your sound will be hollow, weak, tinny, or bassy and unintelligible. What's tricky here is that the sound problem will not occur all the time. You may test for it and not hear a problem, but the president who's a

FIGURE 10-19 Wrong way to use two mikes on one person.

little shorter than you and leaning a little more forward than you did may find just the magic spots where the sound will fade out.

The right way to set up redundant desk, stand, or podium mikes is shown in Figure 10–20. The "heads" of the two mikes should be very close together, almost touching. The left mike should be pointing to the right to pick up the sounds in that area. The right mike would point to the left, picking up sounds in that area. If two CARDIOID mikes are used, their pickup areas will overlap nicely, covering the entire front of the podium or desk area. Because

FIGURE 10-20 Right way to set up two mikes for one person.

MINIGLOSSARY

Phase The *timing* of when electrical or sound vibrations reach a place, like an input or a microphone. When IN PHASE, the vibrations strengthen each other, making a strong signal. OUT-OF-PHASE signals cancel each other out, weakening the result. Electrical and sound signals need to be kept IN PHASE.

WIND or "pop" screen on microphone.
(Courtesy of Radio Shack)

the mikes are close together, they both hear all sounds at the same time, and your signals will never be OUT OF PHASE.

TIE CLIP microphones can also be made redundant by clipping two mikes, one above the other, to the performer's tie or lapel. There are also double clips which will neatly hold two TIE CLIP mikes at once.

Banishing Unwanted Noise from a Recording

Wind. Even a slight breeze over a microphone can cause a deep rumbling and rattling that sounds like a thunderstorm in the background of your recording.
Solutions:

1. Stay out of the wind and don't interview politicians.
2. Buy a WINDSCREEN, a foam boot that fits over the mike and protects it from breezes while letting other sounds through.
3. In a pinch, take off your sock and put it over the mike to deflect the wind. Be prepared for wisecracks like "Your audio stinks."

Hand Noise. The shuffling and crackling of nervous hands holding a microphone can be avoided. Set the mike on a stand or hang a LAV around the talent's neck with the warning "Don't touch it. Don't touch it. Don't touch it." If the performers *must* handle the mikes, tell them merely to grip the mike and not to fidget with or fondle it.

Stand Noise. The mike is on a table stand, and every time the talent bumps the table, it sounds like a kettle drum rolling down a stairwell.

Solutions:

1. Have the talent keep their hands and knees still.
2. Insulate the base of the mike stand from the table with a piece of carpet, a pad, a quiche Lorraine, or anything spongy.

Lav Noises. Too tight a cord results in excessive throat sounds. A very loose cord results in a mike that swings like a pendulum and bumps things. Find a happy medium. *Before recording anything from a LAV, first check to see that there are no buttons or tie clasps for the mikes to clank against.* Murphy's 34th Law states that the clanking starts only when the show starts, not during rehearsals.

Mouth Noise. Performers love to put their lips to the mike. Perhaps they don't trust the wizardry of electronics to sense their feeble sounds from a foot away and amplify them to spellbinding proportions. As a result, two things happen. When the performer speaks loudly, the sound distorts. When the performer pronounces the letters *t*, *b*, and especially *p*, it sounds like bombs bursting in air.
Solutions:

1. Teach performers to trust the mike. Have them keep their distance. Cover the top of the mike with erect porcupine quills.
2. Use professional mikes designed for such abuse. They have a wire screen atop them which puts

Home-made WINDSCREEN.

MINIGLOSSARY

*Windscreen** Foam boot that fits over a microphone to shield it from wind noises.

some distance between the lips and the actual microphone. They also contain built-in WIND-SCREENS and ''pop'' filters. Furthermore, they can withstand the excessive volume found ½ inch from a rock singer's lips.

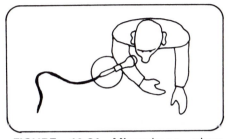

FIGURE 10-21 Microphone placement to avoid "popping p's" at close range.

Room Noise. The closer the mike is to the performer and the louder the performer speaks, the less room noise will be heard. So what do you do if he starts popping his *p*'s, as described above? As shown in Figure 10–21, place the mike at *an angle to the side* rather than directly in front of the performer's mouth. The offensive consonants will fly straight forward, hurting no one yet missing the microphone.

As mentioned before, CARDIOID and UNIDIRECTIONAL mikes are best for rejecting extraneous room noise. They are, however, worse for ''popping p's.''

Try to place your performer in a quiet part of the room, away from windows, fans, loudspeakers, and—not to be forgotten—the whirring video tape recorder.

The echoes in a reverberant room can be quelled by placing a carpet on the floor, drawing the curtains, or hanging blankets on the walls. I recall taping a lesson once in a hospital room which echoed like a giant canyon. I stretched a clothesline between two opposite walls, scrounged 30 doctors' smocks from the hospital cloakroom and hung them across the clothesline. Except for the stage area, the place looked like a dry cleaner, but the smocks succeeded in subduing the echoes . . . echoes . . . echoes.

Feedback. This loud screech or howl (sometimes called ''back squeal'') is very common when loudspeakers or public address systems are in use. It results when sound goes into the microphone, gets boosted by the amplifier for the loudspeaker system, and comes out the loudspeakers only to be picked up by the microphone again and amplified. It goes out through the speakers and into the mike: 'round, 'round it goes, getting louder all the time.

Solutions:

1. For immediate relief, turn down the volume on the amplifier that is causing the FEEDBACK. Disconnecting or switching off the microphone (some

have ON/OFF switches) will also terminate the noise. This solution has the disadvantage of negating the whole purpose of having a loudspeaker system. What good is it if you can't turn it up loud? Solution 2 is the better answer once solution 1 has been employed in order to save eardrums in the interim.

2. Proper placement of speakers and selection of microphones will give long-range relief from FEEDBACK. The whole trick is to keep the microphone from hearing the loudspeakers:
   a. Keep the loudspeakers away from the mikes.
   b. Aim the loudspeakers away from the mikes. Note that some loudspeakers allow sound to project behind them as well as in front of them. So don't get behind the loudspeakers.
   c. Aim the mikes away from the loudspeakers.
   d. Use CARDIOID or UNIDIRECTIONAL mikes.
   e. Keep the performers as close to the mikes as is practical. If excessive movement makes standing mikes inappropriate, use LAVALIERS so the mikes remain close to the performers.
   f. If the FEEDBACK comes as a shrill screech, turn down the amplifier's TREBLE control. If the FEEDBACK is a deep whoop or wail, have the BASS turned down. Figure 10–22 shows a proper public address setup that avoids FEEDBACK. Figure 10–23 shows situations to avoid.

Testing a Microphone

Clunk, clunk, blow, blow, testing—1—2—3—testing—1—2—3—testing—testing. . . .'' Such is the traditional prelude to every sound recording.

The ceremony of testing a microphone has two steps: The first is to find out if it is working *at all*; and the second is to make it work well. The process is easiest if you have a helper at the controls to adjust volumes, to monitor the sound, to check plugs, and to watch the audio meter. After you have plugged the mike into its proper receptacle (mixer, VTR, or whatever) and turned up the volume for that mike, you're ready for the tests.

Test 1. Either tap the mike with your fingernail (while listening for the clunk, clunk over the monitoring speaker) or speak into the mike while your assistant listens for sound and notes whether the audio meter needle wiggles. If it doesn't, read the troubleshooting section of this chapter. One way *not* to test a mike is to blow into it. The ritual of blowing into a mike is like squeezing eggs in a supermarket to test for freshness. Some mikes are too fragile to withstand the ''blow'' test.

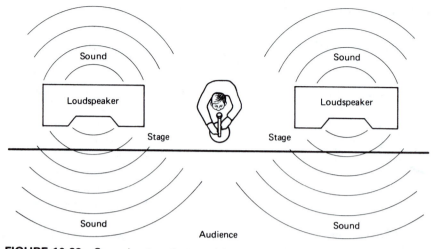
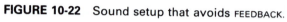

FIGURE 10-22   Sound setup that avoids FEEDBACK.

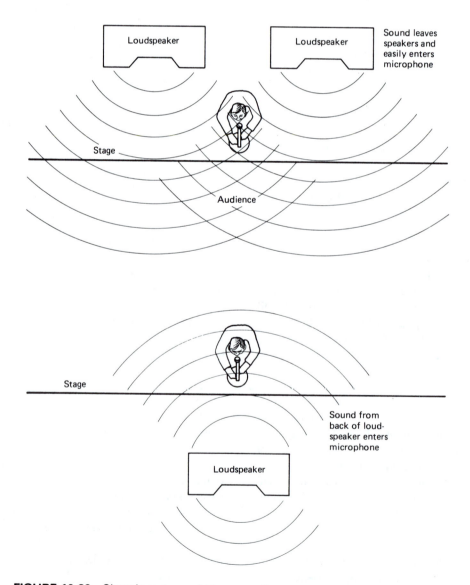

Sound leaves speakers and easily enters microphone

Sound from back of loudspeaker enters microphone

FIGURE 10-23   Situations to avoid because they cause FEEDBACK.

Test 2.   Once you've established that you're getting sound, you make the *audio level* adjustments to ensure that your recording is made at the right volume. This is done by having the talent talk normally into the mike. It is difficult to make novice performers speak normally. So give them something to talk about. Tell them to count to 30 or to say their ABCs. *Do not* allow them to hold their LAVALIERS up to their mouths while you are checking the audio level—you will not get a representative sample of their normal speech volume. *Do not* allow them to stoop down to or lean into their stand mikes. This, too, will yield an unrepresentative volume level. *Do not* let them say just ''testing'' or ''One—two—three . . .'' because people tend to shout these words with unnatural loudness.

If the mike is feeding directly into a VTR with manual audio controls, you observe the AUDIO METER. It should wiggle when people speak, but it should rarely dip into the red area. ''Rarely'' doesn't mean never; loud outbursts are expected to sweep the needle into the red for a moment. If possible, listen to the sound on headphones or a speaker in order to judge the sound quality. Is the sound distorted? Is there a buzz, a hum, or a hiss in the background? If so, something is wrong. Off to the troubleshooting section later in this chapter.

In short, if the meter reads low, turn up the volume. If the meter points into the red area, turn the volume down. The meter should always be wiggling visibly but not be lingering in the red.

PROPER AUDIO LEVEL

Automatic Volume Control

Automatic volume control, sometimes abbreviated AVC or AGC (for automatic gain control—*gain* is another word for *volume*), is present on most portable VCRs and nearly all home VCRs. A few of the better models have a switch on them which allows you to select AGC or MANUAL volume control.

The advantage of AGC is obvious. Recordings are made automatically at the right volume level—no muss, no fuss. It's all done with a circuit that ''listens'' to the audio, and if it gets too loud, it turns the volume down. If it gets too low, it turns the volume up.

So why ever bother with manual controls? There are cases where AGC is not helpful at all. Say you were using an AGC machine to record an interview in a blacksmith's shop. The talent speaks, everything sounds fine, and then somebody's hammer strikes an anvil. The AGC reacts to the loud sound by lowering the record volume drastically

and then slowly raising it again to the level appropriate for speech. The recording could sound like this:

> Under the spreading chestnut tree, the Village Smithy sta—WHANG! . . . ty man . . . He . . . ith . . . and sinewy hands. And the muscles—WHANG! . . . rns . . . strong . . . iron bands.

It would be better if the loud noise came and went in a flash, leaving most of the speech intact, like this:

> Under the spreading chestnut tree, the Village Smithy stands. The smith, a mighty man is he with—WHANG! . . . and sinewy hands. And the muscles of his— WHANG! . . . arms—WHANG! . . . strong as iron bands.

AGC is similarly troublesome in situations where long, quiet pauses occur between sentences. When the talent stops speaking, the AGC circuit ''hears'' nothing and slowly turns up its volume. Still ''hearing'' nothing, it turns the volume up higher and higher. Turned way up, the machine records every little noise in the room, shuffling, sniffing, VTR motor noise, some electronically caused hum or buzzing, automobiles outdoors, and fire whistles in the next town. Then the first syllable out of the talent's mouth, after this long pause, is thunderously loud because the volume is far too high for speech and hasn't yet turned itself down.

In short, AGC is helpful when you expect a fairly constant level of sound. AGC doesn't like long silent pauses or short loud noises. AGC tends to wreck music which is *meant* to have quiet parts and loud parts. There are, incidentally, AGC devices made to overcome these problems. They are somewhat expensive and aren't built into the reasonably priced VTRs.

Manual Volume Control

As described before, you adjust the RECORD VOLUME LEVEL (or whatever they call it) to wiggle the meter without making it loiter in the red area. Also, monitor the sound coming out of the VTR (using headphones or a monitor/receiver) to see if the sound is clear and undistorted.

Once the audio level is set (during checkout while you were setting up), the circumstances of the production will dictate whether it will need to be adjusted again. If the audio is from a professionally prerecorded source or if the audio is from a speech or dictation in which the loudness was relatively even, it is possible to make the entire recording without twiddling the audio knobs and with only occasionally checking the meter. If the sound source changes its volume frequently or drastically, you may have to ''ride

MINIGLOSSARY

*Audio meter   Meter that indicates the loudness of an audio signal.

audio,'' twiddling the knobs and watching the meters intently.

How much do you twiddle? Answer: the least possible to keep the audio level about right. You're twiddling too much if every shriek or cough makes you turn the volume down. Brief noises are loud, yes. They sound distorted when recorded at high volumes, yes. But they are gone in an instant and easily forgotten. Brief pauses and whispers are quiet, yes. But whispers are supposed to be quiet, and silent pauses are not unnatural for us to hear and may even be a refreshing break from the monotony of constant chatter. While playing back a tape, it is irritating for the viewer to have to rise from his comfy seat to readjust rising and falling volume because of some overzealous knob-twiddler who adjusted the record level too often. In short,

1. React quickly to sustained bursts of noise or substantial passages which would be lost if not adjusted for.
2. Don't react to momentary sounds or silences.
3. As conversations ebb and flow in volume, gradually make tiny adjustments in order to compensate. Do it in such a way that no one will notice that the volume is being changed.

MIXERS

A mixer accepts signals from various sources, allows each signal to be individually adjusted for loudness (even adjusted all the way down to no loudness at all), and sends this combination to a VCR or some other recording device. As with all production equipment, you can have small, medium-sized, and enormous mixers which can do basic, moderate, and miraculous tasks. Figure 10–24 shows some mixers.

Inputs to the Mixer

Microphones are generally plugged into the MIC INPUT sockets in the back of the mixer, up to one mike for each volume control knob on the front of the device. On the more professional models, the inputs are the XLR types for BALANCED LINES, and they have a little pushbutton near each socket to release the plug so that it can be removed.

All the microphone inputs are LO-LEVEL INPUTS, which means they accept tiny signals only. It is possible to use (besides microphones) telephone pickup coils, guitar pickups (the signal comes on a wire directly from the guitar itself, not from an amplifier), tape heads (from unpreamplified audio tape decks), and phonograph turntables. In each case, the signals are weak and match the mixer's sensitive inputs. Although turntables and tape heads have the right signal strength for use with MIC INPUTS, these sources may sound tinny or bassy when used this way because they

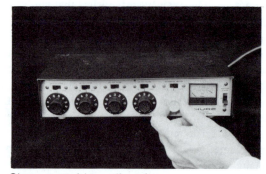

Shure portable audio mixer

Fostex medium-sized mixer

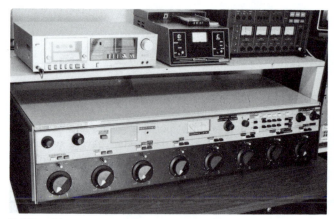

Professional audio console

FIGURE 10-24   Several mixers.

require extra circuits to make the tone right (a process called EQUALIZATION). Some audio equipment accommodates these devices by having special inputs labeled TAPE HEAD or PHONO where the EQUALIZATION is built in.

The MIC INPUTS are not designed for stronger signals like those coming from an FM tuner, a VCR LINE OUT or AUDIO OUT, any earphone output, any speaker output, any preamplifier output, or just about anything which boosts the signal before sending it out. For these LINE LEVEL or HI LEVEL sources, one must use a different input on the mixer (if it has such): the AUX (for AUXILIARY) IN or the LINE IN.

These inputs are less sensitive than LO LEVEL inputs and work well with stronger signals. Use AUX or LINE IN when you have some musical background or sound effects which you wish to mix with the voices on the microphones. The HI-LEVEL or LINE outputs of tape decks, FM radios, record players, cassette players, or similar devices can be connected to the AUX IN for this purpose.

The microphone inputs may have switches next to each of them that say HI Z/LO Z. These switches change the INPUT IMPEDANCE of each mike input. When using a LO-IMPEDANCE mike, set the corresponding switch to LO Z. Meanwhile, if another mike is HI IMPEDANCE, its input gets switched to HI Z. If you don't know the microphone impedance, try the mike and switch from one position to another. One position probably won't work or will sound noticeably terrible.

Once everything is plugged into the mixer and is working, *label each of the mixer's knobs* to tell which source is controlled by which knob. For example, near knob 1, stick a piece of masking tape with the word CASS marked on it. This way, during production, you don't have to be asking yourself, "Let's see, is knob 1 the audiocassette player or is it the record player?"

Outputs from the Mixer

The mixer's output sends the combined signals to the VCR or other device. Just as each microphone's volume is adjustable with a knob on the mixer, the volume of the signal the mixer sends out is also adjustable with the MASTER

REMINDER: IMPEDANCE VERSUS LEVEL

HI Z is different from HI LEVEL, and LO Z is different from LO LEVEL.

LEVEL refers to the *signal strength* of a source. LO LEVEL (weak) sources should feed LO LEVEL inputs. HI LEVEL (strong) sources should go to HI LEVEL inputs.

Z (IMPEDANCE) describes an electrical characteristic of inputs and outputs. HI-Z mikes and sources must go to HI-Z inputs. LO-Z sources should go to LO-Z inputs and may sometimes go to HI-Z inputs.

Although IMPEDANCE has nothing to do with LEVEL, you may find that LO-LEVEL things frequently (but not always) have LO IMPEDANCE and HI LEVEL things often have HI IMPEDANCE.

---

volume control. Usually this knob is a different color, shape, or size from the rest. Turning the MASTER down turns down all the signals coming out of the mixer. This is a very convenient feature, especially at the end of a program that uses multiple mikes—you'd have to be an octopus to turn down all the individual mike volume controls simultaneously.

The mixer may have several outputs. Although one output from a mixer is all that is needed to feed a VCR, the others are there to permit flexibility in setting up and using the audio system.

Mike Out or Mike Level Out or Lo Level Out. This is an audio output that has a tiny signal like a microphone; thus it can be plugged into the MIC IN of a VCR or other device. To this end, the audio cable and plugs are just like microphone cables and plugs. It is as if the mixer were pretending to be a microphone: It is putting out a signal just like a mike, and the signal goes to wherever a mike's signal could have gone.

Line Out or Hi Level Out or Aux Out or Audio Out. A medium-sized signal emanates from this output and is destined to go to a VCR's LINE IN or AUX IN.

Headphone. This is another output for monitoring the audio signal over headphones. This output is a fairly strong signal, one inappropriate for feeding VCRs.

Monitoring Audio

Back in videoland we found it convenient to view a TV signal before it went to the VCR and to monitor it again after it went *through* the VCR. By observing the monitors, we could tell whether a problem was in the original signal or in the VCR. The same is true for audio.

The mikes and other things feed into the mixer and get combined. The audio signal could be fed to the VCR and monitored either on the VCR's speaker (if it has one), or on headphones plugged into the VCR, or on a monitor/receiver connected to the VCR's audio and video outputs. What happens if you get no sound on the VCR monitor/receiver? Is it the VCR's fault? The mixer's fault? The mike's fault? If the mixer has a meter and the meter wiggles when someone speaks into the microphone, the mixer and mike are most likely working. If you can plug headphones into the HEADPHONE jack on the mixer, you can make doubly sure that the mixer and mike are working well. If the head-

---

MINIGLOSSARY

Equalization A tone adjustment for audio frequencies, often needed to boost high or low tones coming from a phonograph cartridge or microphone, or audio tape head.

*Master audio control Mixer knob that adjusts, up or down in volume, all mixer inputs at once. Useful for fading out all mikes

and sounds together.

Headphone Muff-type earphones to fit over your head. Also the socket into which such phones are plugged, either on a VCR, mixer, or other audio device.

phones and the meter are still, then the problem is probably in the microphone or in the mixer. Try another microphone: If it works where the first one didn't, the first mike (or its cable or plug) may be bad, or its ON/OFF switch is off, or perhaps its battery (if it uses one) is dead.

Once you've proven that your signal is playing through the mixer, you turn your investigation to the VCR. Is the mixer's output plugged into the VCR's input? Is the VCR's volume up? Is it in the RECORD mode (remember that many VCRs don't allow you to monitor their signals unless they are in RECORD)? Is the VCR's audio monitoring system (or the monitor/receiver) volume turned up? Are all the cables tight?

Some studios make a big deal of monitoring their audio accurately. They use the LINE OUT from the mixer to feed the VCR. They use the MIC OUT to feed a separate loudspeaker system in the control room so that they may hear exactly what is being sent to the VCRs (not relying solely on the wiggling meter needle). Some places monitor the LINE OUT *from the VCR* by sending that signal to another loudspeaker system. By doing so, they know what the VCR is putting out when it is recording or playing back a tape.

Given a choice, the best place to monitor audio during a recording is the VCR's LINE OUT. Thus, if anything goes wrong with your audio signal *anywhere* in the system, you'll hear the problem there at the final output.

Adjusting Audio Levels

The VCR monitor/receiver has a volume control. The VCR has an AUDIO LEVEL control. The mixer has a MASTER volume and a bunch of knobs controlling the microphones. So what do you turn down if the sound is too loud? The procedure for setting proper audio levels is somewhat involved. The objective is to make each device do its share of work, avoiding the situation where one device is turned way up while another is turned way down. The sound comes out hollow, flat, raspy, or otherwise distorted when one volume control is too low while another is straining at the top of its range.

The simplest way to balance the volumes is to set everything at one-half full volume, try a source, and then twiddle knobs until things sound good. Then look to see what is turned way up relative to something else. Whatever is low should be raised somewhat, while whatever is high should be lowered to yield the same result with all volumes more or less equalized. This is the informal method. For those perfectionists with numerous sound sources to control and a desire to do it *exactly* right, here are some step-by-step procedures.

Mixer Has No Meter and VCR Has Only Automatic Volume Control. In this case, no matter what you do, the VCR is going to try to adjust for it with its automatic controls. Your objective is to make sure the mixer

is doing a good job and is putting out a signal which is neither too weak nor too strong for the VCR.

First, let's get the mixer working.

1. Set the mixer's MASTER volume halfway up.

2. Set one of the mixer's individual controls halfway up and send a fairly steady sample signal through it. One excellent source of sample signals is an FM tuner, tuned to a strong local station. Its output should be sent to the AUX input of the mixer.

3. Monitor the resulting signal:
   a. If the mixer has a HEADPHONE output, listen to it. If the sound is weak and has a lot of hiss, then turn up the mixer's MASTER and individual volume control. If the sound is too loud, distorted, or has poor fidelity, then your volumes may be too high; turn both of them down. Unless your source is sending out too strong or weak a signal for your mixer, you should be able to find a happy medium on your mixer where your sound is loud enough and has good fidelity.
   b. If the mixer has no HEADPHONE output, then you have to monitor the signal some other way. You could send the signal into the AUX INPUT of an amplifier and loudspeaker system or even into a TV monitor/receiver which has an audio input.

4. Once you are happy with the mixer's sound, connect the mixer's output to the VCR's input and see what the VCR does with it. Again, you could use headphones or a monitor/receiver or some other device to check on the VCR's volume. If the sound is flat, clipped, or otherwise distorted, your mixer is overpowering your VCR's input. Even the automatic gain control can't handle signals which are *too* strong. Check to see that you are indeed sending a HI LEVEL OUT to the VCR's AUX or AUDIO IN. Alternatively, you could send the mixer's LO LEVEL OUT to the VCR's MIC IN, or you could use an ATTENUATOR to knock the signal down to size.

5. Once you have good sound, mark the mixer's MASTER volume setting and turn down the individual source control.

6. Leaving everything else the same, try another source into another mixer input set at half volume. Adjust the mixer's MASTER for good sound and mark the setting.

7. Do this with all the inputs. Because some sources are louder than others, the MASTER will have been set at different places. Examining all the marks on the MASTER control, find a compromise setting among all of them. Mark that spot and erase the

others. *This is probably the best setting for the mixer's* MASTER *control.*

8. Using this new MASTER volume setting, try a sample from each of the individual sources again and mark them where they sound good. Now each time you use the mixer, you need change only the individual volume controls to adapt to particular circumstances and you'll already have an idea of about where they should be.

Mixer Has No Meter and VCR Has Manual Volume Control with Meter. Again, we want to get good sound to come through the mixer and then have it be compatible with the VCR.

1. Set the mixer's MASTER volume halfway up.
2. Set one of the mixer's individual controls halfway up and send a fairly steady sample signal through it.
3. Set the VCR's volume control halfway up, press RECORD (if necessary), and observe the meter.
4. If the meter reads low (barely wiggles), turn all three volume controls up equally until the meter shows a proper level (wiggles a lot but stays out of the red). If it reads high, turn the three controls (VCR audio level, individual source control, and MASTER control) down equally. Mark the position on the mixer's MASTER volume control when everything sounds good.
5. Leaving everything else the same, test another input to the mixer. Turn its volume control halfway up. Check the VCR meter and turn the MASTER control on the mixer up or down until the meter is happy. Mark the MASTER knob again.
6. Repeat this process with the various sources. When done, look at the marks on your mixer's MASTER control, average them, and make that your official mark.
7. Set the mixer's MASTER at *this* point and test all the other sources again, turning their individual controls to whatever makes the VCR's meter happy. Mark the controls for later reference.

Mixer Has Meter and so Does VCR. The meters will make this process easier and more exact.
First we adjust the mixer:

1. Set the mixer's MASTER volume halfway up.
2. Set one of the mixer's individual controls halfway

up and send a fairly steady sample signal through it.
3. Observe the mixer's meter and turn both controls up or down until the meter is happy. Mark the position on the MASTER control.
4. Try another source, adjust both the individual and MASTER controls, and when the meter is happy, mark the position.
5. Do this with all the input sources to the mixer and then average the marks you've placed on the MASTER volume control. Set the MASTER at this point and try the other controls again. Mark them for later reference.

Now we adjust the VCR:

6. Press RECORD on the VCR and adjust its volume up or down until the VCR's meter is happy. To work properly, the VCR's volume control should be between one-third and two-thirds of its range. If the volume control is very near the top or the bottom of its range, to make the meter happy, you may need to have a technician weaken or strengthen the mixer's output relative to the VCR. This is often a simple process of installing an audio ATTENUATOR to cut the signal down a little. Or if the signal is too weak, connect the mixer to the VCR's MIC IN and install an ATTENUATOR.
7. When you finally get the mixer and the VCR working well together, mark their knobs for easy reference.

Mixer Has Tone Generator. The TONE GENERATOR is one gadget that can save you a heap of time. It gives you a steady reference signal to work with when adjusting volume controls. Here's the process:
First set up the mixer:

1. Switch on the TONE GENERATOR. Often this means flipping a switch in the back of the mixer to the TONE OSC position, turning the volume of mike 1 up about halfway, and then turning up the MASTER volume to make the meter move.
2. Adjust the MASTER volume so that the mixer's meter reads 0 dB as shown in Figure 10–25.
3. Now switch your VCR to RECORD and using the AUDIO LEVEL control, set *its* meter at 0 dB (or if a nonprofessional meter is used, set the meter a shade below the red).
4. Moving to your audio monitor, adjust its volume so that the sound is pleasant to listen to.

MINIGLOSSARY

*Tone generator Electronic circuit often built into mixers, which can create an even, standardized audio signal. Used for checking volume levels, it provides a handy reference tone.

FIGURE 10-25   Mixer's meter at 0dB.

From now on you need only adjust the mixer's individual volume controls to vary your sound levels. All the other equipment will take care of itself. The mixer's meter and the VCR's meter will read exactly the same, so now you only need to watch the mixer's meter and forget about the VCR's meter.

5. Now switch off the TONE GENERATOR and begin making audio level checks on all your sources, attending only to the mixer's meter and controls. Be careful to choose a MASTER volume level which doesn't force some of the individual controls to be too high or too low.

Controlling Excessive Volume Levels

Sometimes, no matter what you do, you

1. Can't keep the meter out of the red
2. Can't get a volume control to give proper volume unless it is turned almost off
3. Find that the sound is raspy and distorted

These are cases where a VCR or mixer is receiving more sound than it can handle. It happens when you connect a LINE OUT or HI LEVEL OUT to a MIC IN or LO LEVEL IN. It also happens when you try to use the signal from an EARPHONE or HEADPHONE OUTPUT from a radio, or cassette tape recorder, or SPEAKER, or EXT SPEAKER output from a hi-fi or similar device. Such signals are usually too strong even for HI LEVEL INPUTS to handle.

The solution for this is to buy an ATTENUATOR or PAD. This is not like the paper tablet you use for writing home to the folks twice a year a week after Mother's Day or Father's Day. It is an inexpensive little box with a circuit in it that *throws away* most of an audio signal and *passes on* a tiny fraction of it. Plugging a powerful signal into it and then plugging it into your mixer results in the mixer receiving an acceptable signal even though an excessive signal was being put out at the source. A PAD can also be plugged in between the mixer and the VCR to cut down the mixer's volume. (For instance, some home VCRs don't have

---

MORE ABOUT DECIBELS (dB)

A decibel (named after Alexander Graham Bell) is a measure of signal strength.

When describing the signals from microphones, the signal is very weak, yielding a number like −49dB. A stronger signal would be −40dB. These are LO LEVEL audio signals.

HI LEVEL audio signals run about −10 to +10dB, appropriate for AUX or LINE inputs and outputs of audio devices.

The loudness of a sound can also be measured in dB. Silence is 0dB. A whisper is about 30dB. Ordinary speech is about 65dB. Painfully loud rock music punches the eardrum at 140dB.

Audio mixer meters try to relate the electrical dB measurements with the volume of sound measurements all on one scale. Such meters are sometimes called VU meters (the VU stands for volume unit). Here, 0VU or 0dB *doesn't* mean *no sound.* It means *optimal* sound.

The minus dB numbers (like −20dB) are very weak sounds, the slightly minus numbers (like −3dB) are slightly weak sounds, 0dB is the "perfect volume," and the plus dB numbers represent excessive volumes. The more "plus dB" your meter goes, the more distorted the sound will become. More than +3dB is noticeably bad, while between 0 and +3 is not too irritating.

Some audio meters measure sound volume in percent. 0% means silence, and 100 percent means the optimum. Anything over 100% begins to distort the sound.

HI LEVEL audio inputs, and all signals must be sent through the MIC IN.) PADS are cheap to buy and are easy for a technician to make and are a real pal around the studio.

Other Gadgets on Mixers

Mic/Line Input.   Some mixers have a switch on one or more of the microphone inputs that changes the inputs from LO LEVEL to LINE LEVEL to accept stronger signals. By switching the switch to LINE, you can now use a knob to control an audiocassette player, for instance, rather than a microphone.

Multiple Inputs Per Control.   Say you had 5 audio volume knobs on your mixer, but you had 10 sources (four microphones, a couple of audio tape decks, an FM receiver, a couple of VCRs, and a partridge in a pear tree). Instead of plugging and unplugging each source into the back of the mixer, some mixers have multiple inputs. Thus, everything plugs into the back of the mixer, and on the front, over each knob, is a switch. In one position, the switch will make the knob control one source. In another position, that same knob controls another source.

This system usually works out fine because most TV productions—although you may own 10 sources—only use 5 sources during a show. The only problem is that you may not use two sources at once on the same POT (POT is short for POTENTIOMETER, which means the same as audio control knob). For instance, if you wish to use an audiocassette player and mike 1 at the same time, the cassette player must be plugged into one of the *other* inputs controlled by a different POT.

Low Cut Filters. Just as microphones can have LOW CUT FILTERS to reduce the rumble and room echo picked up by the microphone, some mixers have a LOW CUT FILTER built into them to do the same thing. Generally, this is a switch near each volume control which removes deep, bassy sounds from the signal. You might switch these filters on to record speech (where the medium and high tones are most important) and switch the filters off (or switch them to OUT, taking them OUT of the circuit) when you wish to record music with full fidelity.

Peak Level Indicator. This is another device for measuring excessive volume levels. VU meters, though very useful for measuring overall volume levels, do not react fast enough to show brief bursts of sound. So in addition to the traditional VU meter, some new equipment also has an LED (light emitting diode) that flashes when sound transients exceed an allowable level. When you see your PEAK LEVEL lamp flashing frequently, turn your volume down a little. If it only flashes occasionally, pay no attention.

Professional Audio Control Boards

The professional audio control boards are mixers with extra gadgets. They are bigger, cost more, and are very flexible in application, last a long time, and can pass a very high-quality signal. One of the important features on these giant mixers allows you to AUDITION or CUE a selection before use.

Audition or Cue. Say you wish to play the sound effect of a "boing" while someone winds his watch. How do you find the specific groove on your sound effects record that has the "boing" without having your audience hear you search for it? You can't have your listeners hear you play "hee haw," "meow," "cluck, cluck," and "plop" as you seek out the "boing." You would like to play these selections to yourself, find the right one, get it ready to go, and then, at the right time, play it for everybody to hear.

Near the mixer's knob which controls the audio from the record player will be a switch which is likely to say PROGRAM/OFF/CUE. On other models, the knob itself, when turned fully counterclockwise, will click into a CUE position. One way or another, that knob will either work in the PROGRAM mode or the CUE mode or will be switched OFF and will not send signals anywhere.

In the PROGRAM mode, the mixer sends its signal through to the VCR after you have adjusted its volume by the individual volume control on the mixer. In the OFF position, the mixer refuses to pass the signal, and nothing gets recorded. In the CUE position, the mixer passes the audio signal but not to the VCR or to the audience. The signal goes to a separate output or to a CUE SPEAKER built into the mixer. You hear the selections on this speaker as you find the right place on the record. Once found, you are CUED UP and ready to play the record at the appropriate time once you have flipped the switch back to PROGRAM.

AUDITION is much the same as CUE. Sometimes you would like to send a signal to someone else (or yourself) to hear but not the audience. This is easy to do if your mixer has an AUDITION channel. Putting these ideas together, we can see that each POT on the mixer can send its signal to perhaps four places: OFF (no place at all), PROGRAM (to the VCR), CUE (to the audio person for setting up sound effects, etc.), and AUDITION (for the audio person or anyone else who needs to hear a signal passing through the mixer).

Here is an example of how all these controls could be used during a production: Your show is rolling along, and people are speaking into several microphones. Those POTS are turned to PROGRAM, and the sound is being recorded on the VCR. During the show you are trying to set up the music for ending the show. The music is on a record player, so you turn the PHONO POT to CUE and drop your needle into what you think is the right groove. You (and not the audience and not the VCR) listen to the musical selections on the CUE SPEAKER on the mixer. When you've located the selection you want, you stop the record player and switch the PHONO POT to PROGRAM (you may wish to turn the PHONO POT down first so that you can FADE UP the music when the time comes). Meanwhile, the director may ask you if the VCR is recording everything all right. You could get up and walk over to the VCR and check its meter or turn up its monitor, but what if the VCR is in the next room or on another floor altogether, as at CBS? Here's where you use AUDITION. The VCR is probably sending its AUDIO OUT to

MINIGLOSSARY

VU Volume unit—a measure of loudness. A VU meter measures the strength of an audio signal. A 0 VU ("zero V-U") setting is considered optimum sound volume.

Potentiometer Also called a pot, it is a volume control on a mixer or other audio device.

*Peak level indicator Tiny light often built into mixers and audio recorders that blinks when sound volume is too loud.

LED Light emitting diode, a tiny lamp that can blink very quickly, uses little power, and lasts a long time. Often used as an indicator on equipment.

*Audition The act of checking on a sound signal but not recording it. Also, a mixer channel that can be listened to or adjusted but is not necessarily recorded.

*Cue In audio, to "set up" a sound effect or music or narration so that it will start immediately when a button is pushed. Also a mixer channel used by the audio person who listens to the sound effect being set up. The cue channel does not get recorded.

*Program On audio mixers, the sound channel sent out to the VCR.

one of the inputs on your mixer. If you could listen to that input, you could tell how the VCR was doing. You could turn that knob to CUE and listen to the sound. Generally, the CUE speaker is a cheap speaker built into the mixer useful only for finding the right place on records and things. It is not good for measuring fidelity. The AUDITION channel, however, is generally a high-fidelity channel and may feed a HEADPHONE output on the mixer or may feed a speaker in the control room. By switching the POT marked "VCR" to the AUDITION position, you will be able to hear the VCR's output either on headphones or the control room speaker in full fidelity.

Often the HEADPHONE output of the professional audio control board can be switched so that the audio person may listen to the PROGRAM, AUDITION, or CUE channel. This is nice because the control room crew usually suffers enough confusion without hearing a menagerie of "hee haw" and "cluck, cluck" as you CUE UP your sound effects.

Source Switches.   A medium-sized studio may have six microphone lines, an AM/FM tuner, a cassette player, a cartridge player, turntables, an audio tape deck, a 16mm film/sound projector, an 8mm film/sound projector, and sound from three video players. It takes a *very* large audio mixer to have 18 individual source controls. Most semiprofessional mixers have between 5 and 8 knobs to twiddle. So how do you get 18 sources to run through 8 volume controls? The answer is that each control has a source switch that selects which of 2 or more sources may go through it. On channel 1 you may have the choice between mike 1 or a turntable. On channel 2 you might have the choice between mike 2 and the audio tape deck. On channel 5 you might have 6 push buttons so that you may select any one of 6 sources to go through that channel. You never use 18 sources simultaneously, and you rarely need more than 8 at any time during a production, so such a system works out just fine. Think ahead about which sources you can team up together so you don't get stuck wanting to play a taped announcement and a sound effects record at the same time only to find them both sharing the same input on the mixer.

Talkback or Studio Address.   Some professional audio consoles have a TALKBACK system which allows the AUDIO DIRECTOR (or anybody else who gets his or her hands on the switch) to speak into a mike in the control room and have the sound come out of a loudspeaker in the studio. It's good for making announcements like "Stand by" or "Everybody take a break."

Foldback.   Some studios use one mixer for microphones alone and another mixer for all the sound effects, background, music, and so forth. When set up a certain way, the two mixers can permit an audio technique called FOLDBACK. Some mixers have this feature built in.

Say you wanted a surprised look on a performer's face as soon as her watch went "boing." Normally, with a single mixer, the performer wouldn't hear the "boing" (and thus couldn't react to it at the exact time of the "boing") *unless* the sound was piped out to the studio for her to hear. When you send the mixer's sound out to the performer, her microphone's sound gets piped out there too. Unless the sound volume from the speaker in the studio is very low (running the risk that the performer may not hear it), the sound from the loudspeaker will go into the performer's microphone, into the mixer, and eventually out the speaker again, around and around. Instead of a "boing," you get the "wheech" of FEEDBACK.

To avoid FEEDBACK, you wish to send only the sound effects out to the studio, not the microphone sounds. This is done by having all the sound effects and music go into one of two mixers. The output of this first mixer goes to two places: to the studio and to the second mixer. The studio now gets only the output of the first mixer, that output being the sound effects or music. The second mixer also gets a taste of that output on one of its volume controls, while it may mix the sounds from the other mikes. The second mixer sends this total combination to the VCR for recording. Figure 10–26 diagrams this setup.

Some mixers can use the AUDITION channel for FOLDBACK, while others, as mentioned before, have FOLDBACK circuits built into them.

The average TV production doesn't have much call for "boings," but a common use of FOLDBACK involves music. A rock group, for instance, can perform a song wearing street clothes (or rags), reading sheet music, and standing near their carefully placed microphones. Through the magic of FOLDBACK, they can later hear their recording played back and can dance and jump in the studio, *pretending* to be performing the music. The prerecorded music would be video-recorded, along with their "live" screeches, howls, comments, or harmony recorded on camera.

SOUND MIXING TECHNIQUES

There is no substitute for creativity. There are some basics, however, that can help. In fact, your library may have

---

MINIGLOSSARY

*Talkback   A loudspeaker system to allow the control-room crew to speak directly to studio personnel.

*Foldback   Audio mixing system to allow sound effects, music, etc., to be mixed, amplified, and sent to the studio for performers to hear, as well as being recorded, mixed with the sounds of their microphones.

*Segue (pronounced "SEG-way")   A smooth change from one sound, place, or subject to another.

Sting   Short sound effect or a few notes of music or just a chord used to introduce, segue between, or end scenes.

Music under   Music volume is reduced into the background so that narration or something else gets the audience's attention.

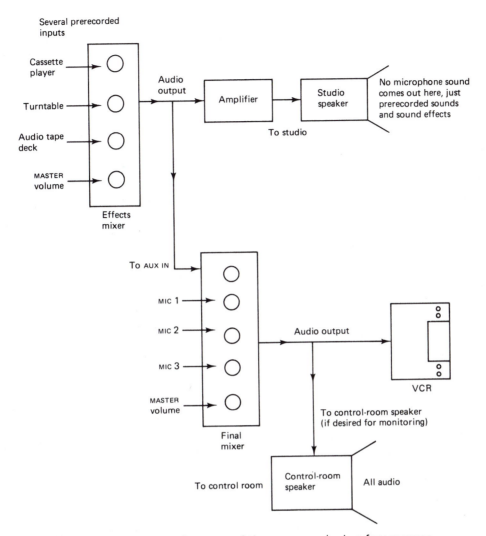

FIGURE 10-26 Two mixers permitting FOLDBACK (only a few sources are shown).

several books written solely on the basics of audio; it gets that involved.

Segue

SEGUE (pronounced ''SEG-way'') is a fade from one sound to another. For instance, the sounds of machinery can be smoothly replaced with music. The machinery's volume control is lowered at the same time that the music volume control is being raised. This is also often done between two pieces of music; as one finishes, the other is being faded up.

A more sophisticated SEGUE uses an intermediate sound when changing from one audio passage to another. For instance, to go from one scene in a play to another, after the last line in the first scene come a few bars of appropriate music. As the music fades out, the first line in the next scene is delivered. Briefer things like jokes or single statements may deserve a sound effect, laughter, applause, or a single note or chord of music (such a musical passage is called a STING) between them.

Some SEGUES prepare the listener for things to come like faint machinery noise before we open the engine room door or the sound of windshield wipers before the actors begin to speak in the car on a rainy night. A famous Hitchcock SEGUE is a woman's scream suddenly changing to the scream of a train's whistle as the train chugs into the next scene. Another popular SEGUE is the dance troupe rehearsing to ''one, two, three, four, one, two, . . .'' as we dissolve the picture and sound to the actual on-stage performance of the number with all the glitter and music.

Music Under, Sound Mix, Voiceover

Your production begins with a snappy musical selection. The title fades in and then dissolves to the opening scene. Someone is about to speak. The music fades down just before the first words are heard. This is a MUSIC UNDER.

The music became subordinate to the speech and is played *under* it.

Sometimes you have to decide whether to fade the music out entirely when the action starts or to MUSIC UNDER holding in the background throughout the scene. If the music is needed for dramatic effect, either to create a mood or just to provide continuity through long gaps in action or conversation, then keep it in. If, however, the action or conversation is very important, then don't distract your audience with background music.

So how loud should the background music be? The answer, of course, depends on the particular situation: There's no hard rule. In general, keep in mind that background music is *background* music. Keep the volume low—lower than your natural inclinations would have you set it. How many amateur productions have you sat through straining to hear the dialogue through that "noise" in the background?

One guide to proper volume setting can be your VU meter. If your narration makes your needle huddle as it should around 0 dB (the 100% mark on the scale), the background should wiggle the needle about −8 dB (about 40% on the scale). If you only have a nonprofessional record level meter, the narration should wiggle the needle just below the red, while the background should wiggle it one-fourth to one-third of its range.

Again, these are just generalities. Some musical selections are inherently more obtrusive than others. For instance, while listening to a narration, the viewer may hardly be aware of instrumental music in the background. Conversely, a song with words competes with the narration for the viewer's attention. Because singing with words is so distracting, it's best to avoid it in favor of instrumentals.

Not all background sound is music. Street sounds, machines, sirens, motors, gunfire—all can be background to your dialogue. Some of these sounds may not be background at all but are interjected between dialogue such as "thud," "crash," and the like. How all these sound effects and backgrounds are woven together is called the SOUND MIX, and it's the AUDIO DIRECTOR's job to mix them effectively. He or she may vary engine background noise up and down to favor the actors when they speak. The AUDIO DIRECTOR may drastically reduce the engine noise to coincide with a change of setting or of camera position. He or she may combine music with barnyard sounds, mix that with the dialogue of the actors, and top it off with a few specialty effects.

Sometimes you are handed a video tape or a film and are told to add narration. The original sound on the tape will be kept but only as background for the narration. Add-

ing narration is called a VOICEOVER. The voice you're adding is imposed over and is louder than the original sounds. Although sound editing will be covered more completely in Chapter 14, here, briefly, is what happens.

Mr. Expert brings in a tape showing his foundry in action. The tape shows the busy machines while you hear them foundering away in the background. Mr. Expert also brings a script that he wishes to read through parts of the recording. To do this, you set up a VCR to copy his original tape from a VCP (videocassette player). The VCP's video goes directly to the VCR. The audio from the VCP goes to a mixer. Mr. Expert's microphone also feeds to the mixer. The mixer combines and regulates the two sources and feeds the combination to the VCR. As the VCP plays, the VCR records, copying the picture and whatever original sound the mixer lets through. Mr. Expert reads his script as he keeps one eye on the VCP's monitor screen. You adjust audio levels, sometimes favoring the background sounds (when the narrator is silent) and sometimes lowering them (when the narrator speaks). That's a VOICEOVER. If he doesn't like the way the final tape comes out, you can erase it and do it over since his original tape was not altered in the process. You can do this over and over again until you get it exactly right. If, after repeated redoing, Mr. Expert still doesn't like it, you may choose to alter his original tape—over his head!

Sometimes the script is narrated by someone far away or long ago. You'll get a videocassette with background sounds and an audiocassette with the narrator's story. Here the process is about the same except an audio tape player is providing the narration rather than a live person. Instead of the narrator slowing down, speeding up, or stopping his reading to coincide with the pictures he sees from the VCP, you, the AUDIO DIRECTOR, must stop and start the audio tape player to coincide with the pictures. If you have the script or memorize the narration, you will be able to choose good places to stop the audio tape from playing without catching the narrator between words.

Several Performers—Each with His Own Microphone

When all the microphones are turned on at the same time, you not only hear the sound of the one person speaking, but you hear the breathing and shuffling of the others in the background. You also get the hollow echo of the speaker as his or her voice is picked up on everybody else's microphone.

To avoid this, turn down all the inactive mikes, allowing only the speaker's mike to be live. This is easy to

MINIGLOSSARY

*Voiceover   Narration added to and louder than background sounds or music.

Sound mix   The process of editing and mixing numerous sounds into the final form heard by the audience.

*Audio director   Studio crew member who handles the microphone placement, sound mix, and other audio responsibilities before and during the show.

do in a scripted production like a newscast or a play, but it is difficult to do in a free discussion. In such cases, you have to suffer the disturbing background sounds resulting from leaving all the mikes live, because if you turn off the unused mikes and then turn them up after a new person has started speaking, you will lose his or her first few words.

One partial solution may be to *lower* by a third or a half the unused microphone volumes, raising them after the person begins speaking. Although the person's first words may be weak, they will be audible and will soon be up to full volume. Allow the dynamics of the discussion to be your guide. If the speech is coming in a crossfire from everyone, keep all the mikes open. If someone seldom speaks, lower his or her mike. If the speakers render long monologues, take a chance on turning down the other mikes until the speech sounds finished.

Good audio requires lightning-fast reactions coupled with a dose of anticipation.

Cueing a Record

You wish to push a button and have that "boing" happen instantaneously, right in sync with the action of the performer as he winds his watch. Once you find the "boing" on the record disc, you need to get it backed up to just before the "boing" so that it will play the instant the switch is thrown. If you leave too much space before the "boing," when you play it during the production, you'll get " . . .boing." That's too late. If you don't back it up far enough, you'll get "ing," the tail end of the sound, or "woooing," the sound of the turntable picking up speed while the effect is being played. Proper CUEING of a sound effect (assuming an appropriate turntable is being used) goes like this:

1. Switch the mixer's "turntable" input to the CUE mode. Adjust volume controls as needed.
2. Turn the turntable motor on and put it in gear so that the disc rotates.
3. Locate the "boing" on the disc by sampling various grooves with a needle. The record label may even tell the contents of various bands on the record.
4. Once you find the "boing," pick up the needle and set it down two or three grooves earlier in the record. You can anticipate that the "boing" will come up in a few seconds. Be ready to stop the disc from turning.
5. Some disc jockeys prefer to stop the disc manually with a finger (touching the edge of the disc and allowing the turntable platter to revolve beneath it). Others shift the turntable to NEUTRAL and use a finger to brake both the disc and platter together. Others turn the motor switch off, brake

the platter with a finger (along the edge), and put the turntable into NEUTRAL afterward. Whatever method you use, somehow you've got to stop the disc when you get to the "bo . . ." in "boing."

6. Once you have the "bo" in "boing," rotate the disc in reverse—with the needle still playing—to find the beginning of the sound. Sometimes it is hard to tell exactly where the sound begins, so with your finger you rotate the record forward and backward, trying to recognize the sound of the beginning of "boing."
7. Once you've got it, back up the record one-eighth of a turn more one sixteenth of a turn for good turntables (the better the turntable, the faster it can pick up speed and the less backing up you have to do to assure it has time to pick up speed before it plays the sound).
8. Put the turntable into NEUTRAL with the motor turned on *or* leave the turntable in gear with the motor turned off.
9. Switch the mixer from CUE to PROGRAM.
10. When you want the "boing," just switch the gearshift to the desired speed *or* turn on the motor switch.
11. When the "boing" is finished, either turn the volume down or switch it off so you don't catch the "cluck, cluck" that follows the "boing."

Two things to watch out for:

1. If you are rotating the disc itself back and forth with your finger while the turntable is still moving, note the following:
   a. Make sure that the turntable's felt platter is clean so that it doesn't scratch the record as it rubs against it.
   b. Try not to shake the disc as you move it, or the needle might jump to another groove.
   c. Once you are CUED UP and if you are in a rush, you may wish to hold the record by its edge with one hand while switching the mixer to PROGRAM and waiting for the performer to wind his watch *while the turntable keeps turning*. When the time comes, just release the record and out comes "boing." If, however, the watch winding isn't for awhile, you may prefer to stop the turntable's motion, release the disc, and at the proper time start the turntable going again for the "boing."
2. Some mixer's meters will display a source's level in the PROGRAM mode but not in the CUE mode. So once your production starts, you can CUE UP your sound effect but you can't tell whether it's

loud enough (using the meter) until you're in the PROGRAM mode and the sound effect is already being aired—perhaps at the wrong volume until you make the adjustment as it plays, which is already too late. Solution: Prior to your actual TV production, be sure to check the proper volume level for your sound effects.

Always make a habit of marking the proper volume settings on the knobs so that you know where to turn them during the show. This preproduction planning can save guesswork, confusion, and precious time during the actual shooting.

Cueing up other devices requires a similar technique. Since some machines don't run backwards, you may be required to play forward to the desired sound, stop, rewind a speck, play again, and by trial and error find the beginning of the desired passage.

Cueing a Reel-to-Reel Tape

The technique will be the same as with the record; only the mechanics are different.

Reviewing the technique: You switch your mixer to CUE and play the tape. When you get to "boing," you stop the tape, back it up to just an inch or so before the sound began, and leave the tape parked there ready to play. Switch the mixer to PROGRAM. Hit PLAY when the time comes.

Mechanically, you do the following:

1. While your mixer is in the CUE mode, you play the tape until you come to the "boing."
2. Find the beginning of the "boing." Depending on the kind of reel-to-reel tape recorder you are using, you may use one of the following two procedures:
   a. Procedure 1
      1. Press STOP to stop the tape.
      2. Throw a CUEING lever, which physically places the tape recorder heads against the tape. In this position, the heads will "listen" to the tape when it moves, even when the machine is on STOP.
      3. Place your left hand on the left reel and your right hand on the right reel and manually back the tape up, keeping it

CUEING A RECORD	
Switch mixer to CUE mode.	. . .
Record plays forward in a search for the sound effect.	*CLUCK, CLUCK . . . PLOP, PLOP . . . BOING, BOING . . .*
Lift needle and back up a couple of grooves.	. . .
Play the record.	*—LOP, PLOP . . .*
Get ready to stop the disc.	. . .
Stop the disc at the first sound.	*. . . BOIN—*
Rotate the disc counter-clockwise to find the beginning of the sound.	*GNIOB . . .*
Rotate the disc clockwise to find the beginning of the sound, then backwards, then forwards again, eventually finding the starting "b" in "boing."	*. . . BOI—* *—IOB . . .* *—BO—* *—B—* *—B—*
Rotate the disc backwards an extra one-eighth turn.	*—B . . .*
Leave the turntable stopped in that position.	. . .
Switch the turntable's input on the mixer from CUE to PROGRAM.	. . .
At the appropriate time, start the turntable.	*BOING*

Lever places audio heads against tape so that you can hear it when tape is manually moved.

slightly taut. You will hear the "boing" playing backwards.

4. If you are unsure of what you are hearing, you may wish to "rock" the tape back and forth, listening to the *boing* a few times until you are sure you've got the *b* in *boing*.

5. Back the tape up an inch more. Good sound effects tapes leave space between the effects so that you should not run into the *hee haw* which precedes the *boing*. Also, some reel-to-reel tape recorders start and stop on a dime. You may need only ½ inch of tape silence before the sound effect to give the machine a running start to play the sound smoothly. Older mechanical tape machines may need a couple inches of tape to get a smooth rolling start. If your sound effects are recorded at 7½ inches per second (as they should be), then 1 inch equals about ⅐ second, a very small delay before your sound effect begins. If, however, your sound effects tape was recorded at 3¾ inches per second, the delay for 1 inch of tape is doubled to about ¼ second, somewhat long.

b. Procedure 2:

1. When you find your *boing*, press PAUSE on your tape player.

2. While in PAUSE, manually back up your tape by turning the reels.

3. When you reach the *b* in *boing*, move the tape just a little farther (about an inch) so that it has some "start-up" time.

3. Switch your mixer to PROGRAM. When the time comes, press PLAY on your audio tape player, and out comes *boing*. Don't forget to kill the sound afterward, or you'll catch a *cluck, cluck* too.

Using a Cartridge Tape Player

There is one machine that CUES itself up—the CARTRIDGE (abbreviated CART) tape player (Figure 10–27).

Especially popular in radio stations, the CART player (which is often a recorder, too) is supersimple to operate. Your sound effect, musical intro, or whatever is on a CARTRIDGE. You simply shove the CARTRIDGE into the player, and when ready to play the sound effect, you press PLAY and it does the rest.

Inside the plastic CARTRIDGE is an endless loop of recording tape (the end of the tape is attached to the beginning of the tape, making a large loop which is compactly wound up to fit into the CARTRIDGE). The tape CARTRIDGES can have as little as 5 seconds of tape to as much as several minutes of tape in them.

When the CART tape reaches its end, a sensor in the machine stops the tape. Pressing PLAY starts the tape at the beginning. This process may be repeated any number of times.

Here's how you would use a CART:

1. Find the CART with the desired sound effect.

2. Slip the CARTRIDGE into the CART player.

3. Unless you wish to test the sound effect first, switch the mixer's input to PROGRAM and turn up its volume.

4. When ready, press PLAY, and your sound effect will be heard.

5. Do not stop the CART machine; it will wind itself automatically to just before the beginning of the sound effect and be ready to play it the next time you press PLAY.

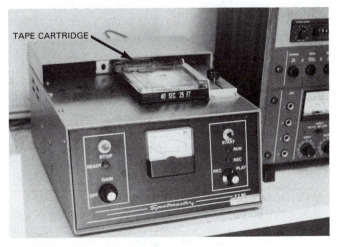

FIGURE 10-27 CARTRIDGE tape player.

When sound effects or music are used frequently or in rapid succession (making it hard to CUE each up in such a short time), the CART machine is a real lifesaver. To use it, however, you must record your own CARTRIDGES ahead of time. When doing so, remember to have your sound effect recorded at the very beginning of the CARTRIDGE (so there isn't a long pause before the sound effect is heard when the CARTRIDGE is played), and after the sound effect is finished, leave blank tape (no sound) to the end of the CARTRIDGE. You don't want any surprises playing from the CART if you forget to turn off the PROGRAM switch on your mixer after the sound effect is over.

Mixing Sound Without a Mixer

Sometimes you have two sound sources you'd love to mix together, but you don't have a mixer to combine them (you're working for Pete's Poverty Productions, Inc.). Here are some alternatives that *may* work but they come with no money-back guarantee that they'll work well.

Acoustical Mixing. Picture yourself driving through Dodge City with the twang of country music on the car radio. A passenger is taping the sights out of the car window. The camera mike picks up the sound of the radio and adds that western flavor to the view. When narration or comment seems appropriate, you just lower the radio volume and talk.

Similarly, you can DUB narration and musical backgrounds over existing visuals. Simply find a record player and select appropriate music. Set up your mike (or the mike built into the camera—it can make AUDIO DUBS even though the camera isn't recording any new picture) close to you and the phonograph. The closer the speakers are to your microphone, the less room echo you'll pick up. Perhaps start with the music at normal volume and then fade it down while you speak. Fade it back up when you're finished speaking. Although the sound fidelity won't be too great, the music and voice mix will add pizzazz to your tape. Also, you're likely to enjoy the thrill of being a true DJ as you attempt to coordinate switching on the VCR, starting the record, fading the volume, and making sense as you try to speak at the same time.

Y Adapter. Figure 5–24 showed how this little widget could take your VCR's one audio output and send it to your stereo's two inputs. A Y ADAPTER with the right plugs can work the same way in reverse, taking two sources and combining their signals to send to one VCR input.

Stereo Amplifier. Most stereo hi-fis today have lots of inputs in the back for tape players, tuners, phonographs, and even microphones. By turning your amplifier's INPUT SELECT switch to one of these positions, your amp will "listen" to that pair of inputs. Often you can plug a microphone into one of the inputs (MIC input works best, but others may work too—experiment) and talk through your hi-fi. Since stereos have two inputs (left and right channels) for every source, you could plug a *second* mike into the twin input and talk through *two* microphones. By adjusting the right and left channel volume controls or by adjusting the BALANCE control on your amp, you can vary the loudness of each mike.

Now to get the signal into your VCR: Find an output on your amplifier that you can connect to the AUDIO IN of your VCR. This output may be labeled TAPE OUT, AUX OUT, LINE OUT, or PREAMP OUT, or if there are no outputs, use the HEADPHONE socket (you may need to buy an ATTENUATOR or special plugs for this). Now you have a stereo output to send to your VCR, but many VCRs aren't stereo. No problem. Either switch the STEREO/MONO switch on the amp to MONO, which will mix the two signals, or use a Y ADAPTER to make the left and right channels mix together. This mixed signal is then sent to the input of your VCR. Thus, you now control the volume of two sources using your amplifier's volume controls and can send the result to your VCR for recording.

Audiocassette Deck. Many stereo audiocassette decks have microphone and auxiliary inputs and independent volume controls. By plugging your mikes into the deck, pressing RECORD and PAUSE, twiddling the controls for the proper volume, and sending the deck's output to your VCR, you can mix audio sources like you could with your stereo amplifier. Some audiocassette decks will allow you to individually vary the volume of the two microphone inputs. Other decks will allow you to do this *plus* independently vary the volume of the two AUXILIARY inputs. With these decks you may be able to control four inputs at once.

The resulting signal will be stereo. If you want it to be monaural, you need to use a Y ADAPTER to combine the deck's stereo outputs into a single signal.

Stereo VCR. Some VCRs record in stereo. It is possible to send one audio source to the left channel and another source to the right channel and either record a stereo presentation or at least get two sources onto the tape at the same time. When playing back the tape, you adjust the volume of the two outputs to perform your sound mix. The

MINIGLOSSARY

Cartridge player Audio tape player (often a recorder as well) that accepts endless loop cartridges containing ¼-inch recording tape. The tape cartridge (cart) is slipped into the machine and is ready to play. After playing, the tape automatically recues itself.

Dub In audio, to replace an old sound track with a new one, leaving the video unchanged. In video, sometimes means to duplicate a tape. To keep things clear, use the term *audio dub* to indicate audio only.

disadvantage in this is that every time you play the tape, you have to adjust the controls manually for the right mix. Also, some decks won't allow you to adjust the volume of each channel; they only allow you to play channel 1 *or* channel 2 *or* an equal mix of both channels.

It would be nicer if the mix were made first and *then* recorded on the tape correctly so you wouldn't have to mess with the machine during playback. Still, if you're in the field with a stereo recorder and no mixer, you may be better off running one sound source into one channel and another sound source into the other so that you have the raw materials for making a better sound mix when you get back to the studio.

AUDIO TRACKS ON A VTR

All ¾U VCRs have two audio tracks. You can record on channel 1 (track 1) or channel 2 (track 2). Upon playback, you have the choice of listening to channel 1, channel 2, or a mixture of both by flipping the AUDIO MONITOR selector from CH-1 to CH-2 or to MIX.

Given a choice, which channel should you record your sound on? If your VCR can record on both at once, then record both channels. That way no one can flip the AUDIO MONITOR switch to the "wrong" channel; they're both good. Also, if you play back two equivalent tracks at once, you have a stronger, less noisy signal.

If recording on only one of the two channels, make channel 2 the main one. That's pretty much the "standard" channel used by everyone.

One-inch B- and C-format VTRs have *three* audio tracks, track 1 being the primary one. Stereo VHS VCRs also have two linear audio tracks with track 1 being the primary. The same is true for Betacam. Hi-fi VCRs (VHS, S-VHS, MII, Betacam-SP) have two additional audio tracks imbedded invisibly in the picture, making a total of four tracks. These, being part of the picture signal, cannot be AUDIO DUBBED after the picture is recorded; nor can the picture be edited without erasing the hi-fi sound.

Regular 8mm and Hi8 VCRs have three tracks: One monaural hi-fi track (20–20 kHz) imbedded in the picture, and two PCM (digital pulse code modulated) tracks with medium high fidelity (30–15 kHz). These PCM tracks can be erased and AUDIO DUBBED independently from the picture.

D2 VCRs record four digital hi-fi (20–20 kHz) audio tracks.

BASICS OF AUDIO TAPE RECORDING

Audio tape recording is a subject unto itself. Complex high-fidelity recordings often require an audio studio with a mil-lion dollars worth of equipment. Such studios may have tape recorders that can record 24 channels at a time. Such recorders allow each audio source to have a separate TRACK on the tape. These sources can be mixed freely for any desired effect. Sometimes these sound mixes are so complex that they require a computer to turn all the controls at once at the proper time.

Such studios also do OVERDUBBING. This is where a performer sings a song on one track and the tape is then rewound and that track is played through the performer's earphones. The performer simultaneously sings harmony with herself, and that harmony is recorded on another track. The tape may be again rewound, and the performer sings a third track.

Another OVERDUB technique is to record the orchestral accompaniment in one studio at one time and later bring in the "star" singer to sing on a separate track while listening to the accompaniment play on the first track. Thus, a large complex recording can be produced over a long period of time using people who have never met each other and who have never worked directly together.

OVERDUBBING also allows a track of "normal" sound to be played back and run through a number of audio devices which change the sound. The result is then recorded on a different audio track. This is called SWEETENING because the sound can sometimes be improved in the process. Commonly, echoes are added, while high or low tones may be boosted or weakened.

Large audio studios may be equipped with well-tuned grand pianos and room for an entire orchestra. They may have the expensive microphones necessary to record the sound in full fidelity. Some studios have DIGITAL RECORDING equipment capable of preserving the sound *exactly* the way it was recorded, adding no circuit noise or hiss or hum to the sound.

Back to the real world, every TV studio will probably have some audio recording equipment for producing audio backgrounds, narration, or fancy sound tracks to go with the video. Remembering the adage that "audio is half of video" reminds us that learning to mix and control sound creatively and effectively is definitely part of the TV process and should not be glossed over.

Reel-to-Reel Recorders

Although it may seem like an ancient format, reel-to-reel audio tape recorders are still preferred by professionals. The better models provide excellent sound fidelity and can record and/or play back two to four separate channels of sound simultaneously. They accept various HI-LEVEL and LO-LEVEL

MINIGLOSSARY

Audio monitor Device that allows you to listen to and check on the quality of a sound signal. Also the switch on a VCR that chooses which channel (or both) is fed to your headphones or in some cases to the VCR's audio output.

*Track A pathway along a tape set aside for a discrete (usually audio) signal.

audio sources and can control their volumes independently. The better ones have two sets of record/playback heads which allow you to listen to your tape recording played back a fraction of a second after it is made. Listening to your tape this way allows you to know for sure that your recording is coming out okay.

Reel-to-reel tape machines make editing easy. The tape travels quickly and can be manually "rocked" back and forth, allowing you to play just part of a sentence and sometimes even split words.

The tape is large (¼ inch) and easy to handle, making it possible to cut it with scissors to remove unwanted passages and to hitch the two ends together using SPLICING TAPE. Thus, the content of an audio recording can be shuffled around in any order, shortened (by cutting out parts), or lengthened (by adding parts from other tapes).

To make editing easier, professional reel-to-reel audio tape recorders have an opening next to the audio heads that allows you to see and touch the tape as it passes over the head. This is handy for when you want to precisely mark (with a grease pencil or felt pen) the beginning or end of a segment which you intend to later SPLICE out. Put another way, you listen to the tape, find exactly the spots where you wish to delete something, mark the tape right where the head is to show the part of the tape you wish to delete, and then remove the tape from the machine and splice out the offending parts.

Tape Speed.   Nearly all professional and semi-professional audio tape recorders move the tape at 7½ inches per second (ips) for recording or playback. This speed yields excellent fidelity and fairly good tape economy. At 7½ ips an 1800-foot reel of tape will last 48 minutes per side. 7½ ips is also a good speed for editing tape. The tape is moving fast enough to leave space between words.

Home and semiprofessional ATRs (audio tape recorders) may also have the speed 3¾ ips. This speed gives poorer fidelity (especially poorer high frequencies) but packs twice as much on a tape, for excellent tape economy. Editing is a bit difficult because sounds are packed closer together on the tape, making it harder to find where one ends and the next begins.

Professionals often prefer the fast tape speed of 15 ips. This speed offers excellent fidelity but uses up tape like crazy (24 minutes per side on an 1800-foot reel of tape).

The high tape speed spreads the sounds out further, making it even easier to edit.

Some ATRs have variable speed controls (sometimes called variable pitch). These controls allow you to record or play back a tape at up to 15% faster or slower than normal. This feature has two applications: If an instrument was tuned slightly high or low during a recording and now another instrument has to be played along with it, one can *tweak* the speed control up or down to raise or lower the pitch (frequency) of the first instrument. Thus, the two instruments can be made to match.

A second application for variable speed control involves timing. Say you were trying to produce an advertisement that had to last exactly 1 minute. No matter how hard you tried, every time you performed the ad, it ran 1 minute and 5 seconds. To get the ad to run exactly 1 minute, you could play this 1-minute-and-5-second ad faster so that it would be over right on time. No one is likely to notice the slight increase in speed.

The same technique can be used for matching music to action on the screen. If the screen action lasts 2 minutes and your best choice of music is 1 minute and 50 seconds long, you could simply slow down the music a few percent and have it finish exactly at the end of your action.

Reel Sizes.   Tape reels come in various sizes: 3, 5, 7, and 10½ inches. The professional ATRs generally use 10½-inch reels. This way they can record up to an hour even at 15 ips.

When using an ATR, it is good to use the same size reel both for TAKE-UP and SUPPLY. If you don't, the machine is unbalanced and has difficulty stopping both reels at the same time after winding or rewinding. You run the risk of spilling or stretching your tape. Reels of the same size work better together.

Making Tracks

Common reel-to-reel audio tape can be recorded as ONE TRACK, TWO TRACKS, or FOUR TRACKS, as diagramed in Figure 10–28.

If you record the entire width of the tape at once, you get better fidelity. It is also easy to slice the tape with scissors (or a razor blade) and cut out segments. ONE-TRACK recording is preferred by professionals.

MINIGLOSSARY

*Overdub   Recording sound on one audio track and then recording a related sound on another track. Each track may have its unique sound recorded or may be a mixture made from the playback of an already-recorded track plus the new sound.

Sweetening   Manipulation of recorded sound to give it echo, filter out a noise, boost a particular frequency, or mix it with other sounds.

*Splicing tape   In audio, special adhesive tape used to attach the ends of audio tape together for continuous playback. A "splice"

is a physical "cut" in a tape followed by attaching the tape to another tape.

*ATR   Audio tape recorder.

Take-up reel   On a reel-to-reel recorder, this is the empty reel that fills as tape is played.

Supply reel   On a reel-to-reel tape machine, this reel contains the program you wish to play or the blank tape you wish to record.

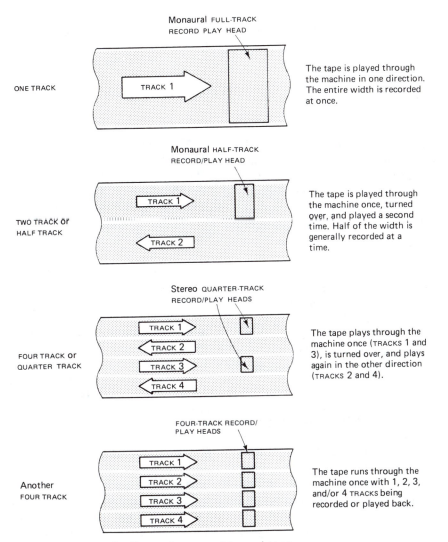

ONE TRACK
Monaural FULL-TRACK
RECORD PLAY HEAD
TRACK 1

The tape is played through the machine in one direction. The entire width is recorded at once.

TWO TRACK or HALF TRACK
Monaural HALF-TRACK
RECORD/PLAY HEAD
TRACK 1
TRACK 2

The tape is played through the machine once, turned over, and played a second time. Half of the width is generally recorded at a time.

FOUR TRACK or QUARTER TRACK
Stereo QUARTER-TRACK
RECORD/PLAY HEADS
TRACK 1
TRACK 2
TRACK 3
TRACK 4

The tape plays through the machine once (TRACKS 1 and 3), is turned over, and plays again in the other direction (TRACKS 2 and 4).

Another FOUR TRACK
FOUR-TRACK RECORD/ PLAY HEADS
TRACK 1
TRACK 2
TRACK 3
TRACK 4

The tape runs through the machine once with 1, 2, 3, and/or 4 TRACKS being recorded or played back.

FIGURE 10-28 Audio tracks on reel-to-reel tape.

TWO-TRACK recordings were most popular some years ago. The ATR would record one TRACK over half the width of the tape. When the tape ran out, the full and empty reels swapped places, and the tape was run through the machine a second time. Automatically, the second half of the tape's width would now be recorded. One problem with this system is that if you physically cut the tape to delete something, you could be cutting two TRACKS at once, accidentally deleting something from track 2. If you plan to record with a TWO-TRACK machine and will edit by splicing, pass the tape through the machine only once; do not record on TRACK 2.

A FOUR-TRACK tape has four paths recorded down the length of the tape. Two paths are recorded in one direction and two in the other. This setup is commonly used in stereo recording where TRACK 1 is the left channel and TRACK 3 is the right channel. When the tape comes to an end, it can be turned over and played through the machine a second time to pick up TRACKS 2 and 4. In short, two TRACKS are

recorded in one direction with a space between them, and two more TRACKS are recorded in the other direction, filling those spaces.

Incidentally, there is always a tiny gap between the TRACKS that never gets recorded. This gap helps to keep the audio head for one TRACK from "hearing" the sound from the adjacent TRACK.

Four-channel ATRs can record one, two, three, or four TRACKS at once. It can record them all in the same direction or can record two in one direction and two others in the other, as described.

By crowding four TRACKS onto a single tape, the sound fidelity decreases somewhat but not enough to affect the common ¾U industrial recording. Only audio gourmets will hear the difference.

As with the TWO-TRACK tape, the FOUR-TRACK tape cannot be edited by splicing. Cutting the tape would cut all the TRACKS at once. This *might* not be a problem with a

FOUR-TRACK recording where all four TRACKS contained exactly the same program. In such a case, the FOUR TRACKS are much like a ONE-TRACK recording.

Channels Versus Tracks

CHANNELS are not the same as TRACKS. You can have a TWO-CHANNEL ATR making a FOUR-TRACK tape. The number of CHANNELS tells you how many independent signals the ATR can record on the tape *at the same time*. The common stereo ATR can record TWO CHANNELS on TRACK 1 and TRACK 3 while the tape is moving in one direction. When the tape is turned over, CHANNEL 1 and CHANNEL 2 will automatically be recorded on TRACK 2 and TRACK 4. Put another way, the audio head for one CHANNEL can record one-fourth of the width of the tape; the audio head for the SECOND CHANNEL records another one-fourth of the tape's width. Turning the tape over places two more quarters of the tape's width in contact with the two heads.

A FOUR-CHANNEL ATR can record one, two, three, and/or four tracks at once. You can also record TRACK 1, rewind the tape, and then record TRACK 2, and rewind the tape and then record TRACK 3, and then rewind the tape and finally record TRACK 4 using a FOUR-CHANNEL machine.

Overdubbing

Say you had recorded scenes of a jeep traveling through the rugged Canyonlands National Park. You wanted to edit these scenes so that the viewers could watch the jeep bouncing over the rugged roads while hearing the sound of the jeep in the background, mixed with some music, mixed with the sound of the narrator telling where the viewers were going.

First, you would need to record the sound of the jeep driving over rough terrain. You might do this while recording your video, or you might do this using a good ATR and good microphone, strategically placed. If the sounds of the jeep hitting water puddles, bumps, and bridges had to synchronize exactly with the picture, you would probably want to use the audio track from the tape. If, as is true in most cases, the sounds don't have to precisely match the picture, you can audio-record them separately using more care for better fidelity. Either way, you arrive home with a video tape or an audio tape with the jeep sounds on it. These sounds may be DUBBED (copied) onto TRACK 1 of an audio tape.

Next, we choose the music and record it on TRACK 2 of the audio tape. Unless the music has to be timed precisely to the action, you can simply record the entire selection on TRACK 2 and just use the parts you want later.

On TRACK 3, you then record your narration. Since each TRACK is separate from the others, if your narrator makes a mistake, all you have to do is go back and rerecord his TRACK. The work you did on the other TRACKS is still safe.

Now that we've used three TRACKS, we have a choice: We can play them all and vary their volumes (from off to maximum) while sending the signal to the video tape recorder to be DUBBED IN parallel with the picture. Or we could mix these three TRACKS and record the result on the FOURTH TRACK. The procedure for both cases is fairly similar.

Picture it working something like this: We see the desert without the jeep. The music (TRACK 2) fades up and runs at full volume for a few seconds. The sound of the jeep (TRACK 1) slowly fades up, and we see the jeep come into the picture. It is almost the time for the narrator to start speaking, so we fade up TRACK 3 in anticipation. Just before the narrator speaks, we fade the music (TRACK 2) down to one-fourth volume and simultaneously fade down the jeep sounds (TRACK 1) to half volume. The narrator's volume is already up to full strength when he speaks. During long pauses in narration, the music and jeep sounds can be brought up higher in volume. To end the segment, the narrator is silent (and the volume for TRACK 3 may be turned off), and as we watch the jeep drive into the sunset, the music fades up to full volume and the jeep sounds slowly fade out. When the picture finally fades out, the music is faded out at the same time. If you're really lucky or if you have timed the music exactly right, you may reach some crescendo or the end of the song at this time. Timing this is hard, but the effect is quite stirring.

Cassette Recorders

Audiocassette recorders (ACRs) give poorer fidelity than reel-to-reel recorders. Editing on an ACR is quite difficult. Except to salvage a broken tape, an audiocassette tape cannot be SPLICED; it is simply too fragile. The tape moves so slowly (1⅞ ips) that it is hard to stop it between words or to CUE it up at a precise point.

Except for specialized models, ACRs generally have TWO CHANNELS and can record FOUR TRACKS on a cassette. TWO TRACKS are recorded in one direction, and then the cassette is turned over (to side 2 or side B) and the next two TRACKS are recorded. Thus, there are FOUR TRACKS on a cassette, but only TWO TRACKS can be played at a time. Figure 10–29 diagrams the TRACKS on an audiocassette.

Some simpler and cheaper audiocassette decks which are monaural (not stereo) have but one CHANNEL and can record only one TRACK at a time. This TRACK covers half the width of the tape.

A FOUR-TRACK tape recorded on a stereo machine will play back fine on a monaural TWO-TRACK machine. You will hear everything; it just won't be stereo. Similarly, a tape recorded on a TWO-TRACK monaural ACR will play back fine on a stereo ACR; you just won't hear stereo.

For best fidelity, if you are making a monaural recording on a stereo recorder, use a Y ADAPTER or some other mechanism to record your sound on both CHANNELS si-

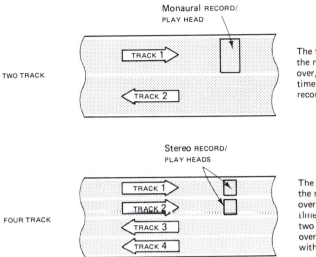

Monaural RECORD/
PLAY HEAD

TWO TRACK

TRACK 1

TRACK 2

The tape is played through
the machine once, turned
over, and played a second
time. Half of the tape is
recorded at a time.

Stereo RECORD/
PLAY HEADS

FOUR TRACK

TRACK 1

TRACK 2

TRACK 3

TRACK 4

The tape is played through
the machine once, turned
over, and played a second
time. Two CHANNELS record
two TRACKS (stereo, generally)
over half the tape's width
with each pass.

FIGURE 10-29 Tracks on an AUDIOCASSETTE tape.

multaneously. Similarly, when playing a monaural tape from a stereo ATR, mix the two channels together (perhaps using a Y ADAPTER) to get the best fidelity. The law here is as follows: *Two signals are stronger and "cleaner" than one.*

Audiocassettes. Most audiocassette tape comes in three types: NORMAL, CHROME, and METAL. NORMAL tape (FeO$_2$) is least expensive and gives satisfactory frequency response. CHROME (CrO$_2$) tape gives better frequency response and is more expensive. METAL tape gives the best frequency response and of course is most expensive. Any of the tapes will play on any ACR, but the better ACRs have a switch which optimizes the machine to play that particular tape. When recording, set the switch to the position which matches the kind of tape that you are using in order to take advantage of the full potential of the tape.

Some tapes absorb high frequencies better than others. To match the machines to the tapes, some ACRs have a switch called EQUALIZATION. "Normal" tape uses a 120μs (microsecond) EQUALIZATION. While recording with this tape, you would throw the EQUALIZATION or (EQ) switch to the NORM position. When recording on CHROME tape (which has an equalization of 70μs), you would throw the switch to the CHROME or CrO$_2$ position. For best results, you would set the EQUALIZATION switch to the appropriate position when playing tapes as well.

Preventing Accidental Erasures. All audiocassettes have tabs in the rear which can be broken off, leaving a small hole. This hole tells the ACR not to try to record (and erase) on the cassette. Popping off that tab protects your audiocassette from accidental erasure, just like videocassettes can be protected.

The tab to protect side A of a cassette is on the left rear as you look down on the side A cassette label with the tape facing toward you. If you turn the cassette over so side B is up with the tape toward you, the corresponding tab will again be in the left rear of the audiocassette.

To render the cassette recordable again, stick some scotch tape over the hole where the tab was.

Cartridge Recorders

Before recording on a CARTRIDGE recorder, you must first select a CARTRIDGE of the proper length. For a single sound effect, a 5-second CART would probably do. For a musical piece to begin or end a tape, perhaps 15 seconds would be fine. In some cases, longer CARTS are appropriate.

It's not hard to imagine reasons why you want to have a CART long enough, but there are some reasons why you don't want the CART to be *too long*. If your sound effect, for instance, lasted only a few seconds and your CART was a minute long, then after playing your sound effect, the CART would have to recycle for a full minute before it came to your sound effect again. If you wanted to use your sound effect twice in the same minute, you couldn't. You want your sound effect to play and then get itself ready to play again as soon as possible, to be really useful. Therefore, always select a CART just a *little* longer than the passage you wish to record.

To make your recording, place the CART in the machine, play your source, and check the RECORD LEVEL on the CART machine. Once you have the right volume for the recording, then carefully CUE UP your sound effect or music. Start your source playing at exactly the same time you start the CART machine recording. This way the sound effect will appear at the very beginning of the tape. When the sound effect is over, turn down the source's volume or the CART's record volume so that it records only silence for the remainder of the tape. Let the CART continue until it stops itself.

When using the CART, simply press the PLAY button,

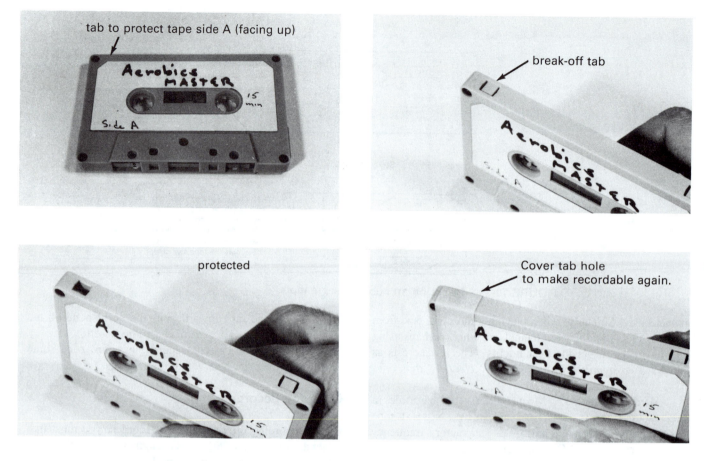

Protecting an audiocassette from accidental erasure.

and your sound effect will play. Do not stop the CART machine when the sound effect is over, or the loop will stop in the middle, and the next time you use the CART, your sound effect won't be ready. Let the CART play to the end, and it will stop itself and be CUED for its next use. This doesn't mean that you can't fade out the sound while the CART is playing. You can fade the sound up or down as you please. What you shouldn't do is stop the CART before its cycle is finished.

RECORDING STEREO

Few educational and industrial television programs today are produced in stereo. Some of the commercial TV stations are beginning to broadcast in stereo, and a few TVs and VCRs are being made which can pick up these broadcasts and play them to you in stereo. Many home VCRs are able to record or play back in stereo.

Although the following discussion will focus on stereo for video recording, the rules for stereo audio recording are about the same. So how *does* one record in stereo?

With Mikes

With microphones it's easy. Simply plug one mike into the left MIC IN and another into the right MIC IN and shoot. Whatever is picked up by the left mike is recorded on the left channel, and whatever goes in the right mike is recorded on the right channel. Figure 10–18 diagramed the process.

Although some VCRs adjust the recording volume automatically, others have two manual controls: one for the left volume and one for the right volume. Set both controls so that the volumes are fairly equal and rarely exceed 0 VU on the meter scale.

From Records and Tapes

To record stereo sound from a record or tape player, simply connect *two* PATCH CORDS from the player's left and right

MINIGLOSSARY

*Patch cord   A short cable that connects the output of one device to the input of another.

MTS   Multichannel television sound, a technique of broadcasting stereo audio on TV.

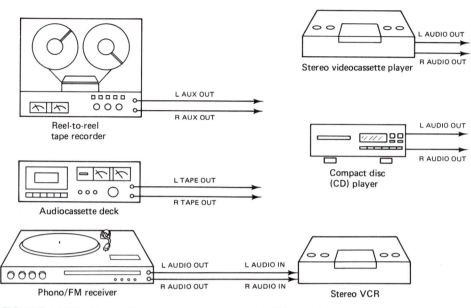

FIGURE 10-30 Recording stereo from records, CDs, and tapes.

outputs to the VCR's left and right audio inputs. Play the record or tape, hit DUB or RECORD on the VCR, and the electric genies do the rest. Figure 10–30 diagrams the hookup.

Off-Air Simulcasts

Although opera, classical, and rock music are more enjoyable in stereo, your TV (in some cases) can't give it to you that way. The only way many people can hear stereo TV is to tune in a SIMULCAST, the simultaneous broadcasting of a TV show over a TV channel and the show's sound in stereo over an FM station. The viewer tunes in the TV station, turns down the TV set's volume, tunes in the corresponding FM station, and listens to that. (You may find it amusing to tune the "wrong" FM station and note how the TV picture and sound almost make sense anyway—try it!)

Some stereo VCRs will record SIMULCASTS but need the help of an FM tuner for the stereo sound. As shown in Figure 10–31, you split your antenna signal and send it to two places: your VCR (through a cable TV converter box, if necessary) and your FM tuner or receiver. The VCR records the picture while your FM receiver tunes in the sound, sending that to your VCR in stereo.

Stereo Broadcasts

Stereo television broadcasts use a system called MTS, which stands for multichannel television sound. It is a system where older monaural TVs and VCRs will still pick up regular television sound; TVs and VCRs equipped with *special tuners* will be able to decode the MTS signal to give you stereo. *Note:* Having a stereo VCR or a TV set with two speakers will not squeeze stereo out of a stereo broadcast; you must have a *tuner* designed to pick up MTS signals. It *is* possible, however, to buy a separate MTS tuner which can make the stereo audio signals for you to feed to your VCR or elsewhere.

MTS is more than just stereo. It contains a third channel (strangely named SAP, for supplementary audio program). It is a monaural channel which the broadcasters can use to send private messages from station to station. SAP can also be useful for sending an additional sound track to homes. For instance, while watching an Italian opera in stereo, folks could switch their TVs to the SAP mode and listen to the English translation.

Some cable TV stations aren't equipped to relay MTS sound. The cable TV sound comes out mono even though the original broadcast was stereo.

SPECIAL AUDIO DEVICES

Noise Reduction Systems

"Noise" is anything you hear in your recording that doesn't belong there. The sounds of cars on the street and jets flying overhead are "noise." We try to get rid of this noise by using soundproof rooms and strategically placed DIREC-

MINIGLOSSARY

SAP Supplementary audio program—a third channel of sound broadcast using the MTS system.

*Noise Unwanted interference that creeps into your signal. Au-

dio noise could be hum or hiss. Video noise could be snow, graininess, or streaks in the picture.

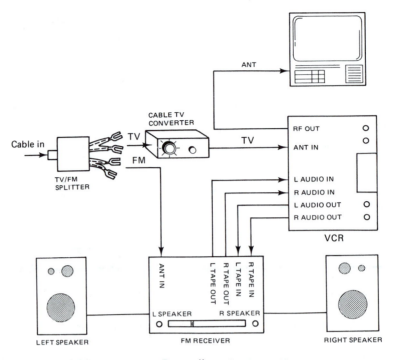

FIGURE 10-31 Recording stereo SIMULCASTS.

TIONAL microphones. Other noises creep into our sound during the recording process. This could be hum or hiss from our wires and circuits. It could be distortion and more hiss from our tape. Nothing is perfect; everything adds a little noise to the signal that passes through it. NOISE RE-DUCTION systems are circuits which endeavor to amplify the desired sound while leaving the noise behind.

Dolby. DOLBY (trademark of Dolby Laboratories) comes in three versions: DOLBY A, DOLBY B, and DOLBY C. DOLBY A and DOLBY C are used on professional equipment, while DOLBY B is used in home audio and video recorders.

DOLBY is one of several types of NOISE REDUCTION systems to diminish background hum, hiss, and noises in recordings. By background noise, I don't mean the kind of sound that gets recorded as you interview someone on a busy street corner. DOLBY won't reduce the unwanted sounds that enter your microphone because of noisy surroundings. What it *will* reduce is the hum and hiss created *internally* by imperfections in your tape and recording equipment.

It would be nice if the sounds you *wanted* to hear were 100 times stronger than the noise. Often this is the case, but sometimes when your music or speech or whatever

is very quiet, maybe 1/100th of its normal volume, the background noise is now as loud as your music or speech. Thus, in quiet passages, the noise becomes quite noticeable.

DOLBY electronically raises the volume of the quiet musical passages during the recording in an attempt to keep the music many times louder than the noise. To make everything come out right, DOLBYIZED recordings must be *played back* through a DOLBY system to de-emphasize the boosting which occurred in the recording process. The result: Quiet and loud passages sound normal through the DOLBY system while background noise is markedly reduced.

DBX. Another system of noise reduction is DBX. Like DOLBY, it reduces the background noise during quiet passages and thus reduces audible tape hiss.

Also like DOLBY, DBX has to be ENCODED (DBX modifies the sound) during recording. During playback, the sound is DECODED (the sound is put back to normal) before you hear it.

For both DOLBY and DBX, the system doesn't help if you don't *both* ENCODE and DECODE. ENCODED tapes sound harsh when played back on normal players. Normal tapes sound muffled when played back through DECODERS.

MINIGLOSSARY

Noise reduction system Electronic device that attempts to reduce electronic noise when something gets recorded and/or played back.

DBX A scheme for reducing audio noise in recordings by encoding and decoding. Effect is more pronounced than with Dolby.

Encode Modification or processing of a signal while it is being

recorded, usually to make it less "noisy" during playback when the signal is decoded.

Decode Reprocessing of a signal to extract the desired part. In audio, a signal is encoded on recording; on playback is decoded so that it sounds normal, but noise is reduced.

Some audiophiles take great pride in the silence of their systems during quiet passages.

Graphic Equalizer

Sound is made up of high, medium, and low tones. Bass and treble controls can boost or diminish these tones. Sometimes it would be nice to boost or diminish *one particular tone*, not messing up the rest of the highs and lows in the process.

A GRAPHIC EQUALIZER (Figure 10–32) does this. It contains selective filters that allow you to pick a sound frequency that you don't like and remove it, while passing the rest of the audio untouched. Hum from a poorly grounded system can be diminished by filtering out the 60-Hz frequency. The rumble of wind can be diminished by filtering out the 30-Hz frequency. The high-pitched whine of an electric motor may require removal of the 10,000-Hz frequency. A dusty record may need removal of the 15,000-Hz frequency.

Audio Patch Bay

Figure 10–33 shows an audio PATCH BAY. It is like a telephone operator's switchboard that allows any phone to be connected to any other phone in the building. What the operator's switchboard did for phones, the AUDIO PATCH BAY does for audio signals. Without a PATCH BAY, if you wanted the CART player to go where the cassette player now goes, you'd have to dig around the spaghetti of wires behind the equipment to make this change. If the CART player had a plug different from the cassette player's, you'd have to find an adapter to make the plug fit in the socket. You may even have to remove the equipment from the console just to *see* the plugs.

What a drag! Solution: Send all the inputs and outputs for *everything* to a place called a PATCH BAY. Here, like the telephone operator, you can connect any device to any other device externally, simply, and with a standardized plug.

Just because the PATCH BAY has lots of plugs and sockets, it doesn't mean that you can connect *anything* to *anything*. If you PATCHED microphone 3 to the place where the audiocassette player went, you'd probably get no sound. The audiocassette player was sending a HIGH-LEVEL signal to a HIGH-LEVEL input. When you plugged the LOW-LEVEL

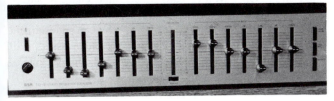

FIGURE 10-32 GRAPHIC EQUALIZER.

FIGURE 10-33 Audio PATCH BAY.

MINIGLOSSARY

*Graphic equalizer Electronic audio device that cuts or boosts particular sound frequencies passing through it.

*Patch bay In audio, several rows of sockets connected to the inputs and outputs of various audio devices. Plugging a patch cable into a pair of sockets connects them so that the signal can travel from one device to the other.

microphone into the HIGH-LEVEL input via the PATCH BAY, the mike didn't have enough oomph to register a signal. In short, only compatible outputs can be PATCHED to inputs if you want sweet sound.

Many PATCH BAYS are NORMAL-THROUGH. This means that if you don't plug anything into the PATCH BAY, whatever is available at the top socket automatically goes to the bottom socket. For instance, if the top socket is labeled VCR OUT and the socket directly beneath it is labeled MIXER 4 IN, then the VCR is connected to that mixer input even though nothing is plugged into the sockets. As soon as you plug a PATCH CORD into *either* of those two sockets, that connection is undone. The socket then gets its signal from wherever the PATCH CORD goes. Disconnecting the PATCH CORD re-establishes the NORMALED circuit.

Limiter

A LIMITER is an electronic device which limits the loudness of an audio signal. It's a box which connects between your mixer and your VCR or between a source and your mixer. It "listens" for loud sounds, and when it hears one, it knocks the recording volume down to keep it from distorting.

Limiters aren't the same as AGC (automatic gain control) found on some VCRs. AGC turns up the volume if it gets too low and turns it down when it's too loud. LIMITERS do half this job. When the sound is loud, they knock it down. When the sound is medium, they do nothing. When the sound is weak, they do nothing. Thus, they don't mess up soft and medium passages of music; nor do they wrongfully raise up the volume during quiet pauses. What they do is protect against loud sounds—sort of "distortion insurance."

This insurance allows you to record a tad louder than you normally would and results in better fidelity and reduced background noise.

PRERECORDED MUSIC AND EFFECTS

There's nothing like a little wind and thunder to add spice to a video show. And some music is a nice way to start off a program or wrap it up. Where do these sounds come from?

Music and Sound Effects Libraries

Nearly all music and sound effects heard in industrial video programs come from music and sound effects *libraries*, sets

of record discs or tapes containing everything from a cuckoo clock to a steamship whistle or from a dripping roof to a raging waterfall. Music is also widely varied. It may be organized by length, instrument type, or style. Styles could befit the stately entrance of a king, the silent stalk of a slithering snake, or the gaiety of a camping trip.

Over 1000 albums are available from Thomas Valentino (New York), Soper Sound (Palo Alto), De Wolfe (New York), and Capital Productions (Hollywood), and others stock music and sound effects producers. Usually, you pay a flat fee for the library of records, and you are free to use them any way you wish. There are others who charge a NEEDLE DROP fee; you pay the company each time you *use* one of its sound effects.

Popular Music

It is also possible to use commercially available prerecorded music from a record album or recorded from the radio. Such music is easily accessible and recognizable. There's just one problem: copyright clearance.

Prerecorded music is generally copyrighted. Someone else owns the right to copy, perform (play in public), and make money from it. Radio stations, record producers, sheet music distributors, and filmmakers all pay royalties to the copyright holder (and sometimes to the performers, too) for the right to distribute the work. You, too, need permission to use someone else's music in your TV production. In most cases, this permission will cost you money—a lot of money if the music is very popular.

Before committing yourself to a popular music selection, find out who produced it (the name will be on the record label, or you could look the data up in music-listing catalogues). Send them a letter telling what song you wish to use, how much of the song will be used, how it will be used in the production, who the audience will be, and how the program will be distributed. The record company (or copyright holder) will either say you may not use the music, may use it for free, or may use it for a fee.

The temptation to use "uncleared" music in a production is sometimes great. It's so easy, and "Who will catch me?" you're thinking. If your TV production is worthy of wide distribution, the chances of being caught are fairly high. Once you're caught, criminal charges can be brought—fines, lawsuits for damages, even jail. What is most likely to happen (unless you were a blatant and frequent infringer) is that your employer will be embarrassed, will settle the damage claim with the copyright holder out of

---

MINIGLOSSARY

*Normal-through or normaled**   In a patch bay, the top socket is automatically connected to the socket directly beneath when no patch cable is plugged into either. The signal "normally" travels through from one to the other.

*Limiter**   Electronic audio device that automatically reduces the

volume of loud audio signals but doesn't change the normal or weak signals.

Needle drop fee   A charge made by record libraries for each sound effect used in a production.

court, and then will turn you into a speed bump in the company parking lot.

If you are unclear as to who owns the copyright to the music you want, try this: Check the record label for the letters ASCAP, BMI, or SESAC. These are three large licensing agencies that act on behalf of many copyright holders. Their addresses follow:

ASCAP (American Society of Composers
    and Performers)
General Licensing
One Lincoln Plaza
New York, NY 10023
Tel.: (212) 870-5756

BMI (Broadcast Music, Inc.)
Vice President of Licensing
320 West 57th Street
New York, NY 10019
Tel.: (212) 586-2000

SESAC (Society of European Stage Actors
    and Composers)
Vice President of Licensing
10 Columbus Circle
New York, NY 10019
Tel.: (212) 586-3450

Commercially available music which will be used as part of a TV production must also have a SYNCHRONIZATION CLEARANCE. This is the right to mate someone's music to your TV show, i.e., an introduction or background. For SESAC music you can obtain SYNC rights directly from SESAC. For all the others, you contact

The Harry Fox Agency
Synchronization Licensing
110 East 59th Street
New York, NY 10022
Tel.: (212) 751-1930

They'll send you a standardized form asking how the music will be used and will calculate the fee.

Home Brew

Your third option is to perform your own music. You still have to pay a copyright fee if you're performing someone else's song, but if you make up your own, it's yours to use as you please. Original music tracks are fun and challenging, especially if you have talented friends willing to ply their skills for free. Often, all that is needed is a drum rhythm or the repetition of a couple chords—nothing fancy. Something nonmelodic and unobtrusive is best. Inexpensive electronic synthesizers are a source for many varied and creative musical and sound effects.

Music Selection

When selecting music, the most important consideration is how fast or slow it moves. Tempo will create a mood or reinforce the nature of the visuals. A fast tempo would support a montage of quick scenes previewing or reviewing a program. Something slower would go better with scenes of a thief sneaking through a building but would speed up upon his/her hasty retreat.

If scoring the music behind a voice, try to match the tempo of the music to the pace of the voice. Slow, serious, reflective narration deserves slow, serious music. Excited happy narration needs bright, up-tempo music.

Avoid recognizable music or music with rhythmic or melodic elements that draws attention to itself. It may distract the audience and interfere with their ability to concentrate on the narrator's message. This is especially true in educational or industrial TV programs.

The musical selection could be short to emphasize a point (*Ta da*, the orchestra rejoices as our hero conquers the mountain peak) or could linger in the background to set a psychological attitude. A rich string sound conveys class and luxury. Electric guitars and keyboards conjure up pop/rock visions associated with youth. Banjos and fiddles promote a "down-home" or southern country feeling.

Match your music to your audience. Do you think the board of directors will appreciate their yearly sales figure supported by a Nashville pickin' fanfare? Would your exercise class enjoy a stately and conservative orchestral introduction?

COMMON AUDIO AILMENTS AND CURES

Troubleshooting audio difficulties is like prospecting for gold nuggets. Putting them in your pocket is easy—it's finding them that is hard. Similarly, finding the source of an audio defect is most of the battle. Once the offending machine or cable is located, it is a simple task to check the obvious connections and switches and then if the device appears defective to find a substitute and to send the dud off for repairs.

MINIGLOSSARY

ASCAP American Society of Composers and Performers—an agency that licenses the use of copyrighted music.
BMI Broadcast Music, Inc.—an agency that licenses the use of copyrighted music.
SESAC Society of European Stage Actors and Composers—an agency that licenses the use of copyrighted music.
Synchronization clearance Permission acquired from music copyright holders to use parts of their music to go along with parts of your TV show.

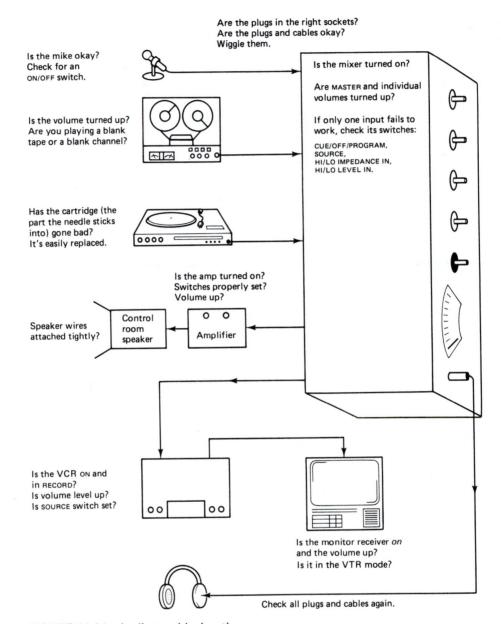

FIGURE 10-34 Audio troubleshooting.

As you prospect for audio problems, keep in mind where the signal is going and where it may be interrupted. Figure 10–34 shows the progression of several audio signals through a system. Sound can be ruined at its source, in the cables, at the mixer, in the mixer's cables either to the control room monitor or to the VCR, at the VCR, in the cables from the VCR to it monitor/receiver, or in the VCR monitor/receiver.

One good place to start prospecting is at the mixer. If the mixer's meter doesn't wiggle and if the control room speaker and mixer headphones make no sound, the problem is *not* likely to be in the VCR or its monitors. The problem is either in the source or in the mixer. To find out which it is, try two sources into the mixer. If something plugged into one of the mixer's inputs gives no sound while a second

source plugged into another input does, the problem is most likely in the first source. It is also possible that a switch or knob controlling the first source's mixer input is misadjusted.

If the mixer's meter does wiggle and if there's a signal over the headphones and control room speaker, you know the source being tested *and* the mixer are okay; your problem, if you are having one, is downstream from the mixer.

Downstream from the mixer is the mixer-to-VCR cable, the VCR, the VCR-to-monitor/receiver cable, and the TV monitor/receiver. If there's no signal on any of these devices, suspect the mixer-to-VCR cable or the VCR.

To test the VCR, plug an audio source directly into the VCR. If its audio meter now wiggles, the VCR is working properly, and your problem was in your mixer-to-VCR cable. If the audio meter does not wiggle, the VCR is either malfunctioning, or its RECORD button isn't pushed, or its controls are misadjusted.

If the VCR's meter wiggles but the VCR monitor/receiver emits no sound, suspect the monitor/receiver or its cables. If substituting another cable between the VCR and its monitor/receiver doesn't solve the problem, suspect the monitor/receiver.

Table 10–6 summarizes the steps taken in prospecting for lost audio.

Here are some specific problems and their solutions in detail:

1. *Mixer passes a signal from all but one microphone*: The problem is in that microphone, cable, plug, or mixer input. Does the microphone have an ON/OFF switch that's off? Is it an ELECTRET CONDENSER mike with a dead battery? Is the mike plugged into the right mixer input? Do the wires look damaged near the mike, or does its plug look like the one in Figure 10–35? Are the mixer's switches properly set? Does the mixer's IMPEDANCE match that of the mike?

TABLE 10-6
Troubleshooting flow chart

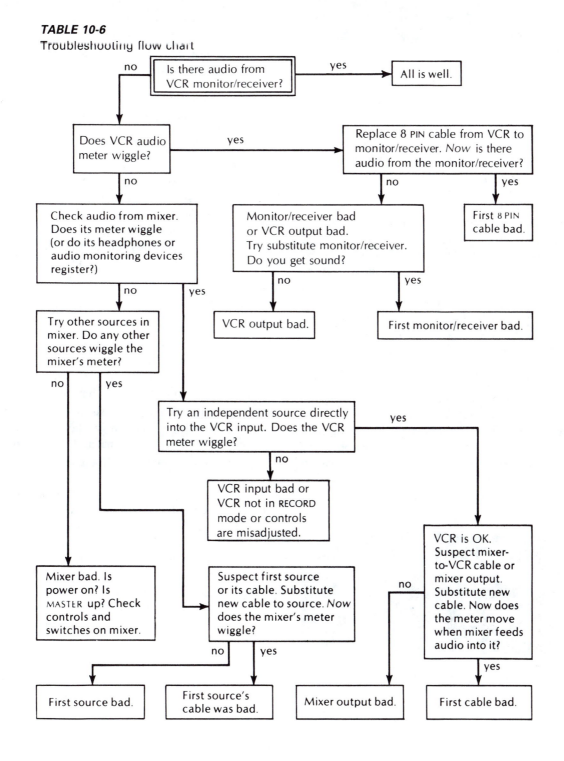

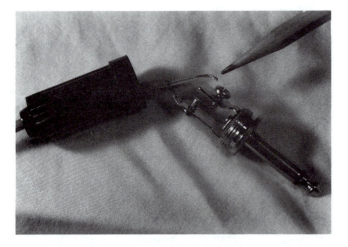

FIGURE 10-35 The handle screws off many plugs so you can see (or fix) the wires.

Try another mike in the same input. If the problem persists, check the mixer once again because that mixer input is probably at fault. If the problem stops, the first mike or its cable is defective.

2. *Mixer passes a good signal from all sources but one, which is very low in volume*: Turn up individual source volume. Check to see if a switch for that input is on HIGH LEVEL when it should be on LOW LEVEL. Try switching the IMPEDANCE switch on the mixer. If it doesn't have a switch, check the microphone to see if *it* has an IMPEDANCE switch and then try switching it. Check the source for loose wires in the cable or plug, wiggling them to see if it makes a difference. Does the source have a volume control of its own? If so, turn it up. Substitute a duplicate source; if it works, your first source has a defect.

3. *Mixer passes a good signal from all sources but one, which is loud and raspy*: Could it be that you are using a HIGH-LEVEL output from a source going into a LOW-LEVEL or MIC INPUT on the mixer? Perhaps your source's signal is too strong for the input. Are you using an earphone or speaker output as your source? That may even overburden a HIGH-LEVEL or LINE or AUX input. If this is the case, put a PAD or ATTENUATOR in the line.

DIGITAL AUDIO TAPE recorder. (Courtesy Technics)

4. *Mixer passes a good signal from all sources except that one has a lot of hum, buzzing, and hiss and sometimes may also have a weak signal*: Hum, buzzing (which is just loud hum), hiss, and a weak signal are usually wiring problems. The troublesome input may be wired with:

a.   A loose connection where the plug plugs into the socket. Disconnect it and plug it in again. Is it in *all* the way?

b.   A broken wire in the cable. Replace the cable.

c.   The wrong kind of audio cable (unshielded). Replace with a shielded cable.

d.   A disconnected wire in the plug itself. (If you unscrew the plug handle, you can see that each wire should be soldered tightly to its corresponding pin in the plug. No wire should have a loose or broken strand touching another wire or another part of the plug as you saw in Figure 10–35.) Resolder the wires or replace the cable and the plug.

e.   An UNBALANCED mike going into a BALANCED line. This should rarely happen because the plugs and sockets are different and don't mate together. But maybe someone found a way to make a naive adaptation which got the plugs to interconnect while the wires still didn't go to the right places. Refer this to a technician if you find a mike with an UNBALANCED line (with a PHONE plug at the end of its cord, for example) connected somehow to a BALANCED input (a female XLR socket—the kind with three holes).

f.   A BALANCED mike going into an UNBALANCED input without *properly* wired ADAPTERS. This can cause the same problem as in (e). Refer the suspect ADAPTER to a technician.

g.   An IMPEDANCE mismatch. Try switching the IMPEDANCE switch on the mixer. The source's IMPEDANCE must match that of the mixer.

h.   A level mismatch. If a mike or other LOW-LEVEL source is plugged into a HIGH-LEVEL input and the volume is turned all the way up, then hiss, hum, and a weak signal will be the result.

i.   Proximity to noise sources. Sometimes if the wire or the microphone is next to an electric device, the electronic field from the device is radiated into your sound system. Keep at least 2 feet away from fluorescent lamp fixtures, electric motors, light dimmers, high-power electric current cables, amplifiers, power supplies, TV sets, and things that use electricity.

If you encounter hum, hiss, buzz, or no sound at all, the quickest remedy is to find out if the culprit is the mike and its cable, the mike extension cable, or the mixer input. Take the following steps to find which is to blame:

a. First disconnect the old mike (and extension cable) from the mixer and substitute a similar mike. If this one doesn't work, suspect the mixer input or associated mixer controls. On the other hand, if this one works, the fault lies with the mike or its extension cable. On to step b.

b. Reconnect the extension to the mixer and plug the substitute mike into the extension. If this works, suspect the first mike or its cable or plug. If this doesn't work, suspect the extension cable. To confirm the extension cable faulty, do step c.

c. Disconnect the extension cable from the mixer and connect the first mike in its place. If the mike works now, the mike is good, and the extension is bad. Off to the repair shop with the extension.

If the problem doesn't seem to be wiring, then it is likely to be the result of a sensitivity mismatch somewhat like that described in step h above. If the source puts out a terribly weak signal and the mixer must be turned *way* up to make it audible, then perhaps the source is just too weak for this kind of mixer. It may need a boost from a preamplifier (turntables sometimes need this). Is it possible that the source needs power to make a hidden preamplifier in it work? Look for an AC plug. If it *does* have a stray plug, the device when plugged in may even make enough signal to power a HIGH-LEVEL input.

Inexpensive preamplifiers, when used on some equipment, may cause hum or hiss. If turning them off (and waiting half a minute for them to "cool off") makes the noise go away, the noise is probably the preamp's fault. If the noise stays, it may be caused by other wiring to or from the preamp.

A weak hum with adequate signal strength is often caused by correct but inadequate wiring between the source and the mixer (the technicians call this a "ground loop" or "floating ground"). To check out this possibility, take a wire, touch it firmly to the *bare metal* chassis of the mixer, and simultaneously touch the other end firmly to the chassis of the source (or the metal body of the mike). Scrape it a few times to assure good contact. Does the addition of this wire decrease the hum? If so, you have a grounding problem. A technician will have to beef up the cables or connect a wire between the chassis of the two devices.

5. *Mixer passes a good signal from all sources but one, which has either flat or boomy sound*: Some mikes, some sources, and some mixers have controls which adjust their tones.

a. Check for a tiny switch on the mike. This may be a "roll-off" filter, which means that it throws

out certain tones and passes the rest. Often, this filter throws out the low tones, thus making the sound flat—which is good for speech but bad for music where we like to hear deep bass notes.

b. Some sources have treble and bass controls. If the controls have a position called "flat," then put them in that position generally. This doesn't mean that the sound will be flat; in this context, the word means that the device will give a true, even response to all sounds both high and low. Some cheaper sources have a single "tone" control. Generally, turn it to its sharpest setting (HIGH or TREBLE) as the other settings may sound muffled. Some turntables have a switch that says RIAA or FOREIGN or 78. Switch it to RIAA for long-playing 33⅓-rpm (revolutions per minute) discs. Switch to FOREIGN for discs made outside of the United States. Switch to 78 if you're playing moldy oldy 78-rpm records. This switch adjusts the electronics to conform to the various properties of the record. Some turntables and amps have "rumble" and "scratch" filters. Unless you hear rumbling and scratches, leave these switches off. The rumble switch filters out some of the low tones, while the scratch switch filters out some of the highs.

c. Some mixers have a LOW CUT filter near each individual volume control. These switches cut the low tones, a move which may be beneficial for speech recording but not for music.

If the problem is very bad, it may be a wiring difficulty. If the sound is tinny and weak, the IMPEDANCE may be mismatched, or the BALANCED LINES may be connected wrong or miswired to an UNBALANCED input. Try a substitute mike or a substitute source. It is not uncommon to come across miswired mike extension cables using BALANCED LINES. There are two ways to wire them, and if one cable followed one standard and the other followed another, they would *look* compatible but wouldn't be.

6. *All audio from the mixer has hum*: Test for ground problems as described before by touching one end of a wire to the chassis of the mixer while, this time, touching the other end to the chassis of the VCR (or wherever the signal is going). If grounding the two devices together reduces the hum, a technician may have to wire the two devices together. Incidentally, if you hear the same hum on the mixer's headphone (if it has such an output), the problem may be in the mixer itself. Try unplugging the AC plug, turning it a half turn, and reinserting it. This sometimes cures hum.

Make sure the VCR volume and the mixer volume are not way up or way down. If one is high and the other is low, try to even them out. If both have to be way up in order to get enough sound, try connecting the mixer to the VCR's MIC input, which is more sensitive than its regular AUX input.

7. *All sound going into the VCR is too loud or is raspy:* Is the mixer master volume up too high? Is the VCR audio level up too high? The meter will show evidence of this by staying way in the red area. (Incidentally, audio people call it "pinning the meter" when the needle reads all the way at the top of its scale.) Is the mixer HIGH LEVEL out feeding into the HIGH-LEVEL or LINE or AUX input of the VCR as it should? Try to listen to the mixer output directly (perhaps through headphones into the mixer). If it is raspy there and it happens to all inputs, the mixer may be defective. If the mixer sounds good, the problem lies between the mixer and the VCR. Try a PAD to attenuate the signal going into the VCR.

8. *The sound is good except when you hear radio stations, police calls, CB radios, the buzz of fluorescent lamps, or the tick, tick, tick of automobile ignitions in the background of your recording:* Presumably, you couldn't hear these sounds with your unaided ears but you could through the audio devices. These signals are electromagnetic radiation inducing a signal in your wires. In a way, it's as if your wires were acting like an antenna picking up these signals. To keep this interference from seeping into your recordings,

    a. Try to keep a distance away from the interference sources.

    b. Ensure that you have good shielding on your source cables (refer this to a technician).

    c. Assure that the cables from the mixer to the VCR are also well shielded.

    d. If you have a choice, run the mixer's HIGH LEVEL OUT to the LINE IN of the VCR rather than using the MIC LEVEL OUT and the VCR's MIC IN. By using the stronger signals, the interference will seem weak in comparison to the signal.

    e. Ground the devices together with a wire connecting their chassis.

    f. Use short cables when possible.

    g. Contact the local FCC (Federal Communications Commission) if you think someone might be broadcasting with too much power. If nothing

else, the FCC may at least send you literature on how to modify your equipment to be less sensitive to this interference.

    h. I saved the best for last: Use BALANCED LINES.

DIGITAL AUDIO

Your steering wheel is ANALOG; you can steer your car slightly left, very right, or anywhere in between. Your headlights are DIGITAL; they're either "off" or "on." Where ANALOG equipment works with varying electrical voltages, DIGITAL equipment uses "on" and "off" signals, or ones and zeroes.

DIGITAL equipment has several advantages over ANALOG:

1. DIGITAL circuits are often smaller, cheaper, and easier to build than ANALOG.
2. Signals are easily copied without degradation.

ANALOG electrical vibrations pick up noise and interference, making audio and video copies worse than originals. DIGITAL copies of a signal are simply copies of ones and zeroes; even if the ones and zeroes are fuzzy, weak, too strong, or slightly distorted, they are still ones and zeroes. Copying machines can "replace" incoming fuzzy ones and zeroes with sharp new ones, keeping the stream of numbers exactly the same and thus reproducing a signal perfectly.

Compact disc (CD) and Digital Audio Tape (DAT)

When sound is recorded DIGITALLY, its electrical vibrations are sliced up into tiny samples (44,100 samples/second in the case of CDs) and the samples converted into numbers. These DIGITS, expressed as binary ones and zeroes, are converted into "pits" and shiny places in a spiral on a CD. When "read" by a pinpoint of laser light in the CD player, the digital data are converted back into ANALOG sound.

Although not as delicate as phonograph records, CDs still need gentle care. Fingerprints and scratches cause data losses heard as clicks or distortions to the sound. Handle CDs by their edges and park them in their jewel cases when not in use.

CD's can't be cued up by rocking them back and forth

MINIGLOSSARY

Analog   Something that varies in infinite gradations. A light dimmer is analog. Analog circuits suffer noise and distortion.

Digital   Something that is either "on" or "off." A light switch is digital. On and off can be represented by the digits 1 and 0. Digital equipment copies signals without introducing noise and distortion.

Compact disc (CD)   Small shiny disc imbedded with micro-

scopic pits representing digital data which can be read by a laser and converted into sound.

Index   Subdivision of a song or track on a CD allowing you to play a particular stanza or phrase.

Dynamic range   A ratio of the softest to the loudest sound reproducible by a device, expressed in dB. A 90-dB dynamic range is more lifelike than a 70-dB dynamic range.

like reel-to-reel tapes or phono records. The best you can do is to play the disc, then PAUSE it just before the desired selection. When ready, UNPAUSE the disc for your sound.

Although all CDs and CD players permit you to jump directly to particular songs or tracks, only certain CDs and professional CD players INDEX the songs into smaller segments. Industrial production music libraries often INDEX their music and sound effects, making it easier to play just one movement of a symphony, verse of a song, or one cock-a-doodle of a rooster.

DIGITAL AUDIO TAPE machines, like CDs, store sounds digitally, permitting wide frequency response, low noise, wide DYNAMIC RANGE, and superb copyability. The tape looks like an 8mm videocassette (slightly larger than an audiocassette) and is recorded by a spinning video head, much like in a VCR. Besides making superb hi-fi original recordings, DAT machines can record the data directly from other CDs or DATs. To prevent people from duping off copies and copies-of-copies of copyrighted music, the manufacturers have equipped DAT machines with copy-thwarting circuits called SCMS (Serial Copy Management System). You can, of course, copy any of your own original sounds as many times as you want.

SUMMARY: GETTING THE BEST AUDIO SIGNAL

Room

1. Use a quiet room, one with thick walls and tight-fitting doors to seal out extraneous noise.

2. Keep the room quiet by turning off fans, noisy air conditioners, and other machines while taping. Move the VCR as far away from the microphone as possible, perhaps by placing it behind something or by moving it to another room.

3. Use an anechoic room. Reduce echoes by hanging curtains, laying carpet, or draping felt or blankets over the walls.

Microphone

1. Use a LO Z mike with a BALANCED LINE to keep electronic "noise" from sneaking into your cable.

2. Whenever feasible, use DIRECTIONAL microphones to reject room noise, especially if the mikes cannot be kept close to the performers.

3. Keep mikes close (6–18 inches) to your performers.

4. When possible, keep the performers from handling mikes or cords during the recording *or* use a specially constructed mike which insulates against hand and cord noises.

5. If the mike has a switch called LOW CUT filter, leave it *off* for music and *on* for speech.

6. Use a WINDSCREEN if miking outdoors.

7. Place carpet or foam under the mike stand to keep floor or desk vibrations from being recorded.

8. For best fidelity, use a DYNAMIC, CONDENSER, or RIBBON microphone instead of a CRYSTAL microphone.

9. If the mike is to be hand-held, choose an OMNIDIRECTIONAL one. CARDIOIDS and DIRECTIONALS are often too sensitive to hand noises and "booming" when held too close to the performer's mouth.

Cable

1. Again use LO Z BALANCED LINES whenever possible. Keep cable runs to 300 feet or less.

2. If using HI Z UNBALANCED LINES, make sure you are using *shielded* cable only. Keep cable length to an absolute minimum, no more than 30 feet *maximum*.

3. Keep people from tripping by tying cables to your mike stand base or in heavy traffic areas by taping cables to the floor.

4. Keep cable extensions from disconnecting by tying them in knots at their plugs.

5. Loop the LAVALIER mike cable through the performer's belt so the cable isn't tugging directly on the microphone as he or she moves.

Inputs

1. Use LINE or HIGH-LEVEL signals whenever available to feed AUX or HIGH LEVEL inputs (i.e., feed the mixer's HIGH LEVEL OUT to your VCR's LINE IN). This reduces the electronic noise that can creep into your cables.

2. If you must feed a HIGH LEVEL signal into a LOW LEVEL INPUT, use a PAD.

3. Keep enough ADAPTERS handy so you can use *what's best* and not have to make do with *what fits*.

Mixers

1. To judge volume levels, trust the mixer's meter. To judge fidelity, trust your ears. Monitor everything.

2. Avoid using mixer filters when recording music. Use them primarily for speech or for adapting to poor room acoustics.

3. Don't run some volume controls high while others are low. Whenever possible, run them all nearest their middle settings.

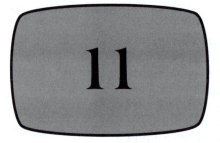

CAMERA SWITCHERS AND SPECIAL EFFECTS

Now we're in the Big Time, using several cameras at once. To select which camera gets "on the air" (or onto the tape), you need a switcher.

Learning to operate a switcher from a book is much like learning surgery from a pamphlet, learning origami over the telephone, or learning sex from a sex manual. As one studies, one must have the equipment in hand in order to visualize the results of each manipulation. Experiment. See what happens. The experience of *doing* will develop skills faster than the experience of *reading*. However, reading is a good starting point, and, besides, Prentice-Hall sells books, not switchers.

Some folks get along just fine without ever having a switcher or more than one camera. They may record all the MEDIUM SHOTS at one time and then move the camera and record all the CLOSE-UPS later. This raw footage is then viewed and logged, and the best scenes are selected. One by one, these scenes are copied onto a master tape. Because the scenes change from close shots to far shots, the viewer never knows that only one camera was used. It looks like the director simply switched from one camera to another. Thus, it is possible to assemble a complete, nice-looking show with but one camera and no switcher.

The problem with this technique is that it is so time-consuming. The performance has to be done and redone for each camera angle recorded. Special attention must be paid to continuity of action so that if the farmer is holding the chicken in his left hand in the long shot, when you shoot the scene again for close-ups, the chicken had better be in the left hand again. The viewing, logging, and selecting of shots and final editing of the raw footage into a master tape takes many hours.

This is where a switcher saves a heap of time. If one camera is taking medium or long shots and another is taking close-ups, an alert director, by pressing a button, can select the shot to be recorded. There are no continuity problems, no reviewing, no logging, and no editing. Unless someone makes a mistake, you end up with a final, nice-looking tape at the end of one performance. What took the single-camera people half a week to produce may take you only a half hour to produce using a switcher.

Producing a whole show "live" with a switcher requires great skill (and a pint of stomach acid). In the real world directors and performers make mistakes. Editing may still be needed, but since most of the shows have already been assembled by the switcher, the task is much quicker.

There are different kinds of switchers, from supersimple to outrageously complex. We are going to start with the simplest and gently work our way toward those multibutton dragons seen in commercial television stations.

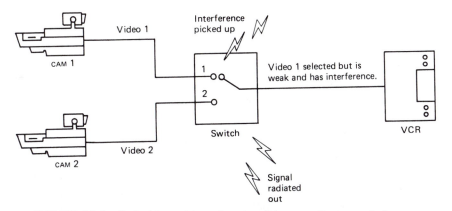

FIGURE 11-1　Switching video signals with an ordinary switch.

PASSIVE SWITCHER WITH UNSYNCHRONIZED CAMERAS

Ordinary Switch

Pretend you have two ordinary TV cameras (not special studio ones) and you wanted to select the picture from one or the other of them to send to your VCR. Why couldn't you simply run their video outputs down a wire to an ordinary hardware store electric light switch as diagrammed in Figure 11–1 and then run the selected signal straight to your VCR? You could. It would work, but your signal would be weak and sensitive to interference (the picture may appear gray, grainy, jiggly, or have wavy lines running through it).

Coax or Shielded Switch

COAX or SHIELDED switches are specially designed for video or RF. They protect the signal from outside interference, keep your video signal from "leaking out," and do other handy things like switching firmly and quickly and not messing up the electronic properties of the signal, such as IMPEDANCE.

VCR's Input Selector Switch

Remember the INPUT SELECTOR or CAMERA/LINE/TV switch you used to select which video input would be recorded? What do you suppose would happen if you somehow connected one camera to the VIDEO INPUT of a VCR and another to the CAMERA input and switched from one camera to an-other using that selector? Answer: The idea would work. By flipping the switch from CAMERA to LINE, you could switch from one camera to another. This is a cheap way to get two cameras to work with one VCR without the fuss and bother of a special switcher.

This method of switching is especially appropriate for mobile, one-person operations which do not require fancy production techniques. For instance, to record a classroom lesson, the system can be set up as shown in Figure 11–2. The camera operator runs the VCR and selects the pictures either from camera 2 (with a stationary wide shot of the whole class) or from her own camera 1 (with a zoom lens for close-ups and follow-the-action shots). One operator can set up, adjust levels, operate one camera, and troubleshoot from the same spot. Although the switching from camera to camera will look a bit rough (and therefore it should be done sparingly), the content of the lesson will suffer very little.

This cheapie system isn't perfect, however. When you play back your tape, you will find a little "blip" or "glitch" that interrupts your nice smooth picture every time you switched (see Figure 11–3). The blip will disappear in a few seconds, and all will be well again.

The blip is the result of using a PASSIVE switcher and UNSYNCHRONIZED cameras. The reason for the blip will be discussed shortly. But first, let's look at a few ways to connect up a simple PASSIVE video switcher.

Connecting a Passive Switcher

Like the built-in switcher, the PASSIVE switcher selects one of the video signals and passes it on to the VCR. The word

---

MINIGLOSSARY

Shield　Braided wire or foil that creates a flexible pipeline surrounding another wire, protecting it from outside electrical interference.

Unsynchronized　Cameras that create their pictures independ-ently of one another, not using an external sync source to lock their electronics in step.

*Line monitor　TV monitor that shows the final signal being broadcast or sent to the VTRs.

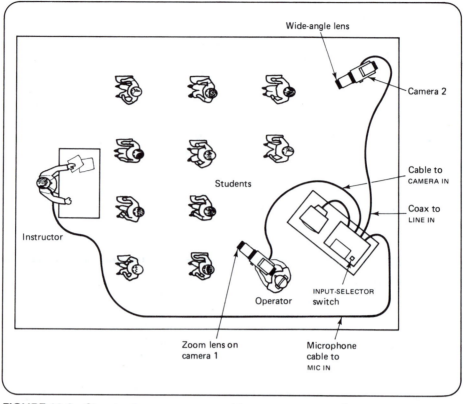

FIGURE 11-2   Simple classroom setup with two cameras and INPUT SELECTOR as switcher.

''PASSIVE'' means that it *doesn't do* anything except select which signal is allowed through: It's just a switch. It doesn't even need power to operate. It can be connected as in Figure 11–4 or better yet as in Figure 11–5 where each camera can be monitored separately and where the switcher output can be monitored before it goes on the VCR.

The monitor that shows the output from the switcher is called the LINE monitor or the PROGRAM monitor. While the VCR is recording, the LINE monitor and the VCR monitor should show the same thing (if all the equipment is working right). If the VCR is playing a tape, the LINE

FIGURE 11-3   Glitch on screen caused by switching from one source to another.

monitor shows the switcher's output while the VCR monitor displays the signal from the tape. Meanwhile, monitors 1, 2, and 3 continually display the pictures from cameras 1, 2, and 3, respectively.

The inexpensive PASSIVE switcher has the same problem the VCR's INPUT SELECTOR had when used as a switcher. Every time you switch from one camera to another, there will be a blip on the screen. When playing back a tape, you will again see the blip appear in the picture and then settle out of your picture in a few seconds. If yours is a small shoestring operation, these blips are no big problem and should concern you only aesthetically. As you become more ''professional,'' however, you may find these flaws to be more and more irritating.

These blips can become serious problems under the following circumstances:

1. You wish to copy the tape—the problem will look even worse on the copy.
2. You wish to broadcast the tape—broadcasting requires excellent sync—without blips.
3. You wish to run the signal through a TIME BASE CORRECTOR or PROCESSING AMPLIFIER or GEN-LOCK device (these machines will be discussed in Chapter 15). These devices may get confused if the sync isn't close to perfect.

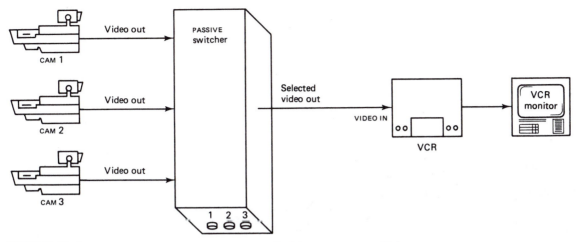

FIGURE 11-4 UNSYNCHRONIZED cameras connected to a PASSIVE switcher.

THE NEED FOR EXTERNAL SYNC TO THE CAMERAS

What causes the blip? For nice blip-free television, all the electronic devices (VCR, camera, and monitors) must have their circuits synchronized. If each camera generates its *own* internal sync, when you switch from one camera to another, the VCR and TV monitor have to resynchronize themselves to the sync of the new camera. It's like two balloons side by side, each with a band playing its own music. As you Lambada through the door of one ballroom and into the next, you discover that everyone there is dancing Electric Glide. It takes a few messy shuffles before you and your date are in step with a new band. Then, as you glide back through the door into the Lambada ballroom again, you might expect a few awkward motions before you are synchronized with the Lambada band. In television these awkward motions create the blip on your screen.

To get rid of the blip, one must synchronize the cameras, which is rather like getting both bands to play the same rhythm simultaneously led by the same band leader. Now you can Lambada from one ballroom to the other without skipping a beat and switch from one camera to another without getting a blip.

The solution to the blip problem is to run EXTERNAL SYNC from some single source to all the cameras so that their electronics are in step. A SYNC GENERATOR will do this.

PASSIVE SWITCHER AND SYNCHRONIZED CAMERAS

Figure 11–6 shows several TV cameras (all in the EXT SYNC mode) connected for use with a switcher. The SYNC GENERATOR synchronizes the electronics of all cameras by send-

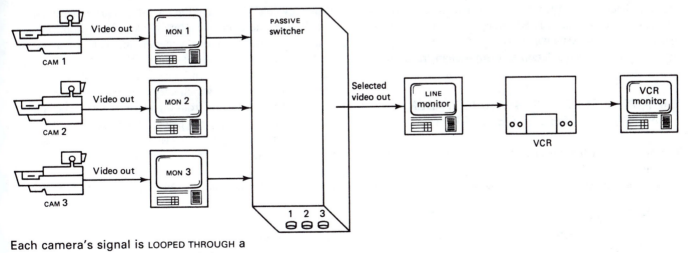

Each camera's signal is LOOPED THROUGH a TV monitor on the way to the switcher.

FIGURE 11-5 UNSYNCHRONIZED monitored cameras connected to a PASSIVE switcher.

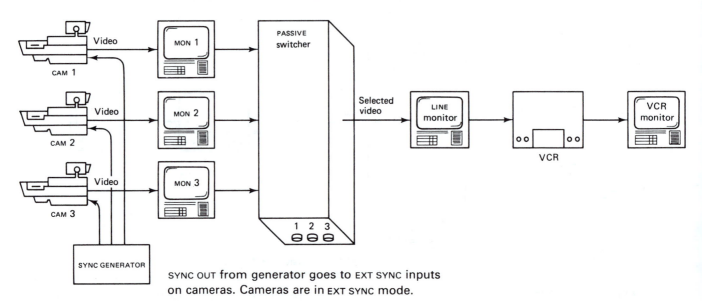

SYNC OUT from generator goes to EXT SYNC inputs on cameras. Cameras are in EXT SYNC mode.

FIGURE 11-6   PASSIVE switcher with SYNCHRONIZED cameras.

ing sync out to them. Each camera sends its video back to the switcher, and a sample of the signal may be sent out to a TV monitor (as shown). The switcher allows one of the camera signals to be passed on to the VCR for recording after the signal has LOOPED THROUGH a LINE monitor.

With the system set up this way, when you switch from camera to camera, you don't get that 2-second blip caused by the cameras not being synchronized with each other—but a *tiny* blink may still remain. As the buttons are pressed on the PASSIVE switcher and its mechanical components snap into place, your video signal is disturbed for just a fraction of a second. This disturbance may appear anywhere in your picture. Such barely perceptible glitches seldom annoy the shoestring educational, industrial, or amateur user, but the pros won't tolerate them. These flaws in the video become more pronounced when recorded or copied or when the signals are passed through TIME BASE CORRECTORS, PROCESSING AMPLIFIERS, GENLOCK, or other highfalutin gadgets. For the solution to this problem, read on.

THE ACTIVE SWITCHER, SWITCHER/FADER, AND SPECIAL EFFECTS GENERATOR

The ACTIVE switcher has circuits in it which help do more than just switch. For one thing, it may (but not always) have a built-in SYNC GENERATOR connected as shown in

Figure 11–7. The sync goes to the cameras through the camera cable, and the video comes back to the switcher through the same multiconductor cable. Note that the ACTIVE switcher may have separate outputs to feed each monitor so that you don't have to go to the trouble of LOOPING the camera signal through each one.

Depending on how elaborate it is, the ACTIVE switcher may handle VERTICAL INTERVAL SWITCHING, FADING, SPECIAL EFFECTS, GENLOCK, and more.

The following is a list of some of the things an ACTIVE switcher may be able to do depending on how elaborate it is.

Vertical Interval Switching

If the device *doesn't* have the VERTICAL INTERVAL SWITCHING feature, there may be a tiny blink or flash on the screen when you switch from one camera to another. It may occur so fast that you can't see it, but it is still there. The flash has nothing to do with sync; that's already taken care of by the SYNC GENERATOR, which keeps everything in step. This momentary flash results from the motion of the mechanical push button as it snaps into place. This minor flaw is cured (at extra expense) by using VERTICAL INTERVAL SWITCHING, which does the switching electronically rather than mechanically. You still push a button on the switcher, but because electronics can do things faster than people and

MINIGLOSSARY

*Vertical interval switching   Electrically controlled switch that toggles from one source to another (i.e., from camera 1 to camera 2) during the brief instant that the TV is *not* making a picture on the screen. It switches during the vertical interval, the black line

below the bottom of the screen.

*Fade   TV picture smoothly turns black (fade-out) or black smoothly turns to a TV picture (fade-in).

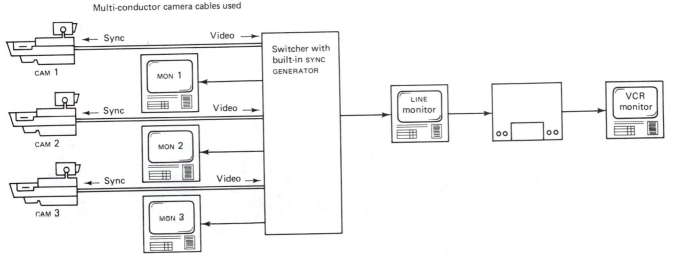

FIGURE 11-7 Several cameras in use at once with ACTIVE switcher and SYNCHRONIZED cameras.

because electronics can also figure out the exact split second to execute the switch, it is done cleanly and perfectly. The name given to that exact split second is VERTICAL INTERVAL.

The VERTICAL INTERVAL is the black line which remains just below the bottom of your TV picture. You see this line only when you misadjust the VERTICAL HOLD control on your TV set (refer to Figure 2–3). In VERTICAL INTERVAL switching, any flashes that might occur happen in this black line out of the viewer's sight.

Fader

Figure 11–8 diagrams a simple FADER. This FADER has two channels, A and B. By pressing (for instance) button 2 on the top row of buttons, camera 2 is now feeding into channel A. By pressing button 1 on the bottom bank, we get camera 1 to feed to channel B. Thus, camera 2 is on A and camera 1 is on B. The two FADER levers to the right control how much of each channel is selected for recording.

As shown in the figure, the A FADER is all the way *on* and the B FADER is all the way *off*, so the device is passing only the A signal, which is what? Camera 2. Now to DISSOLVE to camera 1's picture, pull both levers all the way down. This decreases A's strength while increasing B's.

From here let's DISSOLVE to camera 4. First switch the A channel (which is now *off*) from 2 to 4 by pressing the 4 button. Then slide both levers all the way up. A becomes stronger as B becomes weaker, and what is A now? Camera 4, so 4's picture is now *on*.

To FADE OUT, merely move the FADE lever, which is *on*, to *off* while not touching the other lever. The left lever (A channel) would be down while the right one (B channel) would be up; both are now *off*, and both pictures have disappeared.

Pulling both levers only halfway allows both pictures to be seen at once. This effect is called a SUPERIMPOSITION, abbreviated as SUPER. Figures 11–9 and 11–10 show SUPERS of two images. Here you get one half of A's signal and one half of B's signal, making one whole picture (made of two pictures). It is also possible to leave the A lever all the way up (*on*) and the B lever all the way down (*on*) and get a SUPER, but this method may pose a problem: If you get a whole picture signal from A and add a whole signal from B, that makes your total signal twice as strong as it is supposed to be and messes up the video levels on your VCR. So, don't SUPER two pictures by moving both FADE

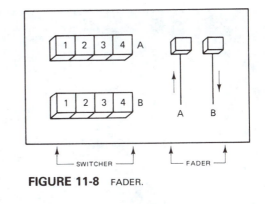

FIGURE 11-8 FADER.

MINIGLOSSARY

*Fader A slider or handle on a switcher that allows you to fade in or fade out a picture or dissolve from one picture to another.

*Dissolve (or lap dissolve) TV effect where one picture slowly melts into another. One picture fades to black while another si-

multaneously fades up from black.

*Superimposition Two pictures shown atop one another. They may look semitransparent or "ghosty." A dissolve stopped halfway.

camera 1

camera 1

camera 2

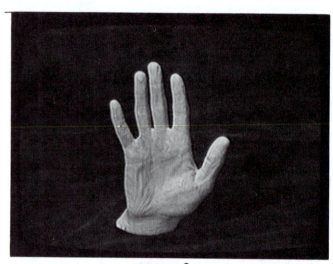

camera 2

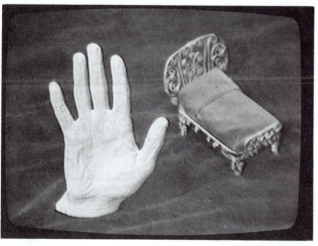

SUPER

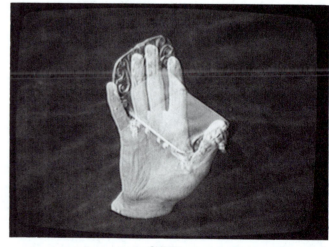

SUPER

FIGURE 11-9 SUPER two images which do not overlap.

FIGURE 11-10 SUPER two images which overlap.

levers all the way *on*; just move each to half way *on*. *An exception to the rule:* If you have two white pictures with black backgrounds (as happens to be the case in Figure 11–9) and the images will not overlap when the SUPER is executed, then both FADE levers may be moved all the way to *on*. In fact the picture will look best when made this way. This particular operation is permissible because you are not adding one signal (one picture) *over the top* of the other signal (the other picture). The two white images occupy different places on the TV screen and don't mix together creating doubly bright spots on the screen.

You will notice from Figure 11–10 that when two pictures are SUPERIMPOSED, they have a "ghosty" effect; you can half see through both of them. The image is confusing and not very appealing but can be useful for special effects (like creating a ghost, for instance).

Words can also be SUPERIMPOSED over a picture. Camera 1 may view the performer while camera 2 views a graphic with printing on it.

Notice what happens, however, if we use normal black lettering on white paper as in Figure 11–11. When SUPERED, the white from the paper mixes with the other picture, causing it to look faded. The words don't stand out very well either.

Figure 11–12 shows the right way to do this. Use black paper with white lettering. The white lettering will show up while the black part of the graphic will disappear and not fog up the rest of your picture.

SUPERS are an imperfect way to put words on a screen because they are so "ghosty" and seem to disappear into light parts of the picture as shown in Figure 11–13. If you must SUPER words, then use the technique similar to what

camera 1

camera 2

Picture faded. Words show
up poorly.

SUPER

FIGURE 11-11 SUPER of picture with a graphic having black lettering on white background.

camera 1

camera 2

SUPER

FIGURE 11-12   SUPER of a picture with a graphic having white letters on a black background.

camera 1

SUPER

camera 2

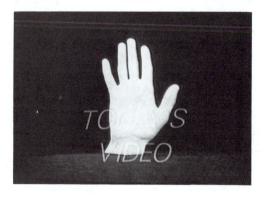

FIGURE 11-13   White words disappearing into parts of a white picture.

camera 1

camera 2

SUPER

FIGURE 11-14 SUPER white letters over dark part of picture.

you saw in Figure 11–9. You place the performer in one part of the picture, leaving a dark area in another part. You position the white lettering on another camera so that it will appear over the dark part of the first picture as shown in Figure 11–14. It is not necessary that the spot where the letters go be absolutely black; just try to make sure that it is very dark—the darker the better.

Special Effects Generator

The simplest SPECIAL EFFECTS GENERATORS (hereafter abbreviated SEGS) have the familiar switches and levers for FADING and also have a few more buttons and levers (see Figure 11–15). The buttons select various CORNER INSERTS, and the levers do WIPES. The buttons are labeled with little pictures to show what the effect will look like, and these pictures may also indicate which part of the final picture is from channel A and which part is from channel B. If you slide the WIPE levers to the middle position, your TV screen

should show an image that looks like the little picture on the push button.

Moving one of these WIPE levers at this point will *widen* or *narrow* the CORNER INSERT. Sliding the other level will *lengthen* or *shorten* the CORNER INSERT height. It is possible to widen the CORNER INSERT all the way across the screen, thereby creating a horizontally SPLIT SCREEN. Now, by adjusting the other lever, you can raise and lower the split in the screen. This action is called a VERTICAL WIPE and is shown in Figure 11–16. You can lower it enough to make it disappear or raise it enough to fill the whole screen.

Learning which lever to move to get the desired effect takes time and experience with the machine. Experiment. There is no button or lever on the SEG that can permanently harm anything if you play with it.

The simplest SEG (such as the one shown in Figure 8–15) may have buttons to select whether the device will create FADES *or* WIPES. More complex SEGS allow you to DISSOLVE *to* a SPECIAL EFFECT and then DISSOLVE back *from*

MINIGLOSSARY

*Special effects generator   Electronic video device that creates effects such as wipes, fades, keys, etc.

*Wipe   Special effect that starts with one TV picture on the screen, then a boundary line moves across the screen (vertically, diagonally, or whatever), and where it passes, the first picture changes into a second picture.

*Corner insert   A special wipe pattern that stops partway across

the screen so that a corner of the TV picture is taken up by part of another camera's image.

*Split screen   A wipe that stops partway across the picture, revealing a section of the original picture and a section of the new picture.

Vertical wipe   A wipe where a horizontal boundary line sweeps vertically through the screen, changing the picture as it goes.

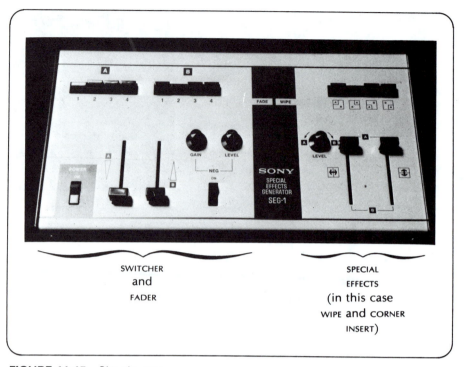

FIGURE 11-15   Simple SPECIAL EFFECTS GENERATOR.

camera 1

camera 2

Horizontally split screen. If the split is moving higher or lower, it is called a VERTICAL WIPE.

FIGURE 11-16   HORIZONTAL SPLIT SCREEN.

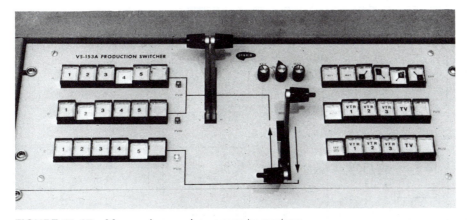

FIGURE 11-17 More advanced SWITCHER/FADER/SEG.

it to a regular camera shot. The more complex SEGS (like the one shown in Figure 11–17) do this by having three channels to work with. Figures 11–18 and 11–19 diagram how they work.

Channels A and B in Figure 11–18 are called the EFFECTS channels or EFFECTS BUS, while C is just a regular channel. You DISSOLVE TO EFFECTS by sliding the FADE levers from the C channel to the EFFECTS channels where A and B do their stuff.

Here are some examples of how to use this device:

1. *To switch (CUT) directly from one camera's picture to another:* Let's start with the FADE levers down so that the C channel is on. Simply press any C button for the desired camera shot. It doesn't matter which buttons are pressed on A or B or where the WIPE levers are.

2. *DISSOLVE from camera 1 to camera 2:* You might start out with the camera 1 button pressed on the C BUS and

the FADE levers *down*. You will eventually push the FADE levers *up* to DISSOLVE to camera 2's picture. To have camera 2's picture waiting for us when we FADE to the EFFECTS BUS (channels A and B), we first have to manipulate those levers. If on the B BUS we press the camera 2 button and slide the WIPE levers all the way down, we will be creating a picture where camera 2 fills the screen. Put another way, the WIPE levers create a CORNER INSERT which looks like the little picture on the push buttons. The A channel fills the little black box on the button, and the B channel fills the white area on the button. By moving the WIPE levers all the way to the B position, the little box will get smaller until it disappears and only the B channel will show. If camera 2 is what the B channel is showing, then your EFFECTS are simply a picture from camera 2.

Back to our exercise: We wanted to FADE from camera 1 to camera 2. To do this, we created an EFFECT which was simply camera 2. We did this by punching the 2 button on channel B and pulling the WIPE levers down. FADING to camera 2 simply required sliding the FADE levers up.

Once the FADE levers are up and the WIPE levers are down, channel B is what is being shown. If you now wish to switch to camera 3, you may do it by punching the 3 button on the B BUS.

Now if you wish to DISSOLVE from camera 3 to camera 4, you would first press 4 on the C BUS and then slide the FADERS down to DISSOLVE from the B BUS (camera 3) to the C BUS (camera 4).

3. *DISSOLVING to a CORNER INSERT and back out again:* Let's let channel A show camera 1, channel B show camera 2, and channel C show camera 3. Figure 11–19 will help you follow along.

The EFFECTS buttons are pressed so that A (camera 1) is in the upper right-hand corner of B (camera 2) when the

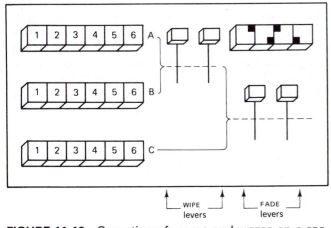

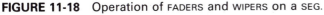

FIGURE 11-18 Operation of FADERS and WIPERS on a SEG.

MINIGLOSSARY

Bus A channel of a group of related buttons on a switcher/SEG.

*Effects bus Group of related buttons on video SEG/switchers to create special effects. A channel on the switcher which you can

dissolve to and from, bringing a special effect onto the screen or taking it away.

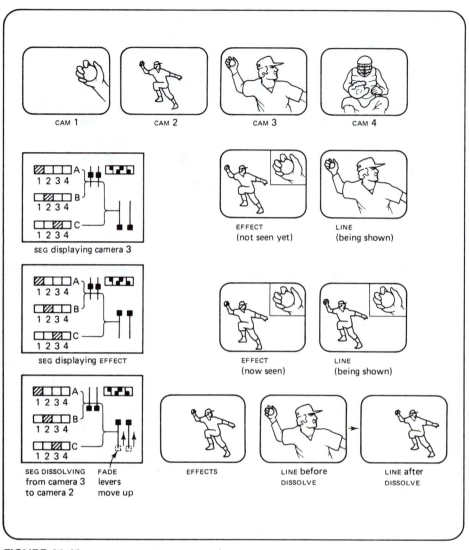

FIGURE 11-19   DISSOLVING to a corner insert.

WIPE levers are halfway down. With the FADE levers all the way down, we are displaying camera 3 only (the C channel).

To show the CORNER INSERT of cameras 1 and 2, just DISSOLVE TO EFFECTS by moving the FADE levers up. To DISSOLVE to 3 again, just move the FADE levers back down.

If you didn't want to DISSOLVE to a fancy CORNER INSERT but just wanted to DISSOLVE to camera 2 from camera 3, you would first have to move the WIPE levers to make the CORNER INSERT (which is channel A's picture) as small as possible until B (camera 2's picture) fills the screen entirely. Now that the EFFECTS channel shows camera 2's picture, you can DISSOLVE from camera 3 to camera 2 by moving the FADE levers up.

Say you wish now to CUT to camera 4, which button do you press? Button 4, right? But on which channel—A, B, or C? Well, we know C isn't the answer because a moment ago we slid the FADER to the EFFECTS mode to show camera 2 (if we were still in the C channel, the answer would have been to press 4 on channel C). What you do to

C at this point won't show until you slide the levers back down again. That leaves A and B channels. Although logic and time could give you the answer, there is a faster way to figure this out. Camera 2 is *on* right now, and you want camera 4. Whichever channel has the 2 button pressed is the channel you want to switch to 4. Thus, you press button 4 on channel B.

4. *DISSOLVE TO A SPLIT SCREEN:* Say you are in the C channel with camera 2 as diagrammed in Figure 11–20. You want to DISSOLVE TO A SPLIT SCREEN with some words (from camera 1) at the bottom of camera 2's picture, like the example shown in Figure 11–16. You want camera 2 to remain on the screen except for the bottom part, which you want to dissolve to the words. How is this done?

Since there is no law that stops you from punching in the same camera number in more than one channel, you punch a 2 into A and a 2 into C. B would get switched to camera 1. The WIPE levers would be manipulated for the

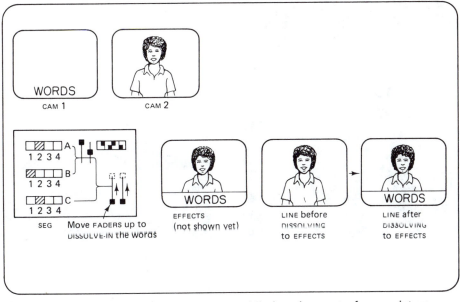

FIGURE 11-20 DISSOLVE to a SPLIT SCREEN while keeping part of your picture.

desired effect. In other words, channel C is showing camera 2. The EFFECTS BUS is showing camera 2 with the words from camera 1 under it.

You now can DISSOLVE from C (camera 2) to the EFFECTS by raising the FADE levers.

On fancier SEGS that offer diagonals, diamonds, squares, circles, and so on, the process is essentially the same. Figures 11–21 through 11–23 show some examples. First punch the button that selects the desired type of EFFECT. Next, create the desired EFFECT on the EFFECTS channels while displaying the C channel. You may then DISSOLVE *to* and *from* for the EFFECTS by moving the FADE levers.

You get out of the EFFECT by moving the FADERS back down to the C channel.

camera 1

camera 2

CORNER INSERT

FIGURE 11-21 CORNER INSERT.

FIGURE 11-22   VERTICAL SPLIT SCREEN.

STUDIO PRODUCTION SWITCHER

The STUDIO PRODUCTION SWITCHER is a complex SEG with goodies galore. These switchers vary greatly in the features they possess. Indeed, if you visit a network control room, you'll see the most elaborate switchers in existence. Figure 11–24 shows one.

This switcher has more buttons than Napoleon's dress uniform, but don't let that throw you. Its basic operation is much like what you've already seen. Refer to Figure 11–24 as we follow these basic exercises:

1. *Switch from camera 1 to camera 2:* There are many ways to do this:

   a. Assume the bottom FADER is up and the PGM (PROGRAM) BUS (second long row from the bottom) has camera 1 punched on. The PROGRAM, or final output from the switcher, is thus camera 1. To switch to camera 2, simply press the camera 2 button on the PGM row of buttons. Done.

   b. If the PGM BUS has the EFF 1 button pressed (four buttons to the left of the bottom FADER), then the EFF 1 (EFFECTS 1 BUS) is in charge. Whatever the top two rows of buttons and their FADER do will make your final picture. To the left and right of the top FADER are buttons which make that lever act as a FADER or a WIPER or something else. Let's assume the MIX (FADE) button up there is pressed so the FADER can DISSOLVE us from channel A to B. If the lever is up and channel A has camera 1 punched on, then camera 1 is what will show. If you press the camera 2 button on A, you then switch (also called TAKE or CUT) to camera 2.

circle with soft border

modulated pattern

box with border

FIGURE 11-23   Other EFFECTS (there are hundreds)

MINIGLOSSARY

*Studio production switcher   A large active switcher/SEG that receives all the video sources (inputs from cameras, etc.) and is used to select the pictures or effects to be shown.

*Program   The final output from a switcher that is broadcast or recorded.

*Program (PGM) bus   The group of buttons on a switcher that

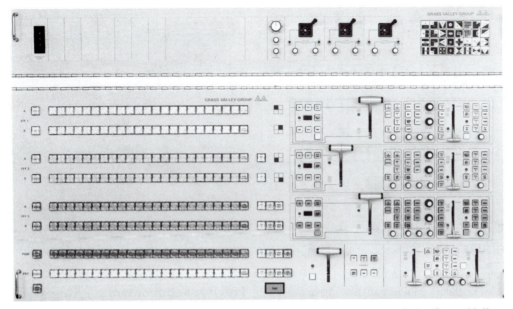

FIGURE 11-24 Professional STUDIO PRODUCTION SWITCHER. (Courtesy of the Grass Valley Group)

Thus, all switching is done on the top row of buttons.

c. If the PGM BUS has the EFF 2 button pressed, then the two EFF 2 rows of buttons (EFF 2 BUS) are in charge. If EFF 2's FADER is in the FADE mode and up and if its channel A has the camera 1 button pressed, then camera 1's picture will show. Simply press the camera 2 button on the EFF 2 channel A row to CUT to camera 2.

d. Similar to example c: If the PGM BUS is in the EFF 3 mode, then the third FADER does the business between its A and B channels. If that FADER says the B channel is on and camera 1 is punched in there, simply punch the camera 2 button next to it to switch to camera 2.

e. If you press the camera 2 button on the PGM BUS, camera 2 will appear, and the EFF 1, 2, or 3 buttons will switch off, and camera 2's picture will appear.

2. DISSOLVE *from camera 1 to camera 2:* We have a choice of carrying out this DISSOLVE effect on EFFECTS BUS 1, 2, or 3. Say we choose EFF 1. The bottom FADER will be up, its EFF 1 button is pressed, the top FADER (let's say)

is up, and camera 1 is punched up on channel A. To get ready to DISSOLVE to camera 2, you press that button on channel B. Double-check the FADER control buttons to make sure that lever will DISSOLVE (not WIPE, etc.) from channel A to B when you move it. When ready, pull the top FADER down to DISSOLVE to camera 2.

3. DISSOLVE *from one effect to another effect:* This is what all those extra buttons and EFFECTS BUSSES are all about. You can set up one effect, say a CORNER INSERT, on EFF 1. You can set up another effect, say a DIAGONAL SPLIT, on EFF 2. On EFF 3, channel A, the EFF 1 button is pressed. On EFF 3, channel B, the EFF 2 button is pressed. Moving the EFF 3 FADER down, assuming it's in the MIX mode, will DISSOLVE the picture from the first effect (CORNER INSERT) to the second effect (DIAGONAL SPLIT). For all this to be recorded, the PGM BUS must have the EFF 3 button pressed.

This process is called CASCADING or REENTRY. You use a FADER to go from one picture to another. *How* you go is determined by the buttons which tell the FADER to WIPE, DISSOLVE, etc. *Where* you go depends on what's selected on its A and B channels. You could punch up just a camera, or you could punch up *another* EFFECTS BUS. That BUS, in turn, could be showing a simple camera shot or a fancy effect determined by *its* FADER and *its* A and B chan-

---

MINIGLOSSARY

directly selects (when pressed) which picture or special effect is broadcast or recorded.

Eff 1 (or effects bus 1)   A video source can be selected on channel A. Another source can be selected on channel B. A combination of these two can be a special effect which is available through a circuit called the effects bus. There may be several effects set up on several effects busses. Eff 1 is the name given to just one of those effects busses.

*Mix   One of the ways of going from one TV picture to another

(as opposed to wipe and key). Mix is often the name on the button that tells the fader levers to dissolve rather than wipe or key from one picture to the next.

Diagonal split screen   A split screen divided diagonally.

Reentry   The process of creating one special effect and then using it as a source (as if the picture had come directly from the camera) that can now be mixed with another picture to create another special effect. Process allows effects to be piggybacked atop one another.

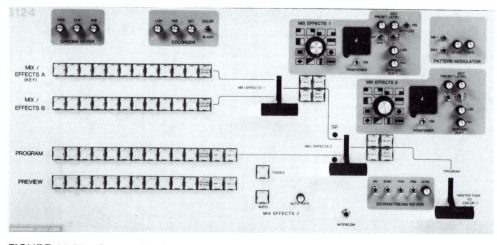

FIGURE 11-25   Crosspoint Latch STUDIO PRODUCTION SWITCHER. (Courtesy of Crosspoint Latch)

nels. *And*, one of *its* channels could be displaying *another* EFFECTS bus, with another effect either happening or waiting to happen.

Although the *professional* STUDIO PRODUCTION SWITCHER is a fun-looking toy, we're going to scale back a bit to a smaller model. Educational, industrial, and business users can generally afford the $4000–$10,000 variety of switcher such as the one shown in Figure 11–25. Let's discuss some of the features found on one of these monsters.

Inputs

STUDIO PRODUCTION SWITCHERS generally have 6–12 inputs. They could come from cameras, videotape players, character generators (electronic titlers), or any other video sources.

If the signals are coming from SYNCHRONIZED cameras or other sources where the video signals are locked in step, they are called SYNCHRONOUS, and the switcher will be able to CUT or DISSOLVE or WIPE, etc., from one to the next smoothly. If, however, the input signal is not SYNCHRONIZED to the others (such as an off-air TV broadcast or the output from a common VCR), the signal is called ASYNCHRONOUS. You could switch to that signal using your switcher, but there would be a blip in the picture. If using these sources causes a glitch, then why use them? Convenience. They allow you to plug everything you've got into this one "master control" to make original recordings through it or copy tapes through it all at the push of a button. During a live show, however, you wouldn't try to switch from one ASYNCHRONOUS source to another.

Preview and Program

Remember a thousand pages ago when you wanted to DIS-SOLVE from camera 2 to a CORNER INSERT between cameras 1 and 2? It would have been nice to have seen the INSERT before we took potluck and DISSOLVED to it. PREVIEW lets you see the EFFECT before you are committed to it. Your PREVIEW monitor will display the effect automatically, assuming that the right buttons have been pressed. On the Crosspoint Latch switcher shown in Figure 11–25, you can PREVIEW any source simply by pressing its button. By pressing the ME1 or ME2 buttons (they're like EFF 1 and EFF 2 discussed earlier), the switcher will PREVIEW automatically whatever scene is ready to come up next.

Once you are satisfied with the picture on the PREVIEW monitor, you can record it by pressing the appropriate button on the switcher's PROGRAM BUS.

PROGRAM means whatever signal or combination of signals is to be recorded by your VCR or broadcast. PREVIEW is like AUDITION or CUE on the audio console: It is for your information only so that you can make sure pictures are satisfactory before you use them.

Using Preview to Show
What Your VCR is Doing

Your switcher selects a signal and sends it out the PROGRAM channel to your VCR. What do you suppose would happen if you took the VIDEO OUT from that same VCR and fed it back into the switcher as shown in Figure 11–26? If you then pressed the PREVIEW button for "VCR," you could

---

MINIGLOSSARY

*Cut   Switch from one picture to another directly, in the blink of an eye.

Asynchronous   Not synchronized. Running independently without external sync circuits locking to an outside clock.

*Preview   A channel on a switcher/SEG sends out to your preview monitor a view of an effect so that you can adjust or perfect the effect before using it. Like *audition* in audio.

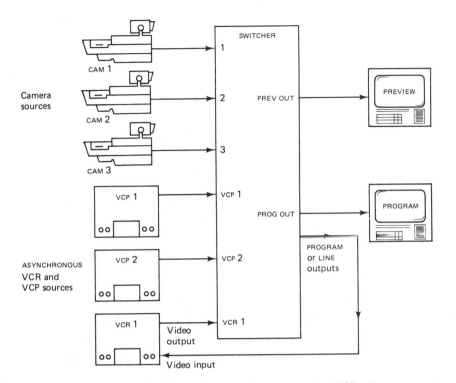

To display VCR 1's signal on the PREVIEW monitor, press the VCR 1 button on the PREVIEW BUS. To preview camera effects, press the appropriate button (could be called ME1, ME2, or AUTO, depending upon manufacturer) and the effects will automatically be displayed on the PREVIEW monitor

FIGURE 11-26 Using the PREVIEW monitor to display the output of a VCR while it is recording.

PREVIEW the output of your VCR *while* it is making its recording. So what's the advantage of PREVIEWING the VCR's output? It shows the same picture as the LINE monitor, right? Right—when the VCR is recording correctly. However, if the VCR's VIDEO LEVEL creeps up too high, the problem won't appear on the LINE monitor; it will show up on the VCR monitor or the VCR's meter. If the VCR and its monitor are in the next room or are hard to see, you may not notice the problem. If however, the VCR is playing its signal through the PREVIEW monitor, which is generally right in front of you, any VCR problems will capture your attention.

Not only can the PREVIEW monitor show you what your VCR is doing while it records, but it can also display your VCR's picture during playback. This is most useful when you need to EDIT OUT a mistake during a show. The PREVIEW monitor will allow you to watch the tape play while you're getting ready to EDIT IN to a new live scene.

Of course the PREVIEW monitor can display only one picture at a time, so when it is showing the VCR's output,

it's no longer previewing EFFECTS. To keep tabs on both, you may switch back and forth between the VCR and the EFFECTS by alternately pressing the VCR button and the AUTO PREVIEW (or ME1 or ME2, or whatever) button on the switcher's PREVIEW BUS.

In short, using the PREVIEW monitor to display VCRs, VCPs, as well as EFFECTS puts many signals at your fingertips without cluttering the console with numerous monitors. One does all the jobs.

Key

KEY is a special effect where part of a camera's picture is cut out, as if with scissors, and replaced with another camera's picture. You can DISSOLVE to and from this effect the same as you would DISSOLVE to a SPLIT SCREEN.

There are two kinds of KEY effects. One replaces all the *black* parts of one camera's picture with parts from another camera. The other, called CHROMA KEY, replaces all the *blue* (or some other selected color) parts of one

---

MINIGLOSSARY

Auto preview Mechanism on a switcher/SEG that automatically displays on your preview monitor any effect *not* being recorded but ready to be shown once selected.

*Key (or luminance key) Special effect where the dark parts of one camera's picture are replaced with parts from another camera's picture.

camera's picture with parts from another camera's picture. KEY can be used to create a special effect or to make something disappear or to make words appear on a screen. Let's first study the black-and-white KEY.

Black-and-White or Luminance Key. Figures 11–27 and 11–28 show examples of a simple black-and-white KEY.

Going to this mode requires some forethought and a little knob twiddling. To create the effect shown in Figure 11–27,

1. Press the KEY button on the switcher.
2. On channel A of the switcher, punch in camera 2.
3. On channel B of the switcher, punch in camera 1.

The camera on the B channel will have some parts of its picture replaced with parts from the A channel. All of B's dark parts will be replaced with A's picture. It is as if all the light parts of B were real and as if all the dark parts were transparent and through them you could see A's picture. Now the question is, How dark must parts of B's picture be before they become transparent? The answer lies with the KEY SENSITIVITY (or KEY SENS) adjustment. This knob determines the threshold at which the equipment will call something black or white. Adjusted all the way in one direction, the KEY SENS will consider *everything* in B's picture dark enough to be transparent. Turned the other way, *nothing* from B will be transparent. With the help from PREVIEW and a little experimenting, the right effect can be perfected and DISSOLVED to when needed.

4. Adjust KEY SENS so that the part you wish to have disappear disappears. Watch carefully to make sure that other dark parts of the picture do not also disappear. Careful adjustment of KEY SENS will keep the edges from your dark object from appearing too grainy and ragged. Smooth shadow-free lighting is often necessary to achieve a good KEY effect.

If the person is moving around as in Figure 11–27, she might slip out from behind the mask you created for her. An alert camera operator may need to aim camera 1 around so that the mask moves with the performer.

In Figure 11–28, the process is similar. You must adjust your lighting and KEY SENS so that only the black sheet of paper disappears. Camera 2 must be carefully aimed so that the desired part of the picture shows when you go to the effect.

camera 1 (channel A)

camera 2 (channel B)

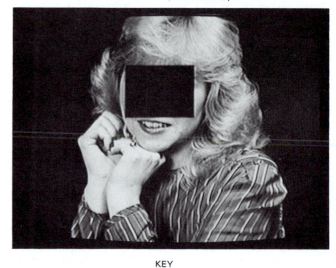

KEY

FIGURE 11-27 KEY over someone's eyes.

MINIGLOSSARY

***Key sensitivity** Control on a switcher/SEG that determines *how dark* something has to be before it disappears and is replaced by another camera's picture.

Keying Words. *The most frequent use for the* KEY *effect is for* KEYING IN *words on the TV screen.* Here a white word on a black background would be fed to channel B of the switcher. In the KEY mode, the black background will disappear and be replaced by the picture from channel A. Thus, it will seem like solid white paper letters were pasted over the picture from channel A.

This is a sharper, more solid effect than the SUPERIMPOSITION you saw earlier. With the SUPER, the words were ghosty. With KEY, the lettering is opaque and solid. Figure 11–29 compares the two effects.

As in the cases described earlier, if you want to DISSOLVE from, say, camera 2 to 2-with-a-word-on-the-bottom using the simple switcher in Figure 11–17, you could do the following:

1. Feed camera 2's picture to the C channel (program) and A channel.
2. Put the word on the B channel.
3. Switch the EFFECTS BUS to KEY.
4. Adjust KEY SENS while examining it on the PREVIEW monitor.
5. DISSOLVE to EFFECTS.

The effect will fade in while the rest of the picture will remain the same. You can get out of the effect by DISSOLVING in the opposite direction from the EFFECT back down to the C channel.

External Key. EXTERNAL KEY is like regular KEY only fancier. Here an additional camera decides, depending on the light and dark *it* sees, which of the two other cameras' pictures will be shown. Figure 11–30 shows examples of an EXTERNAL KEY. There camera 1 is the controlling camera deciding which parts of camera 2 and camera 3's pictures will be shown. When camera 1 sees white, camera 2's picture will show. Where camera 1 sees black, camera 3's picture shows.

To make your switcher work in this mode, you need to find the INT/EXT knob near the KEY knob on the switcher. On some switchers there may not be any button for you to press to select which camera does the controlling. In such cases, the controlling camera signal must be sent into the switcher through a plug in the back of the switcher. But, normally, switching the INT/EXT knob to EXT automatically makes the external camera the controlling factor. Meanwhile the remaining two (or more) cameras may be selected on the A and B channels as usual. You can DISSOLVE TO and FROM this effect also.

No law demands that the EXTERNAL KEY must remain still. It could be a graphic, and your camera operator could

camera 1 (channel B)

camera 2 (channel A)

KEY

FIGURE 11-28 Black area replaced with another picture.

MINIGLOSSARY

*****External key**   Key effect where the dark and light parts of one camera's image determine which of two other cameras' pictures will be shown.

camera 1

camera 2

SUPER

KEY

FIGURE 11-29   SUPER and KEY compared.

zoom in on the graphic, making the black and white parts grow larger. Imagine for a moment that the controlling camera were looking at the word *today* spelled out in big fat letters. The camera operator could start zoomed out on the word and zoom in on it until the "o" in *today* completely filled the screen. The result would be one camera's picture with another camera's picture spelling out the word *today* as if someone had cut the letters out of one photograph and pasted them onto another photograph. As the camera zooms in, the letters get larger, and the second camera's picture takes over the whole screen.

One fancy way to change from one camera shot to another is to make a movie that starts all white, and little by little gets blackened in. Little squares can fill the picture as shown in Figure 11–30 or blobs of paint can be spattered on the picture, drooling until they cover the picture. If this movie (or video tape) is now played into the EXT KEY input, you will see an exotic transition from camera 2's picture to camera 3's picture happening wherever the black appeared on camera 1's picture.

Using black and white paper, cutouts, paint, and other moving methods to change a white screen into a black screen, you can create inexpensive and unique wipes and special effects. Some SEGs cost thousands of dollars because of the wide array of special effects they have programmed into them. Using EXT KEY and a little time, you can recreate or surpass these effects yourself at little cost.

Chroma Key.   In color systems, one may use a color to trigger the KEY process. The technique is the same, only instead of all the black parts of the picture becoming transparent and being replaced with another picture, all the blue (or some other selectable color) parts are replaced with another camera's picture. The process is called CHROMA KEY and is demonstrated in Figure 11–31.

CHROMA KEY is easier to use than black-and-white KEY when you wish to replace part of one picture with part of another. A recurrent problem with black-and-white KEY is that the camera can never quite decide whether something is black or not and tends to take out the wrong parts of your picture. Even with careful lighting, it is impossible not to have the insides of someone's mouth, the shadow of his lapel, or some black part of a prop disappear and be replaced with another camera's picture. This is where CHROMA KEY becomes handy. Pure blue is a relatively rare color. Except for blue eyes, it is not found on the body, and you can easily take precautions to see that costumes and props don't have any blue in them. Thus, placing a person and his props in front of a large blue curtain (often called a CHROMA KEY BLUE CYCLORAMA), you can replace the entire background with a picture from a postcard, slide, movie, or video tape. You could even make your performers disappear from the

MINIGLOSSARY

*Chroma key   Key effect triggered by the color blue (or some other selected color) rather than black.

camera 1

camera 1

camera 2

camera 2

camera 3

camera 3

EXTERNAL KEY

EXTERNAL KEY

FIGURE 11-30 EXTERNAL KEY.

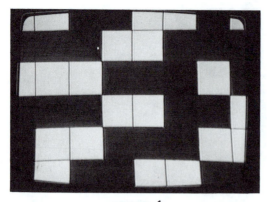

camera 1

camera 1

camera 2

camera 2

camera 3

camera 3

EXTERNAL KEY

EXTERNAL KEY

FIGURE 11-30   Cont'd

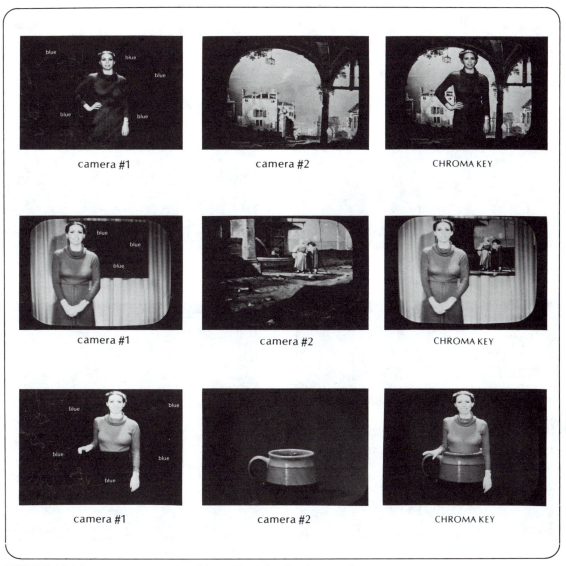

FIGURE 11-31   CHROMA KEY using blue. (See also back cover.)

from the picture by dressing them *completely* in blue. If they carried props around the set, the props would mysteriously float when seen on the TV screen. It is also possible for someone to hold something blue in her hand, like a blue crystal ball. A swirling ghostly effect could then be created in the crystal ball. Zooming in on the crystal ball, that ghostly scene would then become larger and larger until it filled the screen entirely, a nice transition from the real world to the crystal ball world.

CHROMA KEY is not perfect. Care must still be taken in lighting the set. TV cameras tend to turn dark shadowy areas into dark blue instead of pure black. If you're KEYING on blue, your CHROMA KEYER may cut those areas out of your picture. You would counteract this by making sure that your lighting was soft and that there were no deep shadows. Another problem you may have seen on newscasts or weathercasts involves a performer who wears clothing

close to blue. The TV camera sees the aqua or lavender in someone's tie and immediately slices it out and replaces it with a hurricane over the Carolinas.

Sometimes when a performer gets too close to his blue background, the reflection of the blue background on his light-colored clothing makes his clothing look bluish, especially at the edges. The camera, seeing blue-edged clothing, cuts out the shoulders or elbows of your performer, replacing him with a rough and grainy edge. One way to counteract this effect is to use AMBER GELS in your backlights. The amber light tends to counteract the blue, giving your performer sharper edges which are easier for the CHROMA KEY to "see."

CHROMA KEY only works with color TV cameras and SEGS. You can make an EXTERNAL CHROMA KEY using the same technique as black-and-white EXTERNAL KEY. As before, the CHROMA KEY sensitivity to blue is adjusted with a

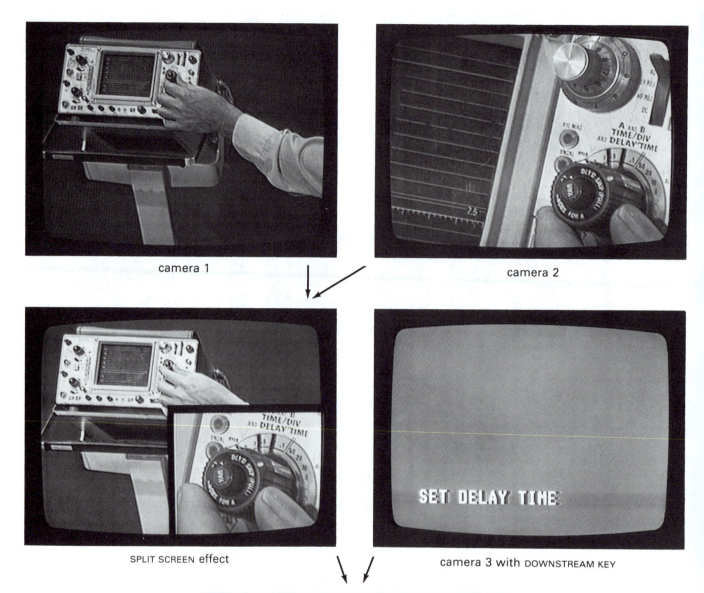

camera 1

camera 2

SPLIT SCREEN effect

camera 3 with DOWNSTREAM KEY

result

FIGURE 11-32   DOWNSTREAM KEY.

KEY SENS control. Another control on the switcher adjusts *which* color the CHROMA KEYER will be sensitive to. You could CHROMA-KEY on yellow if you wanted to replace the leaves on an autumn tree with psychedelic flashing blue and red candy stripes.

Downstream Keyer. Sometimes you'd like to do several effects at once. If your switcher is busy doing a WIPE, how can you KEY a word over that WIPE? The answer is with a DOWNSTREAM KEYER.

First you set up the WIPE, SUPER, or KEY EFFECT, or whatever else you want your picture to be. The next trick is to KEY some words or effect *over the top of* this picture. By switching on the DOWNSTREAM KEYER, you can send a graphic or title from a camera or character generator into the switcher to be added to the picture after, *downstream*, all the other effects have been done. Figure 11–32 shows a SPLIT SCREEN with a DOWNSTREAM KEY added to it.

Matte

MATTE is like KEY only you can adjust the color and brightness of the part being KEYED IN.

On the simple fader, you would SUPER a word on the screen. It would be readable but pale; you could see through it; it would have no zip. KEY added opaqueness to the lettering, but still the zip was lacking. The word would only be as bright as the whiteness in the lettering. If perchance you were KEYING a white word onto a too light scene, the white word would almost disappear. What you would like is the ability to turn that white word black or some other color which would show up well (without recreating a graphic again). This is where MATTE is helpful.

MATTE allows, through the LUMINANCE adjustment on the switcher (probably near the MATTE and KEY knobs), the lettering to appear very white, gray, black, or whatever. The black matting comes in handy when you haven't any dark places on the screen to put a white word. Instead, you find a light part of the picture and emboss a dark word. Another possibility is to make the MATTE *very* white so that the word is brighter than the rest of the picture. Figure 11–33 compares SUPER, KEY, and MATTE.

Except for MATTE's adjustable coloration, it is in all other respects like KEY. It can be DISSOLVED TO, it needs a KEY SENSITIVITY adjustment to decide which grays will be

considered black (or in color systems which colors will be considered the trigger colors), and it can be auditioned on the PREVIEW monitor.

MATTE isn't just good for words; it can be used for artistic effects. Where KEY would allow you to place a person into a picture, MATTE would allow you to take the continuous gray tones out of the person before sticking her on the picture. The effect is cartoon-like and is called POSTERIZATION. Figure 11–34 shows an example.

In color systems, not only can you MATTE in various shades of black and white, but you may also COLORIZE the effect. Take caution, however, when MATTING color subtitles onto a picture—the tinted words may look pretty, but white is usually more legible.

Joystick

The JOYSTICK was invented long before the video game. The JOYSTICK is a lever on the switcher which can move in all directions to position various special effects on the screen. For instance, if you wanted to compose a diamond-shaped insert of somebody's face in the upper right-hand corner of the screen, you'd first select that particular pattern on the EFFECTS BUS of your switcher. Next, you'd adjust the size with the WIPE levers. Then, using the JOYSTICK, you'd position the pattern in the corner of the screen. Don't forget that the camera operator has to position the face in the upper right portion of his or her viewfinder screen as well. Figure 11–35 shows an example.

Soft Key and Soft Wipe

SOFT KEYS and WIPES have soft fuzzy edges. Figure 11–36 shows a SOFT WIPE.

SOFT KEYS are handy for making realistic KEY effects through glass and other semitransparent objects. Instead of the glass having a hard, unrealistic edge, a SOFT KEY allows you to see *through* the glass yet also see its rim and highlights.

Border

Sometimes you'd like to SPLIT a screen and not have anyone realize you did it. This would be handy if you had one performer standing still in front of one camera and a second

MINIGLOSSARY

Downstream keyer A circuit in the switcher/SEG which will key an image (usually a word) over the top of a picture or special effect. This is often the last thing done to the signal before it exits the switcher to be recorded.

*Matte A special kind of key effect where light parts of a picture are removed and replaced with a chosen color.

Posterization Visual effect of reducing a picture's varied brightness levels down to just one or two, giving it a flat poster-like or cartoon-like look.

*Colorize Adding color to something electronically. A matte can be white, black, gray, or colorized; so can wipe borders and backgrounds.

Joystick A multiposition lever on a studio switcher that positions special effects anywhere on the TV screen.

Soft key Key effect with a fuzzy, soft edge.

Soft wipe A split screen or wipe effect with a soft border where the two pictures join.

camera 1

camera 2

SUPER

KEY

MATTE BLACK

MATTE WHITE

FIGURE 11-33 Comparison of SUPER, KEY, and MATTE.

camera 1

camera 2

KEY

MATTE BLACK

MATTE WHITE

FIGURE 11-34 MATTING a person over a scene.

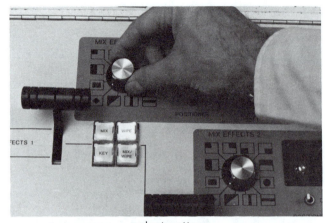

select pattern

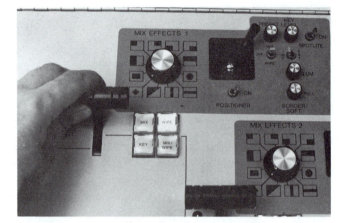

adjust size of pattern

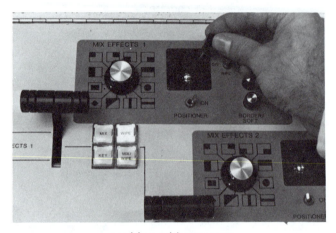

position with JOYSTICK

camera 1

camera 2

EFFECT

FIGURE 11-35 Positioning an effect with the JOYSTICK.

FIGURE 11-36 SOFT WIPE.

performer dancing in front of a second camera. If you horizontally SPLIT the screen at both performers' belt lines, you could have a very silly-looking effect of someone's torso standing still while their legs are dancing like crazy. (For this effect to work, you would also need the same background for both performers.) Using an invisible SPLIT, you are able to have props and pictures occupy part of the screen without letting the audience know that the effect is contrived.

Sometimes, however, you wish to place a border around your INSERT or WIPE to make it stand out better, as in Figure 11–37. Many switchers will allow you to adjust the width of the border as well as its color.

Colorizer

CHARACTER GENERATORS and monochrome cameras aimed at black-and-white titles can give you white lettering on a black background. If your switcher has a COLORIZER, it can change the color of that lettering to any color you like. This is much like the MATTE effect where you could change both the brightness and color of the MATTE. You can use this for strange POSTERIZATION effects as well as for making pretty colored words and borders for your WIPES and INSERTS.

Spotlight

Sometimes you would like to point your viewers' attention to something on the screen. One way would be to have your performer physically point to the object. Another way would be to have the desired part of the picture be brighter than the surrounding parts, as if a spotlight were aimed at it. This is the SPOTLIGHT effect shown in Figure 11–38. It is

quite useful for technical and educational programs where your viewers need to focus their attention on one small part of a larger machine or a particular line of poetry.

Master Fade to Black

It's easy to FADE TO BLACK during a TV show. It's like dissolving from one camera to another only that other camera happens to be capped, turned off, or just not there. Some switchers even have a button called BLACK which you can fade to. Still, some productions get so complex that it is difficult to figure out how to undo the effects that have been set up. MASTER FADE TO BLACK saves us from all this thinking. It fades your picture to black no matter what effects or buttons you have selected on your switcher. It *always* works and therefore is handy when you are in a panic. Think of

FIGURE 11-37 BORDER.

FIGURE 11-38 SPOTLIGHT. (Courtesy of Panasonic)

MINIGLOSSARY

Spotlight Special effect to highlight a portion of the picture as if a spotlight were aimed at it.

Master fade A fade lever that always fades the picture out to black or a chosen color.

it the same way you think of the MASTER VOLUME on your mixer. It fades down all sources at once.

Sometimes when you start using the switcher at the beginning of the day, you may wonder why you can't get a picture out of it. This may be because your MASTER FADER is turned off. Before you go pulling out your transistors or your hair, check your MASTER FADER to see that it's faded up.

Background Generator

A BACKGROUND GENERATOR will turn a black background into the color of your choice. Say you were going to KEY a title onto the TV screen. Either you'd use a CHARACTER GENERATOR or a camera aimed at white letters on a black background to create the title. These letters could be simply KEYED over a black background, but that would be boring. You could substitute the signal from the BACKGROUND GENERATOR for the black background and make the picture more interesting. Don't forget that you can also COLORIZE the letters in your words using another function of your SEG in order to make your letters look pretty too. For best results, you should use a dark-colored background and light-colored letters.

Tally System

All STUDIO PRODUCTION SWITCHERS have a system that simultaneously lights the switcher buttons when they are pressed along with the TALLY LIGHTS on the cameras.

Genlock

Sometimes GENLOCK is a piece of equipment separate from the switcher and sometimes it is part of the switcher. If the switcher and everything else in the studio is driven by a separate external SYNC GENERATOR, then that SYNC GENERATOR is probably in charge of GENLOCKING also. If, however, your switcher manufactures its own sync to drive the cameras and whatnot, then a GENLOCKER may be inside the switcher. What does a GENLOCKER do?

Synchronizing a number of color TV cameras requires that they all receive SYNC and a number of other sync-related signals to keep their electronics in step. These other signals may be called HORIZONTAL, VERTICAL, BLANKING, BURST, and our old friend SYNC. An EXTERNAL SYNC generator can make all these signals to drive the entire system, or your

switcher can make all these signals. By having all your equipment synchronized by these signals, you can CUT, FADE, and do SPECIAL EFFECTS smoothly between all your SYNCHRONOUS sources.

What happens if you wish to CUT or FADE between ASYNCHRONOUS sources? Since they are not in step with the rest of your system, you'll get a glitch. Here is where GENLOCK comes in. GENLOCK is a circuit that will "listen" to the SYNC from an independent video tape player, home video camera, or broadcast TV channel and will manufacture SYNC and other signals for the whole system using this outside SYNC as a guide. Thus, your whole studio system is synchronized to this one source. Put another way, instead of the SYNC GENERATOR or the switcher being "boss" of the system, the outside VTR or camera can be "boss" of the system.

Beware when you are GENLOCKED to a source like a VCP playing a tape that if the VCP stops playing or there is a discontinuity in the tape (a glitch or a bad edit), your whole system will hiccup right along with your VCP. Except in special cases, it is best not to use GENLOCK or to GENLOCK to a very reliable source.

Some smaller switchers use GENLOCK for another reason. They do not manufacture their own SYNC and they don't accept an EXTERNAL SYNC GENERATOR. Instead, one of the cameras (switched to INTERNAL SYNC) manufactures its picture and its own SYNC and feeds this to the switcher. The GENLOCKER in the switcher uses this to send SYNC out to the other cameras (which are in the EXTERNAL SYNC mode) to keep them in step. Thus, camera 1 becomes the SYNC GENERATOR for the whole system.

MINIATURE AND SEMIPROFESSIONAL SWITCHERS AND SEGS

For those who like to travel really light, there are miniature switchers which can be strapped to your hip and connected to portable TV cameras.

Designed for the home video user but also used in shoestring television operations are other switchers that can do special effects. There are $600 boxes of electronics able to fade from a camera or video player's picture, to a chosen color. One can get pretty tired of seeing the picture fade or wipe to green between shots. To dissolve between two independent (non synchronized, i.e., home video) cameras or VCRs, one could buy a small switcher with time base cor-

---

MINIGLOSSARY

*Background generator   SEG circuit adds color to a black background, useful for keying words onto a colored background.

*Genlock   Electronic device which, when fed a video signal, will manufacture synchronized sync signals so that (1) its picture will synchronize with the source's video signal or (2) its sync signals will help other devices (like cameras) synchronize themselves with

the source's video signal.

Blanking   One of th sync signals that determines the size of the black sync bar at the bottom of the TV picture.

Burst   One of the sync signals to control the hue and color accuracy of TV pictures.

director with the task of "calling the shots" (telling everyone what to do), a task requiring only four eyes, three arms, one and a half brains, and ulcers of steel.

SELECTION OF SHOTS AND EFFECTS— THE DIRECTOR'S JOB

Here you are in the big time, facing a bank of silent gray monitors in a chilly darkened control room. The smell of nervous perspiration is in the air—your perspiration. Which shots do you use? Do you CUT or DISSOLVE? Do you WIPE the subtitle up from the bottom or MATTE it over the picture? The answers to these questions are subjective. Composition of shots is an artistic and creative endeavor. Here are some generalizations.

Dissolves Versus Cuts Versus Wipes Versus Fade-outs

CUTS are easiest and quickest to do, requiring only the push of a button. CUTTING from one camera to another is the least obtrusive way of showing *different views of the same thing*. Changing from a close-up to a medium shot is best done with a CUT. Changing from a performer's face to what the performer is seeing or holding is appropriate for a CUT. Switching from a shot of a football kicker to a long shot of both teams on the field and then to a shot of the receiver is best accomplished with CUTS.

DISSOLVES, on the other hand, imply a *change in time or place*. One would DISSOLVE from an interview to a slide or tape of the subject being discussed. DISSOLVES imply

Panasonic WJ-AVE5 video/audio mixer with special effects. (Courtesy of Panasonic)

rectors built into it, such as the Panasonic WJ AVE-5. This $1800 gadget performs dissolves, wipes, mosaics, strobes, posterization, and numerous other effects between two independent video sources (VCRs, cameras, computers with video output, or character generators). It handles composite and S video signals, and mixes audio too. It even genlocks to a third source, such as a camera or character generator used to key in titles. It will also colorize the title and add edging or drop shadow to the letters. Another neat gadget is the Video Toaster (made by Newtek), a $2500 circuit board that fits into an Amiga computer. The likeness of a fancy switcher appears on the computer screen and you move the "controls" with a mouse. You then feed synchronous (or TBC'd) signals into the Amiga which now emulates a broadcast quality $30,000 production switcher with a zillion hi-tech effects.

THE TECHNICAL DIRECTOR

Now that you see what goes into pushing all those buttons on the switcher, it makes sense to have a crew member dedicated to that one task. Trying to direct and do the switching for your own show requires six eyes, five arms, three brains, and nerves of steel. In the meantime, who runs the audio, who watches the video levels, and who follows the script? In productions of any complexity, the TECHNICAL DIRECTOR (abbreviated TD) removes from the director's shoulders the load of *finding* the right buttons and *perfecting* the effects on PREVIEW before using them. This leaves the

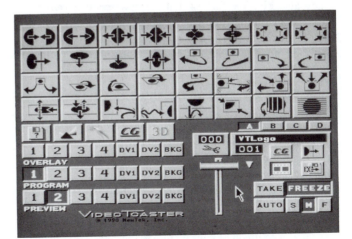

Video Toaster puts a video switcher/SEG/character generator and graphics generator into your Amiga computer. (Courtesy Newtek)

MINIGLOSSARY

*Technical director   The person who pushes the buttons on the switcher/SEG during the show.

Kiss black   Fade to black followed immediately by a fade up on a new picture.

"meanwhile, back at the ranch" or "later that evening" or "see this example of what I am talking about." Like a new paragraph, a DISSOLVE brings an end to one series of thoughts and a beginning to another. DISSOLVES are generally inappropriate for merely changing shots. You wouldn't DISSOLVE from the batter who just hit the ball to the shortstop who is about to catch it. Nor would you DISSOLVE from a medium shot of a performer to a close-up shot of a performer (unless it is for dramatic effect such as a slow DISSOLVE from a long shot of a singer to a close-up of the singer's face).

A montage of rapid-fire shots is best portrayed with CUTS, which in themselves can add tenseness and excitement to the scene. DISSOLVES, on the other hand, are not only difficult for the switcher to handle in rapid succession but in themselves create the opposite impression, one of calm, casual, serene, relaxing entertainment. Beauty, awe, and deep significance are often portrayed with DISSOLVES.

FADE-OUTS or FADE TO BLACK imply an ending, while FADE-INS or FADE-UPS imply a beginning. FADE-OUTS followed by FADE-INS (which the experts call KISS BLACK) indicate a *significant* change of place or time or train of thought. Where the DISSOLVE was similar to going from one paragraph to a new paragraph, a KISS BLACK is more like ending one chapter and beginning a new chapter.

Incidentally, DISSOLVE technically means to melt from one picture to another, whereas a FADE means you are changing from a picture to black or vice versa. Many experts, however, outside of their classrooms, will use these words interchangeably. Perhaps this is because it's easier to say FADE than DISSOLVE when you are in a hurry. So don't be surprised if you hear someone say, "Fade from camera 1 to camera 2."

WIPES like DISSOLVES imply changes in time or place. They also give the feeling that the new scene is *replacing* rather than simply following the previous scene. WIPES are spicier than DISSOLVES and should be used very sparingly lest their novelty wears off and they begin to distract your audience from the content of your production.

Split Screen Versus Matte and Key

Say you have a title, subtitle, or label you wish to have share the screen with your performer. Do you SPLIT SCREEN it at the bottom or do you MATTE it somewhere? In making this decision, here are some things to consider: A split screen intrudes on your picture (taking a section of it), whereas the MATTE is less obtrusive; you can see the scene between the words in the MATTE so that you don't feel that you are sacrificing so much of your picture to the subtitle. However, with a MATTE, you have little control of the background behind the words as the scene moves and changes. What do you do if the bottom of the screen has a lot of white in it? Quickly change your matte to black? Shove your words to the top or to the middle of your screen? Such action

would be more distracting to the viewer than simply SPLITTING the screen so that the words always have the best background for easy reading. The decision about whether to SPLIT SCREEN or MATTE depends on the kinds of scenes expected and how important the words are. If the words are paramount, sacrifice a section of the picture with a SPLIT SCREEN. If the words are secondary and the scenes will accommodate them, use MATTE. Whatever you decide, try to be consistent. Use one method or the other throughout the show lest the audience be distracted by your change in technique.

Transitions

How often should you change shots and what kinds of shots should you change to? The main object is to follow the action, keeping it near the center of the screen (unless, for dramatic effect, you are hiding the action from the audience). If keeping the action on the screen means frequently alternating shots between cameras, then do it. But if you have a good view of the action on one camera and an equal or worse shot of the same thing on another camera, then keep the first shot. Don't change just because the buttons are there on the switcher. As long as the audience sees what they want and need to see, they don't care how many shots were used in the process of displaying it to them.

Some directors change shots just to add variety to the show. The wisdom of this procedure depends on the creativity and savvy of the director, the objectives of the program, and the content of the particular shots. Switching to a side shot of the news anchorperson revealing the busy newsroom in the background adds variety and style to the end of the newscast. Doing this *in the midst* of a news presentation would detract from it. Cutting to a close-up of an interviewee's nervous wringing hands adds variety (and insight) to a show. Displaying this shot while the interviewee is making an important point, however, would be distracting to the audience. Showing this shot while the interviewer was asking the question might even confuse the audience —they might assume they were watching the *interviewer's* hands.

Perching a camera up in the rafters offers a great opportunity to spice up your presentation of a square dance. This shot adds variety. Since you've gone to the trouble of hoisting the camera up there, why not also use it to view the poetry reading coming up next? This unusual shot should add plenty of variety, right? Ridiculous as it sounds, this is one of the toughest decisions facing a fledgling director —whether to get mileage out of something in one's production arsenal or to disregard the dazzling things one *can* do and instead do what is best for the total production. The production should come first. Try not to get pizzazz-happy. Skip the fancy overhead shot if it doesn't add to the drama of the poem.

Sometimes the shots you use imply what's coming next on the screen. The switch from a medium shot to a long shot of a performer indicates that a performer is about to move or be joined by someone. Cutting to a close-up of a performer's face readies the audience to catch her expression. A gesture will be expected if you now switch to a medium shot, and a shot of the door prepares the viewers for an entry. Making any of these shot changes without a specific purpose will not add variety to your show—it will only confuse your viewers. Therefore, change shots for a purpose, not for idle variety.

One outgrowth of the preceding philosophy is the following general rule: Avoid going from a two-shot to another two-shot or going from a long shot to another long shot or from a close-up to another close-up. These are called MATCHED SHOTS. Switch from one thing to *something else*—from a long shot to a medium shot, for instance. The main idea is to offer a *different* view of something when changing shots. The medium shots of the same thing reveal nothing new and are purposeless: In fact, it is often good to assign camera operators a responsibility for specific kinds of shots so that you avoid getting a duplication of shots from two or more cameras. What good is it to have essentially the same picture coming from two different cameras? One is being wasted.

A few exceptions to this rule include the case described earlier where the medium shot of the news anchorperson was followed by another medium shot of the anchorperson from the side with the newsroom in the background. Another exception to the rule relates to the portrayal of a dialogue. Shooting over one person's shoulder, we get a close-up of the second person's face as he speaks. When the first person speaks, we swap everything around, taking a close-up of the first person over the shoulder of the second. In both of these cases, although we are switching from a medium shot to another medium shot or from a close-up to another close-up, the content of the shots is significantly different. The shots thus justify themselves. They have a valid purpose. They display something that needed to be shown and perhaps could not have been shown as well in some other way.

This doesn't mean that over-the-shoulder shots should always be matched in size; it just means that they *may* be matched. In fact, having one person higher and larger than the other person in the over-the-shoulder shot will imply power, importance, assertiveness, and dominance to the larger person. Over-the-shoulder shots between a boss and an employee would generally have the boss positioned higher and larger than the employee.

With the possible exception of the over-the-shoulder shots, you almost never shoot any subject from opposite sides. The opposing shots can easily confuse the audience because what was on the left in one shot is suddenly on the right in the other. The action that a moment ago was flowing to "stage left" is suddenly moving to "stage right." Consider, for example, a shot of a race car crossing the finish line as viewed from a camera in the grandstand. Switching to a shot of the same car viewed from the island at the center of the track results in a picture on the TV screen of a car first speeding right and then speeding left. Did the driver turn around? Is this another car, and is a collision about to occur? Where am I (the viewer)? Similarly, as a performer exits through a doorway walking to the right, when the camera outside the room picks her up, the performer must still be traveling to the right. Even though the performer may be walking mostly away from you in the first shot and mostly toward you in the second shot, still the flow of travel in both cases is essentially to the right. To have the transition appear otherwise would momentarily confuse the viewers.

One way to avoid perplexing camera viewpoints is to think in terms of angles as shown in Figure 11–39. Two cameras can shoot a subject from 60° apart, from 90° apart, or even from 120° apart. Beyond that, as the angle between cameras stretches to 180°, the risk of confounding the audience's sense of direction increases.

Another way to avoid opposing shots is to draw a mental line straight through the center of the stage or performance area. Professionals call this the VECTOR LINE. The cameras can work on one side of that line only, never on the other. Better yet, the cameras shouldn't even come near the imaginary line while shooting the performance.

What do you do in those rare cases when you *have* to shoot a subject from all different angles? First, you could move the camera while it's "on the air," dollying it (and your audience) around to the other side of the subject. Second, you could have the subject move past the camera as the camera pans, keeping the subject in sight. Third, you could use more than two camera angles to shoot the scene. For example, in Figure 11–39, if camera 1 were first shooting the scene, you could switch to camera 3 or 4. From either of those, you could later switch to position 5. From 5 you could later switch to position 6 and then from 6 switch back home to position 1. Each camera angle was within 120° of the last, but taken in steps, the angles total a full 360° sweep.

Two more kinds of shots in the director's arsenal are ESTABLISHING SHOTS and REVELATIONS. An ESTABLISHING SHOT is usually a long shot introducing the viewer to the

MINIGLOSSARY

Matched shots Similar-looking views of a subject from two cameras at the same time.

Vector line Imaginary line dividing the set (action area) in two; all camera angles must be taken from one side of this line to maintain smooth transitions from shot to shot.

Camera angles (as seen from above)

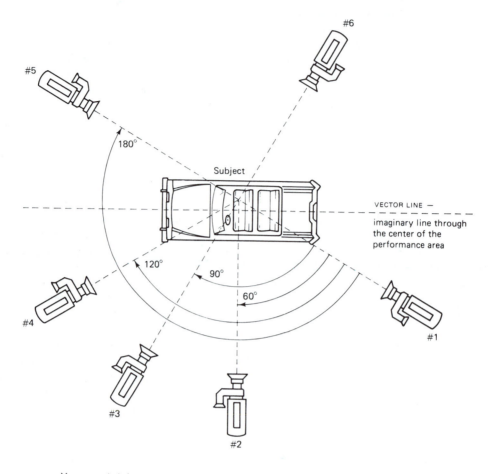

VECTOR LINE —

imaginary line through
the center of the
performance area

You can switch from cameras #1 to #2, #1 to #3, or #1 to #4, but switching
from #1 to #5 or from #3 to #6, would confuse the viewer.

FIGURE 11-39 Camera angles (as seen from above).

setting where the action is about to take place. It may be a shot of the village before we see the inn, or a shot of the inn before we see the lobby, or a long shot of the lobby before we meet the innkeeper. It may be a shot of the patient on the operating table before we see the tonsillectomy. Whatever the ESTABLISHING SHOT, it sets the stage for the program so the viewers have an idea of what to expect. Without the ESTABLISHING SHOTS of the inn or its lobby, the viewers—instead of observing the content of the opening scene—may be asking themselves, Is this a home? Is this a tavern? Is this a city hotel or a country motel? Is this today or 40 years ago? In the case of the tonsillectomy, the audience of medical students may be wondering, Is this patient a child or an adult? What are the operating conditions? The answers to all these questions *may* become apparent as the

show progresses, but it's more efficient to sweep them all out of the way with a moment's establishing shot.

A REVELATION does just the opposite. Something is purposely omitted from the scene and is later revealed to surprise the audience. For example, we are watching the end of the guided tour at the city zoo. The camera slowly zooms out from a close-up of the tour guide's talking face to reveal that he is behind bars and on the cage door the sign reads "Tour guide—please do not feed." Another example would be an opening close-up of a news reporter picking up a teddy bear from beneath a wooden plank. As she begins to speak, the camera draws back to a long shot, revealing earthquake desolation in the background. Here, for dramatic purposes, visual curiosity is piqued and then satisfied with a revelation.

MINIGLOSSARY

*Establishing shot   An introductory shot showing viewers where the scene takes place.

Revelation   A camera angle that hides something important and then reveals it for dramatic effect.

The director plans the shots, considers the camera angles, selects which shots are to be used, and decides how they are to be assembled as a final product. The director must continually judge the quality of his or her product and, when necessary, redo a scene (if possible) until it is perfected. For even the most skillful director, there's one job that is probably hardest of all: Calmly and open-mindedly accepting criticism about the show from a viewer (or reviewer) who hasn't the faintest idea of how much effort went into producing it.

This chapter has focused primarily on how to work the equipment. It has only briefly touched on production techniques. Some of those techniques were covered in Chapter 8. Although camera techniques and picture composition sound like they're the cameraperson's business and not the director's, *everything* in a TV production is the director's business. He or she must be fully aware of the various camera angles and how they will fit together into the final production. It is a poetry of cooperation to see a TV director DISSOLVE from one camera shot to another while the first camera is defocusing its shot and the second camera is focusing on something else. To work smoothly, three things have to happen exactly at the same time and in perfect harmony.

More TV directing techniques will be covered in Chapter 17. On the way there, we will pick up more tips on how to switch from one camera angle to another in Chapter 14 (editing). After all, as far as the viewer is concerned, it doesn't matter if we edited from one camera angle to another or simply switched from one camera to another "live." When done right, it all looks the same.

Yes, it all looks the same—invisible. The good director focuses the viewer's attention on the story, not on production techniques. Camera angles and transitions should happen without the viewer being aware of them. The director uses these tools to communicate to the viewer's subconscious. This is one of the reasons why television is such a powerful medium. You can *show* the viewer something. You can let the viewer *hear* a conversation. You can add a musical background to create a mood. You can adjust the lighting to reinforce the mood. You can add sound effects to make the environment more convincing. Already you're "messaging" the viewer in five different ways and still you have more tricks up your sleeves. Camera angles and transitions from shot to shot give the director two more doorways to the viewer's mind, or heart, or conscience.

No wonder television, when skillfully executed, can make a nation riot, or laugh, or weep.

12

TELEVISION GRAPHICS

The essence of TV graphics boils down to three rules:

1. Make it fit the shape of a TV screen.
2. Keep it simple.
3. Make it bold.

ASPECT RATIO

A TV screen is a box that is a little wider than it is tall. If the screen were 16 inches wide, it would be 12 inches tall. If it were 4 inches wide, it would be 3 inches tall. However wide it is, it is three quarters as tall. This is called a 3:4 (three-by-four) ASPECT RATIO and is diagramed in Figure 12–1.

As a consequence, visuals for television should have a 3:4 ASPECT RATIO if they are to fill the screen evenly. Panoramas don't fill this ratio because they are too wide. Telephone poles don't fit because they are too tall. Strictly speaking, even a square box is too tall to fit perfectly on a TV screen.

When showing a panorama on a TV screen, one must either display a long, long shot of it, showing a lot of sky and foreground, or one must sacrifice some of the width of the panorama, getting just a fraction of it. To display a square box, one must decide whether to cut off its top and bottom in the TV picture or whether to get all of it, leaving an empty space on its left and right.

Making the Picture Fit the TV Screen

Just because your TV screen is a box doesn't mean that you can only shoot box-shaped things. If your scene, or a photograph of your scene, is very tall, as in Figure 12–2, there are three ways to show it:

1. Zoom out to get it all in but with wasted space on the left and right.
2. Zoom in to fill the screen with the most important part of the picture, sacrificing the remainder of the picture.
3. Zoom in on one part of the picture, perhaps near the bottom, and slowly tilt up to take in the rest of the picture.

The first method is somewhat obtrusive and works best when the audience is prepared for it, such as viewing snapshots from an album. Solution 2 is fine if losing part

MINIGLOSSARY

*Aspect ratio   The shape of a TV screen expressed in height compared to width.

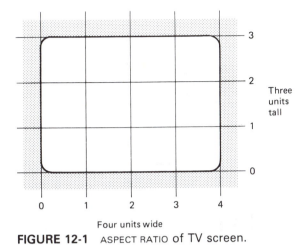

FIGURE 12-1 ASPECT RATIO of TV screen.

Three units tall

Four units wide

of your picture doesn't hurt your message. Method 3 has the advantage of adding motion to your scene, but it is sometimes hard to find a meaningful starting and ending point for your tilt.

For wide scenes, use a similar technique:

1. Zoom out, leaving a margin at the top and bottom of the picture.
2. Zoom in on part of the picture, sacrificing the rest.
3. Zoom in on part of the picture and pan to another part.

Making Words Fit the TV Screen

The composition of words, titles, and logos (a logo is a TV station's or program's symbol or trademark) leaves us more flexibility than pictures. One can arrange the words or whatever to fit the 3:4 dimensions of the screen. Figure 12–3 shows some good and bad graphic compositions.

Sometimes a formula or single line of text must remain intact as a single line and cannot be shaped into a box. In these cases, consider adding a supporting graphic to the empty space on the screen to add balance and help fill the screen. Figure 12–4 shows a good example of how adding graphic symbols to a chemical formula strengthens the instructional concept of the formula while conforming to the TV screen shape.

When accurate workmanship is necessary, it is good to measure out, in faint lines, the 3:4 shape of the TV screen on your blank art paper before you or your graphic artist begin to draw. The lines will act as a guide and can later be erased. When such care is not warranted, one may simply "think boxes" when planning graphic composition.

The Chalkboard Dilemma

Although the old "chalk and talk" presentation is not television at its best, such TV lessons are produced every day in high schools and colleges and at business meetings. Often the instructor, unfamiliar with television ASPECT RATIO, writes across the chalkboard in her normal fashion. This does not televise well as the lettering is often too small when you zoom out to get it all in. But when you zoom in, your camera views only part of the sentence at a time. TV instructors have to be taught to "think boxes." They will probably never conform to the 3:4 ASPECT RATIO you desire, but if they can be convinced to always write in short multiple lines, the results will look much better. Figure 12–5 shows an example.

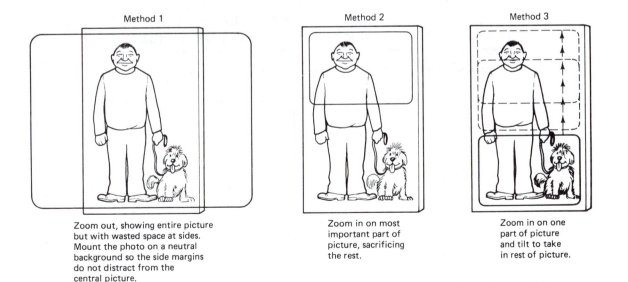

Method 1 — Zoom out, showing entire picture but with wasted space at sides. Mount the photo on a neutral background so the side margins do not distract from the central picture.

Method 2 — Zoom in on most important part of picture, sacrificing the rest.

Method 3 — Zoom in on one part of picture and tilt to take in rest of picture.

FIGURE 12-2 Making a vertical picture fit TV's ASPECT RATIO.

FIGURE 12-3   ASPECT RATIO of words on screen.

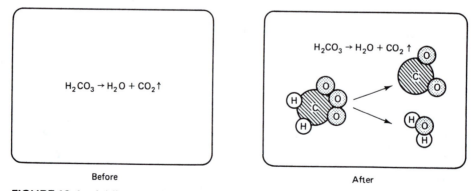

FIGURE 12-4   Adding graphics to a line of text reinforces the concept while maintaining the ASPECT RATIO.

Long sentence is either too small, or part of it is missed.

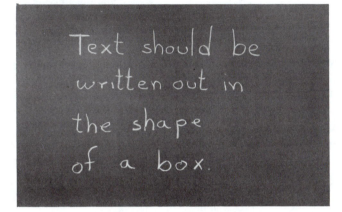

Have instructor "think boxes" when writing.

FIGURE 12-5 Writing on the chalkboard.

SAFE TITLE AREA

Two things you *don't* want to do are

1. Show your audience the edge of your graphic or title sign
2. Have a piece of the title disappear behind the edge of the viewer's TV screen

Both problems can be avoided by

1. Leaving an adequate margin around the title or drawing
2. Shooting the graphic so as to leave a little extra space around all sides of its image on your TV monitor

This extra space allows for the fact that the camera and the studio monitors generally show the *whole* TV picture, while the home viewer's TV cuts off the edges. Some home (and school) TV sets are poorly adjusted anyway, causing even further loss of the TV picture on the edges. To allow for this, the SAFE TITLE AREA is utilized, principally confining all important matter to the middle portion of the TV screen. See Figure 12–6.

The SAFE TITLE AREA holds the essential information which should *always* show on the home viewer's screen. The DEAD BORDER AREA is the blank part of your title card, which *never* shows on the TV screen. In fact, along its edge you'll often find little notes or picture sequence numbers penciled in to help the studio crew arrange and position the cards. Between the DEAD BORDER AREA and the SAFE TITLE AREA is the nether world of the SCANNED AREA. The SCANNED

MINIGLOSSARY

*Safe title area   Central portion of a graphic or a control-room monitor's TV screen that can always be seen when the picture is viewed on misadjusted TV sets; the place where it is "safe" to put a title because you know it will all show.

*Dead border area   Blank margin around a graphic that never shows on TV.

*Scanned area   Part of a graphic "seen" by the studio camera and control room monitor but not necessarily seen on the home viewer's TV set.

Use the SAFE TITLE AREA.

AREA shows on your studio TV monitors and cameras but cannot be trusted to *always* show on home TV sets. Since this part of the picture is untrustworthy, it should not contain anything important. Figure 12–7 shows the SCANNED AREA, SAFE TITLE AREA, SUPPLEMENTARY AREA, and DEAD BORDER AREA of a title card.

Figure 12–8 shows a template you can buy or make. Before creating a TV graphic, you would lay this template over the blank paper and mark along the holes in the template to indicate the edges of the SCANNED AREA and SAFE TITLE AREA. Title cards and templates can be any size (you zoom in on them if they are small or out if they're large, making your graphic always appear the same size on the TV screen). For convenience sake, 11 × 14-inch art cards are the most popular.

The DEAD BORDER AREA is the camera operator's margin. It doesn't matter if the border is an inch or 3 inches in width as long as it's large enough to make it easy for the camera operator to shoot the picture without getting the edge of the title card in the shot.

Some TV control rooms use preview and program TV monitors that are UNDERSCANNED (show the entire scanned picture plus the black sync pulse which borders the picture). To remind them of what part of the TV picture the home viewer can actually see, the control room crew often puts marker lines on their TV screen indicating the SAFE TITLE AREA. Figure 12–9 shows what this would look like. As before, anything important belongs inside the marked-off SAFE TITLE AREA.

When making titles and visuals, it is sometimes convenient to make them all about the same size. Although the titles, regardless of their actual size, may all come out looking the same once the camera operator has zoomed in or out on the title, it is easier on the cameraperson not to have to make this adjustment each time—especially if the visuals come in rapid succession.

Slides

Thirty-five-millimeter slides (also called 2 by 2 slides because the cardboard slide mounts are 2 inches square) are a staple of the video industry. A photographer with a simple lightweight slide camera can be easily dispatched to various locations and come back with pictures (after processing) that look just as real as life when they completely fill the TV screen. Thirty-five-millimeter slides of inanimate objects come out looking especially real on TV. Also, taking photos is usually less expensive than shooting a tape on location with a video crew.

The problem again is that of ASPECT RATIO. Slides, when mounted, make a picture $^{15}/_{16}$ inch tall by $1^{13}/_{32}$ inches wide. That (if you are good at arithmetic) makes a ratio of 2:3. Even if you project the slide, it still comes out with an ASPECT RATIO of 2:3, too wide for our 3:4 TV screens. This means that a piece of the left and right edges of the slide must be omitted in order to make it fit the TV screen. In addition, we mustn't forget Uncle Homeviewer, with his maladjusted overscanning TV set. We must leave a SUPPLEMENTARY AREA around the edge of the slide just for safety.

If you wish to be precise in this matter, you can buy or make a template like that shown in Figure 12–10. The template endeavors to take everything into account—ASPECT RATIO, SAFE TITLE AREA, etc.—to conform to TV's 3:4 ASPECT RATIO. The part of the slide that is actually scanned is $^{15}/_{16}$ inch by $1^{1}/_{4}$ inches. Leaving a supplemental margin around the edge cuts the SAFE TITLE AREA down to $^{5}/_{8}$ inch by $^{7}/_{8}$ inch.

The reader may notice (with pocket calculator a–twitter) that the SUPPLEMENTARY AREA given in this template is one sixth of the picture's height and width rather than the usual one tenth as in Figure 12–7. Why this increased safety space? Because the tiny slides are prone to have tiny mounting inaccuracies. The slides don't always sit evenly in the projector (sometimes because of little bumps in the cardboard mounts), and all these small irregularities add up to make the pictures project inaccurately, sometimes a little high, sometimes too far to the left, and so on. These difficulties may sound inconsequential, but they are not. Be-

---

MINIGLOSSARY

Supplementary area   Outside edge of a graphic or control room monitor's picture "seen" by the camera and control room monitor but not usually seen by the audience because it is just off the edge of their TV screens.

*Slide chain   Slide projector connected to a TV camera for con-

verting slides to video.

*Multiplexer   Mirrored device that selects which one of several projectors shines its image into a TV camera for transferring film to video.

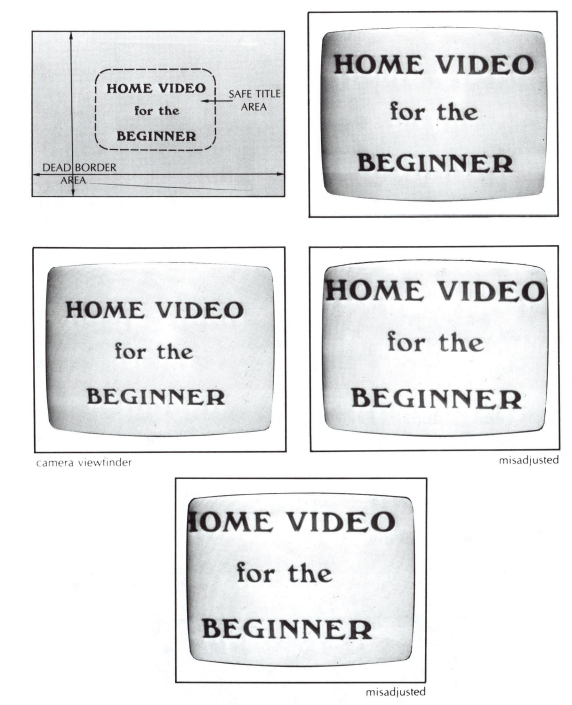

FIGURE 12-6 Picture areas compared.

cause the tiny slide is magnified to giant full-screen proportions, minor mounting and projection flaws are exaggerated. The increased SUPPLEMENTARY AREA allows for the expected flaws.

You can make one of these template slides yourself by starting with a black slide and cutting the appropriate size rectangle out of the center of the slide. Test it on your

TV system for accuracy. Thereafter, you may use the slide to lay over other slides to show how much of their picture will appear on the TV screen.

You may also wish to mark the eyepiece of your 35mm camera to show you where the SAFE TITLE AREA is.

Vertically mounted slides are out of the question for SLIDE CHAIN (or MULTIPLEXER) use; they are just too tall

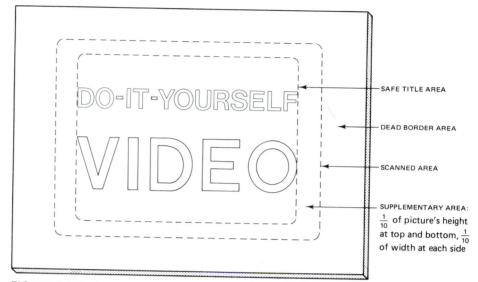

FIGURE 12-7   Areas of a title card.

SAFE TITLE AREA

DEAD BORDER AREA

SCANNED AREA

SUPPLEMENTARY AREA:
$\frac{1}{10}$ of picture's height
at top and bottom, $\frac{1}{10}$
of width at each side

Put unimportant stuff like background outside the ESSENTIAL AREA. (From *Single Camera Video Production*—Prentice Hall)

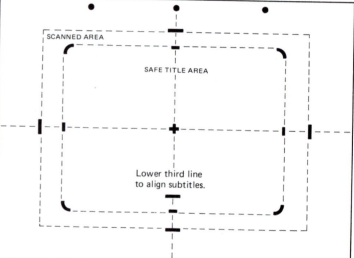

FIGURE 12-8   Template for making title or art cards.

FIGURE 12-9   UNDERSCANNED control room monitor with SAFE TITLE AREA marked in.

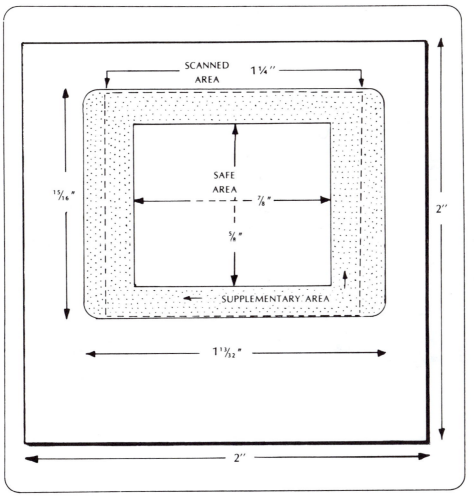

FIGURE 12-10   Framing dimensions for a 2 × 2 slide.

and narrow. If you must use vertically oriented slides, then project them on a smooth, white, dull surface and aim a camera at the projected image. In short, with vertical slides you have three choices:

1. Zoom out and get the whole picture with a blank border on both sides.
2. Zoom in on the essential information, disregarding the rest of the slide.
3. Zoom in on one part of the slide and tilt to another part.

BOLDNESS AND SIMPLICITY

Unlike cinema, slides, photographs, and the printed page, TV is a fuzzy medium. Fine detail turns into blurry grays and hazy shadows. With your eyes alone, look at a newspaper 3 feet away; you can probably read the entire page. Fill a TV screen with that same page and you can read only the main headlines and even they don't jump out and grab you. Figure 12–11 shows what I mean.

Visuals for TV need to be bold, simple, and uncluttered. Figure 12–12 compares some examples of poor and good visuals for TV. Figure 12–13 shows some more good examples of simple, easy-to-read, yet attractive tables and graphs.

Titling for TV needs to be brief, broad, and bold in order to have impact. Wordy subtitles that need to be small and unobtrusive should be limited to *no more than 25 to 30 characters per line* to remain legible. Remember, too, that something that looks pretty sharp in a dimly lit control room and that is seen through your high-resolution video monitor connected directly to your TV camera will lose a lot of oomph once it is recorded, edited, and copied and the copy is played back on an inexpensive video player through RF into a casually adjusted TV set. It's a wonder that there is any picture left at all, much less a sharp one. Figure 12–14 shows what happens to character-generated text which has run the gauntlet from camera to editor to playback of a tape copy.

One way to test for boldness is to step back from your proposed visual, squint your eyes, and look at it through your eyelashes. This is what it will look like when the viewer

FIGURE 12-11 Comparison of regular photograph with same shot played back on a VCR. Notice how TV loses a lot of detail.

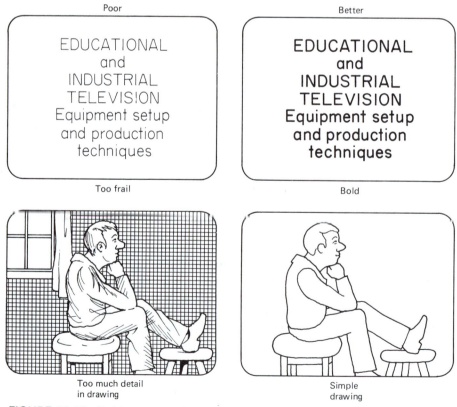

FIGURE 12-12 Boldness and simplicity in TV visuals.

FIGURE 12-12 Cont'd.

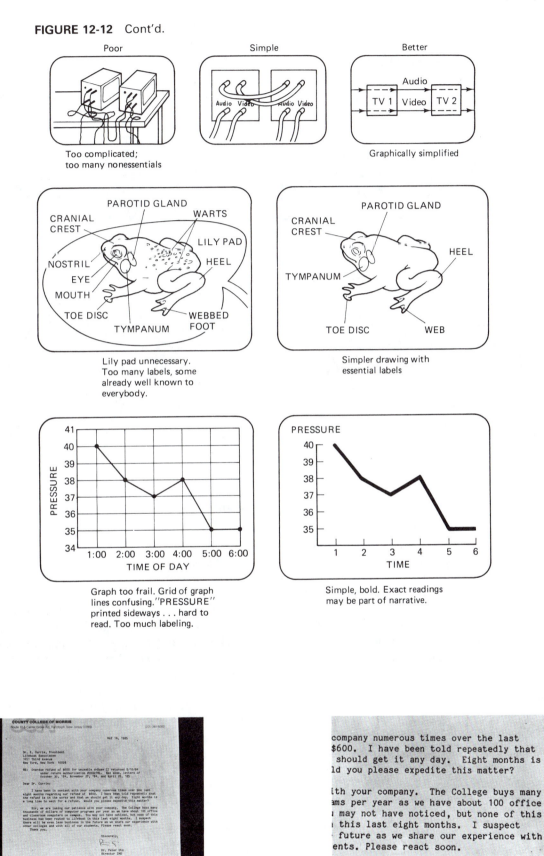

Poor

Too complicated;
too many nonessentials

Simple

Better

Graphically simplified

Lily pad unnecessary.
Too many labels, some
already well known to
everybody.

Simpler drawing with
essential labels

Graph too frail. Grid of graph
lines confusing. "PRESSURE"
printed sideways . . . hard to
read. Too much labeling.

Simple, bold. Exact readings
may be part of narrative.

Too many words, too distant.

Zoom in and limit words.

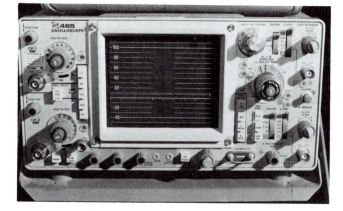

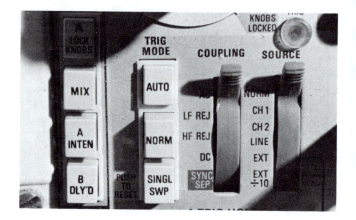

Cluttered photo. Hands hidden. Labels on cans obscured.

FIGURE 12-12   Cont'd.

Simplified display. Unnecessary decorative bowl gone. Action easy to see. (From *Graphics for Television—* Prentice Hall)

sees the image. Give every title, drawing, and photograph the old ''squint test,'' and two things are bound to happen: Your visuals will stand up to the rigors of the TV medium, and your friends will arrange optometrist appointments for you.

One thing to avoid in graphics, photos, and your talent's wardrobe is patterns of fine lines or herringbones. They sometimes vibrate with a shimmering rainbow of colors.

GRAY SCALE

Early in Chapter 9 we learned that a TV camera could accept a LIGHTING RATIO up to 1:30. The brightest thing in a scene couldn't be more than 30 times brighter than the darkest thing in the same scene. Once the cameras, VTRs, and TV sets are finished with your picture, you'll find that the bright-

est white you will ever see on your *TV screen* is only 20 times brighter than the darkest black. This brightness range can be divided into 10 gradations, which is called the GRAY SCALE (Figure 12–15). For one object to *appear* brighter than another *on your TV*, it must be *at least* 1 step brighter on this reference chart, the GRAY SCALE; otherwise, the brightness difference is just too small to register.

What does this all mean to the graphic artist? It means one should not place things which are very close together on the GRAY SCALE close together on a picture. Even though they may have a slightly different brightness, your camera and TV may not pick this difference up, and the two items may merge together into a single mass. To make one object stand out from another, it needs to be significantly lighter or darker than its neighbor (a 1- to 2-step jump on the 10-step GRAY SCALE). For instance, if black were step 10 and white were step 1 on the scale, Caucasian faces would fall

MINIGLOSSARY

*Gray scale   A standard of 10 steps from black to white used to measure contrast ratios. To be visible on TV, objects must be at

least 1 gray scale step different in brightness from their backgrounds.

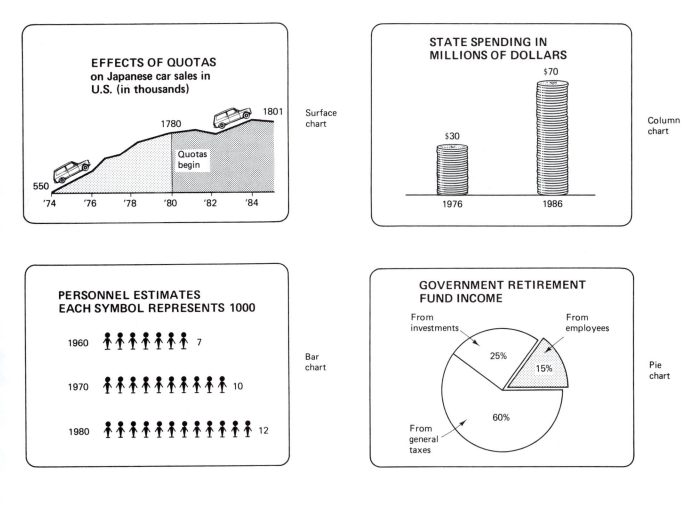

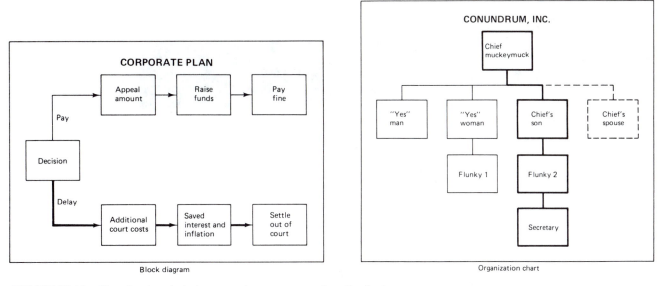

FIGURE 12-13 Simple visuals help your viewers remember the facts.

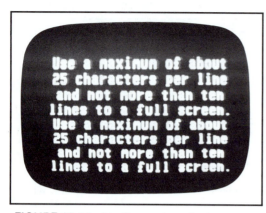

FIGURE 12-14 Limit your long text passages to 10 lines of 25 characters per line.

between 2 and 4 on the scale. If lettering is to be seen across such a face, it should be darker than step 5 or as white as step 1 in order to stand out. Although 1-step differences may show, they will fail the boldness test. A 3-step difference in GRAY SCALE comes with a money back guarantee that your object or lettering will stand out from its background.

Try to avoid using pure whites and pure blacks in your graphics. If some part of the picture is blisteringly white, it will make all the light gray, medium gray, and darker parts of the picture look black in comparison. If, on the other hand, the whites weren't so terribly white when compared with the grays, all the tones will get a fighting chance to be seen. As a general rule, use off-white paper (yellow, light green, or buff) to allow the full range of grays to be seen by the camera. Incidentally, those off-white backgrounds, under most lighting, will still look white when seen on the TV screen.

If there are no grays in your picture (such as a black title on a white background), you can get away with using white paper and black print *or vice versa*. Go easy on the lighting as some TV cameras will smear the letters if the contrast is too great.

Color Compatibility

If you place a light blue card next to a light green card next to a light red card and look at them, you can tell one from another easily (unless you're color-blind). Aim a color TV camera at the three and observe the results on a color TV set, and you will still be able to tell the three apart. If however, you view the result on a black-and-white TV or record it using a black-and-white camera, the three cards may look exactly the same, blending into one mass. This is because, although the colors are different, the cards' relative brightness on the GRAY SCALE is the same.

Since there are over 83 million black-and-white TV sets in use in the United States, it behooves us to consider these viewers when planning our graphics. For graphics to look good both in color *and* black-and-white, important features of a graphic should be 1 or 2 GRAY SCALE steps different from their backgrounds. When in doubt, it is always good to test your graphic on a black-and-white TV to see if it maintains its punch. (See color compatibility example on back cover.)

TYPOGRAPHY

What do you use to print your titles and subtitles? The answer depends on quality, budget, and purpose. Do you have a lot to say? Do you have a lot to spend? How good must they look?

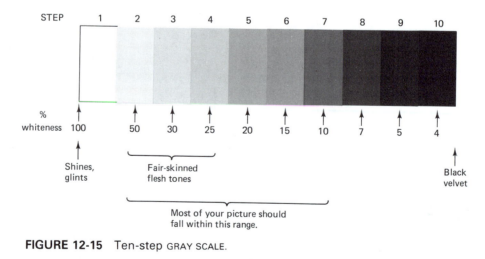

FIGURE 12-15 Ten-step GRAY SCALE.

MINIGLOSSARY

Color compatible An image that can be viewed easily on black- and-white TVs as well as color ones.

Press-On or Rub-On Letters

Press-on and rub-on letters are wax-coated letters that come in a sheet and can be rubbed off onto paper or posterboard (Figure 12–16). The process is quite simple, and the letters are neat and bold. A wide variety of sheets is available, giving you numerous FONTS (type styles) and letter sizes to choose from. The wax sheets are inexpensive, and no equipment is necessary except for a stick with which to rub the letters. The process, though simple, is too slow for doing many words but is great for short titles, credits, short lists, and captions to be affixed to existing photographs, objects, or charts. The letters rub off, however, almost as easily as they rub on, so some care must be taken with the finished product.

Of all the methods, this one gives the most professional-looking results for the least bucks.

Scrabble Board

If your graphics budget is zero, then go home and dig out that old Scrabble game from behind the furnace. You can arrange the letters on the board for a true Scrabble look, as in Figure 12–17, or you can lay the titles on a sheet of paper or across a towel or carpet for varied backgrounds. The letters are easily moved and rearranged. Unless there is a Scrabble theme to your show, this type of lettering may look a bit hokey. Maybe you can think of some other games around the house that come with little letter squares. How about kids' building blocks?

These methods are only appropriate for short titles as you'll probably run out of letters (and patience) after just a few lines.

Letter-On Machine

The letter-on machine (Figure 12–18) cuts adhesive tape into desired letters. The letters may then be stuck to posterboard or whatever. The lettering is bold and fairly durable, and the process is fast enough for making short titles when

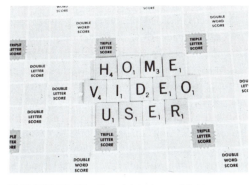

FIGURE 12-17 Scrabble board letters.

necessary but too slow for long credits and text passages. Various FONTS and letter sizes are available, and the tape comes in different colors including black and white.

Professional Typesetting

Modern print shops are equipped with ways of making bold flawless type in seconds through the use of a typewriter connected to a computer that is in turn connected to an instant photographic type-making device. Lettering comes out evenly spaced with justified margins (straight, even margins on both sides, like the type in this book) and can be made in a wide variety of size and type FONTS. The method is used for magazines and newspapers and generates both text and headings. Costs run about $30 per page (higher for special typography), and a lot can be put on a page. With the use of special close-up lenses or lens attachments, columns of lettering can be blown up to fill the TV screen. Choosing the proper column width allows lists, credits, sentences, paragraphs, and other long passages to be moved vertically across the screen for easy reading.

Professional typesetting is visually preferable to regular typing because of its boldness, variety of type style and

FIGURE 12-16 Rub-on letters.

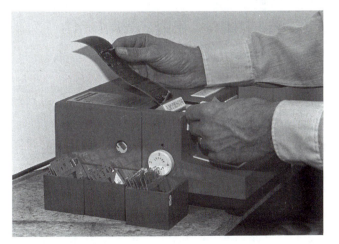

FIGURE 12-18 Letter-on machine.

size, and margin control and especially because of its type spacing. Unlike the typical typewriter, the typesetting machine allows more space for wide letters like ''m'' and ''w'' while using much less space for narrower letters like ''i'' and ''l.'' This seemingly minor attribute makes a big aesthetic improvement over typewritten text when blown up on the TV screen.

Typewriters, Scrabble-board letters, and a few other mechanical type-making machines use MEASURED type spacing—every letter is exactly the same distance from its neighbor. Typesetters, special typewriters which use PROPORTIONAL spacing, and carefully aligned rub-on or press-on letters allow for OPTICAL SPACING—the distance between the letters depends on the shape and size of the letters. Figure 12–19 compares OPTICAL with MEASURED type spacing. Notice how much more professional OPTICAL looks.

When choosing type styles, follow the adage that *less is more*. Keep the type simple. Avoid script, open-faced, or condensed types. The more ornamental type styles (Figure 12–20) are often distracting or hard to read. Use sans serif typefaces such as Optima, Theme, or Helvetica. Other typefaces appropriate for TV are Futura, Metro, News Gothic, Spartan, and Tempo. The bold typefaces usually show best, but don't get them so bold that the holes in the ''A's,'' ''B's,'' ''O's,'' etc., start to close up. On TV they'll close up even more.

Here are a few more terms to help you communicate with your typesetter: Letter sizes are measured in POINTS. A POINT is 1/72 inch. A 36-POINT letter would be 1/2 inch tall from the top of an ASCENDER (the part of the letter that rises above the main body, such as a ''d'') to the bottom of a DESCENDER (the part that extends below the main body as

in a ''g''). The width of a line of type is measured in PICAS. There are 6 PICAS to an inch. Given these facts, you can shape the type to fit your needs. Just remember to limit yourself to 25 to 30 characters per line for television.

Typing

Thrifty but lacking in boldness is our old friend the typewriter. Although it is much maligned as a TV titling device, when used creatively, type can be the budget TV producer's workhorse.

Close-up lenses or inexpensive lens attachments permit the TV camera to take very tight close-ups of the typing. The resulting image is sharp, legible, and even bold when extremely tight close-ups of brief messages are taken (Figure 12–21).

If your typewriter can use a *carbon* ribbon rather than a nylon fabric ribbon, you will get a blacker, sharper print.

Typing on various paper surfaces yields an assortment of background possibilities. Illuminating the titles is simple with a desk lamp or two. Special effects such as a spinning title can be viewed by attaching a card to a phonograph turntable with the TV camera viewing from overhead.

The great advantage of typed titles is that they are quick and cheap. One disadvantage of the typed title method, however, is the likelihood that a minor flaw in the card's preparation will be exaggerated when the small picture area is magnified to large TV screen size. The author remembers

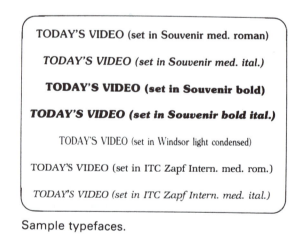

Sample typefaces.

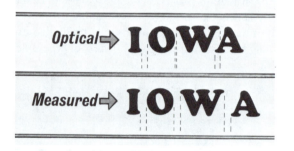

FIGURE 12-19 PROPORTIONAL or OPTICAL type spacing vs. measured.

MINIGLOSSARY

*Font   Style and shape of lettering.

Measured spacing   Typography where the space between letters is always the same.

Optical spacing   Typography where the space between letters depends on the shape and size of the letters.

*Serif   The flare or crook at the ends of some letters, like at the top and bottom of the capital letter ''I.''

Point   A measure of the height of lettering. 72 points equals one inch.

Ascender   The part of a letter that rises above the main body, like the top of the lowercase ''k.''

Descender   The part of a letter that drops below the line, like the bottom of the lowercase ''p.''

Pica   A measure of a line's width. A pica equals 1/6 inch.

ADE	Letters with skinny parts. The skinnys may disappear.
ADE	Letters with a SERIF. Although in this case the SERIF may be fat enough not to disappear, it complicates the legibility of the letters.
EAD	Type with multiple lines. These lines vibrate on TV. Some may disappear.
HANDBOOK	"Tempo black" style here is too bold. Holes in letters close up.
COMICAL ~Care Free CLASSIC·Biblical	Type nicely expresses the mood of the title, but may be too ornate (especially upper case lettering) for TV.

FIGURE 12-20   Type styles that are troublesome on TV.

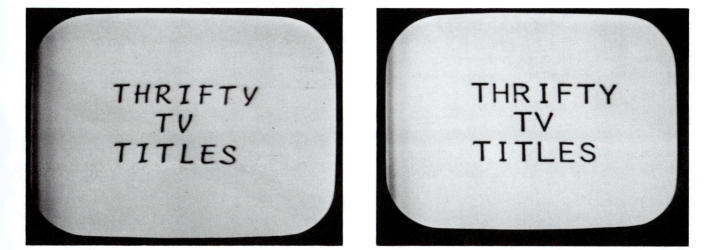

FIGURE 12-21   Typed titles.

Avoid the amateurish "stencil look" by filling in the natural gaps in the letters. (From *Graphics for Video—Prentice Hall*)

once when his associate sneezed near one of the little title cards during a production. The damage went unnoticed until the tape was played back on a 21-inch monitor. The flaw then became grossly apparent, very grossly.

As one might guess, typing is useful for both short and long textual passages. It is perhaps more appropriate for the long ones, while the short frequently reused and more important titles are left for the slower, bolder methods of typography.

Stenciling

Cheapest, and perhaps most familiar, is stenciling. A template with punched out letters guides the sign maker as he or she outlines the letters with pencil. The letters can then be filled in with ink or marker.

Stencils may also be spray-painted, but this method may destroy the templates, and the paint often seeps under the stencil, smudging the letters.

If you use stencils, fill in the breaks in the letters afterward for a more professional look.

Stenciling is a tedious process, good for short titles only.

Character Generator

A CHARACTER GENERATOR is a computerized typewriter which electronically places the words on a TV screen rather than on a piece of paper. Just type, and it's there. Depending on the kind of CHARACTER GENERATOR used, the device may store sentences—even pages—for later display. It may display bulletins crawling horizontally across the TV screen

or whole columns rolling vertically from the bottom to the top of the TV screen. Some machines allow the editing and changing of prepared material in the generator's memory while other parts of the material are being displayed.

For about $700, the CHARACTER GENERATOR will probably just display words and numbers and have perhaps a couple of pages of memory (so that the typing doesn't have to be done "live" when you use it, but, rather, can be prepared and previewed in advance of taping). Many CHARACTER GENERATORS will colorize the lettering and will even generate background colors.

For about $1400 the generator will form better-looking letters in several sizes with edging or drop shadows (shadows at the right and bottom of each letter to make it look three-dimensional).

Depending on the model, the CHARACTER GENERATOR may be able to scroll data like weather reports across the top of the screen (repeating the information when it reaches the end) while also scrolling news across the bottom of the screen. Meanwhile, it can flip from advertisement to advertisement in the middle of the screen as you've probably seen on some of your cable TV channels. Certain words may flash on and off to attract attention. Besides several pages of active memory, some models allow the data to be recorded on floppy discs or computer chip memory so that hundreds of pages may be stored. Figure 12–22 shows a CHARACTER GENERATOR, and Figure 12–14 showed the type style you can expect from this low priced model.

CHARACTER GENERATORS costing $3,000 to $6,000 have more bells and whistles than you can ever ring or blow. Some features:

- Numerous roll speeds (to run the credits quickly . . . until they come to *your* name)

FIGURE 12-22   CHARACTER GENERATOR.

MINIGLOSSARY

*Character generator   Typewriter keyboard that electronically displays letters, numbers, and symbols on a TV screen.

*Anti-alias   The smoothing out of jagged or stair-steppy edges of electronic graphics or generated characters.

Transitions   Ways of changing from one title or graphic to another.

- Automatic centering of a line of text, or a whole page of text
- Several standard FONTS (type styles), upgradable with additional fonts
- Array of more colors (maybe 512 instead of 16 or 64) for text or background
- Hundreds of character sizes
- PROPORTIONAL letter spacing
- Preview channel (so you can type on one channel while the other is being recorded or broadcast)
- User-created fonts (like drippy letters, letters with teeth or cat whiskers)

The CHARACTER GENERATOR is so handy that once you have it connected to your TV system, you'll probably never want to use anything else. Before you throw away your ink and posterboard, however, consider that your titles should look more creative and interesting than the common text of the credits and subtitles in your production. Only the high priced CHARACTER GENERATORS can create fancy logo or angled or glowing type like you can make with the old-fashioned graphic cut, paste, and draw techniques.

In-Camera Titler

Some camcorders come with CHARACTER GENERATORS built into them. They can type letters and numbers or even display the date and time on the TV screen. Most models will allow the words to appear over your camera's picture, or if you just want black for the background, you can simply cap the camera's lens.

Some models like the one in Figure 12–23 will store your typed information in memory so that you can tape a few scenes with your camera and then press the memory button to bring the title onto the screen.

Keyboard with a button for each letter

Image in camera viewfinder

TV screen image

FIGURE 12-23   In-camera titler.

Without ANTI-ALIAS round letters are jagged.

ANTI-ALIAS blurs the "jaggles," making the letters smoother and more natural looking.

ANTI-ALIAS.

Professional Dubner CHARACTER GENERATOR.

The letters on these mini-CHARACTER GENERATORS are quite chunky and perhaps inappropriate for professional television productions. Also, the process isn't quite as fast as touch-typing, but at least it won't dry out like felt pens do three letters before you finish printing your title.

Chapter 6 had a section describing how to operate an in-camera titler.

COMPUTER GRAPHICS

Some computers like the Amiga create a signal which can be mixed (genlocked) with studio camera images or recorded directly on a VCR. Such computers have an NTSC video output, which means they make standard television signals.

Most other home computers, on the other hand, make nonstandard video signals. Although they can send an image to their own viewing screens, this image cannot be mixed with other camera signals or recorded by a VCR. Some of the reasons why:

1. *Character size.* Computers used for word processing and mathematics create 80 characters (letters) per line. The letters are too small and too close together to be seen on regular TV systems. Some home computers, designed to connect to home TV sets, create only 40 characters per line. Although they are sharp enough to be viewed directly on a TV set, if you tried to record them, they would play back so fuzzy you couldn't read them.

2. *Safe title area.* Computers tend to put letters at the very edge of their viewing screens. On most "normal" TVs, the extreme edge of the picture doesn't show, so part of the information gets lost.

A computer, to be compatible with TV, must use only the SAFE TITLE AREA.

3. *Wrong frequency.* It takes 525 horizontal scanning lines per second to make a normal TV picture. Many computers use a different number of lines per second.

4. *Wrong sync.* The sync pulse which holds normal TV pictures steady is a black bar at the bottom of your picture. Computers often create the wrong-sized bar at the wrong frequency, and in some cases the bar is white instead of black. TV systems won't tolerate this.

5. *Signal strength is different.* Normal TV systems are designed to use a 1-volt video signal. Computers often make an overpowering 4-volt signal.

6. *Analog vs. digital.* TV systems are designed to record varying tones of brightness. Many computers make a digital signal which is either on full blast or off entirely.

7. *Colors are RGB vs. NTSC.* Normal NTSC signals carry all the colors on a single wire. Color computers put their signals on three wires, one for red, one for green, and one for blue (RGB).

All these problems are solvable with the help of electronic boxes and computer modifications. Suffice it to say that computers fall into three categories:

1. Those designed to make standard NTSC signals
2. Those that can put a picture on a TV set, but the signal is still too weird to be *recordable*
3. Those designed to work only with specialized monitors

Let's look at computers used for graphics and television.

NTSC-Compatible Computers

Some computers, like the Amiga, can make a recordable video signal. Others, the genlock type, go one step further by synchronizing themselves to your video system just like a camera. This allows you to mix their images with those of your other cameras. If you have a Macintosh or IBM (or compatible) computer, you can slip a $700 to $1,500 NTSC board into it to convert your computer images to NTSC. Some boards even genlock your computer's image with a video signal for mixing.

Remember that making a kosher NTSC signal doesn't guarantee a *pretty* picture. Try recording a standard spreadsheet or word processing document and you'll see barely legible mush. You need a titling or graphics program de-signed for television if you want broad, bold, anti-aliased lettering with edging.

Above $6,000, the CHARACTER GENERATOR starts taking on new tasks, like "grabbing" images with a TV camera (such as a company logo or carnival background), and "painting" new colors or objects over the images. Some will perform simple animation by strobing through two to eight pictures in memory, much like a movie.

Some have an ANTI-ALIAS feature that smooths out the jagged edges of letters. Others move the words around on the screen in tumbles and spins as you change from one title to another. These are called TRANSITIONS. The most common (and unobtrusive) TRANSITION is the simple dissolve from one title to the next (while the background—perhaps a long shot of the newsroom—remains on the screen).

You can load a "titler" program into the computer, which will teach it how to make fancy letters. You may then type your message on the keyboard and have it appear on the computer monitor screen for examination or editing. The letters can be edged and colored, and the background can be independently colored. Various sizes and fonts are available. As you can see from Figure 12–24, some are pleasing, yet affordable.

The titles, once perfected, can be stored on floppy discs for recall later.

Some computer programs also allow you to create straight lines, grids, circles, rectangles, bar graphs, and pie charts and label them using the titler program.

Graphics Tablet. The keyboard isn't the only way to tell the computer what you want to draw. A separate graphics tablet, shaped like a large clipboard with its own electronic pen, can be connected to the computer, allowing one to "draw" pictures on the tablet and have them appear directly on the TV screen. The computer offers a menu of line thicknesses and colors. Some programs allow you to draw a picture, blow it up to larger proportions, comfortably change a small detail in the picture, and then reduce it back down to its normal size. In fact, the picture can be blown up or reduced to whatever size is convenient for your TV production. Figures 12–25 and 12–26 show a graphics tablet in use and the images it can create. The images, incidentally, look a lot more exciting in color than they do here.

Specialized Graphics Computers

There are more specialized (and expensive) computer graphics systems designed to create not only TV pictures, but produce colorful and sharp 35mm slides for educational and business presentations as well as artwork for books and magazines. With practice, an artist can create finely detailed electronic pictures in minutes and then change them radi-

Text without border

. . . with border

. . . with shadow

FIGURE 12-24   Computer generated text using a $50 titler program.

FIGURE 12-25   GRAPHICS TABLET and electronic pen.

MINIGLOSSARY

Graphics tablet   Flat surface connected to an electronic pen or a sliding puck called a *mouse* and a computer that allows you to create video images as you "draw" on the tablet with the pen or mouse.

cally in just a few more minutes. The efficiency of moving electrons instead of pens and erasers makes it possible to create more pictures for fewer dollars.

DESKTOP VIDEO

Now within reach of the industrial user are several camera/computer graphics systems formerly used by television broadcasters. These systems allow a TV camera to make part of the image, a graphics tablet to make another part of the image, and a keyboard to handle the lettering.

Here's how such a system works: A TV camera looks straight down on a work surface. The artist may place an object on the surface and press a button, and that image will be stored in memory. Next the artist places another object on the surface and stores that *on top* of the old image. The final picture is created layer by layer from all the single images taken from the camera. One could, for instance, start with a rainbow background, add graph paper lines to

Desktop Video   Video, made in an office, using a computer and camera.

Image drawn electronically.

Part can be erased electronically.

Image blown up to make it easier to add detail.

Parts of picture can be moved.

FIGURE 12-26 Images created from a GRAPHICS TABLET.

it, add a photograph of a person to that, and then add a title to that. Even a title can be made fancy by keying in a metallic or textured look to the body of the lettering. The final result may then be stored on computer disk, video, or film.

GRAPHIC DESIGN

There are many good books on graphic design. There are a few good books on graphics for television. Here, condensed into a few good (I hope) pages are some of the main points of graphic design as it applies to industrial television.

Title Placement and Background

We saw in Figure 11–13 that a white title disappears when placed over a white background. Figure 11–14 showed how moving the title to a darker part of the picture made the title show up better. Similarly, a dark title should be moved to a light part of the picture to show up well.

Some backgrounds, especially the busy, bright, and contrasty ones, compete with titles no matter what color they are. Figure 12–27 shows an example.

You should avoid busy backgrounds. If you have no choice, then try darkening the background, perhaps with the camera's IRIS or GAIN controls. Another possibility might be to defocus the background. This will make the foreground stand out more. If carefully planned, you can have the title disappear as the background picture comes into focus. If it is the end of a production, you could have the picture defocus as the title or credits appear.

A title should not appear like a bumper sticker wrapped across someone's face as in Figure 12–28. Instead, try to

Busy, bright, contrasty background obscures text.

Darken background so text stands out.

FIGURE 12-27   Text competing with background.

Title fights with face.

Move face and title.

FIGURE 12-28   Title across face.

find a place for the face and a place for the title so they don't fight with each other.

All these warnings should not make you afraid of giving backgrounds to your titles. A drawing or photo can add meaning and pizzazz to a title. In fact the picture may be remembered long after the words have been forgotten, so this powerful visual tool should not be dismissed off-handedly.

If possible, always leave a large border around your title graphic so that you have the choice of zooming in (making the letters large and bold, filling the screen) or zooming out (making the title smaller and easy to fit in a quiet part of the screen). If your title card lacks a large

border, when you zoom out you may overshoot the title card and get part of the studio in your shot. As mentioned earlier, it doesn't matter a whole lot whether the lettering on your title is large or small since you can always zoom in or zoom out to make it the size you want.

Title Clustering

Titles, like good friends, should stick together. Words should be clustered close enough together to be read as a group, to convey a single concept. For instance, read the following program credit:

The Video Club
of
Transylvania
proudly
presents
Transylvania High School's
performance of
Vladimir Dracula's
BLEEPS, BLOOPERS,
and
PRACTICAL JOKES

When these credits roll through the screen, you see only four or five lines at a time. The four or five lines you see never make sense together. Instead, the credits should be condensed and regrouped and shown as three separate screens or rolled across the screen:

Vladimir Dracula's
BLEEPS, BLOOPERS,
and
PRACTICAL JOKES

Produced by
TRANSYLVANIA
HIGH SCHOOL

Sponsored by
TRANSYLVANIA
VIDEO CLUB

Now the TV screen always holds a complete concept or phrase at a time.

At the end of a show, the credits can appear in one of two ways or a combination of both.

1. Single credits can flash on the screen one after another.
2. Credits can roll continuously up through the screen.
3. A few important credits can flash individually, followed by the lesser credits rolling through the screen.

When organizing the text for rolling credits, you have two choices, as shown in Figure 12–29:

1. As shown in example A, blocks of copy are separated by only a line or two of blank screen. Each block follows closely on the heels of the preceding block. This may be the only way to move an immense amount of text through the screen in a short amount of time. Of course, this presentation diminishes the importance of the text and the viewer's retention of the information.
2. Method B in the figure places a large gap between each block of copy so that as one block is leaving the screen, another is entering. Be sure to leave a large blank area before and after the credit roll so that you have a clean place to start and a place to end with no words on the screen. Method B is best for briefer credits or when you wish to place greater emphasis on each block of information.

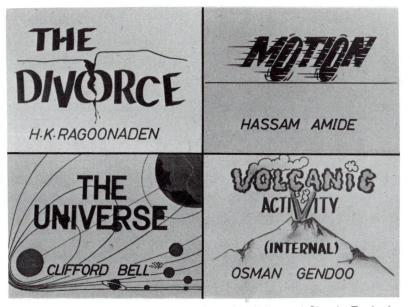

Title backgrounds. (Drawings by Jose Marjolin and Sheela Teeluck, courtesy of Joseph Sauder).

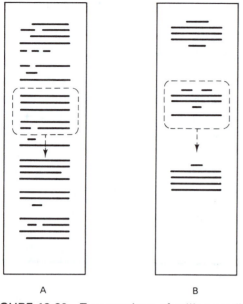

FIGURE 12-29   Two versions of rolling credits.

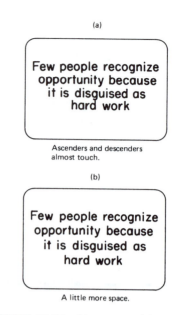

(a)

(b)

FIGURE 12-31   Upper- and lower-case letters require a little more room between lines than half letter height.

Title Spacing and Legibility

When one title is all you want to see, then that is all you want to see. If placing several titles on the same piece of paper, then leave space around each so that the camera operator can focus on it without interference from nearby words.

Titles look best when neither crowded nor sprawled. As shown in Figure 12–30, the general rule is to leave a one-half-letter-height space between lines of text. When using upper- and lowercase letters, you may need a little more space between lines. The capital letters and *ascenders* (parts that stick up from letters like "b," "d," "l," "k") shouldn't almost touch the *descenders* (letter parts which stick down like "g," "p," "q") from the line above. See Figure 12–31.

Long words, especially unfamiliar ones, should be presented in upper- and lowercase. Words printed all in capitals are hard to read. See Figure 12–32.

For the best legibility, letters should be bold but not too bold. For best results, the width of your letters should be one eighth to one fifth of the height of the letters. Too thin a stroke produces an image that is skimpy and faded. Too bold or broad lettering tends to close up, turning into solid bars of text.

Certain letters and numbers, especially when hand-written or printed in unusual fonts, are easy to confuse. Especially difficult are letter groups like

B38R

2Z7

GCOQ

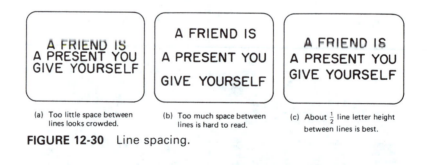

(a) Too little space between lines looks crowded.

(b) Too much space between lines is hard to read.

(c) About ½ line letter height between lines is best.

FIGURE 12-30   Line spacing.

MINIGLOSSARY

*Edging   A dark (or occasionally white) ridge around letters to make them stand out.

*Drop shadow   A dark ridge placed on one side of letters making them look three-dimensional as they cast a shadow. The letters become easier to see because of the edging.

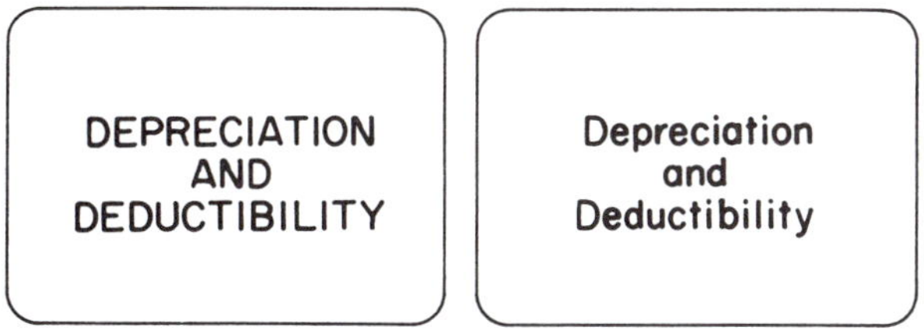

DEPRECIATION
AND
DEDUCTIBILITY

Depreciation
and
Deductibility

FIGURE 12-32 For long words, upper- and lower-case lettering is easier to read than all capitals.

6b

49

Can you tell what the instructor has written in Figure 12–33?

Letter Edging and Color

You've already learned that to make things stand out from each other, they must be two or more steps apart on the GRAY SCALE. To make things stand out even more, they should be EDGED, that is, have a border to further accentuate their edges. This is especially true for lettering. If you can't manage to put a border all around your letters, then try for a DROP SHADOW which gives one edge a dark ridge, also making the letter look three dimensional. Looking back at Figure 12–24, notice how without a border, the letters are visible, but lack ''snap.'' The black EDGING sets the letters off and the DROP SHADOW sets them off further with added dimension.

For lettering coloring, white is a guaranteed winner (except over pure white backgrounds). Yellow and pale colors are good too. Avoid saturated (pure) colors with very little white in them. They will appear fuzzy on screen. In Figure 12–34, the word *Mercury* was white. *Earth* and *Venus* were pale yellow and green, but *Mars* was deep red. When working with character and computer graphics generators, you usually have total control over the colors of

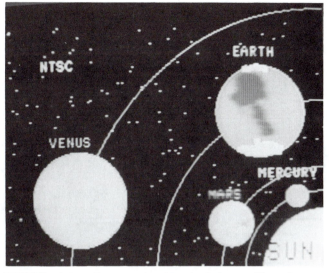

FIGURE 12-34 Saturated colors look fuzzy. (See also back cover.) The word *Mercury* is white, *Earth* and *Venus* pale yellow and green, but *Mars* is pure red. *Mars* is fuzzy even when viewed in color.

your backgrounds and/or lettering. For best visibility, the letters should be pale (high LUMINANCE, low color saturation) and the background should be dark (dark gray or blue works fine).

Drawings

Paints, inks, color tint sheets, line tapes, crayon, brush and ink, and pencil are all old friends of the graphic artist. The problems is that television is such a hurry-up medium that artists seldom have time to produce graphics using all the skills at their disposal. Sometimes there is no graphic artist. You or anybody with a felt pen in their desk drawer becomes the graphic artist.

Felt pen doesn't look too bad on television. A few colored markers can take you a long way toward becoming a video Rembrandt. Here is a tip: With just a little extra attention, it is easy to spice up your plain felt pen pictures. As shown in Figure 12–35, the sketch on the left was done with felt pens and colored markers only. It looks fine but slightly cartoonish. On the right is the same sketch with India ink, crayon, and white pencil. The pen and ink added detail and shadow. The crayon and white chalk added shading and highlighting to make the image stand out with a more realistic look.

Plain line drawings and maps are perhaps best done on 20 percent gray posterboard. Black lettering on the gray will look black. White lettering will look white. To make your black lettering stand out better, first use white chalk to lay down a background for your black letters. Notice in Figure 12–36 how some of the lettering stands out?

FIGURE 12-33 Confusing letter groups. A 13 C

Felt pens and colored markers only.

Same sketch with highlights and detail added from india ink, crayon, and white pencil.

FIGURE 12-35   Drawing with felt markers plus india ink, crayon, and white pencil. (From *Graphics for Television*—Prentice-Hall)

MAKING GRAPHICS COME ALIVE

Those of us blessed with creativity will have no trouble making graphics come alive. Good artists and photographers can find ways to make almost anything look real or interesting.

There is no law which says that a still picture must remain stationary. A camera can pan, tilt, and zoom over a photograph or over a painting as if it were shooting something "live" in the studio. Quick cutting and active movements can make the pictures themselves seem to be moving. Still photographs become movies; battle scenes become the actual battles (with the help of sound effects); birds glide through the sky; ships roll back and forth; earthquake scenes shake up and down; amusement park rides streak by while the lights of the midway grow blurry and dissolve to the next scene.

Cutouts can be placed over visuals and moved, simulating animation. Scenes can have holes cut in them with movement behind the holes simulating running water, snow, vehicles passing by, or whatever. Figures 12–37 through 12–39 show how PULL CARDS can make a graphic move.

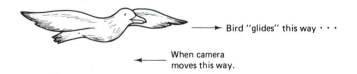

Bird "glides" this way · · ·

When camera moves this way.

Panning across a still picture to create illusion of motion.

The setup in Figure 12–37 requires careful lighting and camera adjustment so that the seams in the black cardboard do not show and just the white lettering shows. Sometimes using the KEY effect will help hide the seams.

In the documentary "The Story of Manhattan Beach," the author recalls finding a black-and-white photograph of an iron gate, a landmark torn down in World War II. By enlarging the photo, gluing it to posterboard, and cutting the posterboard into a gate shape, we were able to open the program with the opening of the gate. By keying the gate over another scene, it appeared that the gate had this other scene behind it. Naturally, the cutout had to be very carefully lit and smoothly opened in order to maintain the illusion.

Detailed photographs and paintings offer an excellent opportunity to zoom in on one part of the picture and slowly pan, taking a trip from one part of the scene to another. Your audience can slowly stroll down a street in Luxembourg or be barraged with close-ups of Civil War cannon fire and soldiers diving for cover. All these shots may be taken from the same picture, just using close-ups of different parts, panning from one area to another, and zooming in on centers of interest.

When you wish to show something complicated, or unfamiliar, it is often best to begin with a wide establishing shot of the item and its surroundings. Then you zoom in on the specific part of the picture, detailing the concept you wish to highlight.

If you know you are going to zoom or pan across a picture, you should draw the picture with that camera move-

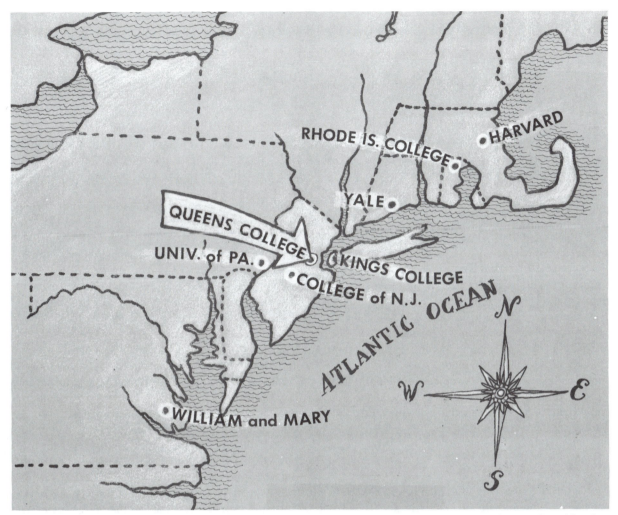

FIGURE 12-36 White highlights make black lettering stand out on gray posterboard. (From *Graphics for Television*—Prentice-Hall)

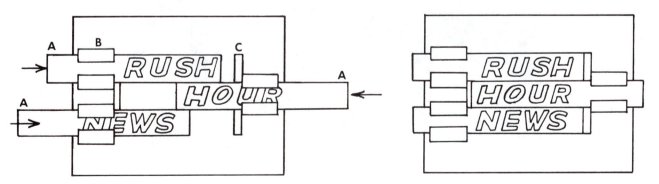

A: Sliding tabs with white letters on black background can be pulled or pushed.
B: Guide.
C: Cardboard "stops" prevent tabs from overshooting when pushed.

FIGURE 12-37 Moving title with PULL CARD. (From *Graphics for Television*—Prentice-Hall)

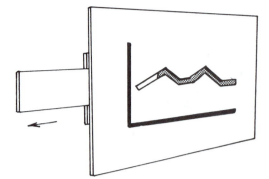

FIGURE 12-38   Moving graph. Graph is actually a slot cut into light-colored front board. As light-colored tab is pulled, darker background is revealed, and the graph appears to draw itself. (From *Graphics for Television*—Prentice-Hall)

ment in mind. Looking at Figure 12–40, we see a simple, well-drawn visual. This drawing is fine if shown in its entirety. But if you plan to zoom in on the hypothalamus, you'll notice that the camera can't help but see parts of words and parts of the surrounding picture. The long shot is good, but the close-up isn't. In Figure 12–41, the graphic is redesigned with the camera movement in mind. As the camera zooms from a wide shot to a close-up of the hypothalamus, there are no overlapping parts of words and diagram to get in the way.

Visual Tricks

There's a bagful of tricks to this trade. Because TV is a two-dimensional medium, there are many tricks that can be

played with the nearness and farness of things. A person standing next to a mural of a house on a mountain may appear to be observing a real house on a distant mountain. A model robot near the camera lens may look like a life-size attacker.

The author recalls having fun with a video tape of a college president giving a speech. The speech was played on a 21-inch TV monitor so that the president's head was nearly life size. A TV camera was carefully focused on the TV screen image. In front of the TV screen a "live" hand (I'll never tell whose) reached up to make unsavory gestures to the audience and pick the president's nose. The hand looked real. The resulting tape was a riot (though never shown publicly). Viewers couldn't tell the TV face from the real hand.

More tricks: Lighting changes can make three-dimensional objects seem to move. Strong backlight with very little key or fill light simulates night scenes. Raising the key and fill lights ushers in the day. Puppets, when not sharing the scene with real people, may begin to look like full-sized people themselves. Visuals may be burned on camera. Title lettering may be blown away. Delicate hands may enter the scene and may turn over a tarot card, revealing titles or credits. Smoke in the foreground can lead to the final fade-out. In the key or super mode, blood (simulated, or course) may drip onto a title. The camera zooms in on the drip, filling the screen as the next visual is revealed in the blackness of the drip. Matted titles can be made to disintegrate into a mushy cloud as the camera shooting the title goes out of focus.

Creativity—it's the fun part of television. Use it. At the same time, however, keep the objectives of your pro-

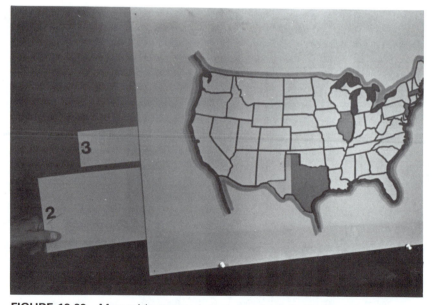

FIGURE 12-39   Map with states changing colors. Pulling tabs reveals background colors behind cutouts of states.

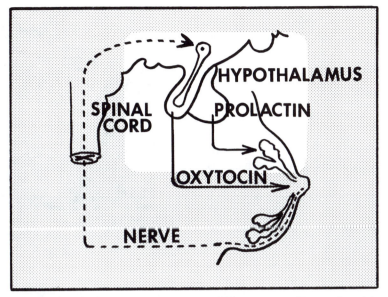

FIGURE 12-40 Graphic not designed for a zoom-in. (From *Graphics for Television*—Prentice-Hall)

duction in mind and don't let the fancy stuff carry you away. How appropriate would it be for the ending credits on a medical research tape to change from one to the other with holes burning in them or blood dripping on them?

DISPLAYING GRAPHICS

Mounting Graphics

Photographs and flimsy paper will always curl under hot television lights. The graphic then goes out of focus and causes uncontrollable reflections. It has to be held stiff and flat.

Masking Tape. Stiff posterboard and foam-filled cardboard sandwiches called FOAM CORE do the job nicely, although you could mount your graphic on tombstones if you wanted to. If your illustration has a large border which will never show in your TV picture, you can tape it down with MASKING TAPE, a beige paper adhesive tape which comes in various widths and is easy to tear.

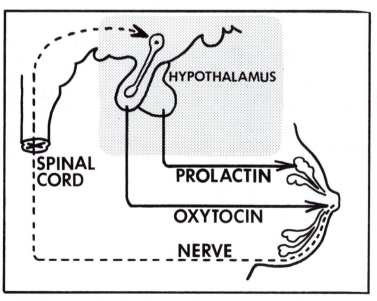

FIGURE 12-41 Graphic designed for a zoom-in on the hypothalamus. (From *Graphics for Television*—Prentice-Hall)

If your graphic is thick or you are working with a photograph whose edges will show, there is a way to use the MASKING TAPE so that it won't show. Tear off a 2-inch strip and bend it into a loop so that both ends stick together, looking like a small sticky wheel. Press the loop against the corners of the back of the picture and flatten them out some. The tape sticks to the picture yet there is a sticky side still facing out. Place the picture on the mounting board, press it down, and it should stay put. One word of warning, though: Valuable pictures should not remain mounted this way. In a couple of years, the MASKING TAPE will bleed its gooey stickum through the paper, leaving a yellow stain on the front of your picture.

Double-Faced Tape.   Instead of making loops out of MASKING TAPE, you can buy special DOUBLE-SIDED or DOUBLE-FACED TAPE that is sticky on both sides to mount your pictures. Unlike the MASKING TAPE, you have to peel the protective paper backing from the tape before using it, as shown in Figure 12–42.

And then there is always chewing gum.

Photographic Tape.   Black PHOTOGRAPHIC TAPE is another useful type of masking tape (also shown in Figure 12–42). It's handy for neatly mounting graphics on black backgrounds, as demonstrated in Figure 12–43. The tape not only holds down your graphic, but it nicely covers the

FIGURE 12-43 Covering the white borders of a photo with black PHOTOGRAPHIC TAPE while also mounting the photo. (From *Graphics for Television*—Prentice-Hall)

white borders of photographs. The tape gives a sharp straight edge to ragged pictures, and several rows of tape can be applied to crop a picture smaller. PHOTOGRAPHIC tape can reshape a 4 × 5-inch or 8 × 10-inch photograph into the TV's 3:4 ASPECT RATIO.

Most types of MASKING TAPE are fairly gentle on photographs and if removed in a couple of hours may cause little or no damage to the photo. After several days, however, the tape hardens and doesn't let go so easily.

Spray-on Adhesive.   There are various spray glues that artists use to mount pictures onto posterboard. You spray the adhesive onto the back of the picture, let it dry a minute, and then position the picture onto the posterboard and press the two together.

Two cautions when using spray glue:

1.   Always work in a well-ventilated area.
2.   Before spraying your photo, place it face down on a very large scrap of cardboard or newspaper. Otherwise, the sticky overspray gets onto everything, is hard to remove, and collects dust.

Rubber Cement.   There are two ways to use rubber cement. The first way is to smear some onto the back of your picture and then press your picture onto the posterboard. Try not to use too much glue or the picture will have bumps and the glue will squash out from under the edge of the picture. If it does, don't try to remove it just yet. Rubbing

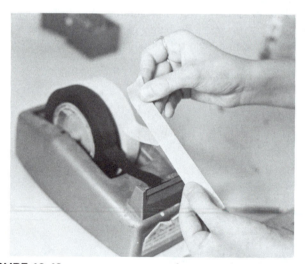

FIGURE 12-42 DOUBLE FACED TAPE for mounting pictures. (From *Graphics for Television*—Prentice-Hall)

MINIGLOSSARY

Foam core   Stiff mounting board made of plastic foam sandwiched between paper.

Masking tape   Paper adhesive tape used for holding things temporarily because it tears easily and removes easily.

Double-faced tape   Adhesive tape sticky on both sides—good for use between pictures and backings.

Photographic tape   Colored adhesive tape (usually black) used for framing pictures.

it with your fingers will leave oily smudges behind. After the glue dries, use a gum pickup or an eraser to rub off the rubbery glue.

The second rubber cement technique creates a stronger bond, but you must be more careful. Here you paint a square of rubber cement on the place where you want the picture to go. Next you paint the back of your picture completely with rubber cement. Let both dry completely. If you have the hands of a surgeon, you can lay the photo onto the rubbery patch exactly the way it should go. It will stick like crazy. If you are a normal human being, a corner of the picture will errantly touch some part of the gum surface, and only Mr. T will ever be able to separate the two.

Mask. When pictures are part of a book or are already mounted in a combination or when you only want one part of a picture, you may need a MASK in order to cover the parts that you don't want. A MASK is simply a black (or gray) cardboard cut to the desired shape (perhaps 3:4 ASPECT RATIO) with a razor knife. The edges of the hole can be blackened with a felt marker. You then lay the MASK over the desired part of the picture as shown in Figure 12–44.

Clothespins, C-clamps, and spring clamps from your hardware store are handy for holding books and magazines flat and holding masks in place.

LIGHTING GRAPHICS

A professional-looking title is generally one which leaves no hint as to how it was constructed. Curly edges on letters and grainy paper fibers in the background make titles look amateurish. Even flat, smooth titles have minor scratches, ridges, and lumps in them which remain hidden until they are revealed by the all-seeing TV camera.

Some of these flaws can be de-emphasized by the use of flat, shadowless lighting. As shown in Figure 12–45, lamps are set up to the left and right of the camera and aimed at the graphic. Each light "washes out" some of the shadow created by the other light, making the image fairly shadowless. The "softer" (more diffused) the lights, the better.

Two 100-watt desk lamps with frosted bulbs and large reflector shades may work very well for black-and-white. Photofloods or professional lights may be necessary for color.

The angle of the lights is not too critical. If, however, they are too close to the camera, glossy or shiny paper or lettering surfaces may reflect light into the camera lens. Placed at too great an angle from the camera, the lights begin to create shadows and may also illuminate the visual unevenly. An angle of 45° from the camera/visual axis, as shown in the figure, is usually satisfactory.

Sometimes, it is more convenient to mount a camera

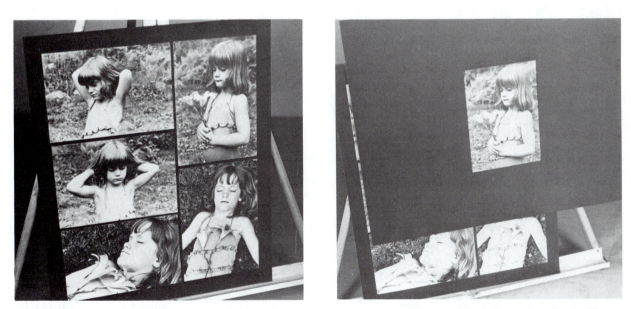

FIGURE 12-44 A cut MASK reveals only part of a picture. (From *Graphics for Television*—Prentice-Hall)

MINIGLOSSARY

Mask A cardboard cutout used to frame a picture with a smooth border or cover unwanted parts of a picture or other background.

*Copy stand A device for holding a camera so it can easily be focused on a graphic.

Handy assorted clamps for holding books, graphics.

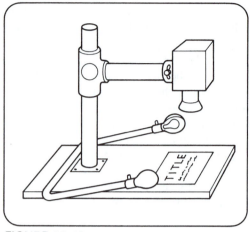

FIGURE 12-46   COPY STAND.

facing straight down using a COPY STAND as shown in Figure 12–46. The camera can be raised and lowered to suit particular needs. Close-up lenses make it possible to blow up small visuals to a large size. Because the base is flat, materials tend to lie flat and are easier to shuffle around than when graphics are placed on a vertical or angled stand. A pane of antiglare glass may be used to hold items flat.

Make sure to turn off any overhead ceiling lights when using the COPY STAND to avoid getting unwanted reflections off the visual from these lights.

FOCUSING ON GRAPHICS

All the focusing procedures in Chapters 6 and 7 also apply to focusing on graphics; however, small things like graphics are harder to focus on than larger things. Although 1 foot may make little difference from 20 feet away, 1 inch makes a big difference 6 inches away from the camera lens.

One way to minimize the focusing problem is to first assure that the graphic is exactly perpendicular to the camera's line of sight, as shown in Figure 12–47. This way, all parts of the graphic are equidistant (almost) from the camera lens and are therefore all in focus at the same time. You can take the guesswork out of this alignment process by laying a small mirror over the graphic. As you look in your TV camera viewfinder, you should see the reflection of your camera's lens. If the center of the lens appears in the center of your viewfinder, you are all set.

Another way to minimize the focusing problem is to flood the visual with light and to "stop down" the camera

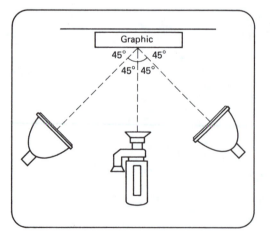

FIGURE 12-45   Lighting graphics (top view).

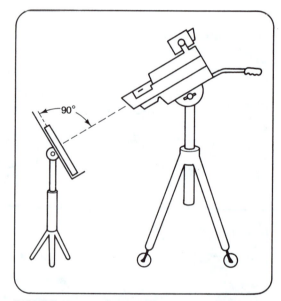

FIGURE 12-47   Graphic kept perpendicular to camera's line of sight.

MINIGLOSSARY

*Keystone   A distorted view of an object (usually a graphic) caused by aiming a camera at it from an angle; the nearer part of the graphic appears larger than the farther part. A square could take on a trapezoidal "keystone" shape.

lens to a higher f-number for maximum depth-of-field. In very tight close-ups, it may be impossible to focus all parts of the visual accurately because the edges of the visual are a shade farther away from the camera lens than the center of the visual. In such cases, stopping down the lens may be your only recourse for an all-around sharp picture. Remember that heat from intense lights, too close to the visual, especially a photograph, for too long, may curl it or its mount. At close range this will throw your focusing off.

SHOWING A SERIES OF VISUALS

If there is plenty of time between visuals, it is no trouble for the camera operator to stack them in order on a stand and change them when her camera is off. If a little less time is available between shots, two cameras may be employed. While one camera shoots its visual, the other is preparing for the next. The series of pictures can thus be shown by switching or dissolving back and forth between cameras.

If just a little time is available between shots, you may need to set up all the shots at once, each on a stand facing a camera, each prefocused. When each camera is off, it will then take only a second to pan it to the next picture and center it. When using such methods, be aware that each visual must be perpendicular to the camera's line of sight. Putting three pictures on a wall as in Figure 12–48 will result in focusing problems and KEYSTONING (a phenomenon where the closer part of the visual looks larger than the farther part).

Given three cameras and the preceding method of setup, it may be possible to change visuals at the rate of once per second. The director merely switches from camera to camera. While each camera is on, it displays a visual. When the camera operator sees his tally light go off, he has 2 seconds to swing to the next visual, center it, and be ready for the time when his camera comes on again. It takes coordination, but it can be done.

If only one camera is available for visuals and it must change visuals before the viewers' eyes, you have a problem. One solution may be to take 2-inch × 2-inch slides of each visual, load them into a DOUBLE-DRUM SLIDE PROJECTOR (if you have one as part of your MULTIPLEXER), and display the images that way. The projector will advance the slides in a blink of an eye.

A less rapid but less expensive way to change visuals is to devise some system of mounting them all on the same size posterboards and finding a way to flip from one board to another. Figure 12–49 shows one such setup that uses a

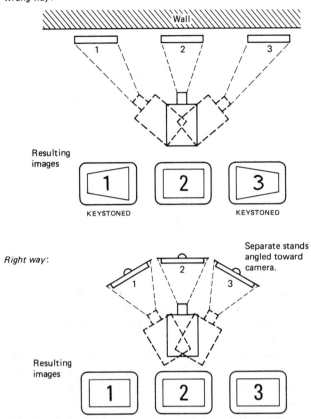

Wrong way:

Right way:

FIGURE 12-48 Setting up several visuals (top view).

ringed loose-leaf binder and graphics mounted on three-hole posterboards. The special stand can even be equipped with its own lighting.

A much slower method of changing titles is to edit from one to the next. This requires a video recorder with editing capability. Unless you have fancy editing equipment, don't expect to pop from visual to visual any faster than one per second. Your fingers and the editing machinery just aren't that accurate.

Crawl and Roll

Technically, a CRAWL is a *sideways* movement of text across the TV screen (like a news bulletin), while a ROLL is the *vertical* movement of words through the screen (like titles, credits, or lists). Regardless of which way the text moves, the machine that does it is called a CRAWL. The text may be printed on long strips of paper and attached to a belt stretched over rollers or affixed to a big drum which is

---

MINIGLOSSARY

*Double-drum slide projector Slide projector with two trays (or drums) of slides that alternately shows a slide from each tray. The switchover from one to the other is very quick.

*Crawl The sideways movement of text across the TV screen

(usually near the bottom of the screen).

*Roll Vertical movement of text across the TV screen (always moving upwards).

FIGURE 12-49 Flip chart method of changing titles.

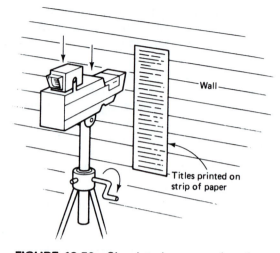

FIGURE 12-50 Simulated CRAWL using the camera's pedestal elevator.

rotated by a hand crank or a variable speed motor. Strangely, this device is called a DRUM-CRAWL, although you'd think the proper term would be DRUM-ROLL (perhaps DRUM-ROLL just didn't sound right).

Nowadays, studios with character generators move the text across the screen electronically. Such machines will vary the speed of the roll or crawl and can stop it at any point.

Small studios without a CRAWL or a character generator can simulate the effect in a couple of cheapo ways. One is to type the desired text in a narrow column on a strip of paper. Then disengage the typewriter's ratchet so that the page can roll smoothly through the typewriter as you twist the roller. Using a close-up lens, aim your TV camera at the text where it exits the typewriter. Try to keep the paper flat in the area you are shooting. To "roll credits," simply hand-turn the roller slowly as the camera looks on.

Another cheapie technique is shown in Figure 12–50. Here the camera is wheeled up to a wall on which is mounted the titles, list, or whatever on a long strip of paper. The strip should be evenly lit from top to bottom. Use a close-up lens if the print is small. Align the camera and list so that the camera may be aimed at the top of the list (or even above the top) and, on cue, slowly and evenly crank the pedestal down, lowering the camera. The words will appear to rise through the screen and disappear at the top. Leaving a blank space at the bottom of the page allows you to crank the last words off leaving a blank screen.

Simply tilting the camera will not work. First, it is very hard to slowly and smoothly tilt a camera across something small like text. Second, since the top and bottom of

the strip are farther from the lens than the middle of the strip, they will be out of focus. Third, for strips of any length, the keystone effect would be quite noticeable.

You can liven up your CRAWL by SUPERIMPOSING or MATTING your text over pictures, a movie, the empty set, or selected scenes from your show. Be sure to subdue the brightness of these backgrounds so your white lettering doesn't get lost in the picture.

SET DESIGN

No rule says performers must have a curtain for a background. Artificial sets and props add to the realism of your production as well as create a mood. As in the case of a newscast, for instance, a busy newsroom background conveys the feeling of late-breaking, somewhat informally presented news. A slick, smooth, well-lit set implies anchor headquarters where the news has been digested and will be presented with professional special effects. A newsroom with paneled walls, stuffed swivel executive chairs, and a wooden desk implies that the presentation will have a corporate or business flavor. In each case the audience *expects* a certain kind of news from a certain kind of set.

The main thing to keep in mind when planning backgrounds is the following: *Keep it simple so that it is not distracting.* Who's going to look at your performer's dreary puss when Mickey Mouse and Pat Benatar are in the background—dancing together—on the edge of a cliff—surrounded by leprechauns—wearing skintight raincoats? The background should be as unbusy as possible, and it

MINIGLOSSARY

*Logo A visual symbol identifying a specific company, organization, or TV station.

should be a few shades darker on the GRAY SCALE than the performer. The performer is usually the central object and must remain the most vivid thing on the screen.

Often the program's LOGO or the sponsor's name is integrated into the scene's background so that it appears during the show. Just as we did earlier, we have to plan the scene design with camera angles and zooms in mind. The LOGO or title should be small enough so that the camera doesn't have to get ridiculously far away to "get it all in." It also should be positioned so that it doesn't disappear behind the talent's head or fight for attention during medium and close shots. It should be positioned so that either all of it shows or none of it shows. It is disconcerting to see just part of a title showing in the background. Figures 12–51 and 12–52 show examples of an awkwardly placed LOGO and a well-placed one.

What is true for LOGOS is also true for props and other interesting parts of the background set. They shouldn't distract or obscure the action or your performers and in most cases should appear completely or not appear at all. Half of a window, half of a person doing something, or half of a TV monitor in the background is distracting. So are pictures or objects which hover next to the talent's head. Review Figure 8–7 for examples of distracting backgrounds.

This doesn't mean that you can *never* show just *part* of something. A large potted plant in the foreground of a

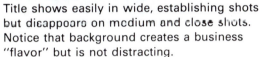

Title shows easily in wide, establishing shots but disappears on medium and close shots. Notice that background creates a business "flavor" but is not distracting.

FIGURE 12-52 Well-designed and positioned LOGO. (From *Graphics for Television*—Prentice-Hall)

picture can help frame the picture. Part of a large tree in the background can give the woodsy atmosphere without requiring you to import a giant sequoia from California.

Altered Perspective

Most TV studios are small, not quite the place to build a house or a city. A good artist may be able to "fake it" by altering the perspective of the background scenery. Figure 12–53 shows what such scenery can look like from the camera's viewpoint and from the bystander's viewpoint. Such methods allow big things to be compressed into small studios.

Most of us lament the fact that television is a two-dimensional medium as opposed to 3-D like our eyes. When it comes to sets and graphics, the two-dimensionality of TV becomes our ally. It allows us to cheat in all sorts of ways. Things which are far apart can look like they are close together. Things close together can look far apart, all depending on lighting, contrast, focus, and size. A person looking to the side can appear to be speaking to an object painted on the background behind him as long as both are in focus and are lit the same (in fact this is how some performers read cue cards while *looking* like they're addressing someone or something directly in front of them).

If multiple camera angles are needed, consider having "holes" in your sets for your camera to look through from behind the set. These "holes" can be camouflaged as windows, corners, or darkly lit areas of visual clutter.

The whole LOGO appears only in super-wide shots, appropriate only for the show's beginning and end. Meanwhile, it hangs there, partly in the picture for all remaining shots, sitting on the talent's head. It unbalances the shot, adding too much weight to the right of the shot. Note also the set seam running through the talent's head.

FIGURE 12-51 Poorly positioned LOGO. (From *Graphics for Television*—Prentice-Hall)

MINIGLOSSARY

*Flat Shallow, lightweight, standing scenery used as background or to simulate walls of a room.

Camera's view

Bystander's view

FIGURE 12-53   Scenery with altered perspective.

If building easels, chalkboards, writing tablets, or maps into your set, consider your talent's height, left- or right-handedness, and where he or she will stand. The talent should be able to easily point to or write on all parts of the graphic while not blocking the camera's view. This means that the graphic should be about shoulder height and remain within range of the outstretched arm. There should be a plain (unbusy) area to the side of the graphic where the talent can stand. The camera should be positioned to easily see the graphic (obscured occasionally by the talent's hand) and, with a small zoom out, catch the performer *and* graphic in a medium shot.

Set Building

If a set is to be used frequently over a long period of time, then it needs to be built like a house, with solid wood, screws, and a firm foundation. Most other sets, however,

are frequently moved, mixed, and matched with other sets and may be hung or temporarily braced. They may spend 90% of their lives in storage. Such sets need to be lightweight and flat (in fact, people in the business call them FLATS).

FLATS may be built out of 1-inch × 2-inch or 1-inch × 4-inch frames covered with muslin or canvas. They are lightweight and cheap.

HARDWALL FLATS use similar frames but are covered with 4-foot × 8-foot sheets of a hard-surfaced material such as:

1. *Cardboard.* Cardboard is cheap, strong, and easy to staple to, but easily creases, punctures, and deteriorates with moisture. Unless carefully painted, it will warp from the dampness.

2. *Foam core.* Foam core is inexpensive, light-weight, smooth, and easy to cut, staple, and paint.

It will bend a little but creases if bent too far. It punctures easily and has a smooth vinyl-like surface sandwiching ¼ inch or more of white foam.

3. *Felt board.* Felt board, also called Upson board or bulletin board, is a pressed fibrous material smooth on one side. It accepts staples and thumbtacks, but nails may pull out too easily to carry any weight. It can be painted, but moisture harms it.

4. *Masonite.* Masonite is very hard and durable, resists punctures and scrapes, and can be nailed firmly. It is smooth on one side and rough on the other and is quite moisture resistant. It is easily painted. Marlite and some other special forms of Masonite come with one vinyl-covered painted surface. Four-foot by eight-foot sheets of many designs are available from lumber stores, usually in the paneling section. The surfaces may look like brick, tile, or smooth bright colors. The material is normally used in kitchens and bathrooms for walls and counters. It is water resistant, strong, and easily nailed. It's shiny surface, however, sometimes causes lighting problems.

5. *Paneling.* This Masonite-like material is strong; resists water, punctures and scrapes; and comes in a variety of surfaces, mostly woodgrain, stone, brick, or tile.

6. *Plywood.* Plywood is inexpensive, strong, and water resistant and comes in various thicknesses and surface qualities. Try not to nail any closer than 1 inch from the edge of plywood because the edges disintegrate easily.

7. *Vacuum-formed plastic.* More expensive and more professional are the assortments of scenic backgrounds available in raised pattern vacuum-formed plastic. You can get sheets that look like Italian provincial wall paneling, library bookshelves, stone walls, Roman columns, etc.

As shown in Figures 12–54 through 12–59, you start building a FLAT by constructing a frame using 1-inch × 2-inch, 1-inch × 4-inch or 2-inch × 4-inch lumber (depending on the strength you need). It is important that the corners be strong so that the frame does not collapse. The frame can then be hung or braced with a wood triangle as shown in Figure 12–56.

When FLATS have to simulate walls or other continuous structures, they have to be connected together. Figures 12–57 through 12–59 show several ways to do this.

When all is done, the FLATS have to be stored. Piling them atop one another not only makes them hard to get to but scratches and punctures them. FLATS should be stored

upright in a rack (Figure 12–60), or they may be hung (Figure 12–61).

SPECIAL EFFECTS

Wipe and Chroma Key

Wipes, inserts, supers, keys, and mattes are other ways to get a performer to look like he or she is part of some gigantic scenery which really isn't there. The technique, however, is exacting, and the results are not always convincing. Figure 12–62 diagrams how a wipe could create such an effect.

For those studios blessed with chroma key, the illusion can be more complex. For chroma key, arrangements are made so that the performer and his immediate props and surroundings are colored "not blue," for instance, while everything else in the scene is blue. The blue in the scene can all be made to disappear, being replaced with a scene from another camera as was shown in Figure 11–31. This could be a scene of the great outdoors, a postcard, a movie, another video tape, a painting, a drawing, or even computer graphics.

Chroma keys are sometimes hard to do. Parts of your picture may shimmer, look grainy, or drop out. Chapter 11 covered some of the techniques of perfecting your chroma key. Here is a brief review of some of the things to look for if your chroma key isn't convincing:

1. Check to see that you are really keying out *blue*. Special effects generators can be adjusted to chroma key on various colors and might not be precisely adjusted to the blue of your background.

2. Adjust KEY SENSITIVITY (or KEY LEVEL) so that your switcher knows how blue something has to be before it disappears and gets replaced with the other camera's picture.

3. Adjust your FILL LIGHTING to reduce dark shadows. Dimly lit areas sometimes appear blue to TV cameras. Avoid dim lighting. Avoid high lighting ratios. Try to keep FILL LIGHTS as low in angle as possible.

4. Use an amber GEL on the backlight. This helps avoid the "fringing" you sometimes get between a performer and his or her background.

5. Keep your performers away from the background (if that is the area being keyed out). The blue from the background will reflect slightly off their clothes and skin, casting a bluish tinge which confuses the chroma keyer. Also, performers' shadows on the blue background may register as black rather than blue and may remain in the picture.

Optimal placement of a writing tablet used by a left-handed performer.

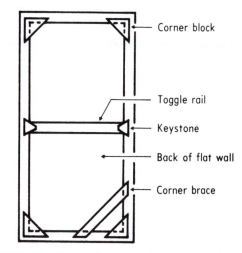

FIGURE 12-54 Frame for a FLAT. (From *Single Camera Video Production*—Prentice-Hall)

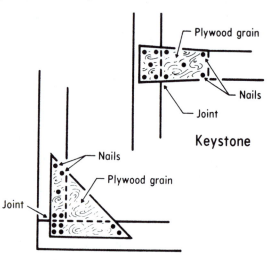

Keystone

Corner Block

FIGURE 12-55 Corner block and keystone. (From *Single Camera Video Production*—Prentice-Hall)

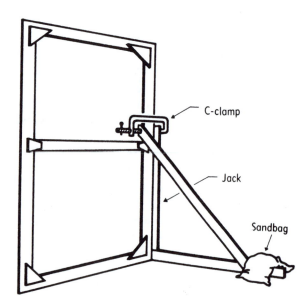

FIGURE 12-56 Floor brace and jack. (From *Single Camera Video Production*—Prentice-Hall)

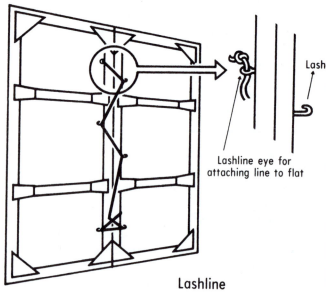

Lashline

FIGURE 12-57 Lashing FLATS together. (From *Single Camera Video Production*—Prentice-Hall)

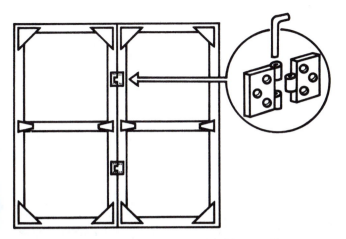

FIGURE 12-58 Fastening FLATS with hinges. (From *Single Camera Video Production*—Prentice-Hall)

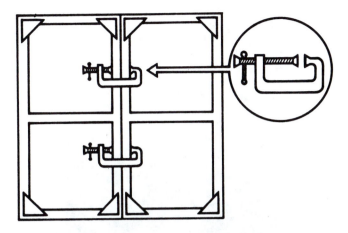

FIGURE 12-59 Connecting FLATS with C-CLAMPS. (From *Single Camera Video Production*—Prentice-Hall)

6. Light your backgrounds (or other key surfaces) as evenly as possible. Be especially careful with SET LIGHTING to ensure that it is even from top to bottom.

7. If your background picture is illuminated from the left (shadows to the right), then your foreground object (or person) should be similarly lit. Sunshine comes from only one place in the sky at a time; don't break the laws of physics.

FIGURE 12-60 Rack to store FLATS.

8. If your background is supposed to look a distance away (e.g., a mountain), then it should not be too vivid. Distant things naturally look grayer and bluer than nearby things.

Projected Backgrounds

CHROMA KEY has taken over most studios, but a few still use FRONT and REAR PROJECTION as a way to create backgrounds simply and cheaply. Here, a movie or slide projector projects the desired background onto a screen behind the performer.

Front Projection

If you project an image against the *front* of a screen, the method is called FRONT PROJECTION. It is the familiar projection technique used in theaters. When this method is used to project a background in a TV studio, however, there is a problem: When the performer stands in front of the screen (his "background"), he casts a telltale shadow. Solution: Hide the shadow behind the performer so the camera can't see it. Moving the projector to within an inch or so of the camera lens will help a lot. The shadow will become a thin outline, hardly noticeable. Better yet, one may use a two-way (half-silvered) mirror as shown in Figure 12–63. This method makes the camera's and projector's point of view exactly the same, thus hiding the shadows completely from the camera. This method has the added advantage of allowing the camera to pan, tilt (within the projected area and within the frame of the mirror), and zoom, whereas with chroma key, zooming in on the performer leaves his background exactly the same size (unless camera 2 zooms at

MINIGLOSSARY

*Front projection Technique of projecting an image onto the front of a screen (or onto a sheet or white wall).

FIGURE 12-61 FLATS hung by hooks.

jection screen and "wash out" the projected image. Meanwhile, the projected image may reflect off the performer's face and clothes. The solution is to use a special, highly directional, reflective projection screen. Such screens are usually made of a foil material on a solid background and are frequently used in classrooms or with home television projectors.

With such a screen properly oriented, the studio lights are reflected from the screen onto the floor while the projector image is reflected back full force to the TV camera.

Rear Projection. If you project an image against the *back* of a special screen so that the image may be seen from the front of the screen, the method is called REAR PROJECTION. Small, self-contained REAR SCREEN PROJECTORS are usually found in travel agencies, schools, study carrels, and expositions and are used on some one-piece home television projectors.

Much larger is the REAR PROJECTION SCREEN used in TV studios. Made of a thin, gray plastic membrane, the screen is stretched over a frame, and a projector is aimed at it from behind as in Figure 12–64. Light from the projected image passes through the screen and can be seen from the other side. A performer, standing in front of the screen, doesn't interfere with the projected image, and thus there is no shadow problem. Some effort must be taken, however, to spill as little studio light as possible on the REAR PROJECTION SCREEN in order to avoid washing out the image on it.

Note: Most rear projection screens have a front side and a back side. Be sure to project against the shiny side (your camera views the dull side) for the sharpest image and least glare.

Projector Tips. Projector bulbs darken as they age. This affects both brightness and color temperature. If your projected image lacks oomph, try replacing the bulb.

REAR-PROJECTED images appear backwards. If you're

exactly the same rate). Likewise, panning and tilting with chroma key causes the performer to move across the screen while his background remains stationary. This effect isn't desirable unless you are making a show about floating people and psychokinetic props.

A second problem with FRONT PROJECTION has to do with lighting. Studio lights are likely to spill onto the pro-

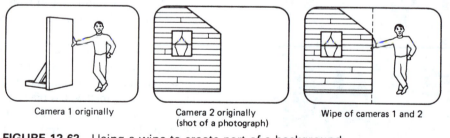

| Camera 1 originally | Camera 2 originally (shot of a photograph) | Wipe of cameras 1 and 2 |

FIGURE 12-62 Using a wipe to create part of a background.

MINIGLOSSARY

*Rear projection Technique of projecting an image onto the rear of a translucent screen so that the image can be viewed from the front.

Dissolve unit Device to dim the lamp on one projector while simultaneously brightening the lamp on another so that the image from the first seems to dissolve into the image of the other.

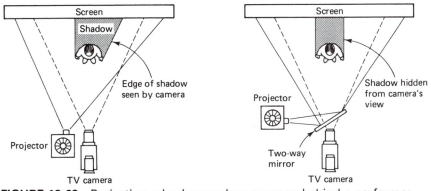

FIGURE 12-63 Projecting a background on a screen behind a performer (top view) using FRONT PROJECTION technique.

using a slide projector, the slide can be reversed so that the image comes out correctly. If you're using a movie projector aim the projector into a mirror as shown in Figure 12–65. Often it doesn't matter whether the picture is reversed (unless it has words or a recognizable scene), so you don't always have to bother reversing the image for REAR PROJECTION.

If you have a small single-camera studio without the capability of doing dissolves, you can often fake it by using two slide projectors and a DISSOLVE UNIT, a device which fades down the brightness of one projector while fading up the brightness of another. You would aim the projectors at a REAR or FRONT PROJECTION screen while your TV camera records the images as they dissolve from one to another. Figure 12–66 diagrams the process using a REAR PROJECTION SCREEN.

Slides which are going to be shown for only a few moments in the projector can use the standard cardboard slide mounts with no problem. If a slide is going to be projected for more than a minute, it will probably heat up

and buckle, going out of focus if anchored in a cardboard mount. Special plastic slide mounts are available which allow the slide to expand with heat without buckling.

PHOTOGRAPHY FOR TELEVISION

Why Use Photos?

If you are surrounded with video equipment, why step back into the Stone Age of slides, prints, and movies?

The pros and cons of video versus film have been debated for years. There are some things film does better than video and others that video does better than film.

Movie and slide cameras are lighter and more compact

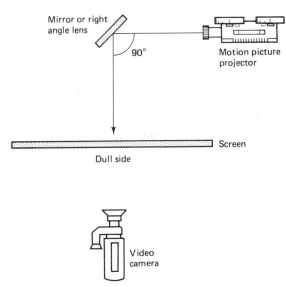

FIGURE 12-65 A mirror or right angle lens can correct image reversal problems on a REAR PROJECTION screen.

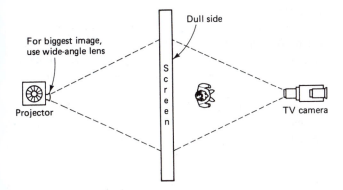

FIGURE 12-64 Projecting a background on a REAR PROJECTION SCREEN behind a performer (top view).

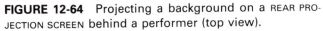

MINIGLOSSARY

*Assemble-edit   Editing shots together onto a video tape in consecutive order.

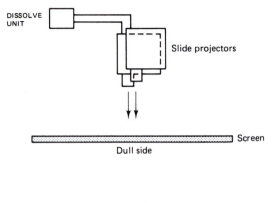

FIGURE 12-66   Using a DISSOLVE unit and two slide projectors.

than VCR/camera combinations. They can be used underwater, skydiving, on frozen Mt. Everest, or on a roller coaster (although the author has taken a VTR on a roller coaster, the gyroscopic effects of the motion caused the tape speed to waver). Film cameras are relatively inexpensive, making them a good choice when there is a risk of damage or robbery. They are easy to operate and break down infrequently. Slide and movie cameras are only slightly affected by extreme cold, and only the film is damaged by extreme heat for extended periods. Film cameras can be aimed directly at the sun or other bright objects without damage.

Setting up a VCR to take cutaways is somewhat cumbersome. A video crew member with an inexpensive slide camera can snap away 20 cutaway shots with very little fuss. Unless there is a lot of motion in the scene, brief slide cutaways can look natural, and their lack of motion will hardly be noticed.

Movie cameras ASSEMBLE-EDIT "as you go." Each scene follows the other without a glitch and without special editing equipment. Simple film editing is cheap, requiring only scissors and glue. Also, segments can be physically added to or deleted from the final program, changing its length, unlike tape, where once a program is recorded, segments may only be erased and *replaced* with new segments. Film is quite standard—you can buy it almost everywhere, even in foreign countries. Movie and slide projectors the world over use the same size film and play at the same speeds. Foreign video standards and tape speeds are different from ours, making their programs unplayable on our VCRs and TVs without the help of specialized equipment.

Silent movie cameras and slide cameras are unaffected by powerful magnetic fields, CB radios, high-tension wires,

nearby radio transmitters, or even radiation; however the film itself may be affected by X-rays or radiation. Video electronics are sensitive to all of these energy fields, and the tape is especially sensitive to magnetism.

Film cameras don't have to warm up before you start shooting. They can sit on the shelf for a year and take a picture 2 seconds after you pick them up.

Film can accept broader contrast ratios than video can. Film can often be shot with less light than that needed for TV equipment, especially color TV equipment. With Kodak VR-1000 and other fast-speed print films, you can take pictures almost in the dark. With the wide variety of high speed 16mm film emulsions, it is true that "If you can see it, you can photograph it."

So why do people use portable VCRs, you ask, while preparing to deposit this chapter in the wastebasket. With tape, you can see your work immediately, without waiting for film processing. Also, you don't have to guess about lighting; you can see what you'll get in your viewfinder. Although video equipment costs a lot more than movie equipment, what goes in it costs a lot less. Video tape costs about 50 cents per minute of play, while movie film with processing costs $4 or more per minute. Besides, video tape can be erased and used over. Video tape can record from 20 minutes up to 8 hours, while movie film cartridges run for only 3 minutes and 16mm film reels become large and cumbersome for more than 20 minutes of running time.

Sound is automatically recorded as part of the video. Sound film equipment is often bulky, and the sound film costs more to shoot and process than silent film. Editing sound film is quite complicated. So is audio dubbing.

TV cameras are silent. There are no motors inside to alert anyone that they are on. TV cameras can be operated some distance from the VCR, which means that monitoring, tape changing, and so on can be handled in one place while the camera is in another (handy for recording shots of Mt. St. Helens without getting yourself blasted off the face of the earth).

Film (except for super-8mm) is much sharper than video. With film it is possible to take a wide shot and blow up part of the scene later without having it become excessively fuzzy. With video, all cropping is done while the picture is taken by the camera.

Special effects such as split screens, mattes, or dissolves are usually cheap and easy to do in video (if you have the equipment) but are expensive or difficult to do photographically. On the other hand, speeded-up motion, stop motion, slow motion, backward motion, and time lapse photography are quite easy to do with film but are next to impossible for standard, moderately priced video equipment. Therefore, do your animation photographically and your split screen and special effects with video.

In conclusion, each medium has its advantages. For quickie shots and rough environments, the film medium is cheaper and easier. For lengthy presentations with sound, video is the answer. In short, film does have a place in the video world, and the two can make a happy marriage.

Shooting Slides and Prints for Television

Contrast. The human eye yields a useful image even when one part of a scene is 1000 times brighter than another, a contrast ratio of 1000:1. A 16mm movie projected on a theater screen can have a contrast ratio of 600:1 for outdoor scenes or 160:1 for indoor scenes. A glossy print offers a contrast ratio of 65:1. A well-maintained, high-quality TV set can work with a contrast ratio of 30:1. So what does this tell the photographer about taking pictures for TV?

Keep the contrast down. The equipment cannot handle the lighting ratios that the photo medium can. The dazzling skies, brilliant highlights, and dim nuances in the shadows that are so pleasing when you see the film projected on the screen will all but disappear on the TV screen. The brilliant whites will merge into a chalky area with no detail, and the black nuances will run together into solid muck.

To avoid this problem, the photographer must try to keep the scene's lighting ratio to within 30:1 or about 4 f-stops on a photographic meter. To keep contrast down, bright lights and dark shadows must be avoided. Beware of scenes which include outdoor sky or indoor windows (with light streaming in from outside). White backgrounds, white shirts, white paper, and shiny objects taking up more than 3% of the picture should be avoided as they will make your performer's skin tones look dark. They will also force automatic TV camera controls to darken the entire picture to keep the whites within a range the camera can handle.

Substitute medium-tone backgrounds, off-white paper, and pastel shirts. White cotton shirts can be washed in tea, shiny objects can be treated with dulling spray, and shadowy areas can be illuminated with lamps or with light from reflectors.

Once a film has been shot without regard for TV, it may still be salvageable. Special low-contrast TV prints can be made from films. Also, there is low-contrast photographic paper for printing your negative.

Television seems to increase the contrast of photos. Shadowy details turn into black blobs. Any grain in a slide will also be increased. To reduce the grain, use a medium film speed such as ASA 64. For 16mm movies, Kodak 7294 high-speed negative film or ECN (Ektachrome negative) are excellent film stocks. If you're shooting 8mm movies for TV, try Kodachrome 40. Otherwise, avoid using Kodachrome as its colors are often too vivid for television and its contrast ratio too great. If you are stuck with Kodachrome, you can keep the contrast down by slightly overexposing your shots.

Whenever a slide is copied, contrast accumulates. Dark areas get blacker, and light areas get whiter. Try to use slide duping film such as Ektachrome 5071 if you expect to transfer the slides to video or duplicate them because it is a lower contrast medium and minimizes contrast buildup.

Film (and video) is better suited to colors that translate into the middle shades of gray. Light, saturated colors appear oversaturated when transferred to video. Especially avoid strong reds and oranges. Slides with dark colors do not reproduce well.

One way to reduce a high contrast ratio in a slide is to shoot it slightly overexposed. When viewed, the slide can be projected with reduced light (perhaps by using a filter), making it normal again, except the contrast is reduced.

White type on slides tends to be too bright for video. If the type is over a plain black or colored slide, a neutral density filter or a gel can be used to darken the lettering. This process doesn't work, however, if the white lettering is over a picture because the filter or gel will darken the picture along with the lettering.

Other Considerations. Here are some more do's and don'ts for preparing photography for video:

1. *Remember the TV* ASPECT RATIO. Keep important information in the middle of the shot and away from the edges. As a rule, always shoot in a horizontal format. Using a vertical slide is like trying to cram an oblong peg into a square hole. It doesn't fit. 35mm slides and photographic prints have roughly a 5:7 ASPECT RATIO. As your TV camera views them, you can expect to lose about 15% of your photos, especially on the sides.

2. *Avoid "busy pictures.* Extreme wide shots are often lost on television. The detail becomes too small and too cluttered. If the background appears too busy, focus selectively on the main object, making the background fuzzy and unobtrusive.

3. *Don't overcrop.* The TV camera can zoom in on a photograph to take just the shot that it needs, but it can't do the opposite. If you've cropped your shot too tightly, then the TV camera will pick up the edge of your photo, or it may be difficult to get the camera to show 99% of the picture without having the edge sneak in by accident. Because film is sharper than video, you can afford to shoot pictures with wider angles than you normally would. Don't overdo the wide angles, or you'll end up with problem 2.

4. *Provide room for movement.* Take pictures which will allow the TV camera to pan, tilt, and zoom across the picture. This *doesn't* apply if you are shooting slides which will be shown in a FILM CHAIN where the camera cannot pan, tilt, or zoom.

5. *Overshoot.* Taking many pictures will allow you more to select from. Also, video can often incorporate more pictures than a slide show normally does.

Film-to-Tape Transfer

Advantages and Disadvantages. Both at home and at school, the inconvenience of dragging out projection screens and projectors, or smashing your shins while crossing your darkened "theater," makes you want to convert all your slides, filmstrips, and movies into video. Perhaps you would prefer the sound of background music and polished narration to the cement mixer roar of your projector fan. Or maybe a travel collection composed of 20 different media, some slides, some snapshots, some super-8 movies, a few audiocassette tapes, a reel-to-reel audio tape, some postcards, and Fibber McGee knows what else are just too difficult to show in a class or business meeting. Furthermore, you'd like to organize and consolidate this mess into a single, neat videocassette that runs itself.

Are you convinced that all your films should be transferred to video? Don't be. Before you start blowing dust off the film cans, take another peek at Figure 12–11. Look at the detail you will lose. Granted, by zooming in and panning across pictures at close range, you can preserve much of the important detail, but in cases where you wish to see the whole picture at once, with all its fine print, all its subtle highlights, and all its true colors and be able to distinguish every feather, every hair, every leaf, and grain of sand individually, then keep your photos as photos. Until HDTV (high-definition television) comes along, video just isn't a sharp enough medium to do the job.

Video is simply a trade-off, convenience for sharpness.

Do-It-Yourself Methods. If you've decided to transfer your slides, filmstrips, and photographs to video tape, here are some methods to do it.

Front screen projection: This method is cheap, flexible, and nearly as good as any other method. You simply aim your projector at a screen and aim your camera at the image to record it, as shown in Figure 12–67.

Use a flat, white posterboard as your screen. Any smooth, matte-finished (not shiny), very white, flat surface will also do. Regular projection screens tend to reflect "hot spots," overly bright areas on the screen, while allowing the edges and corners to look dim.

Position the projector so that the projected image is about 18 inches tall. Larger images start to become too dim. Smaller images accentuate the minor imperfections in the screen's surface. Experiment. Be sure to focus the projector *very well.*

Mount your camera on a sturdy tripod. Position it as near the projector as possible so that you don't get KEYSTONING of your image. Turn out *all* lights (or close curtains) so that *no* light seeps in to "wash out" your projected image.

Focus the TV camera and adjust your camera's iris, color temperature, and white balance controls as usual. The "indoor" setting on the color temperature control may work best. Experiment.

The CONTRAST RATIOS of slides and movie film may be great enough to produce blooming whites, lag, or mild burn-ins. These may be reduced by reducing the projection

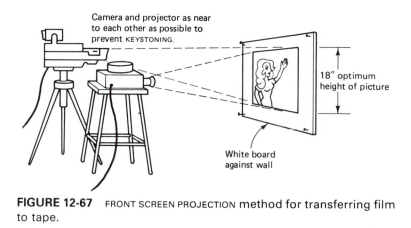

FIGURE 12-67   FRONT SCREEN PROJECTION method for transferring film to tape.

MINIGLOSSARY

Shutter bar Occurs when a TV camera records a movie from a projector—a soft dark band runs through the TV picture when the projector doesn't synchronize its shutter with the TV camera's picture-making frequency.

*Telecine Movie projector/TV camera combination designed for converting movie images to video. Also called a film chain.

bulb brightness, or adding a neutral density filter, or stopping down the camera's iris.

Dark areas of the picture may tend to blend together as a solid mass. To cure this problem, add a small amount of light to the room while watching the result on your camera viewfinder. The room light will add just a little more brightness to the darkest areas, bringing them up to a level that the camera can register.

Adjust your camera's zoom lens so that the projected image fills the TV screen. Because film pictures have a wider ASPECT RATIO than video, this will mean the loss of a little of the left and right borders of your scene. Feel free, however, to zoom, pan, tilt, fade, or defocus your camera in order to focus attention on the main event, or to add movement when needed, or to dress up the transitions between scenes.

If working with home video equipment and a single slide projector, you can omit that blank or black screen you get between slides on most projectors by pausing your VCR at the changeover point. You can also pause while changing movie reels.

If you plan to add an audio track to a silent slide show, don't try to do it while you are trying to record the video. You can rehearse your narration while recording if you want, just to get a sense of timing, but save the actual narration for a later audio dub. You have plenty to handle right now, whereas later, giving audio your full attention, you're likely to do a smoother, more creative job. Also, you won't have projection fans roaring in the background.

Rear screen projection: REAR PROJECTION has one advantage over FRONT PROJECTION: The image is brighter, and that means you can leave a dim light on in the room (your shins will thank you for that). Figures 12–64 through 12–66 showed several REAR SCREEN setups. For transferring slides and movies, the screen only needs to be about 18 inches tall. If you don't have professional Polacoat lenscreen to use for a REAR PROJECTION SCREEN, then try some transluscent or vellum paper to project onto.

Movies: There's one annoying problem faced by anyone trying to transfer a movie to video tape, and that's SHUTTER BAR. Your video image flickers or has a dark horizontal band running through it, perhaps moving, sometime stationary. The bar looks a lot like the HUM BAR shown in Figure 3–38. Here's the cause.

Movie projectors work by beaming light through transparent still pictures and by focusing that beam on a screen. A shutter mechanism in the projector aims the light through one picture, then blocks off the light while changing to the next picture, and then opens to beam the light through the next picture. The process happens so fast that you perceive a continuous moving image. Super-8 projectors display 18 pictures per second (24 per second for sound), and industrial 16mm projectors show 24 frames per second.

Your TV camera scans its sensitive vidicon tube at the rate of 60 pictures per second. Because the numbers don't match, there are times when the TV camera is "looking" for a picture, but your projector isn't making one—it's busy changing pictures, and the screen is black. Thus, the projected image which looks fine to *your* eyes could look bad to the *TV camera*.

Professional TV setups, like studios and commercial film-to-tape transfer shops, have special motors and five blade shutters in their projectors to resolve this problem. This way they always manage to have a picture on the screen when your TV camera is looking for it. If you don't have a professional TELECINE movie projector, you may be able to minimize the SHUTTER BAR if you can increase or decrease your projector's speed a little.

In Chapter 15, you will see the professional solution to this problem. It is called a FILM CHAIN or TELECINE or MULTIPLEXER. It is a device with one camera and several projectors (usually one DOUBLE-DRUM slide projector, one super-8 movie projector, and one 16mm projector) and a mirror mechanism to select which projector focuses its image on the camera. The movie projectors are all modified for TV use.

Commercial Services. If you want to save the hassle, time, and expense of transferring your slides and movies to video, you can always have someone else do it for you (for a price, of course). A commercial service will employ professional cameras and projectors and usually will give better results than you can get aiming your TV camera at a screen. Pay a little more, and you can have a professional transfer shop use a Rank Cintel Flying Spot Scanner, the Cadillac of TELECINES.

Fotomat and other consumer outlets will transfer super-8 movie film or 35mm slides to VHS or beta tape for a reasonable charge. Of course you lose nearly all "creative control" over the result. There would be no panning, zooming, fancy transitions, or mixture of media (a slide, a movie, then a snapshot) that you could do yourself. Also, vertically oriented slides would be recorded "as is" without regard for ASPECT RATIO.

After all this talk, graphics still boil down to three rules:

1. Make it fit the shape of a TV screen.
2. Keep it simple.
3. Make it bold.

COPYING A VIDEO TAPE

By this time, you may already know how to copy a video tape, and this subject may seem too simple to deserve a chapter of its own. But if you consider that nearly every video taped program a viewer ever sees is actually a copy, you will realize that it is our copies by which we are judged, not our originals. A poorly made copy will reflect on the entire production. Thousands of dollars worth of equipment and time will go to waste if the copy of your masterpiece looks bad.

A second reason why copying is so important is that if you visit a TV production studio for any length of time, you will see tapes being copied. A substantial portion of every teleproducer's day is spent cranking out copies of one tape or another.

Copyability is one of the marvels of video tape. Books were not a big deal until the Gutenberg press made the printed page accessible to everyone. The second giant leap in the print industry was the Xerox machine, which made it possible to

1. Make a copy instantly, *when it's needed*
2. Make just one copy if only one is needed

Video tape works the same way. Copies are fairly quick to make (it takes 1 hour to copy a 1-hour tape), and you can make just one if that's all you want. All it takes is two VCRs and some wire to connect them.

To make video copies by the hundreds, you can send your MASTER tape to a DUPLICATION HOUSE (Figure 13–1) where hundreds of VCRs (SLAVES) copy it at the same time.

Another nice thing about video is that you can copy a tape from one format to another. Although different-format tapes may not be compatible, their video signals are. By using a ¾U VCR and VHS VCR, you can make a VHS copy of the ¾U tape. Similarly, using a VHS and an 8mm machine, you can make 8mm copies from the VHS tape or vice versa. Naturally, to make VHS copies of the VHS tape, you need two VHS machines.

SETUPS FOR COPYING A VIDEO TAPE

Using Standard VCRs

Here's how to copy a tape: Get two VCRs. Using separate audio and video cables, connect the audio and video *outputs*

MINIGLOSSARY

*Master tape   The original copy of the finished version of a tape. Could be original footage of a ''live'' show, or could be a program edited together from other tapes. The master is the best-quality copy of this program in existence.

Duplication house   A company that duplicates videocassettes, usually hundreds at a time.

Nation's largest is Video Corporation of America, located outside Chicago. It can make 2000 copies in one hour.

FIGURE 13-1 Videocassette duplication facility (Courtesy of JVC Company of America).

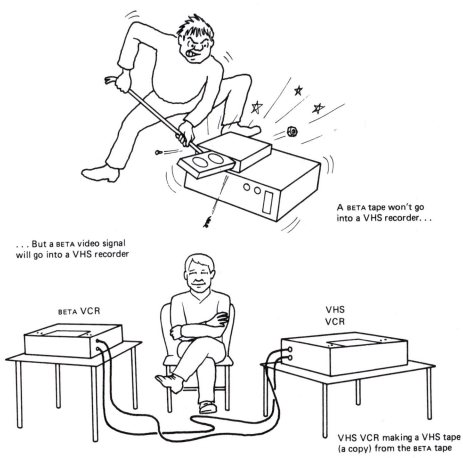

A BETA tape won't go into a VHS recorder...

... But a BETA video signal will go into a VHS recorder

BETA VCR

VHS VCR

VHS VCR making a VHS tape (a copy) from the BETA tape

Different format tapes may not be compatible but their video signals are.

of the player (called the MASTER machine or the VCP for videocassette player) to the audio and video *inputs* of the recorder (called the SLAVE). Connect your TV set to the SLAVE's output so you can monitor the results. Slip your MASTER cassette into the MASTER player and a blank video cassette into the SLAVE. Adjust the INPUT SELECT switch on the SLAVE so it "listens" to the MASTER. Select the desired tape speed; hit PLAY on the MASTER and RECORD on the SLAVE. Let the machines do the rest.

There are several ways to hook up your equipment to carry out this task. Figure 13–2 diagrams some of them. In each case, there must be a way to get the audio and video

signals out of the VCP and into the VCR. It is also handy to have some way to monitor the results, so you need a way of getting your signal out of your VCR and into a TV.

The method you use for getting the signal from the VCR to the TV is not too important. You could use separate audio and video cables or a multiwire 8-pin cable, or you could send RF over an antenna cable to the TV. Although the video cables will give you a slightly sharper picture on your TV than the RF methods, the technique you use will not affect the quality of your recording.

What *will* affect the quality of your recording is the method you use to send the signals from the VCP to the

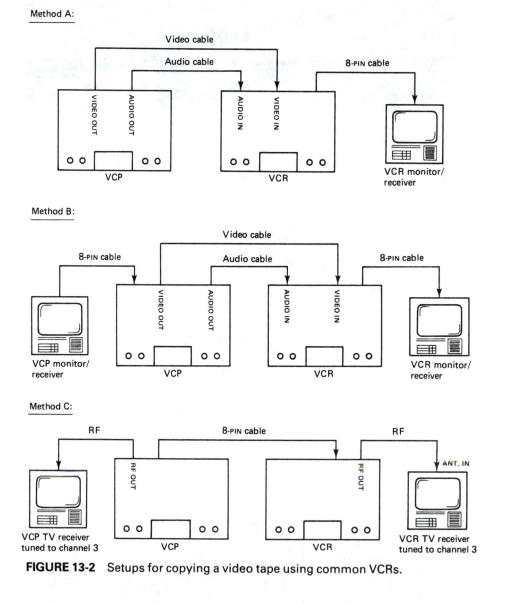

FIGURE 13-2   Setups for copying a video tape using common VCRs.

MINIGLOSSARY

*Slave   In the tape copying process, the videocassette recorder that actually does the recording.

Master recorder   In the tape copying process, the VCR that plays

the tape that the slaves copy. Also called the VCP, videocassette player.

Method D:

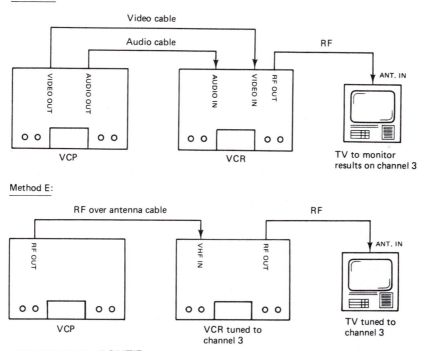

Method E:

FIGURE 13-2 CONT'D

VCR. Separate video and audio cables or an 8-pin cable as shown in methods A–D in Figure 13–2 are the best. Method E, although it employs fewer wires, is the worst method because your video and audio signals must be modulated into a channel number by the VCR (losing a little sharpness) and then demodulated by the tuner in the VCR (losing a little more sharpness).

Methods B and C in the figure have an additional advantage of showing you the picture produced by the VCR as well as the picture produced by the VCP. If the VCP's picture looks good and the VCR's picture looks bad, you have an inkling that something is wrong with the VCR or the cables running to it. If you had only one monitor as in methods A, D, and E, you wouldn't know if it was the VCP or the VCR that was causing the trouble.

Incidentally, you would not want to monitor *just* the signal from the VCP. It would tell you that the VCP was playing okay, but you wouldn't know whether the VCR was getting a good signal. Besides, you always like to play a little of the tape back after it has been recorded to make sure it came out okay. With your monitor attached to the VCR, you can view your results directly when you switch the VCR to PLAY.

All these methods apply to home VCRs as well as industrial ones. They work with reel-to-reel video tape players too. You can even substitute a videodisc player for the

VCP. Although the VCP only needs to be a video player, there is no reason why it can't be a videocassette recorder. As long as it can play the tape, that is all that matters.

VCRs with Dub Feature or S Connectors

There's another method of copying a videocassette that is even better than the ones listed above, but it requires higher-quality video tape machines with a DUB input (or output). Figure 13–3 shows the connection. Incidentally, the word DUB means "to copy." Earlier, you saw the word DUB used in conjunction with a VCR's AUDIO DUB feature whereby a new sound track could be recorded (perhaps "copied" from somewhere), erasing the old sound track. In this new context, however, the DUB feature refers to a special input and/or output on the VCR made especially for efficiently copying video tapes.

This system uses a multiwire cable connecting the DUB OUT of the VCP to the DUB IN of the VCR. Machines capable of connecting this way are able to produce sharper pictures with better color than those using common video signals.

Here's why it's better: To record a color picture, a VCR has to lower the frequency of the color signal using a technique called COLOR UNDER. In doing so, it makes the colors fuzzy. When the tape is played back, the color signal

MINIGLOSSARY

*Dub feature   On better VHS, SVHS, and ¾U VCRs, especially editors, this is an input or output that allows the VCRs to copy

unprocessed color signals directly, yielding a cleaner copy.

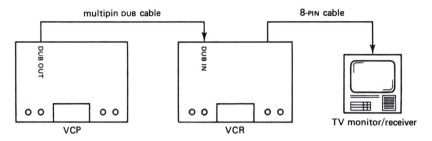

FIGURE 13-3   Setup for copying a video tape using VCRs with DUB feature.

is boosted in frequency, making the color video signal "normal" again; however, the color is still fuzzy from being lowered in frequency that first time. If you take that video signal and rerecord it, the next VCR will *again* reduce the frequency of the color in order to record it on the tape. When the tape is played back, the frequency is boosted again. This cut-boost, cut-boost process makes the color grainy and fuzzy, much like modulating and demodulating a video signal makes it fuzzy. The less meddling you do with the signal, the better. The DUB cable takes a shortcut through the process. It catches the color off the video tape before its frequency has been boosted and allows it to be copied directly on the VCR, saving one boost-cut. Thus, the copied picture has sharper, smoother color.

Some home VCRs have a special DUB feature. You connect the machines up in the usual way, but you throw a switch on the VCP to the DUB position to bypass unneeded enhancement circuits. The result is a cleaner copy. Switch DUB off when using the VCR normally.

SVHS and Hi8 VCRs have inputs and outputs similar to the DUB connectors mentioned above. These S CONNECTORS or Y/C CONNECTORS (described in Chapter 5) can be used in place of the video cables in Figure 13–2. These multiwire cables carry the color signals separately from the luminance signals, thus avoiding color- and sharpness-damaging circuits. Copies made using S CONNECTORS will look better than copies made with plain old video cables.

Home VCRs Connected to Antenna or Cable

Sometimes you may wish to use home VCRs to record shows off the air, copy tapes from one VCR to another, and also allow simultaneous TV viewing of either the VCR's signal or the broadcast signal. Figure 13–4 shows several setups to achieve this.

In each case, let's assume that an antenna signal is used or a cable TV signal is used which has either been UP-CONVERTED to UHF or is being fed to VCRs and TVs which are CABLE READY. Otherwise, we have to face the additional complication of wiring in a cable TV CONVERTER BOX.

In method A the TV signal is split and goes to VCR 1 and VCR 2. Either VCR can record that antenna signal (with its INPUT SELECT on TV) or can record a signal from the other VCR (with its INPUT SELECT on VCR). Either VCR can record from or play back a tape to the other. TV 1 can pick up the antenna signal if VCR 1's OUTPUT SELECT is switched to TV. This will be true no matter if VCR 1 is recording, playing back, or copying a tape. TV 1 can display what VCR 1 is doing by switching VCR 1's OUTPUT SELECTOR to VCR. Similarly, TV 2 can do the same things with VCR 2's signals.

If only one TV is available, then method B can be used. Here an ANTENNA SWITCH selects which VCR the TV will listen to. Either VCR can have its OUTPUT SELECTOR switched to the TV mode, allowing the TV to view broadcast channels even while the VCRs are dubbing a tape.

Method C is the worst method of all. Here only VCR 2 can do the recording. Also, the degenerated RF signal is being copied. The TV can view broadcast programs only while the VCRs aren't occupied duplicating a tape.

MAKING THE COPY

Load your MASTER recording into the VCP and a blank cassette into your VCR. Make sure the VCR's INPUT SELECTOR is in the proper mode to "listen" to your VCP. If using an 8-pin cable between the two, you'll probably switch the VCR's INPUT SELECTOR to VCR. If using separate video cables between the two, you might throw the switch to LINE. If using a special DUB cable, you would switch it to DUB. If, heaven forbid, you were copying via RF, you would switch the VCR's INPUT SELECT to TV and tune the VCR to channel 3 to listen to the VCP.

If you are using a VCR to play the tape, take this one precaution, which will save you apologies and a pint of stomach acid. Place masking tape (or some other reminder) over the RECORD button on the VCR which will do the playing so that you don't accidentally push the button and erase something from the MASTER tape. You don't want to

---

MINIGLOSSARY

Color under   Electronic technique of lowering frequencies of the color information in a video picture making it easier to record.

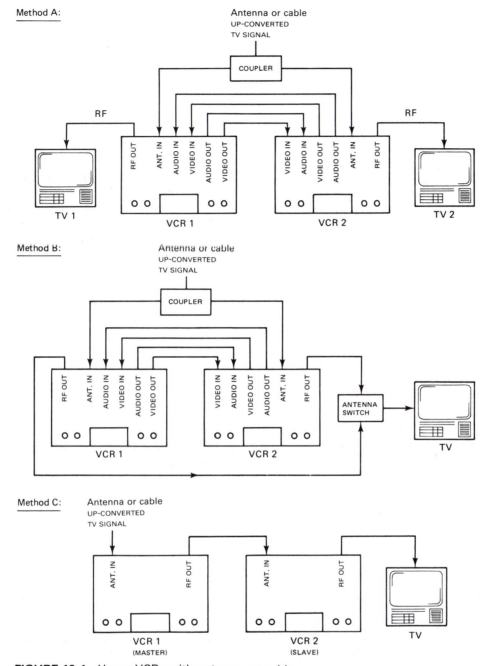

Method A:

Antenna or cable
UP-CONVERTED
TV SIGNAL

COUPLER

RF
RF OUT
ANT. IN
AUDIO IN
VIDEO IN
AUDIO OUT
VIDEO OUT

VIDEO IN
AUDIO IN
VIDEO OUT
AUDIO OUT
ANT. IN
RF OUT
RF

TV 1   VCR 1   VCR 2   TV 2

Method B:

Antenna or cable
UP-CONVERTED
TV SIGNAL

COUPLER

RF OUT
ANT. IN
AUDIO IN
VIDEO IN
AUDIO OUT
VIDEO OUT

VIDEO IN
AUDIO IN
VIDEO OUT
AUDIO OUT
ANT. IN
RF OUT

ANTENNA
SWITCH

VCR 1   VCR 2   TV

Method C:

Antenna or cable
UP-CONVERTED
TV SIGNAL

ANT. IN   RF OUT   ANT. IN   RF OUT

VCR 1   VCR 2   TV
(MASTER)   (SLAVE)

FIGURE 13-4 Home VCRs with antenna or cable

forget which machine is which and suddenly find yourself with a "Rosemary Woods" gap in your original tape.

The next step is to adjust the video and audio levels on your recorder. To do this, first play a sample of tape from the VCP and then press RECORD on the VCR. Set the levels as you would from any other audio or video source.

After setting levels, check the SKEW and TRACKING controls on the VCP to see that the tape plays back as clearly and solidly as it can.

Rewind everything and get ready for the actual copying. When you start up a ¾U VCR or VCP, it takes a few seconds for its motors to stabilize. During those seconds,

the VCP is incapable of producing a good picture, and the VCR is incapable of recording a good picture. To minimize some of the hash you'd record when starting up industrial tape machines, begin the recording procedure in the following way:

1. Press RECORD on the VCR.

2. Switch the VCP to PAUSE or PLAY/PAUSE, allowing the video heads to get up to speed (you can hear them whirring).

3. Switch the VCP to PLAY or FORWARD and then

immediately switch the VCR to FORWARD (holding RECORD down). If you have a helper handy, have the helper switch the VCP to FORWARD at the same moment that you switch the VCR to FORWARD.

By using RECORD and PAUSE in this way (steps 1 and 2), you at least get the video heads spinning so that only the tape has to catch up to speed when you start this process. By starting both machines together (step 3), the following happens: While the VCP is getting up to speed and sending out a bad picture, the VCR is simultaneously getting up to speed and is incapable of recording a good picture anyway.

When using home VCRs, the copying method is slightly different:

1. Select the desired recording speed for the VCR. The fastest speed is best.
2. Press RECORD/PLAY on the VCR and then hit PAUSE.
3. Press PLAY on the VCP.
4. Watch your TV monitor. At the first sign of a clear program, UNPAUSE the VCR to begin recording.

As before, it's always good to make a short test recording to make sure your VCR is recording properly. If starting to record a new tape from its beginning, you may wish to play the blank tape for about 10 seconds first and start your recording from this point. This LEADER gets you past the threading-caused dropouts on the tape. When finished making a copy, view a little of the *end* of the recording. If your video heads clogged during the process, it's the end of your copy that is likely to be bad.

Label Everything

Now that you've successfully copied that video tape, what do you do next? Sit down and have a smoke? *Wrong!* You *label the copy* and make sure to put the word *copy* on that label. Copies and masters look too much alike to be easily sorted out—but they are not *really* alike and should not be treated as equals, so label them and keep them separate. Besides, smoking is bad for your health.

MAKING THE BEST POSSIBLE SIGNAL FOR COPYING

The tape you record can be no better than the tape you're copying from. If your original tape is fuzzy, grainy, or jittery, your copy of this tape will be fuzzier, grainier, and jitterier. Three things are required to make the best video copy:

1. The best VCP and VCR available
2. A good connection between the two
3. The best tape available

The Video Tape Machine

The better the MASTER, the better the copy. A VHS or 8mm tape recorded at the fastest speed (Figure 13–5) will give a better-quality signal than one recorded at the slowest speed. If mastering on VHS or 8mm, use the fastest speed.

The new SVHS (SUPER VHS), Hi8, and ED BETA formats give markedly better quality pictures than their older brothers, making them good MASTERING formats, but these super-recorded tapes are unplayable on regular VHS, 8mm, or beta machines. SUPERBETA and VHS-HQ provide modest improvements over regular beta and VHS, and the tapes are playable (albeit unimproved) on the older VCRs. Similarly, ¾U-SP (Superior Performance) VCRs give better results than regular ¾U VCRs.

Regular VHS and 8mm are not good formats to MASTER on. Although the original tapes may look okay to the eye and their signals are copyable, home VHS and 8mm format tapes are right at the bottom of the heap when it comes to picture quality. You should consider home formats to be one-generation media: They are okay for recording a program off the air and playing it back directly. It is also okay to start with a sharp picture from a professional format tape and copy it onto VHS for distribution. It is a common practice for producers to make an excellent-looking 1-inch or ¾U tape and then to make duplicates on VHS. Although 1-inch and ¾U copies would look better than VHS copies, the expense of distributing large quantities of these large, heavy, and expensive tapes may be prohibitive. VHS tapes are cheap, light, and easy to mail and their quality *as a copy* is "good enough."

A ¾U-SP VCR, SVHS, Hi8, or ED BETA or even a regular ¾U VCR will make a better MASTER tape than a regular VHS or beta or 8mm VCR. The picture will be sharper, smoother, and in the case of ¾U, will be less jittery.

Better than SUPER VHS, ED BETA and ¾U, is 1-inch MII, BETACAM, or D2 formats. These expensive professional machines are used by broadcasters and commercial production houses where the very best picture is required and extensive editing will be involved. Their picture is so good that copies look almost as good as the originals.

Professional COMPONENT recorders record the color signals separately from the black-and-white parts of the signal (hence the word COMPONENT) the color doesn't get mangled in the process. The BETACAM format is the most popular of the COMPONENT recording formats, covering about three quarters of the professional ENG and EFP markets. The remaining one fourth of those markets uses the MII format, where tapes are recorded on high performance VHS video cassettes.

A) Original; As seen by camera

B) First generation; Same scene recorded and played back on VHS machine at 2-hour speed

C) First generation . . . and at 6-hour speed

D) Second generation; Copy of example B made at 2-hour mode

E) Third generation; Copy of example D made at 2-hour mode

FIGURE 13-5   Copies always look worse than originals. Also, home VCRs make better pictures at the fastest speed.

MINIGLOSSARY

Super beta   Beta-compatible video recorder format that yields sharper pictures than regular beta.

VHS HQ (high quality)   VHS-compatible video recorder format that yields sharper *looking* (less grainy) pictures than regular VHS.

SP (superior performance)   ¾U-compatible video recorder format that yields sharper, smoother pictures than regular ¾U.

Component video   Separate color video signals that have not yet been combined into a single video signal. RGB video is an example of component video signals—one for red, one for green, and one for blue.

*Component video recorder   Professional VCR that records separately the distinct color video signals from a camera, offering a high-quality image.

*Betacam   Component VCR format using beta-like cassettes.

A RIDDLE

A teacher shot some video tapes at a school and gave them to her AV department to edit together electronically into one tape. The tape came out so well that the AV director immediately made a SAFE COPY of it in case the precious original ever got damaged. When the school board asked for copies, the AV director wisely used the SAFE COPY as a WORKING MASTER (the MASTER you usually make copies from) to avoid risking the original MASTER. One school-board member copied his copy of the tape and sent it to another school. The teachers liked it and had their librarian make a copy of it for classroom viewing. A student liked it and copied the teacher's copy to play to his folks on his home videocassette recorder. His folks liked it, copied it, and sent the copy off to Grandma, who also had a VCR. The question now is: What did Grandma see? Answer: *crud.*

Why? Every time a tape is duplicated, it loses quality. The more generations you go down, the worse the picture and stability become. The moral of the story is this: *Try to stay as close to the original as possible.* Make copies from the original whenever possible. Make copies from a WORKING MASTER only when it is really necessary to protect the original and when using professional equipment which affords the luxury of going down one more generation.

To review:

Format	Number of Acceptable Generations Possible[a]
VHS (LP, SLP), Beta III	1
VHS (SP), Beta II	1–2
Beta I, Superbeta II, VHS-HQ	2
¾U (using video connection)	2–3
¾U (using DUB connection), super VHS, ED beta	3
¾U-SP	3–4
C	5
Betacam	6
MII	7
Digital	20+

[a] Where 1 equals master only, 2 equals copy of master, 3 equals copy of copy, etc.

Although the COMPONENT recorders are only a little better than the 1-inch machines, they are more popular for EFP and ENG use because the recorders are smaller and the cassettes are easier to load.

Only the larger production houses can afford $10,000+ COMPONENT recorders. One step above them (for the very rich) are DIGITAL VIDEO RECORDERS. Digital video recorders cost $50,000+ and devour tape at wildly fast speeds but can make copies with *no* generational loss. You can make a copy of a copy of a copy 100 generations down if you want without degradation. This is possible because once you have changed the video signal into 1s and 0s, you can copy those 1s and 0s, and they will *remain* 1s and 0s. Even a bad copy of fuzzy 1s and 0s still displays 1s and 0s which can be regenerated back into a picture which looks exactly like the original (within the limits of the equipment).

Almost as good as the DIGITAL VTR is the MII format VCR. It uses metal evaporated tape in cassettes very similar to VHS cassettes, can hold a quality picture for 7 generations of copying, and costs $10,000–$35,000.

If making copies using VHS recorders at their 2-hour speed, try to use four-headed VCRs. Although two-headed machines will work, four-headed machines will use their "fatter" heads at the 2-hour mode, making a stronger recording which will play back smoother and give better still frames and special effects. The better TAPE DUPLICATION FACILITIES employ industrial SLAVES with "fat" heads. As a result, the copies look better than "pirate" copies, which are often made in somebody's basement using home video gear.

The machines you pick to play the MASTER and record the copy should be in tiptop shape. Any abberation in their reproduction will appear even worse as you play your final copy. Maintain your machines and clean your heads once every 100 hours.

If possible, play the MASTER tape on the very same VCR that made the MASTER. Like your own mother, the machine that originated a tape is more "blind" to the recording's minor imperfections and idiosyncrasies and will play it at its best. If you can't play the MASTER on the machine that recorded it, then pay particular attention to TRACKING as you play your MASTER. This may have to be adjusted to yield a smooth, steady picture.

Also watch for TAPE TENSION errors evidenced by FLAGWAVING at the top of the screen. If the machine is working right, the top of your picture should not jitter at all.

MINIGLOSSARY

D2   One type of digital video videocassette recorder.

Digital video recorder   Advanced, professional VCR that records video as 1s and 0s. Digital video tapes can be copied without generational losses.

M2 format   Advanced professional format VCR that uses metal tape in VHS cassettes. Pictures suffer minor generational losses when copied.

Super VHS (S-VHS)   Improved VHS format using special tape and yielding 400 lines of resolution picture sharpness.

If you are copying a VHS or beta tape and you have a choice of which machine to use as MASTER and SLAVE, you might try the tape on both VCRs to see which machine plays the tape the best. Often you will find that the machine equipped with the most visual effects (like NOISELESS SEARCH, REVERSE SCAN, STILL FRAME, etc.) will give the steadiest picture. The machine with the widest video heads (check the VCR's specifications for this) will probably play the sharpest, smoothest picture if the tape was originally recorded with wide heads. Put another way, if a four-headed VCR was used to make the MASTER, then use a four-headed VCR to play it back. In short, whichever VCR plays the tape best should be the MASTER.

If the MASTER tape has been stored at a different temperature than your VCR, allow the temperature to stabilize before playing it. If the tape hasn't been played for a while, wind and rewind it once to "relax" and dry it out.

The Connection

As mentioned earlier, RF is the worst way to send a signal from the VCP to the VCR. If you must do it this way, use a 75-ohm coax cable and route the wire far away from other RF cables. Make sure that the VCR is meticulously fine-tuned. If channel 3 is broadcast in your area, then switch the VCP's RF generator to channel 4 and tune the VCR to channel 4. The object here is to transmit a clean signal with the least amount of outside interference.

The standard way to send signals from the VCP to the VCR is via separate video and audio cables or using an 8-pin cable. For best results, both cables should be less than 8 feet long and should be well shielded. The video cable must use 75-ohm coax (RG-59U is the wire commonly used) and have sturdy connectors at each end. The object here is to convey a strong signal with the least amount of interference from outside signals.

If using SVHS, Hi8, or ED BETA VCRs, use the S or Y/C connection in preference to the video connection between them.

Better than the preceding method is the DUB cable. Try to route this cable away from other electrical interfering signals.

Technocrats have a penchant for dreaming up lofty names for things. Regular video signals are called COMPOSITE. The S or Y/C signals, inputs, and outputs used by SVHS and Hi8 are called Y/C 3.58 (the frequency of the color signal is 3.58 MHz). DUB signals between VHS editors and VHS duplicators are called Y/C 629 (629 kHz is the frequency of their color signals). DUB signals between 3/4U editors and other top-of-the-line U-Matic VCRs use Y/C 688 (688 kHz is the frequency of their color signals).

If copying tapes using component signals, you will use several video cables, one for each of the component signals. These should all be made of 75-ohm coax and should be the same length.

What do you do if the MASTER VCP has a DUB output that's Y/C 688 and your SLAVE records on Y/C 3.58, or Y/C 629? You could always use COMPOSITE video; that always works. Otherwise, you need to use a FORMAT CONVERTER or TBC with the needed inputs and outputs to convert from one type of signal to another.

The Tape

MASTER on the best tape you can get. Although some brands of tape are better than others and some tapes will work on *your* machine better than others, stick with *name brand* tapes for MASTERING. The MASTER is no place to take shortcuts. If editing is expected, there are more expensive, professional, high-grade tapes designed for heavy-duty use. If you're mastering in the home video formats (some people never listen), then at least use high-grade (HG) videocassettes. These tapes will have the *least* dropouts (tiny dots of snow) and will often yield better color.

Superlong thin tape should be avoided. Use the stronger standard-length tapes (T-120, P6-60MP, KCA-60, or UCA-60 or lower number).

Virgin (fresh from the box) tape is best for MASTERING and will also be free from the residual magnetism sometimes left on used tapes which haven't been completely erased.

To guarantee the most complete erasure of a used tape, run it over a BULK TAPE ERASER (described more completely later), which will demagnetize the entire tape at once.

After 30–100 playbacks, the head starts to wear off some of the magnetic oxide in places. The tape also will have had opportunity to pick up dust and scratches. Don't MASTER on old tape.

In short, MASTER on high-grade, professional, name-brand, virgin tape. Avoid old used tape.

The Working Master

Normally you make your copies directly from the MASTER tape, not a copy of it. Copies of a MASTER look better than copies of copies. But what happens if your VCP decides to eat your MASTER tape for lunch? Hundreds of hours of work become the by-products of digestion. Furthermore, the repeated playing of a MASTER tape slowly wears out the tape.

MINIGLOSSARY

*__Working master__ A carefully made copy of a master tape, which is in turn copied. The working master protects the master from damage and wear in the copying process because it is the working master which gets played many times while the master is archived.

For these reasons, some production houses make a copy of the MASTER and use that as their WORKING MASTER from which they strike copies by the hundreds. When the WORKING MASTER gets eaten or worn, they make another from the MASTER.

This process provides safety and assures that the 10,000th copy is as good as the first, but with this insurance comes a one-generation loss.

If you mastered on a 1-inch or COMPONENT format, the quality of the MASTER will be so good that one generation won't appreciably hurt the look of the copies. If you mastered in VHS or 8mm, you definitely can't afford the extra generation. If you mastered in ¾U, the answer depends on whether quality or safety is more important. For all but the most irreplaceable of recordings, I would opt for quality and take the chance. About three generations is all the ¾U format can endure, and your copy would be four generations down if you went

1. RAW FOOTAGE
2. EDITED MASTER
3. WORKING MASTER
4. Circulating copies

To save wear and tear on your MASTER while copying, you could limit the number of passes the MASTER tape had to make through the VCP by taking the MASTER to a TAPE DUPLICATION HOUSE where hundreds of copies could be made at once.

SIGNAL PROCESSING EQUIPMENT

Although a sow's ear can't be copied into a silk purse, there are devices which accentuate desirable attributes of your picture and diminish the negative aspects. There are PROCESSING AMPLIFIERS which stabilize your video signal and IMAGE ENHANCERS which crispen the picture. A TIME BASE CORRECTOR will remove jitter from a picture. These devices, which are connected in the path of the signal as it goes from the MASTER to the SLAVE, are covered in Chapter 15 along with other TV production gizmos.

Also in Chapter 15 we'll study more about UNDERSCAN and CROSS PULSE TV monitors which allow us to view the edges of our TV picture where tracking and skew errors are easily seen. WAVEFORM MONITORS and VECTORSCOPES are more tools for graphically examining the video signal as it is being copied.

In short, we use scopes and monitors to help us "see" what the problems might be in our video signal as we copy it. If the signal needs correction, we send it through a PROCESSING AMPLIFIER, IMAGE ENHANCER, or TIME BASE CORRECTOR for improvement.

A video signal copies best when you don't meddle with it too much. You should put your greatest effort into making a perfect MASTER and then copying it directly (or through DUB cables). Although signal processing equipment may correct defects, they all add a little noise and fuzziness to your picture. Some video signals are so bad that they can't be corrected. If your sync is way out of whack, your PROCESSING AMPLIFIER can't swallow it. If your picture is very grainy or snowy, your IMAGE ENHANCER will make it look even grainier or snowier. If your picture jitters too much, your TIME BASE CORRECTOR may not be able to "lock on" to the signal and will disgorge video vomit rather than a picture.

SCAN CONVERSION

Sometimes you get a tape which is in such bad condition that it barely plays at all. It may be stretched, scratched, or loaded with dropouts, may track poorly, or may have bad sync. Such a tape probably can't be improved by signal processing equipment as it is out of their range. But don't give up yet. There is one more trick in our bag, SCAN CONVERSION (or OPTICAL COPYING).

Often a tape whose signal is too poor to be copied electronically still can be played over a TV set with reasonable results. TVs, especially those designed for use with VCRs, are very forgiving. Some are even more forgiving than others, so you may want to try a few. These sets will "lock on" to a picture even if the sync is bad and will "hide" jitter and FLAGWAVING.

With this in mind, why not aim a camera at a TV screen and record the results? The picture may end up fuzzier and lose some contrast in the process, but in a pinch, something is better than nothing.

What SCAN CONVERSION does is record the picture *optically* from the TV screen. Your camera records what it "sees," which looks good even though the sync, which the camera doesn't see, may be bad. The camera manufactures its *own* sync internally, which is "standard" and good. Your VCR thus records the whole works with fresh sync.

SCAN CONVERSION can be used to display images from computer screens where not only the sync signal is non-

---

MINIGLOSSARY

*Raw footage   Recordings made directly from the camera, intended to be edited into a final program later.

*Edited master   Same as master tape, but created by the editing process.

*Scan conversion   Aiming a TV camera at a TV screen and recording the result, useful for copying the picture from a nonstandard or troublesome tape.

standard but the screen shows too much detail (too many characters per line) to be visible when copied. By zooming the camera in on *part* of the screen, you can blow up the letters to readable size.

SCAN CONVERSION is also a good way to copy foreign video tapes. In fact, that is where the words SCAN CONVERSION come from; you might be *converting* the picture from a foreign 625 *scanning* lines repeated every ⅟₅₀ second to our own standard of 525 scanning lines, 60 times per second. Even though the tapes may have been recorded in an incompatible foreign standard, if you can get your hands on a foreign VCR and TV to display them, you can aim a U.S. camera at that TV screen and make a reasonable copy in your own standard.

The SCAN CONVERSION process is shown in Figure 13–6 and involves aiming your TV camera at a TV screen and recording the image being played from another VCR.

Since audio is pretty much compatible everywhere, you can record that directly.

Here's how to carry out a SCAN CONVERSION:

1. Prepare your tape for playback by finding a good VCP, TV, camera, and VCR and connect them up as shown in Figure 13–6.
2. The camera faces the TV screen and is zoomed to a FULL SHOT. If your camera doesn't have an electronic viewfinder, set up a VCR monitor to observe your final picture quality.
3. Play some of the program and adjust the TV picture controls. Note that the TV screen image that looks best to your eye does not necessarily look best to the TV camera. Adjust the TV screen's brightness and contrast to make a good picture *for the camera* regardless of how it looks to your eye. Use the camera viewfinder or VCR monitor as a guide to the best picture settings. There's a lot of experimentation in this process.

Generally, you get the best results when the TV picture looks a little faded (lacks contrast). You may find that your camera's lens iris needs to be set fairly high to give good depth-of-field. You may wish to set your camera's color temperature filter to sunny daylight (5600°K). A sample tape with flesh tones may give you an idea of where the hue and tint adjustments should be set.

If you noticed faint diagonal or curved lines on the image from the camera, try tilting the camera tripod slightly (lengthening a leg on the left or on the right). The faint lines are called MOIRE (Figure 13–7), and the effect is hard to avoid. Experiment more.

If you notice the picture flashing or pulsing in brightness on your VCR monitor as it shows the picture from the VCP, try plugging all the equipment into the same electrical outlet for power. If using a foreign standard VCP and TV, *they* may have to be run on battery power if they don't take U.S.-type power. Still the picture may pulse some. If it does, try adjusting the TV's brightness and contrast and the camera's iris to minimize it.

4. When ready, turn out the lights (in order to avoid reflections in the TV screen face) and start the VCR recording, start the VCP playing, and cross your fingers.

SCAN CONVERSION is a very useful tool but a tool of last resort. The image always ends up fuzzier and has limited

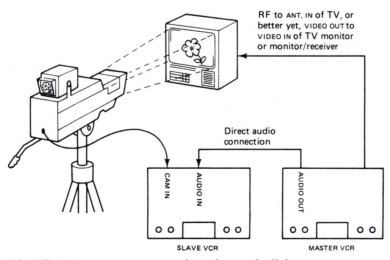

RF to ANT. IN of TV, or better yet, VIDEO OUT to VIDEO IN of TV monitor or monitor/receiver

Direct audio connection

CAM IN

AUDIO IN

AUDIO OUT

SLAVE VCR

MASTER VCR

FIGURE 13-6 SCAN CONVERSION (copying optically).

FIGURE 13-7   MOIRE.
MOIRE pattern of faint diagonal lines, which sometimes appear in camera's electronic viewfinder during SCAN CONVERSION. Tip (rotate) the TV or camera to left or right to reduce lines.

contrast and imperfect color. The process is a compromise between sharp pictures with unstable sync and fuzzier pictures that are stable.

COPYING AND THE COPYRIGHT LAW

Commercially prerecorded videocassettes generally cost $2 and up to rent and $40 and up to buy. Educational programs for schools generally start at a $15 rental and cost $150 and up. Many sell for $400–$600. Wouldn't it be cheaper to just rent a tape and copy it for yourself or borrow a friend's tape and copy that? Yes, it would—but: When the program is copyrighted (most recent ones are), duplicating it without permission from the copyright holder is illegal. It is also immoral, considering the copyright holder spent money producing his or her show to *sell* to you, not to have you simply take it.

The copyright laws and Doctrine of Fair Use are pretty complex. It takes a lawyer to unravel it all. But if you laid all the copyright lawyers end to end, they'd point nowhere. So given vague laws and a diversity of legal opinion, you, the VCR owner, have to decide what you think is safe to do and what isn't.

Not being a lawyer, the author cannot dispense legal advice in a book. You'll have to hire a lawyer to get a solid legal opinion. I *can*, however, render my opinion based on significant research and the legal opinions of others.

The key to copyright law and practice is this: *If what you are copying damages the marketability of someone's program, don't do it. If, however, the copying is minimal, personal, or for limited educational nonprofit use, it may be lawful.*

The following situations are almost always *unlawful*:

1. Copying a popular rental feature film on videocassette and sending it to a friend.

2. Making numerous duplicates of a copyrighted videocassette or off-the-air TV program and distributing them.

3. Copying a TV program off the air for your school and keeping it for use year after year without securing the copyright holder's permission. This is especially wrong when that program can also be rented from an educational film distributor.

4. Buying or renting a home videocassette and showing it in a classroom, at a business meeting, or for entertaining a club or other public group. This is considered a "public performance." Home videocassettes are restricted to private, in-the-home viewing.

5. Renting a movie or videocassette, copying it, and holding the copy for numerous other playbacks in your school.

6. Buying a single copy of a film or videocassette and making duplicates (perhaps for distribution to other schools or for individual use in the library).

7. Charging admission fees to viewers (unless contractually agreed upon with the distributor).

Usually legal practices:

1. Recording an off-the-air TV program in your home for personal viewing.

2. At the request of an instructor, recording an educational TV program off the air for playback to a class which will occur too soon for you to have time to request permission (playback in less than 10 days is a good rule of thumb).

3. Copying or editing out a small portion (less than 10% of the whole and definitely not the "heart" of the program) for playback to classes. Permissible examples might be recording 3 minutes of a 30-minute computer show which describes the differences between read-only memory (ROM) and random access memory (RAM). Another permissible example might be the recording of several minutes of a play or movie to compare acting, writing, editing, or production styles.

4. Recording a TV newscast for research or current events purposes. The newscast may also be archived for historical purposes.

5. Buying a tape and allowing others to view it individually. You can loan your purchased videocassette of "Rocky VI" to a friend. A school that buys a videocassette may place it on their library shelves for individual viewing in study carrels by students.

The most common transgression for schools involves copying TV programs (usually educational) off the air. Typically, the program is not being kept temporarily as a delayed playback but is added to the school's collection and instructional programs for use year after year. Here is a procedure for handling this situation:

1. Don't record a TV program "just in case" a teacher may need it; only record a program that has been *asked for* by an instructor.
2. Have the instructor view the program fairly soon (with his or her class if appropriate).
3. If the program is worthy of keeping, find out who produced it. Usually this information is listed in the credits at the end of the show.
4. Send a letter to the copyright holder or producer requesting permission to keep your copy of the show. Include the following in the letter:
   a. When, where, and how the program was recorded.
   b. Indicate that you wish to keep only one copy of the work.
   c. Explain that the show will be used for educational face-to-face teaching purposes on campus.
   d. Express a willingness to pay a fee if one is charged or to erase the tape if the fee is unacceptable.
5. In most cases, the producer will write back within 2 weeks giving you permission to keep the tape at no charge or will tell you how much the program will cost. Prices range typically from $150 to $350. Alternatively, the producer may tell you where the program can be rented or purchased and may even indicate a price. Some producers may tell you that the program is not for sale at any price. A few producers might not even answer your letters.
6. Either get the copyright holder's permission to have the copy (paying for the privilege, if necessary) or erase the tape.

A clearinghouse for many educational TV programs, typically those shown on PBS (Public Broadcasting Service), is the Television Licensing Center (TLC). You can send for a list of educational programs which can be recorded off the air for a nominal fee. Write to:

The Licensing Center
733 Green Bay Road
Wilmette, IL 60019

The laws are a little tougher on commercial businesses than they are on nonprofit and educational institutions. Before a company can incorporate someone else's material into its advertisements, training programs, or marketing programs, it needs *always* to secure written permission from the copyright holder. If the company is producing a program which *it* intends to sell or rent, it may even have to pay royalties to the original copyright holders or even the performers.

The laws for copying music which is part of a production were covered in Chapter 10. Music is often copyrighted just as video programs are. Permission must be granted before music may be used. This is especially true for recognizable popular songs taken from records. The easiest solution to this problem is to buy a library of musical effects on tapes or records from one of the production houses like Soper Sound, Valentino, or De Wolf. They produce albums of music and, along with the records, sell you the right to copy the music.

Copyrighting Your Own Work

If you, your school, or your business produces a program which you (or they) would like to distribute and profit from, you will need to copyright the work to protect it from duplication and sale by others. Here's how you go about copyrighting something:

1. At the beginning titles or ending credits of your tape, include words like "Copyright 1992, Poverty Productions, Inc."
2. The cassette label and the cassette box label should also have the title of the program followed by "Copyright 1992, Poverty Productions, Inc." (or whatever the year and company).

If you take no further steps, your program will be covered by a weak copyright law that states that you cannot sue an infringer to recover damages but you can make the infringer stop copying your production. To make your copyright ironclad, continue with the following steps.

3. Have the copyright office (part of the Library of Congress in Washington) send you the forms and instructions on how to file a copyright application.
4. File the application, noting whether the production was yours *in its entirety* and the exact date the production was completed.
5. Make a copy of the production and send it with your application and a $10 fee back to the Copyright Office for processing.

With these steps you secure the sole right to duplicate and distribute your production and have the right to sue anyone who duplicates your show without your permission.

Reporting Pirates

The trade associations of both video retailers and Hollywood studios would like to ring the necks of the video pirates—not only are they stealing income from hard-working video producers (one of whom may be you someday), but they tend to produce inferior second and third-generation copies made from poorly maintained VCRs. You can usually tell these purchase and rental videocassettes by their poor labeling (or lack of labels), their unusually low price, and the poor-quality picture. If you'd like to take part in the campaign to stamp out video piracy, here are some places to contact:

1. Video Software Dealers Association Hotline: (800) 257-5259.
2. Recording Industry Association of America Hotline: (800) 223-2328.
3. Motion Picture Association of America, anti-piracy hotline in Sherman Oakes, CA, (818) 995-6600.
4. The copyright division of your closest FBI office.

The MPAA has given rewards to people who have turned in a blatant video pirate.

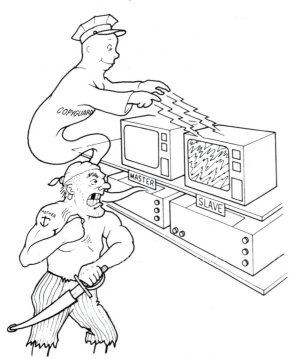

Many rental tape distributors put antipiracy signals on their tapes to discourage copying.

Antipiracy Signals

When you copy a copyrighted tape without permission, the producers call it PIRACY. To thwart the unlicensed duplication of their wares, many distributors do something to the video signal to make it uncopyable. A few years ago, COPYGUARD was the copykiller à la mode. Today it's MACROVISION. Here's how they work.

COPYGUARD reduces your VCR's sync signal enough so that your VCR can *play* the tape, and your TV can *show* it, but another VCR can't "lock on" to the weak signal to record it.

The problem with COPYGUARD is that sometimes it messed up the signal too much; legitimate tape renters and buyers got movies with diagonal lines, vertical rolling, jittering, mistracking, or picture pulsing. Sometimes these problems could be surmounted by adjusting the vertical hold on the TV set, but often, people just brought their tapes back to their vendors and asked for a refund (one good reason for doing business with a reputable dealer).

MACROVISION works a different way. It checkerboards five white and five black squares in the vertical sync pulse. If you misadjust your TV's vertical hold (that black bar between pictures), you can see the boxes flashing on and off. This flashing doesn't bother your TV or VCR as it plays. But if a VCR tries to record the signal, the alternating light and dark boxes "fool" the recorder's automatic gain control circuits, making the recorded picture pulsate dark and light—most unnerving.

Occasionally, MACROVISION messes up the picture for legitimate viewers. Some older TVs have automatic gain controls that are "fooled" by the MACROVISION boxes. If you play your VCR through another VCR (or any other device) on the way to your TV, the MACROVISION signal may trick the second VCR (or device), pulsing your picture. If your TV screen's vertical height control is misadjusted, the sync bar and its MACROVISION sidekick will appear at the top of your TV screen, sometimes causing it to vary in brightness. Adjust your TV's vertical height control to alleviate the problem. Sometimes MACROVISION confuses the auto-color circuits in some TVs. If so, switch them to *manual* and maybe the "MACROVISION blues" will go away.

While some VCRs cannot play guarded tapes, others have such sophisticated circuits that they can record guarded tapes with ease. Tape machines with "special effect" circuits like still frame and fast scan often will do a good job of playing back or recording a COPYGUARDED or MACROVISION videocassette.

So what do you do if (1) you can't even get your guarded tapes to play on your VCR or (2) you wish to

---

MINIGLOSSARY

Piracy   Duplicating copyrighted tapes without permission.

Macrovision   Popular anti-copy signal recorded on a video tape to make it playable but not copyable.

duplicate a guarded tape (copyright holder's wishes not withstanding)? You can run the signal through a PROCESSING AMPLIFIER or VIDEO STABILIZER, which strips off the guarded sync and replaces it with good sync suitable for solid TV pictures, uninhibited "special effects," and (ahem) illicit copying. You'll see more on the PROC AMP in Chapter 15.

PHOTOGRAPHIC COPYING

You were in the audience during the taping of "Family Feud," and the camera panned right by you jumping up and down and waving your hands hysterically in your typical accounting executive behavior. You taped the episode at home and want to show everybody that you were on network television; only nobody wants to watch 30 minutes of Richard Dawson just to see you making a coast-to-coast spectacle of yourself for 2 seconds. They might look at a snapshot, though.

Or you produced a tape of your own that you'd like to distribute, but sending sample cassettes to prospective clients or employers is too expensive. Even the postal charges are prohibitive. Photos, however, are inexpensive to copy and easy to mail.

Or perhaps you would like to share a computer graphics image with a friend and don't have a printer or graphic plotter or other mechanism for storing the data. Photos of the TV screen images would do the trick.

Or maybe you are writing a book, like me, and wish to show authentic examples of TV screen images to your reading public.

Maybe the convenience of carrying around slips of paper appeals to you more than lugging heavy VCRs, TVs, and bulky cassettes. Whatever your reason for wanting to photograph a TV screen, the process is not hard to do.

I'm sure you remember clearly from Chapter 1 how the electron gun of the TV screen creates a picture every $1/30$ second. Every $1/60$ second, it paints the even-numbered lines and then goes back for the next $1/60$ second to paint the odd-numbered lines in between. This every-other-line picture is called a FIELD. This crazy procedure results in a TV picture which has almost no noticeable flicker *to your eyes*. Every $1/30$ second a totally new complete picture (a FRAME) appears.

If a photo camera were to take a picture of a TV screen at, say, $1/100$ second (a good speed for stopping motion, not requiring a tripod), you'd get a picture of a half-scanned TV FIELD. The TV's electron gun would have gotten only about halfway down the screen before the shutter closed. The resulting picture, as shown in Figure 13–8,

would show a dark place where the TV screen hadn't been zapped yet by the electron gun. The resulting diagonal line is called a SHUTTER BAR. To avoid the bar, you have to shoot the TV screen at $1/30$ second or longer, thus getting a complete full-scanned screen. Longer than $1/30$ second is even better, but if things are moving on a screen, a long exposure will make them blur.

How about pausing the playback for a STILL FRAME and recording that? The image (once TRACKING NOISE is adjusted out of the picture) will remain still long enough to shoot with your photo camera on a tripod. There's one problem with this, albeit a small one. When a VCR shows a STILL FRAME, it is really a still FIELD. It is playing only the even-numbered or odd-numbered lines in the picture. (The word STILL FRAME is a misnomer.) It looks like a whole picture, but if you notice carefully, when switching back and forth between STILL FRAME and PLAY, the picture is half as sharp in STILL FRAME.

This difference in sharpness is not awfully noticeable. Go ahead and shoot STILL FRAMES. But if your scene is *not moving anyway*, why lose the sharpness? Wait until a still moment comes up and snap the picture then. You will get a sharper picture (but don't cut yourself).

In short, try to use the slowest shutter speed you can get away with when photographing a TV screen.

Film

For black-and-white prints, medium-speed film such as Plus X gives good results. High-speed films will work, too.

For color prints, use daylight balanced film. The same goes for slides. Use an ASA rating of 50 or higher. Since the color on a TV screen is slightly bluish, perfectionists can compensate by using a CC40R color compensating filter on their photo camera lens. Open your photo camera's iris one stop extra to make up for the additional darkening caused by the attachment.

TV Screen

Avoid small TV screens. They are often not as sharp as bigger ones. Twelve- to seventeen-inch screens are fine.

When taking pictures, first adjust the TV picture so it looks its best to you, then turn the contrast and color saturation down a notch because film tends to exaggerate contrast. Perhaps, turn the brightness up a notch also.

If shooting black-and-white photos, use a black-and-white TV to get a sharper picture. Turn out the room lights to avoid catching reflections in the TV screen.

MINIGLOSSARY

Leaf shutter   A flap inside a photographic camera that moves aside for a moment to let light enter and then snaps closed.

Focal plane shutter   A pair of curtains inside a photographic

camera. One opens to let light reach the film, followed by the other one closing to complete the exposure.

Photo of TV screen taken at 1/120 of a second. SHUTTER BAR could appear anywhere on screen.

Photo at 1/60 second

Photo at 1/30 second

Photo at 1/15 second

Photo at 1/4 second

Photo of STILL FRAME at 1/4 second

Photo at 1/2 second

FIGURE 13-8   Photos of a TV screen.

VIDEO PRINTER (Courtesy Sharp)

Photo Camera

Place your camera on a tripod or other firm surface. Aim it straight into the TV. Almost fill the camera's viewfinder with the TV's image. If the TV screen is small, you may need a close-up lens attachment on your camera to get close enough. Focus very carefully, perhaps trying to see the TV's scan lines on the screen.

Exposure is tricky. Use your camera's built-in exposure control or use an exposure meter aimed at the TV screen. For critical pictures, bracket your exposures by one f-stop either way. Set your shutter for $\frac{1}{8}$ second or longer when possible. Professional photographer Jane Zlotkin gets her best results from shooting $\frac{1}{5}$ second at f5.6 using 400 ASA Ektachrome color slide film in her Nikon FM.

Use a timer or shutter release so that you don't wiggle the camera while pressing the button.

Cameras with LEAF SHUTTERS make less obtrusive SHUTTER BARS than FOCAL PLANE types do. If you *have* to shoot at speeds faster than $\frac{1}{30}$ second, try to use a LEAF-SHUTTERED camera.

For best results, pick recorded scenes which are close-ups or medium-length shots. The detail in long shots is lost in the fuzz.

On the other hand, you can get some offbeat and creative effects by disregarding the rules. Superclose-ups of the TV screen give you a mosaic effect from the phosphor stripes on the screen. Boosting the color or shooting the screen from oblique angles can provide neat distortions to the image.

VIDEO PRINTERS

If you do a lot of off-screen photography, you may want to invest in a VIDEO PRINTER, a device that takes a TV camera or VCR's signal and converts it into a photographic print or slide. The advantages of VIDEO PRINTERS are:

1. They usually make a sharper picture than your photo camera aimed at a TV screen.
2. They're quicker: Some use Polaroid film for instant shots; others crank out thermal prints on special paper in about 15 to 80 seconds.

3. You don't have to darken the room and fiddle with TV and camera controls.

Black-and-white models cost about $400 to $1,400 and color runs about $1,200 to $2,500. Many models digitally "freeze" a frame of the video signal allowing you to capture motion video off-air, from a VCR, videodisc player, or any other source. Higher-priced models accept not only NTSC, but RGB and some computer signals.

Some VIDEO PRINTERS are especially designed to record computer graphics, and produce very high-resolution pictures directly from the computer data. These $15,000+ printers can make slides and prints so sharp that you can project them on a large screen or print them on the cover of a magazine without anyone knowing the picture was totally "electronic." The scan lines in the pictures are so fine, you can't see them.

Printing the Pictures

Retrieving text information stored on tape can get rather expensive if you try to photograph numerous screen pages. Here's a cheaper way out: Take your pictures using high-contrast 35mm film. Have the film developed but not printed or turned into slides. Next, waltz over to your college or public library and sit down at their microfilm reader/printer. But instead of sliding a 35mm *New York Times* microfilm into the machine, thread *your* 35mm film into the printer. Select the desired shots on their big screen. Drop in your dime and press PRINT to get a cheapie copy of your screen shot.

If reading fine print doesn't bother you (and the print is unlikely to end up too fine, considering how few words can appear at once on a TV screen), you can always have your roll of 35mm film developed and *contact-printed*. Then you get 36 TV pages on a single 8 × 10 sheet. You may need a magnifying glass.

Taking super-8 or 16mm movies of your TV screen is quite unmanageable. You can't control most movie cameras' shutter speeds to get long enough exposures to avoid the SHUTTER BARS.

For more information on photographing a TV screen, consult Kodak's publication AC-10, *Photographing Television Images*.

MINIGLOSSARY

Video printer   Electronic device that converts a TV screen image to hard copy.

EDITING A VIDEO TAPE

Editing means different things to different people. In its broadest sense, editing is the organization and the assembly of shots into a logical sequence.

Scripts, movie film, audio tape, and video tape can all be edited. The techniques are similar in that bad material is removed, good material is added, and segments may be moved from one place to another. The methods of editing each medium are different. Scripts may be changed with white-out or a word processor. Movie film can be edited with scissors and glue or a professional MOVIEOLA and FILM SPLICER. Audio tape is similarly sliced up into wanted and unwanted segments and the pieces hitched together with SPLICING TAPE. Audio tape can also be edited by electronically copying desired portions of other audio tapes. In the old days, video tape was edited by slicing it up and SPLICING the desired segments together. The process was tedious and often caused "glitches" in the picture. Nowadays, video tape is always edited electronically.

Some editing methods are quick and dirty. Others are planned and professional. Scenes can be recorded "live" as the camera sees them and assembled on the tape in chron-
ological order. In other cases, scenes may be shot and recorded on one tape and the best scenes selected and copied onto another tape which becomes the final product. Those original scenes on the first tape are called RAW FOOTAGE, or RUSHES, or CAMERA ORIGINALS, or MASTER FOOTAGE. The scenes may be shot days apart, years apart, or miles apart and assembled in any order onto your final tape. The final version of the program, when all these parts are put together, is called the EDITED MASTER.

Notice that in the case of movie film and audio tape the program was physically cut apart and physically reassembled from its pieces. In the end, if the edited program needed to be shortened or lengthened, parts could be removed or added. Video tape, on the other hand, is never physically cut (except to salvage a damaged tape). The program is electronically recorded on a tape either live or from another tape. Unlike film and audio tape, once the edited version is finished, its length cannot be easily altered. A section may be erased with a new section recorded over it (called an INSERT edit), but that doesn't change the length of the whole show. If a show *had* to be lengthened, you

MINIGLOSSARY

Movieola Device for viewing and comparing several reels of film at a time while selecting segments to splice into an edited master film.

Film splicer Mechanical device for clamping and neatly cutting

and holding film steady for gluing.

Splicing tape In video, a thin adhesive tape to attach the ends of a video tape.

would either have to reedit the tape from the beginning or go down one more generation and copy the part of the tape that came before the part you are adding, then add the new part, and then copy the rest of your old edited tape onto this new tape.

As you can see, it pays to make your editing decisions ahead of time. Once performed, some edits are hard to change. For this reason, there are computerized editing systems which allow you to make a list of what the edits should be, preview what the edits look like as listed, and then change the list to refine the program. Finally, the real edit is made from the list.

Why can film and audio tape be cut easily and attached while video tape can't? With movie film, you can slice between the pictures neatly. Audio tape is similar in that you can cut the tape between words. To remove a sentence, merely play the tape to find the sentence, cut the ribbon of tape before the first word of the sentence, and cut again after the last word. Join the ends of the ribbon, and the sentence is gone when the ribbon is played. Video recordings are much more complicated. Instead of being recorded linearly like audio tape, video is recorded as diagonal magnetic stripes (review Figures 5–30 and 5–31) on a tape. Each stripe is a whole picture or a part of a picture (depending on the format of the VCR). Cutting out a segment of tape and joining the ends results in the tape player playing back fragments of one picture and fragments of another. Sync and CONTROL TRACK signals get messed up, causing a horrible-looking glitch. Also, the bump in the tape from the SPLICING TAPE disrupts the picture even more.

EDITING WITHOUT AN EDITING VIDEO TAPE RECORDER

If you desire perfect glitch-free edits, you require special recorders that can edit electronically. If such machines are unavailable, you must make the personal choice between

1. Making a tape with *no* glitches, which means recording the tape all the way through, nonstop. This may limit what kinds of scenes you can put together (no jumping from location to location), and everything must be shot in sequence (no room for mistakes, additions, or deletions).

2. Producing a tape with glitches but with content unbound by the preceding constraints. Shooting could be out of sequence, segments could be done days apart, and parts of a production could be

done over and over in an attempt to achieve greater perfection.

If your tape is ever to be broadcast, shown to large audiences, or copied and used extensively or if it will cost a lot to produce, *no* glitches, not even small ones, are tolerable. Tape copying equipment chokes on glitches, broadcasters are forbidden to transmit glitches, and discriminating audiences will not take glitchy programs seriously.

Whether small-budget productions for nondiscriminating nonbroadcast audiences should have glitches should depend perhaps on how bad the glitches are. Are they wide expanses of snowy picture, or are they little blips gone in a second? The size and obtrusiveness of these glitches range from terrible to imperceptible depending on

1. The method of editing used
2. The kind of VCR used and its condition
3. Luck

Let's start with "terrible" and work our way up to "perfect." Most people consider the word *edit* to imply "good-quality" edit. Since a true electronic editing VCR is not being used in these first primitive cases, what we get hardly deserves the name edit. Pretty or not, for the lack of a better word, let's call them edits anyway. Here are some methods of editing without specialized equipment.

Stop Edits

In STOP EDITING you record a sequence and then hit STOP on the VCR. You practice the next sequence and, when ready to tape it, press RECORD/PLAY and proceed with the recording. When that scene is finished, STOP the machine again. Continue the process until you have assembled your whole show. This method works pretty well with beta VCRs but not with VHS or ¾U recorders. When the latter machines STOP, they unthread themselves, losing your place on the tape. When you hit RECORD/PLAY again, the machines rethread themselves, *but not to exactly the same place where they left off*, and resume recording. You can end up with a pretty big glitch between edits. Beta VCRs don't unthread themselves when you hit STOP, so you don't lose your place, and the glitch in your picture is small.

If you reshoot a scene in the middle of a tape using STOP EDITS, your EDIT-IN point (the beginning of your edit) will have a glitch. Your EDIT-OUT point, regardless of the VCR used, will have a terrible-looking glitch with a band of snow running through it like in Figure 14–1.

MINIGLOSSARY

*****Stop edit** Technique of editing a video tape by stopping the VCR (pressing stop) at the end of one scene and then starting it recording again (pressing record/play) at the beginning of the next.

*****Edit in** Begin recording new material; the beginning of an edit.
*****Edit out** Cease recording new material; the end of an edit.

A band of snow fills the picture and then slides off.

FIGURE 14-1   The end of a STOP EDIT.

Because of the tape rethreading problem, STOP EDITS are hard to accurately place. Add this to the fact that the glitches are hideous, and you have every reason to avoid this type of editing. Unfortunately, most nonediting ¾U VCRs can *only* do this kind of editing.

Pause Edits

In pause editing you record a sequence, hit PAUSE, set up the next sequence, UNPAUSE to continue recording, and so on until the end of the show. This is probably the method you've been using to delete the commercials from TV broadcasts (you hit PAUSE when the commercial began and UNPAUSE to resume recording when the ad was over). Although this trick works on home-type VCRs, it doesn't work on console ¾U VCRs. They won't allow you to PAUSE them while recording.

PAUSE edits are "cleaner" than STOP edits, and they display a barely perceptible blink on the TV screen. On the home VCRs, the edits are usually better when made at the VCR's fastest speed.

Why are PAUSE EDITS better looking? Because PAUSE EDITS, unlike STOP EDITS, don't unthread the tape. You stay put right where you left off—almost. There are exceptions. Some VHS VCRs automatically backspace (run the tape backward) a half second when you hit PAUSE. They do this to ensure a smooth, almost glitchless edit. That's nice for smoothness but a little rough on planning. On such machines, you have to learn to hit PAUSE about a half second late, so when the machine trims off that half second, you'll be right where you want to be.

One problem with PAUSE edits is that you can't leave your VCR in PAUSE very long because the spinning video head wears down the tape in the spot where you PAUSED. Many home VCRs will automatically switch from PAUSE to STOP after a couple minutes to protect the video tape and heads. This limits the amount of time you have to arrange your next scene, plan your next shot, or get your next camera angle.

Some console VCRs and many camcorders allow you to switch the VCR's POWER off (thus stopping the spinning video heads) while remaining in PAUSE. This allows you to take your time setting up between segments. It also saves battery power on camcorders. Hitachi calls this feature Power Saver, while JVC calls it Record Lock.

Camcorders and industrial ¾U portable VCRs perform PAUSE edits when you pull the camera trigger while recording. A light in the viewfinder tells you when the tape is running and when it is PAUSED. Remember that you are using power and wearing down the tape while your VCR is paused between scenes so make your setups quickly.

Recording Something Over

You're recording the Memorial Day parade to send to Aunt Blanche. You assemble fascinating shots of the flags, the bands, and now the horses trotting majestically by . . . uh . . . whoop! You didn't really want to test Blanche's heart pacer with a giant color close-up of Dobbin doin' a dandy on the pavement. PAUSING at this point won't help; the toothpaste is already out of the tube, so to speak. The only choice is to REWIND a ways and replace the unsavory scene with something else, perhaps marching Girl Scout Troop 106.

So you REWIND, PLAY, and watch your viewfinder for a good breaking off point and hit PAUSE. Next hit RECORD/PLAY. UNPAUSE when Troop 106 gets in range.

You may notice that such edits aren't as pretty as PAUSE edits (and it's not the fault of the girls in Troop 106). There may be a few seconds of herringbone lines wiggling through the picture, looking much like Figure 3–33. There may also be a rainbow or smear of colors lasting a few seconds.

This disruption is normal under the circumstances and is not the fault of you or your machine. If you must record over something, two things you can do to minimize the glitch are

1.   Use the fastest tape speed for recording.
2.   Make your edits on pauses in conversations or lapses in action. Then viewers don't feel like they're missing something important.

This technique won't work on ¾U VCRs. They won't let you switch from PAUSE to RECORD.

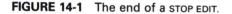

MINIGLOSSARY

*__*Pause edit__   Editing of a video tape while recording by pressing the pause button between takes.

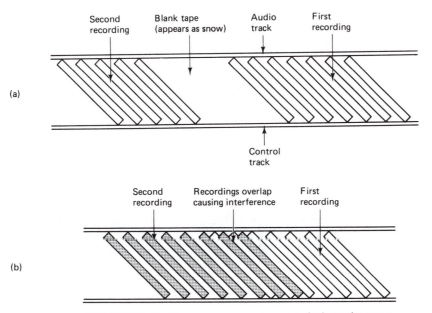

Second recording   Blank tape (appears as snow)   Audio track   First recording

(a)

Control track

Second recording   Recordings overlap causing interference   First recording

(b)

FIGURE 14-2   How a STOP EDIT is recorded on the tape.

HOW VCRS EDIT

Nonediting VCRs

Figure 14–2a shows what happens on a tape when you perform a STOP EDIT. Because the tape doesn't rethread to exactly the same point where you left off, there may be a space of unrecorded tape between the two scenes. This will appear as snow. When they rethread themselves, some machines back up the tape, causing the new recording to be laid over the tail of the old one as shown in Figure 14–2b. Here the old and new recordings interfere with each other causing the effect shown in Figure 3–33.

When you replace a segment in the middle of a recording (INSERT edit), the tape looks like Figure 14–3. While the VCR is recording, its ERASE HEAD is erasing old pictures and sound from the tape several inches upstream from the video and audio record heads as was shown in Figure 5–30. When you stop the edit, the ERASE head has erased some

tape which the record head hasn't caught up to. This gap of erased tape appears as snow when played back. The snow slides off the screen as the spinning video head passes further into the old video.

PAUSE edits, diagramed in Figure 14–4, are much neater. If the tape shifts a little while it's in PAUSE, it may leave a tiny extra gap between the video from scene 1 and the video from scene 2. This will cause a minor glitch. If the tape backs up a little during PAUSE, one or more video tracks may overlap, also causing a brief glitch.

Electronic Editing VCRs

Electronic editing encompasses an assortment of VCRs which at a push of a button will do what is necessary to make a "clean" edit (unlike the edits discussed up to now). How clean is "clean"? As clean as you can afford to buy. The least expensive VCRs make slightly ragged edits; the most expensive ones make perfect edits every time. It may be

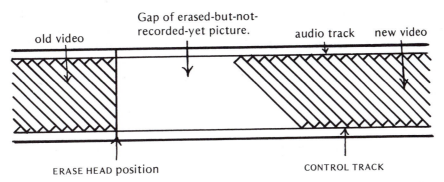

old video   Gap of erased-but-not-recorded-yet picture.   audio track   new video

ERASE HEAD position   CONTROL TRACK

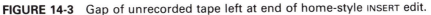

FIGURE 14-3   Gap of unrecorded tape left at end of home-style INSERT edit.

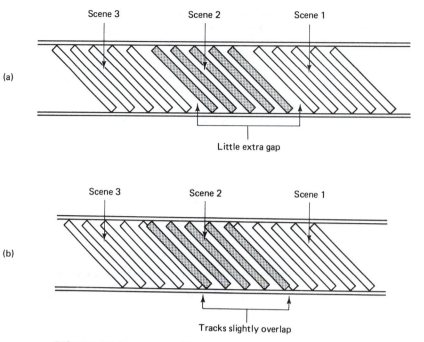

FIGURE 14-4 PAUSE edit.

useful to know a little about how a VCR edits in order to select an editor and to recognize problems with your present editor.

The object is to make a smooth, clear edit. For this, four requirements must be met:

1. The tape must be erased clean as it approaches the record head.
2. The CONTROL TRACK must be continuous and have no interruptions.
3. The switchover must occur invisibly between video pictures.
4. The sync must remain continuous without any aberrations.

The first requirement (clean erasing) is performed by a FLYING ERASE HEAD. It's not a bird and not a plane; it's a super idea. Attached to the spinning video head is a tiny ERASE HEAD, just ahead (a HEAD?) of it. When the EDIT button is pressed, the FLYING ERASE HEAD starts erasing the tape directly before the spinning record head gets to it. Figure 14–5 diagrams the arrangement. Incidentally, the better camcorders come with FLYING ERASE HEADS built in.

The second requirement (continuous CONTROL TRACK) must be met by all video editors. The CONTROL TRACK is a series of pulses recorded on the tape while the VCR is recording video. During playback these pulses guide the tape speed, tracking, and the timing of the video signal. Video recordings are fine when the tape is moving, but when the tape stops, the video and CONTROL TRACK signals become garbled. For this reason, editing VTRs are designed to edit while the tape is still moving. Since the tape doesn't have to stop, start, and then pick up speed again during the edit, the timing of the CONTROL PULSES doesn't get messed up.

Professional editing video recorders will BACKSPACE the tape 2–15 seconds in order to get this "running start" before they perform an edit. This gives plenty of time for the tape speed and machine to stabilize and create new CONTROL TRACK pulses right in step with the old ones.

The third requirement (switchover between pictures) separates the good editors from the excellent ones. Say you start playing a tape in preparation for an edit. When you press the EDIT button, the heads instantly start recording. What if the video heads are right in the midst of playing back a picture? That picture will be interrupted with a brand new one if you didn't happen to press the EDIT button at precisely the moment inbetween pictures. You get edits anywhere, right out there on the screen where you can see them. With the more elaborate VERTICAL INTERVAL EDITORS, the VTR does not execute the edit exactly at the moment you press the EDIT button. It waits a few hundredths

---

MINIGLOSSARY

Flying erase head   A spinning head residing upstream of the video recording head that can erase video tape a split second before the video record head records a new picture.

*Control pulses   Rhythmic signal recorded on a video tape's control track which guides the VCR during playback.

*Vertical interval   The part of a video signal that doesn't show on the TV screen; the black bar (sync) at the bottom of the TV picture when it rolls.

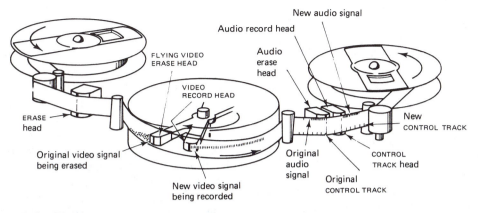

FIGURE 14-5   FLYING ERASE HEAD.

of a second and performs the edit during the VERTICAL INTERVAL, that part of the picture just below the bottom of your TV screen where you can't see it.

Continuous sync, the fourth requirement, is the last piece to the editing puzzle. You would like to edit during the VERTICAL INTERVAL, that split second when the image is invisible to the viewer. But there are two images to be considered here: the old one on the tape and the new one you're about to record. The VERTICAL INTERVAL for one might not come at the exact same time as the VERTICAL INTERVAL for the other. So that the VTR and TV picture do not miss a beat when you edit from the existing taped rhythm to the new incoming rhythm, something must be done to match the two. If we can get both VERTICAL INTERVALS to occur at the same time, we can edit during the invisible parts of *both* pictures. The resulting edit will look superclean as long as the sync and CONTROL PULSES flow smoothly.

To get the sync to match, the VTR must "listen" to the new incoming video signal and try to synchronize it to the signals it is playing. A VTR that listens to an incoming signal while it is playing its own picture is called EXTERNALLY LOCKED. A VTR that doesn't react to incoming video signals while it is playing is called INTERNALLY LOCKED. When a VTR is EXTERNALLY LOCKED, a feature called CAPSTAN SERVO changes the motor speed of the VTR a little so that the tape plays faster or slower until its VERTICAL INTERVALS and CONTROL PULSES match the sync of the incoming video.

In short, to perform their magic, electronic editing VCRs need

1. FLYING ERASE HEADS to clear the tape just before it's recorded
2. BACKSPACING to allow a recording to begin while the tape is moving (to preserve a steady CONTROL TRACK rhythm)
3. VERTICAL INTERVAL EDITING to perform the edit "off screen"
4. EXTERNAL LOCK so that the VCR will synchronize the timing of the pictures it's playing with the timing of the pictures it's about to record.

ASSEMBLE EDITING

There are different strategies for piecing together the parts to a TV show. They go by the names

ASSEMBLE EDIT

INSERT EDIT

AUDIO INSERT

VIDEO INSERT

ASSEMBLE EDITS are produced *in order* from the beginning of the program to the end. You first set up the title and then record it. Next you set up the first scene and record that. Next you set up your second scene, then your third, then your fourth, until you're done.

For each ASSEMBLE EDIT to be "clean," the VCR must be BACKSPACED a proper distance to get a running start. Put another way, you can't simply push a button and begin the

---

MINIGLOSSARY

Externally locked   A VCR that "listens" to an outside video signal and tries to coordinate its own signal to match the other's timing. Such a VCR can synchronize its sync to another source's sync.

Internally locked   A VCR that plays a tape independently, with its sync timed to its own internal clock.

*Insert edit   The recording of a new video segment amidst old,

prerecorded video—unlike assemble edit, which places each new segment at the tail of the last segment.

*Audio insert   An audio dub performed in the midst of an already recorded tape.

*Video insert   Replacing a segment of old video with new video, in the midst of prerecorded tape. Audio is not affected.

scene immediately. You must first get the tape rolling, let it stabilize, and then simultaneously press the button and begin the action.

To perform an ASSEMBLE EDIT, here are the steps in detail:

1. See that your audio and video inputs are connected.

2. Check your audio and video levels. On most VCRs you may do this only after pressing the RECORD button. When finished checking, switch the RECORD button off.

3. Play the tape and learn the exact place where you want to EDIT-IN. It is usually more accurate to use the action in the scene as a guide than to use the tape index numbers.

4. Find the VCR's MODE SELECTOR and switch it to the ASSEMBLE EDIT position. This prepares the VCR's electronics for the upcoming edit.

5. Rewind the tape a ways so that it can play for about 5 seconds (so that the motor speed stabilizes) before the edit is to be executed.

6. Get ready to cue your talent.

7. Switch the VCR to PLAY.

8. When the edit point comes, push the EDIT button (on some editors, instead of pushing EDIT, you press PLAY and RECORD simultaneously) and cue the talent.

9. Switch the VCR to STOP a few moments after the scene is over in order to finish the edit. These few additional moments of extra time after the end of the edit may come in handy later if you decide to attach another scene to the end of this one. If you stop recording abruptly at the end of the scene, your next ASSEMBLE edit will have to be performed with split-second accuracy—if it is too early, it will cut out some of your existing program; if it is too late, you will leave uncovered the messy place where you stopped recording the last scene. Leaving a little "safety" space at the end of your edits leaves you with more leeway to make your next edit.

10. If yours is a mediocre editor, play back your edit afterward to see if it came out. You may have to do it over.

Because you are pushing the EDIT button manually and proceeding with a "live" show, the preceding process

is called a PUNCH-IN ASSEMBLE EDIT. Later we'll see how computerized EDITOR CONTROLLERS can "tell" an editing VCR to perform an ASSEMBLE edit.

The preceding editing technique is the one used most frequently in studio situations when someone flubs a line. You would stop the tape, back it up, and play it, looking for a pause in conversation or some other appropriate place to EDIT-IN. Let your talent know where you will be EDITING-IN so they will know what to say next. Change your camera angle so that you don't get a JUMP CUT (a strange-looking cut where your talent's head "jumps" from one position to another but nothing else changes in the scene). If you can let your talent hear the audio as the VCR plays before it edits, your performers may be able to synchronize their words and pacing with the words on the tape. Furthermore, the words they hear on the tape can cue them more accurately than a pointed finger which they might only see if they were looking at one of your crew members—a look that may appear awkward. Next, back up the tape again, play it, press the EDIT button when you come to the selected spot, and continue recording.

INSERT EDITING

INSERT edits are often made in the *middle* of a program. They are frequently used to correct an error or to change something after your program is finished. INSERT edits *do not* lengthen or shorten your program. They merely replace one part of your program with a new part.

An entire program can be made out of INSERT edits. The INSERT edits may be recorded out of sequence on a prerecorded or BLACKED tape. To edit a 15-minute tape using this mode, you would first record 16 minutes of black using the ASSEMBLE EDIT mode; then you would rewind your tape and switch your VCR to the INSERT EDIT mode and record the beginning title, the first scene, the second scene, and so on in order, or if you wished, you could record the beginning title and skip ahead to the ending credits and then do scenes 3, 4, and 5 and then go back to insert scenes 1 and 2 later.

Notice that an INSERT edit is done over already-existing video (or black). The INSERT edit won't work if there isn't old video to come back to when you end the INSERT. If you expect your new material to run past the end of your old recording, then do an ASSEMBLE edit, not an INSERT.

To perform an INSERT edit, here are the steps in detail:

1. See that your audio and video inputs are connected.

MINIGLOSSARY

*Mode selector   Knob or button on an editing VCR that sets the VCR into the insert edit, assemble edit, video insert, or audio insert mode.

Punch-in assemble edit   An assemble edit executed manually, live, while the actors perform.

Editor controller   A remote control device that can backspace two or more editing VCRs and make them perform an edit.

*Blacked tape   A video recording of black, used to prepare a tape for insert editing.

2. Check your audio and video levels by temporarily pressing RECORD (unless the VCR monitors them in the STOP mode). When finished, switch the RECORD button off.

3. Play the tape and learn the exact place where you want to EDIT-IN. A point where activity pauses is best. Remember this spot and jot down the tape counter number so that you can come back to the spot easily. You may also use the action in the scene as a guide to the EDIT-IN point.

4. Now play ahead to find the appropriate place to EDIT-OUT, that is, to terminate the new recording and go back to the original presentation. Again, a pause in action and conversation is usually a good place to come back to the old material from your edit. Once you find the place, note the number on your index counter. If you wish, you may also use a stopwatch to accurately measure the length of time between the EDIT-IN and the EDIT-OUT points.

5. After learning the EDIT-IN and EDIT-OUT points, find the MODE SELECTOR and switch it to the INSERT EDIT mode. This prepares the electronics for what is about to happen.

6. Rewind the tape a ways so that when the edit is performed the tape speed and motors have stabilized and are running smoothly.

7. Switch to PLAY.

8. When the edit point comes, push the EDIT button (or PLAY and RECORD buttons or whatever the manufacturer instructed) and cue the talent.

9. Pay close attention to your timing. Get ready to press END INSERT, EDIT, or whatever button stops the process. Unless the manufacturer says otherwise, switching to STOP is *not* the way to end an INSERT edit.

10 About 1 second before the INSERT is destined to end, hit the proper button. Depending on the VCR, the EDIT-OUT may take up to 1 second to actuate (because the VCR's electronics sometimes need this amount of time to line up the old sync and CONTROL TRACK that it will be going back to with the new sync that it is now recording; the result is a smooth edit with stable sync).

Essentially what you have done is record a new passage starting with the first index number and ending with the second number. This means that performance must be timed to last *exactly* the length of tape you wish to delete. If the replacement scene is too short, you end up with a long pregnant stare at someone's smiling face while you wait for the final index number to come up. You *must* wait

for the number because if you terminate the edit too soon, you will end up not deleting the tail end of the segment you want removed. If the replacement scene is too long, you'll end up erasing your way into the following material, which you wanted to keep.

This process is not easy. Besides being mechanically difficult, it requires precision timing from the performers and the VCR operator alike. To make things worse, the tape footage counter isn't all that accurate and will throw you off by a second or so anyway. This is why a stopwatch may be a better choice for tracking the length of your edit.

The aspect of INSERT editing that makes it so difficult is the fact that once you've EDITED-IN, you're flying blind. You can't see what you are erasing; you only see what you are recording. You have no visual cue for when to stop, other than your index counter or your timepiece. Add to this the fact that if you make a mistake and EDIT-IN too long a passage, you'll irrevocably erase the next scene as you record over it. For this reason, it is worthwhile to rehearse the edit several times in order to get the timing exact. You may wish to play the tape; *pretend* to edit (as it plays); have the performers dress-rehearse the scene; then *pretend* to stop the edit and, by looking at the screen at this point, determine how far off you were and what should be done about it.

An Important Technical Difference Between Insert and Assemble Edits

When you ASSEMBLE-edit, the VCR records the picture, sound, sync, and CONTROL TRACK pulses. The VCR produces the CONTROL TRACK pulses in step with the vertical sync pulses. You'll remember that the CONTROL TRACK will be used by the VCR during playback to synchronize the movement of the tape through the VCR with the spinning video heads so that the heads retrace exactly the same path the record heads took when making the recording. The CONTROL TRACK pulses are thus the "drumbeat" which keeps the tape and heads moving in step.

If the incoming video (and sync pulses) is smooth, strong, and steady, the VCR will make perfectly measured CONTROL TRACK pulses. If, however, the video (and sync) wavers, then the CONTROL PULSES go into arrhythmia. If you are ASSEMBLE editing, each edit point should have a perfect changeover from the old recorded CONTROL TRACK drumbeat to the newly recorded CONTROL TRACK drumbeat. When edits are perfect, the old and new pulses line up perfectly. If, however, an imperfect edit is performed, there will be a glitch in this smooth train of pulses. This glitch will also appear in your picture.

INSERT editing, on the other hand, doesn't create CONTROL TRACK pulses. These pulses have to be laid down ahead of time (perhaps as you create your BLACKED tape). While INSERT editing, the VCR synchronizes the existing CONTROL TRACK pulses with the incoming video so that the picture is laid down in exactly the right place on the tape. During

CREATING A BLACKED TAPE

INSERT edits can only be made over existing video and CONTROL TRACK. Although you could record your edits over old "Family Feud" reruns, it is generally neater to prepare (and stockpile) a number of BLACKED tapes.

A BLACKED tape is simply a tape recorded from beginning to end (or at least a little longer than your intended edited tape will be) with a stable, glitch-free CONTROL TRACK. You make such a CONTROL TRACK by feeding stable video (and sync) to your VCR, switching it to the ASSEMBLE or NORMAL mode, pressing record, and letting the machinery do the rest.

Where do you get a stable video signal? At Poverty Productions Inc. you would probably connect a camera to your VCR and let it record the image. It doesn't matter what the image is. In keeping with the concept of BLACKED master, you could cap the lens. You may find it a little handier to aim the camera at a clock or stopwatch so that while editing you can tell at a glance how much time you have used.

The stabilitly of your CONTROL TRACK depends on the stability of your sync signal. It therefore behooves us to make the most stable video/sync signal that we can. Feeding your camera external sync from the studio's rock-solid house sync generator gives a more stable signal than the one generated internally by the camera (in its INT SYNC mode). Just make sure that the house sync generator is set to INT (driven by its accurate internal clock) rather than to EXT (driven by some other outside video source which may waver or be interrupted).

If you have a switcher connected to house sync, you don't need to be bothered with the camera at all. Simply fade to black and record the black video (which includes sync). If your studio uses black-and-white and color equipment side by side and you'll be editing a color tape, make sure that your sync generator is set for color, not black and white. Black-and-white sync is slightly different from color sync.

No law says that your tape has to be black. COLOR BARS are pretty. Many sync generators (and cameras too) make COLOR BARS as a handy reference signal. You can tape the COLOR BARS directly from the sync generator and feed it into the VCR without tying up other parts of the studio in the process. Furthermore, the less gadgetry you have in the way, the less can go wrong with the signal as it is being recorded. One extra advantage of COLOR BARS: The beginning of every master tape would have 10 seconds of COLOR BARS before the show starts as a reference signal that technicians may use to adjust the color signal as it comes from the VCR. If the VCR that plays the master tape is out of whack, the bars will be the wrong colors.

CREATING AN SMPTE LEADER

Every important MASTER tape should have a LEADER, a length of tape that comes before the beginning of the program and takes all the threading abuse. Since your program is never manhandled, it is likely to have fewer dropouts, scratches, and stretches.

The LEADER can also contain TEST signals useful to technicians who may need to adjust the video tape player when the program is played back or copied. SMPTE, the Society of Motion Picture and Television Engineers, has standardized a series of test signals that can be recorded on the LEADER of your video tape and will be recognized (even expected) by others in the video industry. It is possible to create your SMPTE LEADER and a BLACKED master tape all at the same time. Here's how:

1. Record 10 seconds of black (for threading).
2. Follow this with 10 seconds of COLOR BARS. During this time, record an audio tone of 1000 Hz at a 0-VU volume level on audio track 2.
3. Follow this with 15 more seconds of black and silence. Later, you may go back and insert a visual ID (called a SLATE) while audibly reading the contents of the SLATE onto the sound track. The SLATE would immediately identify the name of the program, when it was produced, who produced it, where it was produced, what VTR was used to edit the program together (handy to know if the tape someday runs amuck and decides never to play correctly on any other machine but "mother"), and other trivia.
4. Follow this with 10 seconds of leader numbers which go 10, 9, 8, 7, 6, 5, 4, 3, at 1-second intervals. This should be accompanied by eight audio beeps at 1-second intervals.
5. Following this should be 2 seconds of black and silence.
6. At this point the program should start, but you could continue to record black and silence throughout the rest of your tape. As an extra step, you could also record SMPTE TIME CODE (described later) on the tape.

Once you have made a BLACKED tape with an SMPTE leader, do not try to make more such tapes by copying this one. Make each one fresh. Copying this tape would make duplicates with jitter and time base errors recorded onto them. Since your object was to make a pure, perfect CONTROL TRACK, you don't want little timing flaws recorded on your important BLACK master tapes.

editing, the CONTROL TRACK is never touched. If it was laid down correctly at the beginning, it will remain good no matter what pictures you INSERT-edit on the tape. Bad edits and bad sync won't hurt the CONTROL TRACK (they won't make good pictures, but at least they won't hurt the CONTROL TRACK). Figure 14–6 diagrams the differences between ASSEMBLE and INSERT edits.

You are more likely to get a *perfect* CONTROL TRACK throughout your tape if you've recorded it in your studio from beginning to end using your best equipment. Now you can trust your CONTROL TRACK. If instead you ASSEMBLE-EDIT your production, the following hungry gremlins may sneak up and bite you:

1. Your portable VCR may have a weak battery.
2. Nearby electrical interference may mess up your sync.
3. If you move your portable VCR while it's running, gyroscopic aberrations in the spinning video head will mess up your video.
4. Humidity in the field may make the insides of your portable VCR sticky.
5. In the studio, you make a bad (glitchy) ASSEMBLE edit while editing your tape but don't notice it until you finish editing your entire tape. Sometimes in the process of editing you get so involved in the program that you forget to keep a sharp eye on the technical quality of every edit.
6. You ASSEMBLE-edit a piece from another video tape that had a minor flaw in its playback speed

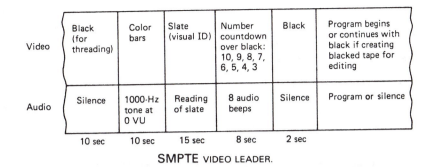

Video	Black (for threading)	Color bars	Slate (visual ID)	Number countdown over black: 10, 9, 8, 7, 6, 5, 4, 3	Black	Program begins or continues with black if creating blacked tape for editing
Audio	Silence	1000-Hz tone at 0 VU	Reading of slate	8 audio beeps	Silence	Program or silence
	10 sec	10 sec	15 sec	8 sec	2 sec	

SMPTE VIDEO LEADER.

or an unseen glitch in its sync. The imperfect material all gets rerecorded onto your edited tape, imperfect CONTROL TRACK pulses and all.

Now that you see the glitch, what can you do about it? The answer is nothing. You can't just redo that one edit on the tape. If you try to use the INSERT mode to cover the bad edit, it won't help. It will change the video but won't change the defective CONTROL TRACK. If you try to make a new ASSEMBLE edit over the old one (starting the edit a few frames earlier so that the bad edit is totally erased), you will have corrected the defective edit at the beginning of that scene, but now, how do you stop the edit? With the VCR in the ASSEMBLE mode, it is making its own pulses and doesn't have any way to line them up with the CONTROL PULSES which will continue when the edit stops. In other words, you get a nice edit at the beginning, but you get a

ragged edit at the end of the scene. Now *that* edit has to be corrected. The only way to fix it is to redo the next scene, and so it goes to the end of your tape. If your original bad edit occurred near the end of the tape, then you lucked out. You'll only have to do a few edits over. If the bad edit occurred in the middle or at the beginning of your tape, you'll experience the joy of creating your masterpiece twice.

In conclusion, it is best to first create a BLACKED tape in the studio and later INSERT EDIT your scenes onto it. But you can still do the edits in order, as if you were assemble editing.

Video Insert Only

In the INSERT edit mode, you can replace the video (and sync) and audio, or you can replace just the video or just

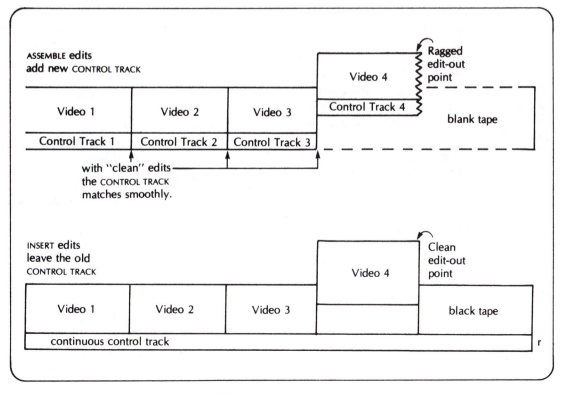

FIGURE 14-6 ASSEMBLE VS INSERT editing.

OVERCOMING THE MURPHY FACTOR WHILE EDITING
Murphy's first law, "Anything that *can* go wrong, *will* go wrong . . . and at the worst possible time," applies exquisitely to video editing. Murphy keeps an office of gremlins right next to every editing console. To keep Murphy from flushing your editing project down the tubes, take the following precautions:
1. Put adhesive tape (or a bandage, or something) over the RECORD button of your *playback* VCR. Pushing that button by mistake could wipe out your original CAMERA MASTER.
2. Double-check your mode. Don't get so wrapped up in what you're doing that you forget to *make sure that the MODE SELECTOR is in the proper position.* Untold numbers of tapes have been ruined by a selector in the ASSEMBLE position during the execution of an INSERT. Hours of audio tracks have been wiped out by an INSERT when a VIDEO INSERT ONLY was called for.

the audio. For audio, you have the choice of replacing audio channel 1 or channel 2 or both.

VIDEO INSERT (or VIDEO INSERT ONLY) permits old video to be replaced with new video in the form of an INSERT edit without touching the existing sound portion of the program. To make a VIDEO INSERT ONLY, do the following:

1. See that your video source is connected to the VCR's VIDEO IN. Also see that you can monitor the VCR's audio as it plays.

2. Check your video record level by temporarily pressing RECORD (unless the VCR monitors this while in the STOP mode). When finished, switch the RECORD button off.

3. Play the tape to learn the EDIT-IN and the EDIT-OUT points. On most machines, you can hear the audio track playing on the monitor while you are performing the video edit. This is an accurate and convenient guide to where you are on the tape so you may wish to use the audio as your cue for when to begin and end your VIDEO INSERT.

4. After learning the edit points, switch the VCR's MODE SELECTOR to VIDEO INSERT ONLY.

5. Rewind the tape a ways.

6. Play the tape.

7. When the edit point comes, press the EDIT button. CUE your talent.

8. Pay attention to your timing. Get ready to press the EDIT or END INSERT button when you hear your cue from the audio track.

9. Remember that you may have to press this button up to 1 second before you want the INSERT to end in order to give the electronics time to carry out the process. Once it has, you may STOP the VCR.

10. If your editing VCR is a poor one, check your edit in case it has to be done over.

As mentioned before, your performers could be fed the VCR's sound track and could learn their cues from that. When they hear a key sentence finish, they begin their action, timed to follow with the sound playing from the VCR.

Audio Insert Only

With AUDIO INSERT ONLY you leave the video untouched and change only the audio. The process is much the same as we saw in Chapter 5 under AUDIO DUBBING.

Since the old material is erased as the new material is put on, a slipup can be dangerous. Once your old sound is erased, it is gone for good. It's wise to rehearse your AUDIO INSERT to make sure your timing is right.

If you can read lips or follow the action closely, you can use your picture as your guide to the EDIT-IN and the EDIT-OUT points. Otherwise, you may have to use your VCR's index counter as a guide.

To perform the AUDIO INSERT edit, do the following:

1. See that your audio inputs are connected.

2. Check your audio levels by temporarily pressing RECORD. When finished, switch the RECORD button off. Be sure to try to match the original sound with the inserted sound. If the AUDIO INSERT changes volume, or a different mike is used, or a different room is used, or the talent is closer or farther from the microphone during the INSERT, or the talent changes the pitch or pacing of his voice, the AUDIO INSERT will stick out like a sore ear.

3. Play the tape and learn the EDIT-IN and EDIT-OUT points.

4. Rewind the tape a ways.

5. Switch the MODE SELECTOR to AUDIO INSERT and pick the appropriate channel to be recorded.

6. Play the tape.

7. When the edit point comes, push the EDIT button (or AUDIO DUB button). Cue talent.

8. Pay attention to your timing. Get ready to press the END INSERT, EDIT, or STOP button, depending on what the manufacturer instructed.

9. Play the result back to see if you got it right.

Incidentally, you can't INSERT new hi-fi audio on a hi-fi video tape. The hi-fi audio becomes part of the picture when it is recorded, and it remains part of the picture forever (or until both are erased). The low-fidelity *linear* audio tracks *can* be edited or replaced.

EDITING TWO-CHANNEL AUDIO

TWO-CHANNEL AUDIO allows two audio tracks to exist on the tape side by side.

When you had but one audio channel, you faced the pressure of making your AUDIO DUB or AUDIO INSERT perfect the first time because while you were DUBBING, you were also erasing the old sound. The original sound couldn't be retrieved if you made a mistake. Also you couldn't mix the new sound with the old sound because the old sound was being erased at the same time that the new sound was being recorded. You got old or new but not both.

With TWO-CHANNEL AUDIO, you can make AUDIO DUBS as before (as if there were no second channel), or you can make a new sound track on the second audio channel without hurting the first audio track.

This is handy if you have, say, the background sounds of people talking and machines whirring on track 1 and you wish to add narration *without losing the background sounds.* Your narrator can be recorded on track 2 without touching track 1. If the narrator flubs his lines, no problem. Just do track 2 over. Track 1 is still safe. Once finished, the tape can be played back with only the original track heard (the background sounds), or with only the narration heard, or a mixture of the two.

Another handy application is the bilingual tape. One track could present the English version of a program while the other has the Swahili version. The person playing the tape can switch the AUDIO SELECTOR to CHAN 1 to hear English, to CHAN 2 to hear Swahili, or to MIX to hear gibberish (both at once).

When planning your audio tracks, it is important to consider who is going to be using your tape and for what purpose. If the complete sound track has components on each of the two channels, a prominent notation must be left on the tape to remind the user that both channels are to be played together. Otherwise, the listener may miss half the audio portion of the program and maybe never realize it. It is distracting, however, for the listener to have to read notes and twiddle knobs to make the sound come out right.

For playback to be totally foolproof, it is best to record the same complete sound track on both channels. Thus, your sound will be heard no matter what position the AUDIO SELECTOR is in. This goes for home video formats, too. If both channels have the same thing on them, the sound will have a little better fidelity, and if a switch or connector gets goofed up by the viewer, nothing will be lost.

On ¾U videocassettes, if you are recording your audio on only one of the channels, make it channel 2. Channel 2 is accepted by the industry as the main audio channel.

Record your finished audio on track 2 (or 1 and 2) when distributing your tape to others.

There are a few occasions when you may wish to keep the audio tracks unmixed and separate. Say you have an EDITED MASTER with English on track 1 and Swahili on track 2. You could make English-only copies by playing track 1 from your VCP into both channels of the VCR. Conversely, for the tapes you send to Zanzibar, you could copy track 2 onto both channels. Thus, one MASTER can spawn two different kinds of copies.

Another example: You expect the music in your production to be changed later. By putting the whole show *except* the music on one channel and the music on the other, you can later change the music without reediting and remixing the original sound track.

I saved the best example for last: TWO-CHANNEL AUDIO is very handy for editing together a final sound track. The technique for making such a finished audio track is called SOUND-ON-SOUND recording or OVERDUBBING. Say you finished making your original recording using channel 1 for your audio. You now wish to add narration and/or music to the original sound in just the proper proportions to make the finished product. To do this, you will need to play back the channel 1 audio, mix it with the music and/or narration sources, and record this combination on channel 2 through the VCR. The process goes like this:

1. Connect the VCP's audio channel 1 output to a LINE LEVEL input of an audio mixer.
2. Also connect your music and narration sources to the mixer.
3. Make an audio level check from all three sources.
4. Connect the mixer LINE OUT to the VCR's channel 2 AUX IN.
5. By putting channel 2 *only* into the DUB or INSERT EDIT mode, get an audio check on the VCR using a sample signal from the mixer.
6. When ready, switch the VCR to FORWARD while the machine remains in the AUDIO 2 DUB mode.
7. The old sound along with the music and narration will all go into the mixer where you can fade, balance, and adjust them as desired. This combination of sounds gets recorded onto track 2 of your tape. If a mistake is made, go back and repeat the process; the track 1 audio is still intact, while the track 2 audio will automatically be erased as you record over it the next time.

Figure 14–7 diagrams the connection.

MINIGLOSSARY

*Two-channel audio   Capability of recording two sound tracks on a tape.

Audio selector   Knob on a VCR that selects whether audio track 1 or 2 or a combination of both will be played back (or recorded upon).

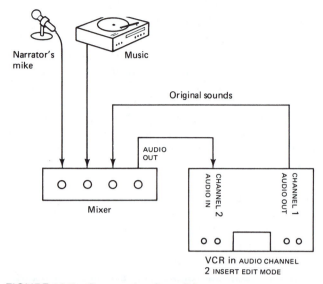

FIGURE 14-7 Connection for editing SOUND-ON-SOUND on a TWO-CHANNEL VCR.

EDITING AN AUDIO TAPE

When the sound portion of a video program is complicated, it is sometimes produced separately on reel-to-reel audio tape and then DUBBED or INSERTED into the video production. Since this is not a three-credit course on audio production, I will spare you most of the gory details of audio editing. But I will share with you a few gory details.

Splicing

If you wanted to cut a 45-minute speech down to a 10-minute speech, you could do just that, cut it. Quarter-inch audio tape is the only format where this is feasible (cassette tape is too flimsy and likely to jam). This is how you would go about editing out parts of a speech:

1. Play the whole tape to determine which parts are to be kept and which ones "cut out."
2. Play the tape again, stopping it at precisely the point where you wish to begin removing the material.
3. Mark the outside of the tape (the side facing *away* from the audio playback head) exactly where it passes over the head.
4. Continue playing the tape until you come to a part you wish to include in your final recording.
5. Mark this spot.
6. Do this until you have finished the tape.
7. Now rewind the tape to the first mark.

8. Remove the tape from the machine and . . .
   a. Using demagnetized scissors, cut the tape at a 45° angle, or, better yet,
   b. Place the tape in a SPLICING BLOCK, and using the groove in the block as a guide, SPLICE the tape at a 45° angle using a razor blade.
9. Leaving the "good" tape still in the SPLICING BLOCK, wind the unwanted tape off the supply reel until you come to the second mark (the end of the "bad" tape and the beginning of the next good segment).
10. As before, slice the tape on the mark. Discard the unwanted tape.
11. Butt the tail end of the first "good" segment up to the head end of next "good" segment. They should touch but not overlap.
12. Place AUDIO SPLICING TAPE over the junction and trim the excess. The SPLICING BLOCK will help guide your hand.
13. Continue this process until all the "good" segments make up a single reel of tape and all the "bad" segments are in the wastebasket.

This process can be done in any order. For instance, you could use the end of the original tape in the beginning of your edited version. In fact, you can mix and match snippets from several reels of tape, making one single production of any length or order. Figure 14–8 shows the process.

Electronic Editing

Video tape is almost always edited electronically. Audio tape is almost always edited by SPLICING, probably because SPLICING is so easy to do and requires so little equipment. Nevertheless, it is still possible to edit audio tape electronically. The technique is essentially the same as making PAUSE edits with a videocassette recorder. The process goes like this:

1. Connect the audio source to the audio input of your audio tape or audiocassette recorder.
2. Press RECORD on your audio recorder and feed it a sample signal and adjust the recording volume level.
3. You record the sound passages you wish to keep while stopping the recorder during the sound passages you wish to delete. There are two popular ways of doing this:

MINIGLOSSARY

Splicing block Grooved metal block that holds tape while it's being spliced (joined and trimmed).

Audio splicing tape Adhesive tape used to join the ends of audio recording tape during the editing process.

To indicate the edit points, mark the tape where it passes over the audio playback head.

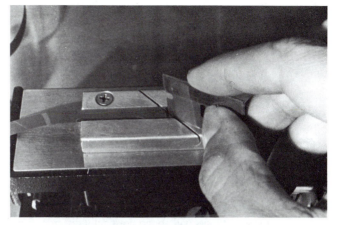

Cut the tape using the SPLICING BLOCK. Pull one piece away.

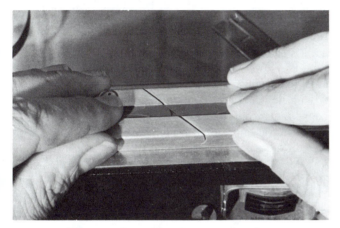

Butt two "good" segments together

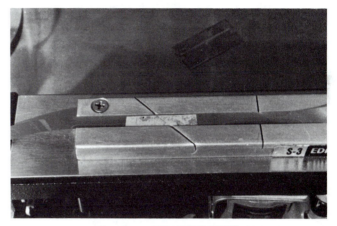

Press on SPLICING TAPE

FIGURE 14-8 SPLICING audio tape.

a. Press RECORD/PLAY and PAUSE simultaneously to get ready to start. When the desired passage comes, simply UNPAUSE your recorder to record the passage. When it's over, PAUSE again. You assemble your tape by PAUSING and UNPAUSING.

b. On other machines, you press RECORD and then let the tape machine sit there until the desired passage comes. When it arrives, press RECORD/PLAY to record the passage. When it's over, press RECORD/STOP, and the tape will stop moving, but the machine will remain in the RECORD mode, ready to grab the next segment that comes along.

Room Tone

The following is true for both audio and video tape recording: The parts of a program that you edit together should flow smoothly. You wish to avoid abrupt changes in volume level, fidelity, and background noise as the tape moves from scene to scene. To make your sound match from one scene to another,

1. Check your audio levels whenever you record to make sure the sound stays the same loudness. Typically, your VU meter should be bobbing around 0 dB.

2. Unless the scenes are drastically different, requiring a different type of microphone to capture the sound effectively, use the same mike in each case. For instance, if your performer is giving a lecture on one day using a lavalier microphone, then use the same microphone on the next day if you plan to edit the 2 days' lectures together.

3. In general, as you go from scene to scene, keep the microphone the same distance from the performer. This way, one scene where the microphone was close to the performer won't sound bassy while the next scene where the microphone

was farther from the talent will sound tinny and echoey. Exceptions: If some of your shots are close-ups and others long shots, it would seem strange for both to sound exactly the same. The close-ups should be accompanied with a close mike to give an intimate sound. The long shots should be miked at a greater distance to get that ''farther away'' sound.

4. Try to keep the ROOM TONE the same. ROOM TONE is the individual character of a sound recorded in a certain room. As sound echoes from the walls and mixes with various background noises, it takes on a color and personality of its own. A recording made outdoors will sound drastically different from one made in the studio, and that will sound different from one made in a living room. All of these will sound different from a sound recorded in a bathroom. Naturally, if your scene moves from one of these places to another, you expect that the ROOM TONE will change. But say your actor flubbed a line during an outdoor scene and you didn't notice the problem until a few days later. If you decide to have the actor read the line again and dub in just the sound (rather than traveling all the way back to Barbados to reshoot the segment), you wouldn't want this AUDIO INSERT EDIT to be obvious to the viewer. Using the same microphone as before (this is where good note taking comes in handy) placed the same distance from the performer as before, have the line reread *outdoors*.

EDITING FROM ANOTHER VIDEO TAPE

Advantages and Disadvantages

The problem with ASSEMBLE editing is that everything must be done in sequence. You progress through your shots in chronological order, unable to shoot the end scene first, then shoot all the airplane shots, and then go to all the bedroom scenes. If you could shoot all the similar scenes in one sitting, you could save a lot of running around locating your performers and setting up lights, audio, and so forth.

The problem with INSERT editing is that if you wish to *add* something to an existing tape, you have to *take something out* to make room for it. If you could alter the content of a show by simply removing scenes or putting in additional scenes, you could exercise uninhibited control over your production.

For these reasons, most professional teleproducers shoot

RAW FOOTAGE or, as the film industry calls them, RUSHES. These are recordings made (on separate tapes, usually) at different locations, at different times, sometimes by different people. These tapes are brought back to the video tape editor, who assembles the best scenes together to make a final tape. One nice thing about shooting RAW FOOTAGE rather than the real thing is that you can take chances shooting scenes which might not work out in the final production. If they don't, well, just don't use them; you're not stuck with them. Also, you can shoot the same scene over and over, perhaps from different angles, until you get it right. Afterwards, you can select the best of the RUSHES to incorporate into your final production.

The disadvantage of editing from RAW FOOTAGE is that your final tape is already one generation down in quality. You can hardly afford this loss if you are using VHS equipment. With ¾U, SVHS or Hi8, the editing flexibility will be worth the loss in picture quality. With the higher formats like 1 inch, MII, or Betacam, there's no question that it's safe to go down one or several generations.

The lower-format VCRs leave you with the difficult decision on which way to go—high picture and sound quality with limited editing, or unlimited editorial control with second-rate technical quality. And to further complicate the decision, you have another alternative: You could ASSEMBLE-edit most of your production (perhaps because it's all shot at one time in one place under easily controlled conditions), yielding a sharp final tape. Then for a few difficult shots (waiting for Junior's ''perfect'' dive or recording a whole movie just to get the juicy car crash) you record them separately and edit them into your final production. This way, most of your show is crisp-looking, and you were able to get the ''hard'' part also, although it's a bit grainy.

How to Do It

The mechanical process of editing from another video tape is the same as for copying a tape, only you're doing little pieces at a time rather than a whole tape.

To make the pieces fit, you must always

1. Find the place on the VCR where you want the edit to occur
2. Find the segment in the RAW FOOTAGE that you wish to use
3. Line both of them up so that when the VCP plays the desired segment, the VCR records it.

Hundreds of thousands of dollars can be spent perfecting the gymnastics necessary to perform the preceding three steps perfectly.

MINIGLOSSARY

Room tone   A character, ''color,'' or individual ''personality'' of a sound recorded in a particular room, caused by echoes and   background noises in the room.

Let's start with the imperfect: how to edit from another tape using home VCRs.

Home VCRs

1. Connect your VCP and VCR for copying a video tape. It's advisable to put masking tape over the VCP's RECORD button so you don't accidentally press it during the confusion of editing. Or, you could pop out the erase-protect tab in the back of your RAW FOOTAGE cassettes.

2. Try a test recording to make sure the VCR is getting a good signal from the VCP playing the RAW FOOTAGE.

3. Play your VCR up to where you want the next edit to begin and hit PAUSE and then press RE-CORD/PLAY.

4. Play your MASTER up to where the first scene to be copied begins and hit PAUSE about 5 seconds before that point.

5. Take a deep breath. Then UNPAUSE the VCP. Immediately position your finger over the VCR's PAUSE button.

6. When you see the desired scene come up, UN-PAUSE the VCR.

7. When the scene ends, PAUSE the VCR.

8. If it won't take you long to locate the next scene on your VCP, just leave the VCR in PAUSE. This will yield the "cleanest" edit. If it will be more than 3 minutes to prepare the next segment, then you have two choices:

   a. Press STOP, and when you are ready to proceed with the next edit, rewind the VCR's tape a ways, then play it, and then PAUSE it at the point of your next edit as described in step 3.

   b. Some VCRs, especially camcorders, allow you to switch the POWER off while the tape is in PAUSE, effectively stopping the video heads from wearing against the tape. When you are ready to roll, power the VCR up again, and you still should be in PAUSE, ready to go.

Industrial and Professional VCRs.

These machines can instantly switch from PLAY to RECORD while the tape is moving. Because the motion of the tape is not disrupted, the edit is glitch-free. Because the editors must BACKSPACE the tape, the process becomes more complicated. Once you know where an edit is to take place, you have to BACKSPACE the tape a certain distance (say 8 seconds). Similarly, once you have found the beginning of the scene you wish to EDIT-IN, you have to BACKSPACE the VCP

8 seconds. Both machines must be started simultaneously and will play together up to the edit point. At this point you (or the VCR if it does this automatically) switch the VCR from PLAY to RECORD/PLAY or EDIT to begin copying the segment from the other tape.

AUTOMATIC EDITING CONTROLLERS which will handle BACKSPACING for you are used almost everywhere and will be discussed shortly. If you don't have such a beast, it is still possible to edit videocassettes.

Manual editing technique for editing videocassettes: The trick is to find a way to BACKSPACE your VCR and your VCP exactly the right amount. One way to do this is to place an audible countdown on an unused audio channel. You might, for instance, read the seconds off your digital watch onto an audiocassette for a total of 30 minutes. You would then AUDIO-DUB this countdown onto an unused audio channel on your VCR and VCP.

To edit using the countdown,

1. Play your VCR to the point where you wish to make the edit. Listen to the countdown as you do this.

2. Jot down the countdown number at the edit point.

3. Subtract 8 from this number.

4. Rewind the tape a ways and play it up to this earlier number and then hit PAUSE.

5. On the VCP, use this same procedure to locate the countdown number, marking the beginning of the desired scene. Subtract 8 from that number. Rewind the tape a ways and play it up to this other number and then hit PAUSE. Now both machines have been BACKSPACED the same amount from the edit point.

6. Switch both machines to PLAY simultaneously and get ready to hit RECORD/PLAY or EDIT on the VCR when the edit point comes. You can use the picture or one of the countdowns as a guide.

This method is accurate to about 1 second. It can also be used for INSERT EDITS and is helpful for timing the EDIT-OUT points.

EDITOR CONTROLLERS

An EDITOR CONTROLLER or AUTOMATIC BACKSPACER is an electronic device which is wired to the editing VCR and VCP and controls the two during an edit. Figure 14–9 shows how one may be connected.

Most industrial models have digital readouts for the VCP and VCR showing exactly where you are on the tape. They create this TIME CODE by electronically counting the CONTROL TRACK pulses on the tape as it plays.

Method 1:

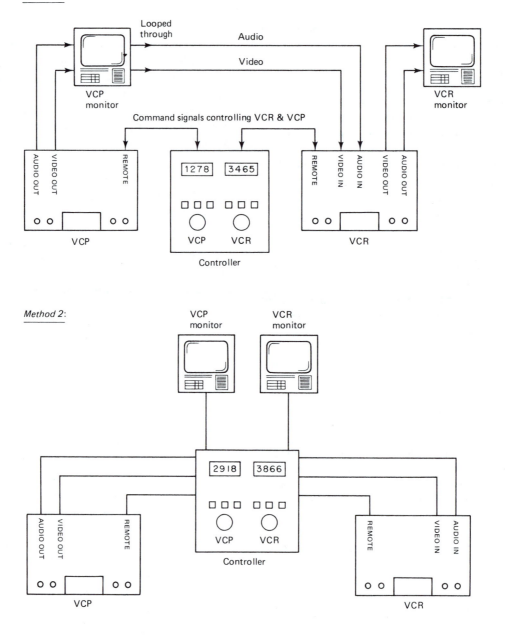

FIGURE 14-9 Connection of an EDITOR CONTROLLER.

When you load a tape into the VCP and start it playing from the beginning, the EDITOR CONTROLLER starts at 00:00:00:00 for the hours:minutes:seconds:frames.

Using this device, you can determine that a particular scene starts at 14 minutes, 32 seconds, and 12 frames from the very beginning of the tape.

These TIME CODE numbers are helpful for locating scenes or for logging events on a video tape for later editing. You could, for instance,

1. List the TIME CODE numbers of the various scenes on the tape

2. Next use those numbers to locate the first scene you wanted to edit

3. Program the CONTROLLER to perform the edit using those numbers

On the other hand, it is also possible to edit tapes together without paying any attention to the TIME CODE numbers. The process might go like this:

1. Find the next desired edit point on the VCR "by eye." Then push an EDIT-IN button to program

MINIGLOSSARY

Time code A way of measuring where (how far from the beginning of a tape) scenes are located. Usually a magnetic pulse recorded on the tape that can be converted into a listing of hours, minutes, seconds, frames.

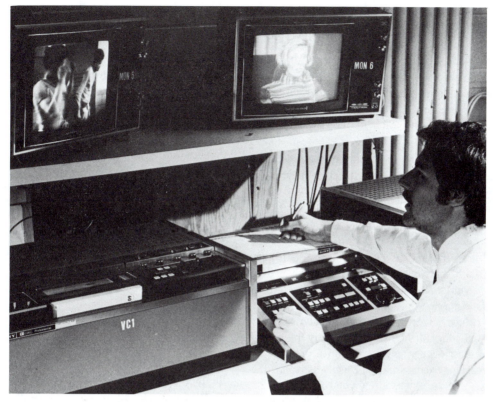

EDITOR CONTROLLER and 3/4U editing VCRs.

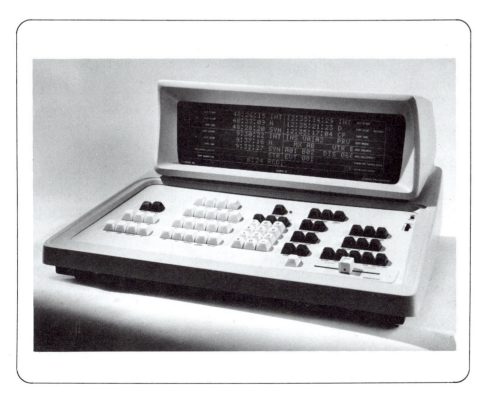

EDITOR CONTROLLER using SMPTE TIME CODE, can perform cuts, wipes, dissolves (Courtesy of Paltex).

the CONTROLLER (the CONTROLLER now automatically keeps track of the *VCR's* TIME CODE number).

2. Browse through your RAW FOOTAGE to find the scene you wish to take from the VCP. At the beginning of that scene, press another EDIT-IN button to program the CONTROLLER (the CONTROLLER now keeps track of the VCP's TIME CODE).

3. Push the EXECUTE button, and the CONTROLLER will automatically BACKSPACE the two machines the proper amount. Next, both machines will start playing, and the VCR will be switched to the EDIT mode at the proper time.

Some EDITOR CONTROLLERS can perform an edit accurate to one-thirtieth second. That's a lot better than the 1-second accuracy you got from using the voice time code described earlier. With one-thirtieth-second accuracy, the user may PAUSE a tape, creep it forwards or backwards a hair (a technique called JOGGING), select precisely the point at which the edit must occur, and then, with a press of a button, have the machine execute the edit exactly as planned.

Some EDITOR CONTROLLERS allow you to rehearse an edit once the edit points are programmed. In this case, instead of pressing EXECUTE, you press PREVIEW, and the machines will go through the motions without actually performing the edit. You can see what the edit would look like by viewing a TV monitor connected to the CONTROLLER as shown in Figure 14–9, method 2. Once you've previewed how the edit looks on the monitor, you can decide whether to execute the edit or to reprogram the edit differently.

Most EDITOR CONTROLLERS have TRIM buttons which allow you to adjust and edit slightly. For instance, imagine you have just programmed the CONTROLLER to execute an edit when the VCR reaches 22:33:14 on its TIME CODE, while the VCP starts *its* scene at 38:49:20. When previewing the edit, you discover that the edit drags too long and you wish to have it be more abrupt and faster paced. In other words, you wish to have the VCP start the scene just a little later, deeper into the action. You could do this by TRIMMING the number 38:49:20 up to 38:49:29 to cut ⅓ second off the beginning of the scene.

You could then rehearse the edit one more time to see if it looked better. If not, you could go back and add some more or trim back some of the time you added.

EDITOR CONTROLLERS usually have a complete set of VCR controls such as RECORD, PLAY, FAST FORWARD, REWIND, and PAUSE on them for both the VCR and VCP. Many also have a SHUTTLE SPEED CONTROL which allows you to run the tape forward or backward at twice, 3 times, or maybe 10 times the normal speed while viewing the picture on the TV screen. The SHUTTLE CONTROL may also allow you to play the tape at half speed or slower. You may even be able to JOG the tape one frame at a time forward or backward while searching for the exact point you wish to edit.

Note: CONTROL TRACK pulse counters can only count pulses when there is video on the tape. If there are blank parts to your tape (no CONTROL TRACK pulses), no additional accounting occurs. The numbers remain unchanged until the machine begins playing video again. For accurate counting, it is good to have continuous video from the beginning to the end of your tape.

SMPTE TIME CODE

CONTROL TRACK pulse counting is not 100% accurate. When the tape stops moving, a few pulses get lost. If the tape is shuttled very quickly, some more pulses may get lost in the shuffle. If you inserted a half-rewound cassette into a VCR and started it playing, the EDITOR CONTROLLER wouldn't know where you were on the tape. Only if the CONTROLLER and cassette *both* start at the beginning, will the numbers be meaningful.

If you *had* to switch cassettes and didn't want to start from zero each time, here's what you have to do (if your CONTROLLER allows):

1. Jot down the TIME CODE before removing the cassette from the machine.

2. Load another cassette and use it (which fouls up your numbers). When done, . . .

3. Reload the first cassette and "dial up" the numbers on the counter so that they match where you left off.

This process is, however, inconvenient and not totally accurate.

The SMPTE (Society of Motion Picture and Television Engineers) TIME CODE is much more accurate. It gives

---

MINIGLOSSARY

Jog   To move a video tape forward or back a very short distance (one or two frames) in search for the "perfect" place to edit.

Trim   Adding or subtracting numbers from the time code at the edit point to make the edit occur earlier or later than originally planned.

Shuttle speed control   A fast scan control allowing video tape

to be played slower or faster than normal—useful when hunting for edit points on a tape.

*SMPTE time code   A time code used to address every frame on a tape with a unique number to aid in logging and editing. The time code format is standardized in the United States by the Society of Motion Picture and Television Engineers.

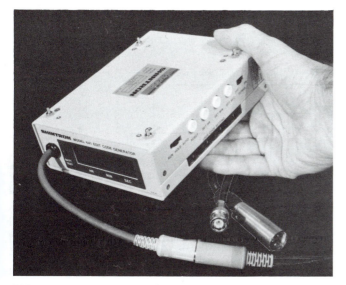

FIGURE 14-10 Portable SMPTE TIME CODE GENERATOR.

a special number to every single frame (picture) on your video tape. Since the code is recorded on your tape, it can be read from the tape at any time, whether the tape has been half rewound or not.

SMPTE TIME CODE is created by a SMPTE TIME CODE GENERATOR (Figure 14–10). This is an electronic box with a special clock in it which is able to create a digital code which is synchronized with the pictures on your video tape. Depending on your VCR, this code can be recorded on one of the audio channels or on a special ADDRESS TRACK or CUE CHANNEL designed for this purpose. Some VCRs can encode the SMPTE TIME CODE into the video (in the VERTICAL INTERVAL where it cannot be seen visually).

SMPTE can be recorded on a tape while the tape is being recorded originally or can be DUBBED onto the tape later.

To "read" the code off the tape, you need an SMPTE TIME CODE READER (Figure 14–11). The READER connects to the appropriate audio channel or ADDRESS CHANNEL out-

put from the VCR and translates the code back into the readable numbers, hours:minutes:seconds:frames.

Connecting SMPTE Time Code Devices

For the SMPTE TIME CODE GENERATOR to give numbers to each video frame, it must be able to "see" them. Thus, the time code generator is always connected to a video wire somewhere to get this information. Figure 14–12 shows a few ways to connect up an SMPTE TIME CODE GENERATOR or READER.

In method 1, a camera sends its signal *through* the TIME CODE GENERATOR on the way to the VCR. The generator samples the video signal and manufactures the time code in step with the video. The code can be fed to one of two places:

1. AUDIO CHANNEL 1.
2. ADDRESS CHANNEL (for VCRs having this third time code channel). Thus, the VCR records audio, video, and time code simultaneously.

Time code can also be recorded in the VERTICAL INTERVAL, the black sync bar just below the bottom of your TV screen. Here, the camera or a videocassette player sends its video signals through the time code generator which embeds the code into the video signal before sending it on its way to the VCR. This method of recording time code is handy if you don't have an ADDRESS CHANNEL or can't spare AUDIO CHANNEL 1.

SMPTE TIME CODE can be added to a tape which has already been recorded. Method 3 shows how. The VCR plays the tape into the time code generator, which creates the code and sends it to AUDIO CHANNEL 1 or the ADDRESS CHANNEL of the VCR where it is dubbed onto the existing tape.

To read the time code (to see where you are on a tape), play the code out of the VCP's address or AUDIO

FIGURE 14-11 SMPTE TIME CODE READER/GENERATOR does both jobs.

MINIGLOSSARY

*SMPTE time code generator   Electronic device that makes the SMPTE time code signal, which may then be recorded on the tape.

*SMPTE time code reader   Electronic device that decodes the time code from a tape on playback and converts it into recognizable

numbers: hours, minutes, seconds, frames.

*Cue channel   Extra (usually a third) audio channel, recorded on an extra track on the video tape—used to carry TV technician messages or time code data, such as the SMPTE time code.

Method 1: To record time code in the
field during production

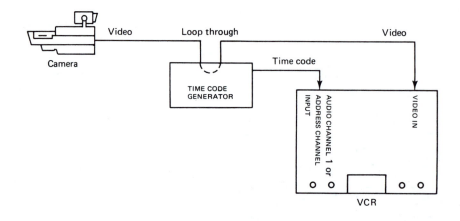

Method 2: To record time code in
the VERTICAL INTERVAL

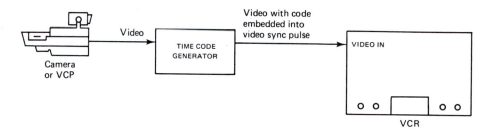

Method 3: To record time code
onto a tape later

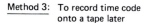

Method 4: To read time code

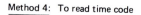

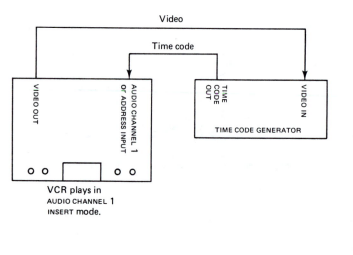

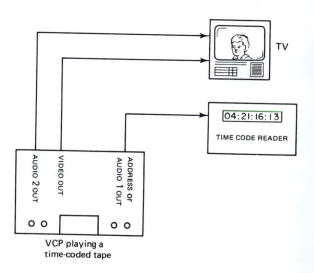

FIGURE 14-12   Connecting up a SMPTE TIME CODE GENERATOR or READER.

Method 5: Reading VERTICAL
            INTERVAL TIME CODE

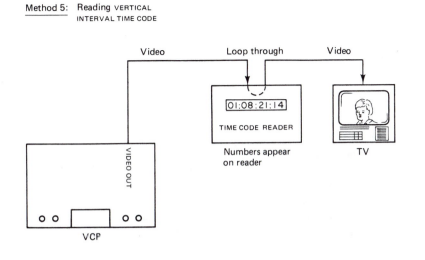

Method 6: To create a WINDOW DUB where the tape
            copy can be played on any VCR without
            a READER because the numbers appear in
            the picture

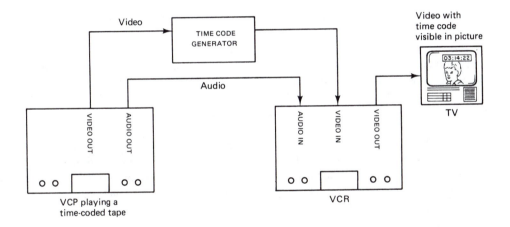

FIGURE 14-12 CONT'D

CHANNEL 1 into the time code reader as shown in method 4. The numbers will appear on the reader.

Method 5 shows how a reader can separate the time code from the VERTICAL INTERVAL to display the time code numbers and pass the video signal onto a TV or another VCR.

Method 6 shows how the time code generator can be used to create a special effect called a WINDOW DUB where the time code numbers are inserted visibly in the picture. By copying a tape this way, you create a tape which can display time code numbers when played on any normal

VCR. Such a tape allows you to review and log the RAW FOOTAGE at your leisure, perhaps using an inexpensive VCR in your home. The WINDOW DUB is never shown to the public; it is just a dummy you can use for reviewing "takes" on a tape or making edit decisions. Your decisions, based on the numbers you see in the WINDOW DUB, are carried out for real later, using the master tape.

SMPTE TIME CODE READER/GENERATORS are not video EDITOR CONTROLLERS. They don't BACKSPACE VCRs or perform edits: They merely tell you where you are on the tape. More advanced EDITOR CONTROLLERS have SMPTE built in

---

MINIGLOSSARY

Window dub Copy of a time-coded video tape with one change: time code numbers are visible on the TV screen, making it possible to log edit decisions while playing the tape on common VCRs not equipped with time code readers.

and *can* use the code to BACKSPACE and perform edits. The most advanced models have video display screens showing a complete EDIT DECISION LIST, the list of edit-in and -out points and transitions you have selected.

STEP-BY-STEP EDITING PROCEDURE

The mechanics of editing aren't too hard; it's mostly a matter of pushing buttons at the right time. The planning and decision making of editing is the hard part. The process will test your patience and your stomach lining. Planning, viewing the RAW FOOTAGE, logging the scenes, and charting the edits are all tedious, unglamorous ordeals. Perhaps that's why most folks skip these steps and end up making uninspiring productions.

Here's a general plan to follow:

1. Before you start, decide how your story will unfold and determine the scenes necessary to tell the story (more on this in Chapter 17). Then shoot your RAW FOOTAGE.

2. View all your RAW FOOTAGE and log each event with some sort of index number as shown in Figure 14–13. For index numbers you could use
   a. The tape footage or time elapsed counter on the VCR (but few counters are accurate)
   b. A verbal reading of the time on your digital watch, recorded on a spare audio channel (but this is hard to do and is only accurate to ±1 second)
   c. Control track pulses (you need an EDITOR CONTROLLER or a VCR with a control track counter; these have to be set at 0 at the beginning of your tape and are accurate to within a few frames)
   d. SMPTE TIME CODE (this most accurate and convenient method requires an expensive reader/generator)

      The SHOT SHEET or LOG/CUE SHEET makes it easy for you to find quickly the desired "take" on the tape when you get around to editing the scene into your final production. SHOT SHEETS are especially helpful when you record your scenes in January but don't edit them until August. In the SHOT SHEET shown, NG stands for "no good," CU for "close-up," and MS for "medium shot." You may wish to put a star next to the "take" that you like the most.

SHOT SHEET		Cassette #3	
Project: Dining Out		Date: 12/7/87	
Take	Counter	Action	Comments
1	0–23	Leader	NG
2	24–44	Testing	NG
3	45–65	CU sandwich hits floor	Excellent
4	66–88	CU sandwich hits floor	Dark
5	89–100	CU sandwich hits floor	OK
6	101–152	MS sandwich hits floor	Fuzzy
7	153–208	Al sits at table	OK
8	209–250	Al sits at table	Glitch
.	.	.	.
.	.	.	.
.	.	.	.

FIGURE 14-13   SHOT SHEET for logging "takes"

3. Plan your editing strategy using an EDITING SHEET like the one shown in Figure 14–14. You lay out the sequence of scenes you wish to assemble, enter the counter numbers of these scenes from your SHOT SHEETS, and you are ready to edit.

Notice that the beginning and end titles can be shot "live" with a camera (as could other scenes) and be assembled along with taped scenes.

If your editing equipment is very simple or your tape index is inaccurate, you may decide to play the edit sequence several times, rehearsing it, and then perform the edit manually using the scene as a guide. If your equipment is advanced, you should be able to "edit by the numbers," selecting the scenes by their TIME CODE numbers and then having the machine execute the edits on the numbers you have selected. If you don't like how the edits are coming out, you can always pick new numbers and redo the edits.

EDITING SOUND OR PICTURE FIRST

Sometimes it is easier to put all the visual scenes together and then go back and DUB in the sound (assuming lip synchronization is not required). This method, called a VOICE OVER, works best when you find it hard to judge (or don't wish to take the time to calculate) how long each scene will be and when you don't wish to adjust any of the visual scenes to make them fit the sound. So you first make a "silent movie" and then go back to DUB in music, narration, or nothing (leaving the original sound track intact). With this method, the picture is most important, and the sound is a slave to the visual timing. The narrator reads, watches, pauses, and reads again, and if the scene is too short for

MINIGLOSSARY

*Shot sheet   An index of all shots recorded on a tape. Includes time code numbers for each shot plus a commentary on the quality of each take.

*Editing sheet   A plan showing which shots will be used to creat the edited master. Usually time code numbers and edit-in and out points are included.

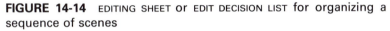

EDITING SHEET		Cassette #27		
Project: Dining Out		Date: 12/15/87		
Segment	Action	Cassette	Edit-in	Edit-out
1	Intro	Camera	275	301
2	Boys meets girl	5	870	880
3	Invites her to dinner	5	40	65
4	Takes her to restaurant	1	153	208
5	Sits at table	3	59	75
6	Orders jelly sandwich	2	422	435
7	Sandwich arrives	2	501	503
8	Boy spills sandwich	2	61	63
9	Sandwich hits floor	3	70	75
10	Girl leaves	1	93	94
11	Boy stunned	1		
12	End	Camera		

FIGURE 14-14 EDITING SHEET or EDIT DECISION LIST for organizing a sequence of scenes

the narrator's text for the scene, the narrator's script must be abbreviated to fit the scene. During gaps in narration, light music or background sounds may be used to fill in the silence.

Sometimes it may be necessary to have lip synchronization, perhaps as a process is explained by a factory worker at his machine. This case calls for a combination of editing techniques:

1. You may wish to shoot the first sequence in "silent movie" style. It may show the titles, an exterior shot of the factory, plus a wide interior shot of the equipment as we dolly in toward the lathe. Once these scenes have been edited on the tape, we can go back and DUB IN music plus the narrator's voiceover introducing the program.

2. The next sequence shows the factory worker describing how to program the computer which runs the lathe. For this scene, we record both picture and sound together.

3. As the lathe goes to work, we zoom in, hearing its "live" sound in the background. As this scene is edited into the program, we can fade out the background sound and fade up the narrator's voice as he describes the dangers of getting one's tie caught in the spinning lathe.

One of the great strengths of AUDIO DUBBING is its ability to tie scenes together. A montage of unrelated pictures become related by a single strain of music, sound effect, or conversation.

Editing scenes with divergent sound levels can be smoothed with an OVERDUB. Say you shoot a birthday party. The long shot of the whole gang has a lot of kids screaming in the background. The close-up of the cake's candles being blown out is fairly quiet. The close-up of someone eating cake has many voices in the background. The visual scenes here are fine, but the disparate sound levels are jarring. To smooth this out, these shots should have the same continuous sound in the background.

This sound could come from the VCR's original sound track during part of the party (as long as the VCR ran continuously long enough to get an adequate stretch of sound—this is called a SOUND BITE), or it could have been made by a separate audiocassette recorder. Except where important words are spoken and must be heard synchronously with the lips speaking them (LIP SYNC), you simply DUB your prerecorded background sound onto your edited tape to create a smooth din of authentic party noise. No one may ever notice that the cacophony of sound doesn't exactly match the bedlam of visual activity.

Many of the TV ads that you see have had their audio DUBBED. Sometimes it is because the best-looking models don't have the best voices or because better sound control is available with someone standing exactly 1 foot in front of a mike in a sound studio rather than moving around an echoey bathroom (sounds of a running shower can be added later). Although you would have difficulty performing an exact LIP SYNC like the pros, you can often use these techniques in many scenes where a close, well-aimed mike is impossible. Redoing the audio later permits you to remove unwanted sound effects (such as jet planes in an old-time western or the roar of electric fans in your "windy" scene). In the studio, you get only the sounds you want, and while you are at it, perhaps you can *add* a few sounds of your own for authenticity (clippity-clop in your western, thunder and gale winds in your storm scene).

After all this buildup, it is only fair to mention that editing audio on a VCR is rather cumbersome. Depending

MINIGLOSSARY

Sound bite   A long enough stretch of acceptable sound to be useful during editing.

on your machine, the beginning of a DUB may have the new sound recorded on top of the old (and not-yet-erased) sound for a moment. The end of the DUB may be followed by a few moments of silence as the erased-but-not-yet-recorded tape comes around. For this reason, avoid complicated DUBBING. Never try to delete just a word or two. Try to start and finish the AUDIO DUB during pauses in conversation. Experiment to see what your machine will do.

Don't forget, AUDIO DUBS, like INSERT EDITS, *replace* material. You don't add or subtract it. The total program always remains the same length. Also, with a true AUDIO DUB, you lose the old sound as you record the new sound. Only if you are copying from one tape to another or recording your new sound on one channel while leaving the old sound on the other channel can you preserve your original sound in case you've made a mistake.

There are times when the sound track carries the main theme of your tape. Here, the visuals are enslaved by the timing of the audio. Such is the case when you have a prerecorded song and you wish to have the visuals coincide with the melody or change to the beat of the music. You can't change the music to fit the visuals; rather the visuals must be stretched or condensed to fit the song.

The first step is to lay down the sound track while recording "black." Next, you make VIDEO INSERT ONLY edits for all the visuals. Here is an example:

Using a single camera, you record a dance number from beginning to end, taking care to feed your VCR good sound from whatever music source is used. Next, you arrange for the dancers to repeat portions of the dance for you. Arrange your tape machine so that it will play back the music which it recorded earlier for the performers to dance to. Change your camera angle to catch a particular move. Back up the tape one-half minute or so, play it, and follow the action with your new camera angle. Have an assistant hit the VIDEO INSERT button on the VCR to cut to a close-up of the desired move. End the insert when the move is over. When the tape is viewed, it will look like two cameras were used, one for long shots and the other for a close-up. This process can be repeated several times for further close-ups until the dancers get tired of repeating the performance and drill a pirouette into the top of your VCR.

A similar technique is useful for music videos. The song may be recorded in a sound studio and copied onto the sound track of a BLACKED tape. The musicians are then gathered in the TV studio under proper lighting with fog props and video effects at hand. Listening to their sound track play from your tape, they LIP-SYNC and perform to the music they hear. It is even possible to take the performance outdoors, into the street, or underwater, just as long as the performers can hear the sound on your tape and act in rhythm with it.

Once you get used to editing both ways (video first/sound later or sound first/video later), you'll discover at your disposal the power to have almost anything come out the way you want. Your flexibility is almost limitless.

For another example, a reporter could memorize the first and last paragraphs of her story and read the rest. The entire report is shot at the scene with the reporter, holding a microphone, speaking to the camera. With action buzzing in the background, everything looks candid and unscripted. All the places where the reporter reads from her notes get replaced with VIDEO INSERT edits of the action, close-ups, and so on. The same voice carries through from beginning to end, and since the viewer never saw the reporter refer to her notes, it appears that this very gifted reporter was able to give the entire report candidly and directly to you, the viewer.

Another example of using a combination of editing methods is employed in places where you are unable to shoot everything you want at a particular time. Say you are trying to show how preschoolers in a day care center deal with some educational toys. With little trouble, you shoot the kids playing in the room, but every time you get in range of the close-up action, the little buggers stop what they are doing and look up at you. So you try sneaking up on them, and you happen to catch a few absorbed in their play. But in the few moments before you're noticed, you can't get focused. The lighting is poor for close-ups; the kids move the toys around so fast that you can't get a tight close-up on important details. Solution to the problem:

1. Assemble with sound as many of the various medium and long shots as possible of the children playing with the toys.

2. Later, either at the TV studio or at the day care center, after the children have gone home, you set up the toys again. This time you light the toys perfectly. This time you choose your shots carefully. This time you can zoom in tightly on every detail. And this time you hire one of the youngsters to lend his hands to move or manipulate something in the picture. These carefully prepared scenes all get included as VIDEO INSERT ONLY sequences scattered among the medium and long shots recorded in step 1 (I hope that you, as a good documentarian, will create INSERTS which represent what really happened in the day care center).

3. Next, add narration and explanation by playing the first audio channel into a mixer and combining that sound with a narrator and music and recording the result on the second audio channel. The original background sounds of the children can play at full volume except where brief explanations are needed. At those times, the background is reduced, and the narrator speaks. Sequences (recorded in step 1) of day care center personnel

or children showing and explaining something would certainly have the original audio boosted and have the narrator silent. In the end, one may slowly fade from the day care center sounds to music and end the tape with the end of a children's song.

4. Now that you know all the people who participated in making the tape, you prepare the CREDITS, and at the end of the tape you edit them in via VIDEO INSERT ONLY.

5. By this time, somebody has thought of a title so you rewind to the beginning of the tape and VIDEO-INSERT-ONLY the title.

The end result is a fairly professional-looking tape with varied shots and varied sound. Everything looked natural, although plenty of it was contrived with painstaking care. That's the object. Every scene is thoughtfully edited for maximum information and maximum impact.

The viewers should be so hypnotized by the content that they are totally unaware of any production techniques and totally unaware of your effort behind those techniques. Like the perfect thief and the perfect spy, you are only successful if no one realizes what you did.

AB ROLLS

So far we've discussed "cuts only" edits, those where a scene abruptly cuts from one to another. These are the only kinds of edits that can be performed using one or two editing videocassette recorders. What does it take to dissolve or wipe from the picture on one tape to the picture on another? Answer: AB ROLLS.

No, AB ROLLS are not puffy buns you get when you put AB dough in the oven. An AB ROLL is the complex technique of coordinating three video machines (two players and one recorder), a pair of time base correctors, and an advanced EDITOR CONTROLLER. To do an AB ROLL, the VCR must first be listening to the signal from one videocassette player and then have that signal fade or wipe to the signal from the other videocassette player. The EDITOR CONTROLLER takes care of BACKSPACING the VCPs and VCR as well as controlling the switcher as it dissolves or wipes from one preselected position to another (such as dissolving from VCP 1's picture to VCP 2's picture). Figure 14–15 shows a simplified diagram of the setup.

For an AB ROLL to work, two very carefully prepared tapes (the A reel and the B reel) must first be produced, like the ones diagrammed in Figure 14–16.

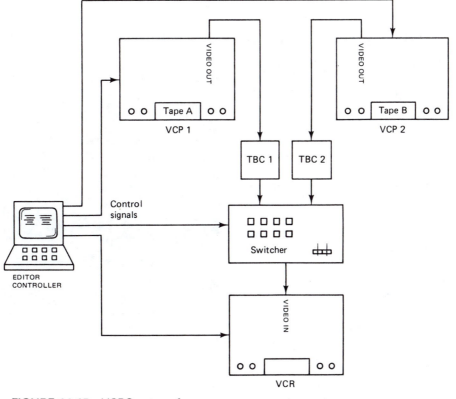

FIGURE 14-15   VCRS set up for an AB ROLL.

MINIGLOSSARY

Credits   The listing, usually at the show's end, of the people who participated in making it.

AB roll   Technique of placing one scene on one video tape (and VTP) and another scene on another and then rolling (playing) both VTPs together, along with the editing VTR, in order to fade, dissolve or do a special effect using both scenes at once.

		Cue tone				Two-second "safety" extension				

FIGURE 14-16 The A and B tapes in an AB ROLL.

To perform an AB ROLL edit,

1. The VCR and VCPs all get BACKSPACED.

2. All three play together up to the EDIT-IN point. Meanwhile the switcher and EDITOR CONTROLLER have been arranged so that only the A tape (VCP 1) can feed its signal to the VCR.

3. The first edit at this point is a standard CUT as the VCR abruptly begins recording the A tape (the B tape is playing to no one).

4. At the desired point, the EDITOR CONTROLLER tells the switcher to dissolve from the A tape to the B tape. Meanwhile all three machines keep rolling. Audio is also faded from the A tape to the B tape, if desired.

When the scene is finished, one could end the edit here with another cut and be done with AB ROLLING. Figure 14–16, however, shows a second special effect where the EDITOR CONTROLLER wipes the picture from the B tape back to the A tape at a certain time.

Back and forth you can go as many times as you want as long as both the A and B tapes are running. You notice, however, that careful planning and timing must go into preparing these two tapes so that when the controller dissolves out of the A tape, there is a picture on the B tape to dissolve to.

ON-LINE AND OFF-LINE EDITING

Elaborate AB ROLL EDITOR CONTROLLERS and time base correctors and computerized EDIT DECISION LIST managers are expensive, often too expensive for the school or industrial user to own. When an intricate program needs to be edited, it is sometimes best to rent time on a fancy editor in a

POSTPRODUCTION HOUSE. Here, for between $50 and $500 per hour (depending on fanciness), you can use the finest of computerized editors and special effects equipment. When you are paying almost $10 a minute to use the machine, you don't want to waste time figuring out how you want to edit the show together. These decision should be made and tested ahead of time so that all you need to do is come in with a list of edits to be performed and maybe make a few minor adjustments here and there.

Editing which is done on the expensive equipment, the computerized controllers, and the 1-inch video tape machines, whether it's done at your own facility or at a POSTPRODUCTION HOUSE, is called ON-LINE EDITING. It is the editing of the final MASTER using the best of equipment, and the result (and process) is called a FINE CUT or FINE EDIT. OFF-LINE EDITING, on the other hand, is done at a more leisurely pace using cheaper, simpler equipment. Here you

Editing suite at a POSTPRODUCTION HOUSE (Courtesy Windsor Total Video).

MINIGLOSSARY

*On-line editing   Editing a video tape with the highest-quality VTRs and editor controllers. Process results in final edited master but costs more than off-line editing.

*Off-line editing   Making a "practice edit" using inexpensive video equipment. Result is a lower-quality "draft" copy used for decision making and to create a list of edits to be performed later for real on-line.

*Rough cut   Approximation of what the edited master will look like. Rough cut is generally performed on off-line editing equipment.

Fine cut   Final edited master, prepared with painstaking care using the best editing equipment available. Fine cut is generally performed in an on-line editing session.

practice various scenes to see how they fit together and make a rough simulation of what the final tape should be. This is called a ROUGH CUT or ROUGH EDIT. It doesn't matter if the edits have a glitch; it doesn't even matter if the time code window shows in the picture. Your whole purpose here is to fit the pieces together, rearrange them if necessary, and "get the bugs out" before going to expensive ON-LINE EDITING to redo the job for real. You may even play your ROUGH CUT to your boss or client for approval (or modification) before spending big bucks on the FINE EDIT.

Editing Strategy

Every editor has a different plan of attack. Here is one editing strategy which is efficient, inexpensive, and fairly popular:

1. Time-code all your RAW FOOTAGE with SMPTE TIME CODE. (Here's a handy trick: Set your SMPTE GENERATOR so that the "hours" number isn't really hours but is your cassette number. The "hours" number really isn't needed anyway because most cassettes are less than an hour long. Now your TIME CODE READER not only will tell you where you are on the cassette but which number cassette you are viewing.)

2. Make WINDOW DUBS of all your RAW FOOTAGE.

3. Using the WINDOW DUBS and an inexpensive videocassette player (perhaps at home), view the RAW FOOTAGE and create your SHOT SHEET for each cassette.

4. Create a draft EDITING SHEET in order to formulate the approximate sequence of events.

5. Begin editing the scenes together OFF-LINE using your inexpensive editing equipment or using PAUSE edits on a pair of home VCRs. It's not important if the edits look glitchy; what's important is that the sequence runs smoothly and that the shots "fit." It is possible that after ROUGH-EDITING half or all of your production, you discover that you wish to add or subtract a few scenes (perhaps the show's length was too short or too long). Rather than redoing all the edits, you would simply copy your ROUGH CUT from the beginning up to the point where you wanted to add the scene and then edit the scene into your copy. Follow that by editing in the remainder of your original ROUGH CUT. You have now made your show longer but

didn't redo a whole bunch of edits. It doesn't matter that this copy is now third generation in quality; you're only using it as a guide.

6. Present this ROUGH CUT to your boss or client for final approval. At this point changes are still easy to make without redoing all the edits by copying the tape once again up to the point where you are making an addition or deletion, making the change, and then copying the rest of the tape.

7. View the ROUGH CUT once again, this time writing down the time code numbers corresponding to each edit you see on the screen.

8. Add to this list any embellishments such as dissolves, character-generated titles, or special effects. The list now becomes your final EDIT DECISION LIST.

9. Bring this list and your original RAW FOOTAGE to the ON-LINE EDITING suite to make the final production. If all of your decisions were perfect, the process can be done "by the numbers" in your absence. Usually a little "tweaking" is necessary to make the edits perfect or to assure that the special effects are the way you want them.

SHOOTING FOR THE EDIT

Planning is the key to editing. It is a pleasure to start with a treasure chest of RAW FOOTAGE, molding it into a dazzling teleproduction. On the other hand, it is really the pits trying to change chicken feathers into chicken salad. By planning your shooting strategy before the edit, you can approach this ordeal with cassettes filled with the right shots, or you can approach it with an armful of chicken feathers.

Here are some hints for shooting RAW FOOTAGE:

1. *Provide sufficient material.* You can never have too much RAW FOOTAGE. If one scene doesn't work, it is always good to have another that you can use in its place. Here are some examples of what an ENG or EFP crew should bring back with them *besides* the footage they were told to get:
   a. CUTAWAYS (described further in the next section), scenes of related or surrounding action which can be inserted to avoid a JUMP CUT. For example, after an interview, you would aim your camera at the interviewer

---

MINIGLOSSARY

*Cutaway The act of "cutting away" (taking a shot of something else) from the main scene for a moment to hide jump cuts. Also the name given to this backup shot, which is generally a long shot of a performer, a host, news reporters, or some other related scene.

*Jump cut An edit from one scene to a very similar scene, causing the picture to "jump" from one position to another. Such edits should be hidden by video inserts of related scenes (cutaways).

reacting to the person's answers, perhaps nodding his or her head. It may also be helpful to have the interviewer ask the same questions a second time with the camera facing the interviewer (make sure to ask the questions the same way as before so that the answers will be authentic).

b. Hold your shots longer than normal. This may only be important in 1 case out of 10, but that extra long-drawn-out shot may be just the thing to end the newscast (while credits roll) or for showing something while the narrator catches up with his or her explanation.

c. As you shoot and reshoot a scene, vary the pacing. This is handy when shooting ads as the editing for a 30-second ad is quicker paced than the editing for a 1-minute ad. The production will look best if the pacing of the performance matches the pacing of the editing. Furthermore, if the action happens quickly, it is easier for the editor to "get it all in" when editing a shorter version of the ad.

d. Keep your OUTTAKES. The bleeps, blunders, and foul-ups that get recorded are called OUTTAKES. These are generally discarded (except when they are used on network TV programs). It is not expensive to keep them until the edit is completed, and you never know when an OUTTAKE may be useful. For instance, in a popular commercial, an actor's laugh was recorded between takes. The laugh looked so natural and spontaneous that it was inserted into the commercial.

e. Shoot ESTABLISHING SHOTS to orient the viewer to locales and surroundings.

f. Throw in an "arty" shot or two such as a shot looking down on the action or up from the ground or through a leafy bush or through the reflection in a window. These unexpected gems help an editor spice up a bland but well-organized program.

g. Shoot a few out-of-focus shots. These may later be used as backgrounds for titles.

h. Record a few minutes of ambient sound. This may later become background sound in the final production or may fill in silent parts between narration or music.

2. *Reframe a shot when fixing an error.* Say some-

one flubs a line. Instead of reshooting the entire scene, you might have him redo the scene starting with the line *before* the flub. This is called a PICKUP because you are "picking up" just before you left off. The problem is, when you edit these scenes together, you will have a JUMP CUT as the performer's head snaps from one position to another. To avoid this, reframe the shot (zooming in or out and arcing the camera to a different vantage point).

3. *Check continuity.* Note the pacing of the scene so that it matches the next scene. Someone who just ran up a flight of stairs should be breathless in the next shot when he reaches the top. Also, try to duplicate facial expressions from shot to shot.

4. *Leave a 15-second preroll time before beginning the scene.* This leaves space for BACKSPACING during editing. Fifteen seconds of good stable video before the scene begins will assure that the EDITOR CONTROLLER and time base corrector (if used) will be able to "lock onto" the signal for a reliable edit. If actors need to be cued, the camera/VCR operator can say to them, "READY," start the VCR rolling, count off 10 seconds to himself, and then count backwards . . . five, four, three, silence, silence, . . . and the scene begins.

5. *Start each tape with 1 minute of color bars.* This gets you past the dropouts which occur at the beginning of a tape due to threading and also allows the technicians to adjust playback levels when setting up the tapes later for an edit.

6. *Provide a SLATE* (Figure 14–17). This will simplify locating particular shots for editing. The SLATE should contain the following information:
   a. Cassette or reel number
   b. Date
   c. Scene
   d. Take number
   By holding the SLATE in front of the TV camera during the leader preceding each shot, you can clearly identify the scene and take. If a SLATE seems too much bother, then at least call out the take numbers so that they are recorded before each shot. Simultaneously, hold your hand in front of the camera with one or more fingers showing to indicate the take number. This is called a FINGER SLATE. If there's no time to SLATE the beginning of a shot, then make a TAIL

MINIGLOSSARY

*Outtake   A shot which for some reason (i.e., a flubbed line) you don't plan to use in the final production.

*Slate   A visible and/or audible cue recorded at the beginning (or end) of a take, identifying the take number for later reference.

Finger slate   A slate made by holding one or more fingers in front of the camera at the beginning of a take.

Tail slate   A slate identifying a scene after it has been recorded (at the tail end of the scene).

FIGURE 14-17 SLATE used by professionals to mark the beginning of each "take."

SLATE at the end of the shot by turning the SLATE upside down (or holding your fingers upside down).

7. *Label your RAW FOOTAGE* so that it doesn't get lost. Mark it as CAMERA ORIGINAL or MASTER FOOTAGE or RAW FOOTAGE so that it doesn't get mixed up with copies which may look about the same.

8. *Remove the record lock buttons.* This assures that your precious RAW FOOTAGE doesn't get erased accidentally.

9. *Overlap lines between scenes.* This will give your actors a "running start" as they say their lines and will make their voices sound more natural when the shots are edited together. This also allows for split audio and video edits where you may edit from one picture to another but the sound (the actor's lines) is still continuing from the first shot. Nine times out of ten this overlap is not needed, but occasionally it comes in handy during fancy edits.

10. *Use many short tapes for your RAW FOOTAGE.* While making your edit decisions and performing your edits, you'll be constantly going back and forth from one recorded scene to another. If the footage for scene 6 is at the beginning of a 60-minute tape and the footage for scene 7 is at the end, you'll be twiddling your thumbs for about 4 minutes as the VCR shuttles from one scene to the other. A hundred edits like this can eat up a lot of time. If each cassette were only 20 minutes long, you would spend more time inserting and ejecting cassettes but much less time shuttling them.

TRANSITIONS

Transitions are the methods we use to get from this shot to that shot. Sometimes you can't get there from here and need in-between shots to smooth the way.

Jump Cut

If you mounted your camera on a tripod and shot an interview and then edited together parts of the interview, you would see the person's head magically "snap" from position to position with each edited sequence. It would be very obvious that the tape had been edited. Such JUMP CUTS (or SNAP CUTS or CAMERA MAGIC as they are called) are obtrusive and disconcerting to the viewer.

When a shot changes, there should be a reason for the change, like

1. A look at something new
2. A look at the interviewer
3. A look at the subject being talked about
4. A different camera angle
5. A closer or farther shot of the talent

By showing something from a different perspective from shot to shot, you provide the viewer with a "reason" for the change in scene, making the edit less intrusive. You also make your program more enjoyable to watch.

To review, *change the kind of shot when you change shots.* Don't go from a long shot to another long shot or from a medium shot to another medium shot. Change from a close-up to a medium shot and then back to a close-up or use some other varied combination of shots.

Besides changing shot size, also change the camera angle when cutting from shot to shot. Besides building a fuller perception of the subject, the change in angle smooths the transition further. Figure 14–18 reviews this concept.

There exist situations where CAMERA MAGIC is desirable. Imagine a scene where the princess kisses her teddy bear and it suddenly changes into a prince. It's an easy feat to perform if you have a tripod. First record your princess kissing the teddy bear. Have her freeze all action while you PAUSE your VCR (or perform your edit). While she remains motionless, you substitute the prince for the teddy bear. Call for "action" and then UNPAUSE. When played back, the prince will suddenly appear in the bear's place. For this to be convincing, you will need a steady tripod so that the camera doesn't move during PAUSE and create a telltale "snap" in the background during the edit.

JUMP CUTS are employed (on purpose) in "hidden camera" commercials such as "taste tests." The SNAP CUTS make the editing obvious but add an air of authenticity and credibility to the scenes.

Cutaways and Cover Shots

When you know ahead of time where you are going to make an edit, you can purposely change the camera angle and shot size to avoid a JUMP CUT. If, however, you don't know where the edit will occur, you have no way of knowing whether the shots will butt together well. Usually they don't,

DON'T   DO

JUMP CUT between long shots and close-ups without changing angle. With a cut, the viewer expects a substantial change in visual information but doesn't get it.

Change angle about 30° when cutting. Adds variety and smooths transition. Builds fuller perception of subject.

FIGURE 14-18   To avoid JUMP CUTS, change camera angle and closeness.

and you'll have JUMP CUTS to hide. The trick here is to use CUTAWAYS or COVER SHOTS.

The COVER SHOT might be a wide view of the talent speaking, one where you can't see their lips moving, or maybe a shot of the audience, or an interviewer, or an object being discussed. This way, you switch from the close shot of the talent to a shot of something else to another close shot of the talent. No one may realize that you edited out something; they may think that you are trying to show them *even more*.

Watch a newscast and notice how the camera jumps from the reporter to the things that he or she is talking about. Observe how presidential press conferences are interspersed with shots of the reporters and photographers. Those folks aren't trying to show you their pretty lights and cameras; they're covering their edits. These press club shots, audience shots, and long shots are called COVER SHOTS or CUTAWAYS and are included as VIDEO INSERTS.

Disparate Light Levels

Unless you are switching scenes from night to day where scene brightness is *meant* to change, keep the light levels in your picture the same. Illuminate scenes to an equal brightness so that the edit doesn't call attention to itself.

Walk-Past

Following people around as they walk from place to place is always hard to condense. One useful trick is to allow the performers to stride toward and past the camera (and out of the picture) and then to edit to another scene of them coming into view from alongside the camera. Also, performers may be allowed to turn corners or pass through doors, leaving the camera viewing an empty set. The next edit begins with another empty set with the performers entering a moment after the edit.

Condensing a long-distance drive into a few seconds is possible by having the vehicle start its journey in one location and drive into or over the viewer. EDIT-OUT (if you survive). The next edit starts with the vehicle going away from (or out from over) the viewer into the new location.

Blank Surface

Here, the angry wife exits a room, slamming the door in the viewer's face, leaving the blank surface of the door

DON'T DO

Change angle without changing shot
size. Twists performer without
apparent purpose.

Change shot size as you change angle
to add variety and interest.

FIGURE 14-18 CONT'D

filling the screen. EDIT-OUT at this point and start the next sequence on another blank surface, perhaps another door which later opens. Better yet, the second surface could be a sheet which disappears as the househusband pulls it from the line. Notice how this transition compresses time as it carries us from the scene of the wife leaving to the scene revealing her husband's plight.

To travel from one place to another, one can tilt up past the treetops into the blue sky and EDIT-OUT. Later, starting with the blue sky, one can tilt down past the tops of city buildings to a new scene.

Similar to the walk-past is the walk-through where the talent walks toward the camera, right into it, obliterating the picture. The next scene can begin with a camera close-up of the talent's back as it recedes.

Swish Pan

To show a move from one location to another (as if to say, "Meanwhile, across town . . ."), pan the camera rapidly to a blur. EDIT-OUT during this pan. EDIT-IN the next scene, starting with a fast pan that stops on the next subject that is to be viewed. The result looks like a hectic pan from one scene directly into another. SWISH PAN edits work beautifully only if they are planned. If the two adjoining scenes are not properly planned and the scene jumps from a pan or a zoom to a static shot, the result will look wretched. So if you can't plan your SWISH PAN edits to come together, avoid even the possibility of getting stuck in one by doing the following:

1. If panning or zooming a camera in a scene, come to a static shot before ending the scene.
2. Start scenes with a static shot before you pan or zoom.

This way all scenes begin or end with static shots and can be edited together in any order.

MINIGLOSSARY

Swish pan   Rapid sideways movement of camera as it goes from one scene to another causing the image to streak.

Still Frame

One transition that you see frequently in movies and TV is used at the end of a scene. The performer looks up and suddenly freezes. It takes special movie or video equipment to make glitch-free STILL FRAMES. If you are blessed with a time base corrector and a videocassette recorder which can make NOISELESS STILL FRAMES (ones which give you clean pictures every time), then you can perform this effect too.

If perfection isn't necessary, you can perform the STILL FRAME on home video equipment. Say you are editing from one tape onto another. To create the STILL FRAME at the end of a scene, merely press PAUSE on the VCP. Your VCR will record the result. If the STILL FRAME happens to be noisy or unstable, your VCR may end up recording useless mush. But if the VCP's STILL FRAME is rock solid, it may copy okay.

Fancier home VCRs and camcorders can make special effects, such as DIGITAL STILL FRAME, at the push of a button. These stills will look a lot better than the PAUSE-type STILL FRAMES mentioned earlier. Some home special-effects gizmos also can make DIGITAL STILL FRAMES. DIGITAL STILLS may copy, edit, and play okay through your TV, but are not always up to broadcast NTSC standards. Avoid these cheapie stills and PAUSES if doing professional work.

Here's an idea for easily working STILLS into your show:

Joe traveler (who won an all-expense-paid one-way trip to Secaucas, New Jersey) is describing his journey and showing his scenic video tapes. The opening shot shows him sitting next to his VCR and his TV. A STILL FRAME of an oil refinery is on the TV. He introduces the travelogue and turns his VCR to PLAY as he finishes his last line while you pan away from Joe and zoom in on the TV screen. Once you have a full shot of the TV, there's a tiny blip on the screen as the picture and sound become sharper. The scene could end with this process reversed, or you could simply cut to a medium shot of Joe switching his VCR to PAUSE (TV set showing in the background of the shot) and turning to your camera to continue his face-to-face discussion. This process makes for very easy transition, easy editing, and trouble-free STILL FRAMES.

What caused that blip and the sharper image? Instead of SCAN-CONVERTING *all* Joe's RAW FOOTAGE, you switched to a direct video (and audio if you wished) feed from his VCR. In other words, his VCP sent its signals to the TV *and also* to your VCR. Your VCR started the scene "listening" to your camera input, but when you reached your

full shot of the TV set, you switched the VCR to "listen" to his VCP directly. Thus, a better copy resulted. The blip is where you switched (and is avoidable if you use a time base corrector). This process also works in reverse, starting with a straight video dub and switching to your camera.

Defocus-Focus

One method of making a CUT look smoother while adding variety to your edits is to defocus at the end of a scene, make your edit at another defocused shot, and then have the next shot refocus. For example, imagine a close-up of a guitar player's hand becoming fuzzy and then magically refocusing into a singer's face.

Or how about a zoom in and defocus on a burning candle, a calendar, a clock, or a baby's bottle? After the edit, zoom out and back into focus with the image of a burnt out candle, an updated calendar, a reset clock, or an old man's wine bottle. Here, the defocus-focus implies the passage of time. It works a lot like the dissolve studied earlier, only it doesn't require a switcher fader or equipment for an AB ROLL. Granted, the dissolve is less obtrusive than this homemade version, but the defocus-focus can create the desired impression—*if not overdone.*

Two things to keep in mind when you apply this method are the following: First, it is far easier to get a close-up way out of focus than it is to get a medium or long shot out of focus. If you want to defocus easily, direct your attention to things you can get close-ups of. Second, use this method sparingly, or else the method won't add variety anymore.

Leading the Action

You are recording a child's birthday party with the usual bedlam. As you tape the proceedings, one child suddenly looks "stage right," and in a few moments all eyes are looking to the right. Now is the time to cut to Mom carrying the blazing birthday cake into the room. The children's looks made a perfect lead-in to your next shot. The viewer *expected* to see a new shot.

Seated performers shuffle preparing to stand. There is your excuse to cut to a long shot of them getting up and strolling off.

Somebody is holding a gem up to his eye. It's time for a close-up of the gem.

The runner lifts her hips in preparation for the gun. It's time for a long shot of the takeoff.

The switch from a medium shot to a long shot of a performer indicates that the performer is about to move or be joined by someone. Cutting to a close-up of his face

MINIGLOSSARY

Defocus-focus A transition from one shot to another by defocusing the first shot, editing (or switching cameras), and following with another defocused shot which then comes into focus.

Digital still frame Electronic method of "grabbing" a still picture on a camcorder or from a tape playing on a VCR.

readies the audience to catch his expression. A gesture will be expected if you now switch to a medium shot. And a shot of the door prepares the viewers for an entry.

In each of these cases the action prepares the viewer for a shot change. The transition from one shot to another becomes natural and comfortable to view. Making any of these shot changes without a specific purpose will not add variety to your show; it will only confuse your viewers. Therefore, change shots for a purpose, not for idle variety.

By building the preceding transitions into your editing strategy, your scenes will introduce each other and flow together.

180° Rule

You almost never shoot any subject from opposite sides. The opposing shots can easily confuse the audience because what was moving left in one shot is suddenly moving right in the other.

Review the section on transitions in Chapter 11 and have another look at Figure 11–39. These remind us about the imaginary VECTOR LINE and how we can't cross it without disorienting our audience.

This mistake is very easy to do when you are shooting scenes hours, days, or weeks apart. If you forget how the flow of the action was moving, you have a 50:50 chance of starting your next scene with the flow going in the wrong direction. If you shot a patient in a hospital bed from one side this week and next week shot some more of him from the opposite side, you'll be unpleasantly surprised when you try to edit these scenes together. To avoid these problems, take notes and draw pictures showing your camera angles and setups. This will assure smooth continuity from scene to scene. And now, more on continuity.

Continuity

Al sits down at the table, picks up his knife and fork, and begins to cut his asparagus. "Cut," you call out, and he feverishly speeds up his sawing until he realizes that the scene is over. You continue the scene with the camera in another position for a close-up of his plate. "Okay, roll," you call out, and when Al realizes you are not calling for his buns, he again picks up his fork and dips into his mashed potatoes. Somebody is in for a surprise when editing time comes—somebody who didn't take notes or didn't examine the tail of the preceding scene to assure matched shots. How many viewers will you entertain with this little slipup?

One way to call the least attention to the exact position of something is to change shots while it is moving. *Cutting on the action* (as it is called) also makes the movement part of *both* shots, fusing the continuity between them.

For example, someone is about to walk out through the door. You shoot the scene first from the inside, hitting PAUSE *as the door begins to open.* Have the talent memorize which hand was on the knob and which foot was forward. Now run outside, compare your door shot, and have the talent repeat the process of opening the door, hands and feet equivalently placed. UNPAUSE as the door opens.

If recording RAW FOOTAGE for later editing, the process is similar. Instead of PAUSING, let the talent complete the exit. Have the talent repeat the process while you shoot it again from the outside angle. Start your tape even before your talent begins to approach the door. During editing, you will now be able to carefully control exactly where you switch from one scene to the next (before the door opens, as it opens, or as she passed through). As long as the action is matched in both "takes," you should have a number of choices of where to make the switch from the indoor view to the outdoor.

Cutting on the action is a science. If you know you will have to edit scenes, contrive action to cut on. Have your actor turn to a blackboard, sit on the edge of the desk, or simply turn to address a new camera angle. If preparing RAW FOOTAGE, remember to repeat one scene's ending action at the beginning of the next scene, overlapping the action. Pay attention to detail—hand, foot, body, and face positions. Maintain screen direction; someone exiting the screen to the left should be moving to the left in the next shot.

As mentioned earlier, each new shot should change *both* the closeness of the shot (image size) and the angle of the shot. This change in size and angle will help cover minor flaws in matching the action in the two sequences. And last, feel free to rehearse an action a few times before recording it. It will help everybody relax and perform smoothly.

This rule of cutting on the action may seem contrary to the earlier dictum of editing on pauses. Let me clarify: If you are doing STOP edits or PAUSE edits or recording a passage over using home video equipment, your edits are likely to have a substantial glitch. If you are editing without the precision of an EDITOR CONTROLLER or doing "live" INSERT EDITS, the edits will not be very accurate. You don't want to lose an important part of your show during a glitch, so you put this messy edit where it will hurt the least, at a pause in the conversation or action. If, however, you are using state-of-the-art professional editing equipment, you don't lose much during the edit, affording you the luxury of applying some of the higher laws of dramatic editing. In short, cutting on the action is a more refined technique requiring accurate editing equipment.

Pacing

Just as camera angles can create a mood, so can the way you edit a program together. Action and excitement call for quick short edits usually showing close-ups or motion. To heighten the excitement further, you can speed the tempo of edits to a staccato of brief scenes, some of which are on the screen hardly long enough to be comprehended. Study the action scenes from *Jaws*, *Psycho*, and *Star Wars*, and you'll see the shots changing at a frenzied pace.

Calm, relaxing love scenes, travelogues, suspense, and drama are edited into longer, more leisurely scenes. Often the change from one scene to another may be a dissolve rather than an abrupt cut.

The beauty of editing is that it leaves an unconscious message. The viewers think that they are only seeing what you are showing them. They don't realize that they are also seeing *how* you are showing them. As your program progresses, you can bring your audience to the edge of their seats and then relax them back down just by varying the pace of your edits. Incidentally, it's good not to maintain the same pace for too long. A sustained slow pace may put your audience to sleep, yet sustained fast pace may jangle their nerves and they will become desensitized. The technique here is to grab the audience by the throat to get their attention, shake them a few times, and then ease them back into their seats and let them breathe.

WHEN YOU'RE FINISHED EDITING

It's 4 a.m. Your masterpiece is finished. Amid a heap of candy wrappers and coffee cups, you stand up for the first time in 12 hours. It's time to call it a night, maybe have a drink, or maybe a nervous breakdown, right?

Oh, no, SMPTE breath!

Have you labeled your EDITED MASTER so some joker doesn't come into the control room tomorrow morning and record a Bugs Bunny cartoon over your precious program? TV facilities are filled with doodleheads who think that any program which isn't theirs is expendable. Label that tape! Put it in a place safe from doodleheads.

Is it possible that your editing VCR is sick but doesn't know it? It can be playing *your* tape back okay tonight (tapes generally play well on ''mother,'' even when she's crazy) but won't play it tomorrow when the TV technician finishes ''fixing'' it. Copy your EDITED MASTER *now*, using the editing VCR to do the *playing* and some other VCR to make the copy. This guarantees that one perfect playback of your tape will occur before something happens to ''mother,'' the VCR who can play your tape best. Besides, even after all those hours, could you sleep knowing that there is only *one copy* of your masterpiece in existence? A second copy kept in another place is excellent insurance and aids sleep better than Sominex.

COMMON VIDEO TAPE EDITING PROBLEMS AND CURES

Many of the headaches faced in editing are the same ones found in simple tape copying, and they stem not from muscle tension but from maintaining a stable clean signal from the VCP. Every flaw in the playback of the RAW FOOTAGE gets compounded in the copy (with the exception of a few problems which can be fixed by the gizmos in Chapter 15).

Practices to Avoid While Editing

1. Don't edit near glitches, blips, video breakup, splices, or bad edits on the RAW FOOTAGE. These aberrations can foul up the speed of the editing VCR so that *its* edit comes out imperfect.

2. Don't change TRACKING or SKEW *on the editing VCR* from edit to edit. Once these controls are set optimally, leave them throughout the editing session. On the VCP, just the opposite holds true. This machine should be optimized for each tape it plays in order to provide the best signal possible.

3. Keep your tapes clean. Leaving unboxed cassettes stacked in disarray around the editing console is not the way to keep dust out of your RAW FOOTAGE.

Other Glitches and Gremlins

1. *Editing VCR refuses to play at the right speed. Speed wanders.* TRACKING *is unstable.* When in the editing mode, the VCR is EXTERNALLY LOCKED; that is, it ''listens'' to the signal from the VCP and uses this signal to guide its motors. If the VCP is turned off, disconnected from the editor, or turned to STOP, it is not sending the signal for the editing signal to ''listen'' to. In most cases, the VCR can disregard this fact and can run satisfactorily. In some cases, it won't. When the VCP is on PAUSE, FAST FORWARD, or REWIND, the signals the VCP sends out cannot be ignored by the VCR, and they drive the VCR crazy because they are not proper signals.

To make the editing VCR happy while it is playing, feed it either a good signal (from a switcher, a camera, or a VCP which is playing a tape) or no signal (turn the VCP to STOP or turn its power off) or switch the VCR to an INTERNALLY LOCKED mode (perhaps to NORMAL if it has such a mode).

If the VCR plays poorly because the VCP is stopped, etc., take note that this will not affect the VCR's ability to make a good edit once the VCP is again running. If you can tolerate the unstable picture while you are preparing for an edit (playing the VCR looking for the edit point while the VCP is also being started and stopped in the process of its edit point being located), then don't worry about it. When both machines are simultaneously started pursuant to the actual edit, the VCR will run smoothly (because it is again locking onto a good signal from the VCP).

When running your VCRs and VCPs through a switcher or video console, it is sometimes possible to accidentally feed a VCR its own output. In other words, the VCR may feed the switcher a signal, and the switcher is sending that signal back into the VCR's input. Such an incestuous loop confuses the VCR, making it play with TRACKING and speed difficulties.

2. *Insert edits don't track correctly: picture plays at wrong speed or tears.* INSERT and VIDEO INSERT ONLY edits must be made over an existing video signal on the tape. Did you forget to record one? And when you recorded that video signal, was it with the VCR in the ASSEMBLE or NORMAL mode as it should have been? Did you use a kosher signal when you recorded that video? Did it have sync?

When INSERTS are done over blank or improperly recorded tape, TRACKING goes to pieces. One way to check the tape for proper video before making INSERTS over it is to play the tape (with the VCR INTERNALLY LOCKED or receiving stable video to it) and watch for a TRACKING problem. If the tape plays a smooth, clean, clear picture (even if it is a picture of black) with no TRACKING problems, it is a good foundation for INSERT EDITING. If the picture plays snow, the tape is blank. If the picture displays an uncorrectable TRACKING error, something is probably wrong with the video signal recorded, or the VCR was accidentally left on INSERT during the recording of this base video.

3. *The end of each INSERT edit looks bad.* INSERT edits are terminated at the press of an END INSERT or an EDIT or some other specialized button (check your instruction manual), generally not by switching to STOP. Switching to STOP amidst an INSERT edit may ruin the end of the edit.

4. *RAW FOOTAGE is of such poor quality that the edit comes out badly or the copy comes out unplayable.* Sometimes a tape may have so many problems that it won't copy, or the copy won't be playable. Still the picture may look fairly good on a TV screen when you play the original. Possible explanation: Perhaps the sync—the part you can't see—is bad even though the picture is good.

Try running the bad footage through a PROC AMP or a TIME BASE CORRECTOR (described later), or try a SCAN CONVERSION. You may find that some TVs may play the picture nicely, while on other TVs the images jitter, tear, or utterly collapse. Find a forgiving TV which shows a stable picture for making your SCAN CONVERSION. The tape which results will have excellent sync, good stability, fair-to-poor contrast, and fair sharpness. Despite its faults, SCAN CONVERSION is a way to edit sequences which otherwise would be unusable.

5. *Edited tape plays poorly on other machines, edits look bad, tape tracking is poor.* Moderately priced editors are seldom perfect—each has a "personality." A tape edited on an editor may play back fine on that editor now but perhaps not a year from now. Nor may that tape always play back well on another player. This problem may be avoided only with the constant care of a video technician who sees to it that all equipment is "tuned up" to specifications all the time.

A way around the problem: When you finish editing a tape, make copies of it immediately *using the editor as the video tape player.* Similarly, when copies must be made of a tape edited earlier, try to find the machine the tape was edited on; then play the tape back from that machine for best results.

OTHER TV PRODUCTION GIZMOS

It has been my experience that for every dollar spent on TV production equipment (cameras, VCRs), another 50 cents eventually is spent on signal testing and signal correcting equipment. After we've checked on our signal and made it perfect, we still may have to send it somewhere, which will cost us a few more ducats for distribution equipment. Then there are "other" accessories, the doodads and widgets we discover we can't live without after we've spent our last cent on cameras and VCRs.

VIDEO MONITORING EQUIPMENT

You might think that a TV set was all you needed to gauge whether you were producing a good picture. Indeed, a glance at a video monitor will tell you a lot about the signal you are recording or sending out. The problem is that TV monitors might not show you a problem when there is one or might not tell you enough about that problem for you to fix it. Many things can go wrong with a video signal on its journey from the camera through the switcher to the VCR and beyond. Video test equipment will save you from the inclination to dim your studio lights because they "seem" too bright, then open your camera lens iris all the way because your viewfinder picture "looks" too dark, then turn down the video level control on your VCR because the meter is reading in the red, and then crank up the contrast on your

TV monitor so the picture "looks good." When measuring your video signal by eye, you're tempted to boost and reduce, boost and reduce. Nearly every piece of video equipment has the ability of boosting or reducing the brightness, contrast, or colors of your TV signal. Just how much boosting or reducing will make an "excellent" TV picture could be anybody's guess—that is, if you are guessing. The trick is to take the guesswork out of the boosting/reducing game and find some instrument that will tell you the *truth* about your video signal.

Underscanned Monitor

The perfect TV monitor for viewers to watch is one which will make the colors look right even when they aren't and will make the picture stable even when it isn't. It will have automatic controls to make the blacks look black and the whites look white. It will cut off the outside edge of the picture so that the viewers don't accidentally see a sync bar, color reference signals, or jitter in the corners.

This is *not* the picture you, the TV person, want to see. You want to see the truth. If the signal is bad, the monitor should show this fact and not cover it up. Special high-performance monitors are made to do this, having circuits that display every little signal flaw on the screen. Such a monitor may have a "long" or "slow" TIME CONSTANT, which means that if your VCR's SKEW is misadjusted, the

452

Note the bend at the top indicating that the VCP's TAPE TENSION control is turned too far to the right.

FIGURE 15-1 Monitor displaying an UNDERSCANNED picture.

top of the picture will FLAGWAVE back and forth. If your video signal is unstable or jitters, your TV picture may twist, squirm, or roll.

Some TV monitors have a switch which allows them to either UNDERSCAN a picture or work normally. In the UNDERSCAN mode, the edges of the picture are visible, making it possible to spot playback flaws such as FLAGWAVING.

Figure 5–23 showed FLAGWAVING which was bad enough to show up on a regular TV set. Figure 15–1 shows minor FLAGWAVING at the screen's edge, a flaw which would not show on a regular TV. It is still a problem that you'd want to catch and correct before it got worse or got copied onto another tape.

The UNDERSCANNED monitor is connected up like any other video monitor. It could make a handy VCR or VCP monitor as shown in Figure 15–2 (once you got used to the smaller TV picture), or you could make your control room preview monitor UNDERSCANNED. This way you could check around the edges of *any* video signal at the push of a button.

Many normal TV monitors can be modified to UNDERSCAN with just a few tender tweaks from your TV technician.

Pulse-Cross Monitor

The PULSE-CROSS monitor (or CROSS-PULSE monitor, as some call it) displays the edge and bottom of the TV picture in the middle of the screen as shown in Figure 15–3. This makes it even easier to study sync, tracking, and tape tension but blows the heck out of trying to watch the TV picture.

Some PULSE-CROSS monitors have a switch which allows the monitors to be used normally or in the PULSE-CROSS mode. The switch simultaneously adjusts the brightness and other circuits to make viewing the sync easier.

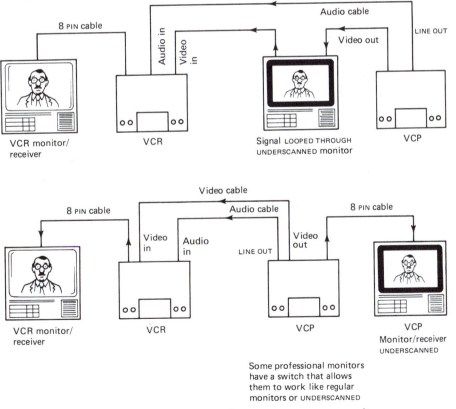

FIGURE 15-2 Possible connections of an UNDERSCANNED monitor.

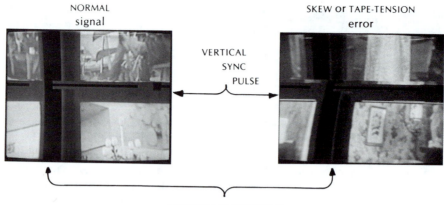

NORMAL signal

SKEW or TAPE-TENSION error

VERTICAL SYNC PULSE

HORIZONTAL SYNC PULSE

FIGURE 15-3 PULSE-CROSS monitor.

There are PULSE-CROSS converter boxes which can be added to existing monitors which will make them display PULSE-CROSS images. Although they save you from buying a separate monitor, they are somewhat inconvenient to use and take up extra space.

PULSE-CROSS monitors are connected into your system just like any other TV monitor. Since they are so handy in displaying how your VCP is behaving, a good place to connect one would be between the VCP and the VCR as shown in Figure 15–4.

Another good place to wire in high-performance PULSE-CROSS monitors is the PREVIEW circuit of your production console. By pressing a button on the PREVIEW bus of the switcher, you can route any signal to the PULSE-CROSS/PRE-VIEW monitor for examination. This way, signals can be evaluated on your special monitor before they are used.

By feeding the video outputs of your VTRs back into the console, you can select one of those outputs to PREVIEW during a recording, thus evaluating on your fancy PULSE-CROSS/PREVIEW monitor the signal the way the VTR sees it.

Waveform Monitor

Even though you are not an engineer, you'll want to keep a screwdriver and a pair of pliers around the studio. They're so handy. Make the third tool you buy a switchable PULSE-CROSS monitor. Make the fourth tool a WAVEFORM monitor. These devices can tell you a whole lot about your TV signal, even if you know nothing about electronics. In the hands of a real engineer, these devices are a necessity. Without them, the engineer is flying blind.

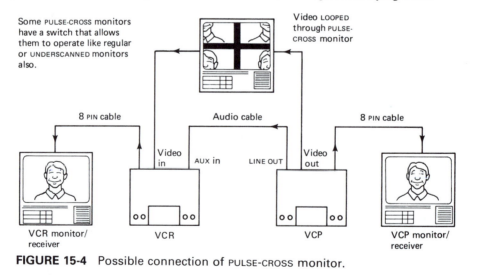

Some PULSE-CROSS monitors have a switch that allows them to operate like regular or UNDERSCANNED monitors also.

Video LOOPED through PULSE-CROSS monitor

8 PIN cable   Audio cable   8 PIN cable

Video in   AUX in   LINE OUT   Video out

VCR monitor/ receiver   VCR   VCP   VCP monitor/ receiver

FIGURE 15-4 Possible connection of PULSE-CROSS monitor.

MINIGLOSSARY

Waveform   A graphic representation of a video signal, showing signal levels (whites and blacks), color, and timing (sync).

*Waveform monitor   A specialized oscilloscope for displaying video signal levels and timing.

IRE   Institute of Radio Engineers, is a measure of video level or ''whiteness'' and is marked off in units of 10 on waveform monitors. A 20 IRE level represents a dark part of a TV picture and 80 IRE, a light part.

Reading the Waveform Monitor. The WAVE-FORM monitor is read like a graph using the little numbers and the scale etched on the screen. The scope displays the picture and sync in terms of signal level. Strong signals are high on the scale; weak signals are low on the scale. The white parts of the picture are the strong parts of the video signal and should appear high on the scale. The dark parts of the picture are weak signals and should appear low on the scale. When creating a picture, one strives for white whites, almost black blacks, and a nice range of grays in between.

You'll notice that the screen of the scope is covered with a graticle, a template with lines and numbers scribed on it in units of 10 from −40 to +100. The units are IRE (which stands for Institute of Radio Engineers) and can be thought of as a percent. 0 on the scale means "no video," and 100 on the scale means "100% white." Dark gray parts of the picture may be at 30 on the scale (30% white), while light gray parts may be at 60 (60% white). Black, according to proper specifications, is at 7.5 on the scale. A little flat-bottomed canyon in the middle shows the horizontal sync pulse, a part of the video signal that holds the picture steady. That canyon should start at 0 and dip straight to −40 for a ways and then jump back up to 0. A proper video signal will have its sync at −40, its blacks at 7.5, and its whites at 100 and have nice sharp peaks and valleys as shown in Figure 15–5.

Using the waveform to adjust cameras, etc.: Use the WAVEFORM monitor constantly when setting the camera

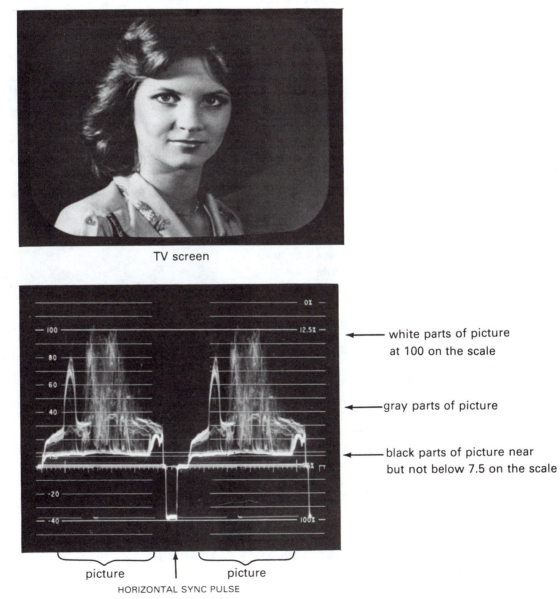

TV screen

white parts of picture at 100 on the scale

gray parts of picture

black parts of picture near but not below 7.5 on the scale

picture   HORIZONTAL SYNC PULSE   picture

FIGURE 15-5 WAVEFORM of a well-adjusted video signal (shown on the WAVEFORM monitor's 2-H scale).

picture controls prior to shooting. First aim your camera at a typical well-lit scene, probably a long shot, something with the full range of brightness in it. Next, switch the WAVEFORM monitor to the HORIZONTAL or 2–H scale, so it shows you the whole picture from left to right (actually, it shows you the picture twice).

The whitest part of the camera's picture should appear on the WAVEFORM as little mountain peaks and should touch the 100 line. If they are too low, the camera's GAIN should be increased until the peaks do touch 100. If no GAIN control is available (such as on cameras with automatic gain controls), open the lens iris (to low f-numbers) or increase the lighting intensity.

If the peaks rise above 100, GAIN should be decreased, as signals over 100 are too strong to be satisfactorily handled by the electronic equipment elsewhere in your video system. *Note*: This setting is appropriate for a standard, well-lit scene, perhaps a typical long shot. Make your camera adjustments for these shots, not the unusual ones which perhaps have no blacks or whites in them. *Note also*: An occasional peak over 100 on the scale probably won't hurt anything. Neither will having the peaks at, say, 80 during part of a performance hurt anything appreciably.

If you plan to MATTE white lettering on a screen, you may wish to adjust the peak brightness of the scene to 80 on the WAVEFORM scale and have the lettering appear as 100 on the scale. This way the white lettering will easily stand out and not get absorbed in the white parts of the scene.

If these little mountain peaks all seem to have flat tops to them at about the same height as in Figure 15–6, this tells you that your camera's internal BEAM circuit is misadjusted. If the peaks appear to be CLIPPING (flattening out) near 100 on the scale, excessive light is probably your problem, or maybe your lens is open too far. If they are clipping some distance below 100 on the scale, there's probably a circuit problem in your camera or amplifier somewhere. The result is a TV picture that looks chalky and lacks highlights.

Sometimes the clipped peak problem results from sending too strong a TV signal to a device, perhaps to a VTR or PROC AMP, which in turn must have its gain adjusted way down to reduce the signal. It's good to check the video levels with a WAVEFORM monitor at both the input and output of a device to ascertain whether the device is receiving and transmitting a proper signal.

The blackest parts of your picture should appear on the WAVEFORM as sharp ravines, each coming to a point. These points should hit 7.5 on the scale as the blackest black acceptable in the picture. If these valleys reside above 7.5, the PEDESTAL (or SETUP or BLACK LEVEL) control on the camera should be turned lower (this may mess up the whites, which will then have to be readjusted). If the crevices dip lower than 7.5 on the scale, raise the PEDESTAL appropriately.

TV screen with chalky picture

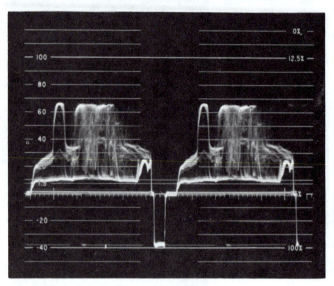

WAVEFORM of a camera's signal with misadjusted BEAM. Note plateaus rather than sharp peaks.

FIGURE 15-6 Camera circuit misadjustment may cause this malady.

If these ravines are allowed to drop past 0 on the scale, they will appear blunt, like flat-bottomed canyons. Under these conditions, the dark parts of the picture merge together. See Figures 15–7 and 15–8.

More light on a subject, lower f-stop numbers, and higher adjustments of the camera's GAIN controls will increase the contrast of a picture. So will increasing the VTR's VIDEO LEVEL control during recording. And if you are running the signal through a PROC AMP or a TBC (described shortly) with such controls, their increased gain will result in a picture with improved contrast.

Figure 15–9 shows too much GAIN in a TV picture. Remember, turning down your TV monitor's contrast control may make this picture look better to you, but it doesn't

TV screen light

TV screen dark, shadowy, muddy

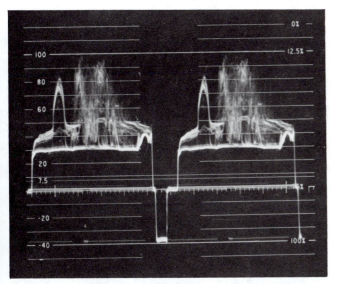

WAVEFORM showing PEDESTAL too high. Notice there are grays and whites, but no blacks.

FIGURE 15-7 PEDESTAL too high.

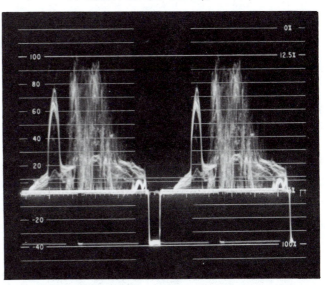

WAVEFORM showing PEDESTAL too low. Valleys hit zero on the scale.

FIGURE 15-8 PEDESTAL too low.

fix the problem. Your trusty WAVEFORM monitor will tell the tale with its peaks rising above the 100 limit on the scale.

GAIN can be decreased by dimming the studio lights, closing the camera iris, reducing GAIN on the camera's CCU, or reducing the VIDEO GAIN controls on any other video gadgets in the circuit. The best thing to do is to test the camera output and make sure *its* WAVEFORM doesn't exceed 100 IRE and then test the other video devices in the circuit, moving downstream one step at a time, to make sure that they are passing a proper signal too.

If your WAVEFORM monitor tells you your video signal is satisfactory but it still looks bad on your TV set, *then* adjust the TV's contrast control to improve the picture.

Figure 15–10 shows the result of too little light on the subject, too high an f-stop number, camera GAIN set too low, or the VCR's VIDEO LEVEL set too low. The picture will look faded, and the WAVEFORM will lack high enough peaks. *Note*: The picture may not look this bad on all TV monitors because some have automatic circuits built in to spruce up bad signals. The picture you see in Figure 15–10 is similar to the one you'll see on a cheap black-and-white TV set lacking automatic controls or a superexpensive control room monitor which shows the truth, warts and all.

In short, VIDEO LEVEL or GAIN stretches the tops and bottoms of the WAVEFORM, increasing contrast. PEDESTAL, on the other hand, raises or lowers the whole works on the scale, making everything lighter or darker. Once the whites

TV screen contrasty

TV screen faded

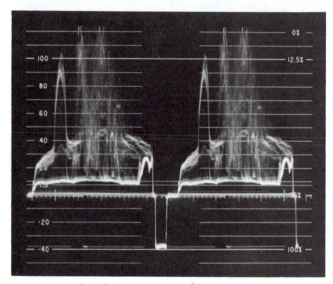

WAVEFORM showing GAIN too high. Peaks exceed 100 on the scale.

FIGURE 15-9   Camera GAIN or VIDEO LEVEL too high.

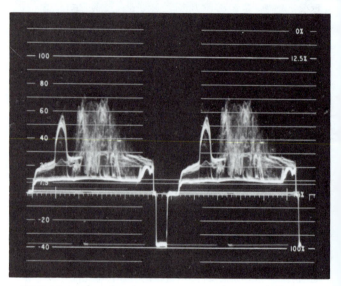

WAVEFORM showing low VIDEO LEVEL (or GAIN). Peaks barely reach 80 on the scale.

FIGURE 15-10   VIDEO LEVEL too low.

are white and the blacks are black, the grays generally take care of themselves.

If you are using three cameras, this entire process has to be repeated with *each* camera. If all cameras are "tuned up" properly, their pictures will match as you switch from one to another. If the cameras are not properly calibrated, then as you switch from one to another, you'll notice that one scene will appear too light, the next too contrasty, and perhaps the next one faded. So it is always good to view each of your camera's signals one by one through your WAVEFORM monitor in order to match levels before you begin your production.

The same goes for editing. Raw footage recorded on different VCRs using different cameras may have different video levels recorded. Before editing your tapes together,

play samples of these rushes through your WAVEFORM monitor to make sure they are kosher.

This is only a taste of some of the secrets a WAVEFORM monitor can show us about our TV pictures. A little later we will come back to the WAVEFORM monitor to discuss the sync and BURST parts of the WAVEFORM.

Operating the Waveform Monitor.   Figure 15-11 shows the controls on a WAVEFORM monitor. Although each manufacturer may do things a little differently, WAVEFORM monitor controls are basically the same. Many of them can be set once and never touched again. Others are useful to the technician when he or she is hunting down elusive video gremlins and are of little interest to the nontechie. Table 15-1 charts the WAVEFORM monitor's controls, what

TABLE 15–1

What the WAVEFORM monitor controls do

Control Name	What It Does	Normal Setting
Response	In the IEEE position, the monitor displays only black-and-white components of the signal. The color gets filtered out. In the chroma position, you see only the color in the signal, not the black-and-white (luminance) part. In flat, you see everything, color and luminance both.	Flat. If using a black-and-white system or if you desire a sharper-looking waveform showing brightness levels, you could also switch to IEEE.
Gain	Adjusts the sensitivity of the device during calibration.	Leave adjustment to technician.
Input select	This waveform monitor has two inputs, A & B. In the A position, the machine listens to its A input, and B listens to B. Here, we've relabeled the position to show what signals we've hooked up to those inputs, the switcher's line output, and the switcher's preview output. Technicians often will switch to CAL (for calibrate) to observe how the machine displays a preselected signal.	Select the input you wish to view.
Position	The left position knob moves the display up and down on the screen. The twin position knob on the right moves the display from side to side.	Adjust display vertically so that the flat plateau just before the sync pulse goes along the zero line. Center the display horizontally so the sync pulse is in the middle of the zero line.
DC restorer	*On:* Holds the display signal steady vertically despite fluctuations in the video signal levels. *Off:* Allows the display to "float," rising and falling with changing dc voltage in the signal.	On
ASTIG	Corrects any slight tilt in the display.	Leave adjustment to technician.
Power/fuse	Turns device on. Has a fuse built inside the switch.	On when in use.
Scale light	Illuminates the graticule so you can see the scale and numbers better.	Wherever comfortable, probably midrange.
Brightness	Makes the green waveform brighter and easier to see.	Wherever comfortable for viewing, probably midrange.
Focus	Makes the green display sharper.	Turn to where the green lines are sharpest.
Sync	*Int:* The scope synchronizes its display to the incoming video. Only one wire carrying a video signal is necessary to feed the device. *Ext:* A separate, external sync source stabilizes the display. Two wires, video, and sync are needed to feed the device.	Int
Sweep	Determines whether the TV picture will be studied from left to right or from top to bottom. *2-H:* Displays the picture from left to right twice on the waveform screen with the horizontal sync pulse in the middle. *2-V:* Displays the picture from top to bottom twice with the vertical sync pulse in the middle. *Mag:* Magnifies either display to make small imperfections easier to see. Notice that in Figure 15–11 the monitor was switched to MAG to observe the sync pulse in detail.	2-H

they do, and where they should be set to achieve the measurements shown in the preceding figures.

Connecting Up the Waveform Monitor.
The WAVEFORM monitor connects to your video system the same as a normal TV monitor would but with one wrinkle. The waveform monitor generally has two inputs, marked A and

B. Each can be looped (and therefore must be terminated if not looped).

If you are using your WAVEFORM monitor to see how your VTRs are operating, then you may wish to connect the scope as shown in Figure 15–12. With the scope's INPUT SELECTOR in position 1, it will display the signal coming from the player. Flipping the switch to the second position

FIGURE 15-11   Controls on a WAVEFORM monitor.

shows how the recorder is handling the signal. If, for instance, the channel 2 display (from the recorder) is bad but the channel 1 WAVEFORM looks good, it is likely that your player is all right but your recorder has run amuck.

It is common in TV studios to connect the switcher's PROGRAM OUTPUT to one channel of the WAVEFORM monitor (perhaps looping it from there to the LINE MONITOR) and to connect the switcher's PREVIEW OUTPUT to the scope's other channel (perhaps looping it to the PREVIEW MONITOR). Thus, anything connected to the switcher can instantly be viewed on the WAVEFORM monitor by punching the right switcher button. Your WAVEFORM monitor, though capable of displaying CHROMINANCE (color) and LUMINANCE (brightness) aspects of your signal, is used primarily to measure the LUMINANCE aspects. The VECTORSCOPE, on the other hand, is used primarily to measure CHROMINANCE.

Vectorscope

Say you were getting ready to record a show and wanted to know if your equipment was faithfully reproducing the proper colors. If a face looked too red, would you adjust the light, adjust the red gain on the camera's CCU, turn the color temperature wheel in the camera, tweak the camera's phase, or simply adjust the hue control on your TV monitor? Would you ask someone else's opinion of the color, adjusting it to suit the majority of observers? What would you do if you dressed for work as I do in a green tie, blue shirt, brown jacket, chocolate pants, and white shoes? Do you think anyone trusts my video color judgment?

Even if you are colorblind, it is possible for you to adjust color cameras perfectly. All it takes is a TEST SIGNAL and a VECTORSCOPE.

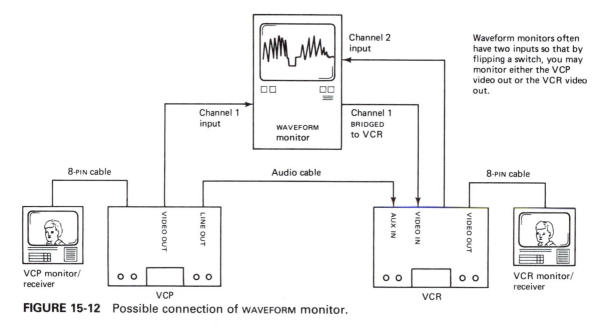

FIGURE 15-12   Possible connection of WAVEFORM monitor.

MINIGLOSSARY

*Test signal**   A video (or other) signal containing certain properties (like color bars) to show whether equipment is working as it should. Test signals meet certain technical specifications useful for calibrating other equipment.

*Vectorscope**   Specialized oscilloscope that graphically displays the color parts of a video signal, precisely showing the colors' strength and hue.

The VECTORSCOPE, like the WAVEFORM MONITOR, samples a little of the video signal and displays components of that signal graphically on a screen. It is generally used before a TV production during the camera setup and adjustment and then left alone for the rest of the show.

Reading the Vectorscope. Figure 15–13 shows a VECTORSCOPE. Figure 15–14 shows the VECTORSCOPE display of COLOR BARS. Figure 15–15 shows COLOR BARS as they'd appear on a black-and-white TV monitor and a WAVEFORM monitor. The COLOR BAR test signal and the VECTORSCOPE are intimately related as you'll see shortly.

COLOR BARS can be created by aiming a TV camera at a TEST CHART, or by switching the camera to the BARS mode (if the camera is equipped with an internal TEST SIGNAL GENERATOR), or by feeding the signal from a TEST SIGNAL GENERATOR to your video system. COLOR BARS consist of seven (sometimes eight) vertical bars of pure white, yellow, cyan (blue-green), green, magenta (reddish purple), red, and blue (and if there are eight bars, a black one is added to the end). The screen of the VECTORSCOPE has little boxes marked Y1, Cy, G, Mg, R, and B, representing these colors (except for the black and white). If a glowing dot falls inside the small crosshatched box marked G, that means

FIGURE 15-13 VECTORSCOPE.

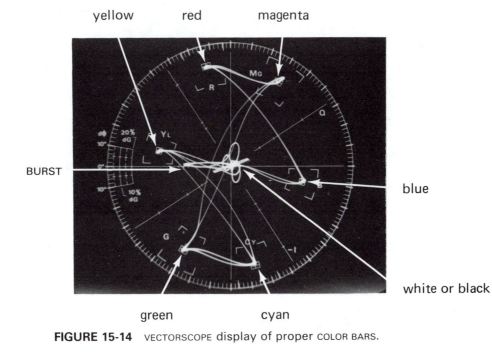

FIGURE 15-14 VECTORSCOPE display of proper COLOR BARS.

that the color is pure green to within 1 percent perfect. If the dot falls somewhere within the larger box outside it, that means the color is green give or take up to 10 percent. The color could also be up to 20 percent too pale or too saturated and still make a dot inside the larger box. Six perfect COLOR BARS should make six dots on the screen, filling the little boxes (plus a few streaks made by the electronic beam tracing from box to box). The splotch of light at the center of the VECTORSCOPE screen represents white (and black) in the picture. The white line that shoots straight left of center represents COLOR BURST, an electronic reference signal from which all the other color signals are measured.

If the six dots of light were further from the center than their boxes, that would mean the colors were too strong. If the six dots didn't reach out far enough, it would mean the colors were weak and needed boosting. If all the dots were rotated clockwise on the screen from their respective boxes, it would represent a total shift in hue. If the blob in the center were skewed off center, that would mean that black or white parts of the picture were contaminated with color.

If you sent a regular TV picture from a camera to the VECTORSCOPE, you would see a mishmosh of colored splotches. They would mean nothing in particular, although if a large red object were on the TV screen, you would see a blob near the R box on the VECTORSCOPE display. The VECTORSCOPE is primarily used with the color TEST SIGNAL in order to calibrate the color circuits in your video system. Put another way, when your VCR or camera is putting out a signal which it thinks is blue, your VECTORSCOPE will verify that the color is indeed blue, not reddish blue, not greenish blue, not pastel blue, and not superstrong cruddy-looking blue. Notice how the VECTORSCOPE eliminates all the guesswork. If the dot is in the box, it's blue—no question about it.

Using the vectorscope to adjust camera color: For this you will need a COLOR BAR TEST CHART or MUNSEL COLOR CHIP CHART like the one shown in Figure 15–16. It is a poster of carefully reproduced colored bars. By aiming your camera at the poster, you can create a COLOR BAR display. Before doing any tests, though, make sure the chart is illuminated adequately. Select the proper COLOR TEMPERATURE filter and perform a WHITE BALANCE. Also, adjust the IRIS and GAIN and make sure the video level stretches from 7.5 to 100 IRE on the WAVEFORM monitor. Next, send the camera's signal to the VECTORSCOPE for the color test.

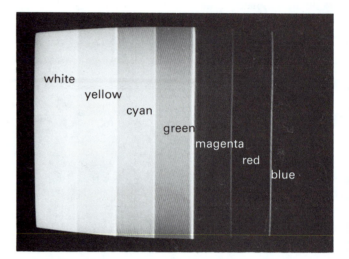

white
yellow
cyan
green
magenta
red
blue

. . . as they'd appear on a black-and-white TV screen

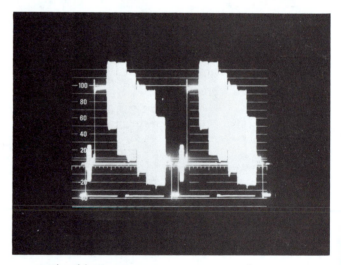

. . . as they'd appear on a WAVEFORM monitor

FIGURE 15-15   COLOR BARS (see also back cover).

MINIGLOSSARY

*Burst   Part of the sync signal controlling the hue and color accuracy of TV pictures. It is a reference signal used by TVs and other video equipment as the benchmark for what all the hues should be.

Test signal generator   Electrical device that makes test signals for calibrating and measuring performance of equipment.

Color bar generator   Electronic device to create color bars for use as a test signal.

Color bar test chart   A carefully prepared poster containing vertical colored bars used for camera testing.

Munsel color chip chart   A particular brand of color bar test chart.

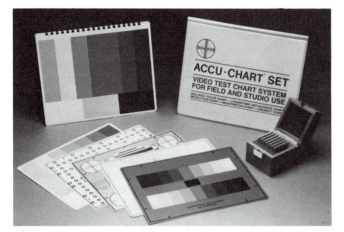

FIGURE 15-16 COLOR BAR and other TEST CHARTS for TV cameras. (Courtesy of Quality/Universal)

If the dots are in the small boxes and if the white blob is centered exactly in the middle of the screen, then you have achieved the VECTORSCOPE-seal-of-approval. If you're using a color camera in the $1000–$3000 range, you can expect the dots to miss the tiny boxes by a little bit but fall inside the larger ones. The dots may be a bit fuzzy, too. If the dots appear to shrink toward the display's center, the CHROMA (color) is too weak: Turn up the CHROMA GAIN or CHROMA LEVEL on the camera's CCU. If the dots fall beyond the boxes, the color is too strong: Turn down the CHROMA controls.

If all the dots seem to have rotated clockwise or counterclockwise from their respective boxes, it means your color hues are off. Adjust the BURST PHASE or SUBCARRIER PHASE control on the camera's CCU.

If the dots in the upper left seem to be pushed to the lower right and the dots on the lower right seem to be pushed toward the upper left but the G and Mg dots look pretty close to normal, tell your TV technician that your camera's I ENCODER LEVEL is *low*. Conversely, if the dots appear to be squashed along the Q axis on the VECTORSCOPE screen, then your camera's Q ENCODER LEVEL may be off. Again, tell your technician.

Sometimes just one or two dots may be off their marks. Double-check your color temperature and white balance. Do you have separate red and blue gain controls on the CCU

that will correct the problem? If it's the red that is weak, boost the R (red) gain control on the camera.

Check the blob in the center of the display. If it lies slightly off center, then you have tinted colors. The white could be tinted, or the black could have a tinge of color. To investigate further, aim the camera at a pure white card illuminated with white light. Recheck the color temperature filter and white balance. If the blob is still off center, as shown in Figure 15–17, check the RED or BLUE GAIN controls on the CCU since the white balance is not perfect.

Next, cap or close the lens to generate a black picture. If the blob is still off center, adjust the RED or BLUE PEDESTAL controls on your camera, a process called BLACK BALANCING.

Essentially what we have done is told the camera not to tint its whites, not to tint its blacks, and to make red things into signals which would be interpreted as red, etc.

Matching two color cameras: When a color TV studio system is first installed, it is necessary for a technician to carefully "time" and "phase" all your cameras and sync generators. If this is not done, when you switch from camera to camera, your picture will shift sideways and/or change color on the TV monitor.

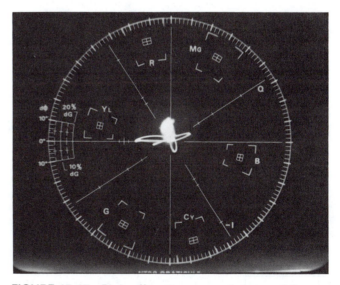

FIGURE 15-17 Blob off center means impure whites or blacks, in this case tinted pink.

MINIGLOSSARY

*Chroma gain   Camera control that boosts the amounts of color in the picture.

*Burst phase   Control on a color camera CCU (or other video gear) that adjusts the timing of the burst signal and thus varies the color hues in the picture.

I encoder   Electronic circuit in a camera that mixes colors into a single color video signal. Responsible for certain colors.

Q encoder   Similar to I encoder but responsible for different colors.

Blue pedestal   Control on a color camera CCU which adjusts the amount of blue signal the camera makes when it "sees" no blue. Similar controls for red and green may exist. Used in balancing black levels.

Delay line   Passive electronic device which, when connected to a video or sync cable, delays (retards the timing of) the signal passing through it; used to slow down "early" signals to keep them synchronized.

Why does this happen? Video travels at the speed of light through the camera cables. Regardless of how fast that video signal zips along, it's going to take longer to go through a 40-foot cable than to go through a 20-foot cable. So the camera 40 feet away gets its signal to the switcher a trifle after the camera that's 20 feet away. In addition, the sync signal which was supposed to keep the cameras in step, gets to the 40-foot camera a little later than the camera that's 20 feet away. That delays the 40-foot camera's signal even more. The technician installs DELAY LINES among the shorter cables to slow down their signals to make everything match up. Once all the camera signals match up, you can switch from one picture to the next without that horizontal shift in your picture or a change in your colors.

Even after this has been done, the colors may still "drift" with age. You may also change cameras or camera cables and again reintroduce delay errors. The VECTORSCOPE is handy for correcting these errors.

If you set up both cameras using a WAVEFORM monitor and VECTORSCOPE and COLOR BARS as mentioned earlier, you probably will have the cameras accurately matched already. If they matched your COLOR BAR standard, then they should match each other.

One way to test if they match is to aim both cameras at COLOR BAR charts and then switch your switcher back and forth from one camera to the other camera quickly while viewing the program monitor. The bars should remain the same color. Next, view your VECTORSCOPE and see if the dots stand still. The less they move as you switch back and forth, the more perfectly matched your cameras will be.

Another technique to compare two cameras is to create a horizontal split screen with one camera's COLOR BARS on the top half of the screen and the other camera's bars at the bottom. The colors should match on your TV monitor. On your VECTORSCOPE you may see double dots for each color, but the dots should be very close together.

If the dots aren't close together, further adjustment may be necessary. If one set of dots is rotated clockwise or counterclockwise from the other, adjust the BURST PHASE or SUBCARRIER PHASE control on one camera's CCU. If one set of dots seems shrunken in from the other set, then adjust that camera's CHROMA GAIN or CABLE LENGTH control.

Once you have matched the second camera to the first, you can now match a third camera to the first and then a fourth camera to the first, continuing until all the cameras in your system are matched together.

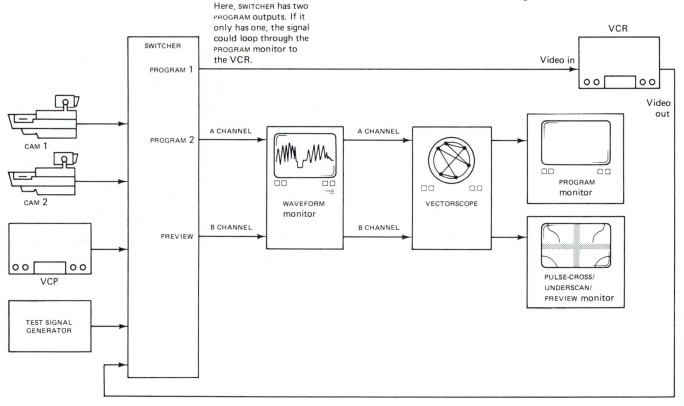

FIGURE 15-18 Ultimate setup of WAVEFORM monitor, VECTORSCOPE, and PULSE-CROSS/UNDERSCANNED monitors in a studio.

You can use a VECTORSCOPE and the COLOR BARS to test not only cameras but switchers, wiring, and VTRs. You'll see later how recording a SMPTE leader or 1 minute of COLOR BARS on your VCR before beginning your production can be useful later when adjusting your VCR's color output.

Operating the Vectorscope. Like the WAVE-FORM monitor, a well-adjusted and calibrated VECTORSCOPE doesn't require much knob twiddling at all. Table 15–2 lists the controls on a typical VECTORSCOPE and what they do.

Connecting the Vectorscope. The VECTOR-SCOPE inputs and outputs are similar to the WAVEFORM monitor's. Connect it just as you would a WAVEFORM monitor. Since most people have both a WAVEFORM monitor and a VECTORSCOPE (usually side by side), it is handy to loop signal A through the WAVEFORM monitor, then through the VECTORSCOPE, and then to its destination. Next, do the same with signal B through the B channels. Again, it is often handy to run the switcher's PROGRAM OUTPUT to the A channel of the WAVEFORM monitor, VECTORSCOPE, and pro-gram TV monitor and run the switcher's PREVIEW OUTPUT through the B channels to the preview TV monitor, as shown in Figure 15–18. In fact, this figure shows the ultimate scenario where the preview monitor is also a PULSE-CROSS/UNDERSCAN monitor. During setup and troubleshooting, you can select any source in your TV system by pressing the appropriate preview button on your switcher and send the signal through your WAVEFORM, VECTORSCOPE, and PULSE-CROSS monitors. When you are ready to begin your TV production, you can switch your PULSE-CROSS monitor back to NORMAL (so that it can be used as a normal preview monitor) and forget about the other two scopes. All this is done with the monitors set to their B inputs. Setting them to their A input would allow you to assess the program signal during production.

Some WAVEFORM monitors have a handy VIDEO OUT circuit which can be connected to the VECTORSCOPE. By using this socket, when you switch the WAVEFORM monitor from input A to B, the VECTORSCOPE will also automatically switch from A to B. Thus, the VECTORSCOPE always displays the same signal as the WAVEFORM monitor. Crafty, those engineers.

TABLE 15–2

Controls on a VECTORSCOPE

Control Name	What It Does	Normal Setting
Gain	Allows you to expand the dots out past their boxes. Good for magnifying the display and making measurements. Switch may also have a factory preset position which displays the dots accurately (not expanded or contracted). Switch may also have a test circle position for calibrating the position of the whole display (you coincide the test circle with the circle etched on the screen).	Preset
Phase	Rotates the entire display.	Set so that burst points straight to the left.
ϕ reference	Determines what reference will be used to compare the phases of other devices. *Position A:* When testing a single source, the phase will be referenced to itself. *Position B:* When testing two items (i.e., cameras) which must be in phase, position B superimposes both their displays. If the corresponding dots line up, both cameras are in phase. If not, one camera must be adjusted (perhaps a phase adjustment) so its dots coincide with the other's.	Position A
Input	Like the waveform monitor, the vectorscope can listen to two sources, A or B.	A or whichever source you prefer
Intensity	The brightness of the dots, etc., in the display.	Where comfortable
Focus	Makes the dots and vector lines sharper.	Make display as sharp as possible.
Vertical position	Moves display up and down. Use in conjunction with test circle for centering.	Once the display is centered, leave.
Horizontal position	Move display left and right. Use with test circle.	Same as above
Scale illumination	Illuminates the graticule.	Wherever comfortable
Sync	Like the waveform monitor, the vectorscope can synchronize itself to the incoming video (INT position) or use a separate sync source entering on a separate wire (EXT position).	INT

Test Signal Generator

Figure 15–19 shows a TEST SIGNAL GENERATOR used for making COLOR BARS and other test signals. The device can be connected up to your switcher like a camera, as was shown in Figure 15–18, or it can be permanently mounted into your studio electronics racks with other signal processing equipment. Often studio SYNC GENERATORS contain TEST SIGNAL GENERATORS and can also perform the job of GENLOCKER (described later). Some make audio test tones, and many have handy switches to create pure blue, red, or green screens, handy for backgrounds behind titles, etc.

Although the TEST SIGNAL GENERATOR is not really video monitoring equipment (it doesn't have any meters or scopes), it does play an essential role in providing a "perfect" color signal for calibrating other video equipment.

Monitor Analyzer

If you don't have TEST SIGNAL GENERATORS, VECTORSCOPES, and other fancy equipment, it is hard for you to adjust your camera colors correctly. You could rely on your TV monitor giving you a picture that "looks good," but what if your monitor's hue is out of adjustment? If your camera were making faces look too red, for instance, you might be inclined to adjust your TV monitor to make the faces look right. The final picture would look fine on your monitor but would look terrible for everyone else in the world.

This is why we need a standard, a set of colors we can trust. If we don't have COLOR BARS, then we try to use flesh tones and adjust our cameras and monitors to make flesh look normal. Of course, as anybody who has viewed their neighbors' and parents' TV sets knows, perfect flesh tone is in the eye of the beholder and varies wildly from person to person. "Perfect" COLOR BARS solve this problem by providing a standard set of colors, but one could still debate how red red should be and how green cyan should be and how rich blue should be. Adjusting the colors on a TV monitor "by eye" is not very accurate.

To make the measurement more objective, you could observe the COLOR BARS through a special dark blue lens

(Figures 15–20 and 21, and back cover). No longer would you see seven varicolored bars but rather four light blue and three dark blue bars. By adjusting the hue or tint controls on your set as you peer through the glass, you'll find a setting where the three dark bars look equally dark and the four light bars look equally light. Your TV colors will now be the *right* colors. You may now adjust the TV's color control to suit your taste, but if you turn the control too far (making the picture too pastel), you'll see the bars begin to combine. Readjust the control so that the bars remain.

Why does this method work? Let's consider the bars one by one. In Chapter 6 we learned the following:

1. White is made up of red, blue, and green, so the white bar has blue in it.
2. Yellow is a mixture of red and green. The yellow bar has no blue in it.
3. Cyan is a mixture of blue and green, so there is blue in the cyan bar.
4. Green is made of green only. The green bar has no blue.
5. Magenta is made of blue and red. The magenta bar has blue in it.
6. The red bar has only red in it, no blue.
7. The blue bar has blue, of course.

Notice the sequence of bars—blue present, blue absent, blue present, blue absent, and so forth. When you look through the blue glass, you see only blue or nothing. Thus, you see blue, nothing, blue, nothing, and so forth on a perfectly adjusted color monitor. If the colors are off a little and leak out from where they should be to where they shouldn't be, the light and dark bars (seen through the glass) become unequal. You might see three dark bars in a row and then a light one. So much for the science lesson.

Kodak sells these blue lenses for about $10 (calling them #47B gelatin filters). A friendly school physics or chemistry department might give you a piece of cobalt glass which is very dark blue. A commercial source for this handy MONITOR ANALYZER is Imero Fiorentino, Inc. Photo supply stores also sell them. Figure 15–21 shows one.

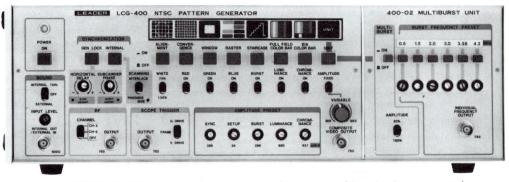

FIGURE 15-19   TEST SIGNAL GENERATOR (Courtesy of Leader Instruments).

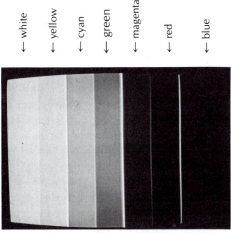

Monitor screen (shown in black and white).

Screen as viewed through the blue lens (shown in black and white).

FIGURE 15-20 COLOR BARS on a well adjusted TV as viewed through the dark blue lens of a MONITOR ANALYZER (see also back cover).

VIDEO IMPROVING EQUIPMENT

Processing Amplifier or Proc Amp

Remember how the color camera's CCU could adjust the camera's brightness (pedestal), contrast (gain), color saturation (chroma gain), and hue (subcarrier phase)? You could make all these changes before the picture was combined with sync and sent to a VCR as composite video.

But what do you do with a program from a VCR which is very dark and off-color (and you're not viewing Eddie Murphy in Concert)? There are no adjustments on the VCR to boost the contrast or change the hue of the picture.

Once the components of the color picture are combined into composite video, they become a package deal. The parts cannot be adjusted individually. However, the package can be taken apart again, readjusted, and recombined into another package. This is what a PROCESSING AMPLIFIER does. It separates sync from the composite video so that its strength can be adjusted and the video's strength can be adjusted. Furthermore, it separates the color from the video so that the brightness and contrast (LUMINANCE) of the picture can be adjusted and the color strength (CHROMA) and hue (BURST PHASE) can be adjusted. Next the PROC AMP assembles the components back together and sends the signal on its way as composite video. Figure 15–22 shows the

front panel of a PROCESSING AMPLIFIER, and Table 15–3 reviews the controls on the PROCESSING AMPLIFIER.

Using the Proc Amp to Adjust the Picture from a Video Tape. The PROC AMP is useful for fixing mistakes in your video signal as well as making unstable pictures stable. It can't make your picture sharper, but it can make your sync sharper.

You'll notice from the WAVEFORM monitor displays earlier that sync always looked the same (unless there was something wrong with it). The picture may change, but sync

FIGURE 15-21 MONITOR ANALYZER. Dark blue lens useful in adjusting HUE and COLOR controls using COLOR BARS on a color TV monitor. (Photo courtesy of Imero Fiorentino Associates, Inc.)

MINIGLOSSARY

Monitor analyzer A deep blue lens used for viewing color bars on a color monitor while adjusting the TV's color hue to obtain proper colors.

Processing amplifier Electronic device that modifies and stabilizes video signals by separating the video from the sync and regenerating brand new, "clean" sync as well as adjusting video and color levels.

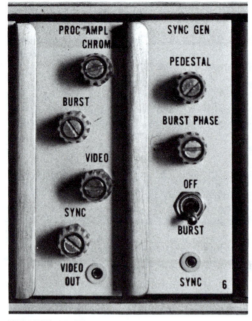

FIGURE 15-22   PROC AMP.

sync always looks the same, hunts for the sync pulse (even if it's a fuzzy one) and removes it, replacing it with new sync which it manufactured itself. Thus, fuzzy sync and fuzzy pictures go into the PROC AMP, and new, clean sync and fuzzy pictures come out. Figure 15–23 shows the WAVE-FORM of a picture before and after it has passed through the PROC AMP. Notice that the sync is straight and clean and that BURST (that little box that follows sync) is less fuzzy.

Here are some things you can do with a PROC AMP:

1. You are copying an old or stretched video tape, and the picture is very unstable, shaking, rattling, and rolling. The top of the picture flagwaves excessively: Sync is probably bad. Run the signal through a PROC AMP which is adjusted to manufacture kosher sync.

2. Somebody made a video recording without enough light. The picture looks very dim: The video level is low. When you run this signal through your PROC AMP, turn the VIDEO LEVEL up.

3. Someone recorded a program off the air from a distant station. The color flashes on and off: The BURST reference signal is probably too weak for the electronics to interpret. Running the signal through the PROC AMP will add new BURST and perhaps solve the problem. If it doesn't, then switch the BURST off, making the picture black-and-white (but at least without color flashing).

4. While playing a tape through your WAVEFORM

always looks the same. When you play a tape, the picture and sync always play back fuzzier than the original signal. Go down two generations, and they both become fuzzier still. At some point, the picture may become so fuzzy and grainy that you wouldn't want to watch it, and the sync would become so fuzzy that your VCP or TV wouldn't be able to "lock onto" the sync and your picture would roll and jitter. The PROC AMP, taking advantage of the fact that

TABLE 15–3

PROC AMP controls

Control Name	What It Does	Normal Setting
Chroma	Varies the chroma, or color saturation in a picture. Color bar dots on vectorscope draw in or expand out when chroma is adjusted.	Where picture looks best, or so that color bars create dots on a vectorscope screen which reach out to the boxes.
Burst	Adjusts strength of burst, a color reference signal. Without burst, there is no color.	+20 to −20 IRE on a waveform monitor.
Video	Varies the strength of the video signal; affects contrast.	On waveform monitor, peaks should reach 100 IRE, valleys reach 7.5 IRE.
Sync	Varies the strength of the sync signal. Without sync, the picture rolls.	On waveform monitor, sync pulse should dip from 0 IRE to −40 IRE.
Pedestal	Adjusts the pedestal, or black levels in the picture; controls picture brightness.	On waveform monitor, valleys (black areas) should be maintained at 7.5 IRE.
Burst phase	Adjusts the phase, or timing of the burst reference signal, effectively changing the hue of all the colors in the picture simultaneously. Dots on vectorscope from color bars rotate clockwise or counterclockwise when burst phase is adjusted.	Where flesh tones look right, or so that color bars, when viewed on a vectorscope, create dots which land in the appropriate boxes.
Burst on/off	Adds burst to the sync signal (even if it didn't have one to start with) or strips burst off, effectively turning a color picture into black-and-white.	On for color, off for black-and-white

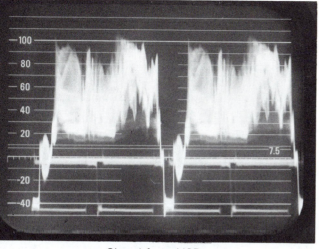

Signal from VCP

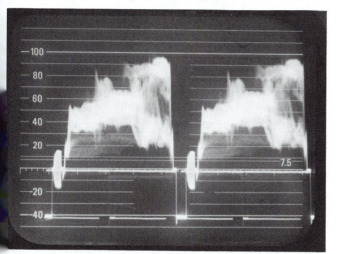

Signal from PROC AMP

FIGURE 15-23 PROC AMP improves sync.

monitor, you notice that the sync is at −20 IRE (rather than at −40 IRE where it belongs): Several things can cause sync to be at the wrong level. If, during the recording process, your signal was DOUBLE-TERMINATED (you screwed a 75Ω terminator into the device *and also* threw the terminator switch into the 75Ω position), the video and sync level both become weak. The same thing also happens if you loop a signal through a device but leave its terminator switch in the 75Ω position. Another example: Someone uses a T connector to split a video signal into two signals. Now each will have a weak video and sync. On the other hand, if someone sends their video signal into a device *without* terminating it, the sync may appear too strong. In each of these cases, you should hunt down the culprit and fix it in the studio before you make your recording rather than try to correct it afterward. However, the PROC

AMP will help correct this video faux pas once you have recorded it.

5. Some clients bring you a video tape with purple faces and magenta trench coats. (Personally, if they looked like that, I wouldn't let them in the door, but in this case it's the *tape* we're talking about). Their camera had a color phase problem which they didn't notice in their black-and-white camera viewfinder or on their misadjusted color TV monitor. When running this signal through the PROC AMP, adjust BURST PHASE to move all the colors back to their correct hues.

6. You wish to copy a videocassette which has MACROVISION or COPYGUARD on it: This is probably an illegal copyright violation. People who put COPYGUARD on their tapes don't want you copying them. Nevertheless, PROC AMPS will strip off the anticopy signals, replacing them with standard sync, resulting in a normal copy of that video tape. This technique does not work with *all* forms of antipiracy signals, just some.

Setting up a Tape for Playback. A properly made video tape will contain 10 seconds to a minute of COLOR BARS in the beginning. Before playing this tape to your audience, play it to your WAVEFORM monitor and VECTORSCOPE to see if the video levels are "up to specs" and your COLOR BARS make dots in the right places on your VECTORSCOPE. If they do, then you are all set; don't bother doing anything to the signal. If, however, these levels are way off, run the signal through a PROC AMP and make the appropriate adjustments.

Figure 15–24 shows the VECTORSCOPE display you might expect from a VCR playing COLOR BARS. The dots

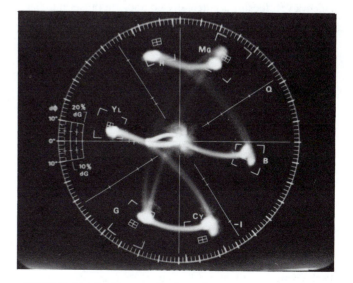

FIGURE 15-24 Vector display of COLOR BARS played back on a VCR. Dots are fuzzy and a little off target. This is normal.

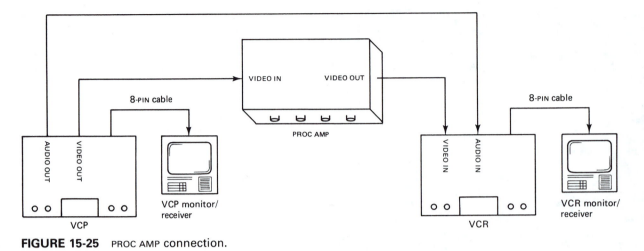

FIGURE 15-25   PROC AMP connection.

appear a bit fuzzy and don't hit their targets exactly, but that is to be expected with ¾U and ½-inch VCRs. As long as the dots are in the larger boxes, you are close enough.

Connecting the Proc Amp.   You place the PROC AMP between the source signal and its destination. If the source is a VCP and you're copying a tape, like in Figure 15–25, you run the VCP's VIDEO OUT to the PROC AMP's VIDEO IN and the PROC AMP's VIDEO OUT to another VCR. The PROC AMP could also be used between cameras, switchers, and any other video devices where the sync, picture, levels, or picture colors needed to be adjusted.

Kinds of Proc Amps.   Industrial PROC AMPS cost $1500 up, depending on their quality and the number of bells and whistles attached to them. Many have empty slots in them where circuits can be added to do extra jobs like DROPOUT COMPENSATION (hiding DROPOUTS when a video tape is played), PEAK CLAMPING (making sure that your video level doesn't accidentally go above a preset number, like 100 IRE), and other enhancements.

Home videophiles can purchase lower-quality PROC AMPS for $100–$250. "Consumer" PROC AMPS are called STABILIZERS and are usually used when copying tapes. In fact, their most frequent use is the illicit duplication of COPYGUARDED tapes.

Image Enhancers

Another device that helps to retrieve the picture is an IMAGE ENHANCER. It makes the picture look crisper and sharper. It doesn't *really* make the picture sharper, as that is im-

possible to do. It only makes it *look* sharper by adding ridges around the edges of surfaces, exaggerating them a little.

The ENHANCER also accents flaws in the picture and will make every wrinkle and blemish on a face stand out clearly. Furthermore, inexpensive video tape players create little flaws of their own. IMAGE ENHANCERS don't know the difference between tape flaws and the regular picture, so they augment both, resulting in a vivid picture with exaggerated flaws—not much of an improvement. See Figure 15–26 for examples.

ENHANCERS are best used in professional applications where the pictures and the VCRs are of high quality to start with. ENHANCERS are often used to sharpen the picture when movie film is being transferred to video tape. They also help sharpen the picture when tapes are copied.

Typically, when a tape is copied, it loses about 4½ dB (about 10%) of its picture sharpness. An IMAGE ENHANCER (or IMAGE CRISPENER or DETAILER, as they are called in the home video field) hides this loss and makes the picture look about 6% sharper.

IMAGE ENHANCERS boost high frequencies. Smooth parts of the picture use low video frequencies. Detailed parts, such as strands of hair, individual leaves on a tree, or the texture of a carpet use high video frequencies. These are rather hard to send through long wires, process through electronic circuitry, and record and play back. The high frequencies tend to become weak, while the low ones remain strong. An ENHANCER amplifies the high frequencies, but our old enemies noise and graininess also occupy those frequencies. If we boost the sharpness, we also boost the graininess.

MINIGLOSSARY

Dropout compensator   Electronic device that hides dropouts by replacing these specks with an adjacent piece of TV picture. Simpler models merely replace dropouts with gray.

Peak clamping   Electronically limiting the maximum video signal level to a certain strength, like 100 IRE.

Stabilizer   Inexpensive processing amplifier made for home video market.

*****Image enhancer**   Electronic device that crispens a TV picture (making it look sharper although it isn't really) by exaggerating the boundaries of parts of the image.

Detailer   Less expensive image enhancer used in the home video field.

Clean signal before ENHANCEMENT

Clean signal after ENHANCEMENT

Grainy signal before ENHANCEMENT

Grainy signal after ENHANCEMENT

FIGURE 15-26   IMAGE ENHANCEMENT.

If the picture has very little noise to start with, it can be crispened without dredging up too much grain. If the picture is grainy, then ENHANCEMENT will only fertilize the weeds along with the crop, as was shown in Figure 15–26.

Professional video ENHANCERS cost $1500 up. Home video DETAILERS cost about $200. The lower-priced industrial ENHANCERS will "sharpen" vertical parts of the picture only. Picket fences will be enhanced but venetian blinds will not. More elaborate ENHANCERS work on both vertical and horizontal parts of the picture, crispening both the fence and the blinds.

When using ENHANCERS, be careful not to double- or triple-enhance your picture. Say you copied a movie onto video tape through your film chain and ENHANCER. Later you make a copy of that tape again using an ENHANCER. You now have a double-enhanced picture. Objects will have vivid outlines, perhaps even double outlines. You may see visible grain in the picture. And if *that* tape ever gets copied through an ENHANCER, the picture will acquire a distinct cartoon effect.

Operating the Enhancer.   ENHANCERS usually have a control called DETAIL allowing you to adjust how much ENHANCEMENT will occur. This is helpful when you want to enhance the picture just enough to make it look sharper but not so much that you bring out the grain.

Another control on some ENHANCERS is *threshold* or CORING. This adjusts the frequencies that will be boosted a

MINIGLOSSARY

Detail   Image enhancer control adjusting amount of enhancement the device will make.

Threshold   Control on an image enhancer selecting which high

frequencies will be boosted (enhanced) and which will be left alone, to reduce graininess in the picture.

ENHANCER/PROC AMP combo available in the home video market.
(Courtesy of Quality/Universal)

lot and those that will be boosted a little. Since most of the objectionable grain in the picture occupies the very high frequencies, you would use this control to reduce the strength of those frequencies. In short, the ENHANCER lets you increase *useful* high-frequency information without boosting the noise too.

Connecting an Enhancer. The ENHANCER is connected the same way as the PROC AMP. Video out from a VCP or other video source goes into the ENHANCER's video in. The ENHANCER's video out then goes to the destination, usually the video input of a copying VCR, as shown in Figure 15–27.

Sometimes you see ENHANCERS in other configurations. When you are copying a movie onto a video tape with a TV camera and film chain, the ENHANCER can sharpen the image coming from the TV camera, making a crisper film-to-tape transfer. ENHANCERS are sometimes found upstream of TV projectors, where a sharp-looking image is necessary.

Time Base Corrector

A TIME BASE CORRECTOR or TBC (Figure 15–28) costs about $1,000 to $14,000 and removes jitter from TV pictures (which usually occur when tapes are played back) and corrects color and sync timing errors.

Small industrial and educational studios generally don't need TBCs. Your tapes will play fine without it. If, however, you're doing A/B roll editing, professional recording and duplication, or broadcast TV work, you will need a TBC.

TRANSCODERS. VCRs can make and receive video signals in different ways, depending on how advanced they are:

- COMPOSITE VIDEO—"Normal" video used by common VCRs. Because the color rides on the same wire as the luminance, both are degraded somewhat.

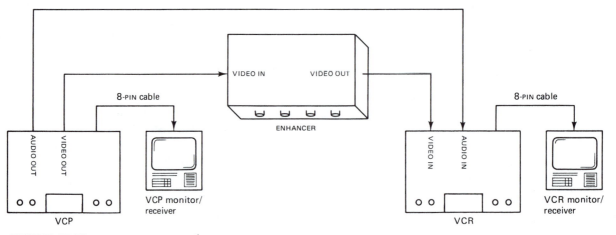

FIGURE 15-27   ENHANCER connection.

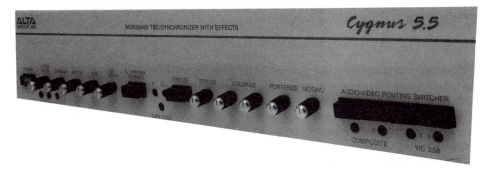

FIGURE 15-28 TIME BASE CORRECTOR (Courtesy of Alta Group, Inc.)

- U-MATIC DUB or Y/688—Higher-quality industrial 3/4U technique of sending color and luminance signals over separate wires.
- VHS DUB or Y/629—VHS method of sending color and luminance signals over separate wires.
- SVHS DUB or Y/3.58 or Y/C—Method of sending separate color and luminance signals from one SVHS or Hi8 VCR to another.
- Y, R-Y, B-Y—Very high-quality method of sending luminance and *two* color signals from one professional VCR (MII or Betacam) to another.
- Y, I, Q—Another method of doing the preceding procedure.

The more advanced TBCs today will TRANSCODE one of these video formats to another, allowing you to edit or copy tapes using dissimilar machines. Even if your VCRs and TBC have COMPOSITE VIDEO inputs and outputs, it is better to use the *separated* color and luminance signals whenever available.

FRAME SYNCHRONIZERS vs. TBCs. A TBC specializes in the job of *correcting* jittery video. A FRAME SYNCHRONIZER, on the other hand, synchronizes two independent video signals without necessarily correcting them. You would use a TBC for correcting a video tape when duplicating it. You would use a FRAME SYNCHRONIZER for mixing a roving reporter's picture with the studio's picture, or synchronizing a VCP with a studio. The more expensive TBCs do *both* jobs but you'll save sheckles by purchasing a plain TBC or plain FRAME SYNCHRONIZER if that's all you'll need.

Built-in TBCs. Some top-of-the-line VCPs have TBCs built in. This not only improves the picture, but makes it directly broadcastable, or makes the VCP connectible to the switcher, like any professional camera.

Some switchers have TBCs or FRAME SYNCHRONIZERS built in. Thus, you can connect cameras, free-running VCRs, or any video inputs to the switcher and have the images mixable.

How TBCs work. When a camcorder or VCR is moved while recording, it creates a GYROSCOPIC ERROR as the spinning video head varies its speed. Also, tape stretches or contracts with temperature and tape tension, slowing down or speeding up the video signals. This pushes the TV picture sideways on the screen or causes hooking or flag-waving of the image (review Figure 5–23). Also, the motors and mechanics of the VCR create tiny amounts of vibration causing jitter in the picture. Smooth, round objects tend to get crusty edges.

All of these problems are due to timing errors in the video signal. The TBC, much like a dam, collects the floods and droughts in the video signal and meters it out smoothly. By using very precise internal clocks, the TBC is able to make perfect sync, burst, and other signals whose timing does not have the jitters and errors that were created by the VCP.

Inexpensive TBCs have small dams capable of correcting small errors. More expensive TBCs have giant dams able to correct large errors perhaps lasting 1/30 second, an entire video frame. These are called FRAME STORES.

The TBC reads the video signal into its memory at the inconstant rate put out by the VCP. The TBC reads the

---

MINIGLOSSARY

*Time base corrector Electronic device to remove jitter and other timing abnormalities from a video signal, usually the signal from a VCP.

*Frame store Electronic device able to store a video picture (a frame) electronically and perhaps manipulate it.

Transcoder Electronic device to convert video signals between Y/688, Y/C, Y/R-Y/B-Y, and others.

Frame synchronizer Electronic device to synchronize two independent video signals so they can be mixed.

Direct Method of time base correction used with professional equipment yielding high resolution.

Heterodyne Method of time base correction used with common and color under VCRs, yielding medium resolution pictures. Also, the type of VCR using color under recording.

video *out* in the same sequence, but at a constant rate, sans error. If the readout is timed in step with the studio sync generator and cameras, then fades, wipes, and special effects are possible between your studio cameras and a video tape or some other free-running video source (like a camera microwaving its pictures from the Goodyear blimp).

DIRECT VS. HETERODYNE *TBCs.*   There are two ways to correct a VCR's signal. One way, called DIRECT, or SUBCARRIER FEEDBACK, or WIDE BAND handles the complete NTSC signal at once and yields about 330 lines of resolution. In order for the method to work, you need professional VCRs or industrial VCRs with SUBCARRIER inputs and TBCs with SUBCARRIER outputs.

The other method is HETERODYNE, which works with any VCR's COMPOSITE VIDEO signal but yields only about 200 lines of resolution.

A little engineering lesson: 3/4U, VHS, SVHS, 8mm and Hi8 VCRs use the COLOR UNDER or HETERODYNE method of recording their colors. The colors are separated from the luminance part of the signal for recording. Also, the colors require more stabilization than the luminance. For economy, the TBC processes the two signals separately. Unfortunately, when a common VCR combines the color and luminance together into COMPOSITE VIDEO, it degrades the sharpness somewhat. The TBC takes in the COMPOSITE VIDEO signal and *reseparates* it, damaging it some more, resulting in fuzzier pictures with slightly degraded color.

Professional VTRs like 1-inch, Betacam, and MII don't use the COLOR UNDER technique, allowing the TBC to process the complete NTSC signal with full resolution; the TBC doesn't have to put more muscle into stabilizing the wavering color signal.

If using VCRs with Y/C, DUB, or other separate luminance/chrominance outputs, use a TBC that can handle these signals separately. And if you have higher quality VCRs and TBCs with SUBCARRIER inputs and outputs, you have a special advantage: The TBC can send a special signal to the SUBCARRIER input of the VCR which is used to *precorrect* the VCR's video signal before it goes to the TBC. This *precorrected* signal can now be handled using the DIRECT method yielding high-resolution images. Without this special connection, the TBC has no choice but to separate the chrominance from the luminance, distorting them somewhat and discarding some picture detail in the process of time base correcting the image.

In short, you get the best result by avoiding HETERODYNE VCRs. Use 1-inch, Betacam, MII, or digital VTRs whose signals can be DIRECTLY processed. If using a HETERODYNE VCR, select one with a SUBCARRIER (SC) input allowing the TBC to precorrect the VCR's signal and handle it DIRECTLY. Also, avoid COMPOSITE video if you can use the separated color/luminance signals among your VCRs and TBCs.

WHAT A TBC DOES

Here are some of the things a TBC can do:

1. Synchronize your VCR with your studio cameras so that the VCR's picture can be mixed with the cameras' signals through your switcher.

2. Synchronize VCRs whose signals need to be mixed during A/B roll editing.

3. Synchronize field sources with the studio cameras so that the signals can be mixed. A portable TV camera in a helicopter with its signal microwaved to the studio is one example of a field source.

4. Correct signals to be broadcast. VTRs create pictures too unstable to be directly broadcast over the air. The tiny jitters and color instabilities which we accept in industrial and educational video do not meet the standards of the broadcast industry and the FCC.

5. Correct VCR signals played out over cable.

6. Perfect VCR signals during duplication. Time base errors add up from generation to generation. Tape duplication houses avoid adding one more layer of jitter by correcting the signal as it is duplicated. Also, video tapes sometimes get so far out of wack that they cannot be copied or their signals won't pass through other equipment. TBCs often fix these instabilities (assuming the original tape isn't so screwed up that the TBC can't "lock onto" it).

7. Process gain and color. Your VCR or studio TV signal may need the brightness or colors adjusted. Most TBCs have a built-in PROC-AMP to do this.

8. Create special effects. A TBC with a FRAME STORE allows you to "grab" a shot with a camera, and while holding it in memory, use the

camera for something else. It's like having an extra camera focused on a stationary graphic. FRAME STORES can also "grab" a picture from a video tape. Many TBCs can do other effects such as STROBE (snapping from picture to picture like a fast slide show), PIXELATION (turning the picture into teeny boxes that can grow large like checkerboard squares), COMPRESSION (squeezing the picture smaller), and EXPANSION (blowing the picture up).

9. Correct VCR signals to TV projectors. Since TV projectors and VIDEO WALLS (many TV sets stacked like building blocks to create a wall of pictures) magnify the jitters and instabilities in a VCR's picture, time base correction becomes necessary.

10. DROPOUT COMPENSATION. A DROPOUT results when bits of tape flake off, or dirt causes your video signal to lose a piece of picture. Some TBCs have DROPOUT COMPENSATORS built into them which will replace the lost piece of picture with a duplicate made from an adjacent piece of picture, so instead of seeing a black spot on your screen, you would see picture and not even notice that it was "fudged."

Connecting a TBC.   Figure 15-29 shows a simple connection of a TBC for copying video tapes. The setup is similar to the PROC AMP and ENHANCER. Video out from the VCP goes to the video in of the TBC. The corrected video out from the TBC goes to the video in of the VCR.

In cases where you wish to mix the pictures from a VCR with the pictures from your cameras during a production, the connection becomes a little more complex. Essentially what happens is this: The house sync generator feeds its signal to the cameras and to the TBC, locking them all to the same drumbeat. The TBC manufactures a new signal called ADVANCED VERTICAL (or VTR LOCK), which it sends to the sync input of your VCP (only the better VCPs have such an input). In this way, the VCP is also locked to the house sync. The VCP plays out its signal with jitter. Put another way, it's "almost" locked to the house sync system. This video signal is sent to the TBC, which swallows it and corrects the jitter, effectively making the signal perfectly locked to the house system. When this video signal leaves the TBC, it looks just like another camera signal and can be switched, faded, or wiped like any camera signal. Figure 15-30 diagrams the setup.

Color Corrector

Say the local March of Dollars asks you to copy a tape for them. The colors are very blue either because the tape was originally copied from an old film or because somebody shot it with the wrong filter on the camera or because the WHITE BALANCE was off. This skew toward one color is called a GAMMA error. You cannot correct this problem by simply running the signal through a PROC AMP and adjusting BURST PHASE, because that will mess up the other colors. You only want to correct *one* color. What do you do?

You could buy a COLOR CORRECTOR. It is an electronic

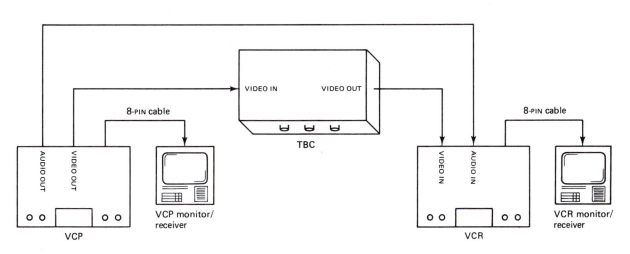

FIGURE 15-29 Simple TBC setup for copying video tapes.

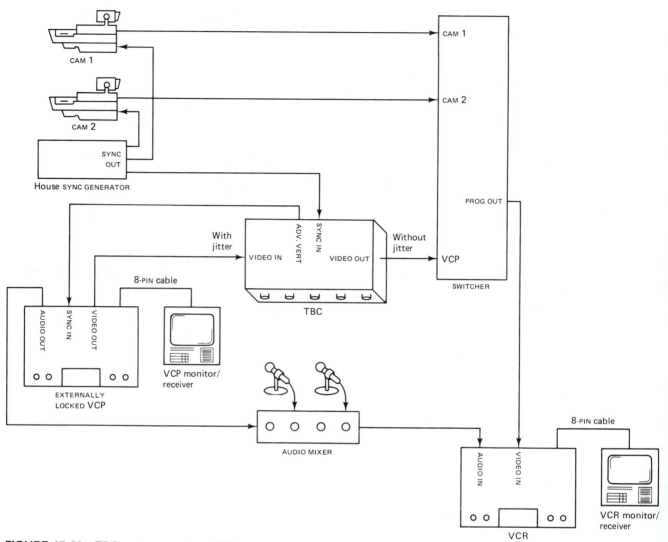

FIGURE 15-30   TBC set up to mix a VTP's signal with other studio cameras.

device that allows you to correct each color separately. Amateur models cost about $500, pro models $9000.

Color Correcting Without a Color Corrector.
If you relish homespun frustration, you can try to make color corrections without a COLOR CORRECTOR. You do it by mixing your video signal from your VCR with some color signal that is the chromatic opposite of the offensive color.

Look at the face of your trusty VECTORSCOPE. The opposite of green is magenta, the opposite of red is cyan,

the opposite of blue is yellow. Other opposites you'll have to figure by guesswork.

In the case described earlier, the offensive color was blue. Yellow is the color we need for neutralizing the blue, so yellow is the color we must mix with the VCP's picture in order to color-correct it.

We first play the original tape through a GENLOCK (described in further detail later) or a TBC so we can synchronize it with our TV system. That way we can eventually mix yellow with the VCP's signal by dissolving *part way* between the tape signal and the yellow signal. How do we make a yellow signal? We could aim the camera at a yellow

---

MINIGLOSSARY

Advanced vertical   Special synchronizing signal sent out by a TBC to a VCP to lock the VCP's video playback to the house (TV system's) sync. Makes the VCP play almost in synchronization with the studio cameras.

Color corrector   Electronic device that dissects the colors of a video signal and allows them to be individually adjusted (i.e., the blues could be changed to aquas without changing anything else).

card, or we could chuck a yellow slide into our color film chain, or we could use a COLOR BACKGROUND GENERATOR device that electronically makes color video signals. Often we find that our COLOR BAR TEST SIGNAL generators will make colors of our choice. Whatever we use, this yellow signal must be synchronized with the tape and partially dissolved (or superimposed) with it to effect the color correction. Expect this process to be loaded with trial and error, mostly error.

The mix can now be rerecorded. Note that this technique works for adding tint to a color *or* black-and-white tape as well.

AUDIO IMPROVING EQUIPMENT

As with video, you can't change a sow's grunt into a silk symphony. Audio improving equipment can be used while making a recording to accentuate the positives and reduce the negatives, or it can be used to enhance recordings as they are copied. It is mostly a process of addition and subtraction—adding desirable sounds, like boosting low tones or adding echoes, or subtracting undesirable sounds, like hums, clicks, and buzzes.

With but a few exceptions, all the devices which will be mentioned connect between the audio output of one machine and the audio input of another. For instance, you could connect the GRAPHIC EQUALIZER between the VCP's audio out and VCR's audio in. Similarly, the device could be connected between a mixer's audio out and an audio tape recorder's audio in. In fact, many of these devices are built into modern audio mixers.

Filter

Just as a cigarette filter stops the tar but lets the smoke (and cancer) through, an audio filter essentially traps certain tones, letting the others through. Generally, these filters are designed to cut high tones or low tones.

The RUMBLE FILTER is a popular filter attached to microphones or their cables to reduce low tones such as the rumble of air conditioners or wind or the hum of machinery. They do throw some of the baby out with the bathwater, however. When heard through a RUMBLE FILTER, a person's

voice will sound shallower and less deep. This often doesn't matter because voices are mostly middle frequencies anyway; the low tones are just for embellishment.

Graphic Equalizer

The GRAPHIC EQUALIZER is a bunch of filters in one box. It contains slider controls, each control affecting a band of frequencies. With a GRAPHIC EQUALIZER, you can reduce the frequencies around 150 Hz while boosting the frequencies around 1000 Hz yet leave the remaining frequencies relatively untouched. The GRAPHIC EQUALIZER was discussed in more detail in Chapter 10.

Parametric Equalizer

The basic idea behind the PARAMETRIC EQUALIZER is to divide the audio spectrum into three ranges: low, mid, and high. Each range has a control that may select any frequency within that range. Another control adjusts the amount of boost or cut for that frequency.

PARAMETRIC EQUALIZERS are handy for when you wish to cut out a very specific frequency and not mess up the neighboring frequencies. You could use it to wipe out the hum of a motor or the screech of a saw. You could also use it to boost the chirp of a bird in the woods without boosting the sound of wind in the trees and overhead jet planes.

The PARAMETRIC EQUALIZER works a lot like the GRAPHIC EQUALIZER except you can very precisely choose the frequency you wish to boost or cut. The GRAPHIC EQUALIZER only gave you a few choices of frequencies.

Dolby and DBX

Dolby and DBX were discussed in Chapter 10. Essentially DOLBY and DBX are methods of reducing hissy noise in a recording. They work by boosting the audio signal (usually the high parts) during recording in a process called ENCODING, and then reducing them during playback in a process called DECODING.

MINIGLOSSARY

Color background generator   Device which electronically creates a screenful of a desired color without the help of a camera. Color could be used as background behind character-generated text.

*Filter   In audio, an electronic device to trap a certain frequency of sound, letting others pass through.

Rumble filter   Audio filter to trap low frequencies (rumble from wind, noisy phono records, hum) and pass the rest.

Parametric equalizer   A tunable equalizer on which you can select a particular frequency or band of frequencies to boost or cut, perhaps to remove much of an unwanted sound from a recording.

Compressor/Expander

Say you had a conversation where a person shouted, then whispered, and then shouted. If you keep your volume control high enough to make the whispering audible, the shouting would be so loud that it would distort. If, however, you turned down the volume when the person shouted, the whispering would be so low that it couldn't be heard. This is a problem of DYNAMIC RANGE, the capacity of an audio system to play very soft sounds and very loud sounds. If you were lightning fast at your audio controls, you could turn up the volume when things were quiet and turn down the volume when the action got loud, thus keeping it all within the limited range of your recorders.

Since no one is lightning fast, enter the COMPRESSOR/ EXPANDER, an electronic circuit which "listens" to sound peaks and turns the volume lower the louder they get. The system reacts quickly and could vary the volume level five times in a single sentence. Thus, the whispers are quite loud, and the shouts are just a bit louder.

You may have noticed that the disc jockey chatter on AM radio stations seems to cut through traffic noise and the whoosh of air with your car window open. Advertisements and DJ talk are often highly compressed, making the station easy to hear (although not necessarily pretty to listen to). If you were to view a VU meter displaying the sound signal, instead of seeing the needle bob up and down between loud and soft words, it would hover around 0 dB in a constant barrage of sound at the 100 percent level.

Although this is a good way to squeeze a wide range of volume levels into the small DYNAMIC RANGE recordable on audio and video equipment, this process gives the sound an unnatural color. Shouts *should* be louder than whispers. To correct this audio compression, the person at the listening end could use an EXPANDER.

An EXPANDER is the opposite of a COMPRESSOR. When sounds get louder, the machine automatically exaggerates them, making them louder still. It adds DYNAMIC RANGE to a sound (assuming your loudspeaker system can handle it).

Putting this all together, here's what happens: The shout is a thousand times louder than the whisper. Your recorder, however, can only record a loudest sound 100 times greater than the softest. The COMPRESSOR knocks the 1000 times louder sound down to a manageable 100 times

louder sound where it is recorded. When the tape is played back, the 100 times louder sound goes into the EXPANDER, which multiplies it back up to 1000 times louder. Thus, the final sound is very similar to the original, yet we didn't exceed the limitations of our recording equipment.

Echo and Reverb

Once an echo is recorded on a tape, it is almost impossible to remove. Therefore, we try to record in echo-free studios and in the field try to reduce the echoes from our surroundings. This guarantees us maximum intelligibility of our sound, especially voices. There are times, however, when we overdo it and our sound is too "dead." We may want to add reverberation to the sound to make it more "normal."

Another case where we wish to add reverberation is when we have moved from one room to another and changed our ROOM TONE. If this change in sound coloration is drastic, it may be too obvious and distracting to the viewer. Since we cannot reduce the echoes from the first room, we may choose to add echoes to the second. For these echoes to match the echoes of the first room, we will need to listen to those first echoes and determine whether they mostly consisted of high or low frequencies. We then filter the sound coming from the ECHO CHAMBER so that only the desired frequencies come out.

REVERBERATION TIME is another part of room echo which varies from room to room. Bathrooms sound different from caverns. ECHO CHAMBERS have an adjustment to make long or short echoes.

Although the words are generally used interchangeably, there is a technical difference between ECHO and REVERB. ECHO is technically the sound which comes back to you over and over again like the hello, hello, hello you get shouting into a canyon. REVERB, on the other hand, is more like the sound you get singing in the shower or shouting into a huge tank. You never hear your words come back; you only hear a general mix of tones which slowly die out. This process is done either electronically or with mechanical springs which vibrate.

Dynamic Noise Reducers

A DYNAMIC NOISE REDUCER endeavors to reduce tape hiss when tapes are played. Unlike Dolby and DBX, which

MINIGLOSSARY

*Compressor   Electronic audio device to reduce the range of volumes in an audio signal down to a range easier to record. Creates a "flat" sound where soft and loud passages are about the same volume.

*Expander   Opposite of a compressor, an electronic audio device that extends the range of volumes in an audio signal, making loud parts louder than they actually were. Undoes the effects of a compressor, making compressed audio sound more normal.

*Dynamic range   A ratio comparing the lowest level of sound audible (above the noise of the machine) with the highest level;

the range of loudness a device can handle without distorting. Wider dynamic range represents truer sound fidelity.

Echo chamber   Device that adds echoes to an audio signal.

Echo   The repetition of a sound like hello, hello, hello, etc.

Reverb or reverberation   The slow decay of a sound when it's finished, like the ringing in the air heard after you clap your hands. Technically not the same as echo. Reverb is often added to music to make it sound fuller.

Reverberation time   The amount of time it takes for a loud sound to fade to silence.

require ENCODING and DECODING, DYNAMIC NOISE RE-DUCERS can be attached to any device at the final stage without any ENCODING necessary. They don't work as well as Dolby and DBX, however.

If you were listening to shouts and whispers and there was tape hiss in the background, you'd notice that the hiss was audible only when quiet passages were playing. You'd hear the hiss during whispers but not during the shouts. You could reduce the hiss by turning down the tone or treble control on your amplifier, but this would kill all the high frequencies, making everything sound muffled. A DYNAMIC NOISE REDUCER will automatically turn down its treble control when your program is very quiet and will stop filtering the highs when sounds become louder. Thus, your shouts will be as sharp as ever; only your whispers will be muffled. By adjusting the controls on the device, you can make it so that only very soft whispers are muffled but loud whispers are not. This way, you are only muffling small parts of your program, allowing 95 percent of your program to pass untouched and undoctored.

TV Stereo Adapters

Stereo adapters are little boxes of electronics which endeavor to make monaural audio sound like stereo. They don't work so hot. Most of them will take an audio signal and separate it into four frequency bands. The low and middle frequencies may be sent to the left speaker, and the high and medium-low frequencies will be sent to the right speaker. This *pseudostereo* may be interesting to listen to, but it is not stereophonic sound.

VIDEO DISTRIBUTION EQUIPMENT

The following devices help us get our signals from place to place.

Video Distribution Amplifier

The VIDEO DISTRIBUTION AMPLIFIER, or VDA, is an amplifier which will take one video input and amplify it and send it out to several places at once, as diagramed in Figure 15–31.

The VDA is handy when you have to send a video signal to many places. Looping the signal through many monitors makes it weak. With a VDA, you can loop one output through two or three monitors, the second output through two or three monitors, and so on, thus feeding many monitors.

The VDA also doesn't care whether you terminate its outputs or not. As shown in the figure, the fact that output 2 isn't terminated does not ruin the signals going to outputs 1, 3, and 4. This is handy when you have several devices

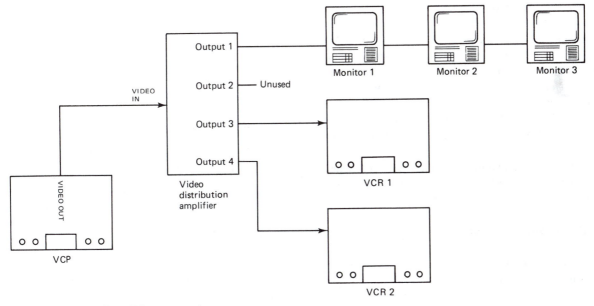

FIGURE 15-31 Possible setup of a VIDEO DISTRIBUTION AMPLIFIER.

MINIGLOSSARY

Dynamic noise reducer Audio filter that "listens" to the sound and turns off when sounds are loud (thus not "coloring" the sound) and turns on when sounds are soft (when hiss would be most noticeable if left unfiltered).

Stereo adapter Device that creates pseudostereo sound (fake stereo sound) from monaural sound.

*Video distribution amplifier Electronic device that splits one video signal into several (often four) and boosts each to make them as strong as the original signal.

to receive a signal from one source, but sometimes you remove a few of the devices and don't want to mess up the signal to the remaining ones.

VDAs are often used to allow the signal from one VCP to feed many VCRs. A VDA may have 4, 6, or more outputs. They may be cascaded so that one VDA feeds 4 VDAs and each of them feeds 4 more VDAs so that there are a total of 64 outputs.

VDAs are often used in schools where one video tape machine feeds many classroom TVs at once.

VIDEO DISTRIBUTION AMPLIFIERS often come in racks with slots for many VDAs to be added to the original one, expanding its capacity.

Multiplexer

If you projected a movie onto a screen and videotaped the results, you would notice a dimness and graininess to the picture. You would also notice a fuzzy bar moving through the picture and making the picture pulse in brightness. This is called a "shutter bar" and results from the projector not making the picture images at exactly the same rate at which the camera "sees" them.

Projectors can be modified with special motors and shutters so that they work well with cameras. To improve the picture, the projection screen is omitted, and the image is projected directly into the camera through a special lens. The setup is called a TELECINE or FILM CHAIN and allows you to convert movies into video tapes using the projector, the camera, and the VCR. When only one projector is used to feed the camera, the device is called a UNIPLEXER. When you have several projectors able to feed the same camera, the device is called a MULTIPLEXER.

A MULTIPLEXER (like the one shown in Figure 15–32) has a mirror which in various positions selects which projector's image will reflect into the camera's lens. Often the projectors and the mirror system are remote-controlled from the video console.

MULTIPLEXERS can be built to handle almost any projection format. Most popular are 16mm movies, regular or super-8mm movies, slides, and filmstrips.

Double Drum Slide Projector

At home, when you show slides on a 2-inch × 2-inch slide projector (the kind that takes standard 35mm slides), you pay little attention to the blank or black screen that appears every time you change slides. On television, this pause during slide changes appears more obvious and may be annoying. The solution is to use a DOUBLE DRUM slide projector. It is really two projectors hitched together as shown in Figure 15–33. When one projector is changing

FIGURE 15-32   MULTIPLEXER.

(a) TV camera
(b) Mirror box selects which projector shines into the camera
(c) filmstrip projector
(d) 35mm DOUBLE DRUM slide projector
(e) 16mm projector

slides, the second is showing a picture. When the second is changing slides, the first is showing a picture. As you change from slide to slide, there's barely a blink as you watch the image on TV.

One complication of working with DOUBLE DRUM projectors is that you have to load your slides as follows:

1. Put slide 1 into projector 1's slot 1.
2. Put slide 2 into projector 2's slot 1.
3. Put slide 3 into projector 1's slot 2.
4. Put slide 4 into projector 2's slot 2.
5. And so on.

It gets confusing, especially when you decide to add or delete one slide; the whole sequence gets off.

Essentially, all the even-numbered slides end up in drum 2, and all the odd-numbered slides end up in drum 1.

Even more elaborate than the DOUBLE DRUM slide projector is a DISSOLVE unit which lowers the brightness of one slide while increasing the brightness of the second. Once the brightness is all the way down, this projector advances to the next slide while the twin projector keeps you busy

MINIGLOSSARY

Uniplexer   Device to couple a film projector to a TV camera, useful for making video copies of movies, etc.

FIGURE 15-33 DOUBLE DRUM slide projector.

looking at its picture. Note that this DISSOLVE unit is built into the projectors themselves and has nothing to do with your switcher. If you have a plain old DOUBLE DRUM slide projector, there is no way that you can dissolve from slide 1 to slide 2. That projector will only switch from slide 1 to 2. If you *have* to dissolve from slide 1 to slide 2, you can always set up a TV camera, projector, and screen in the darkened TV studio and dissolve from the MULTIPLEXER (showing slide 1) to camera 1's image (showing slide 2).

Video Patch Bay

Figure 15–34 shows a diagram of a VIDEO PATCH BAY or PATCH PANEL which, like a telephone operator's switchboard, can route video signals between devices. Without a PATCH BAY, one would have to reach behind the equipment

(sometimes removing it from the console) to get to the cables if a change in wiring were desired. Some of the cable connectors wouldn't mate and would require adapters.

The VIDEO PATCH BAY solves this problem by displaying all the video inputs and outputs in the open so that they are easy to reach. The connections are all standardized, and the sockets can be arranged and labeled in an organized way.

The sockets across the top of the PATCH BAY are the *outputs* from signal sources such as cameras, VTRs, and so on. The sockets across the bottom are *inputs* to monitors to the switcher, to a PROC AMP, to a TBC, to a VDA, or to a VTR. Generally, the panel is wired so that an output is directly over an input that it usually goes to.

Some PATCH BAYS are equipped with a feature called NORMAL-THROUGH. Here, when a socket has no patch cord (a patch cord is a short cable with a plug on each end) plugged into it, the PATCH BAY internally routes the signal through its typical destination (the socket directly below it). When a plug goes into a socket, that socket automatically disconnects from its typical destination and now connects with the patch cord. The signal then goes wherever the patch cord sends it. When the plug is removed from the socket, the socket automatically reconnects with its normal destination.

For example, in the PATCH BAY diagrammed, VTR 1 normally feeds into the switcher. To feed VTR 1's signal to a PROC AMP, plug one end of a patch cord into the VTR 1 socket (an output) and the other end into the PROC AMP input socket.

Another example: The switcher burns out (somebody spilled coffee into it), and you need to make two copies of

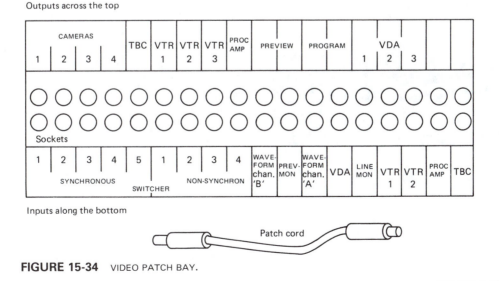

FIGURE 15-34 VIDEO PATCH BAY.

MINIGLOSSARY

Patch bay Several rows of sockets connected to the inputs and outputs of various devices. Plugging a patch cable into a pair of sockets connects them so that the signal can travel from one device to the other.

a tape right away. Also, let's assume that the master tape's sync isn't good enough to pass through the TBC but *will* go through the PROC AMP all right.

First connect the output of VTR 3 (the one with the master tape on it) to the PROC AMP input. Next, patch the PROC AMP output to the VDA input. The VDA automatically sends its signal to VTR 1 and VTR 2. Now the processed signal from the master tape can be recorded on VTR 1 and VTR 2 simultaneously.

Another example: You wish to copy a video tape for a friend, but you have a production to handle at the same time and don't want to tie up the production console with the copying.

First choose which VTR to use for the production, say VTR 1, and delegate the other two VTRs to the copying process. Patch the output of VTR 2 to the input of VTR 3. Now, regardless of what you are doing at the console, VTR 2 can play a tape while VTR 3 copies it.

If during your production you have a moment and would like to look in on the unattended VTR 3 to see how it is recording, you may. Since VTR 3 is NORMAL-THROUGH to the switcher, you may PREVIEW that signal by pressing button 3 on the NONSYNCHRON PREVIEW bus in the switcher. Such a move won't affect your recording of the regular production at all because PREVIEW signals don't get recorded.

Remember that with the standard VIDEO PATCH BAY, only the video is routed. A separate AUDIO PATCH BAY may be necessary to route your audio from device to device.

Routing Switcher

More elaborate and more convenient than the VIDEO PATCH BAY is the ROUTING SWITCHER. The 12 × 13 ROUTING SWITCHER shown in Figure 15–35 sends any of the listed

TV master control PATCH BAY.

FIGURE 15-35   12 × 13 ROUTING SWITCHER.

12 sources to any of the listed 13 inputs. Instead of connecting the inputs to outputs by patch cables, you merely press a button, and the signal is routed. Pressing the button labeled "VT1" directly across from the label marked "WFM" will send the output of VTR 1 to the input of the WAVEFORM monitor.

To copy a tape from VCR 4 into VCR 2, simply press the button marked "VC4," across from where it says "VC2 IN" down in row 11. If you wish to simultaneously monitor the tape as it plays, press "VC4" across from "WFM." Now you're routing VCR 4 to both the WAVEFORM monitor *and* VCR 2. Notice how you can send one signal to two places.

This particular ROUTING SWITCHER is also set up to send its signals to many rooms around a campus (thus the room listings "C100s," "C200s"). No spaghetti of patch cords, just buttons. Too bad it costs about $10,000.

The better ROUTING SWITCHERS have an AUDIO-FOL-LOW-VIDEO feature, which means that when you route the video from one place to another, the audio will also follow. It's like having two switchers in one.

MINIGLOSSARY

Synchronous   Synchronized, running to the same electronic rhythm. Synchronized video sources can be mixed, wiped, or switched without a glitch.

*Routing switcher   Switch that sends the signal from one of several sources to several destinations. A push-button version of a patch bay.

Smaller ROUTING SWITCHERS are also sold to the home video markets for connecting two or three machines to three or four sources. These ROUTING SWITCHERS carry RF and are especially handy for routing antenna and cable signals to various recorders and TVs and allowing different TVs to display the signal from videodisc players, VCPs, or broadcast television.

Modulator or RF Generator

You'll remember from Chapter 5 that all home VCRs and some industrial VCRs have MODULATORS built into them so that they put out channel 3 or 4 as well as separate video and audio. Some video devices don't have MODULATORS, and you must buy a separate one if you wish to use RF.

RF is handy for sending a signal to many places at once or for sending several signals down the same wire. In a school building where classes in different rooms are watching different stations on TVs connected to a cable, the RF in that cable has to come from somewhere. Sometimes it comes from a MASTER ANTENNA that picks up broadcast stations off the air. It is possible to create a station right inside the school building by using a video tape player and a MODULATOR.

The tape player sends out audio and video signals to the MODULATOR, which combines them and codes them into a channel number. Several MODULATORS can be connected to the same RF cable, making it possible to play tapes from three VCPs simultaneously into three MODULATORS which send the signal out over three channel numbers sharing the same RF cable. Any TV receiver can connect to the cable and can, by switching to the proper channel, receive each of the three programs. Figure 15–36 diagrams the process.

Four inputs × 3 outputs (Courtesy of Video Commander)

RF ROUTING SWITCHER used by home videophiles.

Long-Range Cable Runs

No discussion of video distribution would be complete without considering the special problems encountered when sending video or RF signals a long way.

If you are sending TV signals around a campus, for instance, and your TV playback facility is in one building and your classrooms are in other buildings 1000 feet away, you can expect your signal to become very weak by the time it reaches those buildings. Here are some steps to take to ensure a strong signal:

1. Carefully select a distribution cable which has very low resistance. The common RG-59U used around your studio is inadequate for long runs. You may need RG-6U or better. For very long RF runs, you may need to use the same kind of wire that cable TV stations use, HARDLINE. It's expensive but very efficient.

2. Amplifiers may be needed along the way in order to give the signal a "kick," making up for the losses in the cable.

3. Higher frequencies don't travel through a cable as well as lower frequencies. Video, at 4.5 MHz, is a relatively low frequency and travels through long cables easily (but you will need a separate wire to carry your audio). The lower RF channels, like channels 2, 3, and 4, start at 54 MHz and go through long cables with medium difficulty. Channels 7–13 are around 200 MHz and experience greater difficulty going through long cables. Hardest hit are the cable SUPER BAND channels J–W, which run from 216 to 318 MHz. These signals are likely to become very grainy. Almost impossible to send over long wires is UHF TV, 470–890 MHz.

Figure 15–37 shows a typical setup for sending RF or video signals from one point to several buildings over long wires.

Sometimes when you run a long wire through a field or forest from one building to another, you may need to amplify the signal along the way, but you don't have any electricity that you can get to your amplifier to power it. In such cases, you can use an IN-LINE AMPLIFIER, a special amplifier which is able to receive its power over the same video cable that its signal is coming over. The power that feeds the IN-LINE AMP comes from a POWER SUPPLY which

MINIGLOSSARY

Audio-follow-video A special switch that routes an audio signal along with the video signal at the press of a single button, like two switches in one.

Hardline Special low-loss wire used principally by cable TV companies for long cable runs.

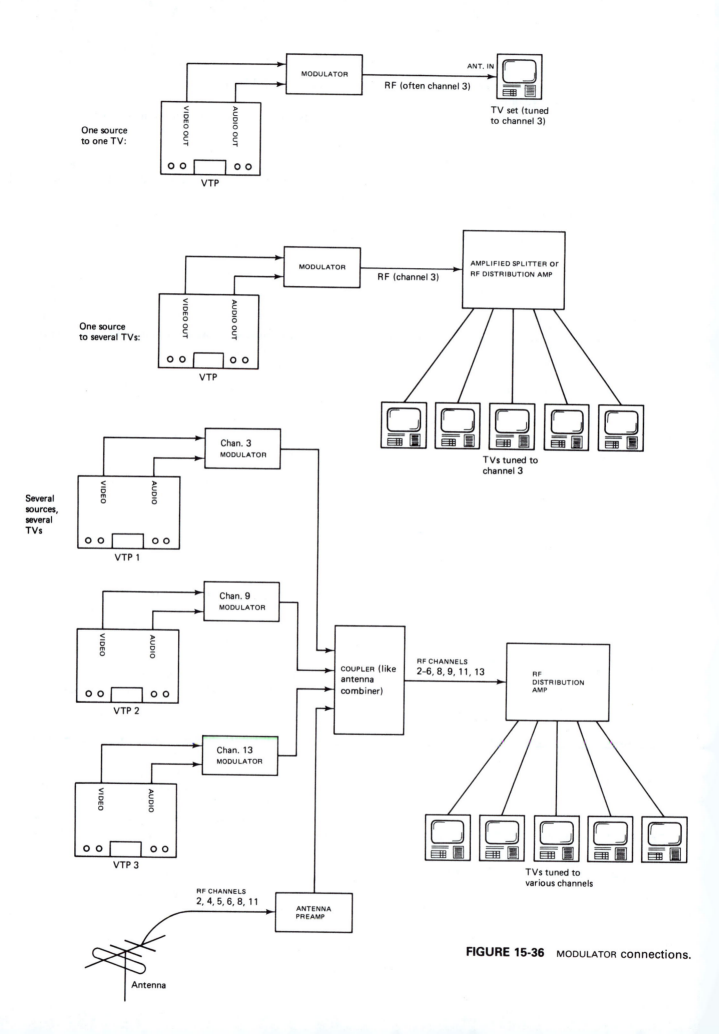

FIGURE 15-36 MODULATOR connections.

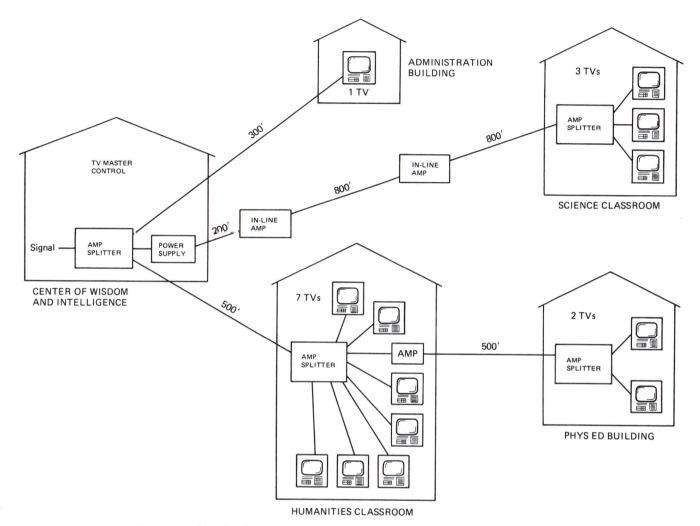

FIGURE 15-37 Long-distance cable distribution.

is also connected to the cable, usually in a handier place near electricity, like in a building.

In short, if you are sending a signal over a long cable, it probably needs amplification before it starts its journey and may need an extra boost along the way if the cable is very long. When the signal arrives at its destination, it may be so weak that it cannot be split, so it will need an additional AMPLIFIED SPLITTER to distribute the signal to various TV sets.

Infrared Transmitters

Just as the INFRARED REMOTE CONTROL on your VCR or TV can send invisible light messages across the room, larger-scale INFRARED OPTICAL TRANSMITTERS can beam strong infrared signals up to one-half mile carrying audio and video information. The devices cost about $4000 per pair (you need a TRANSMITTER and a RECEIVER) and save you the expense of digging up your parking lots, lawns, and build-

---

MINIGLOSSARY

In-line amplifier An amplifier inserted between two wires to boost the signal through them, and gets its operating power through the same wires from a distant power supply.

Infrared transmitter Electronic device to convert electronic signals, such as audio and video, to infrared (invisible to the eye) light beams. Beams travel through air and can be converted back to electrical signals by an infrared receiver.

Infrared receiver Device to convert an infrared light beam into electrical signals.

*Microwave Extremely high-frequency radio waves, about 1 billion vibrations per second (1 GHz, one gigahertz), used to transmit video, audio, RF, telephone, and computer data over long distances.

*Microwave transmitter Device to convert video and audio signals into microwaves for broadcasting to a receiver which can convert the signals back.

*Microwave receiver Device for picking up microwaves and converting them into an electrical signal such as video and audio.

INFRARED TRANSMITTER and RECEIVER (Courtesy American Laser Systems).

ings to lay video cable between them.

An additional advantage of INFRARED TRANSMITTER/ RECEIVERS is that they aren't as permanent as cable. If you decide to do business with building A for a year and then wish to switch to building B, you won't have to waste the cables you buried in the ground going to building A and lay brand new cables in the ground going to building B.

It is possible for one centrally located building to have several TRANSMITTERS feeding RECEIVERS at several buildings. For each link, one TRANSMITTER/RECEIVER pair is needed.

It will still be necessary to use normal wire to send your signals to the TRANSMITTER and also to take the signals from the RECEIVER and bring them to where you want them. But this process generally involves laying small amounts of cable indoors and is easy to do.

The INFRARED signal is disrupted by very heavy fog and tree branches. It is absolutely necessary to have a clear line of sight between each TRANSMITTER and each RECEIVER to get a good signal. Rain and snow do not affect the signal very much.

Figure 15–38 diagrams an INFRARED TRANSMITTER/ RECEIVER setup.

Microwave Transmitters

MICROWAVES are very, very high-frequency radio waves, higher even than UHF. You can put a TV signal (audio and video) on a MICROWAVE frequency and transmit it just like any TV transmitter could.

MICROWAVES are generally used for long-distance

transmissions. Cable companies often use MICROWAVE TRANSMITTER/RECEIVER pairs to send signals 5–30 miles, usually from mountaintop to mountaintop. The process is cheaper than laying cable (with booster amplifiers) all those miles.

MICROWAVES have their weaknesses, too:

1. A MICROWAVE TRANSMITTER/RECEIVER combination will cost over $20,000.
2. You need an FCC license to install one.
3. The signals travel only in straight lines, which means you need to send/receive from the tops of buildings, mountains, or towers.

The setup looks just the same as Figure 15–38 only MICROWAVE transmitters are used instead of INFRARED transmitters and the distance could be up to 30 miles.

MICROWAVE TRANSMITTERS are often used by television news teams doing live broadcasts. Since they are far from the studio, they need to send their signal to a portable MICROWAVE TRANSMITTER, usually mounted on a truck, which is aimed at a receiver at the top of a mountain or at the top of the studio building.

Sometimes a direct line of sight cannot be found between the news area and the studio. In this case, a MICROWAVE RELAY is necessary. The news team sends their signal to a MICROWAVE TRANSMITTER atop their mobile van, and that transmitter sends the signal to a helicopter overhead or a receiver on a mountaintop within sight. These receivers then retransmit the signal using another MICROWAVE TRANSMITTER which *can* "see" the TV station. The MICROWAVE

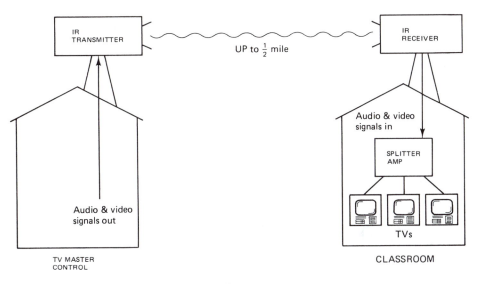

FIGURE 15-38 INFRARED TRANSMITTER/RECEIVER setup.

RECEIVER atop the TV station's building receives the signal from the RELAY and passes it on to the television master control where it can be mixed with other signals. Figure 15–39 diagrams the process.

Fiber Optics

It is possible to send a television signal as a light wave rather than vibrating electromagnetism. Instead of sending your signal down a wire, you could send it down a hair-thin, long glass tube. The process is called FIBER OPTICS and has four great advantages:

1. The signal can be transmitted a mile or two with very little degradation.
2. The signal is insensitive to electrical interference such as lightning, power lines, and nearby radio stations.
3. Many TV channels can be sent over the same fiber at the same time.
4. Fibers may be bundled together so that numerous separate programs can travel over the same umbilical.

FIBER OPTIC cable is not expensive, but the transmitters and receivers which convert the video or RF signals into light and back from light into electronic signals again are expensive. For long cable runs, though, this may be a better

deal than spending money on premium-quality coax cable and numerous amplifiers.

FIBER OPTICS can carry signals 2 miles without a boost. After that, their signals need to be amplified and can then travel another 2 miles.

ITFS

ITFS stands for instructional television fixed service. It is used mostly by schools who need to broadcast their signals throughout a school district with school buildings spread out over a 20-square-mile area. Unlike the other systems discussed, this system uses *broadcasting* which sends its signals in all directions at once rather than in one specific beam or through a cable to a single destination.

ITFS uses frequencies higher than television frequencies, which means your TV signal must be converted to one of the superhigh ITFS channels to be transmitted. At the receiving end the signal is downconverted back into regular TV signals.

ITFS stations are very weak (10 watts or less), covering only a small area like a school district. This way, other ITFS stations can be used in neighboring municipalities.

The ITFS spectrum consists of 28 channels, and it is possible for a broadcaster to use several channels at once to supply several different TV programs to various schools at the same time.

MINIGLOSSARY

Microwave relay Device that receives microwaves, boosts them, and beams them out in another direction.

Fiber optics Technique of converting a signal (such as audio or

video) to a light beam and sending it down a hair-thin strand of glass. Light beams can travel several miles without amplification. The signal is then converted from light back to an electrical signal.

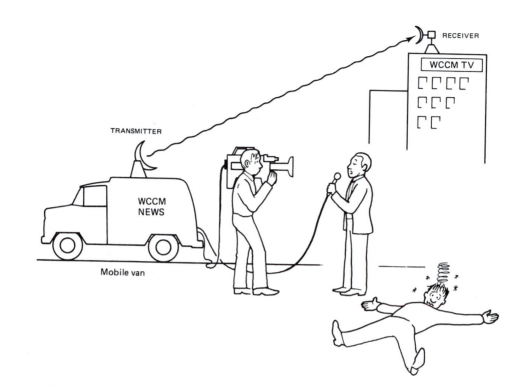

FIGURE 15-39   MICROWAVE links from "live" news crew to station.

To run an ITFS station, you need an FCC license.

ITFS TRANSMITTER/RECEIVERS are somewhat expensive at $50,000 per pair. This compares favorably with stringing a web of cable over an entire city, though.

LPTV

LPTV stands for low-power television. The FCC recently made it possible for schools, municipalities, and private industry to own and operate low-power television transmitters. These transmitters would cover small areas about 12–15 miles in radius. Their advantages are the following:

1. Unlike a regular TV station which may cost $500,000 to build, LPTV startup costs are as low as $25,000. The equipment is also relatively easy to maintain.
2. Because VHF channels are used, everyone with a normal TV set and antenna will be capable of picking up your station in the local area. This makes it possible to present telecourses, local news, or items of municipal interest to residents in an area without resorting to cable TV or harder-to-handle UHF transmission.
3. It is usually possible to find a VHF frequency which is not being used by a commercial broadcaster in your area. Where a high-powered TV station might interfere with some distant broadcaster, your low-power TV station's signal will die out before it reaches distant viewers.

OTHER ACCESSORIES

There are some TV gizmos which are hard to categorize, so I'll put them here.

Genlock

When you wish to neatly switch, dissolve, or do special effects between cameras, you synchronize the cameras with one sync source, the sync generator. Some devices, such as character generators, videodisc players, and portable TV cameras, are designed to work alone *or* work with your TV system. When tied to your TV system, they have to follow the sync from your house sync generator. They can do this easily if they have a built-in GENLOCK capability.

GENLOCK allows a video device to "listen" to video or sync being fed to it from some other source. It uses this signal to synchronize its circuit and send out a video signal in step with the rest of your system. For instance, a character generator that can be GENLOCKED would listen to your house sync system and would create text which you could superimpose or key over the pictures from your other cameras.

Another example: If you had two run-of-the-mill portable TV cameras and wished to hitch them up to a switcher or fader, you would have difficulty. Each camera runs independently and will not be synchronized with the other. Your switches would be glitchy. You wouldn't be able to dissolve. If camera 2 were GENLOCKABLE, then you could send a taste of camera 1's video signal to the second camera to "lock it up," and the second camera will now create a signal that can be mixed or switched with the first camera without a glitch.

Some GENLOCKERS are designed to run an entire TV system. They "listen" to an outside video or sync source and create sync signals which can be sent to the rest of the equipment in your system to drive the whole system.

For example, you wish to do special effects between a VCP and a camera. To do so, they must be synchronized to the same sync source. If you allowed the VCP to play its tape and have that signal go through the GENLOCK device, the GENLOCKER (instead of your normal sync generator) could synchronize the cameras and the rest of your video system. Since now the VCP and cameras are synchronized, their signals are mixable.

The GENLOCKER differs from the normal sync generator in the way that it generates its signal. The sync generator has an accurate clock inside it to make the sync pulses. GENLOCK, on the other hand, uses the signal from the VCP to make its pulses. GENLOCK is a very convenient tool but has a few drawbacks.

Since VCPs don't play back signals with perfect timing, everything in the studio GENLOCKED to these imperfect signals will create imperfectly timed signals. The picture from camera 1, for instance, will be just as shaky as the picture from the VCP. Whether this poses a problem to you or not depends on

1. How much of a perfectionist you need to be
2. Whether the tape will be played on forgiving TV sets
3. Whether the tape will be edited, time-base-corrected, or broadcast

One solution to the jitter problem may be to use a TBC to improve the timing of the signal from the VCP. Some TBCs are designed to perform the GENLOCKING task

---

MINIGLOSSARY

ITFS Instructional television fixed service—a method of broadcasting TV programs throughout school systems via low-power high-frequency transmitters.

LPTV Low-power television—the technique of broadcasting lo-cal programming through a very low-power, inexpensive VHF TV transmitter. Limited signal range keeps LPTV stations from interfering with distant TV stations using the same channel frequency.

while correcting the signal from the VCP at the same time. The result is a very stable final product—even the cameras' pictures will be more stable.

The second drawback to the GENLOCK setup is the following: What happens if the VCP stops, rewinds, or pauses? What synchronizes everything then? You know the sync source that keeps everything afloat is derived from the VCP faithfully playing back a tape. Stopping, starting, rewinding, and pausing the tape will discombobulate the whole sync system in the GENLOCK mode. What is needed is a way to switch smoothly from the studio's sync generator to GENLOCK when it is needed and then back to the sync generator again when the tape segment has been used.

The simplest way to make that changeover is to shoot all the studio scenes using sync from the sync generator. When you come to a scene which mixes both tape and live material, edit to that scene while GENLOCKING your whole TV system to your VCP throughout the scene. When the live-plus-recorded scene is over, stop taping, switch the system back to the house sync generator, edit to the live-only video material, and proceed from there. Continue using the sync generator until you come to the next editable scene requiring GENLOCK.

GENLOCK is often built into house sync generators, which simplifies your wiring. As shown in Figure 15–40, the house sync generator can run the entire system alone, or, by throwing a switch, the house sync generator can then become a GENLOCKER and listen to the sync coming from some other source, like a VTP.

Bulk Tape Eraser

A BULK TAPE ERASER is a black box (Figure 15–41), which plugs into a wall outlet and has a push button on it. When turned on, the device creates a powerful magnetic field capable of erasing any magnetic recording.

It is useful for quickly clearing material off tapes without bothering to run them through recorders (which automatically erase tapes as they record on them).

They are very handy when you have used a tape from beginning to end and now wish to put a new program on the first half of the tape. When your new show finishes, you don't want your audience to see the remaining leftover half of the old program. (Imagine your boss and stockholders catching remnants of ''Lust in the Woods'' which didn't get erased from last New Year's party.) The BULK TAPE ERASER will wipe the entire tape clean so that you may now trust it.

BULK ERASERS *cannot* erase *part* of a tape. They erase the *whole reel* at once.

Some types of tape, such as CHROME (CrO_2) and METAL PARTICLE, used in SVHS, Hi8, MII and Betacam VCRs, are harder to erase than others. They require BULK ERASERS that are bigger, more powerful, and more expensive

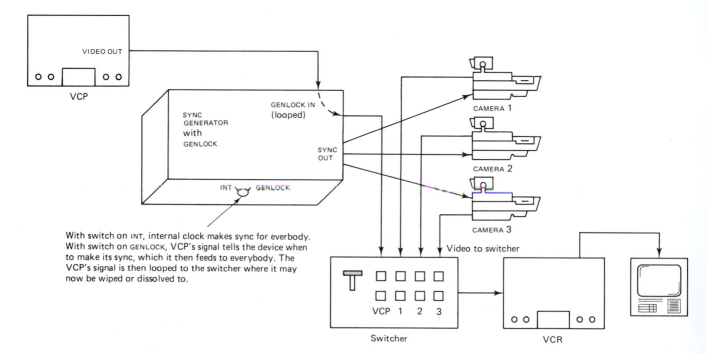

With switch on INT, internal clock makes sync for everbody. With switch on GENLOCK, VCP's signal tells the device when to make its sync, which it then feeds to everybody. The VCP's signal is then looped to the switcher where it may now be wiped or dissolved to.

FIGURE 15-40 TV system GENLOCKED to a VCP.

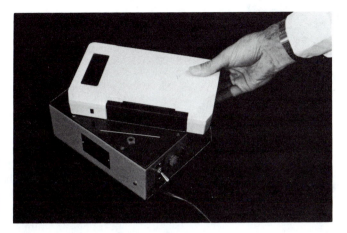

FIGURE 15-41   BULK TAPE ERASER.

(of course) than the erasers used for common VHS cassettes and audiocassettes.

To use a BULK TAPE ERASER,

1. Bring the BULK ERASER and the tape to be erased at least 6 feet from your other tapes so as not to erase them slightly with any stray magnetic fields.

2. Place the tape on the ERASER and press the ON button and hold it down.

3. Rotate the tape a few times so that all of the reel or cassette ends up passing over the device.

4. Just to be thorough, turn the cassette or reel over and repeat the process with the other side facing the machine. This step is especially necessary if you are erasing a thick tape, like ¾U or 1-inch.

5. *With the button still held down*, lift the tape from the ERASER and hold it out to arm's length from the machine and let up on the button.

The reason you hold the tape away from the BULK ERASER when you let up on the button is that when you turn the ERASER off, it creates a surge of magnetism that leaves a small signal on the tape. Holding the tape away keeps this small pulse from being recorded on the tape.

One little caution about using a BULK ERASER: If you wear an electric watch, take it off. The powerful magnetic field may scramble your circuits (something like this could get one ticked off!).

VIDEO MAINTENANCE

An ounce of prevention is worth a pound of aspirin. Today's video equipment, though very complex and expensive to fix, rarely breaks down, if properly maintained. You can save a lot of repairs by caring for your video equipment. I don't mean romantically; I mean like someone who just bought a $1000 piece of artwork. You would probably grace it with a dust cover when it wasn't in use, right? You probably won't set drinks or pizza slices on it either. Although this may seem a bit extreme, you might even forbid smoking in the same room. Smoke film builds up on VCR innards and glass lenses just as it does on windows and curtains.

VCRs and other video gear generally don't like the following things and will find frequent excuses to visit the repair shop to avoid them:

1. *Dust*: chalk dust, plaster, cement, sawdust from renovations, beach sand

2. *Dampness*: rain, sprays (including beach salt sprays), dripping bathing suits, spilled drinks, damp basement storage, being used cold in a warm room (water vapor condenses inside)

3. *Shock*: rough handling during transportation (often due to inadequate padding), falling over after being balanced on something

4. *Heat*: radiator tops in winter, car trunks in the summer, operation without ventilation (sitting on a deep pile carpet, for instance)

PREVENTATIVE MAINTENANCE SCHEDULE

Preventative maintenance is as much a philosophy as it is a skill. Preventative maintenance is the art of cleaning, lubricating, and tweaking things *before* they actually break down. Some people, however, swear by the principle "If it ain't broke, don't fix it." This is not a ridiculous philosophy, either. I've seen equipment sit in one place and work for years without a spot of attention, while other equipment which gets jostled, opened, and probed and has its circuit boards withdrawn and reinjected comes down with major and exotic ailments.

Perhaps the best policy is to consistently *check* the performance of your video equipment to ascertain if it is "on spec." If it is, leave it alone. If it isn't, then it's time to tweak. Furthermore, keep it clean and keep its screws and bolts tight.

There are two times when equipment especially deserves to be checked for proper operation:

Video maintenance.

1. Before a very important shoot, especially a portable production

2. After a portable shoot in a hostile environment

Sometimes the hostile environment walks right into your studio. A person may bring you old, shedding video tapes to be copied. You should expect your VCR heads to clog like crazy. A rock band performs in your studio complete with smoky fireworks and fog effects. Expect the oily film to coat your lenses, studio lamps, and everything. Lightning strikes your building, blowing your TV antenna to smithereens. It is wise to check each device that was plugged in, even if it was turned off, to see if its circuits were damaged. Superdry weather builds up a static charge on everything. Each machine receives a blue spark from your fingertips as you travel from device to device. Most video equipment, if well grounded, will shrug off this static charge, but it would be good insurance to spray antistatic treatment on your carpeting to reduce the problem (or of course you could always ground yourself by wrapping one end of a long wire around your leg and the other end around a cold-water pipe).

Table 16–1 suggests a maintenance schedule for common video equipment used 20 hours per week, 50 weeks

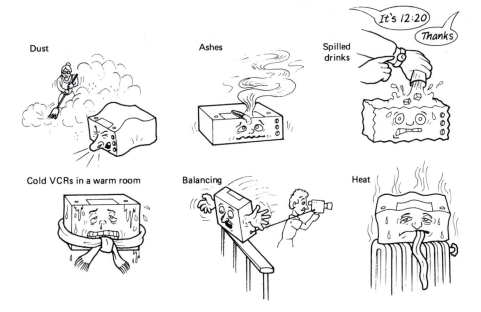

Things VCRs and other video equipment abhor.

TABLE 16–1
Suggested maintenance schedule for common video equipment

Equipment	Weekly	Bimonthly	Yearly
Color camera	Check registration if tubed. Check for loose parts.	Check complete camera setup, levels, color, reproduction.	Check lens. Overhaul optics if necessary.
VCPs	Clean heads	Check brakes, clutches, tape tension. Make a sample test recording of color bars, etc., and examine carefully.	Lubricate. Check for worn guides (look for grooves or flattened places). Vacuum out dust.
Editing VCRs	Same as above	Same, but check video heads for wear.	Same, but check all heads for wear and replace rubber parts, such as pinch roller, belts.
Monitors		Wipe down screen. Check proper cooling.	Check color convergence with test signals. Replace missing knobs.
Audio console and power amps		Check cooling fans and filters. Check loose knobs.	Test meters for accuracy. Do performance check with test signals to examine output levels, frequency response, signal-to-noise ratio.
Audio tape recorders	Clean heads Demagnetise heads	Check heads for wear (grooves, flat spots). Check pinch roller; is it glazed and slipping?	Check tape tension. With test tape, check overall performance, frequency response, noise.
TBCs, sync generators, switchers		Check fans and filters. Check burst frequency.	Using test signals, do a complete performance check.
Tape tension measuring device and other test signal generators			Test calibrations to assure *they* are telling the truth.

per year for a total of 1000 hours annually. The table assumes that your operating environment is clean, the users are gentle, and high technical standards are required of your video signals.

You should expect your yearly repair and maintenance costs for your entire TV operation to equal about 10% of the total value of your equipment. This is for parts only; labor is extra.

Keeping accurate maintenance records from the day the equipment is installed will help you spot repair trends. You will then be able to judge how many video recorder heads, belts, rollers, transistors, and circuit chips to stock up on. When ordered, some of these parts take a long time to arrive. Having the most-likely-to-go parts stockpiled will save you the embarrassment of shutting down your entire video system for 3 weeks while a replacement framus is mailed from Tokyo.

Another use for your maintenance schedule is to help you determine when to ditch ailing old dinosaurs. Tell your boss, ''My equipment is worn out—I need new stuff,'' and your words may go straight into the wastebasket. Tell your boss, ''I have a 5-year-old VCR which has a market value of $1000. It is costing us between $2000 and $3000 per year to maintain. A new one with more features would cost

$5000. I suggest that we replace it as soon as possible.'' Words like these linger in your boss's (and accountant's) ears much longer and are likely to get you the equipment you need.

REPAIR SERVICES

If yours is a big $250,000 operation, then you'll probably have a full-time TV engineer on staff to repair your equipment. The engineer will have test signal generators and tape tension gauges galore and can fix almost anything, save the most complex gadgets which only the inventor and God can understand.

The $25,000 studio is more likely to be maintained by an AV or TV repairperson working from some other department. This person can easily solder broken wires in a connector and replace vidicon tubes and may even be able to replace the heads on a VCR. But deeper difficulties may have to be sent out to the manufacturer or an authorized service center.

The smaller your video operation, the more likely it is that you will be the person maintaining the equipment and most likely it is you who will have to send it to someone else for real repair. Some schools and small video users buy MAINTENANCE CONTRACTS, agreements made with an outside repair facility guaranteeing the upkeep of your equipment for a fixed yearly cost plus parts. Some MAINTENANCE CONTRACTS even include a yearly "tune-up," a very minimal preventive maintenance schedule.

Individuals who buy home video equipment are often pressured by the store salespeople to purchase a MAINTENANCE AGREEMENT which lasts for about 2 years. Such an agreement for a single VHS VCR might cost $125. These are generally a bad deal because

1. If something is likely to go wrong with the VCR, it will happen within the first 5 days after you take it out of the box. Therefore, it is wise to use the heck out of the equipment for the first 5 days to spot such potential problems. Most dealers will take their equipment back within the first 5 days and will replace it with another, assuming you kept the box and packing and didn't damage anything.

2. The manufacturer's warranty lasts for 90 days to a year, so this time is already covered without the help of the SERVICE CONTRACT.

3. VCRs generally don't break down during their first 2 years of use.

Video repair costs start at about $45 per hour plus parts. They end at about $125/hr plus an arm and a leg. Some common expenses are the following:

Item	Labor	Parts
VCR video head	$45	$125
VCR brakes or belts	30	20
VCR motor	35	40
17-inch color TV picture tube	90	175
Integrated circuit replacement	20	35
Power supply	35	35
Tuner	35	60

These rates assume the repairpeople know which parts to replace and do not have to spend hours troubleshooting an elusive gremlin. Arrange to have them phone you if costs exceed a set amount.

CLEANING VIDEO HEADS AND TAPE PATH

The cleaning of the video heads and tape path is the most common video maintenance you are likely to do.

When to Clean

If you play a tape that you know is good and you get a very grainy or totally snowy picture (as in Figure 2–10) but the sound remains okay, then you've most likely got dirty video heads. This problem may come on gradually or all of a sudden.

Be absolutely sure that the tape you are testing your VCR on is indeed good (imagine scrubbing the heads for hours only to find out that the tape was bad). What can you do to *really* be sure your sample tape is perfect? One solution is to buy a TEST TAPE which is made under stringent conditions so it's "perfect." It also contains certain test signals to assist you in calibrating and troubleshooting other VCR and TV problems. One such TEST TAPE marketed by the Society of Motion Picture and Television Engineers (SMPTE) is called aptly *the SMPTE Video Tape Cassette for Receiver/Monitor Setup*.

If you've just cleaned the heads and the problem still persists, clean the heads again. Unlike the overalls in detergent commercials, the heads sometimes require several scrubbings.

If you run the VCR near the sea or near acid or alkali gases or in a dusty or dirty environment, expect to clean the heads once per hour or so. The same is true if you use flaky, wrinkled, or old tape.

If you perform a lot of pausing and editing, the heads get dirty quicker. Under heavy use, clean the heads once a day.

If the VCR is used in a clean, climate-controlled studio, it can probably go for weeks without cleaning. Closing the cover over the VCR between uses also lengthens the interval between cleanings.

When shooting on location, make frequent test recordings and always carry head cleaner with you.

If you make a test *recording* and play it back and find the picture is snowy, you've probably got dirty heads.

If numerous serious head cleanings have not solved your snowy picture problems, it may be that your heads are worn out. Video heads generally last 1500–2500 hours. After that, as the heads wear, your picture becomes grainy. You lose sharpness. Colors smear, especially the saturated reds. Dropouts seem more prevalent. In the bitter end, you finally get snow.

MINIGLOSSARY

*Maintenance contract (or service contract) An agreement with an outside repair company to keep your equipment in good working order for a fixed yearly charge (plus parts, usually).

Test tape A video tape made under "perfect" conditions, used to test the performance of VCRs, VCPs, and sometimes other video equipment. Tape may include special test signals for measuring signal strengths, timing, and purity.

Although head wear is gradual, sometimes heads can be damaged by snagging on a wrinkled video tape or splice or by riding over a few grains of sand.

A dirty tape path can also cause picture problems. If you play a tape that you know is good and the picture is unstable, tracks poorly, or FLAGWAVES uncontrollably or if the VCR motor speed wanders (you can hear the speed changing), it may be due to a dirty CONTROL TRACK HEAD. As one of the components in the tape path, the CONTROL HEAD guides the speed of the motor and keeps the picture stable. The problem could also be the fault of a dirty CAPSTAN. If oily, the CAPSTAN may slip and fail to keep the tape moving at the smooth, constant pace that is necessary for a stable picture. Sometimes the instability problems can be caused by a lumpy or dirty PINCH ROLLER. It, too, can upset the constant speed of the tape as it passes through the machine.

If the tape squeaks or sticks as it passes through the VCR, clean the tape path. (This problem may also be caused by a cold VCR or a cold video tape. In this case, let them warm up to room temperature before you use them.)

What to Use

If you've been reading the previous chapters, you know you don't use Head and Shoulders and a scrub brush to clean video heads. Essentially, you have four choices for cleaning the heads and the tape path.

1. Use a nonabrasive head cleaning cassette.
2. Use an abrasive head cleaning cassette.
3. Open the VCR and manually clean the heads and tape path with a swab and solvent.
4. Send the machine out for cleaning.

Number 1 above is the easiest and least harmful but is ineffective against really tough dirt. Number 2 is easy and works pretty well but wears down your heads. Use this method sparingly. Number 3 is best—*if* you are handy with a screwdriver (or a vodka collins) and a swab. It's thorough and is effective against sticky dirt. Number 4 is expensive and unnecessary but not a bad idea *if* you are having your machines periodically cleaned, adjusted, and lubricated while it's on the shop's operating table.

Bring a head cleaning cassette with you when shooting on location. If your video heads clog, you can fix them in a few seconds by chucking in a head cleaning cassette. Otherwise, you might get stuck trying to disassemble your

portable VCR to the sound of jungle drums or the sway of a sailboat in a storm.

Let's study each cleaning technique in detail.

Nonabrasive Head Cleaning Cassette. These look like regular videocassettes but contain a cleaning fabric instead of normal recording tape. Some brands you simply slip into the cassette compartment and PLAY. Hit STOP after about 15–30 seconds and you are done. Do not rewind. EJECT, and the cassette is ready to use again. Other brands require you to wet the fabric with some head cleaning fluid before you insert the cassette and play it.

Nonabrasive casettes give a very gentle cleaning. This is both good and bad. They are gentle on your video heads, not sanding them down like the abrasive cleaners do. However, they are not very effective against stubborn dirt or head clogs. For this reason, it's good to use them often enough to avoid any major buildup. About once every 100–500 hours should do it. It is also advisable to clean the heads before storing your machine or leaving it idle for a long time. Just as tomato sauce and grape juice have a way of staining your white dinner jacket if not attended to immediately, video dirt has a tendency to "harden" with time, making it more difficult to remove after it has "set."

Some brands of cleaning cassettes should be used only once, while others are used to the end, rewound and used over. Naturally, the more they are used, the dirtier the fabric becomes and the less cleansing occurs. Conceivably, the crud could even come off the fabric and *add* dirt to our heads and tape path.

Head cleaning cassettes cost $15–$25.

Abrasive Cleaning Cassette. Abrasive cassettes are usually made of unfinished video tape, tape which is not polished smooth. The tape acts like very fine sandpaper which cleans the heads while renewing their surface. While buffing the surface, the abrasive is also wearing down the video heads, so use this cleaning method only when actually necessary and apply the treatment only briefly. You might use the abrasive cleaner only once for every 500 hours of playing time and run the cleaner through the recorder for only about 5 seconds for each application.

The abrasive cleaners are quite effective, cleaning out stubborn head clogs that the nonabrasives can't get. The use of abrasive cleaners is somewhat contested among the experts, however, and their overuse or misuse could be hazardous to your video head's health.

Some brands of abrasive head-cleaning cassettes have a message recorded on the cleaning cassette which may say,

MINIGLOSSARY

*Capstan   Shiny rotating wheel inside a VCR to draw tape through the machine at the proper speed.

Pinch roller   Rubber wheel near the capstan that pinches the tape against the capstan, so it can grip the tape as it pulls it through

the mechanism.

*Head-cleaning cassette   A cassette loaded with a ribbon of material (it could be cloth) which cleans the video heads as the cassette is played.

''When you can read this message, your heads are clean. Stop the recorder now.''

Swabs and Solvent. You can use cotton swabs such as Q-Tips, but they tend to leave cotton fibers behind. It is better to use chamois-covered sticks or foam swabs. If you are creative, you could buy a single sheet of chamois from an auto parts store, cut it up into little squares and glue the squares to tongue depressors, and have a lifetime supply for very few dollars.

You can buy VCR cleaning fluid or use ethylene dichloride, or liquid Freon, or ethyl alcohol. Ethyl alcohol is commonly available at drug stores. Its toxic brother methanol or denatured alcohol is sold in paint stores.

Don't use rubbing alcohol. It has water and oil mixed into it.

Don't use water or soap and water to clean heads. You'd be risking residue and rust.

There are some professional spray can solvents on the market for cleaning heads. Do not just blast the dirt away with a squirt from the can. The spray is icy cold and can shatter warm heads. Instead, spray a swab and use *it* to scrub the heads.

If you're shooting on location with actors costing $100 per hour, and the only VCR within 100 miles has a clogged head, and you don't have head cleaner, it's time to take a chance. Open the machine, stick your T-shirt over your finger, spit on the end, and wipe the heads. Inelegant but effective.

Cleaning Heads and Tape Path Manually

One problem with head-cleaning cassettes (besides their expense) is that they aren't thorough. They don't remove stubborn dirt very effectively, and they can't clean places other than their route along the tape path. You can do a better job by hand.

You can also void your warranty with some companies if you disassemble your machine to clean the heads. You'll need a phillips or, better yet, a cross-point screwdriver for removing the screws that hold down the VCR's cover. If you don't have such a screwdriver, go out and buy one that's the right size and of good quality, so it stays sharp and doesn't grind down your screw heads.

To clean the video heads, follow the manufacturer's instructions, if you have them. Otherwise, follow these general procedures and refer to Figures 16–1 and 16–2:

1. Turn off the power to the VCR. Remove the tape, at least from the area where you will be working. You don't want the solvents dripping on the tape. Place the machine on a clean uncluttered worktable. Assemble your tools so they are handy. If your machine has been in use, let it cool down for a half hour before you commence work. Wash

The two (or four for some VCRs) video heads are in the head drum (shown here close-up). Wipe swabs back-and-forth across each head while holding the drum steady with the other hand. *Never wipe up-and-down.*

FIGURE 16-1 Video HEAD DRUM. (Photo courtesy of William Ferreira)

your hands thoroughly and dry them. Extinguish any cigarettes, etc., as many head-cleaning fluids are flammable. Also, ashes dropping into your machine aren't helping you clean it. Besides, smoking is bad for your innards, too.

2. Following the manufacturer's maintenance instructions, loosen and remove the screws holding down the VCR's lid. Use your phillips screwdriver for this. On some machines it is necessary to remove a knob or two or to pop the cassette lid up before you can remove the machine's top.

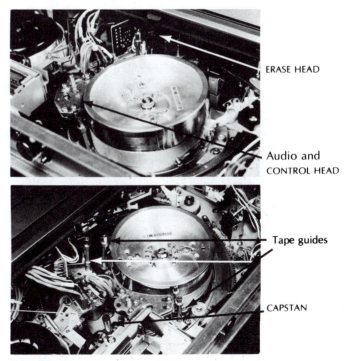

ERASE HEAD

Audio and CONTROL HEAD

Tape guides

CAPSTAN

FIGURE 16-2 Cleaning a VCR's tape path (Photo courtesy of William Ferreira).

3. Locate the HEAD DRUM, a round silver cylinder near the center of the VCR. It will have two small black indentations on the side of it, 180° apart. These are the famed video heads. Some machines have four video heads 90° apart on the drum. On most machines, the head and drum are attached and rotate together. To bring a head around to where you can see it, simply rotate the drum with your hand. Try not to leave fingerprints on the shiny curved outside edge of the drum. On other machines, the drum may be stationary, with the head spinning inside it, peeking out through a gap in the drum. To bring the head into view, locate the VCR's motor and rotate the motor fan shroud. This will move the heads. You'll see them appear somewhere along the gap in the head drum.

4. Moisten a swab with cleaning fluid. If using Freon, cap your bottle immediately as the stuff evaporates like crazy.

5. Wipe the swab *horizontally* back and forth across a video head while holding the head drum stationary with the other hand or by immobilizing the opposite head with your fingertip (covered with a cloth to avoid fingerprints).

   The solvent dries quickly so check your swab from time to time. Be gentle but not too timid. Don't scrub—the heads are delicate. Don't tickle them with a feather touch either—the heads aren't that delicate. About 10 swipes should do the trick; then move on to the opposite head. If there are four heads, do all of them.

Do not, under any circumstances, rub up and down against the heads. They are flimsy in the up-and-down direction and will definitely break. Wipe *only* in the direction the heads move, *horizontally.*

Portable VCRs usually are so crammed with parts that you may be unable to disassemble them for cleaning. Here's a trick which works in many cases: Pop open the cassette lid and peek inside the VCR's mouth. Way inside near the tonsils and wisdom teeth you may recognize the HEAD DRUM. With a long swab and stretched fingers, you may be able to clean the video heads from the outside. Remember, wipe *sideways, not up and down.* Incidentally, some portable VCRs have cassette lids which are easy to remove, gaining you easy access to their innards.

Tape Path. Since you've gone to the trouble of opening your machine, you might as well clean the tape path while you are working on the heads.

You can use the same swabs and head cleaner to clean the tape path that you used to clean the heads, but the swabs will get dirty very quickly. The video heads have to be superclean so clean them first with a fresh swab. The used swab is still clean enough to wipe down the rest of the tape path.

Some cleaning fluids (like ethylene dichloride) shouldn't be used on the rubber PINCH ROLLER in the tape path as it tends to dissolve the rubber. Alcohol, when used lightly, won't hurt the rubber parts.

It is sometimes difficult to determine where the tape path is, there being so many pins and rollers in the machine. It could be instructive to plug in the VCR (definitely keeping your paws out of the machine now) and, using an unimportant videocassette, switch the VCR's power on and turn it to PLAY. Watch how the tape threads through the machine. Memorize the tape path. Note the important parts like the ERASE HEAD, AUDIO and CONTROL HEAD, TAPE GUIDES, CAPSTAN, and PINCH ROLLER. Next, STOP and EJECT the cassette, turn off the VCR's power, unplug the VCR, and proceed with the cleaning (unless you are hopelessly mesmerized by watching the whirring machine do its stuff).

Clean the tape path in this order:

1. AUDIO and CONTROL HEAD
2. CAPSTAN
3. TAPE GUIDES
4. ERASE HEAD
5. PINCH ROLLER
6. Anything else that seems to have accumulated dust or powder

Figure 16–3 shows a simplified diagram of VHS and beta threading patterns. Note the positions of the CAPSTAN, a vertical shiny rod, and the PINCH ROLLER, a black rubber coated wheel. They are important to the smooth motion of tape through your machine.

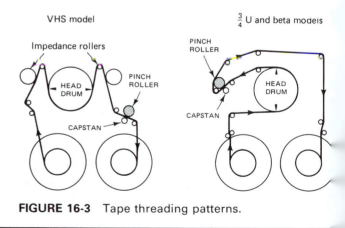

FIGURE 16-3   Tape threading patterns.

MINIGLOSSARY

*Head drum   Shiny cylinder inside a VCR to hold the spinning video heads.

Tape guides   Little posts inside a VCR to guide the tape from place to place as it plays or is threaded.

When finished with the head and path cleaning, let everything dry before rethreading a tape into the machine.

If a snowy picture was what drove you to cleaning the heads and tape path, then before putting everything back together, you may want to check out your work to see if you did indeed unclog the video heads. I must warn you, however, that your machine "open" poses a shock hazard. Don't do this if you are inept at things electrical. Connect up your machine, turn it on, insert a tape, and play it. A clean pictures means a clean head and tape path. Snow and instability mean

1. Maybe the tape is bad or blank.
2. Maybe you weren't thorough enough. Remove the tape and do the whole procedure over. A piece of dirt the size of a particle of smoke is all it takes to clog a video head.

DEMAGNETIZING VIDEO AND AUDIO HEADS

The heads are tiny electric magnets which magnetize a signal onto the tape when electricity is fed to them. The heads, after a period of constant use, may become slightly magnetized themselves. As a result, they magnetize the tape when they are not supposed to or they magnetize the tape the wrong amount while they are recording the tape. When this happens, they need to be demagnetized (or DEGAUSSED).

Video heads generally do not need DEGAUSSING. Demagnetizing won't hurt them, but it's not necessary. AUDIO and CONTROL TRACK heads, however, become magnetized more easily and do need occasional DEGAUSSING.

One sign that your AUDIO HEAD needs demagnetizing is noisy audio or loss of high frequencies. A good time to demagnetize heads might be when the machine is already open for tape path and head cleaning.

What to Use

A special tool called a HEAD DEMAGNETIZER or HEAD DE-GAUSSER (shown in Figure 16–4) is used to demagnetize the heads. A conventional AUDIO HEAD DEMAGNETIZER used on audio tape recorders (they have magnetic head problems too) will work just fine.

Make one modification to the DEMAGNETIZER for using it on the delicate video heads. Stick a layer or two of rubber or plastic tape over the metal end of the probe. Now if the probe touches the video head, the tape will cushion the contact and protect the delicate heads.

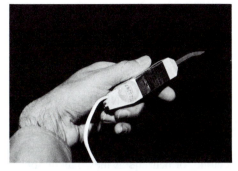

FIGURE 16-4 HEAD DEMAGNETIZER.

What to Do

1. Turn off the VCR.
2. Remove the tape so that *it* doesn't get demagnetized (erased) by the HEAD DEMAGNETIZER.
3. Turn on the HEAD DEMAGNETIZER. Often this is done by plugging it in. Bring the probe very close to the AUDIO, CONTROL, ERASE, and other heads, one at a time. If you have covered the probe with a cushion of tape, you may even touch the heads.
4. Withdraw the probe *very slowly* from each head.
5. It may be a good precaution to pass the DEMAGNETIZER slowly along the tape path in case any of the metal guides have picked up any stray magnetism.
6. When you've finished, slowly withdraw the DEMAGNETIZER from the innards of your VCR and hold it a few feet away from the VCR before turning it off (unplugging it).

Why did you hold the DEMAGNETIZER away from the machine before turning it off? When the DEMAGNETIZER switches off, it sends out a little pulse of magnetism which could DEMAGNETIZE everything near it slightly. Since your object was *de*magnetizing, you don't want to undo any of it. Reel-to-reel and audiocassette recorders need periodic DEGAUSSING too. Do them about once a month. The procedure is the same.

LUBRICATION

Lubrication is best left to the technicians. Too much oil hanging around your machine will collect dust and dirt, making it work worse rather than better.

MINIGLOSSARY

*Demagnetizer (or degausser) Electronic device that makes a fluctuating magnetic field. When the probe is brought near a slightly magnetized object (like an audio record head), the device demagnetizes it.

One place where you might throw *one* drop of oil is the *base* of the CAPSTAN where it disappears into the housing of the VCR. Do not put oil on any shiny surface that touches the tape.

CLEANING VCR AND MONITOR CABINETS

Most cabinets have wood grain, vinyl, or plastic surfaces. Clean with soap and water. Avoid strong solvents like paint thinner, acetone, or benzene as they may dissolve the plastics or mar the veneer.

TAPE CARE

The end of Chapter 5 went on endlessly about how to store tape and how to detect defects in it.

Tape Cleaner/Evaluator

When tape starts to display excessive dropouts, flagwaving, graininess, or hash at the top or bottom of the TV screen, it's time to retire it. If you work with large numbers of tapes, you may try one thing first, before heaving out large numbers of heavily used, suspect tapes: Run the tapes through a TAPE CLEANER/EVALUATOR. This is a machine which winds a tape quickly from beginning to end while sensing edge damage, wrinkles, and dropouts while cleaning the tape. Some models examine the tape by recording a signal on it and playing it back a moment later. This process will destroy any program on your tape (but may save the tape). Another type of TAPE CLEANER/EVALUATOR "looks" at the tape with a light beam and doesn't harm the programs recorded on the tape.

Tapes can be sent out to commercial firms which, for a price, will clean and evaluate your old tapes. As much as 85% of your old tape stock can be salvaged this way.

Splicing a Broken Tape

Halfway through your favorite recording of "*The Galloping Gourmet*," the VCR decides to eat some tape for lunch. What do you do with the leftovers? Folded or wrinkled tape can be flattened out and played if it doesn't look too rough. The image will have lines of snow running through it (like in Figure 5–45) until the bad tape has passed. Using such tape runs the substantial danger of abraiding the spinning video heads and perhaps nicking them. Badly stretched, torn, ragged, or ground-up tape runs a *very high* risk of head damage and should always be avoided. The image

would be unrecognizable when played anyway. What if part of the tape is physically destroyed and this is your only copy? The damaged part will never be reclaimed. To make the remainder payable, the bad part has to be cut out and the two good ends SPLICED together.

Even well-made splices have a bump to them and gooey adhesive can eventually bleed out, contaminating the cassette. Worse yet, the splice itself may snag on a video head and chip it. It is advisable to fix your tape and then copy it right away and thereafter play only your copy. Unfortunately, this leaves you playing a second-generation tape.

Another option may be possible in those frequent cases where the tape breaks near its beginning or end. Here you can cut off the bad tape and attach the new end to the appropriate hub. You lose a little of the beginning or end of your show (often a less important part) and get to keep your original tape without having a nasty splice go through your VCR.

In the preceding case, if a tape breaks and you don't care about the recording, you can always throw away the shortest half of the tape, reattach the longer half to the other hub, and create a shorter-length cassette from the remaining tape. Be sure to label it so the oddball length doesn't surprise you later.

Simple Reconnection Outside the Cassette. The simplest and commonest repairs involve reattaching a tape to its leader or cutting out damaged tape segments and reconnecting the ends together. Often both are protruding from the cassette so you don't have to open the shell to find the ends. Here's how the repair is done:

1. Make a clean work space. A sheet of paper works nicely. Across the bottom of the paper, mark a straight line.
2. Wash your hands thoroughly and dry them.
3. Open the trapdoor (process described in Figure 16–5) and prop it open with a stiff piece of cardboard. This is so it doesn't snap shut, crunching your tape. On beta VCRs, this also releases the tape hubs so that you can withdraw more tape from the cassette. On VHS cassettes, however, you free the hubs by inserting the tip of a narrow screwdriver into a hole in the center of the cassette's underside.
4. Snip off the bad tape and square off the two ends so that they butt together evenly.
5. The outside surface of the tape (shiny side) has contact with the video heads. You don't want any bumps or splicing tape there. Lay your tape down on your paper *inside surface* (dull side) *up*. Using

MINIGLOSSARY

Tape cleaner/evaluator   Electronic device to check video tape for defects while cleaning the tape.

*Splice   To reattach the broken ends of tape together. Also the junction where the tape ends were attached.

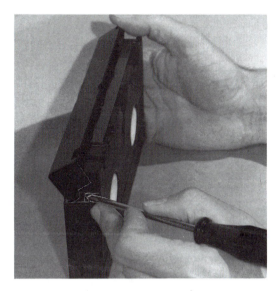

Press lever at bottom left side of VHS cassette to release trap door.

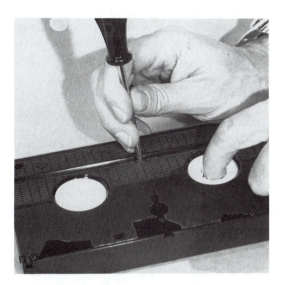

To rotate tape reels (advancing or rewinding tape manually), press button in bottom center of VHS cassette to unlock reels.

FIGURE 16-5 Opening the trapdoor on a VHS videocassette.

that straight line you drew earlier, line up the edges of the two tape ends even with that line. Butt them exactly together. Lay coins on the tape to hold it flat and stationary as you position it.

6. Using aluminum-backed self-adhesive mylar tape, apply the splicing tape across the juncture, anchoring the excess to the paper. Press firmly. Don't try to use the splicing tape lengthwise. It's nearly impossible to line up.

If you plan to recopy the tape immediately, then throw away the original; you can cheat a

little and use Scotch Magic Transparent tape or some other strong, thin, pliable tape. Masking tape, most cellophane tape, and wrapping tape won't do; they are likely to bleed "goo" into the splice or come loose.

Beta owners note: Don't use aluminized tape. Many beta VCRs sense the shiny splice and think they are at the shiny leader at the end of the tape and switch themselves to stop midtape.

7. Trim off the excess SPLICING TAPE with a razor blade or X-Acto knife, making a very slightly crescent-shaped cut as you trim. This will make the spliced portion just a hair narrower than the rest of the video tape so it won't rub against the cassette walls.

8. Wind the tape back into the cassette, unprop the cassette door, and you are back in business.

Figure 16–6 diagrams the process.

For those who expect to do frequent splicing, there are kits that consist of a splicing block to line up your tape and guide your razor blade and a set of gummed tape splices.

Opening the Cassette

As do many sea creatures when disturbed, the tape sometimes retreats into the safety of its shell. Getting it out is a project. You have to take the cassette apart. Do this job right, and you'll only have to deal with 2 or 3 parts and some screws. Do it wrong, and all 36 internal components will spring out at you plus umpty-hundred odd feet of recording tape in a nice neat ball. If you are the kind of person who took clocks apart as a kid—and put them back together so they'd work—then you've developed the kind of skills necessary for "open-cassette surgery."

1. Start with a clean surface and clean hands.
2. If the cassette has a spine label, slit it down the middle along the seam between the two halves (the cassette can't hinge open; it must be completely separated into two parts).
3. Place the cassette upside down and remove the four to six screws holding the halves together. Use a proper-sized, good-quality phillips screwdriver. Keep track of which screw came from which holes so you can put them back in exactly the same places. VHS cassettes usually have two long screws in the front edge and three short ones in the back.
4. When the screws are removed, carefully pick up the cassette sandwich and turn it right-side up. If you are the clutzy type, you may wish to scotch-tape the halves together before you turn the cassettes over, because if you let the halves separate

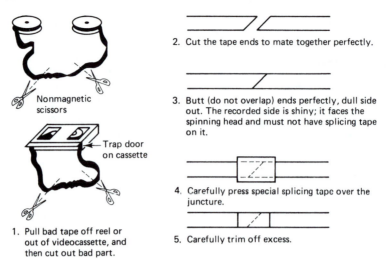

2. Cut the tape ends to mate together perfectly.

3. Butt (do not overlap) ends perfectly, dull side out. The recorded side is shiny; it faces the spinning head and must not have splicing tape on it.

4. Carefully press special splicing tape over the juncture.

5. Carefully trim off excess.

Nonmagnetic scissors

Trap door on cassette

1. Pull bad tape off reel or out of videocassette, and then cut out bad part.

FIGURE 16-6   Splicing a video tape.

at this crucial point, you'll be vacuuming parts out of the carpet for weeks.

5. Lift the top half away from the bottom. You may first have to release the trapdoor manually (Figure 16–5) before the halves will separate.

6. Find the loose tape ends and rethread them through their guides and hold-back pads until they emerge outside the cassette in the proper fashion. See Figure 16–7. For betas, the tape threads between the hold-back pad and guides on each side of the cassette. For VHS, there is only one hold-back pad situated on the left side. On the right side, thread the tape between the metal tape guide and the plastic one just behind it.

7. Once you've gotten the tape ends out (don't try to take the roll of tape out unless you have a degree in neurosurgery), reassemble the cassette halves. The trapdoor may have to be partway open for the halves to mesh. You may wish to reinstall the screws (before you forget to or before you bump this little land mine), or if you are the

type who is likely to lose the tape ends back into the cassette while working on them, just scotch-tape the halves together for now.

8. Splice the tape as shown in the last section and then flip the cassette over and reinstall the screws into their proper holes.

Sometimes the tape or leader doesn't break but pulls free of the reel hub inside the cassette. You can recognize this situation by noting that the free end of the tape still has a leader attached to it. When you open the cassette, you'll probably see a little clamp floating loose in there somewhere. That clamp was supposed to secure the tape to the hub. This is easy to fix.

1. Remove the empty reel from the lower shell. You can release the reel lock by gently pressing a spring in there.

2. From the full reel, draw off just enough tape to work with.

3. Attach the leader to the center of the hub and press the plastic clamp back into its socket. Be careful when applying pressure so as not to bend or break the reel flanges.

4. Rethread the tape through the guides, reinstall the reel, and close the patient.

Handling Wrinkles

When you get wrinkled, you can get a face-lift. When your tape gets wrinkled, you have two choices:

1. Cutting out the bad part (a technique that doesn't work so well on people wrinkles)

2. Ironing the tape, if it is not wrinkled too badly

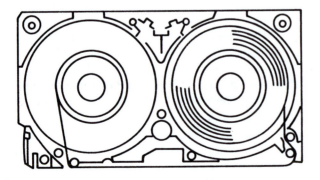

FIGURE 16-7   Inside a VHS videocassette.

(this technique is not recommended for people wrinkles).

The ironing process takes some skill and should not be tried for the first time on your most important piece of tape. Here's the process:

1. Clean the tape of dust and tape fragments.
2. Preheat a household iron to 190°F (this is often the permanent press setting). Make sure that there is no water in the iron and that you don't accidentally leave it in the steam setting.
3. Sandwich the tape between two sheets of white typing paper.
4. Place the sandwich on a hard, smooth, flat surface.
5. Apply the iron with light to moderate pressure for 8–12 seconds.

Even if you manage to eliminate the wrinkles in the tape, chances are that some of the oxide will still have flaked off the plastic tape at the places where the tape was creased. These empty spots will show up as dropouts when you play the tape. With VHS tapes, an automatic light sensor in the VCR may "see through" some of these holes and shut down the machine. To solve this problem, use a black felt marker to "paint out" the pinholes on the back side of the tape.

While we're on the subject, if you discover a cassette tape which seems to eject itself at the same spot always, suspect a pinhole or scratch on the tape to be the blame. When the VCR's sensing light can shine *through* the tape, the VCR coughs it up. (On betas, shiny splicing tape has the same effect.) The solution is to black over the holes with something like a magic marker.

CLEANING LENSES AND IMAGE SENSORS

The cleaning of lenses and image sensors was covered early in Chapter 7. Generally, dust the lens with a camel's-hair dusting brush. To remove fingerprints, fog the lens with your breath and use lens tissue. For stubborn fingerprints, use a soft cloth over your finger dampened with a mild nonabrasive soap and water. Windex—sprayed on the cloth, not directly on the lens—works pretty well. Clean in a circular motion. Wipe off the soap. Wipe the lens dry.

Clean camera tubes the same way as lenses. Be especially gentle with chips, they scratch easily.

CLEANING VIEWFINDERS AND MONITOR FACES

Viewfinders (including the one inside a portable camera) and monitors may be cleaned with a soft, dry, polishing cloth or with a dampened cloth and a little window cleaner. A cotton swab may be helpful for getting into tight corners like in the portable camera's viewfinder. Don't use harsh solvents or lens tissue—these will fog or scratch plastic faceplates.

Wipe TV screens with an antistatic cloth to inhibit dust collecting on them.

Incidentally, you may notice when touching a TV monitor screen that it crackles with static electrical buildup. This is normal. There is nothing wrong with the TV screen.

BATTERY CARE

The rechargeable batteries used with portable VCRs come with instructions for their care and recharging.

The two popular types of batteries used in portable video equipment are GEL CELLS and NICADS. If using both types of batteries, be sure to charge each on its *own* charger. A GEL CELL charger should not charge a NICAD battery and vice versa.

Gel Cells

GEL CELLS use the same technology that starts your car in the morning; lead and acid interact to push electrons out one of the two poles on the battery. By pushing the electrons back into the same pole, you recharge the battery. Unlike your car battery, which has acid which can spill, the GEL CELL contains a gelatinous electrolyte which is sealed in for safe operation in any position.

GEL CELLS cost one-half to one-third what NICADS cost and are very reliable. Many are one-piece units which simply insert into your portable VCR through a trapdoor and will run your machine for up to an hour (depending on how much juice your camera uses).

GEL CELLS have an excellent shelf life—once charged, they retain 80% of their charge for nearly a year if unused and much longer if kept in the cold. As they are used up, their voltage (electrical "pressure") gradually decreases, which means that by looking at a voltmeter (your VCR's battery meter) you can see how much power is left and predict how much remains. GEL CELLS operate over a wide temperature range but give less power the colder they are. GEL CELLS should last 3–5 years or about 200 discharge cycles before they die. As they reach the end of their life, they lose some of their oomph (don't we all!) and can power your VCR for only 45 minutes or less. GEL CELLS can be safely discharged until 100 percent dead without damage, although your VCR will sense the low voltage and shut down long before this happens. If you do discharge a GEL CELL 100 percent, charge it back up right away, or the insides will start to crystallize, shortening the battery's life.

A discharged GEL CELL will take up to 8 hours for an 80% recharge or 16 hours for a total recharge. A brand new uncharged GEL CELL may take 24 hours to charge, and a 100 percent discharged GEL CELL may take 32 hours to totally recharge.

GEL CELLS will not tolerate significant overcharging

as they lose water on overcharging. It is wise to charge the battery in a well-ventilated area because if overcharged excessively, the battery may also leak flammable hydrogen gas.

GEL CELLS don't pack the power of NICADS and are bulkier and heavier. They also can't support as heavy a current draw as NICADS, making them less suitable to power portable TV lights (which use large amounts of current). GEL CELLS usually come encased inseparably in a pack, which means that if one cell from the battery unit goes bad, you ditch the whole package. GEL CELLS can't be discharged/recharged as often as NICADS, making GEL CELLS shorter lived.

If you run your portable VCR on ac most of the time and on batteries for about an hour once in a while, a GEL CELL is probably the best buy for the money. If you are into heavy-duty battery use, then NICADS may be your answer.

Nicads

NICADS are made with nickel and cadmium—hence their name. They cost about twice as much as GEL CELLS but hold about three times the usable power for the same weight and size package. Notice that I said *usable*. As a NICAD expends its energy, its voltage remains fairly constant up there at a level where the VCR can use it. Thus, the NICAD can empty nearly completely before it poops out. GEL CELLS, on the other hand, lose voltage gradually. When your VCR senses that your battery voltage has dropped about 10%, the VCR turns itself off, essentially rejecting the battery even though it has life left. Thus, not all of a GEL CELL's power can be used by a VCR.

Although 1-hour NICAD packs can be bought which can slip into your VCR, 3-hour external models are also available.

The NICAD packs can be charged 500–1000 times, but their lifetime would be better measured in years (3 years, typically) rather than discharges.

Usually, when a NICAD pack starts to give you trouble, a technician can open it, find the bad cell, replace just that cell for a few bucks, and you then go on using the pack.

NICADS can operate over a wide temperature range and, like GEL CELLS, lose their oomph when the temperature drops, but not quite as much. They can operate in any position and do not leak.

NICADS can be "fast-charged" in 2–3 hours using a special charger which senses whether it is overheating the cells in the process. Overheating could cause the batteries to explode. Normally, it takes about 16 hours to recharge a NICAD.

Unlike GEL CELLS, NICADS put out approximately the same voltage until they are nearly out of juice. Then the voltage drops precipitously. In one respect this is good; machinery and circuits will run as well from the battery's fortieth minute as they did from its fourth minute. In another respect, this is bad because a simple meter won't tell you how much charge is left in the battery; it will read full strength for maybe fifty-five minutes and drop on the fifty-sixth as the battery poops out. Since the meter doesn't signal the impending power failure, you have very little advance warning that your shooting is about to screech to a halt. To protect against this unwelcome surprise, many video producers carry several battery packs into the field. You know your shoot is 75 percent over when three of your four batteries are used up.

Although NICADS charge quickly, their shelf life isn't too great. They lose about 1 percent of their charge per day while just sitting, and more if they are sitting in the heat. To keep them ready for action, charge them once per month even if unused. Some users "trickle-charge" their NICADS to keep them 100 percent charged all the time. This process requires only a tiny charger which pumps a small amount of current (a trickle) into the battery every day to make up for the NICAD's 1% daily loss. Constant trickle charging will not harm the NICAD.

Memory. For those who know NICADS, MEMORY has nothing to do with a song from the musical "Cats." MEMORY is a characteristic attributed to NICAD batteries which "forget" their designed capacity and become conditioned to deliver only a small portion of it. This phenomenon is blamed on the battery undergoing short shallow uses and "remembering" how little it had to put out. When called upon to deliver full power, such batteries produce for a few minutes and then run out of power.

If you use your VCR batteries in frequent short spurts, it would be good to exercise them with a deep discharge once or twice every 6 months. You could do this by connecting any TV equipment up to your battery and letting it run until it stops. The process will be faster if you use "power-hungry" items like a TV, lights, or your VCR with a big portable color camera. You would simply run down your battery and then charge it all the way up. Next you would run it down again and charge it all the way up. This exercise should erase the MEMORY.

GEL CELLS do not have MEMORY.

Calculating a Battery's Capacity. How long will a certain battery power your VCR and camera before you run out of juice? To figure this out, you first need to total up your power usage. How many watts does your VCR use *while* recording? Say it's 8 watts. How many watts does

MINIGLOSSARY

*Memory An attribute of nicad batteries whereby they "forget" how long they should be able to provide power if they haven't been worked hard enough. Frequent short duty cycles will eventually make them able to perform only shallow discharges before they need recharging.

your camera use? Say it's 10 watts. The total power drain is 18 watts. Now let's study the battery.

Batteries are rated in ampere-hours or watt-hours. Say your battery has the capacity of 18 watt-hours. Then theoretically your battery should run your 18-watt ensemble for one hour before it blacks out (watts × hours = watt-hours).

If a battery is rated in ampere-hours, you need to make a longer calculation. How many amps does an 18-watt machine use? Since video equipment nearly always runs at 12 volts, we use the following formula: watts/volts = amps. An 18-watt machine uses 1.5 amps (18/12 = 1.5).

A 1.5-ampere-hour battery can run a 1.5-ampere device for 1 hour   theoretically. Batteries often don't perform as well as their ratings, so you should expect to get less than what you calculate.

Table 16–2 summarizes the features of GEL CELLS and NICADS.

Both kinds of batteries lose power rapidly when cold. Neither kind of battery likes to be used as a hammer, immersed in water, shorted out, or disposed of in fire (they may explode).

CABLE CARE

1. *Do* keep a couple of spares around. By substituting a spare, you can determine whether the cable or some other component has failed. Also, the inevitable cable failure is less devastating if you can substitute a good one and get back to business immediately, replacing the broken one at your leisure.

2. *Do* buy heavy-duty name brand cables and connectors like Amphenol or Switchcraft; they'll last longer.

TABLE 16–2
Gel Cells and NiCads compared

	Gel Cell	NiCad
Cost	Low	High
Current capacity	Medium	High
Length of service/charge	Medium	High
Recharge speed	Slow	Fast
Expected lifetime	Medium	High
Shelf life	Long	Medium
Damaged by deep discharge	Yes	Yes
Discharge slope	Gradual	Abrupt
Harmed by trickle overcharging	No	No
Harmed by heavy overcharging	Yes	Yes
Parts replaceable	No	Yes
Operation when very hot	Good	Good
Operation when very cold	Poor	Satisfactory
Memory	No	Yes
Weight	High	Low

3. *Do* keep your connectors off the floor. They can get dirty and flattened underfoot.

4. *Do* clean the contacts on a plug with emery or fine sandpaper or even a pencil eraser when it appears that corrosion is causing you to have an intermittent, weak, or "crackly" connection.

5. *Don't* try to force a plug into a socket if it doesn't want to fit. It will break.

6. *Don't* try to use a plug with a pin missing. It won't work unless all the pins make connection.

7. *Don't* coil cables tightly. *Do* coil them in 15-inch loops to avoid strain.

8. *Don't* let your pets chew on the cords—especially the power cord. Fido might be in for a shock. Chewed video and audio cables don't work so hot either.

9. *Don't* pull a cable out by its cord. Always pull it by the plug. The wires are frail; the plug is strong.

GENERAL CARE

Invest in some tiny jeweler's screwdrivers, a *good-quality* little phillips screwdriver, and some small metric hex wrenches. No, a phillips screwdriver is not a vodka and milk of magnesia cocktail, nor is a hex wrench used by a witch. These are specialized tools (diagramed in Figure 16–8) to keep knobs, screws, bolts, catches, latches, doodads, and thingamahoosies tight. Once the bolts fall out, finding replacement latches and new bolts is next to impossible. This is one situation where an ounce of prevention definitely saves you a pound of hard work—it is so easy to tighten loose screws and so hard to replace lost ones.

Another handy tool may be a can of compressed air. It's useful for cleaning dust out of lenses, tape powder out of VCR innards, cat hairs out of keyboards, and lint off graphics.

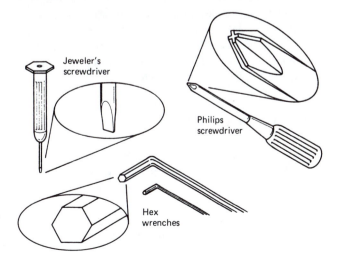

FIGURE 16-8 VCR and camera tools.

PLANNING AND PRODUCTION TECHNIQUES

Television production is 99 percent planning and 1 percent production. The 1 percent part gets the glamour, while the 99 percent part actually makes the show.

Nothing brings good luck like not relying on it. Every step in the production process which hasn't been planned, discussed, reviewed, and made watertight will spring a flood when the tape begins to roll.

Part of the planning process involves assembling the right equipment and creating an environment where it can do its job. Another part of the planning process involves assembling the right people and giving them an environment conducive to performing their jobs. Last, yet most important in the planning process, is assembling the good ideas which will eventually become your program. The finest equipment, people, and production techniques all go to waste if you are producing a show that nobody wants to see or that your sponsor cannot afford. All the pieces to the production puzzle, the equipment, the talent, the crew, the script, and of course the basic plan, must fit together.

SCRIPTS AND STORYBOARDS

First you learn how to write; then you learn how to write between the lines. That's poetry. Script writing gives you many more lines to write between. There's the narration or spoken channel with its statements and nuances. There's the audio channel with its mood-creating musical track or environment-establishing sound effects. There's the visual channel displaying not only the obvious but implying further messages through camera angles and editing.

Planning

State Objective. The French Foreign Legion had a saying: "When in doubt, gallop." Doesn't work in video (didn't work for them, either). Don't start writing until you have a concrete idea of what you want to communicate. Sometimes it's a client or teacher whose fuzzy ideas have to be clarified before you proceed with the treatment. One way to track down a show's elusive purpose is to require the originator (or yourself) to *write, in one or two sentences, what the program is supposed to achieve.* This focuses your energy on a main purpose, an anchor that will keep you from rambling or getting lost in detail.

Keep the plan simple. The secret of being tiresome is to tell everything. The 2-hour treatise showing every detail about something will not be as efficient or memorable as the 10-minute presentation that repeats the main point several times.

If the client's objective is impossible for either of you to grasp at first, try this: Give him or her an audiocassette recorder to dictate the story into. This may help the client to reorganize his or her thoughts while providing you with some insight to what he or she wants. By taking notes from the tape, you can list topics for later reorganization. See ''Instructional Television Techniques—Designing Effective

TV Lessons'' later in this chapter for more on planning objectives and writing for a specific audience.

Determine Audience. Before you can educate, entertain, inform, or persuade your viewers, you need to know who they are and how much they know about your subject. You will be adjusting your time, format, and level of sophistication for your audience. Consider how this program will be seen by the audience. If multiple copies will be distributed, then you need to keep a sharp master tape for making those copies. Use your best equipment and try to reduce multigeneration editing. If the program is to be broadcast, you must produce a glitch-free tape made to high technical standards.

Will you have a captive audience? If so, you have the luxury of spending a few minutes developing your topic or developing the background for your story. If your audience consists of passersby, like in a store, then you have to grab 'em by the throat with something flashy and inject them with a brief message before they wriggle away. Don't let another 20 seconds go by without reaching for their throats again.

Children are also known to have short attention spans. You must reach them with a lively, colorful presentation that covers (and repeats) only a few main topics. Adults, incidentally, aren't much different. They lose interest and daydream just like kids do (only adults are too polite to swing from the chandeliers when a production gets boring).

The only audience that will sit attentively through a detailed, straightforward teleproduction consists of individuals who are deeply interested in the subject matter and intend to use the information right away.

Consider Resources. If you have an editing VCR, then the world of out-of-sequence recording and complex

staging is at your fingertips. If you have a portable VCR, that means that you are not tied to the studio.

What's your budget? How much time and personnel will you have? Usually the budget plays a large role in determining how complex a production can be.

Will you have a content expert or technical consultant there during taping to stop you from making embarrassing mistakes?

Choose a Format or a Combination of Formats. If you draw a blank trying to imagine where to start, Table 17–1 lists a few possibilities.

Get Approvals. When money is involved, it is wise to make sure the sponsor is willing to pay for the work you are doing. It is wise to find *one* individual who has the authority to ''sign off'' on the project each step of the way. Often this approval process is tied to the payment process, as in the following schedule:

Activity	% of Total Payment
1. Prepare general treatment	10
2. Prepare rough script	20
3. Prepare finished script	20
4. Shoot raw footage and prepare rough edit for review	30
5. Complete fine edit	20

In this schedule, the client can stop you before you spend too much money pursuing the wrong angle or creating the wrong mood. At the same time, this protects you from the client willy-nilly changing his or her mind, requiring you to do a lot of the job over. Nothing stops a client from changing his or her mind after signing off; in fact this is commonly done. The client, however, pays extra for this privilege and may think twice before proposing wholesale changes at the last stage.

Writing

Writing is a creative endeavor. Some are blessed with the magic; others write textbooks on video. It's hard to make ''rules'' on writing; the rules are like toys on Christmas: so many are broken but still the fun is there.

Being Visual. Words, words, words . . . is this television? Think pictures. Close your eyes. Imagine telling your story without a single word. Sure, it's not easy, but try anyway. Close your eyes (you're still reading!); what do you see? Can you visualize the story unfolding? The script you are *now* creating is the difference between radio and television. Consider this: The movies *Alien, Close Encounters of the Third Kind,* and *Space Odyssey 2001* ended with a long stretch without words. There was plenty of

TABLE 17–1
Several TV show formats

Format	Uses
Show and tell	Straightforward presentation of facts. You explain an activity as you show it.
Spokesperson	A recognized authority adds credibility to your message.
Interview with man-on-the-street, or victim, or groupie, etc.	Viewers identify with J.Q. Citizen. Adds color to facts and statistics by introducing a human element or drama.
Skits	Visually more exciting. Acted-out situations leave a more memorable impression than a simple declaration of facts.
Animation	Costly but cuts out extraneous visual material and is more entertaining.
Charts and graphics	Simplifies complex ideas. More memorable.

script, plenty of shots and planning, and finally a feast for the eyes, ears, and imagination, but no words. What a relief!

If you look down through your script and find the audio column filled with narration and the video column listing a few spare hints of what pictures will show, then you have a radio script with pictures. Rewrite the script or find somebody who knows how to write for television.

Involving People in Your Subject.

People are most important. How the characters trapped in a situation work their way out involves the audience. Example without people (directly):

> Rusted bolts are sometimes hard to unscrew. No amount of effort seems to break them loose. Before ruining the bolt's head, try applying some penetrating oil to the base of the . . .

But with people:

> "What's the matter, Harry, rusty bolt won't unscrew?" (Harry gets angry, contorting himself and his wrench into ridiculous positions. Harry gets a grip and pulls with all his might.) "Harder, Harry, I think it's coming!" (Harry's grip breaks loose, and he flies across the room.) "Oh, Harry! Before you ruin that bolt and yourself, too, why don't you try some penetrating oil. . . ."

Grabbing the Audience.

"What we want is a story that starts with an earthquake and works its way up to a climax" SAMUEL GOLDWYN. All TV programs need the audience's attention to succeed. Show some action, some mystery, tragedy, beauty, or humor. Or tease them with the sounds of something just around the corner. Whatever, grab the viewer's interest right at the start.

Prose for Television.

Granted, TV can't be all pictures. Even if we try our best to make our shows visual, we often have to provide a narrative. At least if we *have* to use words, let's write them in a visual language. Table 17–2 shows some differences between regular writing and TV writing. Note that these are generalities and not unbreakable rules.

Here's an example of printed prose:

> Affix the mounting wedge to the base of the camera, screwing the mounting bolt into the camera's base plate. Finger-tighten the bolt and then rotate the tightener ring clockwise, finger-tightening the ring. Slip the wedge into the camera head and lock the wedge clamp.

Here's how it would sound if the prose were matched to the video:

> This is the camera mounting wedge, useful for quickly releasing the camera from its tripod. To attach the wedge to the camera, first locate the mounting hole in the base of the camera. Align the fastener bolt with the hole and thread the bolt into the base as you can see here. Notice that the narrow end of the wedge plate is pointing toward the front of the camera. Finger-tighten the bolt and then screw down the tightener ring like this. Now the camera is ready to be placed on the tripod. Slip it into the grooves in the head and slide it forward, like so. Once the camera is snug, find the quick release lever and tighten it down. The camera is now mounted securely to the tripod.

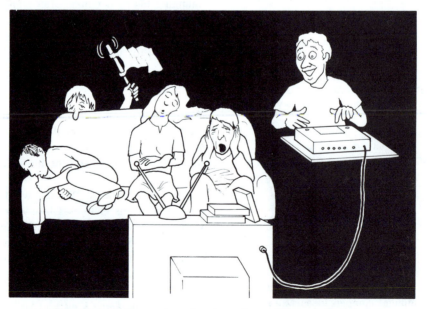

Grab the audience, be visual, and KISS (Keep It Simple, Stupid!)

TABLE 17–2
Regular prose vs prose for television

Regular Prose	TV Style
1. Uses complete sentences.	1. Uses incomplete sentences and phrases and may resemble captions.
2. Is concentrated, compact, and compressed.	2. Is repetitive and more wordy.
3. Uses formal arguments and deductive reasoning, sometimes requiring mental consideration of an abstract thought.	3. Is descriptive. Announces, proclaims, declares.
4. Uses logical transitions between sentences, like *thus*, *because*, *therefore*.	4. Uses adverbial transitions like *here*, *now*, *then*.
5. Presents a logical argument to make a point. Develops a thesis. Ideas are derived through reasoning, surmised.	5. Illustrates the point or narrates a process.
6. Explains verbally.	6. Directs attention.
7. Tends to participal modification such as "the grid lights" rather than "the lights in the grid."	7. Tends to prepositional modification such as "on the dimmer panel" or "at the beginning of the recording."
8. Uses long complex sentences.	8. Uses short simple sentences.
9. Uses variable sentence construction.	9. Uses declarative sentences—subject to verb to object.
10. Uses formal vocabulary and infrequently used words.	10. Uses the common, frequently heard words found in everyday speech.
11. Uses synonymns. Says something only once, figuring the reader can always come back to review the point.	11. Avoids synonymns. Tends to redundantly repeat the same thing over and over again, tautologously duplicating the . . . (you get the idea).

Notice how redundant the TV prose is compared to printed prose. There are several reasons why this needs to be so:

1. A *reader* can concentrate on difficult or complex passages and skim the familiar, skipping back and forth across a sentence. Further, the subject of a sentence can apply to a very complex predicate. Once entangled in the tail of the sentence, the *viewer* may have forgotten the subject but, unlike the reader, the viewer can't zip back to review. Thus, the subjects are repeated, and sentences are kept down to one single thought each.

2. Cues to direct the viewer's attention are critical in order to ensure that important visual elements are not missed.

3. Since the viewer has but one shot at getting your message, it's wise to slow down the presentation rate, especially if the content is difficult to grasp.

4. Most importantly, actions usually proceed more slowly than words. What sometimes takes one moment to describe takes three moments to show so the words have to be spread out to coincide with the action.

This doesn't mean that you must fill every instant of action with constant chatter (don't try to tell this to Howard Cosell, though). Once you've set a tone of continuously narrated action, it may be disconcerting to the viewers to get a burst of information and then have the pictures catch up in silence.

If you find it inappropriate to stretch the narration (you don't want to describe the obvious just to fill time), you can always resort to filling the audio track with the environment's sounds (machines, traffic, office noise) or with light music. The music or sound will bridge the action while letting viewers know their hearing aids didn't go dead.

Narration. A waterfall of words have we and not a drop of visual variety to quench our parched cerebrum. The static boredom of the "big talking head," as the industry calls it, promises to haunt us as doggedly as death and taxes. Whether our excuse is low budget, absence of visual material, or lack of imagination, we will all eventually sink to the depths of the droning voice.

Humankind faces and deals with death and taxes, so shall we bravely deal with "talkie-face." Table 17–3 shows four basic narration formats and their applications. These may be used alone or in combination such as voice-over narration followed by guest expert. Such combinations offer a change of voice, tone, and pace, thus increasing interest.

Using Table 17–3 as a base, we shall now try to squeeze more visual interest out of old talkie-puss.

1. *Shoot talent in their work areas.* Have the designer talk among his drawings, the chemist talk in her laboratory, the pilot talk at the airport, the machinist talk in front of his lathe, or the executive talk at her desk. Such familiar environments relax the talent, providing some insight to their personalities; add credibility to their roles; give you a location to shoot cutaways to hide edits with later; and, most of all, frames Old Talkie-puss with an interesting background.

Taking this idea further, you can *shoot the talent while they perform their jobs or practice their arts.* Not only are the performers in a comfortable environment, but they have something to do with their hands while they speak. Unlike the ''show-and-tell'' format described earlier, the talent aren't necessarily showing or telling *what* they are doing; they are just discussing a topic related to their actions. For instance, an artist could interpret the style of another artist as he himself molds a clay bust. Or the talent could describe the physical benefits of a Swedish massage—while receiving one.

In many situations, the talent can *present their lecture in a stress environment* such as driving a race car, swimming, answering telephones in a busy office, or performing surgery. The surrounding action will add immediacy, naturalism, and excitement to the discourse.

2. *Shoot talent in an atypical environment.* Here we do *just the opposite* of method 1 but for *exactly the same reasons.* The executive explains management strategies while pruning roses and wearing faded overalls. The judge discusses prison reform while playing with her dog in the backyard. The school custodian describes heating systems while tuning his violin for a weekend orchestral performance. The obvious counterpart of such scenes strengthens

TABLE 17–3

Basic narration formats

Format	Description	Example	Speech
Voiceover	As video activity is shown, off-screen narrator describes.	''As oil pollutes the wetlands, waterfowl perish.''	Formal, well enunciated, perfect lines.
Guide	Talks to the camera, sometimes disappearing off screen as an event is displayed. Leads the viewer places. Sometimes interacts with subject.	''We now are entering the shaft where twelve miners died. Here we see. . . .'' or ''Can you tell the folks at home what it's like. . . .''	Off-camera speech is formal. When interacting with others on camera, speech is informal.
Guest expert	An official, celebrity, or expert explains what he or she does or how a task was done. Expert may later join in the action using the eavesdrop format.	''When I first noticed a deviation in the planet's orbit, I thought it was some error in our calculations or an equipment failure.''	Informal, as if talking to a single person, the viewer.
Eavesdrop	Actions unfold as the camera looks on. We overhear conversations between the participants. Interviewing falls into this format or into guide format.	''Hi, Mrs. Pevar, how is your bladder today? Hmmm, your stitches are healing up nicely.''	Very informal.

our respect for the speakers as human beings and adds extra dimensions to their personalities. The shock of the juxtaposition opens our tired eyes.

3. *Shoot talent while they are walking or driving.* Not only do the talent have something to do while they speak (which may also have a calming effect on them), but also the viewer senses that the characters are genuine, doing real-world things. Such shots make excellent bridges from one topic to another, like "So that's how the program should work in theory. I'm taking you now to our field headquarters outside Bakersfield. One reason why this system is so effective. . . ." On our way to Bakersfield we hear introductory descriptions of what we're about to see.

4. *Shoot talent in extreme close-ups.* With the eyes one third down from the top of the screen and the mouth one third up, you get a very intimate and engaging face shot. If you plan to edit, remember that you don't want jump cuts from one ECU (extreme close-up) to another ECU, so provide cutaways or some medium shots. Also, if

you plan to title someone, do it on a medium shot across his chest, not on an ECU across his teeth (for this reason it is good to begin interviews with an introductory "shirt pockets" shot which leaves space for the person's name and title).

Audience reaction research has shown that attention level is much higher when the screen displays full-face ECUs and CUs and there are rapid cuts from one person to another. Many of us assume that a variety of shots including some long shots will maintain interest, but the research does not bear out that assumption. Another thing that the research shows is that you should never keep your camera on a three shot (a long shot) for more than 3 or 4 seconds, especially in an interview. The audience cannot see the participants' facial reactions in a three shot and tends to lose interest.

5. *Shoot the talent out of a cannon.* Only kidding; however, with a telephoto lens and a long mike cord, what an eye-opener you'd have.

6. *Move talent around.* For instance, have Mr. Talkie-

Application	Scripting
No personality superimposed on content. Narrator is the Voice of Truth, representing facts, not controversy. Direct but boring. Requires a constant stream of visual matter.	*Offers maximum control of program. Easiest script to write, shoot, and edit because everything is planned in advance.*
Requires less visual matter. When there's nothing to show, show the guide talking. Guide unobtrusively controls direction of investigation while viewers feel as though they are making "discoveries" on their own. Acting as an audience representative, guide may ask subjects questions. Subjects respond directly to camera.	*Easy script and program control, except when guide interacts with real subjects. Their responses may not fit the preconceived script.*
Lends authority and credibility to content. Must be titled or introduced so viewers know who the expert is. Viewer perceives speaker as a human being and may associate with that relationship. Speaker becomes part of subject.	*Difficult to script because you can only develop a rough outline of material. Preplanning with the expert may aid scripting, but beware of stilted, formal tone if he's not a professional speaker.*
More interesting and memorable presentation. Creates some distance between speaker and viewer. Permits few full-face shots. To be unobtrusive, you must often be able to shoot in confined space using available light, a shotgun microphone, a hand-held or easily movable camera, and only two crew members. And you must usually be able to change focus or f-stop quickly. Skilled interviewer can probe and repeat unanswered questions, thus speeding the reluctant or inarticulate speaker to a brief and informative statement.	*Scripted role-plays require writing finesse and good actors. Using real-life participants affords the scriptwriter little control until the rushes come in.*

Shoot talent in a stress environment

puss refer to models, charts, and other graphics. Instead of placing them together, have them spread around the studio, perhaps in "puddles" of light. The talent strolls from item to item as she talks, perhaps pointing, holding, uncovering, or manipulating the item at each stop. The "puddles" of illumination afford us a change of lighting as she walks through shadows to each well-lit area. Even with the entire set well lit, the change of scene as she strolls and extolls will allure the viewers.

Spreading out the graphics has a side benefit too. It puts into use that old educational axiom that you should reveal material only when it becomes immediately relevant to the topic.

7. *Move the camera around.* If the talent can't move, then move the camera. Arc it around so that the background setting changes. Everyone likes to go on a journey, and moving the camera up, down, or in an arc does exactly that. The change in perspective or background could coincide with the change in topics, thus providing a visual bridge.

You could dolly in or out with the talent (like the walking in example 3). *Do not simply zoom in and out on the talent.* It is not visually exciting. Unfortunately, it is so easy, it gets overdone. It is bad enough that you are accosting your viewers with Talkie-face; don't yoyo their eyeballs with meaningless zooms.

8. *Contrive cutaways.* These could be

    a. A close-up of an object of interest (followed by a zoom out to include the talent).

    b. Just the opposite; start with a medium shot, then zoom in on the object (or symbol of the object) just discussed while the talent moves out of the scene. You could even go out of focus in close-up, dissolve to an out-of-focus shot of the next object, and, while focusing on that, zoom out to include your talent as he or she arrives in the next scene.

9. *Shoot two heads.* Not necessarily on the same performer. They always say two heads are better than one, don't they? Two heads offer a variety of two Talkie-faces interrupting each other perhaps to underscore certain Talkie-points (and if they both talk simultaneously, the show could be over in half the time!).

Like it or not, narration and talking faces are here to stay, probably because they are so cheap and easy to do. This rut of least resistance is addictive, however. Jump at any opportunity you can find to use pictures instead of words, sound effects instead of narration, and lighting, music, camera angles, or editing to make your point.

Script Preparation

Figures 17–1, 17–2, and 17–3 show several sample script formats. Each has been marked with notes and cues by the director.

Generally, the video appears on the left of the page, while the audio and/or narration appears on the right. Sometimes all production information (audio and video) appears on the left, while only narration is on the right.

It is sometimes helpful for complicated audio and video moves to type the page sideways with narration in the middle, video in a column on the left, and audio on the right. Each audio or video cue is keyed to a particular word in a narration by drawing from the word to the cue.

Due to a diversity of production styles, no one set of scriptmarking symbols or methods has become universal. If you invent a system of your own, take care to

1. Make symbols that are clear and unambiguous.
2. Once you have a system, stick to it.
3. Avoid overmarking your script. The clutter will confuse you during the hubbub of production.
4. For "live" productions, place your cue marks *before* the desired action.
5. If the script is clearly marked, simply circle the script cues rather than duplicating them.
6. Feel free to draw little pictures to visualize moves or positions.

Fully scripted shows assist in planning shots, a necessity during complex productions. The director can see everything that will happen and has to contend with few surprises. On the other hand, a detailed script burdens the director with many things to keep track of during the whirlwind of a performance.

Partial scripting is common for interviews, instructional programs, and variety shows where a good deal of ad-libbing is expected. The script would not only contain detail about the production's beginning and end but would

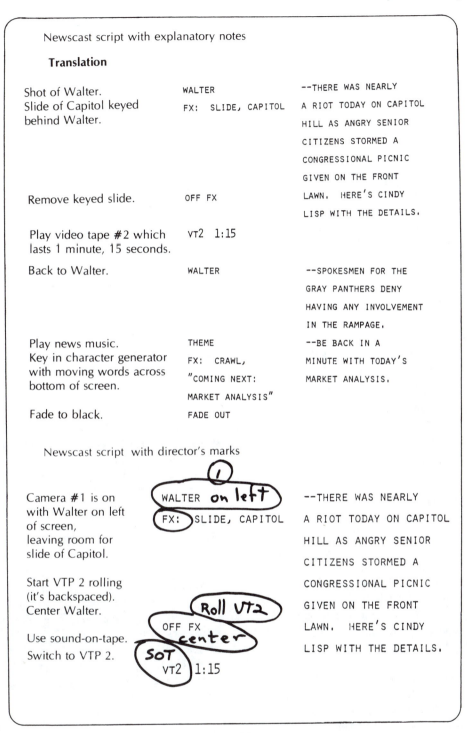

FIGURE 17-1 Newscast script with director's marks.

also include any complicated camera moves involving visuals, films, or sound effects. Otherwise the director just knows the general direction that the program is taking and "wings" it, picking the best shots possible.

A RUNDOWN sheet is the most minimal script. It is a list of things scheduled to happen during the production and the order in which they will occur. For example, the RUNDOWN may say

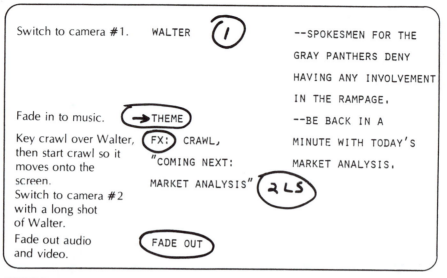

FIGURE 17-1 CONT'D

1. Series theme, title.
2. Instructor introduction, descends ladder onto set.
3. Graphic 1, math puzzle, ladder against wall.
4. Instructor works out problem on chalkboard.
5. Graphic 2; height of tree measured by triangulation.
6. Instructor at board.
7. Sextant described. CU on camera 2.
8. Animated film; "Navigation by Sextant."
9. Instructor conclusion, ascends ladder.
10. Series theme, end title.

Since instructors aren't renowned for their willingness to memorize lines, the RUNDOWN is often as close to a script as you will get. More mistakes and poorer production quality may be the result of such semi-ad-lib programs, but they are relatively cheap to produce and, thanks to video tape, can be done over if flubbed. Incidentally, you'd be surprised at how much preparation goes into a really good ad-lib speech!

Dramatic scripts usually contain no audio or video notes; they read like a play. It is the director's artistry that picturizes the event, indicating shots, angles, and cues.

A good script is especially useful when you are having someone else edit your raw footage. The script acts as an instruction sheet, telling what shots are desired and where. A person who has not been involved with your production could study the script, log the shots, create an edit decision list, and edit your program. Except for a few places where

the taped action doesn't match the script exactly or where editing style plays a substantial role, the final production should come out looking 95% the way you intended it to.

Script Reorganization.   Scripts undergo more updates and changes than any of us would like to admit. We first think our script is perfect and then discover that we left something out or that a scene has become impossible to shoot. A world of kibitzers, including the boss's brother-in-law with golf clubs clanking on his back, will feel free to "suggest" changes in your carefully prepared script. Gratuitous script changes become the norm as you move deeper into public relations and advertising video. You have no choice but to grit your teeth and develop some method of keeping track of the changes.

In some productions, every shot gets numbered. Complex shows may have every line in the script numbered to keep track of them. The revision of shot 25 could be listed as 25a, 25b, or, if you're really lucky, 25z.

The first copy of your script could be on white paper, punched, and placed in loose-leaf notebook. As pages are changed, they are printed in different colors, green for revision 1, pink for revision 2, etc. It is important to distribute script revisions to all interested parties, who would, in turn, dispose of older pages, putting the updates in their place. Eventually, everyone comes to the production with a multicolored script that matches their neighbor's script to the letter.

Actors will generally mark their copy of the script

MINIGLOSSARY

Rundown   List of events planned to happen during a show. The outline of a script.

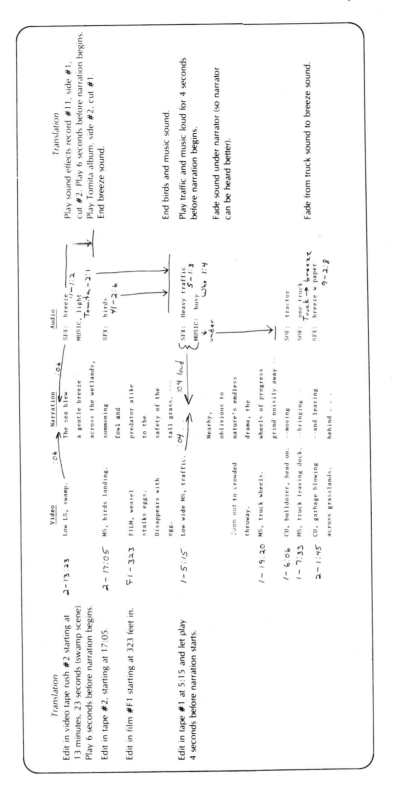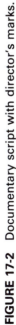

FIGURE 17-2 Documentary script with director's marks.

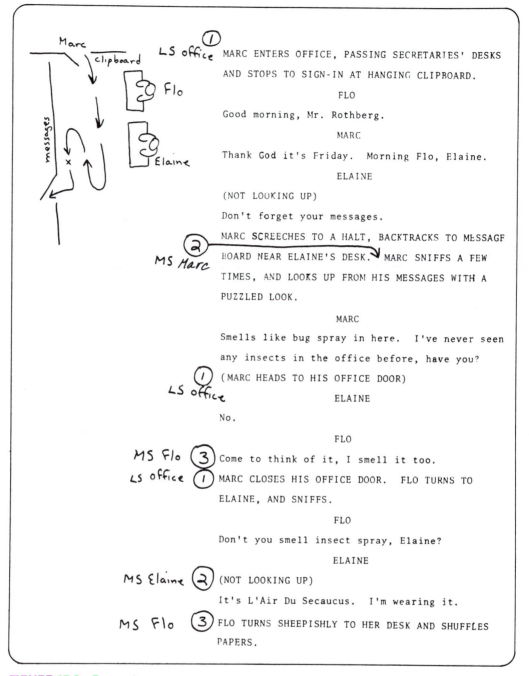

FIGURE 17-3 Dramatic script with director's marks.

with a yellow highlighter, noting only the lines they need to memorize.

To simplify shooting, the director and production personnel need to organize their scripts in another way. Scenes which occur in the same locale need to be grouped together so that they are all shot at once (unless you like taking a dozen flights to Trinidad plus three ferry rides to the Statue of Liberty). This is where scene or shot numbering comes

into play. By going through the script shot by shot, each shot number can be noted on a "location list" like the one in Figure 17–4. The actors would be told what scenes were being shot that day so that they would memorize only the lines needed. The crew would set up the lights and cameras and props in one area and, in one fell swoop, produce all the shots that occur there.

Breaking a script into locations is very exacting. Every

Backyard	Home interior	Driveway	Tug boat LS	Tug boat CU
001	002	005	010	014
003	004		027	016
047	046		029	018
			038	028
			043	030
				042

Bridge LS	Bridge abutment	Barge	Bridge footing
006	007	008	009
011	012	019	017
022	013	023	020
036	015	025	024
045	021	034	031
	026	035	033
	032	037	040
	044	039	
		041	

FIGURE 17-4 Location list.

omission, every mistake, means a complete setup has to be redone, so don't be sloppy.

If doing single-camera teleproduction, it may even be useful to break the shots down further into camera angles. For instance, if two people are speaking together, you would get all the first person's face shots (looking over the other person's shoulder) at once. Then you would move the camera for the second person's face shots. Sometimes it helps the actors maintain continuity if you shoot the scene in its entirety once from one angle and then once again from the other. The scenes will go much faster without starting and stopping the tape for each line, and you will be able to cut back and forth from shot to shot during the editing process.

When only two or three shooting locations are required, it may be simpler to mark locations in the script's margin using a colored magic marker. In the end, your script would have all its margins marked in one of perhaps three colors. You would then shoot all the yellow scenes first, the blue scenes next, and the red scenes last.

Another technique for reorganizing the script for shooting location by location is to simply slice up the script with a razor blade and sort the slips of paper into piles for each location. Each pile then becomes a miniscript to be shot all at once.

To help coordinate the script with the shot sheet or edit decision list, it is sometimes helpful to include the script page number on the shot sheet or edit list. This way, if the brief notations on the shot sheet are confusing, you can look back at the original script to find out what your ancient, cryptic notes really meant.

It will be easier locating shots if you have a separate tape for each location. For instance, all the backyard shots could be on cassette 1; home interior shots, on cassette 2; and barge shots, on cassette 8. If you are shooting the same shot using two isolated cameras (each with its own VCR) simultaneously, you could number these cassettes 1A and 1B, which highlights the similarities in the content of the cassettes.

Word Processing. With the explosion in the use of word processors for handling text, especially text which will undergo numerous revisions, one would think script typing and word processing would be a marriage made in heaven. Unfortunately, the marriage often ends up in the other place, because most common word processing programs handle columns of text poorly. If you use a word processor, you may have to place your video on one page and your audio on another and then cut and paste the pages together. A little more cumbersome but worth the experiment is to use a spreadsheet program like SuperCalc to create columns of text. AVScripter is a neat $40 IBM-compatible computer program available from Tom Schroeppel (4705 Bay View Ave., Tampa, FL 33611) that converts ASCII CODE into dual-column script. Nearly *all* word processors can make ASCII, so with this program you can turn your word processor into a script generator.

Regardless of the method you use, you will not escape the script revising process without a first-class migraine.

Storyboards

STORYBOARD has nothing to do with telling such a boring story that your audience falls asleep. A STORYBOARD is a series of sketches depicting the main pictorial element of a scene (Figure 17–5). The narration usually accompanies each picture. The result is quite similar to a comic strip only the words are outside the picture. STORYBOARDS become especially worthwhile when

FIGURE 17-5   STORYBOARD.

MINIGLOSSARY

*Storyboard   A series of comic-book-like sketches showing what the TV scenes should look like. The corresponding audio is typed at the bottom of each sketch.

ASCII   A universal, standardized code for text and numbers used by computers and word processors.

1. It is difficult to describe a scene with only words.
2. Others will have to carry out your plan. The STORYBOARD clearly shows what you want your audience to see.
3. A production's cost or importance warrants prior approval. The STORYBOARD is a concise way of representing what the final production *may* look like.

If you think that only sissies use STORYBOARDS and that they are amateur and time-consuming, consider this: Alfred Hitchcock drew his own STORYBOARDS—shot by shot—for every one of his movies.

Small educational productions are seldom complicated enough to require a STORYBOARD. As shows move up in complexity, simple hand-drawn STORYBOARDS like the one in Figure 17–5 are quite satisfactory. *You do not have to be an artist to get your point across, so don't let your artistic inabilities hinder you from making your own STORYBOARDS.*

They do not have to be works of art. Advertising agencies and other professionals who have artists on staff make fancier STORYBOARDS like the one shown in Figure 17–6. These are used to obtain sponsor approval before budgeting travel or spending money on production. Figure 17–7 shows how the advertisement finally came out.

When productions are likely to be very expensive, a low-budget version is sometimes produced first, using non-star talent, minimal costumes and props, and industrial-grade equipment and sound. This creates a *moving* STORYBOARD useful in refining a concept or treatment before spending perhaps a million dollars on a full-blown production.

EFFICIENT DIRECTING

The director is responsible for "calling the shots" during a production. There's really much more to the job than this. The director's role begins long before the beginning of the shoot and usually ends in the wee hours after the postprod-

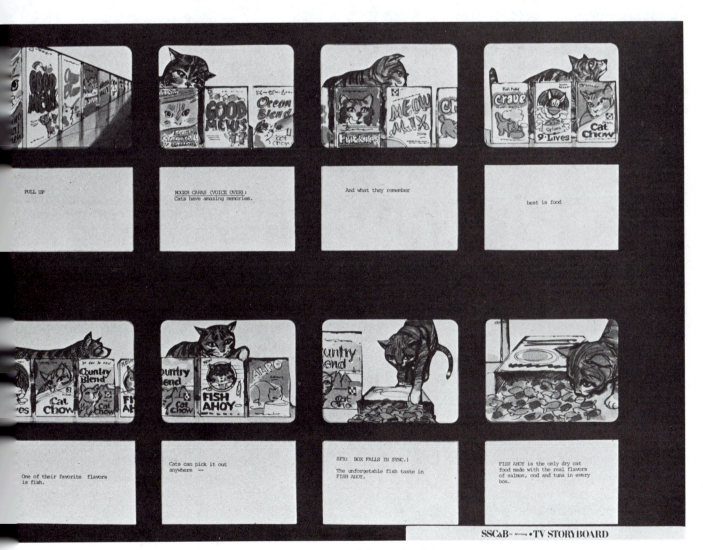

FIGURE 17-6 Advertising agency STORYBOARD. (Courtesy of Carnation.)

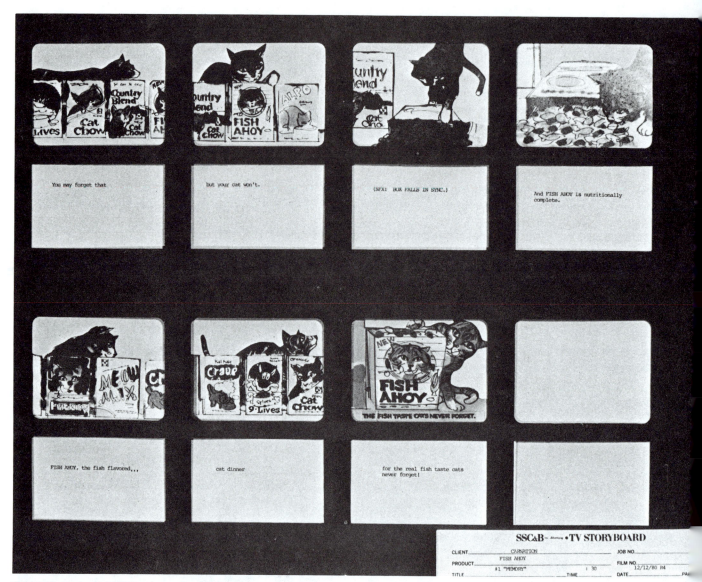

You may forget that

but your cat won't.

(SFX: BOX FALLS IN SYNC.)

And FISH AHOY is nutritionally complete.

FISH AHOY, the fish flavored...

cat dinner

for the real fish taste cats never forget!

SSC&B- ——— •TV STORYBOARD

CLIENT_____CARNATION_____JOB NO._____

PRODUCT_____FISH AHOY_____FILM NO._____

TITLE_____#1 "MEMORY"_____: 30_____TIME_____DATE_____12/12/80 R4____PA

FIGURE 17-6   CONT'D

uction editing. Directors may include in their production planning such things as

1. *Concept planning:* clarifying the objectives of the show and determining the best format for the production
2. *Timing:* making the show and all its segments come out the right length
3. *Visualization:* translating an idea or a script into screen images and mapping out the sequence of images
4. *Lighting and floor plan preparation:* sketching the positions, angles, and types of lights to be used and also diagraming the positions of the set, props, and talent in the scene
5. *Script preparation:* marking the script for camera shots, cues, transitions, and BLOCKING (position and movement of the talent)
6. *Rehearsal:* guiding the actors in their portrayal of the scene as well as in practicing the scene in the studio to familiarize the crew with their moves

A director's postproduction activities may include

1. *Checking the raw footage:* making sure that there is a technically acceptable "take" for each scene (*before* the studio has been cleared and the actors sent home)
2. *Editing the raw footage:* including the insertion of graphics, titles, and music

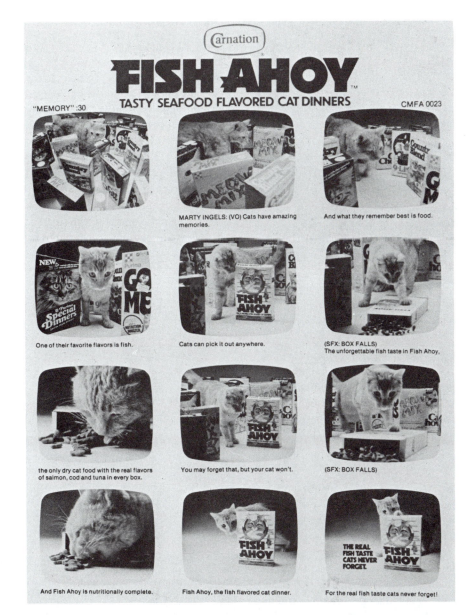

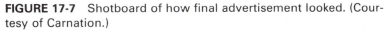

FIGURE 17-7 Shotboard of how final advertisement looked. (Courtesy of Carnation.)

The things a director does before and after production will most likely determine the quality of the show. That's why I've devoted so many precious pages to the subject. Now let's further explore the business the director handles immediately before and during production.

Before the Shoot

Preparing Yourself

1. *Learn the script.* During the show, you won't have time to look at the script for more than a second. Consequently, the script can act only as a reference for you, that is, as a list of reminders and cues. You should already know the content.

2. *Study difficult segments in detail.* The beginning and end of a show generally mix camera moves, music, graphics, and cues all at the same time. Sit down quietly and "talk through" the directions you will give during these difficult parts. Rehearse these segments until you can recite the commands without the script. Thus, when you get to these complex places, you can give your full attention to observing how well they are coming out rather than following your place in the script.

And while you're at it, *make and study a LIGHTING PLOT, FLOOR PLAN, prop list, and perhaps a STORYBOARD*. Doing so will help you plan

and visualize your shots. The prop list reminds you to assemble all the needed props *before* the show.

3. *While setting up and rehearsing in the studio, check your available time.* Where does the time go? Know you're starting to fall behind *before* you're behind. Otherwise, you won't exercise the necessary pressure to stay on schedule.

Unless you're paying an arm and a hoof for studio time, schedule it liberally. TV productions *always* take longer than estimated.

4. *Think through every move.* Every event in a production sequence should be planned out so that there are no "weak spots" or impossible moves to contend with during the actual shoot. For instance, is there time for camera 1 to break away from the title card and get focused on the guest before she's introduced? If you have two sound effects on records and only one turntable, can the audio person get the second cued up in time? Will the camera taking tight close-ups sneak into the other camera's view? Will you have boom microphone shadows? Will the dancers be able to hear the music they're dancing to?

Do you want to simply fade out and roll credits over black at the end or first turn out all the studio lights but one and have the talent exit the lit area and then key the credits over the empty set? Are there enough hands to run the lights? Do you need to leave studio lights on to illuminate the final graphic? Can the white lab coats, cited in the props list, be exchanged for blue ones which won't exceed the camera's contrast limitations?

Preparing Your Crew

1. *Delegate tasks. The most frequent, most costly, most demoralizing error made by amateur directors is failure to delegate.* Time is cheap before a production; time is dear during it. Do you want your crew and talent standing around while *you* adjust the lights, *you* arrange the chairs, *you* look for props, and *you* type the credits into the character generator? No way! In commercial, professional settings, *all* tasks *must* be carried out by specialists, usually long before the shoot. The small-time operator follows a similar but less formal code.

Whip out that FLOOR PLAN you made earlier and give it to a camera operator (in small studios) or to the floor manager (in medium-sized studios). Also give him the prop list and perhaps a copy of the STORYBOARD as a guide in the placement of the sets, furniture, and props. Then go on

LIGHTING PLOT **shows where lights are positioned and aimed.**

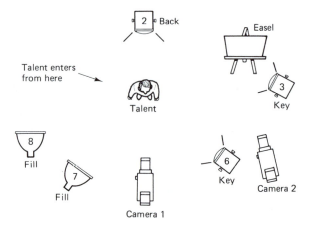

Symbols indicate the kinds of fixtures used. Numbers tell the dimmer circuits used (if known), or the fixture number, or the position on the lighting grid. Due to their similarities, the LIGHTING PLOT and FLOOR PLANS are sometimes shown on the same drawing.

FLOOR PLAN **shows positions of props and cameras.**

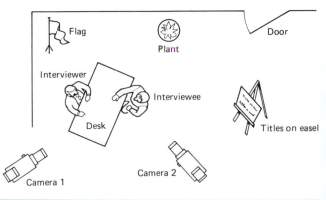

MINIGLOSSARY

*Floor plan   A sketch, seen from above, showing where objects, walls, doors, cameras, etc., are to be positioned on the studio floor.

to something else. Whip out your LIGHTING PLOT and pass it to another camera operator (or lighting director) and then go to something else. Provide a script to the audio director so that she can locate, check audio levels on, and cue up all sound effects. The audio person can then also mike the talent and adjust those levels. With another copy of the script in hand, the switcher can set up special effects and anticipate the kinds of commands to follow. Are there titles to be fed into a character generator? Have somebody do it. Are there slides for the film chain? They should already be sequentially numbered, ready to pass on to someone. The engineer or video tape operator should be told how long a tape to load and whether it's an editing blank to be used in the VCR's insert mode or simply a blank tape. Does the tape need time code? Drop frame or non-drop frame? The VTR operator will also handle the audio and video checks necessary to assure that the VTR levels are set. The STORYBOARD you prepared comes into play as you brief the camera operators and talent on the shots that will be taken.

Here's where your preparations pay off. The studio and control room are abuzz with busy beavers, setting things up and checking them out. Assuming your beavers are somewhat experienced and your written directions are clear, you won't have to spend much time telling people what to do. They'll figure out which lamp goes to which dimmer, which volume control goes to which mike, and which buttons create which special effect. Theoretically, you should be able to leave the area for about 20 minutes and come back to find everything ready to go.

Even if things don't go this easily (they won't), the idea that *you are free to handle special problems will maximize the efficiency of your studio setup. You're free to answer questions, check other people's work, assist the talent, and even (heaven forbid) have time to think.*

2. *Develop a silent response code.* Crew members can communicate with the director over their intercom headphones before a shoot, but during the shoot this conversation will be picked up by the studio mikes. Develop some code where the camera operators can communicate with you (the director) silently. Try having them blow into their mouthpieces—once for "yes," twice for "no," and three times for "I've got a problem." Four blows could mean "I'd like to go to the john," and five blows could indicate "I'd *really* like to go to the john." No telling what six blows could mean.

3. *Direct the studio rehearsal from the control room.* Don't keep running out to the studio to correct things. It wastes time while everyone's concentration lapses. Instead, make your shot judgments based on what you see in your camera monitors and rely on your studio personnel to rectify problems.

4. *Explain what went wrong.* If you stop taping because of a mechanical problem, announce this over the studio address system. Otherwise the crew and actors get uptight, thinking they were at fault. If they did screw up, explain what was wrong and why, so people can correct it. If, for instance, one performer is blocking another or has a hand in the way of something to be shown, don't assume the performer knows it—he or she can't see the camera monitors. Speak up.

5. *Announce return times.* Something breaks down during setup. It will be 15 minutes before you are ready to go. Let your unneeded crew and performers take a break. Tell them *exactly* what time they should be back and ready to go. Otherwise, you'll find yourself waiting and chasing. Meanwhile, since the studio is inactive, shut down those broiling lights.

6. *Do an audio check.* Unless the audio director is handling these details independently, make sure the audio person knows whose mike is on which knob. See that each performer has recited something while the audio person made volume adjustments. Sound effects and music also need to be checked for proper volume.

7. *Do an intercom check.* Call each person on the headphone intercom to make sure everyone can hear you. Chaos results when a crew member loses contact and is flying deaf.

8. *Do a video check.* Unless a video engineer and lighting person have checked this for you, do it yourself just before you start the show. Have the cameras provide long shots of the performers or views of what will be their "typical" shot for the show. Observe the apparent lighting. Are the shadows and highlights appealing? Now check the waveform monitor and, if necessary, adjust each camera's iris, gain, and pedestal one by one to make blacks black and whites white. Switch from one camera to the next to see if the pictures' overall brightness matches. If shooting color, have your cameras aim at the same thing, perhaps a performer, and, while switching between them, observe a color monitor. Is the color matched between the cameras?

Have your camera operators perform a complete and careful focus. Are their pictures sharp?

Equipment electronics "drift," making yesterday's adjustments inapplicable today. The adjustments may even drift during an hour's rehearsal. So the fine adjustments are best made at the last moment, with the lights set the way they'll be, with the talent and props shown the way they'll be, and with the equipment warmed up and ready to go.

If shooting titles on easels, check to see that they are level. A horizontal split across the TV screen will give you a line to use as a reference.

If very precise aiming is needed, adjust all the camera viewfinders, as well as the camera, preview, and line monitors, so that their pictures are horizontally centered and matched. To do so, first calibrate your line monitor, so its picture is accurately centered. Use a test signal, a special wipe pattern, or combination of patterns that displays something in the center of the TV screen. Adjust the horizontal control of your line monitor to center this target. Now get the other monitors to agree with the line monitor. Tell a camera operator to aim at and center something in her view-finder, perhaps a title graphic. While you watch your line monitor, have the operator pan left or right until the title is centered *for you*. When it is, tell the camera operator—without moving the camera—to adjust the horizontal control on her viewfinder to center the title. Now the two of you are calibrated. Next center the image on your camera monitor or any others you wish to calibrate. Once this exercise is over, all of you will be seeing and aiming for things exactly the same way.

9. *Turn off the studio line monitor.* Unless the talent has a special need to see the shots during the show, remove this distraction from their sight. Otherwise, they will tend to look at themselves in the monitor rather than respond to the camera or to each other.

Preparing Your Talent.   Rehearsing and preparing semiprofessional or novice performers is a science deserving a section of its own (coming later). For now, here are a few fundamentals:

1. *Banish the kibitzers.* The well-meaning entourage of PR managers and advisors who follow your company president or chief executive officer around can be a disaster in the studio. They are used to giving their two-cents worth even when that is all their advice is worth. There can be only one director. Be it. With the cackling hangers-on gently removed from the area, the president will be able to listen to *you* without distraction. You may not be used to giving your boss orders, but this is the one time when that's *exactly* your role. And your boss will listen. Senior execs didn't climb the corporate ladder ignoring the advice of the experts they hired. The trick here is to remove the competing voices and take charge as *the expert*, the director.

2. *Rehearse the words first and the BLOCKING (movement on camera) second.* It is difficult for beginners to master both at once. Once your talent is comfortable with their lines, move on to their motions, walking, sitting, turning from one camera to another, or handling props.

Sometimes it is hard to explain to a beginner how to speak forcefully or with excitement. You might try reciting the president's lines into an audiocassette recorder the way *you* think they should sound and then give him or her the cassette the day before shooting.

3. *Try not to overrehearse your talent.* They are likely to lose spontaneity, and you'll hear flattened modulation with less energy and more boredom.

4. *Shoot in short sections.* This gives the novice performer a "breather" and time to master the words and motions of each scene.

5. *Give your talent star treatment.* See that your guests are comfortable. Have someone attach the talents' mikes in the right place for them. This will build their self-esteem and confidence, counteracting their nervousness.

Between takes, offer encouragement, reinforcing the things they are doing well and gently suggesting *one or two* improvements, if needed. Try not to use too many "don'ts." Be positive.

When a production must be stopped, explain to the talent the reason for the "cut." They will be more comfortable knowing what is going on and being "included" as part of the team.

If your talent has been under the lights for a while, have someone bring them a glass of water during a break in the action. This not only soothes parched throats but makes the talent feel important and pampered.

When all is over, if you discover that some of your guests have been "edited out" and will not appear in the final program, send a thank-you letter to the participant, extending your regrets and appreciation.

During the Shoot

1. *Be firm but friendly.* Critical observations that people might accept face to face create tensions when heard over the one-way communications of the studio talk back system.

2. *Continually scrutinize preview monitors.* Is the shot you're taking okay? If so, *there's nothing more for you to do with that shot, so don't waste time looking at it.* Go on to the next shot. Get it set up and ready to go. When you switch to that shot and it's okay, check on the *next* shot, and so on. You end up never looking at what's being used but rather always preparing what's coming up next. *Looking ahead is the essence of direction.*

The director is in charge of quality. This doesn't mean giving commands like "Two, go to graphic 4," and then automatically taking the shot. It means watching the camera (or preview) monitor to check that the shot is steady, sharp, and centered before you call it.

The same goes for special effects. You preview an effect to make sure it's satisfactory before taking it.

3. *Give "ready" cues to switcher, audio, cameras, talent, and others.* Don't just call out "Take 2" or "Kill mike 2." There will be a delay before the crew can carry out these commands. Instead, always say "Ready to take 2—take 2," so the buttons are found and ready to be pressed instantly.

Similarly, warn the cameras of an action about to take place (after all, *you* have the script) such as, "They're going to stand in a second; get ready to follow them up."

Talent and floor manager need ready cues too. Over your studio address system call out, "Standby, we roll in 15 seconds," so they can psych up for action. Say to the floor manager, "Ready to cue talent," so he or she puts a finger in the air, ready to point at the talent when your "Cue talent" comes. Give your "Cue talent" indication a moment

MINIGLOSSARY

*****Blocking**   Planning out everyone's position and movement for the show.

early because there's a lag before the manager can respond. For instance, say, ''Ready to cue talent; ready to fade to 1. Cue talent, fade to 1.'' By the time you've faded to 1, the talent has started.

If you've given a ''ready'' cue and then change your mind, nullify the cue. Don't say, ''Ready to take 1. Take 2.'' Odds are your TD (technical director) will take 1.

4. *Give a ''standby'' cue to sources.* Lighting or film chain operators who have been unoccupied for some time may not be ready when you give your cue. So remind them that their parts are coming up shortly. Thus, they'll have their hands on the controls when you give your immediate cue.

5. *Avoid matched shots.* Say you're ''winging it'' through an unscripted or semiscripted show. Don't let both cameras give you similar shots. You remember how visually poor it is to switch from one long shot to another long shot of the same subject or from a medium to a close-up shot of a subject from the same angle. Since you can't aesthetically switch from one camera to its twin, why have the twin? It's doing you no good. Worse yet, if the shot goes bad from the camera that's on, you have no ''safety shot'' to fall back on, no acceptable place to go.

So delegate certain shots to certain cameras with complementary shots going to the remaining cameras. For instance, say, ''Camera 1, stay with the host; camera 2 stay close on his hands.'' Now while camera 2 is centering or focusing on the tough-to-get shots, camera 1 is giving an acceptable backup and establishing shot.

6. *Give your signals clearly and precisely.* Appear relaxed but alert.

7. *Keep track of which camera is on.* You don't want to call for a cut or dissolve to the camera that's already on. One way to keep track is with your fingers. Thumb on your pointing finger means you're on camera 1, on the second finger for camera 2, and so on.

Some switching consoles have a tally light (or you could easily build one) for each camera monitor. Instead of viewing the program and preview monitors, your eyes are always scanning the camera monitors. You know which camera is *on* by which monitor has its tally lit.

8. *Speak only when necessary.* Too much chatter over the intercom headsets dilutes people's attention.

9. *Call out the camera before giving instructions to it.* For instance, say ''Camera 2, zoom in.'' This avoids having all the cameras start to zoom in until they hear *who* was supposed to do it.

10. *Get cutaways if you plan to do much editing.* These can be produced even after the actual show is over—simple shots of nodding host, attentive guests, audience, or props. Even if you don't expect to use them, get cutaways anyway. You never know when you may have to hide a matched shot or a jump cut or a flubbed shot during postproduction editing.

11. *Cut a beat before the words.* This way the eyes will realize something at the same time the ears hear it. For example, during the line ''At the first sign of smoke or sparks, you should hit the big red panic button,'' you should cut to the CU of the button after the word *sparks*.

12. *Decide whether to start again.* This is the toughest decision a director has to make. While taping a production, an error is made. Do you holler ''Cut'' and start over, or do you let it ride? Factors to consider in this decision are

a. Is the program short and easily ''redoable?''

b. Is this or a series of false starts likely to dampen the talent's performance?

c. Are you near the beginning where very little will be lost by starting over, or near the end of your show?

d. Can you assemble-edit, recontinuing your recording from just before the error?

e. How bad is the error, how discriminating are the viewers, and how hard is it to do the part over?

Rest assured, whatever you ultimately do, Murphy's 308th law will get you: ''If you stopped, you should have kept going; if you kept going, you should have stopped.''

The Director in Action

It's zero hour. The video, audio, and intercom checks are over, the lights are up, and the show is about to begin. Figure 17–8 shows what a partially scripted, simple, 5-minute, two-camera interview show might sound like as heard over Joe Director's intercom.

Ready on black. [To engineer:] *Ready to roll VTR. Over studio speaker: Standby, about thirty seconds. One, tilt up a bit on that title graphic. Good. Hold it. Roll VTR.* [Waits ten seconds for the VTR to get up to speed.] *Ready to start music. Ready to dissolve to one. Two, zoom in on the host a little. . . . There, hold it. Start music. Dissolve to one.*

Camera 1

Figure 17-8 The director's commands during a show, and the resulting shots.

FIGURE 17-8   CONT'D

Ready to cue host. Open host mike. Ready to fade out music. Ready to dissolve to two. Cue talent, dissolve to two, fade out music.

Camera 2

One, get a long shot with host on left. Guest will walk in on right when introduced. Follow them with a two-shot. Open guest mike. Ready to take one. Take one.

Camera 1

Camera 1

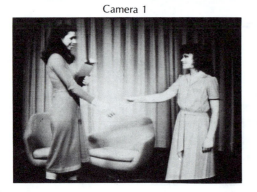

Two, get a medium shot of guest after she sits. One, zoom in a little and follow them down as they sit. Good. One hold that shot.

Camera 1

[Waits for camera 2 to focus on guest.] *Ready to take two. Take two.*

Camera 2

[To switcher:] *I'll be taking one and two with the conversation. Ready one. One . . .*

Camera 1

Two . . . One . . . Susan [the floor manager], *don't forget . . . two . . . to change the graphic* [for the ending]. *. . . One . . .*

[The interview progresses to where an excerpt of the guest's movie is to be shown.] *Two . . . One . . .* [Knowing that the switcher and he will be preoccupied for a moment, the director stays with the TWO-SHOT, an acceptable shot regardless of who's speaking.] *Get ready to roll film. Set up to key CG-one over Film-one on effects.* [CG1 is a title stored in the character generator. It is displayed on the monitor "CG1" and switched with the CG1 button on the special effects bank of the switcher. The film title will be keyed across the bottom of the film. The film appears on monitor F1 and will play through switcher button F1, called "Film-one."] *Ready sound-on-film-one. Ready to kill mikes. I think it's coming soon. . . . Roll film. Ready to dissolve to Film-one.* Waits three seconds for preroll of film watching for it on his F1 monitor. *Dissolve to Film-one, sound-on-film, kill mikes.*

Film-1

How's that key? Good. Ready to dissolve to effects. Dissolve.

CG1 over Film-1

Ready to dissolve back to Film-one. . . . Dissolve. . . .

Film-1

Ready to dissolve to one. Ready to kill sound-on-film. Ready to open mikes. Ready to cue host. . . . Open mikes, cue host, dissolve to one, kill sound-on-film.

Camera 1

Stop film. . . . Ready to take two. Take two. . . .

Camera 2

Ready to take one. . . . Take one.

Camera 1

Set up credits. Set up effects to key CG-one over Camera two. [The character generator operator switches to the next page of stored memory and prepares the machine to roll the credits across the screen.] *Cue up music. Ready to take two. . . . Take two.*

FIGURE 17-8 CONT'D

Camera 2

Signal the host she has one minute left. Two, get ready for a smooth zoom out to a close two-shot. Two, slow zoom out.

Camera 2

One, give me your longest shot. [To lighting person:] Get ready to dim key and fill lights at the end. [This will leave the talent standing in a puddle of backlight and set light, shaking hands and chatting.] Cue host . . . half-minute left. Two, they're gonna stand up; follow them up.

Camera 2

Ready to dissolve to one. Ready to slowly fade up music. Ready to kill mikes. Fade up music. Dissolve to one.

Camera 1

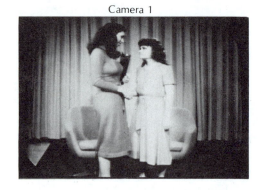

Kill mikes. Kill key and fill.

Camera 1

Two, get graphic. Ready to dissolve to effects. Ready to roll CG. [The credits are positioned on-screen and ready to roll upwards revealing more.] Dissolve to effects. . . . Roll CG.

CG1 over Camera 1

FIGURE 17-8   CONT'D

FIGURE 17-8   CONT'D

Susan, tell them to stroll off the set now. One, don't move.

CG1 over Camera 1

Ready to dissolve to two. . . . Ready to fade out music. . . . Dissolve to two.

Camera 2

Ready to fade to black. Fade out music, fade to black. [Waits ten seconds. To engineer:] *OK, stop tape. Rewind to beginning and play. Bring up houselights and kill studio lights.* [Over studio speaker:] *Very good. Thank you everybody. Please stay put while we check the tape. You want to see it too? . . . Okay.* [To switcher:] *Take VTR-two.* [To audio:] *Play sound-on-tape-two to studio.* [The tape plays through. To studio:] *Looks fine. We're done. Thank you.* [To crew:] *Good work! Thank you.* [Director pulls an Alka Seltzer package from his pocket, opens it, swallows the tablet whole, and staggers from the control room.]

TV TALENT

By *talent*, we mean interviewers, guests, hosts, performers, actors, instructors, anybody on the screen, and not necessarily the talented among them either. As the teleproducer, your job is to help the talent perform at their very best. You may need to dress them, rehearse them, comfort them, educate them, and sometimes pacify them. And just when you get used to the pressure behind the scenes and are questioning the quality of your performers, *presto*, you get dragged into the production and have to perform yourself. This will keep you humble, because performing isn't easy either.

Being the Talent

Whether you happen to be the talent or you've been given the job of transforming your company president from a blob into a smooth-talking, confident William F. Buckley, there are a number of guidelines a performer needs to know in order to "look good" on television. Table 17–4 summarizes the most important of these.

The sections that follow will go into deeper detail about clothing, mannerisms, and self-conduct during an interview. The last subject (interviews) is often taught in expensive workshops around the country. It is given to executives who may be cornered by television reporters and will need to present a good image for their company.

Eyes.   It has been said that 40 percent of the message is verbal and that the other 60 percent is nonverbal. The audience will unconsciously be studying your eyes in search for confidence, credibility, and enthusiasm.

If you are addressing the audience, you should look directly at the camera. You may feel that the unbroken gaze at the lens looks strange, (you wouldn't consider staring at a person that way) but on TV it looks normal and preferred.

If you're being interviewed, you should have 90% eye contact with the interviewer. This will make you appear interested and attentive. Even when the interviewer is talking, you never know if a camera is looking at you. As soon as you look away from the interviewer, you run the risk of being caught by the camera. This also goes for when the interviewer opens or closes the program. Watch the interviewer, not the lighting grid, not the antics of the stagehands. Glue your eyes to the interviewer until the director calls "Cut."

Chairs.   For performers to look comfortable, they need comfortable chairs. A good TV studio will invest in several matching, soft chairs with armrests. These chairs should not rock or swivel because that's what your nervous performers will do in them, making your audience seasick. If swivel chairs are all that are available, the talent should

TABLE 17–4
Basic do's and don'ts of being the talent

1. *Dress right*. Avoid herringbones and small checks (they turn rainbow colors on TV). Don't wear pure white (it dazzles the camera) or pure black or navy (it disappears into a black blot on the screen). Avoid shiny jewelry and digital watches that go "beep" every half hour. Check for combed hair, straight ties, flat collars, zipped flies.

2. *Be punctual*. Time is very expensive in a TV studio.

3. *Be attentive*. Excessive background chatter while the director or others are trying to communicate is punishable by hanging.

4. *Speak with normal volume during the voice check.* Say your ABCs or something. This lets the audio operator know how far to turn up your mike's volume.

5. *Don't acknowledge cues*. When the floor manager points to you (which means "You're on!"), don't nod and say "uh-huh" before you begin. Or when he holds up three fingers to indicate you have 3 minutes left, don't interrupt yourself and look at him with a "uh, 3 more minutes, huh?" Simply act your role, not revealing any awareness of the behind-the-scene activity.

6. *Be aware of the camera*. If you're talking to your audience, talk into the camera lens. Not to the floor director next to the camera. Not to the studio monitor with that big close-up of your smiley kisser. Not into the great televoid of the studio. Not at the floor, your microphone, or the desk. Don't read your notes. Talk to the camera lens.

   Which camera, you ask, flanked by two of the silent beasts? The one with the red tally light that's lit. What if the light goes out, indicating another camera's on? Don't look around until you find the other and say "Oh, there you are!" Simply shift your gaze to something else for a second (such as your notes or another guest) and then look straight to the other camera and continue without skipping a beat.

   If not talking directly to the audience (such as in a drama or interview), *do not* look at the camera or anything else behind the scenes. Ignore the camera if it moves to another position. Be oblivious to the floor manager tripping over a stack of cue cards. Don't look up if a studio light burns out. Disregard the camera operator who starts giggling. *They don't exist.* What's happening on stage is all that exists.

7. *Restrain your movements*. There's no need to project your voice or to dramatize actions as on a stage. The mike will pick up a whisper or a sigh. The camera will reveal a flick of the eyebrow.

   Give warning cues before you move. Jump up from your chair during a close-up, and the audience sees a nice crotch shot. Rather, tentatively move as if to rise and then rise slowly. This the camera can follow. When holding things to be displayed, move them slowly.

8. *Respond immediately to the floor manager's cues*. If the floor manager, for instance, moves a finger across her neck in a slicing move, the show is over, with or without you. Say goodnight, Gracie.

9. *Sit slightly forward in your chair*. So you look interested and not too casual.

10. *Speak clearly and distinctly*. Who wants to listen to a mumble mouth?

11. *Use your hands, but only to express ideas*. The hands will help you communicate. But fussy, nervous movements (finger picking, palm rubbing, pencil tapping, smoking—yes, *smoking*) will catch the viewers' eyes and distract them from your message.

plant one or two elbows on a table or desk in front of them to impede their natural inclination to rock or swivel.

Again, 60 percent of the message is nonverbal body language. As the guest, you should lean forward in your chair to show involvement and interest. This stance also creates an eye-to-eye appearance of equality between you and the interviewer. It also allows a tighter shot because both of your heads are closer together.

Your guest's chair should be at a 45° angle from the interviewer (review Figure 8–13). This causes you to lean slightly to one side, resting your elbow on the armrest, freeing your hands for gesturing rather than white-knuckling the chair.

It is best to sit in the front part of the chair. This stance shows involvement while avoiding the Lincoln Memorial image—or just the opposite, a slouch.

Hands.   Except for wild waving, gestures are constructive. They look natural and illustrate your speech. More important, gestures will take your hands away from twisting rings, tugging at earlobes, gripping your knees, or grasping your chair arms as if you were riding a tilt-a-whirl.

Legs.   Avoid splayed legs. Use the "finishing school" position, crossing your legs at the ankles.

Crossing your legs at the knees may be all right, if room permits. Generally, you are in such close quarters that if you and another person cross your legs normally with the dangling foot closest to the other person, your feet will bump. If you cross your legs so that your foot is aimed *away* from the other person, your body language will appear to be defensive.

Clothes.   Suits should be medium blue, brown, or charcoal gray. Solids look better than pinstripes.

Don't wear a white shirt. Pale blue or tan looks good. Colored ties with conservative patterns look nice.

Vests look a bit formal, constricting TV talent like sausage casings.

Socks should reach halfway up the calf, covering the ankle.

Women appear best in a conservative dress or a good regulation business suit with little jewelry and a scarf for color, perhaps tucked into a jacket so as not to look too busy around the face. Avoid pants and blouse outfits unless you wish to look like a "Wheel of Fortune" contestant. Avoid stripes or loud, bold or vibrant patterns.

Avoid bow-tied blouses—they look too much like a uniform. A blouse should not be bright or white; deep or pale tones are better.

Simple pumps look more feminine than boots on camera.

Jewelry is always a problem because it flashes. Network TV cameras costing $40,000–$100,000 can handle the glint of dazzling diamonds but not the lower-priced educational and industrial cameras. So avoid diamonds and other shiny things. Pearls, gold button earrings, and small gold neck chains should work out okay. Avoid dangling earrings which waggle and distract the audience.

At all costs, avoid black and loud plaids.

Directing the Talent

In our league, most of our TV talent will be semi- or non-professionals, and the less they know, the more you will have to encourage, coach, and calm them. Perhaps the best first step in their education is to present them with Table 17–4, "Basic Do's and Don'ts of Being the Talent." (Don't make a photocopy; just hack out the page. That way, when this page is lost, you will have to buy another copy of the book, enriching Peter Utz and Prentice-Hall.) If interviews are involved, you may also wish to familiarize your talent with a later section of this chapter, "The Interview." Now, Mr. Director, some hints for you.

Prepping Talent.   Don't make your talent work in a vacuum. Tell them about the show as a whole. Let them know how they fit into the overall story or objective. Tell them if you plan to use voice-over (narration), slow-motion, or special effects. Knowing more about the final look of the program will help them "play" to that end.

Nonprofessional talents' biggest problem is usually nervousness. To dispel the "butterflies," invite the performers to the studio before the actual production begins. Talk to them as they sit in the studio getting used to the frenzy of cables, lights, headphones, and people everywhere.

Rehearsal.   Chances are that the rehearsal will not take place in the studio. The TV studio is too expensive a piece of real estate (usually) for people to simply stand and practice lines together. The rehearsals usually occur in an empty room somewhere, and the final run-throughs are done in the studio.

Here's a list of more things to remember:

1. *Give a script to everybody who needs one.* Don't be parsimonious. Be prepared.
2. *Avoid excessive script revisions.* Wrong versions of actions, groupings, and lines have a way of getting remembered and new ones forgotten.

"Body language" is apparent in the way an executive chooses to sit before the cameras.

To relax the performer, invite him to the studio before the actual production begins.

3. *Provide a pronunciation guide to technical language and the names of people.* "Welcome to our show, Mr. Smelleybuns . . . oh, excuse me, Mr. Smel-LAY-bins."

4. *Explain about editing.* Amateurs aren't known for pulling off 20-minute scenes flawlessly. You will probably be shooting in short segments. Explain how the "magic" of editing will assemble the pieces together later.

5. *Let performers know the set* FLOOR PLAN. If they know where everything is, they are less likely to rehearse their moves in a forbidden zone (like on a wall or in a sink). Provide them with reasonable substitutes for props. One actress recalls practicing a phone conversation using just her hand. On live television, the phone rang as scripted. She went to answer it, but instead of lifting the receiver, she automatically picked up her fist to her cheek and held an entire two-minute conversation with her little finger.

6. *Have the moves and identities learned along with the lines.* Executives, teachers, and nonactors will probably have to learn the lines first and the moves second. One problem at a time is all they can handle. Actors, however, even beginners, can generally learn to handle both at once. Include

the moves and identities along with the lines to provide a better "feel" for the action while helping the actors learn the dialogue. Call the actors by their character names to increase their identity and their involvement with the character.

7. *Inform actors of the shots you will be taking.* They will then feel free to wave their hand in long shots, stand very still and include very subtle facial expressions during tight close-ups, and relax when they know that they are not in the shot at all. They should also know to keep still when they're in the background of someone else's shot lest they distract the viewer.

8. *Do not overrehearse.* They'll lose spontaneity.

9. *Praise often:* Never embarrass. Acting is frustrating. Most novices and semipros desperately need feedback and encouragement. Build them up; relax them. Take their minds off their egos and self-consciousness and put them into their role. If you must criticize, do so positively. Be constructive. Besides, if the talent is nonprofessional, you're probably not paying them enough to take *any* abuse—no matter how much they deserve it, no matter how frustrated you become. So be nice, lest they walk out, and then we'll see how frustrated you can *now* become.

Shooting

1. *Shoot the most difficult scenes first when people are fresh.* Go back and do the easy ones later when everyone is tired and cranky.

2. *Remind actors to listen to each other.* Otherwise, they will just recite *their* lines without reacting to other actor's lines in a realistic way.

3. *Walk an actor through a scene when using voice over.* If a narration is going to be dubbed in anyway, you might as well enjoy the freedom of verbally coaching the actor as he performs the scene, like ''Walk a little slower . . . pick up the photograph . . . hold it steady . . . steady . . . good, walk away to your right.''

4. *Give actors something to do.* Unless you want the ''background'' actors standing with their hands in their pockets jingling their keys and driving the sound person mad, give them a prop to hold or something to do. How about a glass of lemonade to drink, a rag to polish with, a bolt to tighten, a book to leaf through, or a dinner date to silently chat with. Anything, as long as it visually doesn't upstage the speaking talent.

5. *Kill the studio lights when not needed between takes.* Don't parboil your actors and crew. Meanwhile, save-a-kilowatt.

6. *Have a pitcher of ice water and cups in the studio.* It's a nice touch and cheaper than Harvey's Bristol Cream.

The Interview

The interview is a quick and inexpensive way to explain something without a disembodied narrator or a lecturer's stilted talking face. The interview program is visually simple to do, though somewhat bland. One way to spice up the image is to carry out the interview at the expert's office, laboratory, workbench, or even outdoors—someplace associated with the expert's or the program's subject. Not only is this visually stimulating, but your guests are more at ease in familiar surroundings. The viewers may also get a better sense of your guests' personalities, associating them with their surroundings. And last, the surroundings offer you the chance to shoot some good cutaways to be used to cover your edits and flubbed shots.

How to Interview Someone. Your object is to systematically extract information from someone while eliminating irrelevant information. Here are some methods for charming an interviewee into willing submission while getting your story. Some of these techniques seem to contradict each other. That's okay. Different situations call for different strokes.

1. *Lead with a question the guest will enjoy answering.* Guests rarely walk out on an interview that's underway, so snag them first with a plum and then work up to the controversial or hard-to-answer questions.

2. *Listen intently and react to your question's answer.* Simply reading a list of prepared questions bores your guest and your viewers alike. To seem alive, the *interview should seem like a discussion.* That means listening, reflecting, restating, and reacting to the guest's response. Try to summarize key points to improve the audience's retention of the facts. Authenticate and substantiate points made by the guest. Illustrate your guest's ideas and include anecdotes, this not only shows interest in his subject but gives him a few seconds to think and relax. A relaxed guest with time to think will give you more reasonable and useful answers, making your interview more instructive.

But don't go too far with this. You don't want to upstage guests by exhibiting superior knowledge of a subject or by performing verbal gymnastics around them. You simply want a stimulating discussion.

3. *Don't interrupt.* Ask only as many questions as are absolutely necessary to get the required information. Then shut up and let the guests talk. If they pause for a moment, don't jump in right away. Given a chance, they'll continue. *Nod your head,* maintain eye contact, and smile to show agreement, fascination, or surprise *in nonverbal ways.* This sort of reaction not only keeps guests talking but gives them confidence and works just as well as words at showing your interest. You won't believe the efficiency and effectiveness of this method until you've tried it.

4. *Set the tone of the interview.* Enthusiasm, like measles, is contagious. The interviewee often assumes the style of the interviewer during the course of questioning. If the interviewer is low-key and serious, the guest may gravitate to a somber, quiet monotone. If the interviewer is bright-eyed and bushy-tailed, the guest is likely to respond similarly with glint of eye and bush of tail.

5. *Be yourself.* If you try to sound like Walter Cronkite or Don Pardo, you're going to sound like a Walter Cronkite or Don Pardo imitation. Your audience will be distracted, if not entertained by your attempted impersonation, and they won't focus on your guest as they should.

6. *Extinguish bad habits.* They quickly become tedious on TV and will drive your audience crazy. First to go should be the phrase ''y'know.'' It not only sounds uneducated; it communicates no useful information. One way to unlearn this habit is to hold a conversation with a friend who's equipped with a dusty vacuum cleaner bag. Tell her to give you a swat in the chin every time you say ''y'know.'' Works wonders. Next, get rid of ''uh huh.'' Replace it with a nod. While you're at it, junk the ''uhhh'' you string between phrases and sentences. Also, don't end sentences with ''and so on and so forth.'' And last, watch out for the ''echo'' syndrome, where you habitually repeat the guest's last phrase:

Guest: " . . . and the empty bottles make nice gifts."

Host: " . . . empties make nice gifts, huh?"

7. *Don't ask questions that require only a yes or no answer.* If you do, that's what you'll get. Then what do you say? Ask, instead, open-ended questions. Replace "Weren't you the first . . ." with "How did you become the first . . ." or "What got you interested in . . ." or "When did you first discover you could. . . ."

8. *Ask simple, not compound questions*

Reporter: "In the light of recent statistical reports indicating a rising trend in manufacturing-related unemployment and economic difficulties for American car dealers, do you oppose discouraging Congress from trying to stop the amendment to abolish the sanctions against foreign imports, and if so, why, and if not, what should be done?

Congressman: "Yes, no, sometimes, and maybe."

Compound questions are likely to have just the first or last part answered. Questions with double and triple negatives will simply boggle the mind. With long lead-ins, the audience is likely to forget the beginning of the question by the time they reach the end.

9. *Do your homework.* Prepare for the interview by studying the guest's background and subject matter. Read his book; see her film. Not only will you be able to hold an intelligent conversation with guests in their areas of expertise, but you and your guests will probably enjoy each other more. And that enjoyment is what makes an interview come alive.

10. *If the guest's answer is unclear, rephrase the question and ask it again.* You must have asked the question for a purpose. What good is a nonanswer or circumlocution?

11. *Save the guest who is in trouble.* The program is a teamwork affair. If the guest goes down the drain, so does the show. Try to present guests in their best light, whether they deserve it or not. For examples, Simon Showoff, a self-proclaimed expert who obviously oversold himself and is now starting to look foolish, needs a bridge to another subject. One escape is to seek out some small thread of truth in Simon's dumb statements (there's usually a thread somewhere) and follow up on it:

Simon: " . . . and the kids today are all hoodlums."

Host: "Some kids seem to be. What do you think society is doing to change that?" Or . . . "Didn't Socrates say the same thing about youth in his generation?"

What do you do with Paralyzed Polly? She was a chatterbox before the cameras went on; now she's speechless. She's probably nervous. Try to lead with a simple question (but not a short-answer one) to make the guest experience some success. Try to center on the guest's pet peeve or some other favorite topic. As she begins to open up, encourage her with nods and smiles.

How about Walley One-Track who insists on pitching his favorite line? Trying to block his attempts will be futile; he'll eventually work his way around to his obsessive topic sooner or later. Perhaps it would be quicker to concede. Let Walley make his point. Then get on to *your* point. Later, you can edit out his soapbox speech. (Hah! The magic of video strikes again!)

Finally, we have Hysterical Harry. Harry may be nervous, just like Polly, but he manifests it as a stream of chatter and bravado apparently aimed at impressing everybody.

You could cut him down with a couple barbs from a Don Rickels Roast. But that's gauche. You could try being the perfect guest and humoring him with laughter, but that may encourage him more. Instead, play it straight. Be cool, slightly unapproving, sincere, and serious. When he suspects he has lost the host as an audience, he may fear that his TV fans may have dropped him as well. Once he's cooled down, you regain control of the interview.

12. *Prep the interviewee.* In some cases, the director may want to edit together several testimonies, leaving out the interviewer's questions. One advantage to this format is that the interviewer, absent from the shot, isn't there to draw attention away from the response. Another is that "stand-alone" responses can be PARALLEL-CUT or JUXTA-POSED (described later in this chapter) without the distractive repetition of the interviewer's questions. A possible side benefit is that anybody can ask the questions off camera; it doesn't have to be the same interviewer throughout the shooting. And even the same interviewer doesn't have to dress exactly the same for each shoot.

Back to prepping. For the preceding format to work, the talent's answers must explain themselves, having the question built into the answer. The talent must be told to *load the first part of their answers with a strong topic sentence* using basic subject-predicate syntax. Each sentence should contain useful information and be able to stand alone. Ideally, it should sound quotable.

For example, answers like "When I was about 15 years old, in school," are useless. You want answers like "I first became aware of my ability to imitate gastrointestinal sounds when I was about 15 years old, in school." Complete responses like this may now carry the show on their own.

Being Interviewed. There's a knock at the door. You answer it in your bathrobe and slippers with the heel that's split apart. Mike Wallace and the "60 Minutes" camera team blind you with photofloods and shove a mike up your nose and ask, "When did you stop beating your wife?" You smile at their impossible-to-answer question and respond, "When she started taking bridge lessons. Now she beats me all the time." Chalk up 5 points for the interviewee. Other helpful hints:

1. *Make certain you understand the question.* It's easier to answer a question you understand, and your response is more credible when it answers the question directly. Have the interviewer rephrase the question if necessary.

2. *Keep one or two good lines in your pocket.* If you expect to be interviewed on the "6 O'Clock News," chances are they're not going to show 5 minutes of you. More likely 30 seconds. Consider, what is a newsperson or interviewer likely to ask? Then think up and memorize a witty, colorful one-line statement that either deals directly with the subject or sidesteps it gracefully. The rest of the interview may end up in the editing raw footage pile, but your pithy one-liner will be too tempting for the director to pass up.

This, in fact, is a way to outmaneuver a manipulative or biased reporter. Despite the reporter's efforts to corner you into making *his* point, you can turn the tables by having a spicy, quotable statement, fact, or anecdote which makes *your* point. How can an editor discard the juicy nuggets while keeping in all the dull drivel which actually did tell the story? In fact, you might be able to gloss over the reporter's question with a plain vanilla answer and then launch into the point *you* want to make using rum mocha chip language. Guess which will end up on the air? Politicians are famous for answering questions nobody ever asked.

3. *Be brief.* NBC News correspondent Michael Jensen once said, "TV news is to journalism what bumper stickers are to philosophy." There is generally no time for follow-ups to your answers, no amplification, no clarification. Whether it's panel discussions or news reporting, TV is hit-and-run. So make a positive point with a short declarative statement; then run.

4. *Don't fake an answer.* There's a saying, "If you can't dazzle them with brilliance, then baffle them with baloney." Well, phooey to that. If you don't know something, say so. No law says you have to know everything. Offer to find out the answer. Suggest someone else who does know the answer. And if you know the answer to a question but don't wish to divulge it, say, "I'm sorry, but that information is confidential right now," or something witty like "Would a gentleman kiss and tell?" Raise your own issues and answers with "I'm not ready to answer that question yet, but I can tell you this. . . ."

Another technique, useful in panel discussions, is to turn a difficult question over to a colleague. To give the colleague some time to think (a much appreciated courtesy) and the director and cameraperson time to move to the other person, you might first announce, "I believe that's a question which Mr. Pevar can answer better than I can," turning toward him while finishing your sentence.

5. *Use "people talk," not industry jargon.* If your viewers are unfamiliar with your field, you won't gain credibility with long words and incomprehensible concepts. Use simple language, analogies, and concrete examples to illustrate your point. Don't call something a "tertiary propulsion module" if you can call it a backup rocket.

6. *Pay attention to the interviewer.* First, a discussion between two people is more interesting than a discussion between one-and-one-half people. Second, your wandering eyes will look shifty and evasive on camera.

7. *Avoid repeating negative words.* It sounds offensive.

8. *If you must interrupt. . . .* Panel discussions sometimes leave you wiggling in your chair, trying to find a way to break in with your two-cents' worth. Here are two ways:

a. *Shirttail.* When the other person says, "*and . . .*," pick up the rest of his sentence. Or start a new sentence using a key word from the other person's sentence to first reinforce his point and then make yours.

b. *Open hand.* During a panel discussion you can get the moderator's attention and the floor by holding up your hand or holding it in a "stop" position toward the moderator. Once you're on, score a brief positive point and then relinquish the floor (or airwaves, or whatever). Don't try this "wild card" interruption more than once per show as it's a bit heavy-handed.

9. *Remain cool when harangued.* Interviewers sometimes, and studio audiences often, can get pretty testy. They can ask venomous questions which are 99 percent complaint and 1 percent query. Now is the time to be cool, courteous, and superpleasant. You'll lose a lot of points (and surely your job as company spokesperson) becoming hot headed and responding in kind.

Panel Discussion. Visually boring but cheap to execute is the panel discussion. Here a host interviews several guests at once, thus offering a variance of expertise and opinion while at the same time taking some of the pressure off the talent. With more people talking, each has more of a chance to think. Here are some tips on how a director can handle panel discussions:

1. *Limit the number of interviewees.* Panel discussions never seem to get off the ground when each member has only a couple minutes' share of the program.

2. *Seat the talent close together.* On TV, gaps between people look large.

3. *Brief the guests.* Guests should know why they are appearing. Tell them what you feel they in particular have to offer. They then become more likely to give you the responses you desire.

It's not necessarily cheating to give the guests sample questions before an interview. It may put the guests at ease

FIGURE 17-9 Over-the-shoulder shot, useful as a cutaway.

and give them time to think up clearer answers to your questions. Lucid answers may be more important to you than catching someone off guard.

4. *Position the host on one side of the guests.* This way, the host and viewers avoid "tennis neck" as their attention Ping-Pongs first to the guest on the left, then to the guest on the right, and back again.

5. *Shoot over-the-shoulder shots for full-face views.* Shooting over the host's shoulder will give expressive full-face views of the guests as well as directing their answers almost into the camera, increasing audience involvement. Shooting over the guests' shoulders (outside the first camera's shot) gives a full-face view of the interviewer, establishing not only the interviewer's singular role but providing possible cutaways of a nodding head for later editing. Figure 17–9 shows an example.

While shooting a two-camera panel show, one camera can show the host while the other is focusing or panning from one guest to the next. This "safety shot" gives the director and camera operator breathing time while setting up shots.

6. *Avoid matched shots.* With matched shots (Figure 17–10), if one shot goes sour, you can't go to the other without committing a *faux pas.* So maintain differing shots between your cameras (that is, a one-shot and a two-shot),

so that you always have something "kosher" to cut to (hence the term "kosher cuts").

7. *Angle the guests.* This maxim applies to any shot with two or more persons in it. People who are faced nose to nose are posed in what seems to be an adversary relationship. It's a great way to portray an argument, but in a panel discussion the shots imply disagreement or debate. Conversely, people lined up facing the camera look like a team of contestants.

8. *Title the participants.* In the opening shots, frame the participants loosely so that you can add their names and titles across their shirt pockets.

Interview Editing Strategies

True-to-life interviews: In the list of no-nos earlier was the jump cut where a person's head snaps from one position to another with each edit. Strangely, the necessary use of same-angle edits in news and documentary television has trained viewers to associate this style with veracity. Using this phenomenon to your advantage, you could shoot a contrived consumer interview, or whatever, loaded with jump cuts. By breaking the "rules" in this case, your interview will look less slick and more credible.

Parallel cutting: If you want viewers to feel as though they've been presented an overwhelming amount of data supporting one particular point of view, link together interviews or testimonials all showing the same responses. This technique, called PARALLEL CUTTING, has a powerful effect in commercials where several individuals discuss their delight in a particular product.

Intercutting: Say you wish to present an idea through interview and testimony. When the raw footage is reviewed, it turns out everybody was essentially in agreement, giving similar stories. What do you do? Show the second person repeating the first person's story? Or just tell the story once with one interview, throwing away the others? One possible answer would be to combine the testimonies to tell the whole story. For example, use one subject's testimony for back-

TWO-SHOT

TWO-SHOT

FIGURE 17-10 Matched shots.

ground, another's testimony to describe the problem, a third's statement on alternatives, and a fourth's testimony to make recommendations. Although each subject probably covered all four areas, each may have a more lucid statement on one particular area, so that's the one to use. This process is called INTERCUTTING.

Juxtaposition: To create an air of controversy, do the opposite of PARALLEL CUTTING. Record the disparate views of several interviewees and then link together their differing responses to the same question. This technique is called JUXTAPOSITION.

Prompting Methods

So far, you've had four choices as a producer/director:

1. Have your talent memorize the script
2. Have them ad-lib loosely from an outline
3. Dub in narration
4. Have your performers read their lines during the performance

The first option is fine when you have actors with the time to learn their lines and the skill to recite them with spontaneity. Option 2 works well with talented subject matter experts working with a RUNDOWN sheet. There's likely to be a lot of chaff in such a production, though. Option 3 needs a lot of visuals and may still be dull. Option 4 allows for careful scripting, little memorization, very little chaff to edit out, and a "live" performer on the screen rather than a disembodied voice. All you have to do is find a way to put words into your talent's mouths so they don't "look like" they're reading.

Teleprompting Systems. Costing about $5000, these devices are generally attached to the cameras and display your script near the lens. You type your script on a roll of narrow paper, and then slip the paper into a motorized feeder that moves the paper under a simple TV camera. The camera feeds the image to little TV monitors attached to the studio cameras (Figure 17–11).

The speed of the text is controlled with a handheld device that can advance the paper slowly, quickly, or whatever is needed. You can even stop the text or run it backwards (not recommended while the talent is trying to read it).

Some models are computerized and, like a word pro-

FIGURE 17-11 TELEPROMPTER.
(Courtesy of Listec TV Equipment Corp.)

cessor, type the words directly on the screen, bypassing the rolls of tangled paper.

To make it possible for the talent to look directly at the camera lens while simultaneously reading script, the video wizards designed a double mirror system (Figure 17–12). Still, when using a TELEPROMPTER, avoid zooming in too close; the viewers will see the talents' eyes zigzagging back and forth as they read from the prompter.

One thing about TELEPROMPTERS—Murphy's 99th law states that your TELEPROMPTER will fail as soon as you're on the air live. The only insurance against such a disaster is to supply your talent with a script to refer to until the prompter catches up.

The script is handy for other purposes too. When the director switches from camera 1 to camera 2, the talent simply looks down at the script, reads a moment, and then looks up at the second camera. It all looks natural. In newscasts, the script becomes a prop. The audience expects a news report to be *read*, not *recited*. The occasional glance down gives the news report a sense of credibility. If you don't believe this fake realism is important, think back to the last editorial reply you saw, with a nonprofessional who read unblinking from the TELEPROMPTER as if hypnotized.

MINIGLOSSARY

Parallel cutting Editing raw footage so that similar or parallel actions are seen one after another, making it look like "everybody's doing it."

Intercutting Editing together several separate events or interviews to tell one story, make one statement, or answer one question

using pieces from each.

Juxtaposition Editing together opposites, like opposing views or conflicting responses to a question.

*Teleprompter Electronic device that shows script or other cues to the talent, who appear to be speaking directly to the TV camera.

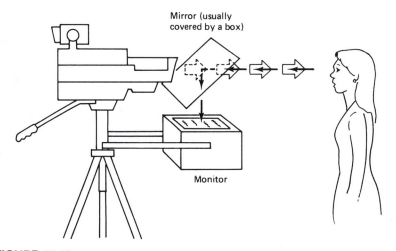

FIGURE 17-12   TELEPROMPTER using two-way mirror and a TV monitor.

Preparing script for teleprompting: The object is to make the script as easy to read as possible. Here are some guidelines:

1.   Type should be upper- and lowercase, sans serif.
2.   The left margin should be justified (even) and the right margin, raggcd.
3.   Aim for 34 characters per line, but . . .
4.   . . .try to make each line a whole phrase.

Good example:

Gorilla warfare broke out today
in the jungles of southern Africa.
Apes from several provinces
joined in the dispute
over treetop territory.

Bad example:

Gorilla warfare broke
out today in the jungles of
southern Africa. Apes from
several provinces joined in the
dispute over treetop territory.

Helping the novice talent: Spoken text usually flows at the rate of about 150 words per minute, but there's a lot of variance to that number. Beginners feel pressured by the TELEPROMPTER because they think they have to keep up with it. Just the opposite should be true, for the prompter is operated by a human being somewhere, rolling it (hopefully) at just the right speed which keeps the spoken lines about one third down the screen. To dispel the talent's fears, tell them to try out the prompter, first speaking slowly and then quickly, and to watch what happens. Eureka, they'll discover that the TELEPROMPTER follows *them*, not the other way around.

A second way to increase a novice performer's self-confidence is to post a crew member next to the TELE-PROMPTER, just inside the talent's field of view. Have this person nod and smile as the talent reads from the prompter. The assurances and positive feedback from the crew member will be seen out of the corner of the talent's eyes and may be comforting. Incidentally, this method works with company presidents, CEOs, and other chief mucky-mucks, just as it does with timid teachers and stunned students.

Cue Cards.   Irreverently aliased "idiot cards" (but don't call them that in front of your talent), these 2-foot × 3-foot posterboards covered with felt pen printing are the backbone of educational and industrial TV prompting. Number the cards to keep them in order. (You can imagine what happens to your show when they are presented out of order.) The CUE CARD HOLDER (catchy job title) holds the cards near the taking camera's lens and to the side so he or she can read along with the talent and know when to change cards (Figure 17–13). Since the cards are not *in front* of the cameras lens, the talent will have that familiar off-center stare during close-ups.

With dialogue, you can play some nifty tricks with CUE CARDS. First, make two sets of cards, one set for the actor facing left and another for the actor facing right. It's possible for actors to read their CUE CARDS over the shoulder of the person they're speaking to without the television audience ever realizing that the two actors aren't looking directly at each other. Here's how: With the two performers facing each other, have one take a half step to the side (toward the camera) but remain looking in the same direction. On the TV screen, they look like they're still facing

MINIGLOSSARY

Cue card holder   Person who holds up the cue cards where the talent can read them.

FIGURE 17-13 CUE CARD held to the side, near camera lens.

carton, and reads something taped to the back while pouring the contents into a bowl. Next he reads from the tabletop while stirring the mixture and then reads a little more of his script among the pages of his cookbook as he checks his recipe. Popping the mixture in the oven, the actor reads a hidden note tucked inside the oven door and turns and reads the final closing from his cuff. It can be done, but it's hard not to have the audience notice.

TV monitors: Some studios have preview and line monitors set up in the studio but aren't using the preview channel for anything important. If this is the case, it is possible to use this idle video channel for cuing purposes. Have someone aim a separate camera at the script typed in narrow columns. Alternatively, type the script into your character generator. Send the signal out to the studio on the preview channel to display on a TV monitor (or monitors) in the studio. The monitors could be placed near the studio cameras to facilitate eye contact. While someone advances the script, the talent reads it off their choice TV monitors.

This method works better with short notes, rundowns, and messages than with long passages of text to be quoted.

Earplug and taped narrative: Some news commentators have used this technique of tape recording their narrative and then playing it back over a tiny earphone plugged into their ear. As they listened, they spoke on camera, performing every inflection just the way it was originally recorded. Since they weren't reading, their eye contact was perfect. The trouble is, some people can do it, while others just don't have the knack for it.

Puppets: The fantastic thing about puppets is that they can look straight into the camera and jabber away while their operators move their hands and read a script (or move their hands to a prerecorded dialogue). This method is so deceptive and so entertaining, there's no excuse for why it

each other. If they react with facial expressions as if they were looking directly at each other, the impression is even more convincing. The performers must avoid looking back and forth between the CUE CARD and each other's face when using this over-the-shoulder method. The change in gaze will be noticeable and will alert the audience to your deceit.

It's also possible for actors to hold a conversation while both are looking in the same direction. Two pilots, people watching a movie or TV, and lovers cuddled near a fireplace could all be reading their CUE CARDS while facing the same direction.

Portable chalkboard: A large portable chalkboard can be placed behind or to the side of the camera and act as a CUE CARD. It has the advantage of being erasable and reusable, not like expensive posterboard. Although the aspect of eye contact still suffers, this method is quite effective for displaying giant notes for an ad-libber.

Overhead projector: As with the chalkboard, you could set up an overhead projector facing a projection screen near the TV camera. An operator could change transparencies as needed or roll a long scroll-type plastic film in sync with the narrative. Use a front or rear projection screen depending on available studio space.

If eye contact with the camera is a problem, hang a white sheet in front of one TV camera and cut a small hole for the lens to stick through. Then project the script onto the sheet for the talent to read, as in Figure 17–14. Sticking a small cardboard cutout onto the overhead's projection surface will cast a shadow on the screen where the TV lens is, thus shielding it from the light.

Crib notes: The actor opens the refrigerator door and secretly reads a little notation there, pulls out a milk

FIGURE 17-14 Prompting with an overhead projector.

Puppets.

isn't used more. Granted you have to buy or make a good puppet (or puppets), find operators who can bring personality to the inanimate characters, and create an excuse for having puppets carry the show, but in cases where a lot of speech has to be learned, puppets are unbeatable for saving hours of memorizing, for providing pseudo eye contact, and for plenty of apparent spontaneity.

TV MAKEUP

Television is a close-up medium. It emphasizes facial characteristics, intensifies colors, and magnifies details—including blemishes.

Males tend to think makeup, even for TV, is "sissy" or false. They may prefer to "be themselves." They may not realize that the hot lights will bring shiny oils and perspiration to their faces. You may have to remind them that the camera accents natural facial shine or discolorations, making the talent *not* look themselves. Makeup helps counteract the camera's infidelity. Without it, the talent doesn't look their best (unless we're making up for *Dracula* or *Night of the Living Dead* characterizations). Makeup is an important element in your talents' metamorphosis from "average street person" to "confident professional."

Still, your attention to makeup will vary with the complexity of your operation. A professional studio will employ a makeup artist. Professional actors will usually know how to apply their own makeup and may come appropriately equipped. Tiny seat-of-the-pants TV studios may never bother with makeup—they have their hands full keeping the recorders running and their wires untangled. Between these extremes we have the situation where you may have to become the "expert," learning the makeup skills and practicing them on your arm and then on a friend and

working your way up to the talent. Don't get discouraged if your results look dismal at first. A few hours of practice will take you a long way.

Even with practice, you may never become as experienced at makeup as a woman who has been applying it to her face since adolescence. She can probably make herself up better than anybody else can. Nevertheless, makeup that looks fine on the street may make your talent look like a hooker on the TV screen. So before she proceeds making up her face, give her these tips so that she can adjust her techniques to the color TV medium:

1. Any greens and bluish tones (and dark reds are bluish) are accentuated by the camera. Avoid them. Light reds, yellows, tans, browns, and dark grays are preferred. Be careful with orange; it sometimes shows up yellow on camera.
2. Avoid heavy shading such as dark eye shadow or dark rouge.
3. Avoid brightly colored lipstick or eye shadow.
4. Avoid lip gloss. Blot lipstick to dull the shine.

And now a word about sex (ah, your undivided attention at last!). Nonprofessional male talent not only have a general aversion to makeup, but they also like being touched by other men even less. If you're a male, you may wish to have some female member of your staff specialize in makeup. First, women are already experienced at applying makeup. Second, and at the risk of sounding sexist, I'll add that if the female makeup artist is cute, the male talent is less likely to resist cosmetics, and they may even find the process enjoyable. Happy talent never hurt a taping session.

Since "straight" makeup (as opposed to character makeup) accounts for 99% of the situations found in the studio, that's what we'll study in most detail here.

Materials

There are three kinds of makeup bases: grease paint, pancake, and a modified grease base or creme stick foundation. Which of the three is best is a matter of taste (but the flavors are awful). Oil-based greasepaint is pretty much obsolete; it's runny under hot lights. Creme foundation has replaced greasepaint in popularity and blends smoothly on the face. Both need to be set with translucent powder. Cake makeup doesn't need to be set and is convenient for most situations. Pancake comes in a dry form and is applied with a damp sponge and washes easily with water. Mehron Star Blend, Max Factor, and Steins brands are found under Theatrical Supply in your Yellow Pages. Mehron Star Blend, for instance, is a solid cake that comes in a jar. It comes in shades of TV4 (the lightest) to TV10. TV4, TV5, and TV6 will cover most situations.

Cake and creme are used differently and don't mix well with each other. If you start with pancake, do the whole job in cake makeup. The same goes for creme stick.

Other basic materials you'll need are

1.   A bottle of witch hazel, alcohol, or other astringent to clean the face before applying makeup
2.   Cotton balls to apply the astringent
3.   A small natural (or sea) sponge to apply whichever type of makeup you choose
4.   A compact of translucent powder used to dull shiny noses, foreheads, and bald pates
5.   Mirror, tissues, hand towels, soap, and water for the removal of makeup

More advanced makeup kits (Figure 17–15) may contain moist/dry eye shadow, eyeliner, eyebrow pencil, tinted or translucent powder, puffs and brushes, blushes or rouges in assorted colors, lipstick, cleansing cream, applicator sponges, crepe wool, mascara, false eyelashes, hair whitener, spirit gum, nose wax, moisturizers, stipple sponges, and facial toners. Distributors will gladly send a catalogue and price list. Student kits cost about $25. Pre-makeup skin care products available from your pharmacy or department store include: shampoo, skin cleansers and toners, baby oil, and after shave.

Before You Start

Find a comfortable place to work, preferably a room with white or neutral-colored walls. The lighting should be even and similar in color temperature to the lighting you will shoot with. If possible, apply the makeup *on camera* while viewing the results in a TV monitor. This way you'll see immediately how the makeup will look to your audience.

FIGURE 17-15   Television makeup kit. (Courtesy of Bob Kelly Cosmetics.)

With a towel or tissues, cover the talent's clothing where it may be soiled by makeup during application.

Place a clean towel on the makeup table and lay out your materials on it. Make sure that your sponges are clean and that your pencils are sharp and chisel-pointed (for sharp or broad lines, depending on how you hold them).

Clean the skin with cleansing lotion (for men) or cleansing cream (for women).

Check the condition of the skin. For dry skin, apply a moisturizer before you apply the base makeup. For very dry skin, use a thin layer of baby oil. For oily skin, administer a facial toner or astringent, such as witch hazel, wiping off the oils with cotton balls.

Applying a Pancake Base

In the simplest situations, you would simply wipe the shiny oils off the talent's face with cotton balls and astringent and then powder the nose, cheeks, forehead, and bald pate to keep the shine away. Base makeup advances us to the next higher level of TV cosmetology.

Select a base makeup color close to the talent's natural skin tone. If performers with divergent skin tones are to appear in the same shot, try to darken the light-toned face a shade.

The base helps even out blemishes, ruddy complexion, paleness, green tinges, creases, shadows, birthmarks, zits, and tattoos.

Cut a foam rubber sponge into a 1-inch × 1-inch × 2-inch rectangle to apply creme stick foundation. Discard it after use. Wipe a corner of the sponge across the makeup and pat a small quantity of base onto the talent's skin, for a nonstreaking finish. Pat and distribute evenly until the entire face is covered. Continue the base up to the hairline and blend down onto the neck area, or else your talent will look like he or she is wearing a mask. Similarly treat any skin that will appear fairly close on camera, such as hands, arms, back, and bustline. To protect talent's clothes, powder the base with translucent powder (or baby powder) where clothing and base come in contact and blot with a damp sponge to take up the excess powder.

Too heavy a layer of makeup will make the talent look like a Sunkist orange. There should be no lumps, streaks, or excessive shine. Pancake can be evened out by patting the area with a damp sponge.

Beards

Even after shaving, most men have a darker tone to their beard line. The camera will perceive this as bluish. A slightly orange base makeup will neutralize the blue tone. Don't use too much, or it's back to Sunkist City.

A thin coat of creme makeup, applied with a stipple

sponge and powdered lightly, with the excess powder removed with a damp sponge, will also dispel the Homer Simpson beardline.

Powder

Powder dulls shine. If applied over base makeup, it sets the base to keep it from smearing while dulling the natural facial sheen. Leave a little sheen on males for a more natural look. Cake makeups have a matte finish already, and so they don't need powder.

When applying powder around the eyes, do it gently so your talent doesn't squint. Squinting creates creases, which the powder may miss.

Shadows and Highlights

It's all illusion. The TV screen is flat. Your talent isn't. To keep dimension in the facial structure, emphasize highlights and shadows.

Apply a base that is one or two shades lighter to prominences such as forehead, cheekbones, jawbone, and nose. Employ the strongest highlight to the center of the prominence and then feather it into your original base tone at the edges. Use darker shades on facial hollows such as under the eyes, nostrils, lower lip, cheeks, and temples.

Light places look wide and dark ones, narrow. To widen a nose, highlight the sides. To slim the nose, shadow the sides with a darker base. Use shadow to recede a too-prominent chin. Shade the top of a forehead to make the forehead look shorter.

Rouge breaks up the flatness of the overall base tone and can be used for men as well as for women. Cream rouge works over cream foundations but won't blend on pancake. Use pink or peach for fair-skinned talent; use coral, tawny, or wine for dark. Between the cheek and jawbone, rub on a strip of cream rouge parallel with the line of the bone. Blend downward from the cheekbone and upward from the jawbone but cover neither. The entire area treated should not be longer than 2 inches along the cheekbone and 1 inch wide. If using pancake, you can get the preceding results by brushing on pressed powder rouge.

Eyes and Lips

Eyes and lips are the most expressive parts of the body and require closest attention.

Highlights around the eyes will make men and women's eyes look wide open and more awake. A light line in the shadowy crease under a baggy eye will de-emphasize the bag. Blend it well into the base. Eyeliner starting at the middle of the upper lid and running close to the lash in a thin line to the outside corner will accent the eyes. For

Makeup artist, Rob Volsky before . . .

. . . and after. Base makeup neutralizes beardline, smooths features, and hides blemishes.

women, apply a similar line below the lower lash, running outward. Blend lines with a brush to soften the effect.

Eyebrows should be brushed smooth and tweezed of stray hairs. To fill in sparse eyebrows, use an eyebrow pencil of a color that matches the hair, using fine strokes in the direction of growth.

Here's how to find where the widest part of the brow should be. Hold a pencil parallel with the nose and just touching the nostril. Where the pencil touches the brow is where the widest part of the brow should start. The arch should peak at a point above the outside of the iris of the eye.

If the lips are too full, cover their edges with base. Draw in a new edge to the lip over the base, using a lip brush or lip pencil, and fill in the rest of the lip using lipstick.

Other Details

You can cover large parts of the body using a mixture of one part cake makeup and three parts water. Pat; don't streak.

Nails should be cleaned and manicured if they appear in close-ups.

Ears tend to look light on camera, and backlights sometimes project through them, making them red. Cover them with a base two shades darker than the face and set with a translucent powder.

Hair tends to get oily quickly under hot lights. Talent who perform often will need to shampoo often. Receding hairlines can be filled in a little with the help of a sharp eyebrow pencil and short feather-like strokes.

Children generally look fine without makeup.

If shooting will continue over several days, you'll want the talent to look the same each day. Take notes on what colors were used and where. Perhaps take a Polaroid snapshot for reference.

Some makeups irritate sensitive skin. You might try a commercial skin preparation designed for use under makeup. Otherwise, see a dermatologist. Hypoallergic base is available at some stores. Avoid makeups which contain perfumes.

Removing Makeup

It's easy for talent to leave the set after a production and march out into the real world forgetting they're still wearing makeup and looking a bit weird to the naked eye. Remind them. Or hang a mirror on the studio exit door.

Cleansing cream or lotion worked into the face with the fingertips and wiped off with tissue will take off the first layers of makeup. To avoid irritation, try not to rub the skin too vigorously. Next, wash the face with warm water and mild soap. Rinse with tepid water several times and use skin toner on a cotton ball as a final rinse. Follow with a dab of moisturizer if needed.

The moment of truth is when you examine your makeup job in the TV monitor. Watch for streaks or hard edges. If the face still looks lousy, put a bag over the talent's head, punch deep holes for eyes using a long sharp instrument, and with a felt pen, draw on a smiley face. The viewers will never notice.

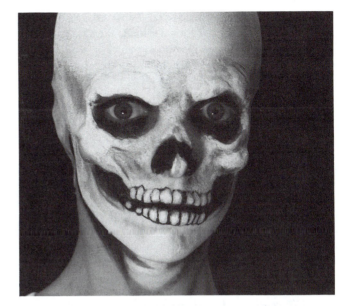

Grotesque makeup using liquid latex, mortician's wax, and bald cap. (Courtesy makeup artist Robin Volsky, Westmore Academy of Cosmetic Arts)

Gruesome Makeup Effects

This is not a way-out topic! Do you believe in the "grab 'em by the throat" theory of television? Why not start your show with a shocker?

If producing the Halloween epic "Rambo Goes Chainsaw Crazy," you could cover your actors with catsup as each falls prey to Rambo's whirring blade, but the result may look more like something to eat than something to cringe at. When creating ghouls, gore, and mayhem, you can get a lot of mileage out of a little creativity and a few kitchen supplies such as corn syrup, gelatin, food coloring or acrylic paint, liquid latex, a bald cap ($5 at theatrical supply stores or department stores at Halloween), and mortician's wax (called Plasto, available from Bob Kelly or Ben Nye).

Latex is widely available from theatrical supply houses. Incidentally, it goes bad (turns dark, won't dry, clumps up) after 5 months on the shelf.

Bald caps and mortician's wax are easy to use: Buy the cap; put it on. Buy the wax, mold it to shape, stick it on.

Blood. You can make blood real looking enough to fool Dracula if you mix red food coloring in clear corn syrup and apply the mixture with a brush or an eyedropper. Since food coloring stains, you may have better luck with a mixture of water color or acrylic paints in corn syrup.

Wrinkles. Stretch the skin and apply liquid latex, artificial eyelash adhesive, or Duo (a surgical adhesive clear latex liquid available in hardware shops) to the stretched skin. When the latex dries, powder it and then stop stretching. A hairdryer will speed the drying process. Voilà, wrinkles. The adhesive is easy to remove; just peel it off.

Bruises.   Blue or blue-green eyeshadow rubbed on the skin makes a nice black-and-blue mark. For a blow to the head or a skinned knee, rub on blush makeup or lipstick.

Swollen Face.   Stuff cotton balls or piece of fruit in the cheeks. Then use your bruise makeup to cover the lumps.

Third-Degree Burn.   Wet a facial tissue with Duo and stick it to the skin while making an air bubble. When it dries, it looks like a big watery blister. Paint the bubble and the surrounding skin area with red acrylic paint to make the burn look more serious.

Charred Flesh.   Using a natural sponge, dab some black acrylic paint onto the affected area of skin (a technique called *stippling*). After the paint dries, the skin will crack when you flex your muscles. If you paint these cracks with red acrylic, it will create the illusion of live flesh beneath the charred skin.

Torn Flesh.   Make a batch of fake blood using red paint and corn syrup. Dab the mixture onto the skin. You may even soak a tissue in the mixture, creating a layer of gnarled "skin" with it. The corn syrup will stick to your skin, forming a "skin" of its own. By twisting and pinching this "skin" into lumps and wrinkles, you can make flesh look like it's been chewed by a shark or a chainsaw. Add a slice of ham if you'd like (but you'll attract cats and dogs).

Wounding the Skin on Camera.   Here the camera actually watches the skin being torn up. Unless you have very dedicated actors, you will want to tear up something other than their real skin. This is done by covering their hide with something that "looks" like skin and then tearing "that" away.

To "peel back" someone's skin, you could first rub a small circle of Vaseline on the skin. Next, paint a larger circle of latex over the jellied area plus the surrounding area and let it dry. Afterwards, anything which rubs over the latex area will tear the latex covering the jellied spot, peeling it back just like skin.

Unflavored gelatin is another way to make fake flesh. Mix the gelatin with water at about 150° using double or triple the proportion of gelatin to water called for on the package. Let the gelatin cool enough so that it doesn't burn your performer but is still thick enough to apply with a paint brush. Because the gelatin is translucent, the actor's skin color will show through, but you can paint, bruise, or modify the actor's skin before applying the gelatin. You could

even mix red and blue strings or yarn with the gelatin to form blood vessels, or you could paint over it using your acrylics. *Note*: Some paints are harsh and may irritate the skin. Look for approved "body paints." Still, it's safest to try a little on your arm, wait a while, and see if your skin is irritated before forging ahead on a larger scale.

If you wanted to create a bloody scratch on camera, you could tint the gelatin with red food coloring before applying it. Next, cover this with base tone makeup so that none of the red can be seen. Now when the actor is being scratched, you'll be tearing away the makeup cover, revealing the red gelatin "flesh" underneath.

Bullet holes can be made by painting a dime-sized red dot onto the actor's skin and then covering it with something that can be pulled off instantly. A dome-shaped sequin with a hole in the center does the job nicely. Attach some fishing line to the sequin and then glue the sequin to the dot, building it up with a little wax. Cover the works with flesh-colored makeup so the skin looks smooth. When the gunfire starts, someone off camera yanks the fishing line and off pops the sequin, revealing your juicy red bullet hole.

A similar technique works for knife gashes. Apply red makeup to an area of skin and lay some fishing line across it. A thin coat of wax will make the fishing line adhere to the skin. Cover the area with flesh tone makeup. Holding the fishing line in the same hand as the knife which is doing the cutting, your actor can pass the knife over (but not near) the victim's skin. As the hand moves past, it will pull the fishing line through the wax coating, and the thin red line which appears will look like a gash.

Skuzzy Eye.   Makeup experts Doug White and Allan Apone (of *Friday the 13th, Part III*, and *Heaven's Gate*) have used the following technique: They paint a new eye on their actor's eyelid. To protect the eyelid, they first apply a cream base makeup on it. Next, they close the actor's eye and glue his or her lashes down with artificial eyelash adhesive. They then attach a false eyelash where the natural lash *used* to be. Now, using acrylics, they paint the lid red, milky, whatever. *Warning*: Nonprofessionals should avoid messing around the eyes. The eyelash adhesive, makeup, or paint could be irritating to the eyes.

Perhaps you could attach a half of a fig, prune, or some other gross fruit over the eye. Ping-Pong balls cut in half also make good eyes when painted with tiny pupils.

Broken Nose.   On a broken nose the nostrils are flared, and the nose is crooked. You can flare the nostrils by finding some ⅜-inch see-through plastic tubing, cutting it into a ¼-inch length, and stuffing it into the nose far enough so that it can't be seen. Opposite the flared nostril, on the bridge of the nose, create a lump with some wax, covering it over with some base makeup.

SHOOTING ON LOCATION

Location shooting brings with it all the pleasures of camping out. You must remember to bring everything imaginable, or you'll end up with coffee but no coffee pot, a flashlight but no batteries, and perfume instead of insect repellent.

Preparations

The northwest loggers had a saying: ''Take time to sharpen your saw.'' If you ever expect to need your equipment in a hurry, take the time to pack it ready to go. Store things together, ready to carry away. Charge the batteries *first thing* after you return from an outing so that they will be ready if your next mission comes sooner than expected. Repair loose or broken parts or other equipment defects right away rather than ''learning to live with them.'' Repair cables and plugs if they malfunction intermittently. This way, you'll avoid having to run around wiggling and testing cables during a production. Keep the lenses and video heads clean. Leave a blank cassette in the machine at all times, ready to go.

In short, be prepared.

What to Take with You

What you take is largely dictated by what you'll be doing. The watchword, nevertheless, is the same: *Be prepared.* For shooting about 2 hours of tape in the next town, one might bring the items listed in Table 17–5.

Some of these items may not apply to all productions and may be left behind. Location shooting just outside the studio may require bringing only a VCR/camera ensemble. The farther you stray from home base, however, the surer you must be that you have everything.

Before packing equipment for a journey of any importance, take this added precaution: *Connect all equipment together and make a 1-minute sample tape. Next, play the tape to make sure everything works.* This is perhaps the most important step prior to going on location. This superfluous-sounding routine pays off in the long run! Most of the time, this testing procedure reveals no problems. About 20% of the time it will. It's better to face your gremlins at the outset rather than getting bit by them on location.

Preparations for an Expensive or Distant Shoot

Things to Bring. If you're really going some distance or if the shoot is a substantial commercial venture, you may wish to bring the following items (in addition to those listed earlier):

1. The location owner's 24-hour phone number
2. Permits (parking, road closings—obtainable from police)
3. Location releases (owner's permission to shoot on his or her property)
4. Insurance certificates (holding the property owner not liable if a crew member is injured)
5. Petty cash (quarters for phone, bribes to silence noisy lawn mowers, tips for favors, or $20 to instantly convert an onlooker into a ''production assistant'' who will carry your gear up three flights of stairs for you)
6. Your best 10:1 zoom lens (for maximum shooting flexibility)
7. A +2 diopter lens attachment (for close-ups)
8. Polarizing and neutral density filters

''The Distant Shoot''

TABLE 17–5

What to bring on location

Items to Bring	Bring These as a Backup in Case Something Fails on Location
1 VCR and camera, zoom lens, cassette (threaded), and carrying case.	1 additional VCR/camera/lens ensemble if the shooting is very important. Shoot with both machines simultaneously. This way, if the camera operator goofs up or the heads clog on one VCR during a shoot, the other will still catch the scene. Bring a complete set of accessories (batteries, tripod, tape) for the second machine. If the scenes aren't rare enough to require two-camera coverage but you're traveling a long way at some expense to do the shooting, bring a second camera anyway and store it on the site. If the first machine fails, you'll have a backup with which to keep shooting.
1 3-hour battery.	1 additional 3-hour battery or 1 ac power supply. Either is in case the first battery dies prematurely or the shooting runs longer than expected.
6 ½-hour cassettes of tape in boxes.	4 extra ½-hour cassettes in boxes. It doesn't cost anything to return with unused tape, but it is inexcusable to run out during a production. Having extra cassettes also makes it easier to categorize your shots during editing.
1 roll of masking tape and a *good* felt pen to label the tape boxes. Keeping track (and not accidentally erasing) of what you have is just as important as shooting it. The adhesive tape is also handy in unpredictable ways.	
1 portable tripod.	
1 lavalier and 1 shotgun microphone with 50 feet of mike cable and an appropriate plug for the VCR.	1 extra length of mike cable in case the first conks out.
1 pair of headphones with the proper plug for the VCR. The headphones will permit an accurate monitoring of audio during taping.	
1 30-foot headphone extension cord if the shotgun mike is to be aimed by a sound person who needs to hear what's being recorded.	1 spare earphone (the tiny one that fits in the VCR carrying case).
2 portable lamps with tripods and barn doors.	2 spare bulbs for the lamps.
3 heavy-duty, grounded, multiple outlet extension cords. Two are for the lights; the third is for the VCR if ac is used. The multiple outlets make it possible to power other accessories near the VCR (a TV monitor, a mixer, a lamp, or a battery charger).	2 extra extension cords. In case the first ones don't reach, these can be connected in series.
1 TV monitor/receiver with an 8-pin-to-VCR cable. This allows you to play back, with sound, the raw footage on site for you and others to evaluate.	RF generator (built into your VCR) and RF cable with a 75–300Ω adapter, in case you need to use someone else's TV set to view footage.
1 flashlight if auditorium or auditorium-based shooting is necessary. The flashlight will help you see to find switches and sockets and to label the tape.	
1 set of close-up lens attachments, if appropriate.	
3 grounded ac plug adapters to allow you to use your 3-prong ac plugs with wall sockets having only two holes.	
1 head-cleaning kit.	
1 audio kit, *if needed*. Kit includes a mixer, mixer batteries, mikes and cables, a cable going to the VCR with the proper plug, assorted audio adapters, audio cables, and an attenuator (in case you must record from someone's loudspeaker system).	

TABLE 17–5 CONT'D

Items to Bring	Bring These as a Backup in Case Something Fails on Location
1 pad and pencil to take notes.	
1 copy of the script.	1 extra copy of the script.
20 model release forms.	
1 enormous two-handled box to carry it all in.	Hernia insurance.

9. Plywood (for bases, ramps)
10. Cue cards, magic markers
11. Rain tarp, raincoat, umbrella
12. Walkie-talkies
13. Garbage bags, broom, shovel (for cleaning up —before or after shooting)
14. Card table, folding chairs (for base of operations)
15. Clipboard
16. Ladder
17. First aid kit
18. Prop kit, makeup, hairdryer, iron with board
19. Hand truck
20. Rubber mat (to lay over cables in high traffic areas)

Things to Check on Before the Shoot. Once you know the kind of setting you want, send a scout around to find an appropriate location for the shoot. Professional location services will provide (for a small fee) pictures of various locations, and if you select one, you pay a referral fee. There are also free-lance scouts who charge about $200 a day to find locations.

Next, you need to pay the property owner for use of the location. Sometimes the charge is nominal, but commercial producers generally pay $300 for an exterior, $800 for a house interior, $900 for an office interior, or $1500 for a small store.

In conjunction with selecting the site, your scout makes a REMOTE SURVEY to determine which sound, visual, power, communications, and crowd problems need to be addressed. The scout should bring a Polaroid camera to take pictures of various aspects of the site. The REMOTE SURVEY should be carried out at the same time of day as the actual shooting so you don't get surprised by a 3 o'clock whistle, a rush of traffic, or unexpected sun reflections.

The scout should also check on the following things.

Location

1. Where is the exact shooting location? Draw up concise directions to the site.

2. Will there be any large distractions in the background (like a competing sponsor's billboard or 99 kids screaming "hello" to Mommy)?

3. Is there a safe place to route your cables? Route them over doorways and tape them down to avoid trip-ups. Don't run your cables alongside electric power cables. The power may interfere with your signal. If you cross high traffic points with your cables, put a carpet or mat over them. If you run cables a long distance through the air, first run a rope; then tape the cables to the rope. This way the cables are less likely to stretch and break.

4. Do you have parking permits for your production van and other vehicles? Do you need to have parking reserved? Do you have admittance waivers (for sporting events, concerts, and the like)?

5. Can you get your switcher and VCRs close enough to your cameras? The umbilical cord between a nonprofessional color camera and its VCR should not be over 40 feet. The coax cable from a professional color camera to the switcher shouldn't exceed 2000 feet.

6. Are there machines nearby that generate audio and video interference, such as an X-ray machine, radar, ham radios, or high-tension power lines?

7. Do you need crowd control, "no parking" signs, areas roped off, traffic stopped? Get police permits. Also, hire an off-duty police officer for crowd and traffic control. It costs about $20 per hour, and officers often come with uniform, gun, radios, and squad car.

8. Has the company's or school's chief of security been informed that you're coming? Do so personally. Perhaps the chief can arrange convenient parking spaces for you. Have security sign in your equipment if you want to avoid conflicts as you leave with it.

9. Who is the plant's maintenance chief? This per-

MINIGLOSSARY

*Remote survey   Visit to a distant shooting location to determine production needs, strategy, and resources.

son is invaluable for knowing where power, ducts, turnoffs, and props are.

10. Who is the chief secretary? Introduce yourself and your crew. This person knows the names (and idiosyncrasies) of the maintenance, cafeteria, custodial, and electrical crew and can cut red tape getting their help.

11. When do the employees change shifts? Will there be a giant turnover during your shoot?

Lighting

1. Where will the sun and shadows be during the shoot?

2. Do you have to shoot indoors with windows in the background? Can they be covered?

3. Where will the lighting instruments have to go? Is there power enough for them? Before you answer that question, consider the following: Who else, besides you, will be drawing power from those circuits at the same time?

4. Include the building's electrician in the details. If shooting in a factory, make sure it has 120-volt outlets.

5. Fluorescent lights tend to produce greenish faces on color cameras. Aim a couple of quartz lights at a white ceiling or wall to balance the color temperature.

Audio

1. Is there excessive ambient noise (machines, air conditioners, jet planes, typewriters, traffic)? Can any of this extraneous sound be silenced?

2. Can you get the mikes close enough to the talent? Lavaliers are preferred in noisy environs. Special close-talk, noise-canceling mikes are used in very noisy places, like helicopters during traffic reports.

3. If using a wireless microphone, is there interference from any electrical sources or competing transmitters? Try it out.

Prentice-Hall, Inc.
Englewood Cliffs, N.J. 07632

Telex No. 13-5423

MODEL RELEASE

I hereby give Prentice-Hall, Inc. the absolute right and permission to copyright and/or publish, or use photographic portraits or pictures of me, or in which I may be included in whole or in part, or composite or distorted in character or form, in conjunction with my own or a fictitious name, or reproductions thereof in color or otherwise, for art, advertising, trade or any other lawful purpose whatsoever.

I hereby waive any right that I may have to inspect and/or approve the finished product or the advertising copy that may be used in connection therewith, or the use to which it may be applied.

I hereby release, discharge, and agree to save Prentice-Hall, Inc. from any liability by virtue of any blurring, distortion, alteration, optical illusion, or use in composite form, whether intentional or otherwise, that may occur or be produced in the taking of said pictures, or in any processing tending towards the completion of the finished product.

DATE_____   MODEL _____

ADDRESS _____

PARENT OR
WITNESS _____   GUARDIAN _____
(Required only if model is a minor)

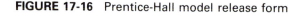

FIGURE 17-16 Prentice-Hall model release form

Traveling with video.

Communications: The program usually falls apart when the director loses contact with the cameras, so

1. Can the director communicate with the cameras? Rock bands and cheering crowds often hinder listening to director's commands over earphones. Use double earphones (both ears are covered) in such cases. Noise-canceling close-talk mikes on intercom headsets ensure good communication from the cameras to the director.

2. Do you need a line monitor or program audio going to the talent so that they can coordinate their presentation with the rest of the show (such as conversing with the news desk or singing along with a prerecorded musical track)?

Release forms: If your show is to be distributed, especially commercially, you had best get written permission from everyone involved before you take their pictures. Figure 17–16 shows a sample release form used by Prentice-Hall.

Generally, you're free to shoot company employees in roles related to their jobs for in-house educational or corporate projects. Nevertheless, it's still wiser to have a signed release from everybody. Some political upheaval may later come up, turning today's friendly faces into tomorrow's lawsuits.

Public gatherings on the street don't require releases from the participants, but an individual interview with a "man-on-the-street" does.

If attending a sports event, you may wish to post a few notices warning that TV shooting is going on. Actually, the signs and cameras add an extra tinge of excitement to less popular contests such as shuffleboard matches or poker competitions.

Most cities have policies regarding filming in public (and even private) places. Check with the city motion picture office or recreation department to secure permission to shoot. They are helpful at providing police escorts, traffic control, and crowd control and for notifying neighbors so they won't be terrified when your six helicopter gunships arrive to attack the farmhouse you rented in the country.

One or more days before the shoot, check on lunch arrangements, equipment, talent, police, everything. Assume nothing. Trust no one.

Traveling Hints

Motels. Use a ground-floor motel room for your headquarters, for storing equipment, and for cleaning, recharging, etc. Also try to get a motel with a parking lot right outside your door. It's worth an extra 5-mile drive to save hauling a half ton of stuff up and down corridors and stairs.

Cars. To avoid subjecting your equipment to extremes of hot and cold, don't store it outside in a car. Besides, it may get stolen, especially if it's in a rented car or station wagon.

Air Travel. Anything that can crunch suitcases into tiny bits can misalign a VCR. So pack everything with plenty of foam. Excellent, sturdy, foam-cushioned carry cases are available for video equipment (Figure 17–17).

FIGURE 17-17 Sturdy cases for carrying portable TV equipment. (Photo courtesy of Fiberbilt.)

Try to book your nonstop flights so that the baggage crunchers don't juggle your bags twice.

Smaller, delicate things like cameras might best be carried on board with you for safekeeping.

When arriving at the airport, put all your baggage in the hands of one seasoned skycap. Your things will get personal attention and better treatment, and they are likely to all stay together. Once the skycap has all your equipment, don't forget the tip!

Food. If you and your crew are very busy, you'll all need a rest. So which is better: racing off to a distant diner, waiting in line for a steamed hamburger, wolfing it down, and racing back to the set, or having the food brought in to you, leaving you time to eat peacefully and digest it? Catering costs more, but it may be worth it in saved time and warmer crew relationships.

If you have a large crew that will be eating in the company cafeteria, contact the cafeteria manager to arrange speedy service and a place to sit without disturbing the cafeteria's normal operation.

Public Relations. This is your big chance to win points or to lose face. Here are some tips at making a good impression on location.

1. *Make sure your equipment works.* This has been said before, but it bears repeating. What kind of impression do you leave behind when you go away empty-handed (or empty-taped) because your equipment failed?

2. *Fly your colors.* The chief mucky-mucks in your

corporation or school may hardly know you exist. When they see your efficient crew scurrying about with cameras displaying your department logo, they'll get that favorable impression that buys you crew members and equipment. Inform the crew members of highly visible or "executive suite" shoots and insist on proper attire. No jeans. You're professionals.

3. *Assign crew duties.* Jobs should be delegated so the production details don't bottleneck with the director. Besides the obvious efficiency in having all the crew knowing what they're supposed to do, you have *impressions* to make. Do you want to look like a "Mission Impossible" team or a Laurel and Hardy comedy?

4. *Function as a team.* If somebody's in trouble, help out. If somebody has screwed up, help out—don't argue or recriminate; save that for after the shoot. Neither director nor crew should panic under pressure. Act calm. Hysteria spreads easily and loses points with the outside world. Besides, going around in circles makes no one a big wheel.

5. *Don't intrude.* Arrange not to have any incoming calls for you or the crew while you're imposing on someone else's office. Ask permission before calling out.

When swarming over an office for a production, take careful notes on the exact position of all the objects in the room so that you can replace them correctly. This restoration of the "office ecology" shows you care and understand corporate office psychology and wins more points than you'd ever guess.

Try to stage your equipment somewhere other than the president's office during your shoot there. If an anteroom is available, set up there, so that only the essential lights and cameras will clutter Mr. Bigshot's inner sanctum.

6. *Clean up.* Although crew members are probably exhausted at the end of a remote production, it pays off in the long run to leave the area clean before you all leave. This includes removing masking tape and its residue, string, food leftovers and wrappers, and promotion posters.

7. *Show appreciation.* Thank the secretary who located that prop you needed, and the custodian who routed your cables over the ceiling, and the security guard who opened doors and watched over your goodies.

Don't forget to thank the police officer who slowed the parade long enough for you to get that "perfect shot" or the grandmother who fed your crew 6 gallons of lemonade on the hottest day. And if you can get people's names, just imagine the uplift in community morale when your helpers see your show's credits: "Special thanks to Sergeant Alan

MINIGLOSSARY

Carnet   A customs document listing your tools and their origin and destination. It guarantees to a country that they were not bought nor will be sold in that country.

*EFP   Electronic field production, producing TV shows outside the studio. Usually involves studio-quality equipment, techniques,

and editing.

*ENG   Electronic news gathering, portable video production for the news. Often quick-and-dirty techniques are used with minimal equipment and crew.

Pevar, NYPD, and Mrs. M. Rothberg.'' You'll have no trouble getting invited back.

Last, but far from least, *thank your crew* when the shoot's over. A good ration of grog might be in order, too.

Customs Registration.   Leaving the United States with your foreign-made VCR is easy. Getting it back into the United States without paying about 5½% duty on it is difficult. A receipted U.S. bill of sale showing serial numbers may act as proof that you bought the gear in the United States. Better yet, fill out a ''Certificate of Registration for Personal Effects Taken Abroad'' with U.S. Customs before you go, listing make, model, and serial numbers of all foreign-made possessions you're taking abroad.

If you're traveling through several countries on your way to a shoot, you may wish to obtain a CARNET, a cash bond and a document listing the tools of your trade and testifying to their origin and where they'll be used.

SHOOTING STRATEGIES FOR ENG

No matter what kind of production you are doing, EFP, ENG, sports, or advertising, the basic laws still apply: Try to tell a story and make the shots flow smoothly from one to another.

Television professionals come to each shooting situation with a bag of tricks up their sleeves (which must make it very hard to move). These are handy production strategies which work very well in certain situations. Before covering them here, let's review a few of the basics which work under *all* circumstances. They deserve this emphasis because they are the foundation of all the shooting strategies.

Shoot Cutaways

When an editor is putting together a series of shots, he or she wants to avoid jump cuts which disturb the continuity of the scene. Wherever you are, whether cutaways are in the script or not, it is wise to shoot about 1 minute of cutaways, general scenes which can look appropriate when inserted in the story.

The cutaway is used often in action sequences where action needs to be compressed. For example, if you were trying to show an exciting 3-minute roller coaster ride in one-half minute, the parts you edited out would create jump cuts. The editing would become obtrusive. If you took the ride a second time and shot the faces of the participants, then took a few close-ups of the tracks zipping by, followed by an under-the-track shot of the roller coaster hurtling by, these shots could be interspersed throughout the segment to cover the jump cuts. These shots will add drama and excitement to the scene and distract the viewer from the fact that the ride was shorter than a real one.

Transition Shots

The cutaway is one type of transition shot, used often to hide an edit. An edit that is well planned for doesn't need to be hidden. This is where the transition shot is built into your raw camera footage so that your edits can melt seamlessly together. For example, consider the following:

1. The scene finishes and the talent walks out of the picture. You can start with your next shot of an empty scene and allow the talent to walk into the picture. This could be a long shot of a building with a street out front and someone walking in the front door. You could then cut to an interior shot of the building as the person reaches her apartment door, *or* you could be inside the apartment watching the door open or hearing the doorbell ring as someone else answers. In each of these cases, we have established the location of the next shot.

2. Close-ups also make good transition shots. For instance, when someone makes an outrageous decision at a board meeting, the next shot could be a close-up of a telephone with all the lights blinking plus commotion and other ringing in the background. This sets us up for the next scene of the bedlam that results. Another alternative would be to have the next shot be a close-up of a pay phone being dialed. We then cut to a medium shot of the corporate spy spreading the news of the unpopular decision.

With these basics properly in mind, let's go on with some specifics.

The Interview Shot

Used heavily in news and documentaries, the interview shot is a scene of a person talking. Here are some guidelines on how to shoot the scene.

1. If the scene is dramatic, use tight close-ups. You are searching for emotion in the face. Try to keep the eyes in the upper third of the picture and the lips in the lower third.

2. If the scene is not emotional, you can get by with medium shots and medium close-ups. In the medium close-ups, keep the eyes in the upper half of the screen. In the medium shots (waist shots), keep the face in the upper half of the screen.

3. Center the interviewee in your picture unless that person is speaking to someone off camera to the side. In this case, allow them to turn toward that

person and leave them a little extra space to "speak into."

4. Always be aware of the background—never forget that it is part of the picture. Make the interviewee's background work for you and not against you. If the background is distracting or runs contrary to the story (such as kids screaming, waving, laughing, and giving the finger to the camera in the midst of a tragic auto accident), arrange the camera and interviewee so that the background is a nearby bush (nobody can get in there to wave at you) or a busy street (nobody will stand in traffic to wave at you) or so far away that it can be put out of focus as you zoom in on the interviewee's face. Zooming in not only reduces the depth-of-field of your lens but will also leave less background to be seen.

5. Try to use the background if you can. If a woman's house just burned down, we should see the smoldering wreckage over her shoulder. If a man is inviting you to visit his new car dealership, then try to work it so that his sign is over his shoulder. *Note*: Don't ruin your foreground shot in order to make your background shot come out. If the car dealer's sign is too high, don't just tilt up to put the sign in the upper left-hand corner of the screen and his head in the lower right-hand corner. Also, don't squat down to raise the person's face in the picture as you will now have a shot looking up at the subject, making him look authoritative and dictatorial, not what you'd call your friendly car salesman. Instead, move the camera and the subject farther from the sign so that it can be placed over the shoulder and good framing can still be maintained.

Stand-Up Reporter

A stand-up is an interview with the reporter or the interviewee standing, speaking directly to the camera.

1. To maintain good eye contact, shoot no tighter than a necktie shot and no looser than a waist shot.

2. If including an important background in the shot, frame the talent off center to leave more space for the background. Walking shots will add interest to the scene. Either move with the talent or use your zoom lens to maintain a waist shot as the talent moves toward you.

3. If the reporter is talking *to* the audience while walking, he or she should be walking generally toward the camera.

4. If the reporter is taking the viewer to another location or wants to show something behind him or her, then it's okay to walk away from the camera.

5. Sometimes your background gets into your foreground and gets in the way to boot. Typically, it's a bystander waving and trying to get his or her puss into your picture while you interview somebody. If you can't cordon off an area or get some other crew members to block for you, the next trick is to use a wide-angle lens and shoot very close to your subject. People are less likely to intrude when the camera and subject are only 3 feet apart. Be aware that when you do this you may encounter special lighting problems. Be careful not to cast a shadow on your subject. On the other hand, if your camera has a light attached to it, either dim, diffuse, or bounce the light off something else so that you don't melt a hole in your poor subject from 3 feet away.

Spot News

Spot news are news events with the potential of becoming history. They could happen in a second like the explosion of the Hindenburg or could be one of a series of related events such as burning children running from a village that has been napalmed.

Shooting spot news, you usually end up in one of three situations:

1. In the middle of the action as it happens
2. In the middle of the aftermath
3. On the perimeter of the action or aftermath

In the Midst of the Action.   If you are in the middle of the action, *start your camera rolling immediately.*

1. Double-check; is it really rolling?

2. Zoom to a medium wide shot and focus for about 30 feet. This way most of your shots will be usable even if they are made by accident.

3. Shoot the scene as you would like to see it. If something catches your eye, look there with your camera. Try not to zoom in on one action if it excludes other actions nearby.

4. Save your zooms for when things are better under control. Do mostly panning.

5. Keep the camera rolling no matter what. White-balance while it's rolling. Walk while it is rolling. Run from your car with the camera rolling. This is a second reason for keeping a wide shot: The picture bounces around less when you move if you have a wide shot.

6. Avoid panning wildly, trying to get everything in. If too much is happening to ''get it all in,'' select just one thing, then hold it as a static shot—as still as you can—and count off *10 seconds* to yourself before leaving that shot and going to the next. Then repeat the process with the next shot. Fight your natural inclination to take a bit of this action and a snippet of that action. Such shots are too short to be useful. It is better to have one useful shot of half the action than two useless shots of all the action.

7. Once you feel that you have recorded the essentials, check your equipment. Are you *really* getting sound? Is the picture *really* being recorded? Go over your mental checklist to verify and assure a good recording, and then once you know that every switch is in the right position and that the tape is rolling again, continue shooting the event.

In the Midst of the Aftermath. Things are under control, and the authorities are in command.

1. Start with an opening wide shot to establish the locale of the event. Hold it steady for a good 10 seconds.

2. Feel free to look for action, medium shots, and close-ups. Look for tight shots of faces that tell the story of the individuals who were there during the actual event.

3. Shoot anything that moves because action is interesting, even if it is after the fact.

4. Try to determine what the story is and shoot the elements of the story. If the story started in one place and moved to another, get shots of both places.

At the Perimeter. Here the authorities got there first, and you are restricted to shooting from quite some distance from the action.

1. Here you will need a tripod and a long focal length lens. A 2X lens extender is a handy accessory to include in your camera bag for such occasions. Since your camera is acting like a telescope, every little wiggle is going to show. Try to set your tripod on solid ground.

2. If a tripod is out of the question, then rest the camera on the ground, a tree, or any solid object to give you a steady picture.

3. Again, you are looking for the action; shoot anything that moves: people, police cars, fire engines, smoke rising, stretchers being loaded, people using walkie-talkies, or unusual equipment being set up.

4. Ascertain what the story is and, again, try to shoot the elements.

5. If you came too late to get any action at all, *shoot something*. It could be a police line holding the crowd back or anything moving.

Perimeter shots are neither exciting nor good television, but don't pack up your equipment prematurely in disappointment. The story may still be important, and the reporter may do an excellent job of telling it. All you need to do is provide 1 minute of *something* to show while the story is being told. Nobody will forgive you if you can't find one lousy minute of *something*.

General News

General news includes the common stories about city council, playground openings, a public bus strike, or a drought. Because the subject matter is so dull, it requires even greater finesse on the part of the photographer and the reporter to tell the story and keep it interesting.

1. Grab 'em by the throat with the first shot. Don't start with the reporter's talking head but rather with a picture of something moving or the sound of something unusual happening. This will get the audience's attention.

2. Try to have the story tell the story, not the reporter. The reporter should be the glue that holds the story together, connecting the interviews or explaining the meaning of scenes shown.

3. If the story is essentially an interview with a person, say the builder of a housing project, show that person for about 10 seconds, which is long enough to establish the person's name and face with the viewers. Use a medium shot, leaving room for his name and title to be keyed over his shirtpockets. Then, with the person's voice edited in the background, go on to scenes of what is being talked about. Come back to the face once again to finish up the scene.

Feature News

Feature news stories are usually lighthearted samples of unusual subjects that entertain, touch, or fascinate the viewer. Usually the schedule is looser, allowing the photographer and reporter to be wild and imaginative. Music, fancy camera angles, and fancy sound effects are all appropriate here.

Incidentally, if assigned such a story, treat it as an opportunity, not as a low-priority time waster. This is a chance to be creative, to have fun, and to really please the viewers, especially after they've experienced 25 minutes of rape, robbery, government corruption, and other bad news.

1. Infiltrate the subject. Gay, fun-loving people turn into zombies when the camera lights go on, and this destroys the warmth and enthusiasm you were trying to show. Mix with the crowd and stay there a long time so that people become used to you and less self-conscious. Soon they will begin to laugh, dance, and interact in their natural ways, and you will get good shots.

2. Feel free to ask the participants to do something. Here, journalism or newsworthiness is not the driving force; entertainment is. You are not telling lies with your pictures; you're just dressing the story for maximum beauty and impact.

Sports

Sports breaks down into two categories: features and competition. Each uses a different production strategy.

Sports Features. Like a news feature, a sports feature requires maximum creativity for viewer involvement. If you are doing a piece on a football star, consider moving the camera into the quarterback's position and take the hike. Stand back a few steps as you (and the audience) see what it's like to be chased by a wall of enormous grunting arms, shoulders, and helmets. Or maybe see what it is like to have a pass thrown to you.

These are experiences the viewer would never get during the real competition, but they make excellent companion pieces to the competition as they involve the viewer with the real flavor of competing.

Competition. Here the object is to catch all the action. Shoot from the best seat in the house. For basketball, this is about one-third up the seating and on the half court line.

For football, shoot from a high vantage point midfield.

Baseball really requires two shooting angles to cover everything. One is good for following the ball, and the other is good for following the runners. Center outfield with a long zoom lens is one good place to shoot from. Another is from home base. Try to follow the ball as the pitcher throws it and it is hit by the batter. Follow the ball a ways after it is hit, but expect it to disappear from view after a few seconds. When it does, tilt down to the player(s) preparing to catch the ball and stay with them until it is caught (or dropped). Then come back to the runner in a medium-wide shot to see who gets to the base first, the runner or the ball.

Golf and hockey are tough to shoot because the ball or puck is so small and moves so fast that it nearly disappears on the screen. If you zoom in close enough to see it, it may zip out of view, and you'll lose it. If you zoom out too far, you'll never lose it, but you'll never see it.

If you're a sports beginner, the best bet is to stay fairly wide and try not to lose the ball or puck. With experience, you will learn to anticipate moves and will be able to zoom in closer as you follow the action. If too many plays are getting away from you, this is your clue that you need to zoom out a bit.

Since the best vantage points for sports are high and far from the action, you'll need to zoom in a great deal to bring it in close. Use a solid tripod, a fluid head, and a long handle in all the preceding cases.

Live News

Live news is reported as it happens. Instead of recording the story on tape and editing it together back at the studio, the story is broadcast as it happens.

To get the audio and video from the news site to the station news room, a microwave transmitter might be used (Figure 15–39). This is a high-frequency radio transmitter which broadcasts its signals from a dish-shaped antenna, usually mounted on a truck. The signals can only travel in a straight line so the truck must find a place where its transmitter can "see" the station's receiving microwave antenna. If such a connection is impossible, some stations take the journey in two "hops." They send their signal from the news site to the top of a mountain or to a nearby helicopter which receives the signal and retransmits it to the station antenna. Because the helicopter or mountain is high, it can "see" both the station and many news sites at once. Such a microwave relay is called a REPEATER.

When an ENG crew tape-records the news, it can spot check its tapes in the field to make sure they are coming out. When the news is live, there are different methods to check on their signal.

Larger stations may stay in radio or microwave contact with their news crew, giving them cues or playing them the off-the-air television signal. Another method that is cheap but cumbersome is to make telephone contact with the news crew. The TV director may be in phone contact with one member of the crew who is standing at a telephone booth. This member may then signal the others via walkie-talkie or by waving his hands. The most popular method of monitoring one's signal is to use a regular battery-operated television set tuned to the station's channel. Here, the news

---

MINIGLOSSARY

*Repeater   A receiver/transmitter that picks up a radio signal (or TV or microwave signal) and retransmits it. Repeaters are usually placed up high where their signals can reach farther than portable transmitters.

crew watches the newscast as it progresses and takes their cue when their own reporter appears on the screen. If suddenly the picture goes to pieces, they know there's something wrong with their signal transmission (although there is hardly time to do anything about it).

Here are a few more strategies for producing live newscasts:

1. Keep your camera near the talent so that there isn't space for someone to walk between you and the talent.

2. Try to anticipate what your shooting conditions are going to be like. At 3 p.m., the bus terminal may be quiet, but at 5 it's rush hour, and you will have crowds and noise to contend with.

3. Plan your lighting with attention to how it will be in a few hours when you are on the air. If it's light at 5, will it be dark at 6? Will the distant shot you can get now by the light of the sun be totally lost when it is illuminated only by the lone bulb of your 100-watt portable studio lamp? Be aware also that backgrounds which are dark but not black will probably appear black to the camera once you have illuminated the talent.

4. Don't let "live" television make you afraid of having motion on your TV screen. If your reporter wants to walk from place to place, then follow him or her (stay close so that no one gets in the way).

5. As soon as your microwave transmitter is sending a signal, the station may start to use it totally out of schedule in the show, without warning. If you don't want to be embarrassed about the things you said and did while your mike and camera were on while you thought they were off, do the following:

   a. Stay alert and pretend that you are on the air every moment as you wait for your cue from the station. That cue is guaranteed to come while you are picking your nose or making a wise crack.

   b. Unplug your mike and switch your camera to color bars so that the station cannot accidentally switch you on without your knowing it.

EFP STRATEGIES

EFP (electronic field production) is essentially studio recording without the studio. Unlike ENG and newscasting, electronic field production is typically well planned, scripted, and shot with intensive editing in mind.

Corporate Video

Corporate video is usually information-oriented rather than entertainment (although some of the wiser corporations have discovered that a little sugar helps the corporate medicine go down). Corporate TV programming may be educational (how to do something), informative (company benefits explained), sales-oriented (how to carry out a company-wide marketing strategy), or company news (stories about management and employee achievements, awards, or company growth). The "news" is usually more public relations than it is hard news. Its intent is often to make large organizations more personal and familiar to the widely scattered members.

Marketing and Motivation

Marketing and motivation shows are intended to inform salespeople about new products and services as well as create a positive attitude and enthusiasm about certain products. Such shows are shot much like an advertisement. They are tightly scripted, fast paced, and very dynamic. The colors and lighting are bright, and the music is upbeat. Quick cuts are used to enhance the dynamic energy of the program.

Commercials

Large-scale advertising is beyond the scope of this book. Small companies, however, sometimes do their own. The local car dealer's or restaurant's budget may be small, requiring you to use a maximum of creativity and a minimum of equipment and time.

If you are working for a corporation, financial creativity may be as important as scripting and production creativity. For instance, if your corporation can't afford a helicopter for a fancy shot, then perhaps the local police station or fire service would let you ride in theirs in return for your corporation producing a free ad for them.

Multicamera Productions

Orchestras, dance troupes, circuses, and car races don't fit in the TV studio (you needed this book to tell you that). Such shows are performed once, live, and if a shot is missed, it is gone forever. For this reason, it is good to employ several cameras at once; the more, the merrier. There are three strategies for doing this:

1. Tie all the cameras to a switcher which feeds the selected picture to a video tape recorder. Similarly, selected microphones are mixed and the sound is recorded along with the picture. This procedure is quick and quite acceptable, especially if the director is familiar enough with the show to make wise shot selections. If the director

goofs up, however, the tape will have a bad part. Although it is possible to edit the tape later to remove the bad part, this will shorten the program and may cause a discontinuity in the action.

2.   As a partial solution to the preceding, two video tape recorders instead of one could be used. One records the switcher's program output while the other records the switcher's preview output. If the shot the director selected goes sour, then the shot the director *didn't* select will go on the second tape recorder. The shot can be inserted later over the first, assuming *that* shot was good. For best results, camera operators should be assigned zones or tasks so that each one is watching for something different. This will avoid the needless duplication of matched shots and will give the director something different to switch to when his main shot goes astray. A popular shooting technique is to assign one camera to take all the general long shots which get all the action in. That camera should always keep a good image on its screen so that the director has a ''safety shot'' to jump to when the other cameras are moving, focusing, or simply giving blurry close-ups of the curtains where the talent *was* standing. These other cameras are assigned the more active tasks of following the action, getting close-ups, taking creative shots, and moving around among the crowd.

3.   Safest but most tedious is to give each camera its own video tape recorder or use camcorders. These independent, free-running cameras are nicknamed ISOs, short for ISOlated. The resulting tapes are then brought back and edited together to create the final show. One handy technique is to start all the cameras at the same time and let them roll nonstop throughout the show (assuming there is enough tape). The tapes are then played back in several video tape players and are synchronized precisely together. This way the director can switch from tape to tape as if switching from camera to camera live. If a mistake is made, the whole process is stopped, the tapes are backed up a ways (equally), and the director makes a different selection. This process, however, requires the use of SMPTE time code and time base correctors for the video tape players in order to make their pictures mixable.

Music Video

Unless you've just arrived from Mars, you have seen MTV (Music Television) and know what music video is all about.

The professional stuff is usually shot at great expense on film under very controlled conditions. You may, however, have an opportunity to make a low-budget music video for someone. Here are a few strategies.

1.   Simply recording the musicians performing their song in the studio and using a few camera angles, special effects, and lighting tricks are not likely to be stimulating enough. The music video fans expect more. You may have to go on location, perhaps one suggested by the song. The Amazon may be farther than your budget will take you, but a railroad station, a garbage dump, a supermarket, or a baseball field may do just fine.

2.   After establishing the group in a long shot, move to tight close-ups for most of the show to keep the energy high.

3.   Use low camera angles to accentuate the presence of the performers.

4.   It is hard to get good sound when producing in the field. It may be best for your band to perform the song in its entirety under controlled conditions with good microphones, proper acoustics, echo, and other enhancements. Next, you travel to your locations and play that music *to* the performers while they lip-sync to it. Essentially, you are shooting video without audio. All the pictures are later edited together to match the original sound track. One technique that makes the process a little easier is to lay down the complete sound track on a ''blacked'' video tape (one which has color bars or a black video signal recorded on it). Next, you make video inserts to go along with the music. You might do this by taking the tape machine out to the field and playing its prerecorded sound track through a loudspeaker system. After a few rehearsals, you press the VIDEO IN-SERT ONLY button on the VCR to begin recording this part of the performance. Hit the button which ends the insert when this scene is completed. When finished, you will have inserted video scenes in sync with the sound all on the same tape. You don't even have to shoot the scenes in sequence; you can checkerboard them.

INSTRUCTIONAL TELEVISION TECHNIQUES

Knowledge keeps no better than fish. The objective of instructional television is to dispense fish and keep it fresh. You've accomplished little unless your viewers *remember* your message. So besides all the laws of editing, camera angles, and scripting, we add a new set of axioms for designing programs with easy-to-grasp, memorable content.

Between half to all of an ITV (instructional television) program's content is normally forgotten by viewers after merely 2 weeks. You didn't buy all that video equipment, read this giant book, and spend weeks producing that show just for your viewers to forget. Using proven techniques in media psychology, you can carefully design your program to improve the impact and memorability of your ITV programs.

Reasons Why Viewers Fail to Remember

1. The facts aren't of any interest to them.
2. The content goes by too fast.
3. The subject is not understood or doesn't make sense.
4. The program is boring.

Designing Effective TV Lessons

Step 1. Know your specific audience.

a. *What age are they?* Youngsters and teenagers have a shorter attention span than adults and require short, to-the-point, fast-paced productions.

b. *How educated are they?* Write up or down to the level of your audience. When there's a wide range of ages, write for a ninth-grade level.

c. *How new is the material to your viewers?* Your language should reflect their experience. If you're describing a variable resistor to electrical technology majors, you'd call it a potentiometer and simply point to it on a circuit diagram. If describing the variable resistor to audiovisual education students, you might avoid the fancy terms and describe it using an analogy such as "Here the variable resistor controls the flow of electricity, much like a water valve adjusts the flow of water in a pipe."

d. *How interested is your audience?* If they don't give a flying quack about the subject, you're in trouble. Either convince the viewers that the message is vital to their future, their grades, their health, or their careers (intrinsic motivation) or trick them into being interested through humor, mystery, drama, or other creative design (extrinsic motivation). The term *edutainment* might apply to this endeavor. You could also promise them a test (more extrinsic motivation).

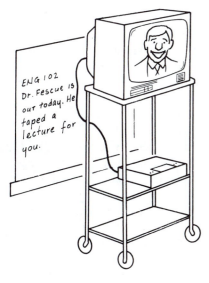

Instructional television techniques.

e. *What do they expect?* An engineer who chucks a videocassette into a player in the laboratory is expecting unadorned facts and isn't overly concerned with production quality. Executives who sit in dimly lit conference rooms watching a color video projector with high-fidelity sound expect a glossy, technically perfect production before they buy that $2 million teabag stapler you're trying to market.

f. *Will viewers get a chance to use this information soon?* If so, you can deal with more complex messages because the viewers' memories will be refreshed through application of their new skill. If not, your message should be short, very simple, very clear, and extremely hard to forget.

Step 2. Establish objectives for your audience. What will your audience be able to do after watching this tape? Make a list. Be specific. Don't list "The student will understand how to operate zoom lenses." Instead, list "The student will attach a zoom lens to a given camera. The student will adjust the iris for a satisfactory picture in the viewfinder. The student will focus on an object so that it remains in focus throughout the lens' zoom range. The student will identify the iris, focus, and zoom adjustments of a lens."

Don't establish unnecessary objectives. Clutter works against you. If a fact or skill is not germane to understanding a topic, leave it out.

Step 3. Tell your viewers what is coming, tell them what they are seeing now, and tell them what they have just seen. This is an old tried-and-true radio technique. It doesn't apply too well to humor, mystery, and drama, but it's inherent to effective instruction. When viewers don't know the purpose of the details you're spewing at them, they will remain unrelated details—not very noteworthy. If you first tell how these details will fit together, then they have meaning and become memorable.

Amidst the trees of your presentation, the viewer may forget the forest, your ultimate objective. So during the body of your presentation drop a reminder of your overall point, such as the following: "Here's another equation to solve. Remember, whatever we do to one side of an equation, we must always do to the other side too." "What is the molarity of this solution? Remember, the molarity clues us as to how much reagent to use for our chemical reaction."

At the end of the presentation, offer a review that links all the points together into a cohesive structure. Restate the overall goal and show how the facts or skills built up to support that goal. Much of the learning will get "locked in" during the closing review.

Step 4. Associate new material with things viewers already know. Viewers may comprehend your new facts and ideas as they are presented, but unless they get connected with something familiar, they're easily forgotten. For example, "A transistor modulates a current that passes through it dependent on the voltage at its base" is useless to someone who is unfamiliar with electronics yet who understands the words. Instead, try the following: "A transistor is like an electrically operated valve in a pipeline. Open it farther, and more water goes through. Feed more electrical pressure to a transistor's base, and it will allow more current to pass through." Associate new ideas with things around the home (such as the water valve), parts of the human body (such as male plugs), or concepts previously studied ("just as Shakespeare's Lady Macbeth suffered a haunting guilt which drove her toward madness, Dostoevski, in *Crime and Punishment* reveals . . .").

Step 5. Use visuals, analogies, and mnemonic devices. Pictures are more easily remembered than words. At every opportunity, draw a picture. Show how ideas relate to each other by making a flowchart or pyramid with labeled boxes. Depict Russia as a bear or the navy as a ship. Use caricatures, cartoons, photographs, pie charts, line graphs —anything that will visualize your concept (review Figure 12–13).

If the concept is ideological, visualize an analogy. Don't say, "Attain the viewer's attention." Instead say, "Grab 'em by the throat," and include a picture of your viewer being yanked into the TV screen, his eyeballs spinning around, his feet off the floor.

Use jingles, songs, and mnemonic devices when possible. They're not kid stuff—they work. For nurses to remember which foot their patients with broken legs should lead with when climbing or descending stairs, they remember "Angels go up (the good foot leads up the stairs); devils go down (bad foot steps down first)."

Do any trigonometry students remember "Indian Chief SOH CAH TOA?" The Sine of an angle is the Opposite side of a right triangle divided by the Hypotenuse. Cosine is Adjacent over Hypotenuse. Tangent is Opposite over Adjacent. Does "HOMES" help you recall the names of the five Great Lakes? Jingles and songs, if you're creative enough to dream them up, work just as well. Your author remembers teaching the bones in the human skeleton using the song "Dry Bones" ("leg bone connected to the thigh bone," and so on), only we sang the proper words for the proper bones while showing them on the screen. Great fun. To this day, most of those students can tell you the names of their bones (but you might catch them humming a bit before they answer).

Step 6. Use repetition. Assuming your facts are mean-

MINIGLOSSARY

Strike   To clear props and set pieces from the studio.

ingful so that the repetition isn't simply rote learning (which is quickly forgotten), hammer them in with relentless restatement. Your monotonous refrain may sometimes irritate your audience, but they'll remember your message.

Besides the obvious fact that repetition improves recall, there's a not-so-obvious fact that *spaced* repetitions are better than massed reiterations. For example if you see the phone number for the National Poison Control Center, 1-800-962-1253, you can repeat it to yourself a few times, holding it in your mind long enough to dial it. Will you remember the number 5 minutes from now? An hour from now? The massed repetitions stored the number only in your short-term memory. Increasing the interval between practices will shift the information into your long-term memory So present an idea, wait a while, repeat it, and review it again at the end of the tape. If the tapes are in a series, bring up the idea again in the next tape and maybe again three lessons later.

To avoid monotony and to help file a concept in *both visual and aural memory, repeat the concept visually if it was first introduced orally and vice versa. Now there will be two mental routes to the concept rather than one. Better yet, devise a tactile, verbal, or physical response to reinforce the fact or idea: The more "handles" a fact has, the easier it is to grab.*

Step 7. KISS (Keep It Simple, Stupid). People will remember one, two, or three bite-sized concepts. Give them five, and they'll remember one, two, or three. Give them ten and they'll remember one, two, or three.

SCHEDULING A SHOOT

A TV studio is an expensive place to make a mistake. Depending on the size of the studio, hourly charges could run between $50 and $500. Whether renting a studio or trying to efficiently schedule your own studio, it is wise to estimate how long it will take to do your thing there.

Most commercial studios work a 10-hour camera day before they charge overtime. From that 10 hours, subtract 1 hour at the start of the day and ½ hour after lunch for technical setup and camera alignment. Next, deduct another hour to STRIKE at the end of the day. No, STRIKE doesn't mean carrying signs and demanding higher wages; it means to disassemble and remove the set and props and clean up the studio. Thus, out of a 10-hour day, you end up with 7½ usable hours.

Expect not to work longer than 5 hours straight without a break. Union and most nonunion commercial studios charge a substantial penalty for working longer than 5 hours in a row.

Unless you are in the middle of a "roll," plan to give your crew a 5-minute (which always turns into 10 minutes) break every hour.

If your lighting and sets are simple, you can probably have that work done during the TECHNICAL SETUP (the period where technicians tweak the cameras). If lights and sets are elaborate, they can be done the day before the shoot (probably at a reduced studio rental rate).

Although sets, lighting, and TECHNICAL SETUP can usually be done all at the same time, it is unwise to try to cram the rehearsals in there too. There will be too much chatter, and your actors will be bumping into ladders and tripping over cables.

One last dress rehearsal should start out your studio production day. This will give the audio and lighting people a last chance to adjust their instruments while showing the rest of the studio crew, especially the camera operators, what the production is all about. Once the crew understands the production, they are better able to participate creatively in its execution. Table 17–6 suggests a studio schedule appropriate for a 20-minute drama performed in a three-camera studio. Of course, every show is different so Murphy's 103rd law applies without mercy: "Estimate the amount of time a job will take, double it, and you'll still run out."

THE BLACK ART OF BUDGETING

When your outgo exceeds your income,
your upkeep is your downfall.

Life would be far more pleasant if
you could make money first,
then make it last.

Most educational and industrial teleproducers live in a sheltered world where they earn a salary, receive a reasonable budget for materials, and sometimes are granted the funds for new or replacement equipment. Their purpose is to provide a service, and their efficiency and productivity are loosely if ever measured in terms of cost versus benefits.

Larger television departments with $500,000 worth of equipment or more may have to pay closer attention to budgets. They may need to justify their existence to their deans or upper management. Often this is done by comparing the total number of dollars spent to the benefits provided to the institution. These benefits could be measured in several ways as shown in the following list:

1. Total number of shows produced
2. Total number of students taught (or tuition earned)
3. Number of grant dollars attracted to the college

MINIGLOSSARY

Technical setup Adjusting the video equipment prior to a show. Also the time period for this process before a show.

TABLE 17–6
Sample shooting schedule

Day 1

9:00 a.m.– 1:00 p.m	Rehearse outside studio Acquire props and set pieces
1:00 p.m.– 2:00 p.m.	Lunch
2:00 p.m.– 5:00 p.m.	Rehearse outside studio

Day 2

1:00 p.m.– 4:00 p.m.	Assemble sets and light the studio Rehearse outside studio
4:00 p.m.– 5:00 p.m.	Rehearse on set (lighting director present)

Day 3: production day

8:00 a.m.– 9:00 a.m.	Finish sets and lights Technical setup Makeup for actors
9:00 a.m.– 9:30 a.m.	Dress rehearsal on set (crew present)
9:30 a.m.–12:00 p.m.	Shoot scenes
12:00 p.m.–12:30 p.m.	Shoot cutaways, reaction shots
12:30 p.m.– 1:30 p.m.	Lunch
1:30 p.m.– 2:30 p.m.	Review raw footage
2:30 p.m.– 3:30 p.m.	Redo bad shots
3:30 p.m.– 4:00 p.m.	Record narrator announcements
4:00 p.m.– 5:00 p.m.	Runover
5:00 p.m.– 6:00 p.m.	Strike

Day 4: editing day

9:00 a.m.–10:30 a.m.	Special effects, character generator, 3 VCRs with TBCs for AB rolls, titles, and dissolves. Audio sweetening and music intro and outro.
10:30 a.m.– 1:00 p.m.	Straight editing with 2 VCRs.

4. Number of dollars saved by producing training or advertising materials in-house rather than going to outside producers

5. Number of dollars earned by the organization renting its services to other companies

6. Number of dollars "earned" by serving various departments in the same company (a system called CHARGE BACK)

7. Amount of prestige earned by the institution which can proudly display state-of-the-art equipment and communications technology to prospective clients or students

The commercial production houses live a tougher life. When they buy a piece of equipment, they must earn back its cost plus some profit. To remain solvent, commercial producers, postproduction houses, and video equipment rental shops must keep their people and equipment busy all the time.

Whether you will be renting someone else's facility or will be creating a facility to be renting to others, your expenses will fall into two main categories:

1. ABOVE-THE-LINE or *out of pocket* costs. These are one-time costs to produce a particular show. They would include the cost of actors, writers, travel, video tape, and the rental of special equipment such as helicopters.

2. BELOW-THE-LINE or fixed costs. This is OVERHEAD, the studio's costs for the privilege of being in business. They include rental of the building, electricity, insurance, salaried staff, interest payments on equipment, and AMORTIZATION (the cost of paying for an expensive piece of equipment over the number of years of its useful life). The company pays these expenses whether it has a customer or not. If a company has a lot of business, these expenses are spread over a greater number of clients. If, however, the company is idle for three days and produces a show on the fourth, that company has to build in charges to pay for the three idle days.

The Role of the Producer

The television PRODUCER is usually responsible for the budget. He or she supervises the production personnel in an effort to get the best possible performance from every individual on the production team. The PRODUCER also

MINIGLOSSARY

Above-the-line costs   Production expenses related only to a particular show. Examples: special talent, writers, travel, charges for special effects.

Below-the-line costs   Ongoing costs realized whether a production company is doing a show or not. Overhead. Examples: staff engineering and production personnel, equipment amortization, telephone, taxes.

Amortization   Splitting up the cost of an expensive item over the number of years the item is used.

*Producer   Creator and organizer of a TV show, usually responsible for budgets, salaries, etc.

*Director   Person in charge of shooting and editing a show, the

actual "builder" of the show.

*Producer/director   Combined job title for a person in charge of undertaking a TV show, handling financial matters, and carrying out TV production details.

*Executive producer   A business manager for a TV production company; a higher-level authority dealing with policies, corporate posture, and money raising; not generally involved with production details.

Associate producer   Lower-level production assistant who handles program details; a bookkeeping/clerical position requiring specialization in TV production.

1. Plans and solicits production bids.
2. Schedules and confirms production facility time and services. This also includes entering production contracts and setting requirements.
3. Organizes and makes talent calls for casting. This may also include supervising music selection and narrators.
4. Supervises and guides the creative interpretation of the production.
5. Follows through with budget/billing, expense reports, payments, and legal matters.

The television producer may work very closely with the TV director, whose job it is to supervise the studio and location and edit production sessions. The director essentially is responsible for transforming the script into video and audio images.

Sometimes both jobs are merged, creating a PRO-DUCER/DIRECTOR who handles both the money and the creative production details.

Some organizations have an EXECUTIVE PRODUCER who takes care of the entire budget and handles station management, advertising agencies, financial supporters, and salaries for principal actors. The EXECUTIVE PRODUCER may have overall responsibility for a complete series of programs.

The ASSISTANT or ASSOCIATE PRODUCER is a very busy but lower-level person on the team. He or she assists the producer by handling details such as telephoning talent, confirming schedules, worrying about deadlines, acquiring artwork, and coordinating other details. The ASSOCIATE PRODUCER may also keep track of expenses.

Creating the Budget for a Production

Although $1000 per finished minute is a common rule of thumb for fairly good-quality educational and industrial television production costs, actual productions may range much higher or lower than this figure. Naturally, a teacher's talking head in the studio for 15 minutes is not going to cost the same as Michael Jackson narrating a 15-minute helicopter tour of Angel Falls in South America.

Budgeting involves making predictions based on past experience, discussion with the production staff, and plain down-to-earth hunches, often based on your track record with the same client or shop. The object is to make your budget low enough not to scare everyone away but high enough not to have a budget overrun. Explaining a budget overrun is about as pleasant as explaining a crease on a master tape.

There are three basic steps to putting together a production budget:

1. Establish the needs.
2. Write the budget.
3. Track the budget.

Establish the Needs. This step determines the tone of the program for the remainder of the production. If the show needs to have remote locations, special effects, and fancy graphics, these costs must be added. Sometimes trade-offs are necessary, resulting in a few still photographs of a location being used rather than shooting actual motion footage there.

Writing the Budget. Table 17–7 shows a sample budgeting sheet useful for listing the costs of a show. Each facility, depending on its needs, will make budget sheets differently. Some companies throw in an extra column so that they may track estimated costs versus actual costs. Other companies break down their budget sheets into separate budgets for manpower (producer, director, costume designer), production equipment (cameras, VCRs, tape), editing, clerical help, and technicians.

Using the sheet in Table 17–7, one would multiply the estimated number of units of each service (like lighting technician) times the cost per hour and come up with an estimated total cost for that service. By adding up all the costs, you end up with a total cost for the production. More complex productions may utilize numerous services not listed on the form. In such a case, you would simply start another page listing those miscellaneous services, such as

Producer's secretary	Director's secretary
Performer's personal attendants	Assistant directors
	Script clerks
Casting office salaries	Special effects directors
Technical directors	First aid
Dance directors	Wardrobe
Camera operators	Hairdresser
Makeup	Insurance
Electrician	Stock shots
Social security	Film
Music clearance	Musicians
Photography	Arrangers
Singers	Location scouts
Royalties	

On the budget estimating sheet, many blank spaces can be filled in with equipment and personnel charges, rental and free-lance rates, and quotes for graphics. Next, based on experience and an understanding of the production process, you estimate how much of various other services will be needed.

Sometimes a large part of a production involves one or two entries in the budget. Instead of lumping a whole

TABLE 17–7
Budget estimating sheet

Program Title: _____ Date: _____				
Function	*Est. Units*	×	*Unit Cost*	=	*Total*

Function	Est. Units	×	Unit Cost	=	Total
Producer	____	×	$ ____ /	=	$ ____
Director	____	×	$ ____ /	=	$ ____
Writer, visualization	____ hr	×	$ ____ /hr	=	$ ____
Graphics:					
Titles	____ (qty)	×	$ ____ /ea	=	$ ____
Animations	____ (qty)	×	$ ____ /ea	=	$ ____
Flats, staging, etc.	____ (qty)	×	$ ____ /ea	=	$ ____
Narrator, announcer	____	×	$ ____ /	=	$ ____
Actor(s)	____ days	×	$ ____ /day	=	$ ____
Audio director	____ hr	×	$ ____ /hr	=	$ ____
Lighting director	____ hr	×	$ ____ /hr	=	$ ____
Remotes:					
Equipment	____ days	×	$ ____ /day	=	$ ____
Labor (_____)	____ days	×	$ ____ /day	=	$ ____
Labor (_____)	____ days	×	$ ____ /day	=	$ ____
Materials (tape, etc.)	____ (qty)	×	$ ____ /ea	=	$ ____
Support expenses _____ crew ×	____ days	×	$ ____ /day	=	$ ____
Transportation (truck, van, air, auto, etc.)				=	$ ____
Video studio:					
Setup time	____ hr	×	$ ____ /hr	=	$ ____
Production time with ____ cameras	____ hr	×	$ ____ /hr	=	$ ____
Film chain	____ hr	×	$ ____ /hr	=	$ ____
Character generator	____ hr	×	$ ____ /hr	=	$ ____
Titling camera	____ hr	×	$ ____ /hr	=	$ ____
Off-line editing (_____)	____ hr	×	$ ____ /hr	=	$ ____
On-line editing (_____)	____ hr	×	$ ____ /hr	=	$ ____
Duplication:					
Tape to film (_____)	____ min	×	$ ____ /min	=	$ ____
Tape to tape (_____)	____ (qty)	×	$ ____ /ea	=	$ ____
Misc. rentals (_____)	____ day	×	$ ____ /day	=	$ ____
Miscellaneous (_____)	____	×	$ ____ /	=	$ ____
Miscellaneous (_____)				=	$ ____
Miscellaneous (_____)				=	$ ____
			Total cost	=	$ ____

bunch of dollars under one entry (such as "remotes: 20 days times $1500 per day = $30,000"), an entry can be broken down into smaller parts so that each can be itemized and tracked separately.

Once the total budget has been established, it is time to make adjustments. If you are way under budget, perhaps you can be lavish with sets or lighting. If you are over budget, perhaps some services can be cut back or deleted entirely.

Once the budget total has been determined, it is wise to add a 10–15% contingency to that number to cover "unexpecteds." This contingency could be as low as 5 percent

MINIGLOSSARY

*Per diem An additional daily living expense payment made to employees working away from home.

Pickup An individual temporarily hired to perform a specific task. Often a free-lance camera operator brought in for a single show.

for simple studio shoots where very little is likely to go wrong . . . go wrong . . . go wrong. Location shooting, on the other hand, runs the risk of lost equipment shipments, uncooperative weather, and various other delays which add to the costs. These estimates perhaps should have a 20 percent contingency added.

Tracking the Budget.   When work begins, someone must track expenditures and match them against the budget, noting discrepancies. If an item goes substantially over budget, another item may have to be reduced if possible.

On-Location Budgeting—an Example

Let's try preparing a budget for an on-location shoot, just for fun (a pleasure which I'm sure rates up there with changing your engine oil and washing out your trash barrel).

First, some information about the production: We will budget for a complete video production, from shooting on location to postproduction through an on-line edit and mastering on 1-inch tape. This is to be a 15-minute educational minidocumentary. Except for one segment, it will be shot single camera and edited together.

We will deal with three locations, two in town and one out of state. Assume further that our script is detailed and specifies locations and types of shots needed. The program will utilize nonprofessional talent, workers in a factory.

The object of the production is to show how a product is manufactured.

Some Decisions.   The 2 days of shooting in town is fairly easy to estimate. You simply rent the necessary equipment and pay the crew. The out-of-state production is more complicated. Should you pay about $1200 in airfares, lodging, and PER DIEM (a daily allowance in addition to regular salary) for a crew to travel to the remote site, or should you hire a ''free-lance'' production crew who lives near that location and also rent the equipment out there too? This is always a trade-off. It generally costs more to send a crew out of state to shoot, but if you give the job over to someone you are unfamiliar with, you don't know if the work will be done well or even whether the equipment will be assembled correctly by the rental company.

The director needs to scout the out-of-state location, which adds airfare, transportation, and lunch to the pre-production charges. The director also spends time procuring crew and equipment.

To keep things interesting, let's assume that the director discovers that the out-of-state shoot requires two cameras at once and a switcher. Our choice (so that you can see an example of the costs of an on-location multicamera shoot) is to send one production team (the director, one camera/lighting person, one technical director/VCR operator, and one audio person) out of state to shoot while we hire a PICKUP (a free-lance camera operator) to run the second camera.

The equipment chosen will be ¾U portable VCRs. The raw footage will be rough-cut in that format, but the tapes will be duplicated to 1 inch for the on-line final cut in order to preserve quality.

The Calculation.   Given these data, we start filling in the estimated costs of equipment and services. The result should look something like Table 17–8.

The Cost of Services

How do you know how much to list on your budget sheets for VCR rentals or salaries for TV directors? Nearly all television production facilities and equipment rental shops publish a RATE CARD. It is a list of what they charge for various services, from the rental of ¾U portable VCRs to the rental of an entire studio with crew and equipment. Postproduction houses have a RATE CARD listing off-line and on-line editing charges. A simple off-line editing closet with two VCRs could cost as little as $50 per hour. A big editing suite with overstuffed chairs, professional editing technicians, special effects, time base correctors, AB rollability, and 1-inch VTRs, plus other bells and whistles, could cost $500–$800 per hour.

People have RATE CARDS too. Some charge UNION SCALE, an hourly rate determined by a union contract, while others free-lance at whatever rate the market will bear.

There are too many services to list what they cost here, and besides, the prices would be out-of-date before this page leaves the author's typewriter. But, to show the difference in prices for some services, consider editing equipment:

MINIGLOSSARY

Rate card   A published listing of charges for services.

Union scale   A salary negotiated between a union and larger producers, covering specific job titles. Scale sets a standard used by others in the industry, even if not members of the union.

Amateur VHS	2 VCRs and an edit controller	$ 25/hr
Low end industrial	¾U editing VCR, ¾U player only, automatic editing controller	$ 40/hr
¾U dissolve	3 ¾U VCRs, 2 time base correctors, computerized edit/controller	$175/hr
¾U 1-inch editing	2 ¾U VCRs, 1 1-inch VTR, 2 TBCs, 1 computerized edit controller	$225/hr
½-inch component professional VCRs	2 component (MII or Betacam) VCRs, 1 1-inch VTR, computerized edit controller, 2 TBCs, and audio equipment	$235/hr
Broadcast CMX editing	3 1-inch VTRs with TBCs, CMX advanced computerized edit controller, full audio sweetening, digital effects	$360/hr

TABLE 17–8
Sample on-location budget

Function	Est. Days	×	Cost/Day	=	Total
Preproduction					
Producer/director (scout location)	1	×	$ 300	=	$ 300
Travel expenses					250
Producer/director (script analysis, production plan, & sched.)	1.5	×	300	=	450
			Subtotal	=	1,000
Production (local)					
Producer/director	2	×	300	=	600
Camera/lighting person	2	×	250	=	500
VCR operator/audio person	2	×	200	=	400
Local transportation for 3	2	×	100	=	200
			Subtotal	=	1,700
Production (out of state)					
Producer/director	2.5	×	300	=	750
Camera/lighting person	2.5	×	250	=	625
Technical director/VCR operator	2.5	×	200	=	500
Audio person	2.5	×	200	=	500
Free-lance camera operator	1	×	150	=	150
Air fare at $250/person × 4 (including excess weight)				=	800
Lodging & per diem at $100/person × 4	2	×	400	=	800
Local transportation for 4	2	×	100	=	200
			Subtotal	=	4,325
Equipment (local)					
Camera	2	×	300	=	600
VCR (with time code generator)	2	×	100	=	200
Lights (4K kit)	2	×	50	=	100
Misc. (monitor, cables, tripod)	2	×	50	=	100
			Subtotal	=	1,000
Equipment (out of state)					
2 Cameras	2	×	600	=	1,200
VCR (with time code generator)	2	×	100	=	200
Lights (6K kit)	2	×	75	=	150
Portable production console (monitors, cables, switcher, waveform monitor, vectorscope, audio mixer)	2	×	600	=	1,200
			Subtotal	=	2,750

TABLE 17–8 CONT'D

Videocassettes

15 cassettes (20 minutes each)				=	150
3 cassettes (60 minutes each)				=	100
Insurance	6.5	×	100	=	650
			Subtotal	=	900
		Total production cost		=	11,675
		10% contingency		=	1,168
		Production budget		=	12,843

Postproduction

Copy ¾U tapes to 1 inch				=	800
Time code burn-in on VHS				=	400
Producer/director (for rough cut)	4	×	300	=	1,200
Off-line editing (for rough cut)	4	×	200	=	800

Function	Est. Days	×	Cost/Day	=	Total
Producer/director (for fine cut)	4	×	300	=	1,200
On-line editing (for fine cut)	.75	×	1,600	=	1,200
Sound mix	2	×	500	=	1,000
Video tape					
2 1-inch 30 min at $50				=	100
4 ¾U cassettes at $25				=	100
15 VHS for rough cut at $7				=	105
Graphics (4 charts)				=	200
			Subtotal	=	7,105
			5% contingency	=	355
		Total postproduction budget		=	7,460
		Total program budget		=	$20,303

VIDEODISCS

A videodisc is like a phonograph record with a TV program on it. You can buy it already recorded, and you play it on a special kind of phonograph called a VIDEODISC PLAYER (catchy name) and send the signal to a TV set.

Figure 18–1 shows a videodisc. Unlike movie film, the pictures that make up the videodisc program cannot be seen with the eye. Unlike a video tape, a common videodisc cannot be erased and rerecorded (but that feature may be coming soon). The video signal is encoded as microscopic pits arranged in a long spiral and then covered with a protective clear plastic surface. A common 12-inch videodisc can hold up to 1 hour of recording on each side.

Movies, slides, audio recordings, or any audiovisual material can be used to create a videodisc; however, video tape is the most common medium. A film or video producer brings a master film or tape to a MASTERING PLANT where a MASTER DISC is made from which duplicates by the dozens or by the thousands are pressed.

The user places his or her videodisc copy on a machin resembling a high-speed phonograph. A beam of laser ligh is focused on the disc, and its reflection off the pits in th disc's surface creates the signal which is converted int audio, video, sync, color, and even computer data.

KINDS OF VIDEODISCS

There are several ways to put pictures, sound, and comput data on a disc. LASERVISION (LV), sometimes called OPTICA VIDEODISC or LASER DISC (LD), is the type found in mo schools and industry, although other types are emerging.

LV—Laservision. A laser reads coded inform tion off a 12-inch disc to produce analog composite vid plus two analog audio tracks and two digital audio track CAV discs play 30 minutes (or 54,000 still frames) per sid CLV discs play 60 minutes per side.

MINIGLOSSARY

*Videodisc A record-shaped disc encoded with a TV program.

*Videodisc player Phonograph-like machine that plays a video-disc, sending the video and audio (or RF) signals to a TV for display.

*Mastering plant A company that converts a video tape or other media into a master videodisc and makes copies of it.

Master disc A specially made original videodisc from whi distribution copies are reproduced.

*LV Laservision. A videodisc read by a laser, the light of whi reflects off microscopic pits in the disc. Not the same as CE videodiscs, which use a groove.

Optical videodisc Same as laservision (LV) videodisc.

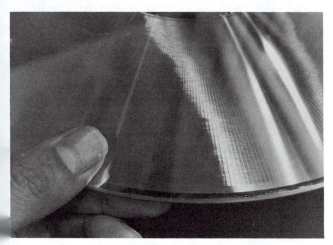

FIGURE 18-1   Videodisc.

CD—Compact Disc.   A laser reads digital audio data off a 5-inch disc yielding 74 minutes of high-fidelity music. Mini 3-inch discs yield 20 minutes of music.

CD-ROM—Compact Disc, Read-Only Memory.   A laser reads coded data from a 5-inch disc holding 650 megabytes (equalling 1,700 floppy disks or 300,000 typed pages) of computer data. The data can include illustrations, but CD ROMs access the data too slowly for motion video.

CD-I—Compact Disc, Interactive.   Improved CD able to access hi-fi music, computer text, still pictures, animation, or even 72 minutes of limited motion video.

CD-ROM-XA—CD ROM Extended Architecture.   Hybrid of CD-ROM and CD-I, handling computer data like a CD-ROM, but audio like a CD.

CDV—CD Video (also known as VIDEO LASER-DISC).   Gold-colored 5-inch disc holds 20 minutes of audio or 5 minutes of video or 9,000 still frames or a combination of these.

DVI—Digital Video Interactive.   A 5-inch laser disc holds 72 minutes of full-motion video. Disc is compatible with IBM PS/2 computer.

Many videodisc players will play only one kind of disc, but a few models will play several of the preceding formats. Since LV is the most popular interactive video format, we'll study LV in more detail.

CLV vs CAV

Videodiscs can be produced in one of two ways: CLV (constant linear velocity) or CAV (constant angular velocity). A CAV disc rotates at 30 revolutions per second (the angular velocity is constant) and can hold programs up to ½ hour in length. These programs can be still-framed, fast-scanned, reverse-scanned, and searched with the help of a computer. CLV discs vary their rotation speed, playing the longer outer spirals slower than the inner ones. This way you get more on a disc, allowing up to a 1-hour TV program to be recorded per side. What you gain in long play, you lose in features: The CLV discs can't still-frame, fast-scan, etc. In short, CLV discs are best for showing movies from beginning to end. CAV discs are appropriate for educational use where particular pictures may be searched out (by a human or a computer) and displayed as still frames.

MULTI-FORMAT videodisc player swallows laser videodiscs (LDs), compact discs (CDs), and CD videos (CDVs). (Courtesy Radio Shack)

MINIGLOSSARY

CLV   Constant linear velocity, the 1-hour mode of an optical videodisc and player. Special effects are not available.

CAV   Constant angular velocity, the half-hour mode of an op-

tical videodisc and player. Special effects are available.

Videodisc recorder   Machine that records a TV signal onto a videodisc.

Disc Recorders

Normally discs are recorded and duplicated by the hundreds at a MASTERING PLANT (Figure 18–2) under tightly controlled superclean conditions. For those who only need a couple discs, or change program content frequently, there are disc recorders costing 12 to 40 kilobucks each. These discs are *not* compatible with regular CD or LV discs and must be played on special players costing 3 to 6 kilobucks (as opposed to about $2,500 for "standard" LV players). The blank discs cost about $350 each.

Videodisc recorders are great for recording animation. Instead of tediously backspacing a VTR thousands of times to record single frames of computer animation, the disc recorder simply zaps one spiral of the disc with its recording laser, then when the next picture is ready, zaps the next spiral.

HOW IT WORKS

A rigid 12-inch LASERVISION disc is spun at 10–30 revolutions per second in the disc player. Pressed into the disc are narrow spirals, like in a record, and each spiral contains a complete video picture (and sound), coded as tiny pits in the disc. A narrow beam of laser light is directed at the spiral. The disc's shiny surface reflects the laser light back to a sensor. The pits vary the amount of laser light reflected. This varying light signal changes into electrical vibrations which are converted into a TV signal. Figure 18–4 diagrams the process.

At a MASTERING PLANT, a video tape is played into a master laser recorder which focuses a laser beam on a glass master disc. The beam creates tiny dots corresponding to the audio and video signals. A photographic process creates the pits, which are then covered with a fine layer of metal.

FIGURE 18-2   LV videodisc MASTERING PLANT (Courtesy of Pioneer Electronics Corp.).

FIGURE 18-3   Videodisc recorder. (Courtesy of Panasonic)

From this master disc, a stamper is made which is used in a replicating machine to crank out videodisc copies.

If you look down at a CAV disc, you might even notice a pattern formed from the billions of tiny pits in the disc. It may look something like Figure 18–5, which diagrams the way the pictures and sync pulses are recorded on the disc. As you can see, if the laser sits in one spot while the disc rotates, it will illuminate one field of video, followed by sync, followed by a second field of video, followed by sync, followed by a repeat of the first field of video, etc. The two fields combine to make a sharp and stable still frame on your TV. Unlike a VCR, which displays only a ''still field'' (only the even or only the odd lines in a picture), the video disc creates a true still frame with *all* the lines, making the picture look twice as sharp as a VCR still frame.

If the laser moves ahead to the next spiral every time the disc rotates, a new picture will be shown every ¹⁄₃₀ second, resulting in smooth motion. The laser can even be told to skip every other spiral, resulting in motion which is twice as fast as normal. A superfast play is possible if the laser plays 1 spiral and skips 10 and then plays another spiral and skips 10 more every ¹⁄₃₀ second. Conversely, slow motion is possible if the laser looks at every spiral twice before advancing to the next. Backward motion is achieved by having the laser jump to the previous spiral at the end of each rotation. Thus, the laser disc can be read forward or backward at numerous speeds, including still frame. All this is done without any harm to the disc because nothing touches the disc except laser light. The disc can rotate all day long, showing just one picture with no wear at all. This was not possible with video tape players. When they showed a still frame, the heads were constantly wearing against the tape.

It only takes a couple seconds for the laser mechanism to jump from the first spiral on a videodisc to the last spiral, making it possible to jump from scene to scene in any order without a long wait.

In disc *recorders* only one copy of a disc is made at a time. There, a video signal is sent to the recorder from a camera or VCP. The signal modulates the beam of a powerful laser which ''burns'' pits into a special sensitive disc. To read back the pictures, a weak laser beam (which doesn't make pits) is used.

FEATURES OF VIDEODISCS AND PLAYERS

Like video tape, videodiscs can have stereo (or dual-channel) analog audio. Recent models also offer two *additional* tracks of digital audio. The sound quality is excellent. Although the discs are breakable, they are fairly rugged. Minor dust and scratches do not affect the program. Dirt and fin-

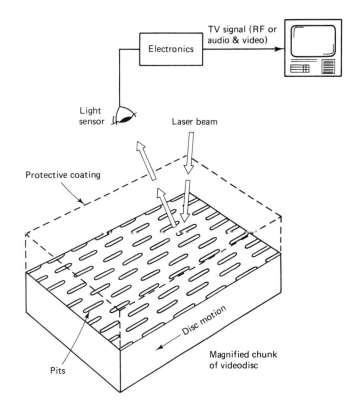

Laser beam is aimed at disc as it rotates. How it's reflected depends on the spacing and length of the pits in the disc. A light sensor converts the reflected beam into a video and audio signal.

FIGURE 18-4 LV system.

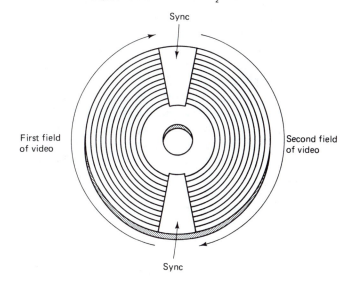

One revolution = 2 fields = 1 frame
54,000 frames = 1 disc side = ½ hour motion

FIGURE 18-5 Where the pictures are recorded on a CAV disc.

MINIGLOSSARY

CAI Computer-assisted instruction, lessons and questions managed by a computer.

gerprints can easily be wiped off using a soft cloth and some window cleaner (try that with a video tape!). They give excellent pictures with the qualities just a little poorer than 1-inch video tape and a little better than ¾U. Industrial videodisc players can be connected to computers to provide pictures, motion, and sound for computer-assisted instruction (CAI). You'll see more on this feature later.

Data Storage

Videodiscs and their CD-type brothers can store more than pictures and sound. Those trillions of tiny pits can represent the 1s and 0s of computer data. Where the typical 5¼-inch floppy computer disc can hold about .36 megabytes of data (⅓ million characters) and a professional "hard" disc can hold about 20 megabytes of data, a videodisc can hold about 300–500 megabytes of data. That's about the same as 1200 floppies! That's about ⅕ million pages of typewritten text.

The Sony VIEW system takes advantage of this fact by combining a computer with the videodisc player and having the computer program stored on the videodisc along with the pictures.

Still Frame Audio

Normally, sound runs along with the picture just as it does with video tape. When the tape stops, the sound stops. When a videodisc is still-framed, the sound stops. If you wish to hear a narration while showing a still picture, you normally have to record a motion sequence of that still picture. It seems a shame to waste 300 frames of picture, all showing exactly the same thing, just to provide 10 seconds of explanation. Two solutions:

1. STILL FRAME AUDIO. It is possible to record 40 seconds of sound as a single video frame. The sound is first compressed and coded into a signal which "looks" like a picture. This "picture" is then recorded on the disc. When the disc is played back, a computer tells it to play the "sound" frame first and store the data in a special device called a STILL FRAME AUDIO ADAPTER. The adapter converts the data back into sound and plays it to the viewer while the videodisc player jumps to the desired picture still frame, which appears now on the TV screen.

2. DIGITAL FRAME STORE. LV players like the Pioneer LDV-8000 can "grab" a frame to display while playing live audio from elsewhere on the disc.

Thus, with the help of a computer to search out the sound and pictures, one can watch a videodisc slide show which lasts for hours. The first system makes it possible,

for instance, to display 9000 historic paintings on a disc along with 500 hours of narration!

ADVANTAGES AND DISADVANTAGES OF VIDEODISCS

Unlike video tape, when you get tired of your videodisc, you can use it as a Frisbee (although catching one may be hard on your hands). Here are some real reasons for choosing the videodisc medium over tape.

Advantages of Videodiscs

1. You can pack a lot on a disc so your programs take up less space than with video tape.
2. Prerecorded discs cost about half of what prerecorded videocassettes cost. Discs can be duplicated by the thousands quite cheaply (in the neighborhood of $10 each), making programs available for $15–$25. By comparison, the *blank* stock for 1 hour in each of these other formats would cost about
    a. $300 for a 16mm color film
    b. $200 for an 8mm color film
    c. $15 for an industrial ¾U videocassette
    d. $4 for a ½-inch home videocassette
    Still you have to record something on the tape or film. To purchase a *prerecorded* show, the costs for a 1-hour color sound program could be
    a. $700 for a 16mm film
    b. $250 for a ¾U videocassette
    c. $60 for a VHS or beta videocassette
3. The picture and sound quality from discs is excellent and typically exceeds the quality found on all but 1-inch video tape.
4. Locating something on a disc takes only a second. You don't have to wind tediously through a long ribbon of tape in your search for a particular program segment.
5. The LV discs don't come in contact with the sensing stylus and thus receive negligible wear even when left in still frame for a long time.
6. CAV discs can be played backward or forward at various speeds while yielding an excellent picture. Some may randomly access any of the 54,000 single pictures that make up a ½-hour show and do this in just a second.

MINIGLOSSARY

*Still frame audio   Technique of turning audio into data that can be stored like a picture on a disc. With the help of a decoder, the "picture" can be "read" and converted back into 10–40 seconds of sound.

Still frame audio adapter   Device to convert a still frame of encoded audio from a videodisc into several seconds of sound.

7. A few videodisc manufacturers are producing videodisc players to play CDs, CDIs, DVIs, etc. These multifunction machines can thus play movies or high-quality digital sound recordings.

8. Most videodisc players can be controlled by a computer.

9. Videodisc *recorders* do a great job of recording single frame digital animation.

Disadvantages of Videodiscs

1. Except for special recorders you can't simply create your own discs. Discs are geared for being produced commercially, duplicated, and sold to you like an educational film or an LP record. If you decide to produce a videodisc, you must first produce a properly formated video tape and send it to a DISC MASTERING FACILITY that will produce the disc from your tape for about $2000.

2. LV players are not *completely* standardized. Although most discs manufactured to play on a Pioneer disc player will also play on a Sony, you'll find a few that won't. Disc recorders make discs totally incompatible with everyone else's LV discs.

3. VCPs cost about ⅕ as much as videodisc players, and the VCRs can both play *and* record.

4. Video tape can be erased and reused. When you buy a disc, the show is yours forever.

5. Although a videodisc can store 54,000 pictures, these are fuzzy *video* pictures, not true pages of a book. It would take about 7 frames of video to hold one readable page of this book. Thus an LV disc can hold 8000 book pages, not the 54,000 intimated by some advertisers. Incidentally, we're talking video pictures here, not computer data. When those videodisc pits are converted directly to computer data which then can be turned to text, the number of text pages that fit on one side of a disc jumps from about 8,000 to about 100,000.

6. For consumers, buying a videodisc is not a good bargain. Usually a movie is viewed once, maybe twice, but seldom more than that. TV programs cannot be played like records in the backgrounds of our lives as we do other things. They require our undivided attention. Few of us have the time or interest to watch movies over and over again.

For this reason, renting a videocassette for $2 a night is a better deal than paying $20 for a videodisc and owning it forever.

INTERACTIVITY

INTERACTIVITY is a reciprocal dialogue between the user and the videodisc system. The viewer may watch a sequence, be asked a question, select an answer by typing something into a keypad or computer, and then see another sequence based on his or her response. INTERACTIVITY is a very efficient way of instruction. Unlike a lecture or a TV show where the viewer sits passively, absorbing only a fraction of the information offered, INTERACTIVE instruction forces the viewer into an active participatory role. When the viewer is forced to respond to questions on the screen, he or she has no choice but to pay attention.

The old adage "practice makes perfect" applies especially well to INTERACTIVE instruction. A student views a lesson and immediately must apply the new knowledge.

INTERACTIVITY offers other possibilities besides instruction. A disc on house construction, for instance, could have sections on laying the foundation or installing the rain gutters. To see more detail on a specific topic, the viewer could jump to that topic and skim through the topic until he or she stumbled across the particular area of interest such as curing the concrete or attaching the rain gutter brackets. The disc may also have menus which act as a table of contents, directing the viewer to various sections on the discs. An auto manufacturer may display and name all the parts of a car along with their part numbers. A person knowing a particular part number could look it up on the disc to see what the part looked like. Conversely, a person finding a part on the disc could find out what its part number was.

One of the more elegant videodisc demonstrations is a tour through Aspen, Colorado. People may view the disc that shows them riding down a street in Aspen. The viewers may press a button and look sideways as they go down the street or look behind them as they travel. At each intersection, the viewers may choose to "turn left," "go straight," or "turn right." The videodisc then displays *that* street. The viewers may even stop to visit a particular hotel and, by pressing buttons, be escorted into the hotel and even have a look at the dinner menu.

INTERACTIVITY allows great flexibility in simulations. A student pilot, for instance, may be offered a number of possible solutions to a flying problem. Upon making his or her selection, the student views the result. In this way it is

---

MINIGLOSSARY

Interactivity   The ability of a machine to react to the responses of its user. An interactive videodisc system may ask the viewer a question, wait for a response, and then display a certain sequence keyed to that response.

Interactive videodiscs help you learn.

possible to experience a real-life situation without the danger or expense of smashing a Boeing 747 to smithereens.

INTERACTIVITY is the one area where the videodiscs beat video tape hands down. A videodisc can randomly access any part of a ½-hour program in just a few seconds or less. Typically, it takes a videodisc ½ second or so to locate a sequence. Video tape, on the other hand, has to wind through hundreds of feet of tape in order to jump from segment to segment. This winding could take as much as 3 minutes. This long wait kills the spontaneity of the human/machine relationship.

It is possible to arrange related segments of a lesson close to each other on a tape, making the search much shorter, but still the search will take 5 seconds or more.

Levels of Interactivity

Not all machines are created equal. Some do more than others.

There are four primary levels of videodisc INTERACTIVITY: LEVEL 1, LEVEL 2, LEVEL 3 and LEVEL 4. To work together, the level of the videodisc *and* the disc player (and computer program, if included) must *all* match. A LEVEL 1 machine may play only a LEVEL 1 disc (you could play other discs on a LEVEL 1 machine, but all you'd see is a montage of random scenes). A LEVEL 2 machine can play either LEVEL 1 or LEVEL 2 but not 3. A LEVEL 3 machine can play LEVEL 3 discs, with the right computer program loaded into the external computer, but can't play LEVEL 2 discs. Many industrial videodisc manufacturers are making players which can do several jobs. Figure 18–6 reviews some of these levels.

Basic Level. This is hardly a level at all but doesn't fit into the regular categories. BASIC LEVEL disc players are used mostly by consumers in the home video market. This home player runs in the CAV or CLV mode and progresses

---

MINIGLOSSARY

Basic level videodisc player   Like a movie, this videodisc and player can only start at the beginning and play to the end of a program. There is no interactivity.

*Level 1 videodisc player   Often a consumer model player with features such as freeze frame, picture stop, chapter stop, scan, and two-channel audio but without the computer memory or the ability to select on its own which sequences to show.

*Level 2 videodisc player   An educational/industrial videodisc player, usually with all the level 1 capabilities plus a small computer built in. Computer programs (instructions) coded in the audio track of the disc are loaded into this computer when you start the disc playing, telling the player which sequences to play as a result of viewer responses.

*Level 3 videodisc player   Videodisc player linked to an external computer. The computer program, perhaps from a floppy diskette, controls the sequences the videodisc player will play. The computer may also display graphics and questions of its own. The disc serves mostly as a storehouse of still and moving pictures and sound.

Level 4 videodisc player   Videodisc player with computer, able to read a large computer program from the disc, and thereafter runs like a level 3 player.

*Level 1/2/3 videodisc player   A combination level 1, level 2, level 3 videodisc player.

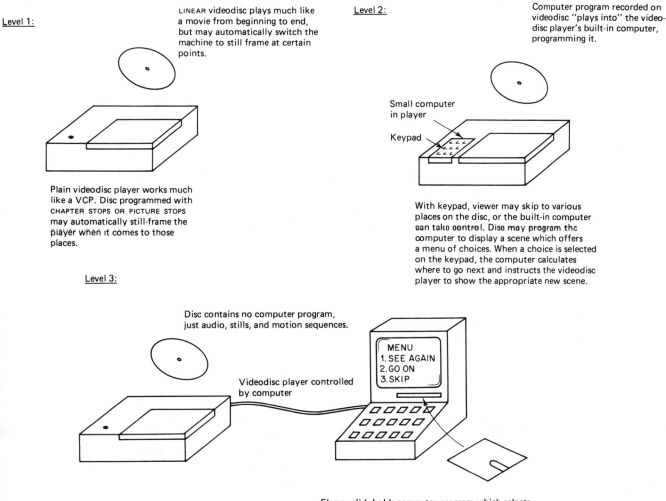

Level 1:

LINEAR videodisc plays much like a movie from beginning to end, but may automatically switch the machine to still frame at certain points.

Plain videodisc player works much like a VCP. Disc programmed with CHAPTER STOPS OR PICTURE STOPS may automatically still-frame the player when it comes to those places.

Level 2:

Computer program recorded on videodisc "plays into" the video-disc player's built-in computer, programming it.

Small computer in player

Keypad

With keypad, viewer may skip to various places on the disc, or the built-in computer can take control. Disc may program the computer to display a scene which offers a menu of choices. When a choice is selected on the keypad, the computer calculates where to go next and instructs the videodisc player to show the appropriate new scene.

Level 3:

Disc contains no computer program, just audio, stills, and motion sequences.

Videodisc player controlled by computer

MENU
1. SEE AGAIN
2. GO ON
3. SKIP

Floppy disk holds computer program which selects scenes which videodisc player will show. Floppy may also contain text or graphics which can also instruct (like standard computer assisted instruction — CAI). Computer reacts to answers typed into keyboard, selecting what questions or scenes will be shown next. Floppy disk, computer, videodisc player, and video-disc must all be designed to work together.

FIGURE 18-6 Levels of INTERACTIVITY.

LINEARLY through the program, much like a video tape or movie. It is not INTERACTIVE.

Level 1. LEVEL 1 is the same as BASIC LEVEL with a few extra features:

1. *Freeze frame.* Useful for showing a table of contents or a menu.

2. *Step motion* in forward or reverse. Movement from one still frame to another, useful for paging through stills, be they slides, charts, questions, or other text.

3. *Scanning.* Fast motion, allowing you to locate certain actions or segments quickly.

4. *Slow motion*

MINIGLOSSARY

Chapter stop A code, embedded in the level 1 videodisc flags where each new chapter or section begins. While scanning fast forward, the player will sense the code and will still-frame at this point. This feature speeds the process of locating segments on the disc.

Picture stop A coded instruction on a level 1 videodisc telling the player to freeze-frame when it comes to a particular picture. Works while the machine is in the play mode.

5. *Chapter stops*. Machine can skim along, automatically stopping at chapter heads to display a still frame showing that chapter's contents or a list of the other chapters the viewer could now jump to.

6. *Picture stops*. While playing, the machine automatically pauses on a still frame (a test, menu, chart) which you may study before continuing.

As you can see, LEVEL 1 machines and discs are hardly INTERACTIVE at all. All the searching and jumping must be done manually by the viewer. At least the viewer can skip along quickly to a particular chapter, with the automatic CHAPTER STOPS halting the machine at each one while *scanning* across the disc. Further, the PICTURE STOPS will halt the machine on a specific still frame as the disc *plays*. Note also that the above features can be enjoyed only while playing a CAV disc. Machines cost $1000.

Level 2. Industrial or educational videodisc players can generally play LEVEL 1 and LEVEL 2 discs. LEVEL 2 machines have a small (2–7 kilobyte) microprocessor in them which can be programmed to control the sequence of what is played. The little computer will also note your responses to menus or questions and will, according to its computer program, branch you hither and thither.

This computer "learns" its program from the videodisc itself. Audio channel 2 contains the program in a 10-second burst of data at the beginning of the disc. If you listened to the signal, it would sound like a wooden chair being scraped across a cement floor.

The disc can be programmed to play, stop on a particular frame (making the LEVEL 1 PICTURE STOPS no longer necessary), and then branch somewhere, depending on the response entered into the player's keypad. All responses must be numerical ("answer 1, 2, or 3") to be entered on the numerical keypad.

If the program is too large to fit in the computer's tiny memory, it can be broken into smaller bite-sized "digital dumps" whereby every so often the videodisc will tell the computer to erase its memory and load some more of the program. Players cost $2000–$3000.

Level 3. Here the videodisc is controlled by an external compatible computer and programmed software (usually stored on floppy diskettes). Here the computer does the branching, asks the questions, stores the answers (which no longer must be numerical but can be anything that is typeable into a keyboard or created with a light pen, joystick, paddle, touch screen, mouse, etc.), and really carries out a conventional CAI (computer-assisted instruction) job. The videodisc is sort of a "slave," providing TV pictures, either motion or still, upon request from the computer. The computer could generate a menu directly, or it could call up, say, disc frame number 1085, which could be the menu.

In short, the computer makes instructional contributions of its own and uses the videodisc to provide the things it can't supply efficiently itself such as sound, motion pictures, true-to-life still pictures, and visual effects. Such workstations cost about $8000.

Level 4. A hybrid of LEVEL 2 and LEVEL 3, the disc feeds a big program to a big computer associated with the videodisc player. The program then takes over, selecting scenes on the videodisc and overlaying text and graphics on the screen, as well as asking the viewer questions. Cost for the ensemble: $8000.

Which Level Is Best? Each level has its advantages and disadvantages. LEVEL 1 is best for simple linear programs which don't interact except for an occasional stop on a menu or still frame. LEVEL 2 allows simple INTERACTIVITY and saves you the money of buying a separate computer and interface. LEVEL 2 programs are fairly well standardized and can be arranged so that they will run on any LEVEL 2 videodisc machine. LEVEL 3, on the other hand, requires your program to be compatible with the specific computer being used to drive the videodisc player. LEVEL 2 is limited in the types of answers it can accept (numerical only) and is also limited in its ability to keep track of your score as you answer questions. LEVEL 3 systems can store quiz scores and typed data in detail.

LEVEL 3 *players* are cheaper than LEVEL 2 *players* because they are "dumb." They contain no computer; they just follow directions from an outside computer.

For LEVEL 3 systems, since the program is on a re-recordable floppy disc, it is easy to change the program to make updates and corrections. Conversely, on LEVEL 2 discs, the computer program is "burned in" and unchangeable. A computer program error on a LEVEL 2 disc will "hang up" the viewers forever (until new discs are mastered). With LEVEL 3, you simply change the program on the floppy. You can even change the program to "skip over" some bad or out-of-date videodisc pictures. Perhaps the omitted segment could be replaced as a computer graphic's scene.

LEVEL 3 and 4 systems have *two* ways of presenting information to the viewer:

1. Text or pictures played from the *videodisc*.
2. Text or computer graphics generated by the *computer*. When the computer makes text on the screen, the text is very sharp, 80 characters of

---

MINIGLOSSARY

Premaster The video tape sent to the mastering facility and transformed into videodiscs. Also, the act of making such a tape.

text pcr line, as opposed to 25 characters per line for fuzzy NTSC video.

In short, it seems that LEVEL 2 is best for simple INTERACTIVE programs which will be distributed widely to people who might not have external computers. LEVEL 3 would seem best for complex programs requiring a lot of text on the screen, a lot of data entry, and frequent updates or debugging of the program.

PRODUCING A VIDEODISC

The only people who actually *produce* their own videodiscs are those with a disc recorder. Panasonic makes a $40,000 model that records 30 minutes of motion on an $800 disc that can be erased and rerecorded 1,000 times. Teac makes a recordable-but-not-erasable WORM (stands for Write Once, Read Many times, but sounds fishy to me) recorder costing $20,000. The $100 discs hold 60 minutes. Sony and Tandy also have videodisc recorders.

Those of us without recorders can produce videodiscs by first producing a video tape and then sending it to a MASTERING PLANT, which then makes the disc. The rest of this section will deal with PREMASTERING, the process of making a proper video tape from which a videodisc can be made.

So Many Alternatives

The videodisc medium allows so many different ways to present information that it boggles the mind. INTERACTIVITY presents a whole new dimension to the range of options available to the teleproducer. (See next page.)

All these features are at your disposal, but to use them, you must be equipped with the right computer, videodisc player, and peripherals. If a light pen is to be used, it must plug into the computer somewhere. When TOUCH SCREENS are used, the television must be equipped with a special overlay which senses touch. This device must then be connected to the computer. You already know that for STILL FRAME AUDIO you need an adapter attached to your videodisc player (or built into it).

If producing a LEVEL 1 disc with CHAPTER or PICTURE STOPS, you will need to rent or buy a CUE INSERTER which places a special signal on your tape which can be read by the disc mastering equipment and converted into a code

which automatically still-frames the videodisc player when it reaches that point.

One consideration when producing a LEVEL 3 videodisc is whether it will be played on a one-monitor or two-monitor system. Figure 18–7 shows the difference between the two systems. In a two-monitor system, the computer screen shows the computer information while a separate TV screen shows the images from the videodisc player. Although twice as much can be seen at once with two monitors, it is often confusing for the viewer to decide which monitor to look at and respond to. Sometimes things are missed when the viewer is watching the computer monitor while the videodisc monitor is displaying some important point.

The solution is to use a single TV monitor to show both computer information and videodisc scenes. Special computers are used which can SUPERIMPOSE (sometimes with the help of an optional SUPERIMPOSER circuit) computer text and graphics *over* the videodisc image. The device may even select whether audio track 1 or audio track 2 is played through the TV.

The solution is to use a single TV monitor to show both computer information and videodisc scenes. Special TV monitors are used which can SUPERIMPOSE (sometimes with the help of an optional SUPERIMPOSER circuit) computer text and graphics *over* the videodisc image. The device may even select whether audio track 1 or audio track 2 is played through the TV.

Such systems work in color and allow very sharp computer text and graphics to be displayed when needed, along with the slightly fuzzy motion sequences available only as video. Incidentally, although the word SUPERIMPOSE is used here, the effect is more like a video KEY effect. The computer text can be opaque, blocking out the videodisc image, or can be semitransparent, like our old video friend, the SUPER.

Figure 18–8 shows an example of how a computer and videodisc player can team their pictures up together: The video from the videodisc player shows a computer keyboard having a special "help" key. While the narration tells what this key is for, the computer, using computer graphics, creates a highlighted window showing where the key is. Such a demonstration would not be possible using a two-monitor system.

Another example: The videodisc displays a still frame of an actual kidney. The computer superimposes the instruction "Point to the area that creates insulin." Using a

---

MINIGLOSSARY

*Touch screen A touch-sensitive TV screen whereby viewers can point to or press their fingers against the screen in response to computer questions rather than using a keypad or computer keyboard.

*Light pen Pen-shaped device connected to a computer that the viewer points or touches to a TV screen in response to computer

questions rather than typing the answers on the computer's keyboard.

Cue inserter Device that puts a coded signal on the premaster tape. At the mastering plant, this cue is transformed into a level 1 chapter stop or picture stop.

Videodisc feature	Uses
1. Still frame picture from disc	Allows a single nonmoving picture to be studied in detail.
2. Motion from videodisc	Shows things that move along with music, sound effects, and narration.
3. Still frame text from disc	Shows menu or instructions which can be studied leisurely.
4. Still frame text from computer	Allows text to be read at a leisurely pace. Computer-generated text has high resolution, making it possible to place more text on a screen at a time. Computer-generated text can be revised, debugged, and updated easily. It is generally easier to produce computer-generated text than it is to create a graphic of the text, tape-record the graphic, and then have it become a permanent part of the disc. Computer-generated text is just a one-step process.
5. Still frame graphics from computer	Here the computer can create a sharp, colorful chart, graph, or line drawing. The advantages of computer-generated graphics are the same as for text.
6. Computer animation	Here, the computer strobes through three or four still graphics in order to simulate motion, or the computer makes calculations to move a figure across the screen (like in video games). The computer-generated animation is sharper than videodisc pictures and can easily be updated or changed. The computer animation can be more INTERACTIVE as it responds directly to the viewer's commands. In this way, an object can be moved a million different ways around the TV screen yet not use up valuable videodisc time.
7. Step frame	If there is too much text to appear in one picture, the viewer may step the player forward (or backward) to see other "pages" of text. Besides paging through text, the viewer could page through pictures, backward or forward, skipping over unimportant ones and studying certain ones at length.
8. Stop motion	Allows the student to study in detail something which is too complicated to comprehend in motion. A medical student, for instance, could view a dissection and stop the disc at a certain point to view the position of arteries and veins.
9. Slow motion	Allows the viewer to slow down and study actions which happen quickly such as a golf swing, a triple somersault, or a magic trick.
10. Still frame with audio	This feature requires extra equipment but greatly expands the capability of showing many still frames with audio. For instance, hundreds of lunar photographs can be studied in detail, each with a narrative of what to look for. Artwork, circuit diagrams, and other nonmoving subjects are well suited for still frame audio.
11. TOUCH SCREEN or LIGHT PEN	Instead of typing an answer on a keyboard, the viewer touches the TV screen or presses a LIGHT PEN somewhere on the screen in response to a question. The question could be as simple as "Touch the trapezoid" or as complex as "Point to the part of the X-ray showing the arterial blockage."
12. Two- or four-track audio	Track 1 could contain the English version of instruction, while track 2 could be in Spanish. Track 1 could contain the music, while track 2 could contain an explanation of the style. Track 1 could play the background sounds and conversations of a nurse examining a patient, while track 2 could narrate the nurse's observations and thoughts. In each case, the student could listen to track 1, track 2, or perhaps both. A computer might even select which track the student hears depending on the student's responses to past questions. Some models support two additional digital audio tracks.

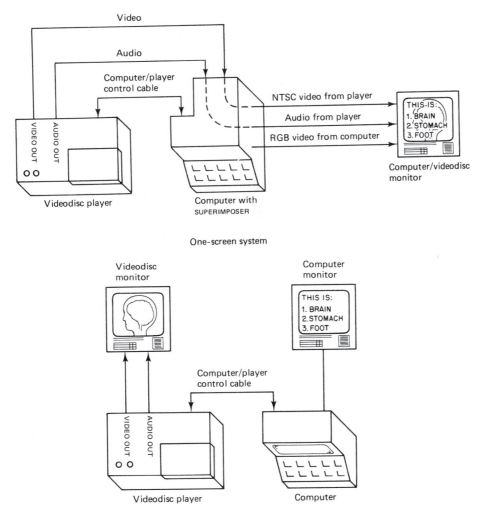

FIGURE 18-7 One-screen and two-screen LEVEL 3 systems.

TOUCH SCREEN or a LIGHT PEN, the student then points to some part of the kidney. The computer would sense which area was being pointed to and would then SUPERIMPOSE a circle around that area, and if the answer were correct, the computer would also display the words "Correct; you have identified the islets of Langerhans, the area of the kidney which produces insulin."

In both of the preceding examples, the videodisc producer has to know what equipment is going to be used to display the program. In the latter example, if a TOUCH SCREEN or LIGHT PEN were not going to be used, the example might be done like this: The videodisc produces a still of the kidney. The computer SUPERIMPOSES text asking "Which of these areas produces insulin?" and draws four circles, labeled A, B, C, and D, on the kidney. Here the student

types in his or her choice, and the computer erases the unselected choices from the screen while telling the student whether he or she is correct.

If a two-monitor system is used, then the image from the videodisc player must already contain the four circles with the letters A, B, C, and D. The computer monitor then asks the question "Which area of the kidney creates insulin?" The student then looks over to the videodisc monitor, makes a choice, and types the appropriate letter into the computer keyboard.

Costs

When you produce a video tape, your expenses are finished when you have edited your master and made your copies.

MINIGLOSSARY

Superimposer   Circuit that lays computer text over videodisc scenes so that both can be viewed together on the same TV monitor.

FIGURE 18-8 Computer graphic SUPERIMPOSED over videodisc image. Overlay highlights "help" key.

The same is also true for record-it-yourself videodiscs. When producing a regular videodisc, the kind that is made in a MASTERING PLANT, you face the additional mastering expenses:

Service	Cost
Levels 1 and 3	
Conversion of tape to master disc (CAV)	$1800 per side
Conversion of tape to master disc (CLV)	2500 per side
Disc copies (1–99)	18 per side
Disc copies (1000–1999)	10 per side
Level 2 (price includes two digital dumps)	
Conversion of tape to master disc	$2500 per side
Disc copies (1–999)	18 per side
Disc copies (3000–4999)	6.75 per side
Additional digital dumps (keypunching of computer data which are to be recorded on the disc; when the disc plays, the computer program plays into the videodisc player's built-in computer, programming it)	225 per dump

As you can see, it is a significant hardship to produce only a few copies of a videodisc. The mastering expenses are substantial. However, if you can spread those costs over many inexpensive copies, the videodisc then becomes cheaper to distribute than video tape.

Production Strategy and Instructional Decisions

In Chapter 17 you learned the first steps to planning an instructional TV program. Those steps still apply:

1. *Assessment.* What problem are you trying to solve? What are your goals?
2. *Audience.* Who will watch this program? Will they be beginners or experts, a captive audience or passersby?
3. *Media selection.* Do you show a simplified diagram or a photo of the real thing? Do you use motion or stills?
4. *Script and storyboard.* What will be said? What will be shown?

Whether you are producing a simulation, a lesson, or a game, the preceding steps of production planning still apply. If you stop there, however, you will end up making a LINEAR videodisc, something that plays like a plain old video tape. You have two more dimensions to work in when using videodiscs:

1. INTERACTIVITY
2. COMPUTER ASSISTANCE

If your disc is going to be INTERACTIVE, then make it INTERACTIVE from the start. After a few seconds of splashy action (the old "grab 'em by the throat"), ask your viewers a question which they must answer. This immediately wakes them up and puts them in the right frame of mind to *participate* in the program. This first question should be very easy, giving your viewers confidence both in themselves and in the fact that the machine can actually work.

Continue the INTERACTIVITY throughout the show. Bite-sized chunks of information may only take 20 seconds to present, a minute at the most. The great power in INTERACTIVITY is that you feed your viewers a new situation or piece of information *and then let them respond.* Don't try to feed them 25 pieces of information. This is not a video tape where everything had to be taught in a single unified program followed by a comprehensive test. With videodiscs you can do one thing at a time.

No law says that videodiscs have to be serious or dull. Surprise and spontaneity are enjoyed just as much by vi-

MINIGLOSSARY

*Flowchart   A diagram, mapping out the events, actions, and branches a program can take.

*Branch   A step in a flowchart or CAI program where a choice is made, and the viewer follows one of several alternate routes through the program.

*Menu   A list of choices, possible answers to a question, or a table of contents. The option selected is the one the player goes to next.

deodisc users as by moviegoers and classroom students. In a chemistry simulation, if the student chooses to add a beaker of ground-up sodium metal directly to water, then some juicy fireworks should occur, followed by a shot of a smoking lab coat draped near a hole in the roof. If a nurse miscalculates and selects an intravenous solution with too much water, the result should be a cartoon of a bulging patient running for the bathroom. Just the opposite calculation could show a cartoon of the patient all shriveled up, on a desert, lapping the last drop from a canteen.

Not only the mistakes should be humorous (the student may choose the wrong answer just because it is more interesting to see), but the right answer should also show some flair. Imagine succeeding through a complex maneuver and after the last correct decision being greeted with the blare of trumpets, the pop of skyrockets, and a handshake from Dan Quayle awash in confetti. At the bottom of the screen, the computer might then add the tiny words "Need we say more?" Such a human-like response from the machine may warm the cockles of your viewers' hearts, making them feel that there is indeed a human behind all of this.

Flowcharts.   Unlike a LINEAR video tape which is a single string of events, the videodisc can go every which way. This is called BRANCHING. A BRANCH is an instruction telling the computer/disc player to diverge from its present sequence in the program and take another path. Like squirrels running along the branches of a tree, your viewer may select this route or that depending on his or her responses to questions.

Sometimes the BRANCHING is the result of a MENU, a list of choices like

Which vital signs would you like to assess?
1. Temperature
2. Pulse
3. Respiration
4. Blood pressure
5. I'm finished assessing vital signs

Sometimes MENUS are followed by other MENUS, allowing the viewer to narrow down a particular topic step by step. The auto mechanics student may, for instance, see the MENU containing

1. Engine
2. Cooling
3. Electrical
4. Exhaust
5. Suspension
6. Emission controls

Upon choosing item 3, the student may then see the MENU

1. Fuses
2. Gauges
3. Ignition
4. Alternator
5. Starter
6. Lights and buzzers

After choosing item 4, the viewer gets the final MENU for the alternator:

1. Rotor
2. Stator
3. Bearings and lubrication
4. Tensioning adjustments
5. Measurements and specifications
6. Troubleshooting

Thus, by making three decisions, the student may view the troubleshooting procedure for an automobile alternator.

Some MENUS can take you around in circles. You may choose to see one thing and afterward come back to the same MENU and have the opportunity to choose another thing.

All these routes have to be planned, and the planning is much more complex than simply writing a story that starts at the beginning and ends at the end. An INTERACTIVE videodisc is a story that starts at the beginning and can go anywhere and everywhere. Keeping track of all this is where the FLOWCHART comes in.

Figures 18–9 and 18–10 show examples of FLOWCHARTS. A FLOWCHART is perhaps the most important step you take in planning your program. It is the skeleton upon which all other components will hang. This is where you choose what will be shown and to whom. Here you must decide what MENUS will be shown, what answers accepted, what segments reviewed, and what data collected from the viewer. You'll also choose whether a segment is to be a videodisc motion, videodisc still, videodisc text or graphic, or computer text or graphic and whether there will be sound and which of the two possible sound tracks you will use. Decisions made at the FLOWCHART stage are far-reaching and therefore require close attention from all members of the production team.

Remember that this FLOWCHART is not meant to list the *order* of segments recorded on the disc (except for BASIC and LEVEL 1). The order of scenes on the disc bears little resemblance to the order in which they will be viewed. *Viewing* order is what your FLOWCHART is establishing.

FLOWCHARTING is an analytical process. You have to

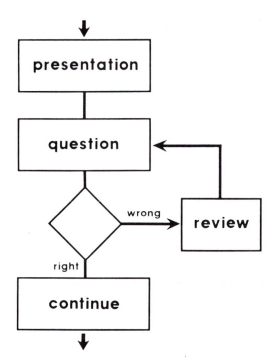

FIGURE 18-9 A FLOWCHART maps out the events, actions, and branches that a program can take.

think of *everything*. You have to test every option, twice. The chart must explain itself so that the computer programmer can translate it correctly and unambiguously into a computer program.

In Figure 18–10, you will notice that different symbols are used for various segments. There are standardized symbols (decision diamonds, computer screen output rectangles, etc.) established by computer programmers, and you may have to invent some of your own (videodisc motion, stills, record the student's response, and digress to subroutines).

Let's read the FLOWCHART in Figure 18–10 together to see what it is telling us: Each step on the chart is numbered for easy reference. The student enters at step 3007 and is shown a videodisc still frame having a MENU with nine choices on it. If the student chooses item 4, "Examine the patient's dressing," the computer tallies this response ("input choice") and then sends the student to step 3016, a motion sequence showing the patient's dressing being checked. When the sequence is over, the student is brought back to the MENU again and asked if there was something else he would like to see. If the student chooses item 2, "Examine patient's vital signs," another videodisc still frame appears with a MENU showing four vital signs to be examined. If the student selects item 1, "Temperature," he will see a motion sequence of the patient's temperature being taken. After that, the student will see the "vital signs" MENU again and at this point may choose "Pulse." This takes the student to segment 3022, which is a videodisc still

of the patient's pulse being taken and the notation on the screen "Pulse 96." The student is then taken back to the MENU and asked if he wishes to see more vital signs. If finished, the student selects choice 5 and exits the loop, going back to the main MENU 3007. Notice that every time the student chose to examine something, the response was recorded by the computer ("input choice"). This information will later help the instructor determine if the nursing student took the most efficient route in caring for the patient. If the student feels he is finished assessing the patient's condition, he may choose item 9 on the MENU and begin a new series of segments starting with 3019. Here the videodisc still frame asks the student to make a nursing diagnosis. The computer stores the student's typed response. The computer in step 3029 then asks the student to substantiate his or her choice and again stores the data. Next comes another videodisc still frame containing another MENU. The student is asked to choose an action. Note that two of these choices result in the student seeing the patient breathing nicely at 2 p.m. followed by a congratulatory still frame which exits the student from this part of the program. The other answers are not so rewarding. Students who choose "Call the doctor" see a videodisc still frame of a golf course with a subtitle "The doctor is out; you can handle this yourself." (Nurses are taught that they can't go running to the doctor for every decision, especially ones that they should make themselves.) All the "wrong" choices lead to a motion sequence of the patient coughing, followed by a still frame telling the student nurse that he goofed. The student then returns to the MENU at the top and starts all over, perhaps this time searching out the right information, making the right assessment based on the clues given, and providing the best care for the patient.

Notice that the INTERACTIVITY of the videodisc in this example simulated a real-life situation. No patient had to be harmed by a student nurse's mistake. The student's errors wasted only time and electricity.

Care must be taken in transitions from scene to scene. Unlike a LINEAR teleproduction where the director knows which scene precedes and follows the scene at hand, with INTERACTIVE videodiscs, the viewer can enter a scene from a multitude of places. Will this movement from scene to scene always make sense? In the nursing videodisc, the first time the nurse enters the patient's room, the nurse says to the patient, "Hello, Mrs. Morris. . . ." What happens when this scene is followed by a decision, an action, and a route back to the patient's room? Is it silly to repeat "Hello, Mrs. Morris . . ." if all the nurse did was to digress a moment to check the patient's chart?

Another continuity trap occurs when time progresses during a simulation. If something happens at noon and later something happens at 2 p.m., is there an unnoticed loop which might accidentally take the student back to noon the same day? If so, we've invented a time machine. If the

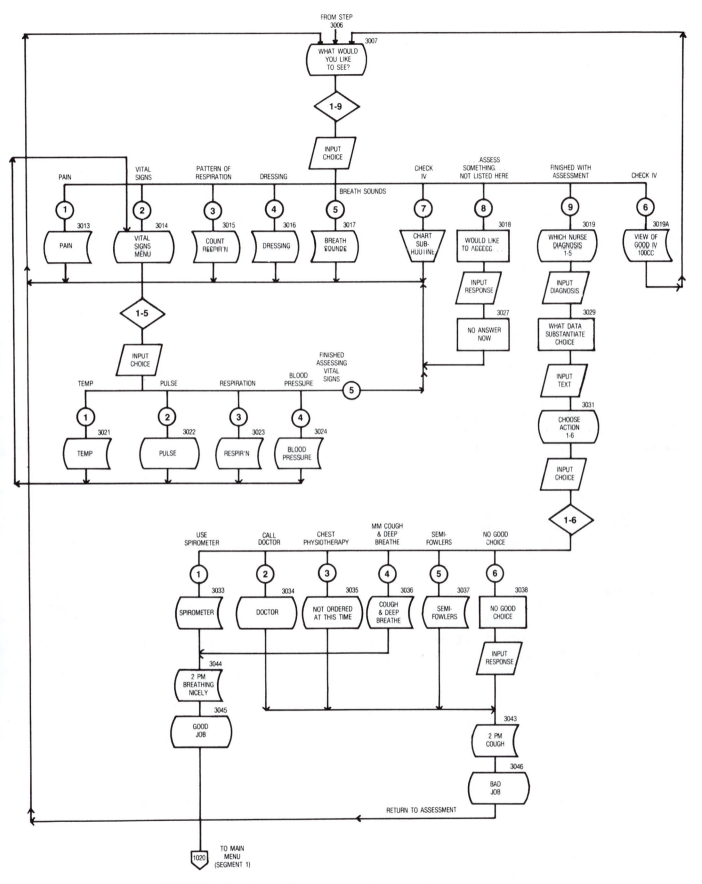

FIGURE 18-10 Part of the FLOWCHART for a LEVEL 3 nursing videodisc.

student is to look at the patient's chart, he had better not see entries for tomorrow when looking at it today. The intravenous fluid flowing in the left arm in one scene had better not be in the right arm in the next scene—unless you changed it, and then it darn well had better stay there. Details, details.

Scripts and Storyboards.   This is the flesh we attach to the FLOWCHART's skeleton. For each little box on the FLOWCHART, we now make a page of script (and maybe a storyboard).

Storyboard the scenes with computer text or graphic superimposition in mind. Will the text be in the way? Will it show up? (Avoid white letters on white backgrounds.) Will the MENU of videodisc function controls (still, backward, play, scan) be showing? These prompts may be necessary if the student is to take control of the videodisc playback at any time. Such control offers the viewer freedom to skim the well-understood, review the misunderstood, and slowly observe the hard to understand or play a second audio channel with detailed explanations of the scene being shown. Notice how you, the producer, must be thinking in five dimensions.

Where a television script had generally 3 columns (audio, video, narration), your multidimensional INTER-ACTIVE videodisc script may have 5 or 10 columns, certainly more than will fit on a normal page. The videodisc producer must develop a script format which will have a place for the sounds on audio channel 1 (perhaps the sounds of the action taking place), the sounds on audio channel 2 (perhaps a narration explaining what is happening), the text or visual coming from the videodisc player, the text or graphics coming from the computer, plus any other useful information like whether the videodisc is in motion, still-framed, or still-framed with audio. The script is now so multifaceted that it needs a special format like the one shown in Figure 18–11.

Notice how those segment numbers now come in handy: Each page of the script is clearly keyed to one of the steps on the FLOWCHART. While the FLOWCHART shows the over-all view of what's happening, the script's pages show specific detail.

Technical Production Decisions

What follows are some guidelines on how to produce a tape that becomes a good videodisc.

Organizing and Preserving Real Estate.   Videodisc REAL ESTATE has nothing to do with buying and selling swampland in Florida. REAL ESTATE is simply the usable space on the disc.

Making the most out of a little space: Since a CAV disc runs only 30 minutes, you have to be careful how you use the time. If something doesn't move, for instance, you will want to use a still frame if possible rather than using 30 frames per second to show the same thing.

If sound is necessary, you may have no choice but to use motion (even if the picture is still) to get the sound to play out of your machine. If you expect to use a lot of still frames with sound, then you may need to employ the still frame audio techniques described earlier which ''grab'' a still frame or compress the sound into just a few frames.

Although you can put text on videodisc still frames, the process requires a lot of time-consuming editing. Would it be better to have the computer generate the text? The advantages of this technique would be the following:

1.   The text is sharper.
2.   It takes up only a minor amount of REAL ESTATE on the disc.
3.   The text is easily changed or updated simply by changing the computer program.
4.   Editing together dozens of still frames onto your video tape becomes no longer necessary.
5.   You can have sound with your text. If a videodisc still frame were used, you might not be able to have audio. However, if computer text is shown on the screen, there is nothing that stops you from having the videodisc play audio. Where does this audio come from? It comes from a sly trick: If you have a motion sequence somewhere on the disc which uses only one audio track to go with it, then you have a spare audio track not being used. Fill this spare audio track with the sound that you would like to have go with your computer text. When the ''real'' motion sequence is accessed, the computer is told to listen only to channel 1 (the real sound). When the segment showing the computer text appears, the computer goes back and plays only audio channel 2, the desired narration (and not the picture from the disc). In this way, no space is wasted. Every scrap of sound track on the disc gets used in one way or another. The sound track does not have to be narration; it could be music to accompany computer-generated graphics or text.

MINIGLOSSARY

Real estate   The space on a videodisc available for your program.

PROGRAM TITLE <u>Recognizing Appendicitis</u>   SEGMENT NUMBER <u> 67 </u>

SEGMENT TITLE <u>Probing</u>

✔Audio 1
✔Audio 2
— Video motion
— Video still
— Computer text
— Computer graphic

Doctor's voice: I'M GOING TO PRESS DIFFERENT PLACES ON YOUR ABDOMEN NOW. YOU TELL ME IF IT HURTS ANYWHERE.

Silence during probing, then...

Patient: OH, OOH. YES, THERE!

— Audio 1
— Audio 2
✔Video motion
— Video still
— Computer text
— Computer graphic

Shows patient's abdomen as doctor probes.

Patient flinches when probing reaches left of groin.

— Audio 1
— Audio 2
— Video motion
— Video still
✔Computer text
— Computer graphic

Text at bottom of screen:

PRESS R TO REPEAT, Q TO QUIT NOW, C TO GO ON.

Notes to programmer: End motion on still frame 5944

Start frame <u>5306</u>

End frame <u>5944</u>

FIGURE 18-11 Sample LEVEL 3 INTERACTIVE videodisc script.

For best results, use white or yellow lettering with edging over a gray or pale blue background. If you are working with a large title which will not be hard to read, you may even have the title superimposed over a picture from the videodisc. Soft, out-of-focus shots of generic things like a hospitals or laboratories make nice backgrounds for computer-generated text.

Where to put things on the disc: The order that you put things on the disc doesn't really matter. The computer can access them in any order. You can have the end first and the beginning last if you'd like. However, careful organization of your REAL ESTATE can save *search time* as the player hunts for the desired sections.

Waiting for an INTERACTIVE program to hunt out a desired section is like waiting for a bus. It seems to take forever. You want your videodisc to have fast response time. To do this, things which are likely to go together should be positioned next to each other on the disc. Where it may take 1.5 seconds for the player to jump from frame 100 to frame 53,000, it may take only .1 second to jump from frame 100 to frame 500. If a certain part of the disc is to be referred to frequently, like a chart or sound effect, it may be wise to place it in the middle of the disc, close to everything else.

If you expect to make revisions to your program *and*

you are not using the whole 30 minutes available on the disc, it may be wise to leave one or two "islands" of black on the disc. These are places where you have recorded pure black and no sound for a minute or so. Later, when you need to revise the disc, you can go to your master *tape* and insert the new sections into these "blacked" areas. You don't have to reedit the entire tape. Because these areas are interspersed throughout the disc, you'll be able to place the new material fairly close to associated material so search time will be minimized.

Premastering. MASTERING was the process of making the first videodisc from your original edited video tape. PREMASTERING is the process of making that edited video tape. Your PREMASTER will include motion and still sequences, audio, coded information numbering all the frames, PICTURE STOPS, CHAPTER STOPS, LEVEL 2 computer "dumps," etc., ready to be transcribed into a videodisc.

Although various media can be used for this PRE-MASTER, ¾U and 1-inch video tape are most common. Of course, 1-inch tape gives a sharper picture than ¾U. If you expect to be going down several generations in your edit, 1-inch tape may be preferable even though it is more expensive. If however, you are editing only to the second generation, you might get by with ¾U.

Some people shoot their raw footage on ¾U and bump

it to 1-inch tape during editing. Others may edit on ¾U and make a 1-inch copy. The advantages here are the following:

1. The cheaper ¾U portable VCRs can be used for the shooting.

2. The 1-inch final version has a higher quality than a ¾U final version.

3. With 1-inch tape, you have a wider selection of MASTERING FACILITIES (a few don't accept ¾U tapes).

4. You can keep your master tape at home while sending a high-quality 1-inch dub to the MASTERING FACILITY.

This last step is important. If working in ¾U, you may be inclined to send your original master tape to the VIDEODISC MASTERING FACILITY in order to minimize generational losses. Sending them a copy of your master would bring you down one more generation, probably your third generation. Sending your master tape anywhere is always risky business. If you spend tens of thousands of dollars producing a tape, you don't want someone smashing it in the bottom of a mail pouch. However, this is a chance you usually end up taking in order to preserve picture quality. Using 1-inch tape, however, makes the generational losses small, allowing you to keep your master and send the dub (or vice versa).

There's another difference between 1-inch and ¾U PREMASTERS. All PREMASTERS must include SMPTE time code. One-inch tapes have two audio channels *plus* an extra cue channel for the time code. Many ¾U VCRs, however, must use audio track 1 for the time code, leaving only one audio track. If you wish to use two audio tracks, then that time code has to go somewhere else. Professional ¾U editors will allow you to place the time code on a separate cue track on the tape, solving this problem.

SMPTE: Your PREMASTER must be encoded with SMPTE full-frame, nondrop frame time code. You'll remember that SMPTE works two ways: drop frame and nondrop frame. The drop frame method keeps perfect track of *time* by throwing away an oddball frame number once in awhile. Nondrop frame SMPTE, however, doesn't tell time perfectly well (it gets off by about a second per hour), but it *will* keep track of every single picture on your tape. Because every picture is important (your computer may search for that particular picture someday), it has to be numbered. Thus, your PREMASTER has to be encoded with the nondrop frame time code.

The time code numbers you place on your tape do not turn out to be the same numbers that the MASTERING PLANT puts on the disc. They use your numbers to assign their numbers, but these numbers are not the same numbers. The

numbers are easily correlated, however. If you look at one frame from your tape and the same frame from the disc and find that the disc's frame number is 125 greater than your tape's frame numbers, that's the way it will be throughout the rest of the disc. Simply add 125 to your tape's SMPTE time code, and you'll get the disc's frame numbers.

Leader: Most MASTERING PLANTS require you to

1. Start with 2 minutes of color bars with 100-Hz tone

2. Follow this with 40 seconds of black

3. Follow this with your program

4. Follow this with 30 seconds of black

Notice that 3 minutes and 10 seconds of your 30 minutes has been used up by the leader. In actuality, most people find that they can get 28 good minutes of program on a 30-minute disc.

Video: It is important that the PREMASTER tape have a perfect control track with no dropouts. You may wish to use a carefully "blacked" master tape before you begin editing.

Video luminance levels in the program should be at 100 IRE. Graphic fonts should be at 80 IRE (this keeps them from smearing and buzzing).

Try to avoid red and high-chroma (saturated color) backgrounds. On videodiscs, they tend to show up noisy and grainy. Use pastel backgrounds for text or, better yet, use gray.

Still frames: Producing still frames for a videodisc can often be a problem:

1. Most ¾U editing VCRs and controllers cannot edit 100% accurately. It is difficult to record *one single frame*.

2. If you are trying to show one frame of something in motion, there may be a blur. The batter's arms may be blurry as he or she swings at the ball. With a little careful searching, however, you may find one frame which is much less blurry than the others and use that as your still frame.

3. FIELD DOMINANCE mismatches may occur when you do single frame editing. Take a look at Figure 18–12. In the top examples, the video tape was edited with FIELD TWO DOMINANCE. For each single frame, field 2 (the even-numbered TV lines) came first. If the videodisc is also arranged with FIELD TWO DOMINANCE, when it searches out a frame, it will always use field 2 of the picture first. Thus, each videodisc frame will be exactly the same as each tape frame.

Now look at the bottom examples. Here, as before, the master tape used FIELD TWO DOMINANCE. Each frame edited on the tape started with field 2. If the videodisc, however, chooses to show a still frame which starts with field 1 (FIELD ONE DOMINANCE, starting each TV picture with the odd-numbered lines), field 2 may be from another

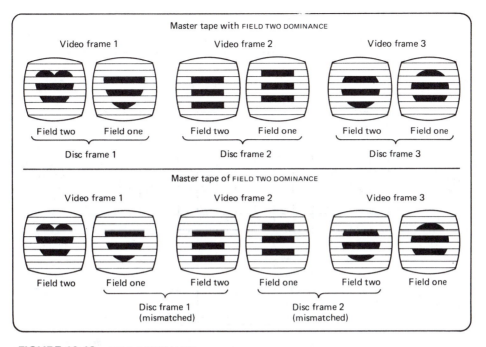

FIGURE 18-12   FIELD DOMINANCE.

picture. The result will be a flickering picture showing two things at once. Disc frame 1, for instance, may show a heart superimposed over a box. Disc frame 2 may show a box superimposed over a circle.

To avoid this problem, one must tell the DISC MASTERING PLANT to produce the videodisc using the same FIELD DOMINANCE which you used to produce the video tape. Include this information with your PREMASTER.

4. Movies can be a big problem. If you are shooting a movie, shoot it at 30 frames per second, and things will be easy because the pictures can be transferred to tape frame for frame (there are 30 movie frames per second and 30 video frames per second).

If, however, the movies are shot at the standard 24 frames per second, when they are transferred to video tape, an adjustment has to be made. Your television film chain makes this adjustment by using a special projector with a 3/4 PULLDOWN mechanism. The projector will show film picture 1 three times for 1/60 second each time. It then shows picture 2 two times for 1/60 second each. It then shows picture 3 three times at 1/60 second each. Thus, it takes 5/60 second to show two movie pictures, or 1 second to show 24 frames

of a movie. Figure 18–13 diagrams what happens next. Notice that video fields 1 and 2 show the same picture from the movie film. These two fields will become a good videodisc frame. Video fields 3 and 4, however, are made up of two different movie pictures. When they become a videodisc frame, they will flicker, superimposing movie pictures 1 and 2. Video fields 5 and 6 do the same thing, also creating a bad videodisc frame. Fields 7 and 8 show the same movie picture and thus create a good videodisc frame.

In short, when a 24-frame-per-second movie is transferred to video tape and then to videodisc, the frames are good, good, good, bad, bad, good, good, good, bad, bad, and so on. You don't notice the problem when the frames play quickly; the motion looks normal. But when you show a still frame, the bad ones will be blurry or show two images superimposed. There is no problem, of course, if the motion picture shows something that isn't moving, but this is seldom the case. To avoid the problem, you have to do some fancy calculations and pick exactly the right still frame to show.

All the preceding still frame problems can be solved with one trick: *never record just one still frame.* Always

MINIGLOSSARY

Field dominance   A determination of which field (the odd or the even) is used first when a videodisc player creates a still frame from two video fields.

Field one dominance   Attribute of a still frame using the odd field as the first of two fields which comprise the whole picture frame.

Field two dominance   Attribute of a videodisc still frame which

uses the even field first, and then the following odd field to create a still frame.

Three-two pulldown   The motion taken as a movie projector plays a 24-picture-per-second movie into a 30-frame-per-second TV system. One movie picture is scanned three times by the TV system and pulled down, and the following picture is scanned twice; then the process repeats.

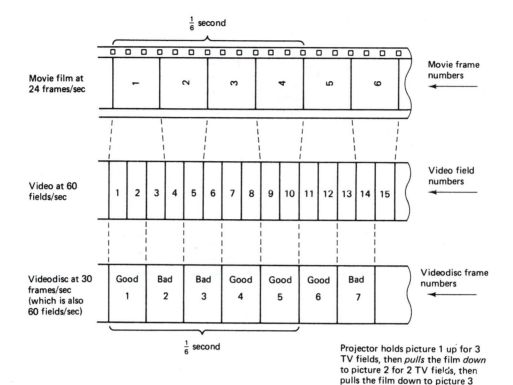

FIGURE 18-13   3/2 PULLDOWN.

make it five single frames. This way you'll always get a good one, no matter what. When your videodisc comes back from the MASTERING PLANT, you can examine it for just the "right" still frame and tell your computer the frame number to access when the time comes.

Of course this method greatly reduces (by a factor of 5) the number of still frames you can get on a disc. Instead of 54,000 frames, you now get only 10,800. For most videodisc producers, this is not a problem. They would rather waste a little space than have something go wrong . . . go wrong . . . go wrong.

Transitions: In a normal video tape, you know which scenes will follow what. You can plan your transitions from scene to scene. In the videodisc, however, the order of the scenes depends on the choices of the viewer. For the scenes to always mate smoothly together, they need some generic beginning and ending which can mate with any other scene easily. Here are some ways to do this:

1. Plan your scripting so that hellos, good-byes, and one-time-only salutations are at the true beginnings and ends of sequences and not in the middle where they are accessible from other parts of the program.

2. Leave a second of audio silence before each scene begins. Viewers don't mind seeing the image on

the screen cut from one scene to the next, but they are jarred by hearing the speech start immediately. Without this moment for the viewers to orient themselves, the series of scenes seems choppy. This tiny pause makes things comfortable and "natural."

3. If you decide to use a fancy transition such as a wipe, dissolve, or defocus-focus, then use that same transition between all the scenes in that section of the program. Put another way, make it so that no matter how the viewer branches, they will always see the same transition going out of one scene as they see coming into the next.

4. Remember that the computer can help make transitions too. It can cover the screen with blue (handy while the player is searching for the next segment) and then wipe the blue off the screen. With the help of certain programs, it can do fancy effects between scenes. Since it is the computer that is doing the effects, the effects are not part of the disc; the disc simply cuts from scene to scene every time. The computer, on the other hand, could wipe a logo or title across the screen and then wipe it off when the next scene starts.

Scheduling: Once you've made your FLOWCHART, script, and storyboards, you can plan your shooting much the same as you always have. If you can predict when the shooting and editing will be complete, you may want to reserve a date for your videodisc to be created at the MAS-TERING PLANT. If you waited until your production was entirely finished and you had a PREMASTER in your hand ready to go, you would find that the MASTERING PLANT would probably not be able to start working on your order for 3 weeks or so. You can get a jump on them by placing a reservation early (if you can finish your tape on schedule).

For almost double the normal MASTERING fee, many plants will speed up the process, giving you next day service.

CONNECTING A VIDEODISC PLAYER

Home videodisc players connect between your antenna and your TV set like in Figure 18–14. When not in use, the player routes the antenna (or cable) signal straight to your TV (or VCR if that is part of your system). When turned on, the player will cut out the incoming antenna signal and substitute its own, usually channel 3 or 4. Many disc players have separate audio outputs to go to your hi-fi for improved sound. Some models also have video outputs for improved picture quality.

Industrial videodisc players are connected as in Figure 18–7. Sometimes the wiring is more complicated than is shown in the figure. The RGB video might be three coax cables, or it might be a single computer cable. The audio from the videodisc player is likely to be carried on two audio cables (left and right channels), *through* the computer, and to the TV monitor. The computer has the job of selecting whether channel 1 or 2 or both are played through the monitor. On some systems, an additional cable between the

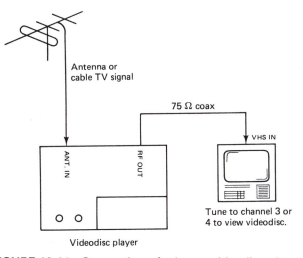

Antenna or cable TV signal

75 Ω coax

VHS IN

ANT. IN

RF OUT

Tune to channel 3 or 4 to view videodisc.

Videodisc player

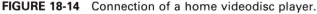

FIGURE 18-14   Connection of a home videodisc player.

computer and videodisc player will carry a special sync signal which helps the computer position its text over the TV screen while the disc is playing an image beneath it.

OPERATING A VIDEODISC PLAYER

To play a disc, the player first has to be loaded. With LV players, the power must be *on* before you can open the lid to insert the disc. The REJECT/OPEN button stops the platter and unlatches the lid.

Most industrial players have several speeds forward and reverse. PLAY is the only speed that has sound (except for the optional STILL FRAME AUDIO). Like VCRs, the players have a switch allowing you to listen to channel 1, channel 2, or stereo or a monaural mix of both channels.

At the back of most videodisc players, there is a CPU/MAN switch. In the CPU position, the external computer takes control of the videodisc player. The function controls on the player are inactive. In the MANual mode, the player responds to your commands directly.

If you discover that your TV screen rolls and twists into diagonal lines, there is probably another switch in the back of the videodisc player called INT/EXT SYNC which can solve your problem. In the INT mode, the player makes its own sync, and in the EXT mode, the computer controls it. If the computer isn't being used and the player is left in the EXT mode, there may be no sync to stabilize your TV picture. LEVEL 2 videodiscs and players have a switch on the keypad allowing you to display the videodisc frame number on the screen. If you have an index of where things are on the disc, you can use these frame numbers to find your place. Players with SEARCH features allow you to punch in the frame number of the desired spot on the disc (say frame 16624 or chapter 5), and the player will skip to that point in a few seconds and begin playing. This FRAME SEARCH feature is only available on CAV discs.

If using a CLV disc, you may find CHAPTERS indexed on the disc jacket. You can use this number to approximately access various parts of a program.

CARE OF VIDEODISCS AND PLAYERS

LV discs are pretty durable. They are stiff and quite insensitive to fingerprints, dirt, and *minor* scratches. They are brittle, however. Whack them, and they'll break. Mail them *only* in a protective sandwich of stiff cardboard.

If you lay your videodisc on an uneven surface for a long time, it may warp, making the disc unplayable.

Videodiscs and CDs can absorb dampness over time, making them unplayable. Keep them in a dry place.

MINIGLOSSARY

Chapter   One section of a level 2 videodisc program, like a chapter of a book.

Videodiscs are not harmed by magnetic fields.

If you *do* have trouble playing your LV disc, remove it from the player and wipe it down using a soft cloth and glass cleaner. Wipe CDs and LV discs *radially* (from the center outwards), not in the direction of the spirals. This way, any accidental scratches you make won't obliterate a whole spiral, but may only create a tiny ''tick'' or ''pop,'' easily hidden by the error-correction circuitry.

The fast-spinning disc inside the player acts something like a gyroscope. Have you ever tried to tilt a gyroscope? They resist movement mightily. *Don't ever attempt to move the videodisc player while the platter is spinning.* Its delicate innards are very likely to self-destruct.

VIDEODISC AND CD ROM DIRECTORIES

The *Complete Interactive Video Courseware Directory* from Convergent Technologies Associates (516-248-5984) lists about 250 courses. Cost is $60.

The *Videodisc Compendium* available from Emerging Technology Consultants (612-639-3973) lists available videodiscs.

The *Interactive Video Directory* available from Future Systems (800-323-DISC) contains 800 videodisc titles. Price is $65.

Future Systems also markets lists of CD ROM titles such as: *CD ROM Yearbook* published by Microsoft Press

for $79.95. Also there is *CD ROMs in Print* published by Meckler at $37.50.

NEBRASKA VIDEODISC DESIGN/ PRODUCTION GROUP

If you were looking for the center of the videodisc universe, you would point yourself toward Nebraska. There at the Nebraska Videodisc Group on the campus of the University of Nebraska—Lincoln, leading videodisc developers, educators, and users demonstrate their projects, techniques, and current research and evaluation, as well as discuss the implications and applications of video technology for the future. They have yearly symposia with tutorials, workshops, and demonstrations of CD-I, DV-I, computer graphics, and other interactive technology research. They also maintain a videodisc publications library with books, papers, catalogs, and periodicals devoted primarily to videodiscs and interactive technologies. For information contact:

Videodisc Design/Production Group
1800 North 33rd Street
Lincoln, NE 68583
Telephone 402-472-3611
FAX 402-472-1785

TV PROJECTORS

No little box can command one's attention like a big screen can. Also, no little box can be viewed as easily by as many people as a giant projected image can. Imagine gathering your group before a huge screen in a dimly lit room to see the NFL playoffs, a flight through the Grand Canyon, or alien invaders being blasted into hyperspace as laser blasts sputter from all directions. Imagine, also, somebody's giant dentures covered with stickum or dodging a torrent of blood as doctors try to control a hemorrhaging spleenectomy. Yes indeed, giant screen TV offers us a whole new television experience.

HOW TV PROJECTORS WORK

Essentially, TV projectors work like TV monitor/receivers except that they project the TV image onto a screen, much like a movie projector does. Since the images from TV projectors are usually not very bright, special projection screens are often used to efficiently bounce most of the light straight back to the audience.

The simplest and cheapest TV projector consists of a TV set with a bright picture tube. A lens gathers the light from the tube and focuses it on a projection screen. More expensive TV projectors usually have three tiny TV sets (one for each primary color—red, blue, and green—the mixture of which makes all the other colors), with three lenses, perhaps a mirror, and the projection screen.

TV projectors designed for theaters with audiences of more than a 1000 use a totally different system for creating the image. The GE and Eidophor models, for instance, form the TV image on a thin film of hot oil. A powerful projection lamp bounces its beam off the oil, through the projector lens, and onto the screen to make the picture.

Several laws of physics apply to all TV projectors:

1. When you make a TV projector's image larger, it becomes dimmer, because you are diluting the brightness over a larger area. Some TV projectors may start out brighter than others, but as you make any of their images larger, they will always become dimmer.

2. When images get bigger, they look fuzzier. A magazine picture which looks sharp from far away looks very fuzzy through a magnifying glass. TV projectors, since they magnify an image, also make it look fuzzier. Some projectors may start with a sharper picture than others, thus yielding a relatively sharper projected image, but no matter what you do, if you make any image larger, it will look fuzzier.

3. Different projection screens will affect how much light will bounce back to the viewers. A white MATTE screen, similar to a smooth white sheet, has a GAIN of 1. The projected light bounces in

589

all directions equally so that a viewer from the side will see as bright an image as a viewer standing near the projector. BEADED and LENTICULAR projection screens, commonly used in classrooms, have a GAIN of 2 or 3, which means that the brightness bouncing straight back toward the viewers near the projector is two or three times as strong as the brightness reflected to the sides. Special rigid screens covered with a curved foil-like surface may have a GAIN of 9, meaning that the image is nine times brighter along its axis than it is to the side. In short, the kind of screen used affects the brightness of the image you see and the viewing area covered.

KINDS OF TV PROJECTORS

Now let's study in more detail the different kinds of TV projectors you could buy or rent. Some come with built-in TV tuners for receiving broadcast TV. Others accept only straight video signals. Most consumer models work both ways, RF or video.

Single-Tube Projectors

Figure 19–1 shows a SINGLE-TUBE TV projector. They consist mainly of a television set (or monitor), a lens, and a box. Figure 19–2 diagrams a SINGLE-TUBE TV projection system.

Advantages of Single-Tube TV Projectors

1. They are cheap.
2. They're small and lightweight. One model is designed to be carried to small meetings.
3. By a simple adjustment of the projection lens, you can vary the distance between the projector and the screen, also making the image smaller or larger.

Disadvantages of Single-Tube TV Projectors

1. The images are quite dim, so dim that the viewing room has to be completely darkened.
2. The images can seldom be more than 3 feet across. As they get bigger than this, they become too dim to view.

FIGURE 19-1   SINGLE-TUBE TV projector (Courtesy of Sony Corp. of America).

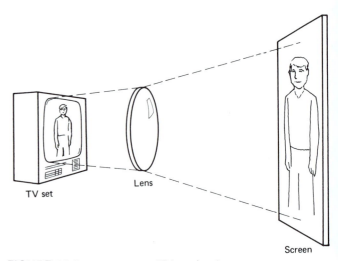

FIGURE 19-2   SINGLE-TUBE TV projection system.

3. The images are somewhat fuzzy. The projected picture can be no sharper than the picture from the color TV screen. Since the screen is made up of scanning lines and colored phosphor stripes, so is the projected picture. These lines and stripes are magnified when the image is projected. Also, depending on the quality of the lens used, the projected image may be a little fuzzy in the corners.

MINIGLOSSARY

*Matte screen   A nonglossy smooth white projection screen. Like a white sheet, it reflects light equally in all directions.

*Gain   A projection screen's reflectivity. The higher the gain number, the brighter the picture, because more light is reflected back toward the projector (but less light is reflected to the sides).

Beaded screen   Projection screen covered with tiny glass beads (looks like white sandpaper); has a gain of 2 or 3.

Lenticular   Gray, metallic-looking projection screen with tiny vertical grooves (looks like smooth corduroy) to distribute most reflected light straight back and to the sides (where the audience sits) and reflects very little on the ceiling and floor.

High-gain screen   Rigid, curved, foil-covered projection screen with a gain of 5 or more which yields a bright projection image, even in a well-lit room.

*Single-tube TV projector   TV projector with only one TV picture tube, usually recognizable because it has only one lens snout.

For the preceding reasons, SINGLE-TUBE TV projectors are used primarily by experimenters and home video consumers.

Three-Tube TV Projectors

TV projectors are a bright idea but not a bright medium. Considerable effort and investments have gone into making brighter (yet sharp) screen images. One trick is to use not one tiny TV screen inside the machine but three. Each little TV displays a sharp, superbright black-and-white picture which is then beamed through a colored lens which turns the image into black-and-blue, black-and-red, or black-and-green. These three primary colored images converge on the projection screen to create the entire spectrum of colors and a single picture. Figure 19–3 diagrams the process.

Since there are now three TV projectors in one box, the system costs about three times as much as a SINGLE-TUBE system. Prices start at about $2500 and range to $4500 for consumer models and up to $8000 for industrial models. Prices depend, to a large degree, on screen size, image brightness, and sharpness.

Advantages of Three-Tube Projectors

1. The image is fairly bright, satisfactory for a dimly lit room.
2. Fairly large pictures are possible, from 5 to 10 feet in width.

3. Images are quite sharp, unlimited by the phosphor stripes on the TV screen face.
4. There are numerous models to choose from with widespread availability.
5. A few models are portable (but still quite large and heavy).
6. With the help of HIGH-GAIN projection screens, images are viewable under average room lighting.
7. Some models can be connected to computers to display computer graphics and color.

Disadvantages of Three-Tube TV Projectors

1. Projectors are generally designed to work at only a certain distance from the screen, yielding only a certain image size. Moving the projector closer or farther from the screen makes it "cross-eyed" whereby the three colored images don't land atop each other to create a single image. Some models, however, do have optional lens systems allowing the projector to be used at a *second* unadjustable distance from the screen, yielding another image size.
2. If the projector is moved, it has to undergo a 2-minute setup procedure whereby the images must be focused and electronically CONVERGED (the red, green, and blue images are moved so that they land atop each other).

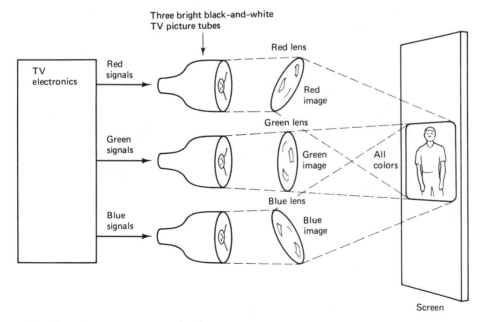

FIGURE 19-3 THREE-TUBE projection system.

MINIGLOSSARY

*Three-tube TV projector TV projector with three TV picture tubes for a brighter, sharper picture than one-tube models. Three-tube projectors are recognizable by their three lens snouts.

Convergence On a three-tube TV projector, focusing and aiming the three colored pictures so that they overlap, producing all colors accurately, without ridges along edges of objects.

Kloss Novabeam Model Two portable THREE-TUBE TV projector (Courtesy of Kloss Video Corp)

ages projected in a partially lit room will look dim unless a special HIGH-GAIN screen is used. This screen adds problems of its own. As shown in Figure 19–4, the rigid HIGH-GAIN screen bounces most of the light straight back to the viewers in the center of the audience. The image gets drastically dimmer as you move up, down, left, or right of center. The colors even change a little, becoming bluer or redder as you move from side to side.

Light-Valve TV Projectors

For theater, conference, and other professional applications, a heavy-duty LIGHT-VALVE TV projector is used. The Cadillac of LIGHT-VALVE projectors is the Ediphor, which costs about $400,000 and can make a bright 30-foot-wide picture large enough for 20,000 viewers. General Electric sells a $60,000 projector good for a 25-foot picture and maybe 5000 viewers.

3. The image, though brighter than a SINGLE-TUBE projector's image, is still fairly dim compared to movie projectors and regular TV sets. A 4-foot image shown on a MATTE screen in a totally darkened room will look fine. Larger images or im-

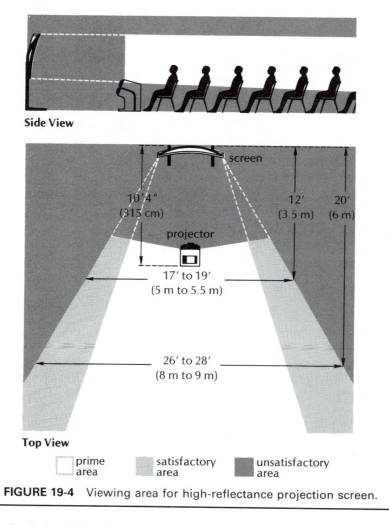

Side View

Top View

□ prime area   ▨ satisfactory area   ▧ unsatisfactory area

FIGURE 19-4   Viewing area for high-reflectance projection screen.

MINIGLOSSARY

*Light-valve TV projector   Professional TV projector capable of very bright, sharp, full-color pictures. Uses a projection lamp rather than TV picture tubes to create the light.

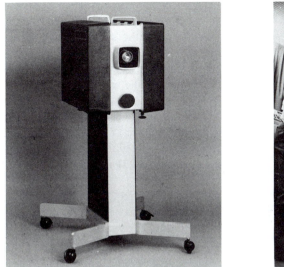 

LIGHT-VALVE projectors. GE P-J7000 (left) and Ediphor 5070. (Photos courtesy of General Electric, and Ediphor.)

Advantages of Light-Valve TV Projectors

1. Excellent resolution.
2. Excellent color rendition and registration.
3. Excellent brightness.
4. GE model is relatively compact and portable.
5. Images are sharp enough for computer graphics and text.

Disadvantages of Light-Valve TV Projectors

1. All LIGHT-VALVE projectors are expensive; even renting one can cost over $1000 per day.
2. The Ediphor projector (the brightest) is not readily available and is very large and difficult and expensive to transport and install.

One-Piece Versus Two-Piece TV Projectors

Anyone with a super-8 movie projector knows that you set up the projector at one end of the room and the screen at the other. The same goes for TV projectors; they must be some distance from the screen in order to project on the screen.

Two-piece units work as projectors and separate screens. The screen is usually against a wall, and the projector is usually in the middle of the room (or sometimes hanging from the ceiling). Although two-piece systems provide the best overall performance at the most reasonable cost, many people don't care to sacrifice a good chunk of classroom or conference room space for the projector. Some manufacturers have housed their projectors in fine cabinetry or shaped them into useful coffee tables while others have hung them from the ceiling.

In the single- and multiple-tube units, it is possible to bounce the image off a mirror and up against a screen which sprouts from the rear of the projection unit, making the projector/screen a single appliance. The mirror and screen sometimes fold into the cabinet, making the device smaller and less conspicuous when not in use (this also keeps kids' greasy fingerprints off the mirrors and lenses). Figures 19–5 and 19–6 show two-piece and one-piece projectors. (I tried to fix one once and ended up with a 187-piece projector).

FIGURE 19-5 Two-piece TV projector. Kloss Novabeam. (Courtesy of Kloss Video Corp.)

FIGURE 19-6   One piece, THREE-TUBE TV projector (Courtesy of Sony Corp. of America).

Two-piece projectors can easily get out of alignment and out of focus if the screen or projector is moved. Not only do each of the three projection lenses have to be accurately focused on the screen, but their images must converge precisely (otherwise you get colored ridges around contrasting objects in the picture). Some users mark their floors with tape so that when their projector gets bumped, they know where to move it back to.

One-piece projectors have to be aligned and focused too, but once this is done, it pretty much stays done because all the parts are fastened together or are internal.

Rear-Screen Projectors

A REAR-SCREEN TV projector (Figure 19–7) is another type of one-piece projector. Instead of reflecting the TV image off the front of a white or silver screen, the projector focuses the image on the rear of a smoky screen.

COMPUTER PROJECTION

How many times have you seen a class of 30 students trying to learn computer programming or database management from an instructor on a computer having a single 9-inch television monitor? Sixty eyeballs all crowd around one tiny

TV screen trying to view a miniscule cursor dancing about leaving itty-bitty dashes of text behind.

Now imagine the same computer display blown up on a screen 5 feet wide. Not only can everyone see without standing on their tippy-toes, but they can actually read what is being typed onto the screen.

In Chapter 2 you undoubtedly remember that there is a difference between normal TV monitors and computer monitors. Normal video monitors reproduce a fairly fuzzy NTSC picture compared to computer monitors, which produce a very sharp RGB image. The same is true for TV projectors.

TV projectors costing around $4,000 are designed primarily for the home and don't take well to computer signals. If you're blessed with an Amiga or some other computer that makes NTSC composite video or RF, then you're home free. Otherwise, things get complicated. For your TV projector to work with your computer, the following must happen:

- *Connectors*. The projector needs an input socket matching your computer's monitor output. Usually this is a 9-pin D connector.

- *Scan frequency*. The projector must be able to scan at the same frequency as the computer. Ordinary NTSC composite video sweeps 15,735 scan lines across the screen per second. Many computers,

FIGURE 19-7   REAR-SCREEN TV projector (Courtesy of NEC Home Electronics, USA).

especially the expensive ones that do high-resolution graphics, use frequencies much higher than that. Industrial multiscan projectors can handle the higher frequencies.

- *Resolution.* The resolution of the projector must be high enough to match the computer. A word processing computer puts 80 sharp characters on the screen, but your NTSC-resolution projector will display only about 40. Some computer programs can switch their output to a lower resolution 40-character mode.

- *Aspect ratio.* The computer-screen aspect ratio must match the projector's aspect ratio. Some computer pictures measure 3:2 rather than NTSC's 4:3. Others make tall pictures, like an 8½″ × 11″ page.

- *Safe title area.* Some computers print their text all the way out to the very edge of the RASTER (the active part of the TV picture). If your TV projector can underscan, this should cause no problem; all the text will appear on the screen. If the monitor or projector cannot underscan, then you are likely to lose the first and last characters in a line and perhaps the top and bottom lines of the computer display.

- *Interlace.* NTSC TV pictures are comprised of 525 scan lines each 1/30 second. First, 262½ even-numbered lines are "painted" on the screen, and 1/60 second later 262½ odd-numbered lines are scanned in between them. This is called 2:1 INTERLACE. To make sharp, jitter-free images, computers often skip this even-odd step, and scan all the lines onto the screen at once. To display an image, the computer and projector must match INTERLACE.

- *Component vs. composite.* Computers generally make RGB video signals; a separate wire carries the signal for each primary color. Common TV projectors and monitors accept composite video; the colors all travel on the same wire. For the two to work together, either the computer (or an external device) must ENCODE the RGB signals into composite video (making it somewhat fuzzy in the process) or the projector must accept RGB signals directly.

- *Signal strength.* The projector must accept the signal voltage put out by the computer. Many computers use TTL (Transistor-Transistor Logic) which creates a limited number of bright, pure colors using 4- or 5-volt RGB signals that are either all the way "on" or "off." ANALOG RGB, on the other hand, varies the voltage between 0 and .7 volts, permitting an infinite variety of colors.

Other things such as sync and voltage also have to work out for your TV projector to be compatible with your computer. If the preceding list hasn't totally discouraged you, take solace that converter boxes made by Extron, Covid, Inline, and Shintron can usually conform one signal into another making the computer/video marriage possible. Table 19–1 reviews some of the computer/video signal differences.

Computer Data Displays

There is an easier way to blow up computer images (without using dynamite), the COMPUTER DISPLAY PANEL, or LCD PANEL. About the size of a book and costing $700 to $2,500 (depending on monochrome or color) this LCD (liquid crystal display, like the numbers on a digital watch) device is placed atop an overhead projector. It plugs into the computer's monitor output and a window in the device displays the computer data. The overhead projector shines through the little window projecting the image onto a screen.

Advantages of LCD PANELS:

1. They are cheaper than video projectors.
2. They are designed for computer use, so they work in many computer-type appliations.
3. They are small and lightweight.
4. They make a brighter image than most TV projectors. (Image brightness depends on the overhead projector brightness.)
5. Like an overhead projector, they can be focused at nearly any distance, permitting various-sized images.

Disadvantages:

1. Image lag makes them poor for motion video.
2. Color models display limited colors.
3. Some have to work with certain computers or boards (ie., VGA, EGA, CGA, and so on).

MINIGLOSSARY

Raster   The scanned, illuminated part of the TV picture.

TTL   Transistor-transistor logic, an all-on or all-off signal put out by some computers and usable by some TV projectors.

Interlace   Method of scanning video lines onto a TV screen, either consecutively, or first odd-numbered lines then even-num-

bered. NTSC uses the latter method, called 2:1 interlace.

Color encoder   Device that converts R, G, and B color signals into composite video.

Computer display panel   LCD device which when placed on an overhead projector can project computer data.

TABLE 19–1

Computer versus video signals

Video system or card	Format	Horizontal	Vertical	Interlace	Aspect ratio
Apple Macintosh:					
High resolution mono or color	Digital	35 kHz	66.7 Hz		4:3
2-page mono	Digital	68.7 kHz	75 Hz		4:3
portrait	Digital	68.85 kHz	75 Hz		3:4
Apple IIGS	Analog composite & RGB	15.7 kHz	59.5 Hz	2:1	1:1
Commodore Amiga	Analog composite & RGB	15,734 kHz	59.94 Hz	2:1 or 1:1 switch-able	1:1 or 4:3 switchable
IBM PC, XT, AT, and PS/2 compatible systems:					
Monochrome adapter	I*	18.4 kHz	49.8 Hz	1:1	1:1
CGA (Color Graphics Adapter)	Digital RGBI	15.7 kHz	59.92 Hz	1:1	1:1
EGA (Enhanced Graphics Adapter)	Digital RGBI	21.8 kHz	59.7 Hz	1:1	1:1
VGA (Video Graphics Adapter)	Analog RGBI	31.5 kHz		1:1	1:1
PGA (Professional Graphics Adapter)	Digital RGB	31.5 kHz	60.0 Hz	1:1	1:1
Composite NTSC and RGB video signals:					
RGB	Analog RGB	15.75 kHz	60.0 Hz	2:1	4:3
NTSC	Analog encoded	15.734 kHz	59.94 Hz	2:1	4:3

*I—intensity

4. Their resolution is sometimes limited.
5. Recent $5000+ models solve the above problems.

Superimposers

In Chapter 18, we saw that some TV monitors could be connected to a computer and a videodisc player at the same time and display the videodisc player's NTSC signal and the computer's RGB signal over the top of it. This way, computer text could be shown on the same screen as a videodisc picture. Some TV projectors are equipped th[e] same way, allowing you to display either NTSC video[,] computer text, or both at the same time. Optional conversio[n] circuits, costing about $200 each, can be purchased whic[h] will mix NTSC video with many types of computer signal[s] and send the mixed result to the projector for display.

WHEN TO USE A TV PROJECTOR

A television monitor can generally serve an audience of [up] to 20. Larger-screen models may serve up to 30 viewers[.] If you have more than 30 viewers, you have two options[:]

1. Use several TV monitors.
2. Use a TV projector.

Multiple TV monitors are fairly easy to set up. Yo[u] merely feed power to them and loop a video and audi[o] signal from one to the other. Similarly, you could send R[GB] to several monitors at once. In each case, you still end u[p] routing cables all over the place, dividing the attention [of] your audience among several TV monitors. TV monitor[s,] however, show very bright pictures, even in well-lit clas[s-] rooms. TV projectors, however, generally require a dim[ly] lit or darkened room.

TV projectors have the advantage of providing o[ne]

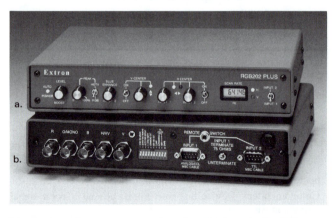

CONVERTER changes computer signals to RGB for TV projectors. (Courtesy of Extron)

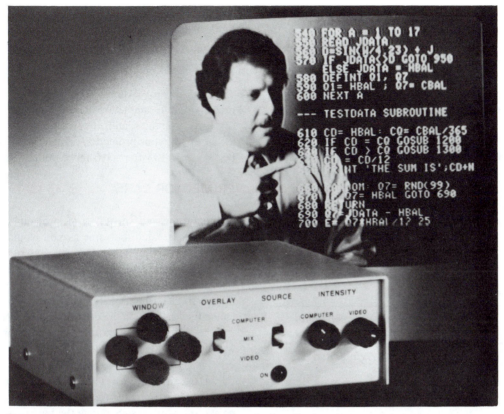

Computer/video combiner mikes both signals for display on one screen. (Courtesy of Quality/Universal)

CD PANEL displays computer data using overhead projector. (Courtesy Telex Communications)

central, easy-to-see image for the viewers. The audience's attention is not divided or distracted.

Computer displays are easier to see when projected on a large screen rather than squeezed onto a tiny TV monitor. If selecting a projector for displaying computer programs and text, a TTL model will probably be satisfactory. If displaying computer graphics or RGB signals from a camera or other source, a wide range of colors may be necessary, requiring an ANALOG model.

TV projectors don't make silk purses out of sows' ears. The semifuzzy picture coming from a ¾U VCP or the fuzzier picture coming from a ½-inch VCP will appear noticeably fuzzier when projected onto the big screen. The jitter and grain of a third-generation video tape also will be noticeable on a giant screen. Naturally, any program recorded off the air from a weak station or a misaligned antenna will appear grossly grainy. The tiny specks of snow that you never noticed on your 17-inch TV will appear the size of nickels when blown up on the large screen. In short, watching a big, jiggly, snowy image will give your viewers

---

MINIGLOSSARY

Analog A smoothly varying signal (the opposite of digital signals, which vary in discrete steps). Normal video signals are analog and require analog TV projectors to interpret the signal. Computers which generate hundreds of colors would need to feed an analog signal to a TV projector.

headaches. Provide your projector with a clean, sharp, stable picture to get the most out of the medium.

In an effort to get the sharpest image possible, always use video in preference to RF. When showing computer text, always use RGB in preference to NTSC.

Screen Size and Audience Size

The bigger your projection screen, the larger your audience can be and still see clearly. To estimate how big a screen is necessary for various size audiences, see Table 19–2. The table assumes that a MATTE screen is at one end of a rectangular room and that the tall lady with the sombrero sits in the *last* row.

TABLE 19–2
Screen size versus audience size

Audience Size	Necessary Screen Width (Feet)
15	1.5 (TV monitor)
30	3
50	4
75	5
100	6
150	8
300	10
450	12
700	15
1,500	20
5,000	25
10,000	30

Seating

The bigger the projected image, the tougher it is on the eyes of those sitting closest to it. Viewers should not be seated closer than 1½ times the width of the screen. If the screen is 4 feet wide, the nearest row of seats should be 6 feet from the screen (1½ times 4).

The best TV seating is at about four times the screen's width away from it. Thus, a 4-foot-wide TV image looks best from 16 feet away (4 times 4).

Unless your audience brought binoculars, they probably wouldn't want to be more than six times the screen's width away. For a 4-foot-wide screen, this calculates to 24 feet (6 × 4).

To avoid obstruction of the screen by the seated audience, the bottom of the screen should be at least 4 feet above the floor.

Figure 19–8 diagrams these screen and seating arrangements. Remember that higher-GAIN projection screens,

although they reflect a brighter picture, do it over a narrower field of view, decreasing audience size.

Lettering Size

Tests have shown that to be easily seen, the height (in inches) of the characters should equal $\frac{1}{150}$ the audience's distance from the screen. For example, if your audience sits 20 feet from your projection screen, your characters should be $\frac{20}{150} = .133$ ft = 1.6 inches tall to be readable.

PROJECTOR CARE

Monochrome and SINGLE-TUBE projectors are fairly hardy and can be treated the same as you would treat a TV set.

THREE-TUBE projectors get out of alignment when bounced around and therefore need to be handled more like babies.

THREE-TUBE projectors require careful positioning because they need to be a certain distance from the projection screen. Also, a CONVERGENCE procedure must be followed to "tune up" the image and make the colors overlap properly.

MATTE, BEADED, and LENTICULAR projection screens are fairly hardy, requiring little care. Use special care when projecting outdoors. Sandbag the base of your tripod screen so it doesn't sail away in the wind. Before rolling up a retractable screen, check for moths and other insects clinging to its surface. Rolling them up with your screen will do neither them nor your screen any good.

The rigid, curved, silvered HIGH-GAIN projection screens are more delicate (and more expensive than their common counterparts, about $350). Fingerprints show up as dark blotches on these screens. On some, the fingerprint can be removed with mild soap and water. On others, fingerprints will never come off. The metallic surface is also easy to scratch, and the scratches show up as a permanent scar.

ONE-PIECE projectors, which have a mirror that folds out, should be kept closed when not in use. The mirror attracts dust and fingerprints. These mirrors are much more delicate than normal mirrors; they scratch very easily. When wiping dust or fingerprints from them, use a soft cloth and, if necessary, a mild window cleaning solution.

The same is true for projector lenses; they also attract dust and fingerprints. They can be dusted with a soft cloth or wiped with a mild window cleaning solution. Some lenses are coated and scratch easily. Others are plastic and scratch easily. In short, everything associated with TV projection scratches easily.

HIGH-DEFINITION TELEVISION

High-definition television (HDTV) is a new, experimental method for broadcasting and displaying wider, sharper T

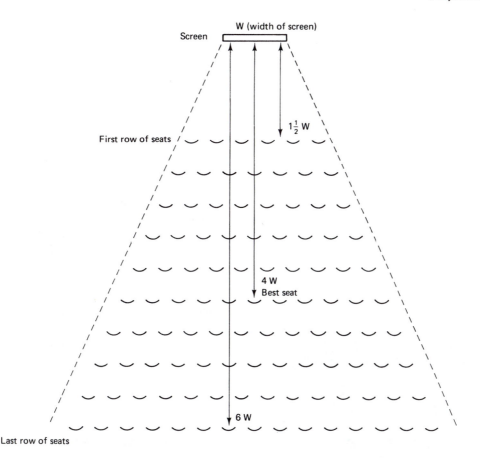

FIGURE 19-8   Screen size and seating arrangements.

ictures. Normal TV pictures are made of 525 horizontal
canning lines. It is sort of like painting a picture on a
Venetian blind having 525 slats. Pictures could look sharper
f we used more slats that were skinnier. HDTV proposes
naking the TV picture out of 1,125 scanning lines.

HDTV pictures are not only sharper in the vertical
irection, but also in the horizontal direction. Where normal
V pictures are composed of about 440 PIXELS (picture
lements, or dots), HDTV will use 1,920 tinier, sharper
IXELS to make a line.

Proposed HDTV pictures will be wider than today's
V screens. Today's TV screens are proportioned in a 4:3
SPECT RATIO. HDTV is likely to be 16:9, or ⅓ wider (see
igure 19–9).

When you see an HDTV demonstration, you are im-
nediately impressed by how *smooth* the picture looks and
ow realistic the wide screen image appears. After watching
DTV for a while, when you go back to normal NTSC, it
els like you are going back to a manual typewriter after
sing a word processor.

Engineers, manufacturers, and even nations are having difficulty selecting one worldwide standard for HDTV.
Many still disagree on the screen shape, number of pixels
per line, number of lines per picture, and number of pictures
per second, as well as more technical details about the color
and contrast. In the United States, the FCC requires that
any HDTV system must be compatible with the existing
NTSC broadcasting standards; you must be able to play an
HDTV broadcast on your regular, old TV.

One of the problems with HDTV is that there is so
much detail in the picture that it requires very high frequencies to transmit the information. It could require two
to five TV channels to broadcast just one HDTV image.
Since it is unlikely that this number of unused TV channels
will become available, it is probable that the first HDTV
will be transmitted by satellite or sent over cable TV or
fiber optics directly to the home. These technologies are
capable of handling the high HDTV frequencies and have
the extra channel space without knocking someone else off
the air. Further, these technologies do not involve *broad-*

MINIGLOSSARY

HDTV (High-Definition Television)   New method of displaying
harper, wider TV pictures than the present NTSC system. Pictures
will probably be shaped into a 16:9 aspect ratio, composed of
1,125 scanning lines, each line having 1,920 pixels.

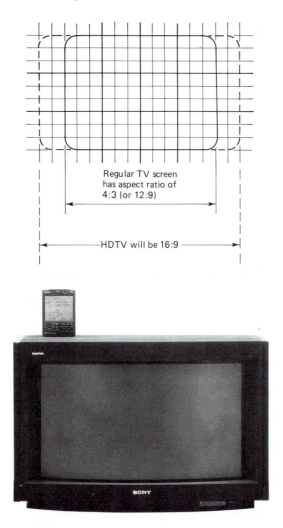

Regular TV screen
has aspect ratio of
4:3 (or 12:9)

◄———HDTV will be 16:9———►

FIGURE 19-9   HDTV vs. regular NTSC.

casting the TV signal and therefore won't be limited by the FCC's requirement that HDTV signals be totally compatible with the old NTSC TV signals.

Future TV sets, to be HDTV compatible, are likely to have removable circuit boards much like the circuit cards that go into a computer. In the future you may buy an advanced TV set and then slip in the circuit boards that make your set receive the kind of HDTV signals available in your area. If you are not interested in HDTV, you buy the *standard* NTSC board. If the FCC selects a compromise broadcast method where HDTV requires only two TV channels in your area, you buy a circuit board that can receive such a signal. If you are receiving a higher-quality HDTV signal over cable, fiber optics, or satellite, a different circuit board would configure your TV to receive the special signals.

TV production companies would also need to modify their cameras, VTRs, and transmitters to handle the sharper wider pictures. You would need a souped-up VCR at home to record the HDTV signals. The manufacturers will, I'm sure, enjoy selling all new equipment to us.

HDTV is likely to be viewed on TV projectors before it is seen on regular TV sets. NTSC video, because it is not very sharp, looks okay on TV sets but doesn't look very good on a large screen. HDTV, on the other hand, will look as good as a movie when blown up. Also it is difficult right now to make the extra wide TV picture tubes, but the technology is already in place for making HDTV projectors. It is quite possible that the first HDTV images you see will be played from an HDTV video tape recorder directly into an HDTV projector.

Hey, guess what, folks—this finishes the last chapter. We're done!

APPENDIX

VTR FORMATS COMPARED

	VHS	8mm	3/4U	3/4U-SP	SVHS	ED BETA
Edit capable deck cost ($)	2,500	1,950	8,000	13,125	4,900	3,300
Tape type	oxide	metal evaporated or metal particle	oxide	oxide	oxide	metal particle
Tape coercivity (oersted)	720	1,450	450	720	900	1,450
Tape cost/min ($)	.13	.40	.30	.31	.20	.16
Studio record time (min)	120, 240, 360	120, 240	60	60	120, 360	120, 180
Field record time (min)	120	120, 240	20	20	120	120, 180
Video luminance bandwidth (MHz)	3	3.2	3.2	4.2	5.0	6.2
Video luminance deviation (MHz) (larger number is white peak)	3.4–4.4 (1)	4.2–5.4 (1.2)	3.8–5.4 (1.6)	5.0–6.6 (1.6)	5.4–7 (1.6)	6.8–9.3 (2.5)
Luminance S/N (dB)	45	45	46	47	48	47
Color S/N (dB)	45 (AM) 40 (PM)	43 (AM)	46	48	50 (AM) 44 (PM)	
Color carrier (kHz)	629	743.44	688	688	629	688
Horizontal lines of resolution	240	250	260	340	400	500
Audio tracks*	2 linear 2 hi-fi	1 linear (rare) 1 AFM 2 PCM 12 PCM (in audio only mode)	2 linear	2 linear	2 linear 2 Hi-Fi	1 linear 2 Hi-Fi
Audio S/N (dB)	45 linear 72 hi-fi	70 on PCM	48	52 (72 w/ Dolby)	44 linear 91 Hi-Fi	70 Hi-Fi
Audio noise reduction	varies	proprietary	none	Dolby C	Dolby B	proprietary
Audio dynamic range (dB)	50 linear 80 hi-fi	80 on AFM 88 on PCM	55	55	90 Hi-Fi 52 linear	90 Hi-Fi
Audio frequency response for linear track (Hz + 3dB)	50–12,000	50–12,000	50–15,000	50–15,000	50–12,500	50–11,000
Audio frequency response for hi-fi track (Hz + 3dB)	20–20,000	20–20,000 mono AFM 30–15,000 PCM	NA	NA	20–20,000	20–20,000
Signal type	composite	composite	composite	composite	composite, Y/C	composite, Y/C
Control track	yes	none needed	yes	yes	yes w/indexing	yes w/indexing
Separate time code channel	no	yes	yes on pro models	yes on pro models	no	no

PCM—Digital pulse code modulation of audio (editable, separate from picture)
AFM or Hi-Fi—Audio modulated within video signal (not editable separate from picture)
AM—Amplitude modulation affecting color saturation

					Type C	D2
Edit capable deck cost ($)	2,300	17,000–30,000	22,000–36,000	9,500–32,000	36,000–100,000	75,000
Tape type	metal particle or metal evaporated	oxide	metal particle or metal oxide	metal particle	oxide	metal particle
Tape coercivity (oersted)	720	720	1,200	1,500	720	1,500
Tape cost/min/ ($)	.45	.30	.30	$1	$1	$1.25
Studio record time (min)	120, 240	94	94	90	180	32, 94, 208
Field record time (min)	120, 240	30	30	20, 90	NA	NA
Video luminance bandwidth (MHz)	5.0	4.1	4.5	4.5	4.2	5.5
Video luminance deviation (MHz) (larger number is white peak)	5.7–7.7 (2)	4.4–6.4 (2)	5.7–7.7 (2)	5.6–7.7 (2.1)	7.1–10 (2.9)	not applicable
Luminance S/N (dB)	47	48	51	49	48	54
Color S/N (dB)		50	53	50		54
Color carrier (kHz)	743.44	time compressed	time compressed	time compressed	high band	not applicable
Horizontal lines of resolution	400	330	470	470	330	440
Audio tracks*	1 linear (rare) 1 AFM 2 PCM 12 PCM (in audio only mode)	2 linear	2 linear 2 Hi-Fi	2 linear 2 Hi-Fi	3	1 linear 4 digital
Audio S/N (dB)	60 PCM	50 (72 w Dolby)	54 (72 w Dolby)	56	56	70
Audio noise reduction	proprietary	Dolby C	Dolby C	Dolby C	none	none
Audio dynamic range (dB)	88 Hi-Fi	55	85 Hi-Fi	80 Hi-Fi	55	90
Audio frequency response for linear track (Hz + 3dB)		50–15,000	50–15,000	50–15,000	50–15,000	100–12,000 linear
Audio frequency response for hi fi track (Hz + 3dB)	20–20,000 AFM 30–15,000 PCM	NA	20–20,000	20–20,000	NA	20–20,000 digital
Signal type	composite, Y/C	Y/R-Y/B-Y component	Y/R-Y/B-Y component	Y/R-Y/B-Y component	composite	composite
Control track	none needed	yes	yes	yes	yes	yes
Separate time code channel	no	1	1	1	1	2

INDEX

(Number in **bold** indicates page where term is defined.)